Progress in Mathematics

Volume 247

Hans-Joachim Baues

The Algebra of Secondary Cohomology Operations

Birkhäuser Verlag
Basel · Boston · Berlin

Author:

Hans-Joachim Baues
Max-Planck-Institut für Mathematik
Vivatsgasse 7
53111 Bonn
Germany
e-mail: baues@mpim-bonn.mpg.de

2000 Mathematics Subject Classification: Primary 18G10, 55T15, 55S20

A CIP catalogue record for this book is available from the Library of Congress, Washington D.C., USA

Bibliographic information published by Die Deutsche Bibliothek
Die Deutsche Bibliothek lists this publication in the Deutsche Nationalbibliografie; detailed bibliographic data is available in the Internet at <http://dnb.ddb.de>.

ISBN 3-7643-7448-9 Birkhäuser Verlag, Basel – Boston – Berlin

Printed on acid-free paper produced of chlorine-free pulp. TCF ∞
Printed in Germany
ISBN-10: 3-7643-7448-9 e-ISBN: 3-7643-7449-7
ISBN-13: 978-3-7643-7448-8

9 8 7 6 5 4 3 2 1 www.birkhauser.ch

Contents

Result

This book computes the Hopf algebra of secondary cohomology operations which is the secondary analogue of the Steenrod algebra.

Preface

Primary cohomology operations, for example the squaring operations Sq^i and the pth-power operations P^i of N.E. Steenrod, supplement and enrich the algebraic structure of the cohomology ring $H^*(X)$ of a space. The Steenrod algebra consisting of all (stable) primary cohomology operations was computed by J.P. Serre and J. Adem in terms of the generators Sq^i, P^i, $i \geq 1$. Using H. Cartan's formula J. Milnor showed that the Steenrod algebra $\mathcal{A}$ is a Hopf algebra with the diagonal:

$$\delta : \mathcal{A} \longrightarrow \mathcal{A} \otimes \mathcal{A},$$

$$\delta(\mathrm{Sq}^n) = \sum_{i=0}^{n} \mathrm{Sq}^i \otimes \mathrm{Sq}^{n-i}.$$

The computation of the Hopf algebra $\mathcal{A}$ led to progress, both in homotopy theory and in specific geometric applications. In fact, in the decades after Steenrod's discovery the Steenrod algebra $\mathcal{A}$ became one of the most powerful tools of algebraic topology. We refer the reader to the survey of R.M.W. Wood [W] concerning applications in topology and the rich algebraic properties of the algebra $\mathcal{A}$.

It is, however, an intrinsic feature of homotopy theory (in contrast to algebra) that primary operations always give rise to secondary and more general higher-order operations. The understanding of higher operations leads to knowledge of homotopy groups of spheres via the Adams spectral sequence. J.F. Adams in solving the Hopf invariant problem and H. Toda in computing low-dimensional homotopy groups of spheres exploited secondary operations in the solution of fundamental problems in topology. This demonstrates that enriching cohomology with both primary and secondary operations, yields a powerful algebraic model of a space.

Though there is a large amount of detailed information on secondary cohomology operations in the literature, the algebraic nature of the secondary theory remained a mystery. We clarify the algebraic structure by showing that secondary cohomology operations yield an algebra $\mathcal{B}$ with an associative bilinear multiplication in the category of pair modules. This crucial fact is missing in the extensive studies of L. Kristensen and his students A. Kock, I. Madsen, and E.K. Pedersen on secondary operations.

The algebra $\mathcal{B}$ and its multiplication are defined in this book topologically in terms of continuous maps between Eilenberg-MacLane spaces. Topologically we also introduce a diagonal

$$\Delta : \mathcal{B} \longrightarrow \mathcal{B} \hat{\otimes} \mathcal{B}$$

which induces Milnor's diagonal δ on the Steenrod algebra $\mathcal{A}$. Moreover we show that $\mathcal{B}$ with the diagonal Δ is a (secondary) Hopf algebra and we compute algebraic invariants L and S of $\mathcal{B}$. We prove that up to isomorphism there is a unique secondary Hopf algebra $\mathcal{B}$ with invariants L and S. For p odd the invariants are trivial, $L = S = 0$, and for $p = 2$ we give explicit formulæ for L and S. This uniqueness theorem yields the algebraic determination of $\mathcal{B}$ as a Hopf algebra leading to an algorithm for the computation of $\mathcal{B}$ in Chapter 16. The author is very grateful to M. Jibladze for implementing the algorithm as a Maple package.

As an application one obtains the computation of triple Massey products in the Steenrod algebra $\mathcal{A}$. In fact, we show that the Milnor generator

$$\mathrm{Sq}^{(0,2)} = \mathrm{Sq}^6 + \mathrm{Sq}^5\,\mathrm{Sq}^1 + \mathrm{Sq}^4\,\mathrm{Sq}^2 \;\in \mathcal{A}$$

yields a non-trivial triple Massey product

$$(*) \qquad \langle \mathrm{Sq}^{(0,2)}, \mathrm{Sq}^{(0,2)}, \mathrm{Sq}^{(0,2)} \rangle \neq 0$$

containing $\mathrm{Sq}^{(0,1,2)}$. This is the first non-trivial triple Massey product in the literature of the form $\langle \alpha, \beta, \gamma \rangle$ with $\alpha, \beta, \gamma \in \mathcal{A}$. We show that in degree $|\alpha| + |\beta| + |\gamma| \leq 17$ all triple Massey products $\langle \alpha, \beta, \gamma \rangle$ vanish. Our algorithm computes also all matrix triple Massey products in the Steenrod algebra $\mathcal{A}$.

A fundamental tool for the computation of the homotopy groups of spheres is the Adams spectral sequence $(E_2, E_3, \dots)$. Adams computed the E_2-term and showed that

$$E_2 = \mathrm{Ext}_{\mathcal{A}}(\mathbb{F}, \mathbb{F})$$

is algebraically determined by Ext-groups associated to the Steenrod algebra $\mathcal{A}$. It is proved in [BJ5], [BJ6] that the E_3-term is, in fact, similarly given by secondary Ext-groups

$$E_3 = \mathrm{Ext}_{\mathcal{B}}(\mathbb{G}^{\Sigma}, \mathbb{G}^{\Sigma})$$

which are algebraically determined by the secondary Hopf algebra $\mathcal{B}$ computed in this book. The computation of E_3 yields a new algebraic upper bound of homotopy groups of spheres improving the Adams bound given by E_2. Computations of the new bound are described in [BJ6]. The author is convinced that the methods of this book also yield a new impact on the computation of E_n for $n \geq 3$ and finally this might lead to the algebraic determination of homotopy groups of spheres in terms of the Steenrod algebra and its higher invariants like L and S above.

The topological construction of both the multiplication and diagonal in $\mathcal{B}$ and the proof of the uniqueness theorem constitute a substantial amount of work. Essentially all the material in this book is required in the proof of the main result. In order to provide the reader with a quick introduction to the new algebraic concepts in secondary cohomology, we state the definitions and more important results in the introduction. Further algebraic properties of the Hopf algebra $\mathcal{B}$ are discussed in [BJ7] where in particular the dual of $\mathcal{B}$ is described extending the Milnor dual of the Steenrod algebra $\mathcal{A}$.

Bonn, October 2003 H.-J. B.

Introduction

The introduction consists of several parts. Part A gives a topological description of the *algebra $\mathcal{B}$ of secondary cohomology operations* and compares $\mathcal{B}$ with the Steenrod algebra $\mathcal{A}$ of primary cohomology operations. Part B introduces the algebraic notion of a *secondary Hopf algebra*. We show that a secondary Hopf algebra structure of $\mathcal{B}$ exists which induces the Hopf algebra structure of $\mathcal{A}$. The uniqueness theorem for secondary Hopf algebras yields an algebraic characterization of $\mathcal{B}$. Part C discusses the concept of *secondary cohomology* of a space X and describes its structure as a secondary algebra over the secondary Hopf algebra $\mathcal{B}$. This result on secondary cohomology can be viewed as an enrichment of the cochain functor $C^*(\ ,\mathbb{F})$ adding fundamental algebraic insight to the recent concept of $C^*(X,\mathbb{F})$ as an algebra over an E_∞-operad, see part D.

Part A. Primary and secondary cohomology operations

Let p be a prime and let $\mathbb{F} = \mathbb{Z}/p\mathbb{Z}$ be the field with p elements. Then $H^*(X,\mathbb{F})$ and $\tilde{H}^*(X,\mathbb{F})$ denote the cohomology and reduced cohomology.

A primary (stable) cohomology operation of degree $k \in \mathbb{Z}$ is a linear map

$$\alpha : \tilde{H}^n(X,\mathbb{F}) \longrightarrow \tilde{H}^{n+k}(X,\mathbb{F}), \tag{A1}$$

defined for all spaces X and all $n \in \mathbb{Z}$, which commutes with suspension and with homomorphisms induced by continuous maps between spaces. The graded vector space of all such cohomology operations is the *Steenrod algebra* $\mathcal{A}$. Multiplication in $\mathcal{A}$ is defined by composition of operations.

It is also possible to define the Steenrod algebra $\mathcal{A}$ by stable classes of maps between Eilenberg-MacLane spaces $Z^n = K(\mathbb{F},n)$. A *stable class* $\bar{\alpha}$ of degree $k \geq 0$ is a homotopy class

$$\bar{\alpha} : Z^n \longrightarrow Z^{n+k},$$

defined for each $n \geq 1$, such that the following diagram commutes in the homotopy category.

$$\begin{array}{ccc} Z^n & \xrightarrow{\quad\bar{\alpha}\quad} & Z^{n+k} \\ \simeq\downarrow & & \downarrow\simeq \\ \Omega Z^{n+1} & \xrightarrow{\quad\Omega\bar{\alpha}\quad} & \Omega Z^{n+k+1} \end{array} \tag{A2}$$

Here Ω is the loop space functor and the vertical arrows are the canonical homotopy equivalences. Since the set $[X, Z^n]$ of homotopy classes $X \to Z^n$ satisfies

$$\tilde{H}^n(X, \mathbb{F}) = [X, Z^n] \tag{A3}$$

we see that a stable class induces a primary cohomology operation and vice versa. Therefore the graded vector space of all stable classes coincides with $\mathcal{A}$. Multiplication in $\mathcal{A}$ is given by the composition of stable classes.

Steenrod constructed the set of generators of the algebra $\mathcal{A}$ given by

$$E_{\mathcal{A}} = \begin{cases} \{Sq^1, Sq^2, \dots\} & \text{for } p = 2, \\ \{P^1, P^2, \dots\} \cup \{\beta, P^1_\beta, P^2_\beta, \dots\} & \text{for } p \text{ odd.} \end{cases} \tag{A4}$$

Here Sq^i is the squaring operation, P^i is the pth-power operation, and β is the Bockstein operation. Moreover P^i_β denotes the composite $P^i_\beta = \beta P^i$. We have to introduce the additional generator P^i_β in order to deal with secondary instability conditions. Adem obtained a complete set of relations for these generators so that $\mathcal{A}$ is algebraically determined as an algebra.

Milnor observed that the algebra $\mathcal{A}$ of primary cohomology operations is actually a Hopf algebra with the diagonal

$$\delta : \mathcal{A} \longrightarrow \mathcal{A} \otimes \mathcal{A} \tag{A5}$$

given by the Cartan formula. The multiplication $\mu : H \otimes H \longrightarrow H$ of the cohomology algebra $H = H^*(X, \mathbb{F})$ is compatible with the Hopf algebra structure δ of $\mathcal{A}$, see (C2) below.

In this book, we offer similar results on *secondary cohomology operations*. Classically a secondary operation

$$\varphi : S_\varphi \longrightarrow Q_\varphi$$

is defined on a subset S_φ of all cohomology classes and has values in a quotient set Q_φ of all cohomology classes. This concept, however, is not suitable in order to study the global algebraic structure of all secondary operations. For this reason we introduce below the new object

$$\mathcal{B} = (\mathcal{B}_1 \xrightarrow{\partial} \mathcal{B}_0)$$

termed the *algebra of secondary cohomology operations*. Secondary operations can be derived from $\mathcal{B}$ in a similar way as Massey products are derived from the structure of a differential algebra. The algebra $\mathcal{B}$ is the secondary analogue of the Steenrod algebra $\mathcal{A}$. For the definition of $\mathcal{B}$ we need the ring

$$\mathbb{G} = \mathbb{Z}/p^2\mathbb{Z}$$

and the ring homomorphism $\mathbb{G} \to \mathbb{F}$ which shows that an $\mathbb{F}$-vector space is a $\mathbb{G}$-module. Let

(A6) $$\mathcal{B}_0 = T_{\mathbb{G}}(E_{\mathcal{A}})$$

be the $\mathbb{G}$-tensor algebra generated by $E_{\mathcal{A}}$ in (A4) and let

$$q : \mathcal{B}_0 \longrightarrow \mathcal{A}$$

be the surjective algebra map which is the identity on generators in $E_{\mathcal{A}}$. The definition of $\mathcal{B}_1$ and $\partial : \mathcal{B}_1 \to \mathcal{B}_0$ is more complicated and relies on the notion of track.

Given pointed maps $f, g : X \to Y$ a *track* $H : f \Rightarrow g$ is an equivalence class of homotopies $f \simeq g$. Here homotopies $H_0, H_1, : f \simeq g$ are equivalent if there is a homotopy H_t from H_0 to H_1 where $H_t : f \simeq g$ for all $t \in [0,1]$. Hence a track is the same as an arrow in the fundamental groupoid of the function space of pointed maps $X \to Y$.

A *stable map* α of degree k is a representative of a stable class $\bar{\alpha}$ in (A2), that is, α is a diagram

(A7)

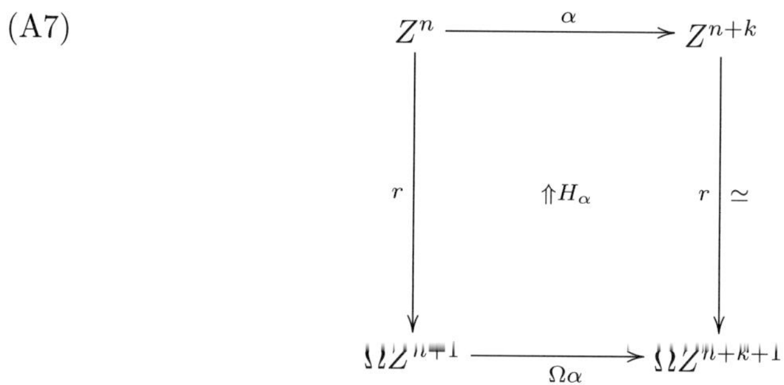

defined for each $n \geq 1$, where α is a pointed map and $H_\alpha : r\alpha \Rightarrow (\Omega\alpha)r$ is a track. Given a stable map $\alpha = (\alpha, H_\alpha)$ a *stable track* $a : \alpha \Rightarrow 0$ of degree k is a diagram

(A8)

0

$\Uparrow a$

$Z^n \xrightarrow{\alpha} Z^{n+k}$

defined for each $n \geq 1$, such that the pasting of tracks in the following diagram yields the trivial track.

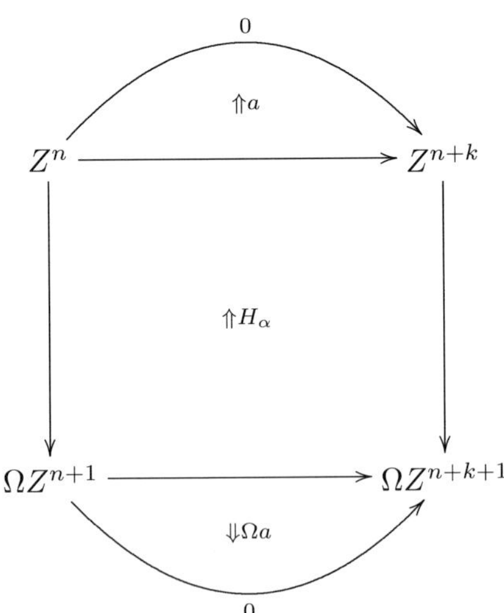

We can choose the Eilenberg-MacLane space $Z^n = K(\mathbb{F}, n)$ to be a topological $\mathbb{F}$-vector space and the homotopy equivalence $Z^n \longrightarrow \Omega Z^{n+1}$ to be $\mathbb{F}$-linear. This implies that stable maps form a graded $\mathbb{F}$-vector space $[\![\mathcal{A}]\!]_0$ and stable tracks form a graded $\mathbb{F}$-vector space $[\![A]\!]_1^0$. There is the linear boundary map

$$\partial : [\![\mathcal{A}]\!]_1^0 \longrightarrow [\![\mathcal{A}]\!]_0 \tag{A9}$$

which carries $a : \alpha \Rightarrow 0$ to α. One now gets the following pullback diagram with exact rows which defines $\mathcal{B}_1$ and ∂ on $\mathcal{B}_1$.

(A10)
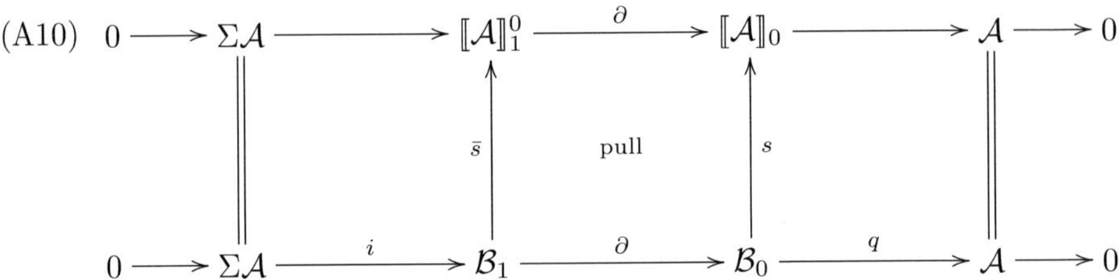

Here Σ denotes the suspension of graded modules. We define the function s in the diagram by choosing for $\alpha \in E_{\mathcal{A}}$ an element $s\alpha \in [\![\mathcal{A}]\!]_0$ representing α. Since $[\![\mathcal{A}]\!]_0$ is a monoid this induces the monoid homomorphism

$$s : \mathrm{Mon}(E_{\mathcal{A}}) \longrightarrow [\![\mathcal{A}]\!]_0$$

where $\mathrm{Mon}(E_{\mathcal{A}})$ is the free monoid generated by $E_{\mathcal{A}}$. Since $[\![\mathcal{A}]\!]_0$ is an $\mathbb{F}$-vector space one gets the $\mathbb{G}$-linear map

$$s : \mathcal{B}_0 \longrightarrow [\![\mathcal{A}]\!]_0 \tag{A11}$$

using the fact that the tensor algebra $\mathcal{B}_0$ is the free $\mathbb{G}$-module generated by $\mathrm{Mon}(E_{\mathcal{A}})$.

It is also possible to define the graded $\mathbb{G}$-module $\mathcal{B}_1$ by the near-algebra of cochain operations introduced by Kristensen [Kr1]. Let

$$C^*(X, \mathbb{F})$$

be the singular cochain complex of a space X. A *cochain operation* Θ of degree k is a function between the underlying sets

$$\Theta : C^n(X, \mathbb{F}) \longrightarrow C^{n+k}(X, \mathbb{F})$$

defined for all $n \in \mathbb{Z}$ and all spaces X, such that Θ commutes with homomorphisms induced by maps between spaces. Let $\mathcal{O}$ be the graded vector space of all cochain operations with addition defined by adding values in $C^{n+k}(X, \mathbb{F})$. Composition of cochain operations yields a multiplication $\Theta \cdot \Theta'$ for $\Theta, \Theta' \in \mathcal{O}$ which is left distributive but not right distributive so that $\mathcal{O}$ is a near-algebra. There is a linear map

(A12) $$\partial : \mathcal{O} \longrightarrow \mathcal{O}$$

of degree $+1$ defined by the formula $\partial\Theta = d\Theta + (-1)^{|\Theta|}\Theta d$ where d is the differential of $C^*(X, \mathbb{F})$. One readily checks that $\partial\partial = 0$.

Kristensen shows that the homology of $(\mathcal{O}, \partial)$,

$$\mathcal{A} = \text{kernel}(\partial)/\,\text{image}(\partial),$$

coincides with the Steenrod algebra. For this reason we get the following pull back diagram with exact rows which also defines $\mathcal{B}_1$.

(A13) $$\begin{array}{ccccccccccc} 0 & \longrightarrow & \Sigma\mathcal{A} & \longrightarrow & \Sigma\,\text{cokernel}(\partial) & \xrightarrow{\partial} & \text{kernel}(\partial) & \longrightarrow & \mathcal{A} & \longrightarrow & 0 \\ & & \Vert & & \uparrow & \text{pull} & \uparrow{\scriptstyle s} & & \Vert & & \\ 0 & \longrightarrow & \Sigma\mathcal{A} & \xrightarrow{i} & \mathcal{B}_1 & \xrightarrow{\partial} & \mathcal{B}_0 & \xrightarrow{q} & \mathcal{A} & \longrightarrow & 0 \end{array}$$

Here s is defined by multiplication and addition in $\mathcal{O}$ similarly as s in (A12). This construction of $\mathcal{B}_1$ in terms of cochain operations yields the connection of the theory in this book with Kristensen's theory of secondary cohomology operations in the literature, see [Kr1] [Kr2].

We point out that for $\alpha, \beta \in \mathcal{B}_0$ the element $s(\alpha \cdot \beta)$ does not coincide with $(s\alpha) \cdot (s\beta)$. But we show that there is a well-defined element $\bar{\Gamma}(\alpha, \beta) \in [\![\mathcal{A}]\!]_1^0$ satisfying

(A14) $$\partial\bar{\Gamma}(\alpha, \beta) = s(\alpha) \cdot s(\beta) - s(\alpha \cdot \beta), \text{ see (A12).}$$

Here it is of crucial importance that we define the tensor algebra $\mathcal{B}_0$ over the ring $\mathbb{G}$ and not over $\mathbb{F}$ since only for elements α, β, defined over $\mathbb{G}$, the term $\bar{\Gamma}(\alpha, \beta)$ is well defined.

According to the pull back in (A11) an element $x \in \mathcal{B}_1$ is a pair $x = (\xi, u)$ with $\xi \in \mathcal{B}_0$ and $u : s\xi \Rightarrow 0$ a stable track. We define stable tracks $\alpha \bullet u : s(\alpha\xi) \Rightarrow 0$ and $u \bullet \beta : s(\xi\beta) \Rightarrow 0$ by pasting tracks in the following diagrams where $\Gamma(\alpha, \beta) = \bar{\Gamma}(\alpha, \beta) + s(\alpha\beta)$ is given by the operator $\bar{\Gamma}$.

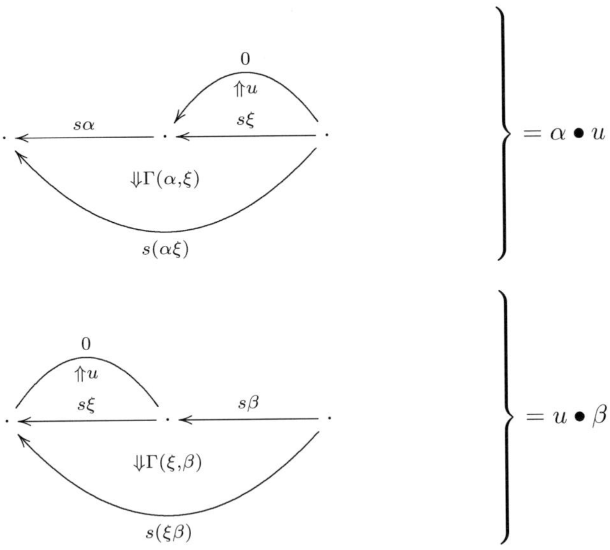

A15 Theorem. *Defining the left and right action of $\alpha, \beta \in \mathcal{B}_0$ on $x = (\xi, u) \in \mathcal{B}_1$ by*

$$\alpha \cdot x = (\alpha\xi, \alpha \bullet u),$$
$$x \cdot \beta = (\xi\beta, u \bullet \beta),$$

one obtains a well-defined structure of $\mathcal{B}_1$ as a $\mathcal{B}_0$-bimodule. Moreover $\partial : \mathcal{B}_1 \longrightarrow \mathcal{B}_0$ satisfies the equations $(x, y \in \mathcal{B}_1)$,

$$\partial(\alpha \cdot x \cdot \beta) = \alpha \cdot \partial(x) \cdot \beta,$$
$$\partial(x) \cdot y = x \cdot \partial(y).$$

The theorem shows that $\mathcal{B} = (\partial : \mathcal{B}_1 \to \mathcal{B}_0)$ is a pair algebra, see Section (B2) below.

We now indicate how elements $x \in \mathcal{B}_1$ are related to secondary operations φ_x. Given $\alpha, \beta \in \mathcal{A}$ with $\alpha \cdot \beta = 0$ in $\mathcal{A}$ we can choose $\bar{\alpha}, \bar{\beta} \in \mathcal{B}_0$ with $q\bar{\alpha} = \alpha, q\bar{\beta} = \beta$ so that $q(\bar{\alpha}\bar{\beta}) = 0$. By exactness there is $x \in \mathcal{B}_1$ with $\partial x = \bar{\alpha} \cdot \bar{\beta}$. Then x induces the associated *secondary operations*

(A16) $$\varphi_x : \{g \in H, \beta g = 0\} \Longrightarrow H/\alpha H$$

of degree $|\alpha| + |\beta| - 1$ with $H = H^*(X, \mathbb{F})$. If g is represented by $\gamma : X \longrightarrow Z^n$ then $\varphi_x(g)$ is represented by the map $\varphi : X \longrightarrow \Omega Z^{n+|\alpha|+|\beta|}$ obtained by pasting tracks in the following diagram.

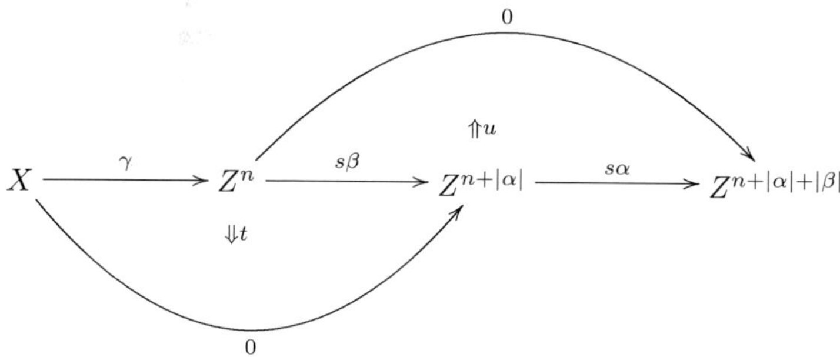

The track t exists since $\beta g = 0$ and $u = \bar{s}x$ is given by x. If $X = Z^m$ is an Eilenberg-MacLane space (with m large) then φ yields an element in $\mathcal{A}$ and the collection of all such elements is the *Massey product*

(A17) $$\langle \alpha, \beta, \gamma \rangle \subset \mathcal{A}.$$

defined for all $\alpha, \beta, \gamma \in \mathcal{A}$ with $\beta\gamma = 0$ and $\alpha\beta = 0$. The Massey product $\langle \alpha, \beta, \gamma \rangle$ can be computed in terms of $\mathcal{B}$ as follows. Let $\bar{\alpha}, \bar{\beta}, \bar{\gamma} \in \mathcal{B}_0$ be elements representing α, β and γ respectively. Then there exists $u, v \in \mathcal{B}_1$ with

$$\partial u = \bar{\beta} \cdot \bar{\gamma}, \qquad \partial v = \bar{\alpha} \cdot \bar{\beta},$$

since $\beta\gamma = 0$ and $\alpha\beta = 0$. Hence we get the element

(A18) $$x = \bar{\alpha}u - v\bar{\gamma} \ \in \mathcal{B}_1$$

by the $\mathcal{B}_0$-bimodule structure of $\mathcal{B}_1$. Since $\partial x = \bar{\alpha}(\partial u) - (\partial v)\bar{\gamma} = 0$ we see that $x \in \Sigma\mathcal{A}$. The element x represents the Massey product $< \alpha, \beta, \gamma >$. This shows that the algebraic determination of $\mathcal{B}$ in this book solves an old problem of Kristensen and Madsen [Kr4], [KrM2].

We derive from the pair algebra $\mathcal{B}$ a *derivation* of degree -1,

(A19) $$\Gamma[p] : \mathcal{A} \longrightarrow \mathcal{A}$$

as follows. Let $[p] \in \mathcal{B}_1$ be the unique element of degree 0 with $\partial[p] = p \cdot 1$ where 1 is the unit of the algebra $\mathcal{B}_0$. For $\alpha \in \mathcal{B}_0$ the difference $\alpha \cdot [p] - [p] \cdot \alpha = x$ is defined by the bimodule structure of $\mathcal{B}_1$ with $\partial x = \alpha \cdot p - p \cdot \alpha = 0$ so that $x \in \Sigma\mathcal{A}$. Moreover x depends only on the image $\tilde{\alpha}$ of α in $\mathcal{A}$ so that $\Gamma[p](\tilde{\alpha}) = x$ is well defined.

A20 Theorem. *The derivation* $\Gamma[p]$ *coincides with the derivation* $\kappa : \mathcal{A} \longrightarrow \mathcal{A}$ *defined on generators as follows.*

$$\begin{aligned} \kappa(Sq^n) &= Sq^{n-1} \ \text{ for } n \geq 1, \ p \text{ even. Moreover} \\ \kappa(P^n) &= 0 \ \text{ for } n \geq 1, \ \text{ and} \\ \kappa(\beta) &= 1 \ \text{ for } p \text{ odd.} \end{aligned}$$

Here $Sq^0 = 1$ is the unit of $\mathcal{A}$.

Part B. Secondary Hopf algebras

The concept of algebra and Hopf algebra is based on the monoidal category of modules with the tensor product as monoidal structure. Primary operations in homotopy theory lead to such algebras in contrast to secondary operations which lead to pair algebras defined in the monoidal category of pair modules like the pair algebra $\mathcal{B}$ of secondary cohomology operations in (A16). Moreover, $\mathcal{B}$ has the structure of a secondary Hopf algebra inducing the Hopf algebra structure of the Steenrod algebra $\mathcal{A}$.

(B1) Modules and pair modules

Let R be a commutative ring with unit and let $Mod(R)$ be the category of R-modules and R-linear maps. This is a symmetric monoidal category via the tensor product $A \otimes B$ over R. A *pair* of modules is a morphism

$$X = (\partial : X_1 \longrightarrow X_0)$$

in $\mathbf{Mod}(R)$ and a map $f : X \longrightarrow Y$ between pairs is the following commutative diagram.

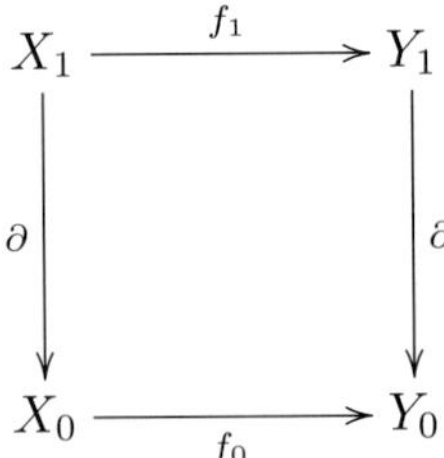

A pair in $\mathbf{Mod}(R)$ coincides with a chain complex concentrated in degree 0 and 1. Let X and Y be two pairs of modules. Then the tensor product $X \otimes Y$ of the underlying chain complexes is given by

$$X_1 \otimes Y_1 \xrightarrow{d_2} X_1 \otimes Y_0 \oplus X_0 \otimes Y_1 \xrightarrow{d_1} X_0 \otimes Y_0$$

with $d_1 = (\partial \otimes 1, 1 \otimes \partial)$ and $d_2 = (-1 \otimes \partial, \partial \otimes 1)$. We truncate $X \otimes Y$ and we get the pair

$$\begin{cases} X\bar{\otimes}Y = (\text{cokernel}(d_2) \xrightarrow{\partial} X_0 \otimes Y_0), & \text{with} \\ \partial : (X\bar{\otimes}Y)_1 = \text{cokernel}(d_2) \longrightarrow X_0 \otimes Y_0 = (X\bar{\otimes}Y)_0 \end{cases}$$

induced by d_1. One readily checks that the category of pairs in $Mod(R)$ together with the tensor product $X\bar{\otimes}Y$ is a symmetric monoidal category. The unit object is $R = (0 \to R)$.

A graded module is a sequence $A^n, n \in \mathbb{Z}$, of R-modules with $A^n = 0$ for $n < 0$. We define the tensor product $A \otimes B$ of two graded modules as usual by

$$(A \otimes B)^n = \bigoplus_{i+j=n} A^i \otimes B^j.$$

We define the *interchange isomorphism*

$$T : A \otimes B \cong B \otimes A$$

depending on a prime p by the formula

$$T(a \otimes b) = (-1)^{p|a||b|} b \otimes a.$$

Here $|a|$ is the degree of $a \in A$ with $|a| = n$ if $a \in A^n$. Hence the interchange of the graded elements a, b always involves the interchange sign $(-1)^{p|a||b|}$ depending on the prime p. Let ΣA be the suspension of A defined by $(\Sigma A)^n = A^{n-1}$ and let $\Sigma : A \to \Sigma A$ be the map of degree $+1$ given by the identity. One has the canonical isomorphisms

$$(\Sigma A) \otimes B = \Sigma(A \otimes B) \overset{\tau}{\cong} A \otimes (\Sigma B)$$

where $\tau(a \otimes \Sigma b) = (-1)^{p|a|}\Sigma(a \otimes b)$. We call τ the *interchange of* Σ.

A *graded pair module* X is a sequence of pairs $X^n = (\partial : X_1^n \to X_0^n)$ in $\mathbf{Mod}(R)$ with $X^n = 0$ for $n < 0$. We identify a graded pair module X with the underlying map ∂ of degree 0 between graded modules

$$X = \partial : X_1 \longrightarrow X_0.$$

The tensor product $X \bar{\otimes} Y$ of graded pair modules X, Y is defined by

$$(X \bar{\otimes} Y)^n = \bigoplus_{i+j=n} X^i \bar{\otimes} Y^i$$

and the interchange isomorphism

$$T : X \bar{\otimes} Y \cong Y \bar{\otimes} X$$

is induced by the interchange isomorphism for graded R-modules depending on the prime p above. Given two maps $f, g : X \to Y$ between graded pair modules a *homotopy* $H : f \Rightarrow g$ is a morphism $H : X_0 \to Y_1$ of degree 0 as in the diagram

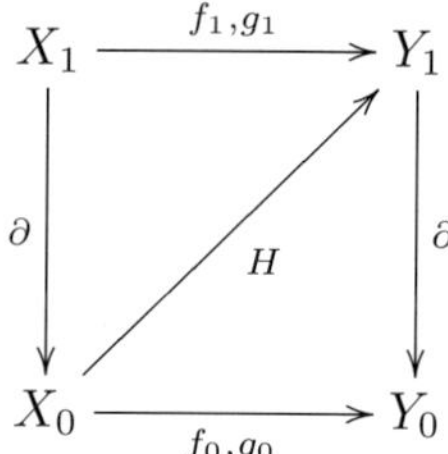

satisfying $H\partial = f_1 - g_1$ and $\partial H = f_0 - g_0$.

(B2) Algebras and pair algebras

An *algebra* A with multiplication

$$\mu : A \otimes A \longrightarrow A$$

is the same as a monoid in the monoidal category of graded R-modules. Moreover an *A-module* is a graded module M together with a left action

$$\mu : A \otimes M \longrightarrow M$$

of the monoid A on M. Similarly one defines a right A-module by a right action and an A-bimodule by an action

$$\mu : A \otimes M \otimes A \longrightarrow M$$

from the left and the right. For example the suspension ΣA of an algebra A is an A-bimodule by using the interchange of Σ.

A *pair algebra* B is a monoid in the monoidal category of graded pair modules with multiplication

$$\mu : B \bar{\otimes} B \longrightarrow B.$$

A left *B-module* M is a graded pair module M together with a left action

$$\mu : B \bar{\otimes} M \longrightarrow M$$

of the monoid B on M. One readily checks:

Lemma. *A pair algebra $B = (\partial : B_1 \longrightarrow B_0)$ consists of an algebra B_0 and a B_0-bimodule map ∂ satisfying $\partial(x) \cdot y = x \cdot \partial(y)$ for $x, y \in B_1$.*

This shows that $\mathcal{B}$ in (A16) is, in fact, a pair algebra.

(B3) Hopf algebras

For graded algebras A and A' the tensor product $A \otimes A'$ is again an algebra with the multiplication

$$(A \otimes A') \otimes (A \otimes A') \xrightarrow{1 \otimes T \otimes 1} A \otimes A \otimes A' \otimes A' \xrightarrow{\mu \otimes \mu} A \otimes A'.$$

In the same way the tensor product $B \bar{\otimes} B'$ of two graded pair algebras is again a pair algebra with the multiplication

$$(B \bar{\otimes} B') \otimes (B \bar{\otimes} B') \xrightarrow{1 \otimes T \otimes 1} B \bar{\otimes} B \bar{\otimes} B' \bar{\otimes} B' \xrightarrow{\mu \otimes \mu} B \bar{\otimes} B'.$$

Here T is the interchange isomorphism above depending on the prime p. Hence the category of algebras, resp. pair algebras, is a symmetric monoidal category.

A *Hopf algebra* A is a comonoid in the monoidal category of graded algebras, that is, A is a graded algebra together with augmentation $\varepsilon : A \longrightarrow R$ and diagonal

$\Delta : A \longrightarrow A \otimes A$ such that the following diagrams are commutative; here ε and Δ are algebra maps.

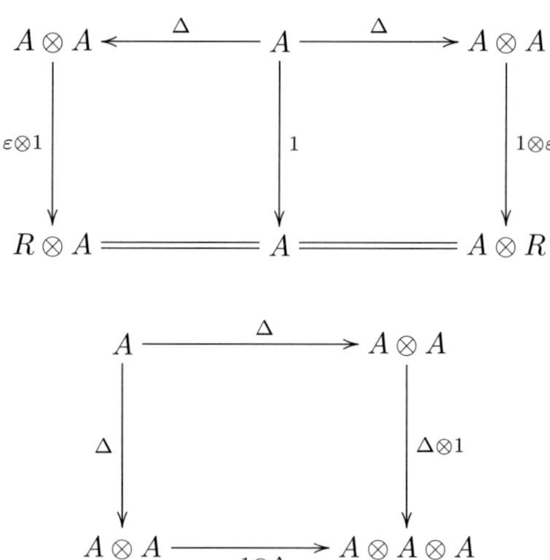

The Hopf algebra is *co-commutative* if in addition the following diagram commutes.

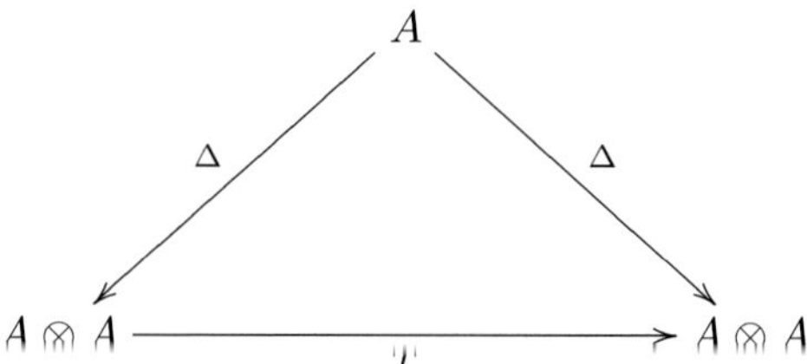

In a similar way it is possible to define a "Hopf pair algebra" as a comonoid in the monoidal category of pair algebras. Secondary cohomology operations, however, lead to a more sophisticated notion of secondary Hopf algebra, defined below in terms of the folding product $\hat{\otimes}$ which is a quotient of the tensor product $\bar{\otimes}$ of pair modules.

(B4) Examples of Hopf algebras

The tensor algebra $\mathcal{B}_0 = T_{\mathbb{G}}(E_{\mathcal{A}})$ is a Hopf algebra with the diagonal

$$\Delta_0 : \mathcal{B}_0 \longrightarrow B_0 \otimes B_0$$

which is the algebra map defined on generators by $(i \geq 1)$

$$\Delta_0(Sq^i) = \sum_{k+l=i} Sq^k \otimes Sq^l \text{ for } p = 2,$$

and

$$\left.\begin{aligned}\Delta_0(\beta) &= \beta\otimes 1 + 1\otimes\beta,\\ \Delta_0(P^i) &= \sum_{k+l=i} P^k\otimes P^l,\\ \Delta_0(P^i_\beta) &= \sum_{k+l=i} (P^k_\beta\otimes P^l + P^k\otimes P^l_\beta)\end{aligned}\right\}\text{ for } p \text{ odd.}$$

One readily checks that (B_0, Δ_0) is a well-defined Hopf algebra which is co-commutative for p odd and for p even, since the interchange isomorphism T depends on the prime p. Moreover the tensor algebra $(T_{\mathbb{F}}(E_{\mathcal{A}}), \Delta)$ and the Steenrod algebra $(\mathcal{A}, \delta)$ are Hopf algebras with the diagonal Δ, resp. δ, defined by the same formula as above so that the canonical surjective algebra maps

$$\mathcal{B}_0 = T_{\mathbb{G}}(E_{\mathcal{A}}) \longrightarrow T_{\mathbb{F}}(E_{\mathcal{A}}) \longrightarrow \mathcal{A}$$

are maps between Hopf algebras. Here $T_{\mathbb{F}}(E_{\mathcal{A}})$ and $\mathcal{A}$ are also co-commutative Hopf algebras.

(B5) The folding product of $[p]$-algebras

Let $\mathcal{A}^{\otimes n} = \mathcal{A}\otimes\cdots\otimes\mathcal{A}$ be the n-fold tensor product of the Steenrod algebra $\mathcal{A}$ with $\mathcal{A}^{\otimes n} = \mathbb{F}$ for $n = 0$. Hence $\mathcal{A}^{\otimes n}$ is an algebra over $\mathbb{F}$ and $\Sigma\mathcal{A}^{\otimes n}$ is an $\mathcal{A}^{\otimes n}$-bimodule. A *$[p]$-algebra of type n*, $n \geq 0$, is given by an exact sequence of non-negatively graded $\mathbb{G}$-modules

$$0 \longrightarrow \Sigma\mathcal{A}^{\otimes n} \xrightarrow{\;i\;} D_1 \xrightarrow{\;\partial\;} D_0 \xrightarrow{\;q\;} \mathcal{A}^{\otimes n} \longrightarrow 0$$

where D_0 is a free $\mathbb{G}$-module and an algebra over $\mathbb{G}$ and $q : D_0 \longrightarrow \mathcal{A}^{\otimes n}$ is an algebra map. Moreover D_1 is a right D_0-module. Using the algebra map q also $\Sigma\mathcal{A}^{\otimes n}$ is a right D_0-module and all maps in the sequence are D_0-linear. Since $(\Sigma\mathcal{A}^{\otimes n})^0 = 0$ we have the unique element $[p] \in D_1$ of degree 0 with $\partial[p] = p\cdot 1$ where 1 is the unit in the algebra D_0. As part of the definition of a $[p]$-algebra we assume that the quotient

$$\tilde{D}_1 = D_1/[p]D_0$$

of $\mathbb{G}$-modules is actually an $\mathbb{F}$-module. For this reason we get for a $[p]$-algebra D a commutative diagram

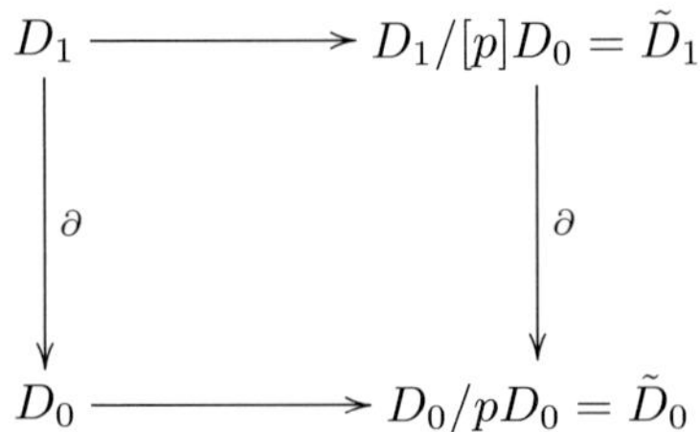

$$\begin{array}{ccc} D_1 & \longrightarrow & D_1/[p]D_0 = \tilde{D}_1 \\ \downarrow{\scriptstyle\partial} & & \downarrow{\scriptstyle\partial} \\ D_0 & \longrightarrow & D_0/pD_0 = \tilde{D}_0 \end{array}$$

which is a push out and a pull back of right D_0-modules. We call $\partial : \tilde{D}_1 \longrightarrow \tilde{D}_0$ the *pair module over* $\mathbb{F}$ *associated* to D. Now let D and E be $[p]$-algebras of type n and m respectively. Then a morphism $f : D \longrightarrow E$ is a commutative diagram

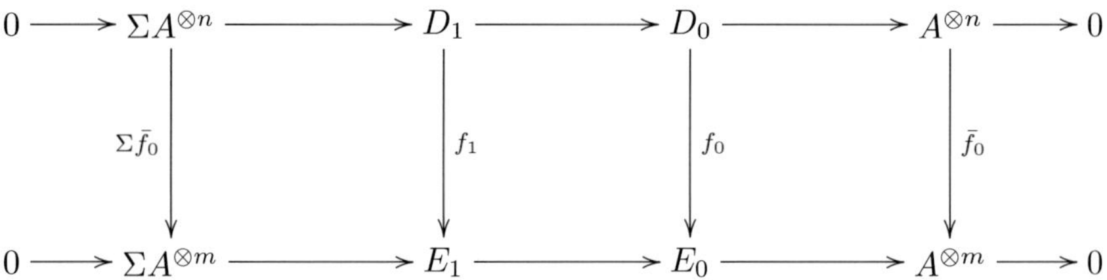

where f_0 is an algebra map and f_1 is an f_0-equivariant map between right modules. We point out that f_1 has the restriction $\Sigma \bar{f}_0$ where $\bar{f}_0$ is induced by f_0. Let

$$\mathcal{A}lg^{[p]}$$

be the category of $[p]$-algebras of type $n \geq 0$ and such maps.

The *initial object* $\mathbb{G}^\Sigma$ in $\mathcal{A}lg^{[p]}$ is the $[p]$-algebra of type 0 given by the exact sequence

$$\begin{array}{ccccccccc} 0 & \longrightarrow & \Sigma\mathbb{F} & \longrightarrow & \mathbb{G}_1^\Sigma & \xrightarrow{\partial} & \mathbb{G}_0^\Sigma & \longrightarrow & \mathbb{F} \longrightarrow 0 \\ & & & & \| & & \| & & \\ & & & & \mathbb{F} \oplus \Sigma\mathbb{F} & & \mathbb{G} & & \end{array}$$

with $\partial|\mathbb{F} : \mathbb{F} \subset \mathbb{G}$ and $\partial\Sigma\mathbb{F} = 0$. For each $[p]$-algebra D there is a unique morphism $\mathbb{G}^\Sigma \longrightarrow D$ carrying 1 to 1 and $[p]$ to $[p]$. We call a morphism

$$\varepsilon : D \longrightarrow \mathbb{G}^\Sigma \;\text{ in }\; \mathcal{A}lg^p$$

a *secondary augmentation* of D.

For $n, m \geq 0$ we obtain the *folding map* φ by the commutative diagram

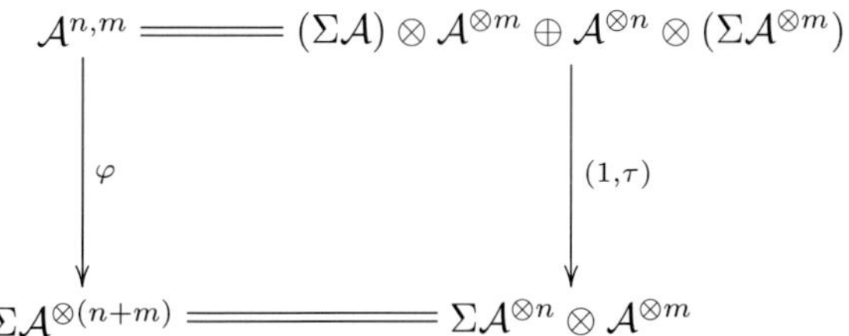

where we use the interchange of Σ . Let D and E be $[p]$-algebras of type n and m respectively and let $\tilde{D}, \tilde{E}$ be the associated pair modules. Then the *folding product*

$D\hat{\otimes}E$ is defined by the following diagram in which the top row is exact.

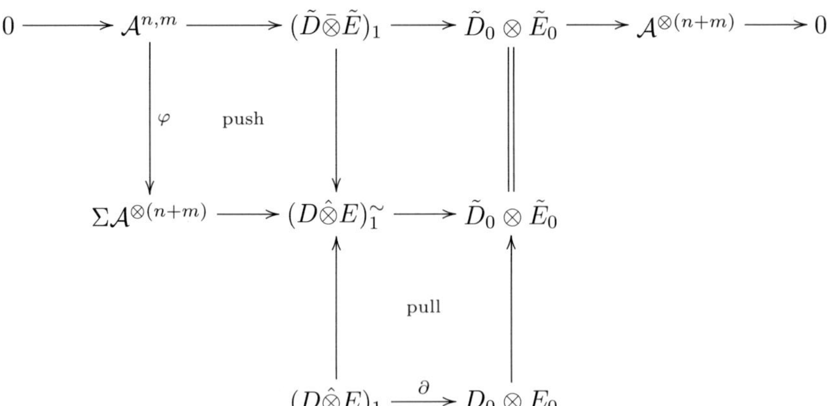

The bottom row defines the pair module $D\hat{\otimes}E$. The algebra $D_0 \otimes E_0$ acts from the right on $(D\hat{\otimes}E)_1$ since D_0 acts on $\tilde{D}_1$ and E_0 acts on $\tilde{E}_1$ and φ is equivariant. This shows that $D\hat{\otimes}E$ is a well-defined $[p]$-algebra of type $n+m$.

Lemma. *The category $\mathcal{A}lg^{[p]}$ of $[p]$-algebras with the folding product $\hat{\otimes}$ is a symmetric monoidal category. The unit object of $\hat{\otimes}$ is $\mathbb{G}_\Sigma$.*

The pair of maps

$$\begin{cases} (D\bar{\otimes}E)_1 \longrightarrow (\tilde{D}\bar{\otimes}\tilde{E})_1 \longrightarrow (D\hat{\otimes}E)_1^{\sim} \\ (D\bar{\otimes}E)_1 \longrightarrow D_0 \otimes E_0 \end{cases}$$

induces by the pull back property of $(D\hat{\otimes}E)_1$ the map

$$q : (D\bar{\otimes}E)_1 \longrightarrow (D\hat{\otimes}E)_1$$

which is a *quotient map* of right $D_0 \otimes E_0$-modules.

The interchange map T for $D\bar{\otimes}E$ depending on the prime p induces via q the corresponding *interchange map*

$$T : D\hat{\otimes}E \longrightarrow E\hat{\otimes}D$$

in the category $\mathcal{A}lg^{[p]}$ of $[p]$-algebras. We are now ready to define secondary Hopf algebras associated to the Steenrod algebra $\mathcal{A}$.

(B6) Secondary Hopf algebras

As in (B4) we have the map

$$q : \mathcal{B}_0 = T_{\mathbb{G}}(E_{\mathcal{A}}) \longrightarrow \mathcal{A}$$

between Hopf algebras. We consider a pair algebra $\mathcal{B} = (\partial : \mathcal{B}_1 \longrightarrow \mathcal{B}_0)$ together with an exact sequence of $\mathcal{B}_0$-bimodules

$$0 \longrightarrow \Sigma\mathcal{A} \longrightarrow \mathcal{B}_1 \xrightarrow{\partial} \mathcal{B}_0 \longrightarrow \mathcal{A} \longrightarrow 0$$

such that $\mathcal{B}_1/[p]\cdot\mathcal{B}_0$ is an $\mathbb{F}$-vector space. Then $\mathcal{B}$ is also a $[p]$-algebra of type $n = 1$ as defined in (B5) and the folding product $\mathcal{B}\hat{\otimes}\mathcal{B}$ together with the quotient map

$$q : (\mathcal{B}\bar{\otimes}\mathcal{B})_1 \longrightarrow \mathcal{B}\hat{\otimes}\mathcal{B}$$

is defined. Moreover we assume that $\mathcal{B}$ induces the derivation $\Gamma[p] = \kappa$ as in (A19). For example the pair algebra $\mathcal{B}$ of secondary cohomology operations in part A has these properties.

Since $\mathcal{B}$ is a pair algebra also $\mathcal{B}\bar{\otimes}\mathcal{B}$ is a pair algebra so that $(\mathcal{B}\bar{\otimes}\mathcal{B})_1$ is a $\mathcal{B}_0 \otimes \mathcal{B}_0$-bimodule which via the algebra map $\Delta_0 : \mathcal{B}_0 \longrightarrow \mathcal{B}_0 \otimes \mathcal{B}_0$ in (B4) is also a $\mathcal{B}_0$-bimodule. One can show that there is a unique $\mathcal{B}_0$-bimodule structure of $(\mathcal{B}\hat{\otimes}\mathcal{B})_1$ for which the quotient map q is a morphism between $\mathcal{B}_0$-bimodules. Here one needs for the existence of the left action of $\mathcal{B}_0$ on $(\mathcal{B}\hat{\otimes}\mathcal{B})_1$ the assumption that $\Gamma[p] = \kappa$ satisfies $\delta\kappa = (\kappa \otimes 1)\delta$.

We say that the pair algebra $\mathcal{B}$ is a *secondary Hopf algebra* (associated to $\mathcal{A}$) if an augmentation

$$\varepsilon : \mathcal{B} \longrightarrow \mathbb{G}^{\Sigma} \text{ in } \mathcal{A}lg^{[p]}$$

and a diagonal

$$\Delta : \mathcal{B} \longrightarrow \mathcal{B}\hat{\otimes}\mathcal{B} \text{ in } \mathcal{A}lg^{[p]}$$

are given such that the following diagrams commute.

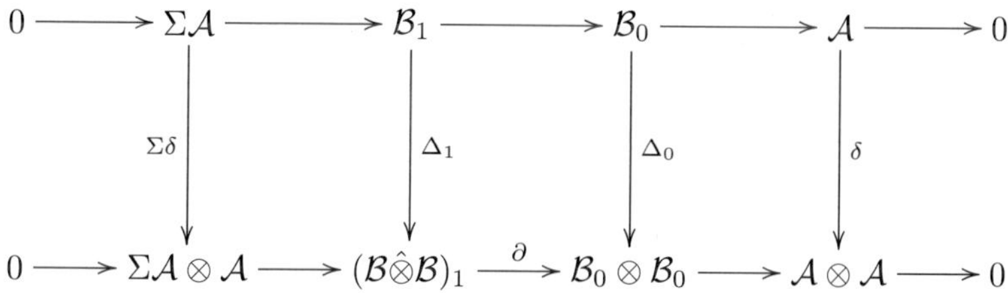

Here Δ_1 is a map of right $\mathcal{B}_0$-bimodules.

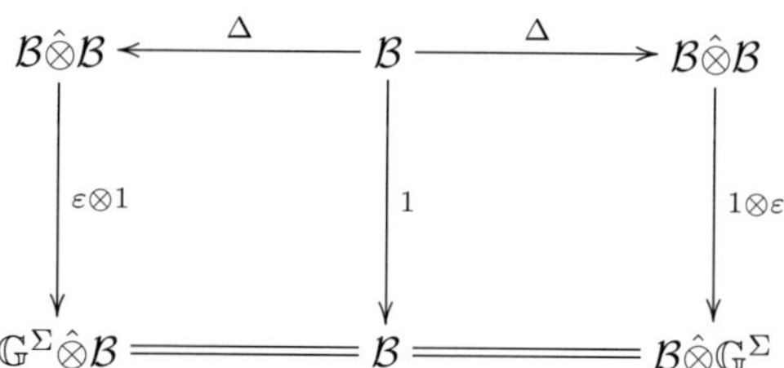

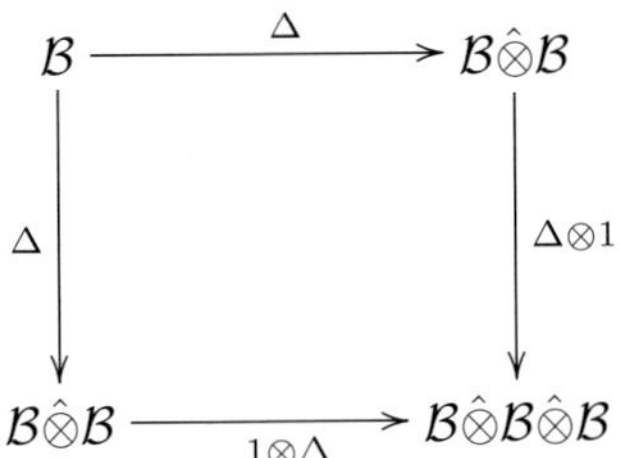

These diagrams show that $\mathcal{B}$ is a comonoid in $\mathcal{A}lg^{[p]}$.

Existence theorem. *The pair algebra $\mathcal{B}$ of secondary cohomology operations has topologically defined augmentation $\varepsilon : \mathcal{B} \longrightarrow \mathbb{G}^{\Sigma}$ and diagonal $\Delta : \mathcal{B} \longrightarrow \mathcal{B}\hat{\otimes}\mathcal{B}$ such that $\mathcal{B}$ is a secondary Hopf algebra.*

Given a secondary Hopf algebra $\mathcal{B}$ let

$$\begin{aligned} R_{\mathcal{B}} &= \mathrm{kernel}(q : \mathcal{B}_0 \longrightarrow \mathcal{A}) \\ &= \mathrm{image}(\partial : \mathcal{B}_1 \longrightarrow \mathcal{B}_0) \end{aligned}$$

be the *ideal of relations* in $\mathcal{B}_0$ with $p \in R_{\mathcal{B}}$ in degree 0. We associate with $\mathcal{B}$ a *symmetry operator*

$$S : R_{\mathcal{B}} \longrightarrow \tilde{\mathcal{A}} \otimes \tilde{\mathcal{A}}$$

defined by the formula in $(\mathcal{B}\hat{\otimes}\mathcal{B})_1$,

$$T\Delta_1(x) = \Delta_1(x) + \Sigma S(\xi)$$

for $x \in \mathcal{B}_1$ with $\partial x = \xi \in R_{\mathcal{B}}$. Moreover we define a *left action operator*

$$L : \mathcal{A} \otimes R_{\mathcal{B}} \longrightarrow \tilde{\mathcal{A}} \otimes \tilde{\mathcal{A}}$$

by the formula in $(\mathcal{B}\hat{\otimes}\mathcal{B})_1$,

$$\Delta_1(\alpha \cdot x) = \alpha \cdot \Delta_1(x) + \Sigma L(\alpha \otimes \xi)$$

with $\alpha \in \mathcal{B}_0$, $x \in \mathcal{B}_1$ and $\partial x = \xi$. Here $\tilde{\mathcal{A}} = \mathrm{kernel}(\varepsilon : \mathcal{A} \to \mathbb{F})$ is the augmentation ideal of $\mathcal{A}$. In case $\mathcal{B}$ is the algebra of secondary cohomology operations we compute S and L explicitly, see Chapter 14. If p is odd we have $S = 0$ and $L = 0$. But if $p = 2$ we get non-trivial S and L.

An *isomorphism* between secondary Hopf algebras $\mathcal{B}$, $\mathcal{B}'$ is an isomorphism $\mathcal{B}_1 \cong \mathcal{B}'_1$ which is compatible with all the structure described above.

Uniqueness theorem. *Up to isomorphism there is a unique secondary Hopf algebra $\mathcal{B}$ associated to $\mathcal{A}$, the derivation $\Gamma[p] = \kappa$, the symmetry operator S, and the left action operator L.*

The existence theorem and the uniqueness theorem are the main results of this book. Based on the uniqueness theorem we describe an algorithm for the computation of $\mathcal{B}$. In low degrees $\mathcal{B}$ is completely determined by the tables at the end of this book. The author is very grateful to Mamuka Jibladze for implementing the algorithm on a computer. His computer calculations are a wonderful manifestation of the correctness of the new elaborate theory in this book. Also the result of Adams [A] in degree 16 is an example of such calculations. Moreover a table of triple Massey products in the Steenrod algebra is obtained this way.

Part C. Secondary cohomology

Cohomology of a space X can be derived from the singular cochain algebra $C^*(X,\mathbb{F})$ or from Eilenberg-MacLane spaces $Z^n = K(\mathbb{F},n)$ by

$$\tilde{H}^n(X,\mathbb{F}) = \tilde{H}^n C^*(X,\mathbb{F}) = [X, Z^n].$$

In a similar way we derive in this book "secondary cohomology" $\mathcal{H}^*[X]$ either from $C^*(X,\mathbb{F})$ or from Z^n. The secondary cohomology $\mathcal{H}^*[X]$ has a rich additional algebraic structure. In particular, $\mathcal{H}^*[X]$ is a secondary permutation algebra over the secondary Hopf algebra $\mathcal{B}$, generalizing the well known fact that $H^*(X,\mathbb{F})$ is an algebra over the Hopf algebra $\mathcal{A}$, see (A3).

(C1) Secondary cohomology as a $\mathcal{B}$-module

We first introduce the concept of secondary cohomology of a chain complex. Let C be an augmented cochain complex with the differential

$$d : C \longrightarrow C \quad \text{of degree} \quad +1$$

and augmentation $\varepsilon : C \longrightarrow \mathbb{F}$. Let

$$\tilde{C} = \text{kernel}(\varepsilon).$$

The *cohomology* of C is the graded module

$$H^*(C) = \text{kernel}(d)/\,\text{image}(d).$$

The *secondary cohomology* of C is the graded pair module $\mathcal{H}^*(C)$ defined by

$$\begin{array}{ccc} \Sigma(\tilde{C}/\,\text{image } d) & \xrightarrow{\ \partial\ } & \text{kernel}(d) \\ \| & & \| \\ \mathcal{H}^*(C)_1 & & \mathcal{H}^*(C)_0 \end{array}$$

Here ∂ is induced by d. Hence we obtain the exact sequence of graded modules:

$$0 \longrightarrow \Sigma\tilde{H}^*(C) \longrightarrow \mathcal{H}^*(C)_1 \xrightarrow{\ \partial\ } \mathcal{H}^*(C)_0 \longrightarrow H^*(C) \longrightarrow 0$$

Lemma. *If (C,d) is an augmented differential algebra, then $\mathcal{H}^*(C)$ is a pair algebra and all maps in the exact sequence are $\mathcal{H}^*(C)_0$-bimodule maps. Here $\mathcal{H}^*(C)_0$ is the algebra of cocycles in C.*

The lemma shows that pair algebras are just the secondary truncations of differential algebras. Now let X be a pointed path connected space. A *topological cocycle of degree* n in X is a pointed map

$$\xi : X \longrightarrow Z^n, \quad n \geq 1.$$

Let $\mathcal{H}^*(X)_0$ be the graded module which is $\mathbb{F}$ in degree 0 and which consists of topological cocycles in degree $n \geq 1$. Here addition of cocycles is induced by the topological vector space structure of Z^n. Moreover let $\mathcal{H}^*(X)_1$ be the graded module consisting of pairs (a, ξ) where $a : \xi \Rightarrow 0$ is a track and ξ is a topological cocycle and let

$$\partial : \mathcal{H}^*(X)_1 \longrightarrow \mathcal{H}^*(X)_0$$

be the boundary map which carries (a, ξ) to $\partial(a) = \xi$. Then $\mathcal{H}^*(X)$ is a pair module termed the *secondary cohomology of* X. One has the exact sequence:

$$0 \longrightarrow \Sigma\tilde{H}^*(X) \longrightarrow \mathcal{H}^*(X)_1 \longrightarrow \mathcal{H}^*(X)_0 \longrightarrow H^*(X) \longrightarrow 0$$

Here $\tilde{H}^*(X)$ is reduced cohomology since tracks are defined by pointed homotopies. The pair algebra structure of $\mathcal{H}^*(X)$ is induced by the multiplication maps

$$\mu : Z^n \times Z^m \longrightarrow Z^{n+m}$$

which are associative. One can check that there is a weak equivalence of pair algebras

$$\mathcal{H}^*(X) \sim \mathcal{H}^*(C)$$

where $C = C^*(X, \mathbb{F})$ is the augmented differential algebra of cochains in the pointed space X.

A stable map $\alpha : Z^n \longrightarrow Z^{n+k}$, as defined in part (A), acts on $\mathcal{H}^*(X)$ by composition of maps. But this action of $[\![\mathcal{A}]\!]_0$ on $\mathcal{H}^*(X)$ is not bilinear. For this reason we have to introduce the *strictified secondary cohomology* $\mathcal{H}^*[X]$ which is a pair module obtained by the following pull back diagram with exact rows.

$$\begin{array}{ccccccccc}
0 \longrightarrow & \Sigma\tilde{H}^*(X) & \longrightarrow & \mathcal{H}^*(X)_1 & \longrightarrow & \mathcal{H}^*(X)_0 & \longrightarrow & H^*(X) & \longrightarrow 0 \\
& \| & & \uparrow & \text{pull} & \uparrow s & & \| & \\
0 \longrightarrow & \Sigma\tilde{H}^*(X) & \longrightarrow & \mathcal{H}[X]_1 & \longrightarrow & \mathcal{H}^*[X]_0 & \longrightarrow & H^*(X) & \longrightarrow 0
\end{array}$$

Here $\mathcal{H}^*[X]_0$ is a suitable free algebraic object generated by $\mathcal{H}^*(X)_0$ which, in particular, is a $\mathcal{B}_0$-module. The function s is similarly defined as the function s in the pull back diagram (A11) defining $\mathcal{B}$. Generalizing the well-known fact that the cohomology $H^*(X)$ is an $\mathcal{A}$-module we show

Theorem. *The strict secondary cohomology $\mathcal{H}^*[X]$ is a $\mathcal{B}$-module inducing the $\mathcal{A}$-module structure of $H^*(X)$.*

Compare the definition of $\mathcal{B}$-modules in (B2). The action of $\mathcal{B}$ is defined similarly as in (A16).

(C2) Algebras over Hopf algebras

The algebra structure of cohomology $H^*(X)$ and the $\mathcal{A}$-module structure of $H^*(X)$ are connected by the *Cartan formula* showing that $H^*(X)$ is an algebra over the Hopf algebra $\mathcal{A}$. More explicitly let μ_2 be defined by the following commutative diagram with $H = H^*(X)$.

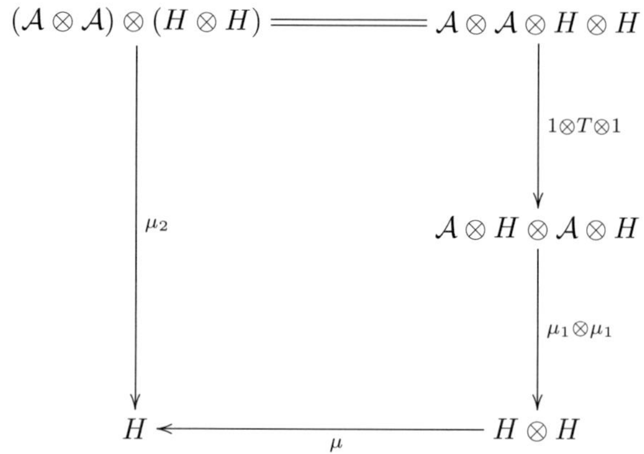

Here μ_1 is the action of $\mathcal{A}$ on H and μ is the multiplication in the algebra H. Then the following diagram commutes where δ is the diagonal of $\mathcal{A}$. This diagram is termed the *Cartan diagram*.

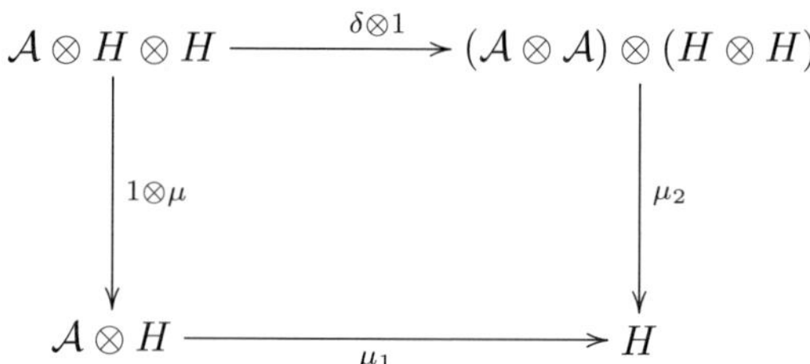

An algebra H, which is also an $\mathcal{A}$-module, is termed an *algebra over the Hopf algebra* $\mathcal{A}$ if the Cartan diagram commutes.

In the next section we describe the corresponding property of the strictified secondary cohomology $\mathcal{H}^*[X]$.

(C3) Pair algebras over secondary Hopf algebras

We have seen that the secondary cohomology $\mathcal{H}^*[X]$ is a pair algebra and a $\mathcal{B}$-module, see (C1). Moreover $\mathcal{B}$ is a secondary Hopf algebra. Now the pair algebra structure of $H = \mathcal{H}^*[X]$ and the secondary diagonal of $\mathcal{B}$ are connected by generalizing the Cartan diagram (C2). There is a unique map μ_2 between pair modules for which the following diagram commutes

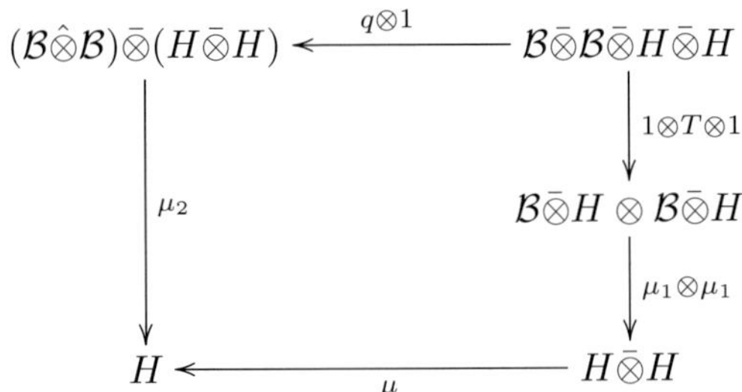

Here μ_1 is the action of $\mathcal{B}$ on H and μ is the multiplication of the pair algebra H and q is the quotient map in (B5). Generalizing the Cartan diagram (C2) we get the following result.

Theorem. *The strictified secondary cohomology $H = \mathcal{H}^*[X]$ is endowed with a* Cartan homotopy C *as in the following diagram.*

$$\begin{array}{ccc} \mathcal{B}\bar{\otimes}H\bar{\otimes}H & \xrightarrow{\Delta\otimes 1} & (\mathcal{B}\hat{\otimes}\mathcal{B})\bar{\otimes}(H\bar{\otimes}H) \\ {\scriptstyle 1\otimes\mu}\downarrow & \overset{C}{\Longrightarrow} & \downarrow{\scriptstyle \mu_2} \\ \mathcal{B}\bar{\otimes}H & \xrightarrow[\mu_1]{} & H \end{array}$$

Here C is a $\mathbb{G}$-linear map natural in X,

$$C : \mathcal{B}_0 \otimes H_0 \otimes H_0 \longrightarrow H_1,$$

satisfying the following properties. First $C : \mu_1(1\otimes\mu) \Rightarrow \mu_2(\Delta\otimes 1)$ is a homotopy between pair maps, see (B1). Let $\alpha, \beta \in \mathcal{B}_0$ and $x, y, z \in H_0$. Then the associativity formula

$$\begin{aligned} &C(\alpha\otimes(x\cdot y, z)) + (C\otimes 1)(\Delta(\alpha)\otimes(x, y, z)) \\ &= C(\alpha\otimes(x, y\cdot z)) + (1\otimes C)(\Delta(\alpha)\otimes(x, y, z)) \end{aligned}$$

is satisfied where $x\cdot y,\ \ y\cdot z$ are given by the multiplication in H. The maps

$$C\otimes 1,\ 1\otimes C : \mathcal{B}_0\otimes\mathcal{B}_0\otimes H_0\otimes H_0\otimes H_0 \longrightarrow H_1$$

are defined by

$$\begin{aligned} (C\otimes 1)(\alpha\otimes\beta\otimes(x, y, z)) &= (-1)^{p|\beta|(|x|+|y|)} C(\alpha\otimes(x, y))\cdot(\beta\cdot z), \\ (1\otimes C)(\alpha\otimes\beta\otimes(x, y, z)) &= (-1)^{p|\beta||x|}(\alpha\cdot x)\cdot C(\beta\otimes(y, z)) \end{aligned}$$

with $(x, y) = x\otimes y$ and $(x, y, z) = x\otimes y\otimes z$. On the right-hand side we use the action of $\mathcal{B}_0$ on H_0 and the action of H_0 on H_1.

Moreover the Cartan homotopy C satisfies further equations with respect to the symmetry operator S and the left action operator L, see Chapter 14.

A pair algebra H which is a $\mathcal{B}$-module together with a Cartan homotopy C satisfying these properties is termed a *pair algebra over the secondary Hopf algebra* $\mathcal{B}$. This is the secondary analogue of an algebra over the Steenrod algebra in (C2).

We know that the cohomology algebra $H^*(X)$ is a *commutative* graded algebra. The corresponding result for secondary cohomology $\mathcal{H}^*[X]$ shows that $\mathcal{H}^*[X]$ is a secondary permutation algebra, see Section (6.3) and Section (14.2). Only by use of this additional structure are we able to compute the operators S and L above.

(C4) Instability

The cohomology $H^*(X)$ is an *unstable algebra* over the Steenrod algebra $\mathcal{A}$ in the following sense:

If the prime p is even and $x \in H^*(X)$ then for $i \geq 1$,

$$(*)\qquad \begin{aligned} Sq^i x &= 0 \ \text{ for } \ i > |x|, \ \text{ and} \\ Sq^i x &= x^2 \ \text{ for } \ i = |x|. \end{aligned}$$

If the prime p is odd then for $i \geq 1$,

$$(**)\qquad \begin{aligned} P^i x &= 0 \ \text{ for } \ 2i > |x|, \ \text{ and} \\ P^i_\beta x &= 0 \ \text{ for } \ 2i+1 > |x|, \ \text{ and} \\ P^i x &= x^p \ \text{ for } \ 2i = |x|. \end{aligned}$$

An algebra H over the Hopf algebra $\mathcal{A}$ is called unstable if these conditions are satisfied and if H is a commutative algebra.

Let $\mathcal{M} = \mathrm{Mon}(E_\mathcal{A})$ be the free monoid generated by $E_\mathcal{A}$. We define the *excess function*

$$e : \mathcal{M} \longrightarrow \mathbb{Z}.$$

For a monomial $a = e_1 \dots e_r \in \mathcal{M}$ with $e_1, \dots, e_r \in E_\mathcal{A}$ put, for $p = 2$,

$$e(a) = Max_j(|e_j| - |e_{j+1} \dots e_r|).$$

Moreover put for p odd

$$e(a) = \mathrm{Max}_j \begin{cases} 2|e_j| - |e_{j+1} \dots e_r| & \text{for } e_j \in \{P^1, P^2, \dots\}, \\ 2|e_j| + 1 - |e_{j+1} \dots e_r| & \text{for } e_j \in \{P^1_\beta, P^2_\beta, \dots\}, \\ 1 & \text{for } e_j = \beta. \end{cases}$$

One readily checks for $x \in H^*(X)$ the equation:

$$\alpha x = 0 \ \text{ for } \ e(\alpha) > |x|.$$

We now consider the *secondary instability* condition of the secondary cohomology $\mathcal{H}^*[X]$ corresponding to the primary instability above. For this let

$$\mathcal{E}(X) \subset \mathcal{M} \times \mathcal{H}^*[X]_0$$

be the *excess subset* given by all pairs (α, x) with $e(\alpha) < |x|$, $\alpha \in \mathcal{M}$, $x \in \mathcal{H}^*[X]_0$. Then there are *unstable structure maps*

$$\begin{aligned} v : \mathcal{E}(X) &\longrightarrow \mathcal{H}^*[X]_1, \\ u : \mathcal{H}^*[X]_0 &\longrightarrow \mathcal{H}^*[X]_1 \qquad \text{for } p = 2, \\ u : \mathcal{H}^*[X]_0^{\text{even}} &\longrightarrow \mathcal{H}^*[X]_1 \qquad \text{for } p \text{ odd}, \end{aligned}$$

which are natural in X with $|v(\alpha, x)| = |\alpha| + |x|$ and $|ux| = p|x|$ and

$$\begin{aligned} \partial v(\alpha, x) &= \alpha \cdot x, \\ \partial u(x) &= \alpha_p \cdot x - x^p \end{aligned}$$

with $\alpha_p = Sq^{|x|}$ for $p = 2$ and $\alpha_p = P^{|x|/2}$ for p odd. Moreover the properties in (13.3.3) and (13.3.4) in the book are satisfied. The existence of v and u corresponds to the instability equations $(*)$ and $(**)$ above.

Part D. Algebraic models of spaces

Sullivan showed that the cochain algebra $C^*(X, \mathbb{Q})$ of a simply connected space X determines the rational homotopy type of X, [S]. In rational homotopy theory, as developed in Félix-Halperin-Thomas [HFT], one has good knowledge of how topological constructions are transformed to algebraic constructions by the cochain functor. Such constructions can be transported to the secondary cohomology functor $\mathcal{H}^*$ as well. We consider the functor $\mathcal{H}^*$ (which carries X to a secondary permutation algebra over $\mathcal{B}$) as an intermediate step between the following models:

- the cohomology $H^*(X, \mathbb{F})$ as an algebra over $\mathcal{A}$,
- the cochain algebra $C^*(X, \bar{\mathbb{F}})$ as an algebra over an E_∞-operad as studied by Mandell [Ma].

Here $C^*(X, \bar{\mathbb{F}})$ determines the p-adic homotopy type of X yielding the p-adic analogue of Sullivan's result. The great advantage of secondary cohomology $\mathcal{H}^*[X]$ is the fact, that the $\mathcal{B}$-module structure of $\mathcal{H}^*[X]$ has a direct connection to the $\mathcal{A}$-module structure of the cohomology $H^*(X, \mathbb{F})$. Representing the action of the Steenrod algebra in an algebra over the E_∞-operad is a lot more involved, [May]. It would be interesting to see, how the action of $\mathcal{B}$ on $\mathcal{H}^*[X]$ can be deduced from the structure of $C^*(X, \bar{\mathbb{F}})$ as an algebra over the E_∞-operad. This, in fact, requires the secondary enrichment of May's "General algebraic approach to Steenrod operations", [May].

Concerning the theory of secondary cohomology operations in the literature we refer the reader to the recent book of John Harper [Ha]. Our approach is new and mainly concerned with the algebraic nature of the theory. All the results in the literature on secondary cohomology operations can be considered as properties of the structure of $\mathcal{H}^*[X]$ as an algebra over $\mathcal{B}$.

Part I

Secondary Cohomology and Track Calculus

In the first part of this book we study classical primary and secondary cohomology operations and we show that a "global theory of secondary operations" is obtained by the track theory of Eilenberg-MacLane spaces. As pointed out by Karoubi [Ka2] Steenrod operations can be defined by "power maps"

$$\gamma : Z^1 \times Z^q \longrightarrow Z^{pq}$$

for $q \geq 1$ and $Z^n = K(\mathbb{F}, n)$. Karoubi uses only the homotopy class of γ.

Since we are interested in the secondary structure we study the properties of the map γ and not of the homotopy class γ. We observe that power maps γ are well defined up to a canonical track. The power maps are part of the homotopy commutative diagrams related to

- the linearity of the Steenrod operations,
- the Cartan formula, and
- the Adem relation respectively.

We show that such diagrams not only admit a homotopy but, in fact, are homotopy commutative by a well-defined track which we call the *linearity track*, the *Cartan track* and the *Adem track* respectively. These tracks can be considered as *generators* of the secondary Steenrod algebra in the same way as power maps generate the classical Steenrod algebra.

We shall work in the category **Top** of compactly generated Hausdorff spaces and continuous maps, compare the book of Gray [G]. Let **Top*** be the corresponding category of pointed spaces and pointed maps.

Chapter 1

Primary Cohomology Operations

In this chapter we show that for the prime 2 the category $\mathcal{K}^0$ of connected algebras over the Steenrod algebra $\mathcal{A}$ is isomorphic to the category **UEPow** of unitary extended power algebras. A similar result holds for odd primes. Power algebras are algebras together with power operations γ_x. The cohomology algebra $H^*(X)$ of a space is naturally a power algebra with power operations induced by the power maps γ in Chapter 7. Our approach of studying secondary cohomology operations in Part II is based on properties of these power maps.

1.1 Unstable algebras over the Steenrod algebra

We recall from the book of Steenrod-Epstein [SE] or Schwartz [Sch] the following facts and notations on the Steenrod algebra.

Let p be a prime number and let $\mathbb{F} = \mathbb{Z}/p$ be the field of p elements. The mod p *Steenrod algebra* $\mathcal{A} = \mathcal{A}_p$ is the quotient of the "free" associative unital graded R-algebra generated by the elements

$$\begin{array}{lll} Sq^i & \text{of degree } i,\ i > 0, & \text{if } p = 2, \\ \beta & \text{of degree 1 subject to } \beta^2 = 0 & \text{and} \\ P^i & \text{of degree } 2i(p-1),\ i > 0, & \text{if } p > 2; \end{array}$$

by the ideal generated by the elements known as *Adem relations*

$$Sq^i Sq^j - \sum_{k=0}^{[i/2]} \binom{j-k-1}{i-2k} Sq^{i+j-k} Sq^k$$

for all $i, j > 0$ such that $i < 2j$ if $p = 2$;

$$P^i P^j - \sum_{t=0}^{[i/p]} (-1)^{i+t} \binom{(p-1)(j-t)-1}{i-pt} P^{i+j-t} P^t$$

for all $i, j > 0$ such that $i < pj$ and

$$\begin{aligned} P^i \beta P^j \quad & - \quad \sum_{t=0}^{[i/p]} (-1)^{i+t} \binom{(p-1)(j-t)}{i-pt} \beta P^{i+j-t} P^t \\ & - \sum_{t=0}^{[(i-1)/p]} (-1)^{i+t-1} \binom{(p-1)(j-t)-1}{i-pt-1} P^{i+j-t} \beta P^t \end{aligned}$$

for all $i, j > 0$ such that $i \leq pj$ if $p > 2$.

In these formulas Sq^0 (resp. P^0) for $p = 2$ (resp. $p > 2$) is understood to be the unit.

The mod p cohomology $H^*(X; \mathbb{Z}/p)$ of a space X will be denoted by H^*X and the reduced mod p cohomology will be denoted by $\tilde{H}^*X$.

1.1.1 Theorem (Steenrod, Adem). *For any space X, H^*X is in a natural way a graded $\mathcal{A}$-module.*

Classically, β (Sq^1 if $p = 2$) acts as the Bockstein homomorphism associated to the sequence $0 \to \mathbb{Z}/p \to \mathbb{Z}/p^2 \to \mathbb{Z}/p \to 0$. N.E. Steenrod constructed the operations Sq^i and the operation P^i, and J. Adem showed that the Adem relations above act trivially on the mod p cohomology of any space. We shall prove these facts by use of power algebras in the next section.

The next theorem is a consequence of the computation by H. Cartan and J.-P. Serre (see [C][S1]) of the cohomology of the Eilenberg-MacLane spaces.

1.1.2 Theorem. *The Steenrod algebra is the algebra of all natural stable transformations of mod p cohomology.*

Here "stable" means "commuting with suspension".

1.1.3 Proposition. *The operations Sq^{2^h}, $h \geq 0$, for $p = 2$ constitute a system of multiplicative generators for $\mathcal{A}$; so do the operations β and P^{p^h} for $p > 2$.*

In fact, this system of generators is a minimal one.

We now describe an additive basis for the Steenrod algebra.

1.1.4 Definition. Let p be 2. For a sequence of integers $I = (i_1, \ldots, i_n)$, let Sq^I denote $Sq^{i_1} \ldots Sq^{i_n}$. The sequence I is said to be *admissible* if $i_h \geq 2i_{h+1}$ for all $h \geq 1$, ($i_{n+1} = 0$).

Let $p > 2$. For a sequence of integers $I = (\epsilon_0, i_1, \epsilon_1, \ldots, i_n, \epsilon_n)$, where the ϵ_k are 0 or 1, let P^I denote $\beta^{\epsilon_0} P^{i_1} \beta^{\epsilon_1} \ldots \beta^{\epsilon_n}$. The sequence I is said to be *admissible* if $i_h \geq p i_{h+1} + \epsilon_h$ for all $h \geq 1$, ($i_{n+1} = 0$).

The operations Sq^I (resp. P^I) with I admissible are called admissible monomials.

1.1.5 Proposition. *The admissible monomials Sq^I (resp. P^I) form a vector space basis for $\mathcal{A}$.*

Proof. It is a consequence of the Adem relations that the operations Sq^I (resp. the operations P^I), I admissible, span the graded vector space $\mathcal{A}$. In fact, let $I = (i_1, \ldots, i_n)$ (resp. $I = (\epsilon_0, i_1, \epsilon_1, \ldots, i_n, \epsilon_n)$) be an admissible sequence. Its moment is defined to be $i_1 + 2i_2 + \cdots + ni_n$ (resp. $i_1 + \epsilon_1 + 2(i_2 + \epsilon_2) + \cdots$). If I is not admissible there exists h, $1 \leq h \leq n-1$ such that $i_h < 2i_{h+1}$. Using the Adem relations one gets:

$$Sq^I = \sum_{0}^{[i_h/2]} \epsilon_t Sq^{I'} Sq^{i_h + i_{h+1} - t} Sq^t Sq^{I''},$$

where $\epsilon_t \in \mathbb{F}_2$, $0 \leq t \leq [i_h/2]$, $I' = (i_1, \ldots, i_{h-1})$ and $I'' = (i_{h+2}, \ldots, i_n)$. The moment of any sequence occurring on the right (i.e. $I', i_h + i_{h+1} - t, y, I''$), $0 \leq t \leq [i_h/2]$) is strictly lower than the moment of I. By induction on the moment we see that admissible monomials generate $\mathcal{A}$ as a graded vector space. The case of an odd prime is proved in the same way.

The admissible monomials are linearly independent. To see this one looks at their action on $H^*(B(\mathbb{Z}/p)^{\oplus k}) \cong (H^*B\mathbb{Z}/p)^{\otimes k}$. Here BV is the classifying space of the abelian group V. □

The mod p cohomology of a space X has, as $\mathcal{A}$-module, a certain property called *instability*:

- if $x \in H^*X$ and $i >\mid x \mid$, then $Sq^i x = 0$ for $p = 2$;
- if $x \in H^*X$ and $e + 2i >\mid x \mid$, $e = 0, 1$, then $\beta^e P^i x = 0$, for $p > 2$.

Here $\mid x \mid$ denotes the degree of x.

1.1.6 Definition. An $\mathcal{A}$-module M is *unstable* if it satisfies the preceding property. In particular, this implies that an unstable $\mathcal{A}$-module M is trivial in negative degrees (recall that one identifies Sq^0, resp. P^0, with the identity operator). Let $\mathcal{U}$ be the category of unstable $\mathcal{A}$-modules. This is an abelian category with enough projective objects. We obtain free objects in $\mathcal{U}$ as follows. The *excess* of an admissible sequence I is defined to be

$$(1)\quad \begin{cases} (i_1 - 2i_2) + (i_2 - 2i_3) + \cdots + (i_{n-1} - 2i_n) + i_n & \text{if} \quad p = 2, \\ 2(i_1 - pi_2) + 2(i_2 - pi_3) + \cdots + 2i_n + \epsilon_0 - \epsilon_1 - \cdots - \epsilon_n & \text{if} \quad p > 2, \end{cases}$$

and it is denoted by $e(I)$. Note that, if $p > 2$, an admissible sequence I such that $e(I) \leq n$ contains at most n entries $\epsilon_i = 1$ because $e(I) = \epsilon_0 + \cdots + \epsilon_n + 2(i_1 - pi_2 - \epsilon_1) + \cdots + 2(i_n - \epsilon_n)$.
Let

$$(2)\quad B(n) \subset \mathcal{A}$$

be the vector subspace generated by admissible monomials of excess $> n$. An $\mathcal{A}$-module X satisfies $B(n) \cdot X^n = 0$ for all n if and only if X is unstable. We have the *suspension functor*

$$\Sigma : \mathcal{U} \longrightarrow \mathcal{U} \tag{3}$$

defined by setting $(\Sigma M)^n = M^{n-1}$. Let $\Sigma : M^{n-1} \to (\Sigma M)^n$ be the map of degree 1 given by the identity of M^{n-1}. Then the $\mathcal{A}$-action on ΣM is defined by $\theta(\Sigma m) = (-1)^{|\theta|}\Sigma(\theta m)$ for $m \in M$, $\theta \in \mathcal{A}$. We obtain the $\mathcal{A}$-module

$$F(n) \;=\; \Sigma^n(\mathcal{A}/B(n)) \tag{4}$$

which is the *free unstable module on one generator* $[n]$ in degree n. Here $[n] = \Sigma^n\{1\} \in F(n)$ is defined by the unit $1 \in \mathcal{A}$. A basis of $\mathcal{A}/B(n)$ is given by admissible monomials of excess $\leq n$. Free objects in $\mathcal{U}$ are direct sums of modules $F(n)$, $n \geq 0$.

The mod p cohomology of a space X is also, in a natural way, a graded commutative, unital $\mathbb{F}$-algebra which is *augmented* by $H^*(X) \to \mathbb{F}$ if X is a pointed space $* \in X$. Let $\tilde{H}^*(X)$ be the kernel of the augmentation $H^*(X) \to \mathbb{F}$. The algebra structure is related to the $\mathcal{A}$-module structure by the *Cartan formula*

$$(\mathcal{K}1)\begin{cases} Sq^i(xy) &= \sum_{k+l=i} Sq^k x Sq^l y, \\ P^i(xy) &= \sum_{k+l=i} P^k x P^l y, \\ \beta(xy) &= (\beta x)y + (-1)^{|x|} x\beta y, \end{cases}$$

where $x, y \in H^*(X)$ and by the following formulas:

$$(\mathcal{K}2)\begin{cases} Sq^{|x|}x &= x^2 \quad \text{for any } x \text{ in } H^*X \text{ if } p = 2, \\ P^{|x|/2}x &= x^p \quad \text{for any } x \text{ of even degree in } H^*X \text{ if } p > 2. \end{cases}$$

This leads to

1.1.7 Definition. An *unstable algebra K over the Steenrod algebra $\mathcal{A}$* or an *unstable $\mathcal{A}$-algebra* K is an unstable $\mathcal{A}$-module provided with maps $\mu : K \otimes K \to K$ and $\eta : \mathbb{F} \to K$ which determine a commutative, unital, $\mathbb{F}$-algebra structure on K and such that properties $(\mathcal{K}1)$ and $(\mathcal{K}2)$ hold. We shall denote by $\mathcal{K} = \mathcal{K}_p$ the category of unstable augmented $\mathcal{A}$-algebras, morphisms being $\mathcal{A}$-linear algebra maps of degree zero compatible with the augmentation.

Hence the cohomology H^* is a contravariant functor

$$H^* : \mathbf{Top}^*/\simeq \longrightarrow \mathcal{K}$$

from the homotopy category of pointed topological spaces to the category $\mathcal{K}$ of augmented unstable algebras over $\mathcal{A}$. For pointed spaces X, Y let $[X, Y]$ be the set of homotopy classes of pointed maps $X \to Y$. This is the set of morphisms $X \to Y$ in the homotopy category $\mathbf{Top}^*/_\simeq$.

The axiom ($\mathcal{K}1$) can be reformulated as follows. There is an algebra map δ (diagonal) from $\mathcal{A}$ to $\mathcal{A} \otimes \mathcal{A}$ such that

$$\delta(Sq^i) = \sum_{k+l=i} Sq^k \otimes Sq^l \text{ if } p = 2,$$

$$\begin{cases} \delta(\beta) = \beta \otimes 1 + 1 \otimes \beta, \\ \delta(P^i) = \sum_{k+l=i} P^k \otimes P^l \text{ if } p > 2. \end{cases}$$

This map determines a co-commutative Hopf algebra structure on $\mathcal{A}$, and it can be used to provide the tensor product $M \otimes N$ of two $\mathcal{A}$-modules M and N with an $\mathcal{A} \otimes \mathcal{A}$-module structure, this structure being determined by the formula

$$(\theta \otimes \theta')(m \otimes n) = (-1)^{|\theta'||m|} \theta m \otimes \theta' n$$

for all $\theta, \theta' \in \mathcal{A}$, $m \in M$ and $n \in N$. Then $M \otimes N$ is an $\mathcal{A}$-module by restriction via δ. Axiom ($\mathcal{K}1$) is equivalent to the $\mathcal{A}$-linearity of the map $\mu : K \otimes K \to K$ in (1.1.7). The structure of the dual of $\mathcal{A}$, as a commutative Hopf algebra, was determined by Milnor [Mn].

As an example of an unstable algebra, recall the structure of the mod p cohomology of the classifying space $B(\mathbb{Z}/p)$.

For $p = 2$ the mod 2 cohomology $H^*B(\mathbb{Z}/2)$ is the polynomial algebra $\mathbb{F}[x]$ on one generator x of degree 1. The action of $\mathcal{A}$ is completely determined by axioms ($\mathcal{K}1$) and ($\mathcal{K}2$) and one finds that

$$Sq^i x^n = \binom{n}{i} x^{n+i}. \tag{1.1.8}$$

If $p > 2$, $H^*B(\mathbb{Z}/p)$ is the tensor product $\Lambda(x) \otimes \mathbb{F}[\beta x]$ of an exterior algebra on one generator x of degree 1 and a polynomial algebra on one generator βx of degree 2. The action of $\mathcal{A}$ is determined by axioms ($\mathcal{K}1$) and ($\mathcal{K}2$) and the fact that β is the Bockstein homomorphism. We obtain

$$\beta(x) = \beta x \ , \quad P^i(\beta x)^n = \binom{n}{i} (\beta x)^{n+i(p-1)}. \tag{1.1.9}$$

1.1.10 Definition. For an unstable $\mathcal{A}$-module X let $U(X)$ be the *free unstable $\mathcal{A}$-algebra in $\mathcal{K}$* constructed as follows. Let

$$T(X) = \bigoplus_{i \geq 0} X^{\otimes i} \tag{1}$$

be the tensor algebra generated by X. Then $T(X)$ is an $\mathcal{A}$-module since the i-fold tensor product $X^{\otimes i} = X \otimes \cdots \otimes X$ is an $\mathcal{A}$-module by the Hopf-algebra structure of $\mathcal{A}$. Let

$$D(X) \subset T(X) \tag{2}$$

be the two-sided ideal generated by the elements

$$(3)\qquad \begin{cases} x \otimes y - (-1)^{|x||y|} y \otimes x, & \\ Sq^{|x|}x - x^{\otimes 2} & \text{for } p = 2, \\ P^{|x|}x - x^{\otimes p} & \text{for } p \text{ odd and } |\, x \,| \text{ even} \end{cases}$$

with $x, y \in X$. Then the quotient algebra

$$(4)\qquad U(X) = T(X)/D(X)$$

is the *free unstable* $\mathcal{A}$-algebra in $\mathcal{K}$ generated by X. We call $U(X)$ *completely free* if X is a free object in $\mathcal{U}$. In particular let

$$(5)\qquad H(n) = U(F(n)) = U(\Sigma^n(\mathcal{A}/B(n))$$

be the completely free object in $\mathcal{K}$ generated by one element $[n]$ in degree n.

Due to a result of Serre [S1] and Cartan [C], see Steenrod-Epstein II.§5[SE], we know:

1.1.11 Theorem. *For an Eilenberg-MacLane space $Z^n = K(\mathbb{Z}/p, n)$, $n \geq 1$, one gets an isomorphism in $\mathcal{K}$,*

$$H^*(Z^n) = H(n).$$

For $n = 1$ we have $Z^1 = B(\mathbb{Z}/p)$ so that in this case $H^*B(\mathbb{Z}/p) = H(1)$ can be described by (1.1.8) and (1.1.9).

1.1.12 Definition. An augmented graded algebra A is *connected* if the augmentation is an isomorphism in degree 0, that is $A^0 = \mathbb{F}$. Let

$$\mathcal{K}^0 \subset \mathcal{K}$$

be the full subcategory of connected unstable $\mathcal{A}$-algebras.

There is an obvious forgetful functor

$$\mathcal{K} \longrightarrow \mathcal{K}^0$$

which carries A to $A^{\geq 1} \oplus \mathbb{F}$. Here $A^{\geq 1} \subset A$ is the (non-unital) subalgebra of elements of degree ≥ 1. Of course $A^{\geq 1}$, in addition, is an A^0-module and Sq^j, P^j, β restricted to $A^{\geq 1}$ are A^0-linear. Moreover the multiplication in $A^{\geq 1}$ is A^0-bilinear. This leads to the following characterization of objects in $\mathcal{K}$.

1.1.13 Proposition. *Let A^0 be an augmented commutative algebra concentrated in degree 0 satisfying $a = a^p$ where a^p is the p-fold product in the algebra A. Moreover let $A^{\geq 1} \oplus \mathbb{F}$ be an object in $\mathcal{K}^0$. Then $A^{\geq 1} \oplus A^0$ is an object in $\mathcal{K}$ if and only if $A^{\geq 1}$ is an A^0-module and Sq^j, P^j, β on $A^{\geq 1}$ are A^0-linear and the multiplication in $A^{\geq 1}$ is A^0-bilinear.*

Proof. Given an object A in $\mathcal{K}$ we see that $x \in A^0$ satisfies $Sq^j x = 0$, $P^j x = 0$, $\beta x = 0$ for $j > 0$ since A is an unstable $\mathcal{A}$-module. Moreover ($\mathcal{K}1$) implies that Sq^j, P^j, β for $j > 0$ are A^0-linear. Finally ($\mathcal{K}2$) implies for $\mid x \mid = 0$ that $x = x^p$. □

The proposition shows that the category $\mathcal{K}$ can be easily described by A^0-objects in the category $\mathcal{K}^0$ where A^0 is an algebra as in (1.1.13). In the next section we use power algebras to describe a category isomorphic to $\mathcal{K}^0$. Using (1.1.13) we thus also get accordingly a category isomorphic to $\mathcal{K}$.

1.2 Power algebras

We here introduce the algebraic notion of a power algebra. Using power maps γ in Chapter 3 we show that the cohomology ring $H^*(X)$ of a path connected space X has the natural structure of a power algebra. Power algebras can be used to define the action of the Steenrod algebra on $H^*(X)$. In this section we describe precisely the algebraic connection between power algebras and unstable algebras over the Steenrod algebra. This clarifies the role of the power maps γ.

Again let $\mathbb{F} = \mathbb{Z}/p$ be the field of p elements where $p \geq 2$ is a prime. If V is a $\mathbb{F}$-vector space with basis $x_1, \ldots, x_n$ we write

(1.2.1) $$V = \mathbb{F}x_1 \oplus \cdots \oplus \mathbb{F}x_n$$

Let **Vec** be the category of $\mathbb{F}$-vector spaces and $\mathbb{F}$-linear maps. The zero-vector space is denoted by $V = 0$. Let

$$S(V) = \mathbb{F}[x_1, \ldots, x_n]$$

be the *polynomial algebra* generated by (1.2.1) and let

$$\Lambda(V) = \Lambda(x_1, \ldots, x_n)$$

be the *exterior algebra* generated by (1.2.1).

A (unital) *graded algebra* $A = \{A^n\}$ is a graded $\mathbb{F}$-vector space concentrated in degree ≥ 0 with an associative multiplication

(1.2.2) $$\mu : A^n \otimes A^m \longrightarrow A^{n+m}$$

denoted by $\mu(x, y) = x \cdot y$ and a unit $1 \in A^0$ with $1 \cdot x = x \cdot 1 = x$. The algebra A is *connected* if $A^0 = \mathbb{F}$ and A is *augmented* if an algebra map $A \to \mathbb{F}$ is given. In particular each connected algebra is augmented. Of course A^0 is a subalgebra of A and all A^n, $n \geq 1$, are A^0-bimodules. Moreover A is *commutative* if $x \cdot y = (-1)^{|x||y|} y \cdot x$. Here $|x|$ is the *degree* of $x \in A$ with $|x| = q$ iff $x \in A^q$. A *map* between graded algebras is an $\mathbb{F}$-linear map $f : A \to B$ of degree 0 with $f(x \cdot y) = (fx) \cdot (fy)$, $f(1) = 1$. Let $\mathbf{Alg}_0$ be the category of connected commutative graded algebras and such maps. The category $\mathbf{Alg}_0$ has coproducts given by the *tensor product* $A \otimes B$ of algebras. Multiplication in $A \otimes B$ is defined by $(a \otimes b) \cdot (a' \otimes b') = (-1)^{|b||a'|} a \cdot a' \otimes b \cdot b'$.

1.2.3 Definition. A *β-algebra* $A = (A, \beta_A)$ is an object A in $\mathbf{Alg}_0$ together with an $\mathbb{F}$-linear map $\beta_A : A^1 \to A^2$. A β-map $f : (A, \beta_A) \to (B, \beta_B)$ is a map in $\mathbf{Alg}_0$ compatible with β; that is $\beta_B f^1 = f^2 \beta_A$. Let β-$\mathbf{Alg}_0$ be the category of β-algebras and β-maps. If $\mathbb{F} = \mathbb{Z}/2$ then we always insist that $\beta_A : A^1 \to A^2$ is given by

(1) $$\beta_A(x) = x \cdot x \text{ for } x \in A^1.$$

Hence for $p = 2$ the category β-$\mathbf{Alg}_0$ coincides with $\mathbf{Alg}_0$.

The tensor product of algebras yields a coproduct in the category of β-algebras by defining

$$\beta_{A\otimes B} : (A \otimes B)_1 = A^1 \oplus B^1 \longrightarrow A^2 \oplus B^2 \subset (A \otimes B)_2$$

via $\beta_A + \beta_B$.

The forgetful functor $\phi : \beta$-$\mathbf{Alg}_0 \to \mathbf{Vec}$ which carries (A, β_A) to A^1 has a left adjoint

(2) $$E_\beta : \mathbf{Vec} \longrightarrow \beta\text{-}\mathbf{Alg}_0$$

which carries V to the free β-algebra $E_\beta(V)$ generated by V concentrated in degree 1. Let βV be given by V concentrated in degree 2 and let $\beta : V \to \beta V$ be defined by the identity of V. Then the free β-algebra $E_\beta(V)$ is given by

(3) $$E_\beta(V) = \begin{cases} S(V) & \text{if } \ p = 2, \\ \Lambda(V) \otimes S(\beta V) & \text{if } \ p > 2. \end{cases}$$

For a β-algebra (A, β) and $x \in A^1$ we define the element $\omega_i(x) \in A^i$, $i \geq 0$ by

(4) $$\omega_i(x) = \begin{cases} (-\beta x)^j & \text{if } \ i = 2j, \\ x \cdot (-\beta x)^j & \text{if } \ i = 2j + 1. \end{cases}$$

Hence we have $\omega_0(x) = 1$ and $\omega_1(x) = x$ and $\beta\omega_1(x) = -\omega_2(x)$. Here we follow the convention of Steenrod-Epstein in V.5.2 [SE]. Moreover we set $\omega_i(x) = 0$ for $i < 0$. If $p = 2$ we see by (1) that $\omega_i(x) = x^i$ for $i \geq 0$.

It is easy to see that the elements $\omega_i(x)$, $i \geq 1$, yield a vector space basis of the free β-algebra $E_\beta(Rx)$. Therefore each element $y \in E_\beta(Rx) \otimes A$ with $q = |y|$ can be written uniquely in the form

(5) $$y = \sum_{i\geq 0} \omega_i(x) \otimes y_i$$

with $y_i \in A^{q-i}$ termed the *coordinate* of y in degree $q - i$ where $|y| = q$.

The *cohomology* H^*X with coefficients in $\mathbb{F}$ is a commutative graded algebra which is augmented if X is pointed and connected if X is path connected. Let

$\mathbf{Top}_0^* \subset \mathbf{Top}^*$ be the full subcategory of path connected pointed spaces. Then cohomology determines a contravariant functor

$$H^* : \mathbf{Top}_0^*/\simeq \longrightarrow \beta\text{-}\mathbf{Alg}_0. \tag{1.2.4}$$

Here $H^*(X)$ is a β-algebra by use of the Bockstein homomorphisms $\beta : H^1(X) \to H^2(X)$ associated to the exact sequence $0 \to \mathbb{Z}/p \to \mathbb{Z}/p^2 \to \mathbb{Z}/p \to 0$.

A finitely generated $\mathbb{F}$-vector space V yields the Eilenberg-MacLane space $K(V,1)$ of the underlying abelian group of V. This is also the classifying space $B(V)$ of V. Let $V^\# = Hom_R(V,\mathbb{F})$ be the dual vector space. Then it is well known that one has a natural isomorphism of β-algebras

$$H^*(K(V^\#,1)) = E_\beta(V). \tag{1.2.5}$$

The isomorphism (1.2.5) holds for all primes $p \geq 2$ since we use the convention (1.2.3)(1) for $p = 2$. We show in Chapter 3 below that the cohomology algebra $H^*(X)$ of a path connected space has naturally the following structure of a "power algebra".

1.2.6 Definition. A *power algebra* (H,γ) over $\mathbb{F} = \mathbb{Z}/p$ is a β-algebra H (i.e. an algebra in β-$\mathbf{Alg}_0$) together with $\mathbb{F}$-linear maps

$$\gamma_x : H^q \to H^{pq} \text{ for } x \in H^1 \text{ and } q \geq 1.$$

The following properties (i), (ii) and (iii) hold:

$$\gamma_0(y) = y^p. \tag{i}$$

Here $0 \in H^1$ is the zero element and y^p is the pth power of $y \in H^q$ in the algebra H. Let $\bar{p} = p(p-1)/2$ be given by the prime p. Then

$$\gamma_x(y)\cdot\gamma_x(z) = (-1)^{|y||z|\bar{p}}\gamma_x(y\cdot z) \tag{ii}$$

for $x \in H^1$, $y,z \in H$ with $|y| \geq 1$ and $|z| \geq 1$ and

$$\gamma_x(\gamma_y(z)) = (-1)^{|z|\bar{p}}\gamma_y(\gamma_x(z)) \tag{iii}$$

for $x,y \in H^1$ and $z \in H$ with $|z| \geq 1$.

A map $f : (A,\gamma^A) \to (B,\gamma^B)$ between power algebras is a map $f : A \to B$ in β-$\mathbf{Alg}_0$ for which the following diagram commutes,

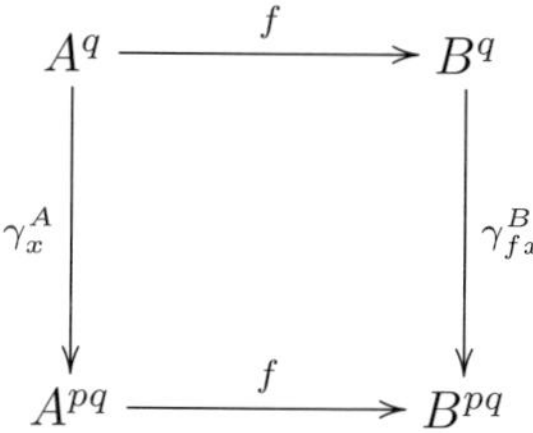

with $x \in A^1$ and $fx \in B^1$. Let $\mathbf{Pow}$ be the category of power algebras and such maps.

This book is mainly concerned with the following example.

1.2.7 Example. Let X be a path connected space. Then the cohomology $H^*(X,\mathbb{Z}/p)$ is a power algebra with power maps γ_x, $x \in H^1X$, defined as follows. In Chapter 3 we obtain for Eilenberg-MacLane spaces $Z^n = K(\mathbb{Z}/p, n)$ the power map

$$\gamma : Z^1 \times Z^q \longrightarrow Z^{pq}$$

which induces via $\tilde{H}^n(X) = [X, Z^n]$ the map

$$\gamma_x : H^q \longrightarrow H^{pq}$$

by setting $\gamma_x(y) = \gamma(x,y) : X \to Z^1 \times Z^q \to Z^{pq}$. We show in section (3.2) that $(H^*(X), \gamma)$ is a well-defined power algebra with properties as in (1.2.6). This yields a contravariant functor

$$H^* : \mathbf{Top}_0^*/\simeq \longrightarrow \mathbf{Pow}$$

enriching the structure of the functor (1.2.4). In particular we get by use of the isomorphism (1.2.5) the next result.

1.2.8 Proposition. *For a finitely generated $\mathbb{F}$-vector space V the power algebra*

$$H^*B(V^\#) = (E_\beta(V), \gamma)$$

is the unique power algebra satisfying (i) and (ii) respectively:

(i) *Let $p = 2$. Then $x, y \in V \subset E_\beta(V)$ satisfy*

$$\gamma_x(y) = y^2 + x \cdot y.$$

(ii) *Let p be odd and $m = (p-1)/2$. Then $x \in V \subset E_\beta(V)$ satisfies*

$$\begin{aligned} \gamma_x(y) &= m!(\beta x)^{m-1}((\beta x)\cdot y - x \cdot (\beta y)) \text{ for } y \in V, \\ \gamma_x(y) &= y^p - (\beta x)^{p-1} \cdot y \text{ for } y \in \beta V. \end{aligned}$$

We prove this result in (1.2) below.

The power algebra $(E_\beta(V), \gamma)$ is *natural* in V, that is, an $\mathbb{F}$-linear map $\varphi : V \to V'$ between finitely generated $\mathbb{F}$-vector spaces induces a map $\varphi : (E_\beta(V), \gamma) \to (E_\beta(V'), \gamma)$ between power algebras. This type of naturality is used in the following notion of an "extended power algebra".

1.2.9 Definition. For $p \geq 2$ an *extended power algebra* H is defined by a power algebra $(E_\beta(V) \otimes H, \gamma)$ for each $\mathbb{F}$-vector space V of dimension ≤ 2. The following properties hold:

(i) The inclusion $E_\beta(V) \to E_\beta(V) \otimes H$ which carries y to $y \otimes 1$ is a map between power algebras $(E_\beta(V), \gamma) \to (E_\beta(V) \otimes H, \gamma)$ where $(E_\beta(V), \gamma)$ is defined in (1.2.8).

(ii) The power algebra $(E_\beta(V) \otimes H, \gamma)$ is natural under H, that is, for $\mathbb{F}$-vector spaces V, V' of dimension ≤ 2 any map $\psi : E_\beta(V) \otimes H \to E_\beta(V') \otimes H$ under H in β-$\mathbf{Alg}_0$ is also a map $\psi : (E_\beta(V) \otimes H, \gamma) \to (E_\beta(V') \otimes H, \gamma)$ between power algebras. Here ψ is determined by the R-linear map $\psi : V \to (E_\beta(V') \otimes H)_1 = H^1 \oplus (V' \otimes H^0)$.

For the trivial vector space $V = 0_\mathbb{F}$ we have $E_\beta(0_\mathbb{F}) = \mathbb{F}$ so that by (1.2.9) the algebra $E_\beta(0_\mathbb{F}) \otimes H = H$ is a power algebra. Proposition (1.2.8) shows that the trivial algebra $\mathbb{F} = H$ is an extended power algebra. A *map* $f : A \to H$ between extended power algebras is a map in β-$\mathbf{Alg}_0$ such that for all vector spaces V, V' of dimension ≤ 2 the map

$$1 \otimes f : (E_\beta(V) \otimes A, \gamma) \longrightarrow (E_\beta(V') \otimes H, \gamma)$$

is a map between power algebras. Let **EPow** be the category of extended power algebras and such maps. The cohomology functor (1.2.4), (1.2.7) yields a contravariant functor

$$\text{(1.2.10)} \qquad \mathbf{Top}_0^*/ \simeq \longrightarrow \mathbf{EPow}$$

which carries X to the extended power algebra H^*X given by the power algebras

$$(E_\beta(V) \otimes H^*X, \gamma) = (H^*(B(V^\#) \times X), \gamma)$$

where we use the product space $B(V^\#) \times X$ and (1.2.7) and the Künneth formula $H^*(Y \times X) = H^*(Y) \otimes H^*(X)$.

1.2.11 Definition. Given an extended power algebra H we obtain for the $\mathbb{F}$-vector space $V = \mathbb{F}x$ the diagram

$$\text{(1)} \qquad \begin{array}{ccc} (E_\beta(\mathbb{F}x) \otimes H)^{pq} & \xrightarrow{pr_i} & H^{pq-i} \\ \uparrow \gamma_x & & \uparrow D_i \\ (E_\beta(\mathbb{F}x) \otimes H)^q & \xleftarrow{j} & H^q \end{array}$$

where pr_i carries an element to the coordinate of degree $pq - i$ as defined in (1.2.3)(5) and where j is the inclusion with $j(y) = 1 \otimes y$. The operator D_i is the composite $D_i = pr_i \circ \gamma_x \circ j$, or equivalently we have for $x = x \otimes 1 \in E_\beta(\mathbb{F}x) \otimes H$ the equation:

$$\text{(2)} \qquad \gamma_x(1 \otimes y) = \sum_i \omega_i(x) \otimes D_i(y).$$

We set $D_i = 0$ for $i > pq$ and for $i < 0$. Moreover we say that the extended power algebra H is *unitary* if (3) and (5) hold.

(3)
$$\begin{cases} D_i(y) = 0 \text{ for } i > (p-1)q \text{ and} \\ D_{(p-1)q}(y) = \vartheta_q \cdot y \end{cases}$$

for $y \in H^q$. Here $\vartheta_q \in \mathbb{Z}/p = \mathbb{F}$ is given by the formula

(4)
$$\begin{cases} \vartheta_q = 1 & \text{if } p = 2, \\ \vartheta_q = (-1)^{mq(q+1)/2} \cdot (m!)^q & \text{if } p \text{ odd} \end{cases}$$

with $m = (p-1)/2$. We point out that $(m!)^2 \equiv (-1)^{m+1}\ mod(p)$, see 6.3 page 112 [SE]. If p is odd then

(5)
$$\begin{cases} D_j(y) = 0 \quad \text{if} \quad |y| \text{ is even and} \\ \qquad\qquad j \notin \{2m(p-1), 2m(p-1)-1; m \geq 0\}, \\ \\ D_j(y) = 0 \quad \text{if} \quad |y| \text{ is odd and} \\ \qquad\qquad j \notin \{(2m+1)(p-1), (2m+1)(p-1)-1; m \geq 0\}. \end{cases}$$

Let

$$\mathbf{UEPow} \subset \mathbf{EPow}$$

be the full subcategory of unitary extended power algebras.

Recall that for $p = 2$ the Steenrod algebra $\mathcal{A}_2$ is generated by elements Sq^i, $i \geq 1$, and that H^*X is an unstable algebra over $\mathcal{A}_2$ as described in Chapter 1. Let $\mathcal{K}_2$ be the category of augmented unstable algebras over $\mathcal{A}_2$, see (1.1.7). The next result shows a new fundamental relation between power algebras and the Steenrod algebra.

1.2.12 Theorem. *For* $p = 2$ *the category* **UEPow** *of unitary extended power algebras is isomorphic to the category* $\mathcal{K}_2^0$ *of connected unstable algebras over the Steenrod algebra* $\mathcal{A}_2$.

Proof. If H is a unitary extended power algebra we define the action of $\mathcal{A}_2$ on $y \in H^q$

(1)
$$Sq^j(y) = D_{q-j}(y) \in H^{q+j}.$$

Now one can check that H is a well-defined object in $\mathcal{K}_2^0$. Conversely if H is an object in $\mathcal{K}_2^0$ we see that H is also an object in **UEPow** as follows. The tensor product of algebras is also a coproduct in $\mathcal{K}_2$ and $E_\beta(V) = H^*B(V^\#)$ is an object in $\mathcal{K}_2$. Hence also $A = E_\beta(V) \otimes H$ is an object in $\mathcal{K}_2$. We now define for $x \in A^1$ the power operation $\gamma_x : A^q \to A^{2q}$ by the formula

(2)
$$\gamma_x(y) = \sum_{j \geq 0} \omega_{q-j}(x) \cdot Sq^j(y).$$

Here we have $\omega_i(x) = x^i$. Now one can check that $E_\beta(V) \otimes H$ is a power algebra natural for algebra maps under H in β-$\mathbf{Alg}_0$ so that H is a well-defined unitary extended power algebra. Further details for the proof are given in Section (1.3), (1.4) below. □

A similar result is true for odd primes p if we use the following Bockstein operators. Let H be an object in β-$\mathbf{Alg}_0$. Then a *Bockstein operator* β on H is given by R-linear maps

$$\beta : H^q \longrightarrow H^{q+1} \text{ with } q \geq 1 \tag{1.2.13}$$

satisfying $\beta\beta = 0$ and $\beta(x \cdot y) = \beta(x) \cdot y + (-1)^{|x|} x \cdot \beta(y)$ for $x, y \in H$. Moreover for $q = 1$ the map β coincides with the β-algebra structure of H. For example there is a unique Bockstein operator β on $E_\beta(V)$ which extends $\beta : V \to \beta V$. The cohomology $H^*(X)$ has a Bockstein operator induced by the extension $0 \to \mathbb{Z}/p \to \mathbb{Z}/p^2 \to \mathbb{Z}/p \to 0$. If β is defined for H and A in $\mathbf{Alg}_0$ then β is also defined for $H \otimes A$.

1.2.14 Definition. For p odd a *Bockstein power algebra* (H, γ, β) is a power algebra (H, γ) together with a Bockstein operator β satisfying $\beta\gamma_x = 0$ for all $x \in H^1$. Let **BPow** be the category of Bockstein power algebras. Morphisms are maps in **Pow** which are compatible with β. In particular $(E_\beta(V), \gamma, \beta)$ is an object in **BPow** defined by (1.2.8) and (1.2.13); this object is natural in V. An *extended Bockstein power algebra* H is defined by Bockstein power algebras $(E_\beta(V) \otimes H, \gamma, \beta)$ for each $\mathbb{F}$-vector space V of dimension ≤ 2. The following properties hold:

(i) The inclusion $E_\beta(V) \to E_\beta(V) \otimes H$ which carries y to $y \otimes 1$ is a map $(E_\beta(V), \gamma, \beta) \to (E_\beta(V) \otimes H, \gamma, \beta)$ in **BPow** where $(E_\beta(V), \gamma, \beta)$ is defined in (1.2.8).

(ii) The object $(E_\beta(V) \otimes H, \gamma, \beta)$ is natural under H, that is, for $\mathbb{F}$-vector spaces V, V' of dimension ≤ 2 each map $\psi : E_\beta(V) \otimes H \to E_\beta(V') \otimes H$ under H in β- $\mathbf{Alg}_0$ is also a map in **BPow**. Here ψ is completely determined by the R-linear map $\psi : V \to H^1 \oplus (V' \otimes H^0)$.

We define maps between extended Bockstein power algebras in the same way as in (1.2.9).

Let **EBPow** be the category of extended Bockstein power algebras and let **UEBPow** be the full subcategory consisting of unitary objects; see (1.2.11).

Recall that for odd primes p the Steenrod algebra $\mathcal{A}_p$ is generated by elements β and P^i, $i \geq 1$, and that H^*X is an unstable algebra over $\mathcal{A}_p$. Let $\mathcal{K}_p^0$ be the category of connected unstable algebras over $\mathcal{A}_p$, see (1.1.7). The next result is the analogue of Theorem (1.2.12) for odd primes.

1.2.15 Theorem. *For odd primes p the category* **UEBPow** *of unitary extended Bockstein power algebras is isomorphic to the category $\mathcal{K}_p^0$ of connected unstable algebras over the Steenrod algebra $\mathcal{A}_p$.*

Proof. If H is a unitary extended Bockstein power algebra we define the action of $\beta, P^j \in \mathcal{A}_p$ on $y \in H^q$ by

$$(1)\qquad \begin{array}{rcl} \beta \cdot y & = & \beta(y), \\ (-1)^j \vartheta_q P^j(y) & = & D_{(q-2j)(p-1)}(y), \end{array}$$

where ϑ_q is defined as in (1.2.11)(4). Now one can check that H is a well-defined object in $\mathcal{K}_p$. Conversely if H is an object in $\mathcal{K}_p^0$ we see that H is also an object in **UEBPow** as follows. The tensor product of algebras is also the coproduct in $\mathcal{K}_p$ and $E_\beta(V) = H^*B(V^\#)$ is an object in $\mathcal{K}_p$ see (1.1.9). Hence also $A = E_\beta(V) \otimes H$ is an object in $\mathcal{K}_p$. We now define for $x \in A^1$ the power operation $\gamma_x : A^q \to A^{pq}$ by the formula

$$(2)\qquad \begin{array}{rcl} \gamma_x(y) & = & \vartheta_q \sum\limits_j (-1)^j \omega_{(q-2j)(p-1)}(x) \cdot P^j(y) \\ & + & \vartheta_q \sum\limits_j (-1)^j \omega_{(q-2j)(p-1)-1}(x) \cdot \beta P^j(y). \end{array}$$

Here $\omega_i(x)$ is defined by $\beta : A^1 \to A^2$ as in (1.2.3)(4). Now one can check that $A = E_\beta(V) \otimes H$ is a well-defined Bockstein power algebra natural for maps under H in β-$\mathbf{Alg}_0$ so that H is a well-defined unitary extended Bockstein power algebra. Further details of the proof are given in Section (1.3), (1.4) below. □

Proof of (1.2.8). Let $p = 2$. Then we have for $A = E_\beta(V) = H^*(BV^\#)$ and $x, y \in A^1 = V$ the formula (see(1.2.11))

$$\begin{array}{rcl} \gamma_x(y) & = & \sum\limits_j x^{1-j} \cdot Sq^j(y) \\ & = & Sq^1(y) + x \cdot Sq^0(y) \\ & = & y^2 + x \cdot y. \end{array}$$

Now let p be odd and $m = (p-1)/2$. Then we get for $A = E_\beta(V) = H^*(BV^\#)$ and $x, y \in A^1 = V$ the formula (see (1.2.15))

$$\begin{array}{rcl} \gamma_x(y) & = & \vartheta_1 \sum\limits_j (-1)^j \omega_{(1-2j)(p-1)}(x) \cdot P^j(y) \\ & & + \vartheta_1 \sum\limits_j (-1)^j \omega_{(1-2j)(p-1)-1}(x) \cdot \beta P^j(y) \\ & = & (-1)^m m!(\omega_{p-1}(x) \cdot P^0(y) + \omega_{p-2}(x) \cdot \beta P^0(y)) \\ & = & (-1)^m m!((-\beta x)^m \cdot y + x \cdot (-\beta x)^{m-1} \cdot (\beta y)) \\ & = & m!(\beta x)^{m-1}((\beta x) \cdot y - x \cdot (\beta y)). \end{array}$$

Moreover for $x \in A^1 = V$ and $y \in \beta V \subset A^2$ we get by the formula in (1.2.15):

$$\begin{array}{rcl} \gamma_x(y) & = & \vartheta_2 \sum\limits_j (-1)^j \omega_{(2-2j)(p-1)}(x) \cdot P^j(y) \\ & & + \vartheta_2 \sum\limits_j (-1)^j \omega_{(2-2j)(p-1)-1}(x) \cdot \beta P^j(y). \end{array}$$

Here (1.2.11)(4) shows

$$\vartheta_2 \;=\; (-1)^m \cdot (m!)^2 \equiv (-1)^m \cdot (-1)^{m+1} = -1 \quad \bmod\ p.$$

Hence one gets for $y \in \beta V$,

$$\begin{aligned}\gamma_x(y) &= (-1)\cdot(\omega_{2(p-1)}(x)\cdot P^0(y) - \omega_0(x)\cdot P^1(y)) \\ &\quad + (-1)\cdot(\omega_{2(p-1)-1}(x)\cdot \beta P^0(y)) \\ &= -(\beta x)^{p-1}\cdot y + y^p.\end{aligned}$$

Here we use (1.1.9) and the fact that $\beta\beta = 0$ implies $\beta y = 0$ for $y \in \beta V$. □

1.3 Cartan formula

Let $p \geq 2$ and let H be an extended power algebra so that for $y \in H$,

(1.3.1) $$\gamma_x(1 \otimes y) = \sum_i \omega_i(x) \otimes D_i(y) \ \in E_\beta(\mathbb{F}x) \otimes H$$

as in (1.2.11). We set $D_i(y) = 0$ for $i < 0$.

1.3.2 Lemma.

$$D_0(y) \;=\; y^p$$

Proof. We can choose the algebra map under H,

$$f : E_\beta(\mathbb{F}x) \otimes H \longrightarrow H$$

which carries x to the zero element in H^1. Then compatibility of γ with this map shows

$$D_0(y) = f_*\gamma_x(1 \otimes y) = \gamma_{fx}(f(1 \otimes y)) = \gamma_0(y) = y^p$$

by (1.2.6)(i). □

We define $Sq^j(y)$ and $P^j(y)$ in an extended power algebra by

(1.3.3) $$\begin{cases} Sq^j(y) &= D_{|y|-j}(y) & \text{if } p = 2, \\ (-1)^j \vartheta_{|y|} P^j(y) &= D_{(|y|-2j)(p-1)}(y) & \text{if } p \text{ odd.}\end{cases}$$

1.3.4 Lemma. *For $p = 2$ the power operation γ in a unitary extended power algebra H satisfies for $x \in H^1$, $y \in H^q$,*

(*) $$\gamma_x(y) \;=\; \sum_j x^{q-j} \cdot Sq^j(y).$$

For p odd the power operation γ in a unitary extended Bockstein power algebra H satisfies, for $x \in H^1$, $y \in H^q$,

(**) $$\begin{cases} \gamma_x(y) &= A_x(y) + B_x(y), \ \textit{with} \\ A_x(y) &= \vartheta_q \sum_j (-1)^j \omega_{(q-2j)(p-1)}(x) \cdot P^j(y), \\ B_x(y) &= \vartheta_q \sum_j (-1)^j \omega_{(q-2j)(p-1)-1}(x) \cdot \beta P^j(y).\end{cases}$$

Proof. For $p = 2$ the equation for $\gamma_x(y)$ holds in $(E_\beta(\mathbb{F}x) \otimes H, \gamma)$ by definition (1.3.3), (1.3.1). Now the element x defines a unique algebra map $E_\beta(\mathbb{F}x) \to H$ carrying the generator x to $x \in H^1$. This algebra map yields a map under H in β-$\mathbf{Alg}_0$,

$$E_\beta(\mathbb{F}x) \otimes H \longrightarrow H \tag{1}$$

which by assumption on extended power algebras is compatible with γ. This implies that $(*)$ holds in H. For p odd we have to prove $(**)$ only in $E_\beta(\mathbb{F}x) \otimes H$. Then the map (1) shows that $(**)$ also holds in H. We know by the assumption on Bockstein power algebras that $\beta\gamma_x(y) = 0$. Hence we have for (1.3.1) the formula in $E_\beta(\mathbb{F}x) \otimes H$,

$$\begin{aligned} 0 &= \beta\gamma_x(1 \otimes y) = \beta(\textstyle\sum_i \omega_i(x) \otimes D_i(y)) \\ &= \textstyle\sum_i (\beta\omega_i(x) \otimes D_i(y) + (-1)^i \omega_i(x) \otimes \beta D_i(y)). \end{aligned} \tag{2}$$

By (1.2.3)(4) we know (since $\beta\beta = 0$)

$$\beta\omega_i(x) = \begin{cases} 0 & \text{if } \ i \text{ is even}, \\ -\omega_{i+1}(x) & \text{if } \ i \text{ is odd}, \ i \geq 1. \end{cases}$$

Hence (2) implies

$$\begin{aligned} 0 &= \textstyle\sum_{i \ \text{even} \geq 2} ((-1)^i \omega_i(x) \otimes \beta D_i(y) - \omega_i(x) \otimes D_{i-1}(y)) \\ &\quad + (\textstyle\sum_{i \ \text{odd}} (-1)^i \omega_i(x) \otimes \beta D_i(y)) + 1 \otimes \beta D_0(y). \end{aligned} \tag{3}$$

Here we have $\beta D_0(y) = \beta y^p = 0$. Now (3) implies

$$\begin{cases} \beta D_i(y) = 0 & \text{if } i \text{ is odd}, \\ \beta D_i(y) = (-1)^i D_{i-1}(y) & \text{if } i \text{ is even} \geq 2. \end{cases} \tag{4}$$

Hence the definition of P^i in (1.3.3) and the assumption that H is unitary imply that $(**)$ holds in $E_\beta(\mathbb{F}x) \otimes H$ and hence in H. □

We now show that the "global Cartan formula" for $\gamma_x(y \cdot z)$ in a power algebra H is essentially equivalent to the classical Cartan formula $(\mathcal{K}1)$ in (1.1.7). For this we need the following formula in $E_\beta(\mathbb{F}x)$:

$$\omega_i(x) \cdot \omega_t(x) = \begin{cases} 0 & \text{if } i \cdot t \cdot p \text{ is odd}, \\ \omega_{i+t}(x) & \text{otherwise}. \end{cases} \tag{1.3.5}$$

Compare (1.2.3)(4). Here we use the fact that $x \cdot x = 0$ if p is odd and $|\, x \,|$ odd.

1.3.6 Lemma. *Let H be a connected unstable algebra over the Steenrod algebra $\mathcal{A} = \mathcal{A}_p$, $p \geq 2$. Then γ_x defined by (1.3.4)$(*)$, $(**)$ satisfies the global Cartan formula* (1.2.6)(ii),

$$\gamma_x(y \cdot z) = (-1)^{|y||z|\bar{p}} \gamma_x(y) \cdot \gamma_x(z).$$

Proof. For $p = 2$ we have $(q = \mid y \mid + \mid z \mid)$,

$$\begin{aligned}
\gamma_x(y \cdot z) &= \sum_j x^{q-j} Sq^j(y \cdot z) \\
&= \sum_j x^{q-j} \sum_{k+t=j} Sq^k(y) \cdot Sq^t(z) \\
&= \sum_j \sum_{k+t=j} x^{|y|-k} Sq^k(y) \cdot x^{|z|-t} Sq^t(z) \\
&= \gamma_x(y) \cdot \gamma_x(z).
\end{aligned}$$

For p odd a similar argument holds. In fact, $(\mathcal{K}1)$ implies

$$A_x(y \cdot z) = (-1)^{|y||z|\bar{p}} A_x(y) \cdot A_x(z).$$

Moreover since β is a derivation $(\mathcal{K}1)$ also implies

$$B_x(y \cdot z) = (-1)^{|y||z|\bar{p}} (A_x(y) \cdot B_x(z) + B_x(y) \cdot A_x(z)).$$

Finally we observe that

$$B_x(y) \cdot B_x(z) = 0$$

since $x \cdot x = 0$ in H, $x \in H^1$, p odd. □

1.3.7 Lemma. *For $p = 2$ let H be an extended unitary power algebra and for p odd let H be an extended unitary Bockstein power algebra. Then Sq^j and P^j defined as in* (1.3.3) *satisfy the Cartan formula* $(\mathcal{K}1)$.

Proof. We use the same equation as in the proof of (1.3.6) in the algebra $E_\beta(\mathbb{F}x) \otimes H$. Then comparing coordinates yields $(\mathcal{K}1)$. For example, for $p = 2$ we have

$$\begin{aligned}
\sum_j x^{q-j} \otimes Sq^j(y \cdot z) &= \gamma_x(1 \otimes y \cdot z) \\
&= \gamma_x(1 \otimes y) \cdot \gamma_x(1 \otimes z) \\
&= \sum_j x^{q-j} \otimes (\sum_{k+t=j} Sq^k(y) \cdot Sq^t(z)).
\end{aligned}$$

Similar arguments hold for p odd. In fact, in this case we get for the first summand $A_x(y)$ in (1.3.4)(∗∗) the formula in $E_\beta(\mathbb{F}x) \otimes H$,

$$A_x(y \cdot z) = (-1)^{|y||z|\bar{p}} A_x(y) \cdot A_x(z)$$

and this implies $(\mathcal{K}1)$ by comparing coordinates. □

1.4 Adem relation

We show that the formula

(1.4.1) $$\gamma_x\gamma_y(z) \;=\; (-1)^{|z|\bar{p}}\gamma_y\gamma_x(z)$$

in a power algebra corresponds to the classical Adem relation. We use the following convention on binomial coefficients.

(1.4.2) $$\binom{r}{j} = 0 \text{ if } r < 0 \text{ or } j < 0, \quad \binom{r}{0} = 1 \text{ if } r \geq 0.$$

Moreover Sq^j and P^j are zero for $j < 0$. All summations run from $-\infty$ to $+\infty$. We have in $E_\beta(\mathbb{F}x)$ the formulas ($i, j \in \mathbb{Z}$)

(1) $$Sq^j\omega_r(x) = \binom{r}{j}\omega_{r+j}(x),$$

(2) $$\begin{aligned} P^j\omega_{2r}(x) &= \binom{r}{j}\omega_{2r+2j(p-1)}(x), \\ \beta P^j\omega_{2r}(x) &= 0, \end{aligned}$$

(3) $$\begin{aligned} P^j\omega_{2r-1}(x) &= \binom{r-1}{j}\omega_{2r+2j(p-1)-1}(x), \\ \beta P^j\omega_{2r-1}(x) &= -\binom{r-1}{j}\omega_{2r+2j(p-1)}(x). \end{aligned}$$

For this compare (1.1.9), (1.1.8), (1.2.3).

1.4.3 Lemma. *Let H be a connected unstable algebra over the Steenrod algebra $\mathcal{A} = \mathcal{A}_p$, $p \geq 2$. Then γ_x defined by (1.3.4)(*), (**) satisfies the global Adem formula (1.4.1) above.*

Proof. Let $p = 2$. Then we get for $q =| z |$,

(1) $$\begin{aligned} \gamma_x\gamma_y(z) &= \gamma_x(\textstyle\sum_i \omega_{q-i}(y)Sq^i(z)) \\ &= \textstyle\sum_i \gamma_x(\omega_{q-i}(y) \cdot Sq^i(z)) \\ &= \textstyle\sum_{k,i} \omega_{2q-k}(x) \cdot Sq^k(\omega_{q-i}(y) \cdot Sq^i(z)) \\ &= \textstyle\sum_{i,k,j} \binom{q-i}{j}\omega_{2q-k}(x) \cdot \omega_{q-i+j}(y) \cdot Sq^{k-j}Sq^i(z). \end{aligned}$$

Here we write $q - i + j = 2q - l$, so that

(2) $$\gamma_x\gamma_y(z) = \sum_{k,l} \omega_{2q-k}(x) \cdot \omega_{2q-l}(y) \cdot D_{2q-k,2q-l}(z),$$

(3) $$D_{2q-k,2q-l}(z) = \sum_i \binom{q-i}{q-l+i} Sq^{k+l-i-q}Sq^i(z).$$

Hence (1.4.1) is a consequence of the following equation in H^{k+l} for all $k, l \in \mathbb{Z}$, $q \geq 1$, $z \in H^q$.

$$D_{2q-k,2q-l}(z) = D_{2q-l,2q-k}(z). \tag{4}$$

In fact, this equation holds in each unstable algebra over the Steenrod algebra $\mathcal{A}_2$. A direct proof of (4), however, is highly sophisticated based on Adem relations in $\mathcal{A}_2$ and the unstable structure of H. We therefore give in (1.5) below a proof of (1.4.3) relying on a result of Serre [S1].

For p odd one uses similar arguments as above though formulas are more involved. We have

$$\gamma_x\gamma_y(z) = A_xA_y(z) + B_xB_y(z) + A_xB_y(z) + B_xA_y(z). \tag{5}$$

Hence (1.4.1) is a consequence of

$$A_xA_y(z) = (-1)^{|z|\bar{p}}A_yA_x(z) \tag{6}$$

and

$$\begin{aligned} &B_xB_y(z) + A_xB_y(z) + B_xA_y(z) \\ &= (-1)^{|z|\bar{p}}(B_yB_x(z) + A_yB_x(z) + B_yA_x(z)). \end{aligned} \tag{7}$$

Using the Cartan formula ($\mathcal{K}$1) and the equations in (1.4.2)(1), (2), (3) we see that (6) and (7) correspond to equations in H (similarly as (4) above). These equations are consequences of the Adem relations and the unstable structure of H. This can be proved directly requiring highly tedious computations. We therefore give in (1.5) below a proof of (1.4.3) as in the case of $p = 2$. □

1.4.4 Lemma. *For $p = 2$ let H be an extended unitary power algebra and for p odd let H be an extended unitary Bockstein power algebra. Then Sq^j and P^j defined as in* (1.3.3) *satisfy the Adem relations.*

Proof. We use the same equations as in the proof of (1.4.3) in the algebra $E_\beta(\mathbb{F}x \oplus \mathbb{F}y) \otimes H$. Then comparing coordinates yields for $p = 2$ the equation

$$D_{2q-k,2q-l}(z) \;=\; D_{2q-l,2q-k}(z).$$

This shows that for $p = 2$ the Adem relation is satisfied, see page 119 [SE]. Here we use the fact that the maps under H in β-$\mathbf{Alg}_0$,

$$E_\beta(\mathbb{F}z) \otimes H \longrightarrow E_\beta(\mathbb{F}x \oplus \mathbb{F}y) \otimes H$$

which carry z to x and y respectively both are maps of power algebras by the condition (ii) in (1.2.8). This implies that the squaring operations Sq^j, P^j on H defined by γ_x and γ_y respectively coincide. For p odd we use (1.4.3)(6), (7) in $E_\beta(\mathbb{F}x \oplus \mathbb{F}y) \otimes H$ and we compare coordinates. This yields the Adem relations in the same way as in VIII.1.8, 1.9 [SE]. □

1.5 The theory of Eilenberg-MacLane spaces

A *theory* $\mathbf{T}$ is a small category with products $A \times B$ for objects A, B in $\mathbf{T}$. Let $\mathbf{Set}$ be the category of sets. A *model* M of the theory $\mathbf{T}$ is a functor $M : \mathbf{T} \to \mathbf{Set}$ which carries products in $\mathbf{T}$ to the products of sets,

$$M(A \times B) \;=\; M(A) \times M(B).$$

Let $\mathbf{model(T)}$ be the category of such models; morphisms are natural transformations. Since the work of Lawvere [L] it is well known that many algebraic categories (like the categories of groups, algebras, Lie algebras, etc.) are such categories of models of a theory $\mathbf{T}$. See also Borceux's book [Bo].

1.5.1 Definition. Let p be a prime ≥ 2 and let $\mathbf{K}_p$ be the following theory. The category

$$\mathbf{K}_p \subset \mathbf{Top}^*/_{\simeq}$$

is the full subcategory of the homotopy category of pointed spaces consisting of finite products

$$A \;=\; Z^{n_1} \times \cdots \times Z^{n_r}$$

with $n_1, \ldots, n_r \geq 1$ and $r \geq 0$ where Z^n is the Eilenberg-MacLane space

$$Z^n \;=\; K(\mathbb{Z}/p, n).$$

Products are defined in $\mathbf{K}_p$ so that $\mathbf{K}_p$ is a theory, termed the *theory of Eilenberg-MacLane spaces.*

1.5.2 Theorem. *There is an isomorphism of categories*

$$\mathbf{model}(\mathbf{K}_p) \;=\; \mathcal{K}_p^0$$

where $\mathcal{K}_p^0$ is the category of connected unstable algebras over the Steenrod algebra $\mathcal{A}_p$.

This result relies on the computation of Serre [S1] and Cartan [C] of the cohomology of Eilenberg-MacLane spaces.

Proof of (1.5.2). We have the forgetful functor

$$\phi : \mathcal{K}_p^0 \longrightarrow \mathbf{Vec}^{\geq 1}$$

where $\mathbf{Vec}^{\geq 1}$ is the category of graded R-vector spaces concentrated in degree ≥ 1. The functor ϕ carries A to the underlying vector space of $\tilde{A}$. Let H be the left adjoint of ϕ which carries V in $\mathbf{Vec}^{\geq 1}$ to the free unstable algebra $H(V)$ generated by V. Let x_i be an element of degree n_i. Then it follows from (1.1.11) that

$$\begin{aligned} H^*(Z^{n_1} \times \cdots \times Z^{n_r}) &= H(\mathbb{F}x_1 \oplus \cdots \oplus \mathbb{F}x_r) \\ &= H(n_1) \otimes \cdots \otimes H(n_r). \end{aligned}$$

Let

$$\mathbf{H}_p \subset \mathcal{K}_p^0$$

be the full subcategory generated by objects $H(\mathbb{F}x_1 \oplus \cdots \oplus \mathbb{F}x_r)$ with $n_1, \ldots, n_r \geq 1$ and $r \geq 0$. Then the cohomology functor yields an isomorphism of categories where $\mathbf{H}_p^{\mathrm{op}}$ is the *opposite category* of $\mathbf{H}_p$,

$$H^* : \mathbf{K}_p = \mathbf{H}_p^{\mathrm{op}}.$$

Since $\mathbf{model}(\mathbf{H}_p^{\mathrm{op}}) = \mathcal{K}_p^0$ we see that (1.5.2) holds. Compare also [BJ4]. □

We have the following commutative diagram of functors corresponding to the well-known equation

$$\tilde{H}^n(X) \;=\; [X, K(n)]$$

for $n \geq 0$.

(1.5.3)

$$\begin{array}{ccc} & & \mathbf{model}(\mathbf{K}_p) \\ & \overset{[X,-]}{\nearrow} & \| \\ \mathbf{Top}_0^*/\simeq & & \| \\ & \underset{H^*}{\searrow} & \| \\ & & \mathcal{K}_p^0 \end{array}$$

Here $[X, -]$ is the model which carries an object A in $\mathbf{K}_p$ to the set $[X, A]$. This model obviously satisfies

$$[X, A \times B] \;=\; [X, A] \times [X, B]$$

so that the functor $[X, -]$ in (1.5.3) is well defined. On the other hand H^* is the classical cohomology functor, see (1.1.7). The isomorphism of categories in (1.5.3) is given by (1.5.2).

We are now ready to prove (1.2.12) and (1.2.15) by the following result.

1.5.4 Theorem. *For $p = 2$ there is a commutative diagram of functors.*

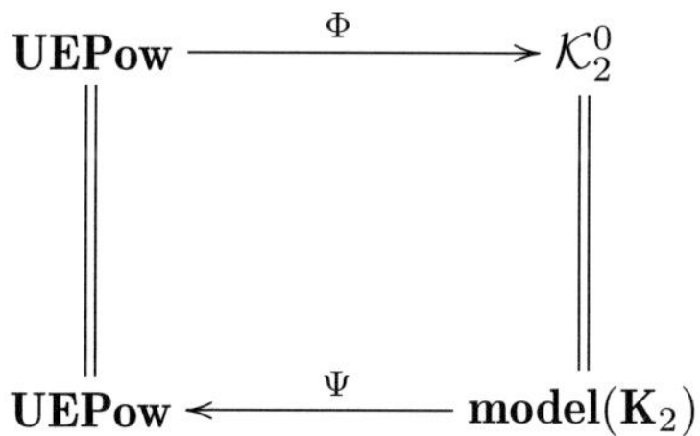

For p odd there is a commutative diagram of functors.

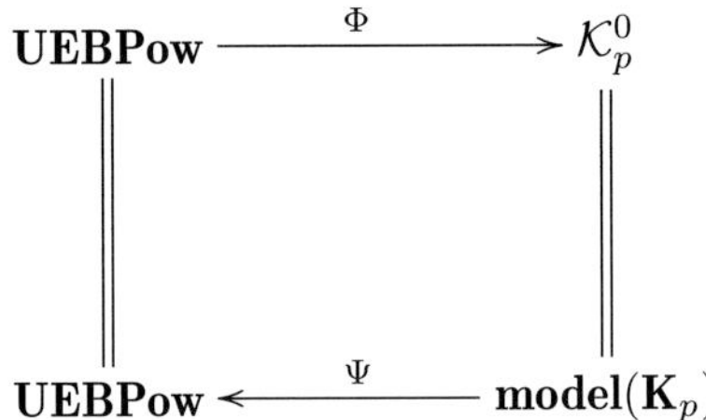

Proof. The functor Φ carries (H, γ) to (H, Sq^i) for $p = 2$ and to (H, P^i) for p odd. By (1.3.2), (1.3.7) and (1.4.4) we see that Φ is a well-defined functor. We shall construct in Section (8.5) the inverse functor Ψ of Φ. □

Chapter 2

Track Theories and Secondary Cohomology Operations

In Chapter 1 the theory $\mathbf{K}_p$ of Eilenberg-MacLane spaces was defined by homotopy classes of maps between products of Eilenberg-MacLane spaces. We now consider all maps in $\mathbf{Top}^*$ between such products and homotopy classes of homotopies between such maps termed tracks. For this reason we choose in section (2.1) Eilenberg-MacLane spaces Z^n with "good" properties. For example they are $\mathbb{F}$-vector space objects in $\mathbf{Top}^*$. Many results in this book will rely on these properties. In Part I we use the additive structure of Z^n as an $\mathbb{F}$-vector space object. In Part II we need the multiplicative structure and the action of permutation groups on the spaces Z^n.

2.1 The Eilenberg-MacLane spaces Z^n

Let R be a commutative ring with $1 \in R$; for example for a prime p let $R = \mathbb{F} = \mathbb{Z}/p$ be the field of p elements. In (2.1.4) below we introduce an Eilenberg-MacLane space ($n \geq 1$)

$$Z^n = Z^n_R = K(R, n)$$

for the underlying abelian group R. The space Z^n is defined in the proof of (2.1.4) by the free simplicial R-module generated by the non-basepoint singular simplices in the n-sphere $S^n = S^1 \wedge \cdots \wedge S^1$. The symmetric group σ_n acts on S^n by permuting the S^1-factors of S^n and hence σ_n acts on $K(R, n)$. We point out that Z^n is an Eilenberg-MacLane space as above only for $n \geq 1$, see also (1.5.1).

The space Z^n is a topological R-module (with R having the discrete topology) and Z^n is a σ_n-space with the symmetric group σ_n acting via R-linear automorphisms of Z^n. Moreover the following additional structure is given.

The cup-product in cohomology is induced by R-bilinear *multiplication maps*

$$\mu = \mu_{m,n} : Z^m \times Z^n \to Z^{m+n} \tag{2.1.1}$$

which are associative in the obvious sense and which are equivariant via the inclusion $\sigma_m \times \sigma_n \subset \sigma_{m+n}$. Let $\tau = \tau_{n,m} \in \sigma_{n+m}$ be the permutation of $\{1, \dots, n+m\}$ exchanging the block $\{1, \dots, n\}$ and the block $\{n+1, \dots, m+n\}$. Then the diagram

$$\begin{array}{ccc} Z^m \times Z^n & \xrightarrow{T} & Z^n \times Z^m \\ \downarrow{\scriptstyle \mu_{m,n}} & & \downarrow{\scriptstyle \mu_{n,m}} \\ Z^{m+n} & \xrightarrow{\tau} & Z^{m+n} \end{array} \tag{2.1.2}$$

commutes where T is the interchange map and $\tau(v) = \tau \cdot v$ is given by the action of $\tau \in \sigma_{m+n}$.

Let sign : $\sigma_n \to \{+1, -1\}$ be the homomorphism which carries a permutation to the sign of the permutation. For example we have $\mathrm{sign}(\tau_{n,m}) = (-1)^{nm}$. The kernel of sign is the alternating group. For the σ_n-space Z^n and for $\sigma \in \sigma_n$ we have the map $\sigma : Z^n \to Z^n$ which carries x to $\sigma \cdot x$. This map induces a homology $H_n Z^n = R$ the sign of σ, that is

$$\sigma_* = \mathrm{sign}(\sigma) : H_n(Z^n) \to H_n(Z^n). \tag{2.1.3}$$

These properties of Z^n are crucial for the definition of the power maps in Part II.

2.1.4 Proposition. *Eilenberg-MacLane spaces Z^n with properties described in* (2.1.1), (2.1.2) *and* (2.1.3) *exist.*

Proof. We shall need the following categories and functors; compare the Appendix of this section and Goerss-Jardine [GJ]. Let **Set** and **Mod** be the category of sets and R-modules respectively and let **ΔSet** and **Δ Mod** be the corresponding categories of simplicial objects in **Set** and **Mod** respectively. We have functors

$$\mathbf{Top}^* \xrightarrow{\mathrm{Sing}} (\mathbf{\Delta Set})^* \xrightarrow{|\ |} \mathbf{Top}^* \tag{1}$$

given by the *singular set* functor Sing and the *realization* functor $|\ |$. Moreover we have

$$\mathbf{\Delta Set} \xrightarrow{R} \mathbf{\Delta\, Mod} \xrightarrow{\Phi} (\mathbf{\Delta Set})^* \tag{2}$$

where R carries the simplicial set X to the *free R-module* generated by X and where Φ is the *forgetful* functor which carries the simplicial module A to the underlying simplicial set. Moreover we need the Dold-Kan functors

$$\mathbf{Ch}_+ \xrightarrow{\Gamma} \mathbf{\Delta\, Mod} \xrightarrow{N} \mathbf{Ch}_+ \tag{3}$$

where $\mathbf{Ch}_+$ is the category of chain complexes in **Mod** concentrated in degree ≥ 0. Here N is the *normalization* functor which by the Dold-Kan theorem is an equivalence of categories with inverse Γ. For a pointed space V the chain complex $C_*(V) = NR\,\mathrm{Sing}(V)$ is the *normalized* chain complex of singular chains in V. We now define

$$K(V) = |\Phi S(V)| \text{ with } S(V) = \frac{R\ \mathrm{Sing}(V)}{R\ \mathrm{Sing}(*)}. \tag{4}$$

Hence $K : \mathbf{Top}^* \to \mathbf{Top}^*$ carries a pointed space to a topological R-module. We define the binatural map

$$\bar{\otimes} : K(V) \times K(W) \to K(V \wedge W) \tag{5}$$

as follows. We have

$$\mathrm{Sing}(V \times W) = \mathrm{Sing}(V) \times \mathrm{Sing}(W)$$

and this bijection induces a commutative diagram in $\Delta\,\mathbf{Mod}$.

$$\begin{array}{ccc} R\ \mathrm{Sing}(V) \otimes R\ \mathrm{Sing}(W) & = & R\ \mathrm{Sing}(V \times W) \\ \downarrow & & \downarrow \\ S(V) \otimes S(W) & \xrightarrow{\wedge} & S(V \wedge W) \end{array}$$

The vertical arrows are induced by quotient maps. For R-modules A, B let $\otimes : A \times B \to A \otimes_R B$ be the map in **Set** which carries (a, b) to the tensor product $a \otimes b$. Of course this map $\otimes$ is bilinear. Moreover for A, B in $\Delta\,\mathbf{Mod}$ the map $\otimes$ induces the map $\otimes : \Phi(A \times B) \to \Phi(A \otimes B)$ in **Set** and the realization functor yields

$$|\otimes| : |\Phi A| \times |\Phi B| = |\Phi(A \times B)| \to |\Phi(A \otimes B)|.$$

Hence for $A = S(V)$ and $B = S(W)$ we get the composite

$$|\Phi S(V)| \times |\Phi S(W)| \xrightarrow{|\otimes|} |\Phi(S(V) \otimes S(W))| \xrightarrow{|\Phi\wedge|} |\Phi S(V \wedge W)|$$

and this is the map $\bar{\otimes}$ above. One readily checks that $\bar{\otimes}$ is bilinear with respect to the topological R-module structure of $K(V), K(W)$ and $K(V \wedge W)$ respectively. Moreover the following diagram commutes.

$$\begin{array}{ccccc} K(V) \times K(W) & & \xrightarrow{\bar{\otimes}} & & K(V \wedge W) \\ \uparrow{\scriptstyle h \times h} & & & & \uparrow{\scriptstyle h} \\ |\mathrm{Sing}\, V| \times |\mathrm{Sing}\, W| & = & |\mathrm{Sing}(V \times W)| & \longrightarrow & |\mathrm{Sing}(V \wedge W)| \end{array} \tag{6}$$

Here the *Hurewicz map* h is the realization of the map in $\Delta\mathbf{Set}$,

$$\mathrm{Sing}(V) \to \Phi R\,\mathrm{Sing}(V) \to \Phi S(V)$$

which carries an element x in $\mathrm{Sing}(V)$ to the corresponding generator in $R\,\mathrm{Sing}(V)$.

Let $S^n = S^1 \wedge \cdots \wedge S^1$ be the n-fold smash product of the 1-sphere S^1. Then the symmetric group σ_n acts on S^n by permuting the factors S^1. It is well known that this action of σ_n on S^n induces the sign-action of σ_n on the homology $H_n(S^n) = R$. We define the Eilenberg-MacLane space Z^n by

(7) $$Z^n = K(S^n) = \left|\Phi \frac{R\,\mathrm{Sing}(S^n)}{R\,\mathrm{Sing}(*)}\right|.$$

Since K is a functor we see that σ_n also acts on $K(S^n)$ via R-linear automorphisms. We define the multiplication map $\mu_{m,n}$ by

$$\mu : Z^m \times Z^n = K(S^m) \times K(S^n) \xrightarrow{\bar{\otimes}} K(S^m \wedge S^n) = Z^{m+n}$$

where $S^m \wedge S^n = S^{m+n}$ and where we use (5). Diagram (6) implies that μ induces the cup product in cohomology. □

Let $V \wedge W = V \times W/V \times * \cup * \times W$ be the *smash product* of pointed spaces V and W. Since the multiplication $\mu_{m,n}$ is bilinear we obtain the induced map

(2.1.5) $$\mu : Z^m \wedge Z^n \to Z^{m+n}$$

which is $\sigma_m \times \sigma_n \subset \sigma_{m+n}$ equivariant. We define the *product* $x \cdot y$ by

$$\mu(x, y) = x \cdot y \in Z^{m+n} \text{ for } x \in Z^m \text{ and } y \in Z^n.$$

Moreover for maps $f : X \to Z^m$ and $g : Y \to Z^n$ let

$$f \boxtimes g : X \times Y \to Z^{m+n}$$

be the map which carries (x, y) to $f(x) \cdot g(y)$. If $X = Y$ and Δ is the diagonal $X \to X \times X$ then

$$f \cdot g : X \to Z^{m+n}$$

is the map $(f \boxtimes g)\Delta$ which carries x to $f(x) \cdot g(x)$.

For pointed spaces X, Y let $[X, Y]$ be the set of homotopy classes of pointed maps $X \to Y$. It is well known that for $n \geq 1$ the set

(2.1.6) $$[X, Z^n] = H^n(X, R)$$

is the nth cohomology of X with coefficients in R. A pointed map $f : X \to Z^n$ is therefore considered as a "cocycle" representing a cohomology class $\{f\} \in H^n(X, R)$. Clearly the cohomology class

$$\{f \cdot g\} = \{f\} \cup \{g\}$$

is given by the cup product in cohomology. The associativity and graded commutativity of the cup product can be derived from the properties of the multiplication maps μ in (2.1.1).

Let ΩX be the *loop space* of the pointed CW-space X. The map μ in (2.1.5) with $m = 1$ induces a homotopy equivalence

(2.1.7) $$r_n : Z^n \xrightarrow{\sim} \Omega Z^{n+1}$$

with the following properties. We choose a map $i_R : S^1 \to Z^1$ which in homology induces the homomorphism of rings $\mathbb{Z} \to R$. Then i_R yields the composite $\mu(1\wedge i_R)$:

$$Z^n \wedge S^1 \longrightarrow Z^n \wedge Z^1 \longrightarrow Z^{n+1}$$

and the adjoint of this map is r_n. Since Z^{n+1} is a topological R-module also the loop space ΩZ^{n+1} has the structure of a topological R-module. Moreover the bilinearity of μ implies that r_n is an R-linear map between topological R-modules. This fact is of main importance in Part I of this book. In addition r_n is equivariant with respect to the action of $\sigma_n \subset \sigma_{n+1}$.

Remark. Using composites of maps $Z^n \wedge S^1 \to Z^{n+1}$ we get a map $Z^m \wedge S^n \to Z^{m+n}$ which is $\sigma_n \times \sigma_m \subset \sigma_{m+n}$ equivariant where σ_m acts on S^m as above. This shows that Z^n is a *symmetric spectrum* as used by Hovey-Shipley-Smith [HS] 1.2.5.

We need the homotopy equivalence $r_n : Z^n \to \Omega Z^{n+1}$ above for the definition of stable maps in the secondary Steenrod algebra, see Section (2.5). The use of r_n for stable maps turns out to be appropriate in Section (10.8); see also (2.1.9)(7) below.

On the other hand we obtain the homotopy equivalence

(2.1.8) $$s_n : Z^n \longrightarrow \Omega Z^{n+1}$$

which is the adjoint of $\mu(i_R \wedge 1)$:

$$S^1 \wedge Z^n \longrightarrow Z^1 \wedge Z^n \longrightarrow Z^{n+1}.$$

Then s_n and r_n are related by the formula

$$\tau_{1,n} s_n = r_n \text{ with } \operatorname{sign}(\tau_{1,n}) = (-1)^n$$

where $\tau_{1,n}$ is the interchange permutation, see (2.1.2). We use s_n (and not r_n) for the following definition of the Bockstein map β. As we shall see in Section (8.6) the use of s_n implies that the Bockstein map is a derivation.

We define the *Bockstein map*

(2.1.9) $$\beta : Z^n_{\mathbb{F}} \longrightarrow Z^{n+1}_{\mathbb{F}}$$

as follows. For $\mathbb{F} = \mathbb{Z}/p$ and $\mathbb{G} = \mathbb{Z}/p^2$ we have the short exact sequence

(1) $$0 \longrightarrow \mathbb{F} \xrightarrow{i} \mathbb{G} \xrightarrow{\pi} \mathbb{F} \longrightarrow 0$$

which induces a fiber sequence

(2) $$F^n \xrightarrow{\partial} Z_{\mathbb{F}}^{n+1} \xrightarrow{i} Z_{\mathbb{G}}^{n+1} \xrightarrow{\pi} Z_{\mathbb{F}}^{n+1}.$$

The space F^n is the fiber of the inclusion i, namely

(3) $$F^n = \{(x,\sigma) \in Z_{\mathbb{F}}^{n+1} \times (Z_{\mathbb{G}}^{n+1}, 0)^{(I,0)};\ ix = \sigma(1)\}.$$

Here we use the path space and we set $\partial(x,\sigma) = x$. Now we have homotopy equivalences

(4) $$Z_{\mathbb{F}}^n \underset{\sim}{\xrightarrow{s_n}} \Omega Z_{\mathbb{F}}^{n+1} \underset{\sim}{\xleftarrow{\pi}} F^n$$

with $\pi(x,\sigma) = \pi\sigma$. Let $\bar{\pi}$ be a homotopy inverse of π. Then β is the composite

(5) $$\beta : Z_{\mathbb{F}}^n \xrightarrow{s_n} \Omega Z_{\mathbb{F}}^{n+1} \xrightarrow{\bar{\pi}} F^n \xrightarrow{\partial} Z_{\mathbb{F}}^{n+1}.$$

The Bockstein map is compatible with r_n in (2.1.7) since there is the commutative diagram $(Z^n = Z_{\mathbb{F}}^n, \Omega_1 = \Omega_2 = \Omega)$.

(6) $$\begin{array}{ccccccc}
Z^n & \xrightarrow{s_n} & \Omega_1 Z^{n+1} & \xleftarrow{\pi} & F^n & \xrightarrow{\partial} & Z^{n+1} \\
\downarrow{\scriptstyle r_n} & & \downarrow{\scriptstyle r'} & & \downarrow{\scriptstyle r''} & & \downarrow{\scriptstyle r_{n+1}} \\
\Omega_2 Z^{n+1} & \xrightarrow[\Omega s_{n+1}]{} & \Omega_2\Omega_1 Z^{n+2} & \xleftarrow[\Omega\pi]{} & \Omega_2 F^{n+1} & \xrightarrow[\Omega\partial]{} & \Omega_2 Z^{n+2}
\end{array}$$

Here r' carries $(\sigma : S^1 \to Z^{n+1}) \in \Omega Z^{n+1}$ to $r'(\sigma) : S^1 \wedge S^1 \to Z^{n+2}$ with $r'(\sigma)(t_2 \wedge t_1) = \sigma(t_1) \cdot \hat{t}_2$ where $\hat{t} = i_{\mathbb{F}}(t)$ is given by $i_{\mathbb{F}} : S^1 \to Z^1$. Similarly r'' carries $(x,\sigma) \in F^n$ to $(r''(x,\sigma) : S^1 \to F^{n+1}) \in \Omega_2 F^{n+1}$ with $r''(x,\sigma)(t_2) = (x \cdot \hat{t}_2, \sigma \cdot \hat{t}_2)$. One readily checks that diagram (6) is well defined and commutative. Diagram (6) shows that the following diagram homotopy commutes.

(7) $$\begin{array}{ccc}
Z^n & \xrightarrow{\beta} & Z^{n+1} \\
\downarrow{\scriptstyle r_n} & \overset{H_{\beta,n}}{\Longrightarrow} & \downarrow{\scriptstyle r_{n+1}} \\
\Omega Z^{n+1} & \xrightarrow[\Omega\beta]{} & \Omega Z^{n+2}
\end{array}$$

We now use notation explained later in this book. For pointed spaces X, Y we have the groupoid

$$[\![X, Y]\!].$$

The objects in this groupoid are the pointed maps $f, g : X \to Y$ and the morphisms termed tracks are the homotopy classes of homotopies $f \simeq g$.

The map s_n in (2.1.7) depends on the choice of $i_R : S^1 \to Z^1$. The full subgroupoid $\mathbf{s}_n$,

$$\text{(2.1.10)} \qquad \mathbf{s}_n \subset [\![Z^n, \Omega Z^{n+1}]\!]$$

consisting of maps s_n as in (2.1.8) is a contractible groupoid by (3.2.5) below. Moreover there is a well-defined contractible subgroupoid

$$\text{(2.1.11)} \qquad \underline{\underline{\beta}} \subset [\![Z^n_{\mathbb{F}}, Z^{n+1}_{\mathbb{F}}]\!]$$

consisting of all maps β as defined in (2.1.9). In fact, let

$$G_\pi \subset [\![\Omega Z^{n+1}_{\mathbb{F}}, F]\!]$$

be the full supgroupoid given by all homotopy inverses $\bar{\pi}$ of π. Then G_π is a contractible groupoid by (3.2.5) below. Therefore the image of the functor

$$\mathbf{s}_n \times G_\pi \longrightarrow [\![Z^n_{\mathbb{F}}, Z^{n+1}_{\mathbb{F}}]\!]$$

carrying $(s_n, \bar{\pi})$ to $\partial\bar{\pi}s_n$ is a contractible groupoid and this image is the subgroupoid $\underline{\underline{\beta}}$ above. The subgroupoid $\underline{\underline{\beta}}$ does not depend on choices.

This shows that two different Bockstein maps β, β' as defined in (2.1.9) are connected by a unique track $\beta \Rightarrow \beta'$ in $\underline{\underline{\beta}}$. We also say that the Bockstein map β in **Top*** is *well defined up to canonical track*. Moreover there is, in fact, a canonical track $H_{\beta,n}$ in (2.1.9)(7) which can be derived from the commutative diagram (2.1.9)(6). This shows that β is a stable map in the secondary Steenrod algebra, see Section (2.5). In general the Steenrod operations $\alpha = Sq^i, P^i$ considered as maps

$$\text{(2.1.12)} \qquad Z^q_{\mathbb{F}} \longrightarrow Z^{q+|\alpha|}_{\mathbb{F}}$$

are not well defined up to a canonical track. In Part 2 we deduce the Steenrod operations from a power map

$$\gamma : Z^1_{\mathbb{F}} \times Z^q_{\mathbb{F}} \longrightarrow Z^{pq}_{\mathbb{F}}.$$

It is a crucial observation in this book that also the power map γ is well defined up to a canonical track.

Appendix to Section 2.1: Small models of Eilenberg-MacLane spaces

The Eilenberg-MacLane spaces Z^n defined in (2.1.4) are very large spaces since they are defined by singular sets of spheres. They have the advantage of good symmetry properties like the commutative diagram (2.1.2).

In this appendix we discuss small models of Eilenberg-MacLane spaces which are frequently used in the literature, see Eilenberg-MacLane [EML] and Kristensen [Kr1]. The small models are directly related to chain complexes of simplicial sets.

We first recall the following notation.

Let R be a commutative ring and let **Mod** be the category of R-modules. Let gr**Mod** be the category of graded R-modules. For A and B in gr**Mod** a morphism $f : A \to B$ of degree k is given by morphisms $f_i : A_i \to B_{i+k}$. We write

(1) $$A_i \;=\; A^{-i}.$$

For $x \in A_n$ we say that $\mid x \mid = n$ is the *lower degree* of x and for $x \in A^m$ we say that $\mid x \mid = m$ is the *upper degree* of x. A chain complex is a map $d : A \to A$ of lower degree -1 with $dd = 0$ and a cochain complex is a map $d : A \to A$ with upper degree $+1$ and $dd = 0$. Using the rule (1) a chain complex is a cochain complex and vice versa. The *tensor product* of chain complexes is defined by

(2) $$(A \otimes B)_n \;=\; \bigoplus_{i+j=n} A_i \otimes B_j$$

where $i, j \in \mathbb{Z}$ and $A_i \otimes B_j$ is the tensor product over R in **Mod** and $d(a \otimes b) = (da) \otimes b + (-1)^{|a|} a \otimes (db)$.

Let Δ be the simplicial category. Objects in Δ are the sets $\underline{n} = \{0, \ldots, n\}$ and morphisms are monotone functions $\underline{n} \to \underline{m}$. A *simplicial object* in a category $\mathbf{C}$ is a functor $X : \Delta^{\mathrm{op}} \to \mathbf{C}$ where Δ^{op} is the opposite category. We set $X_n = X(\underline{n})$, $n \geq 0$. Let $\Delta\mathbf{C}$ be the category of such simplicial objects in $\mathbf{C}$. Morphisms in $\Delta\mathbf{C}$ are natural transformations. We have the well-known *Dold-Kan equivalence* of categories

(3) $$\mathbf{Chain}_+ \underset{N}{\overset{\Gamma}{\rightleftarrows}} \Delta\mathbf{Mod}.$$

Here $\mathbf{Chain}_+$ is the full subcategory in **Chain** consisting of chain complexes A with $A_i = 0$ for $i < 0$. Compare Goerss-Jardine [GJ]. The functor N is the normalization and Γ carries A to a simplicial object in **Mod** with

$$\Gamma(A)_n \;=\; \bigoplus_{\underline{n} \twoheadrightarrow \underline{k}} A_k$$

where $\underline{n} \twoheadrightarrow \underline{k}$ denotes surjections in Δ.

Let **Set** be the category of sets. Then the *forgetful* functor $\phi : \mathbf{Mod} \to \mathbf{Set}$ induces

(4) $$\phi : \Delta\mathbf{Mod} \longrightarrow \Delta\mathbf{Set}.$$

Moreover we have the *realization* functor

(5) $$|-| : \Delta\mathbf{Set} \longrightarrow \mathbf{Top}$$

where **Top** is the category of topological spaces. We also use the *free module functor*

(6) $$R : \Delta\mathbf{Set} \longrightarrow \Delta\mathbf{Mod}$$

which carries X to RX where $(RX)_n$ is the free R-module generated by X_n. We have the natural map $[-] : X \to \phi RX$ which carries $x \in X$ to the corresponding generator $[x] \in RX$. Moreover for $A \in \Delta\mathbf{Mod}$ and $f : X \to \phi A$ in $\Delta\mathbf{Set}$ we have the unique map $\bar{f} : RX \to A$ in $\Delta\mathbf{Mod}$ for which the composite $(\phi\bar{f})[-]$ coincides with f.

For a module M in **Mod** let

(7) $$M[n] \in \mathbf{Chain}$$

be the chain complex given by M *concentrated* in degree n; i.e., $M[n]_i = 0$ for $i \neq n$ and $M[n]_n = M$. Given a chain complex (C, d) in **Chain** we define the *cochain complex* $C^*(M) = \mathrm{Hom}(C, M)$ with

(8) $$C^n(M) \;=\; \mathrm{Hom}_{\mathbf{Mod}}(C_n, M)$$

and differential $\partial = \mathrm{Hom}(d, 1_M)$. Let

(9) $$Z^n(M) = \mathrm{kernel}\{\partial : C^n(M) \to C^{n+1}(M)\}$$

be the module of cocycles in degree n. One has the canonical binatural isomorphism

(10) $$Z^n(M) = \mathrm{Hom}_{\mathbf{Chain}}(C, M[n])$$

which we use as an identification.

For M in **Mod** we define the *Eilenberg-MacLane object* in $\Delta\mathbf{Set}$, resp. $\mathbf{Top}^*$, by

(11) $$K(M, n) = \phi\Gamma(M[n]) \in \Delta\mathbf{Set}^*,$$

(12) $$\mid K(M, n) \mid = \mid \phi\Gamma(M[n]) \mid \in \mathbf{Top}^*.$$

This is the *small* model of an Eilenberg-MacLane space. The construction shows that (11) and (12) are R-module objects in the category $\Delta\mathbf{Set}^*$ and $\mathbf{Top}^*$ respectively. Sine $\Gamma M[n]$ is a simplicial group we know that $K(M, n)$ is a Kan complex in Δ**Set**. Moreover $K(M, n)$ is pointed by

$$* = K(\{0\}, n) \longrightarrow K(M, n).$$

Here $\{0\}$ is the trivial module in **Mod** and $*$ is the point object in Δ**Set**. Of course the realization $\mid * \mid = *$ is the point object in **Top**.

(13) **Definition.** Let X be a simplicial set. Then

$$C_*(X) \;=\; NRX$$

is the (normalized) *chain complex* of X and

$$C^*(X,M) = \mathrm{Hom}(C_*X,M) = (NRX)^*(M)$$

is the *cochain complex* of X with coefficients in $M \in \mathbf{Mod}$. Let c be a cocycle of degree n in $C^*(X,M)$ so that $c : C_*X \to M[n]$ is a chain map which induces the composite

$$c_\# : X \xrightarrow{[-]} \phi RX \cong \phi\Gamma NRX \xrightarrow{\phi\Gamma(c)} \phi\Gamma M[n] = K(M,n)$$

in $\mathbf{\Delta Set}$. If X is pointed by $* \to X$ then $c_\#$ preserves the base point for $n \geq 1$. Moreover for a map $f : X \to Y$ in $\mathbf{\Delta Set}$ we have

$$c_\# \circ f \;=\; (f^*c)_\#.$$

(14) **Lemma.** For $K = K(M,n)$ there is a fundamental cocycle $i_n^M \in C^n(K,M)$ for which $(i_n^M)_\# : K \to K(M,n)$ is the identity.

Proof. We have the equation $(\Gamma M[n])_n = M$ so that $C_nK = RM$ and $i_n^M : C_*K \to M[n]$ is defined in degree n by the homomorphism $RM \to M$ in $\mathbf{Mod}$ which carries $[m]$to m. □

Let $\tilde{Z}^n(X,M)$ be the *module of cocycles* in degree n of

$$\tilde{C}^*(X,M) = C^*(X,M)/C^*(*,M).$$

Then (14) implies that one has a canonical bijection

$$\mathbf{\Delta Set}^*(X,K(M,n)) \;=\; \tilde{Z}^n(X,M) \tag{15}$$

which carries $f : X \to K(M,n)$ to $f^*i_n^M$. The inverse carries c to $c_\#$; see (13). Moreover the bijection is natural in X and M. By (15) we see that the definition of $K(M,n)$ above coincides with the definition of Eilenberg-MacLane [EML].

For a simplicial set X and $M,N \in \mathbf{Mod}$ we have the *Alexander-Whitney cup product* of cochains

$$\bigcup : C^*(X,M) \otimes C^*(X,N) \longrightarrow C^*(X,M\otimes N) \tag{16}$$

which is natural in X, M and N and which is associative. Now let $X = K(M,m) \times K(N,n)$ and let $p_1 : X \to K(M,n)$, $p_2 : X \to K(N,n)$ be the projections. Then the fundamental cocycles $i_m^M \in C(K(M,m),M)$ and $i_n^N \in C^n(K(N,n),N)$ yield the cocycle

$$p_1^*i_m^M \cup p_2^*i_n^N \in C^*(X,M\otimes N)$$

which by (15) gives us the map

$$\mu_{m,n} : K(M,m) \times K(N,n) \longrightarrow K(M\otimes N, m+n), \tag{17}$$

$\mu_{m,n} = (p_1^* i_m^M \cup p_2^* i_n^N)_\#$ in **ΔSet**. The map $\mu_{m,n}$ is again natural in M and N respectively and $\mu_{m,n}$ is associative in the obvious way. Naturality implies that $\mu_{m,n}$ carries $* \times K(N,n)$ and also $K(M,m) \times *$ to the base point of $K(M \otimes N, m+n)$. Hence $\mu_{m,n}$ defines an induced map

$$\mu_{m,n} : K(M,m) \wedge K(N,n) \to K(M \otimes N, m+n).$$

Here $X \wedge Y = X \times Y / X \times \{*\} \cup \{*\} \times Y$ is the *smash product* of pointed simplicial sets. The maps $\mu_{m,n}$, however, do not allow a commutative diagram as in (2.1.2) since the permutation group σ_n acts only by sign on $K(M,n)$.

We finally compare for $R = \mathbb{F} = \mathbb{Z}/p$ the small model $K(\mathbb{F},n)$ and the big model Z^n in (2.1.4). For this we choose a homomorphism

(18) $$\varphi : \Delta^n/\partial\Delta^n \approx S^n$$

which defines a generator φ and a cycle in the chain complex $NS(S^n)$ with $S(V) = R\,\mathrm{Sing}(V)/R\,\mathrm{Sing}(*)$ as in (2.1.4)(4). We thus obtain chain maps

(19) $$\mathbb{F}[n] \xrightarrow{i} NS(S^n) \xrightarrow{r} \mathbb{F}[n].$$

Here i is the inclusion with $i[n] = \varphi$ and i is a cofibration and a homotopy equivalence in the category of chain complexes. Hence we can choose a retraction r of i with $ri = 1$. By applying the functor $\mid \phi\Gamma \mid$ we get the $\mathbb{F}$-linear maps i, r between $\mathbb{F}$-vector space objects in **Top*** with $ri = 1$.

(20) $$\begin{array}{ccccc} \mid \phi\Gamma\mathbb{F}[n] \mid & \longrightarrow & \mid \phi\Gamma NS(S^n) \mid & \longrightarrow & \mid \phi\Gamma\mathbb{F}[n] \mid \\ \| & & \| & & \| \\ & & \mid \phi S(S^n) \mid & & \\ \| & & \| & & \| \\ K(\mathbb{F},n) & \xrightarrow{\quad i \quad} & Z^n & \xrightarrow{\quad r \quad} & K(\mathbb{F},n) \end{array}$$

Moreover i is a homotopy equivalence in **Top***. Using i, r we see that each map $\alpha : K(\mathbb{F},n) \to K(\mathbb{F},m)$ yields the map $\bar{\alpha} = i\alpha r : Z^n \to Z^m$ with the property $\bar{\beta\alpha} = \bar{\beta}\bar{\alpha}$ and $\bar{1} = ir$. Using (15) we get for a simplicial set X with $Y = \mid X \mid$ the map $(n \geq 1)$

(21) $$\tilde{Z}^n(X,\mathbb{F}) \to \mathbf{Top}^*(Y, Z^n) = [\![Y, Z^n]\!]_0$$

which carries the cocycle c to the map $i \mid c_\# \mid$. This shows the connection between the algebraic cocycles $Z^n(X,\mathbb{F})$ and the topological cocycles **Top**$^*(Y, Z^n)$.

2.2 Groupoids of maps

We here recall some basic notation and facts on groupoids. A groupoid **G** is a category in which all morphisms are invertible. The morphisms of **G** are termed *tracks*. The set of objects of $\mathbf{G}$ will be denoted by $\mathbf{G}_0$, and the set of morphisms of $\mathbf{G}$ will be denoted by $\mathbf{G}_1$. We have the canonical source and target maps

$$\mathbf{G}_1 \rightrightarrows \mathbf{G}_0. \tag{2.2.1}$$

Let **Grd** be the category of groupoids. Morphisms are functors between groupoids.

Tracks in a groupoid $\mathbf{G}$ are denoted by $H : f \to g$ or $H : f \simeq g$ or $H : f \Rightarrow g$. Let $\mathbf{G}(f, g)$ be the set of tracks $H : f \Rightarrow g$ in $\mathbf{G}$. Composition of tracks $H : f \Rightarrow g$ and $G : g \Rightarrow h$ is denoted by

$$G \square H : f \Rightarrow h.$$

The *identity track* or *trivial track* of f is $0^\square : f \Rightarrow f$. The inverse of a track $H : f \Rightarrow g$ is $H^{\mathrm{op}} : g \Rightarrow f$ so that $H \square (H^{\mathrm{op}}) = (H^{\mathrm{op}}) \square H = 0^\square$.

The set of connected components of $\mathbf{G}$ is

$$\pi_0(\mathbf{G}) = \mathbf{G}_0/\sim . \tag{2.2.2}$$

Here $f, g \in \mathbf{G}_0$ satisfy $f \sim g$ if there is a track $f \Rightarrow g$ in $\mathbf{G}$. Let

$$\pi_1(\mathbf{G}, f) = Aut(f) \tag{2.2.3}$$

be the group of automorphisms of the object f in $\mathbf{G}$ with $0^\square$ the neutral element in $Aut(f)$.

The groupoid $\mathbf{G}$ is *connected* if $\pi_0(\mathbf{G})$ is a point $*$. Moreover $\mathbf{G}$ is *contractible* if $\pi_0(\mathbf{G}) = *$ and $\pi_1(\mathbf{G}, f) = 0$ is the trivial group for $f \in \mathbf{G}_0$. For two objects f, g in a contractible groupoid $\mathbf{G}$ there is a unique morphism $f \Rightarrow g$ in $\mathbf{G}$. If all automorphism groups in $\mathbf{G}$ are trivial then all connected components in $\mathbf{G}$ are contractible.

The groupoid $\mathbf{G}$ is *discrete* if all tracks in $\mathbf{G}$ are trivial tracks. In this case $\pi_0(\mathbf{G}) = \mathbf{G}$.

A groupoid $\mathbf{G}$ is *abelian* if all automorphism groups $\pi_1(\mathbf{G}, f)$ with $f \in \mathbf{G}_0$ are abelian groups.

2.2.4 Example. Given a topological space X one obtains the *fundamental groupoid* $\Pi(X)$. Its objects are the points of X and morphisms $x_0 \to x_1$ with $x_0, x_1 \in X$ are homotopy classes rel. ∂I of paths $\omega : I \to X$ with $\omega(0) = x_0$ and $\omega(1) = x_1$. Here $I = [0, 1]$ is the unit interval with boundary $\partial I = \{0, 1\}$. Composition in $\Pi(X)$ is given by addition of paths. It is well known that $\Pi(X)$ is an abelian groupoid if X is a topological group or more generally if each path component of X has the homotopy type of an H-space. Moreover $\Pi(X)$ is connected if X is path connected and $\Pi(X)$ is contractible if X is 1-connected. Now let $(X, *)^{(A,*)}$ be the *mapping space* (with the compactly generated compact open topology, see [G]) of all pointed maps $A \to X$. Then the fundamental groupoid of this space is denoted by

$$\Pi((X, *)^{(A,*)}) = [\![A, X]\!]. \tag{1}$$

Objects in $[\![A,X]\!]_0$ are the pointed maps $f,g : A \to X$ and tracks $H : f \Rightarrow g$ in the groupoid $[\![A,X]\!]$ are homotopy classes of homotopies $f \simeq g$. We call $[\![A,X]\!]$ the *mapping groupoid*. The trivial map in $[\![A,X]\!]_0$ is $0 : A \to * \to X$. Let $[A,X]$ be the *set of homotopy classes* of pointed maps $A \to X$. This is the set

$$[A,X] = \pi_0[\![A,X]\!] \tag{2}$$

of connected components in the mapping groupoid. Now let $\Omega X = (X.*)^{(S^1,*)}$ be the *loop space* of X with $S^1 = [0,1]/\{0,1\}$. Then one readily checks that a track $0 \Rightarrow 0$ in $[\![A,X]\!]$ can be identified with the homotopy class of a pointed map $A \to \Omega X$ in $[A,\Omega X]$. Hence we have the equation of sets

$$[\![A,X]\!](0,0) = [A,\Omega X]. \tag{3}$$

2.2.5 Definition. Let $\mathbf{C}$ be a category with products $A \times B$. An *abelian group object* A in $\mathbf{C}$ is given by maps $+_A : A \times A \to A$, $-1_A : A \to A$, $0_A : * \to A$ satisfying the usual identities. Here $*$ is the final object in $\mathbf{C}$ which is considered to be the empty product. A map $f : A \to B$ between abelian group objects is linear, if $f0_A = 0_B$, $f(-1_A) = (-1_B)f$ and $f{+_A} = {+_B}(f \times f)$.

An abelian group object A is an $\mathbb{F}$*-vector space object* in $\mathbf{C}$ with $\mathbb{F} = \mathbb{Z}/p$ if the composite

$$A \xrightarrow{\Delta} A^{\times p} \xrightarrow{+} A$$

is the trivial map $A \to * \to A$. Here $A^{\times p} = A \times \cdots \times A$ is the p-fold product and Δ is the diagonal map and $+$ is defined by $+_A$.

The *category of pairs* in $\mathbf{C}$ denoted by $\mathbf{pair}(\mathbf{C})$ is defined as follows. Objects are morphisms $f : A \to B$ in $\mathbf{C}$ and morphisms $(\alpha,\beta) : f \to g$ in $\mathbf{pair}(\mathbf{C})$ are commutative diagrams in $\mathbf{C}$.

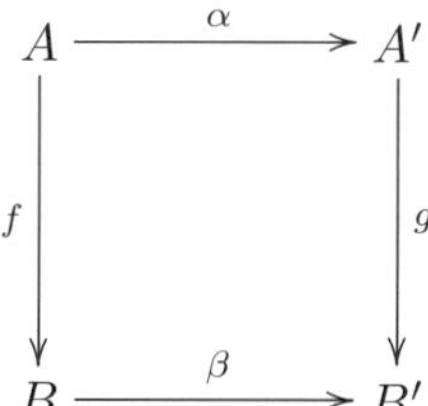

Let $\mathbf{Ab}$ be the category of abelian groups and let $\mathbf{Vec}_{\mathbb{F}}$ be the category of $\mathbb{F}$-vector spaces.

2.2.6 Proposition. *The category of abelian group objects in* $\mathbf{Grd}$ *and linear maps is equivalent to the category* $\mathbf{pair}(\mathbf{Ab})$. *The category of* $\mathbb{F}$*-vector space objects in* $\mathbf{Grd}$ *and linear maps is equivalent to the category* $\mathbf{pair}(\mathbf{Vec}_{\mathbb{F}})$.

Proof. Given an abelian group object G in $\mathbf{Grd}$ we obtain the object

$$\partial : G_1^0 \longrightarrow G_0$$

in **pair**(**Ab**) as follows. Here G_0 is the set of objects of G which is an abelian group since G is an abelian group object in **Grd**. Let $0 \in G_0$ be the neutral object in the abelian group G_0. Then G_1^0 is the set of all morphisms $H : a \Rightarrow 0$ in G with $a \in G_0$ and we define ∂ by $\partial H = a$. The abelian group structure of G_1^0 is defined by

$$(H : a \Rightarrow 0) + (G : b \Rightarrow 0) = (H + G, a + b \to 0 + 0 = 0)$$

where the right-hand side is defined since G is an abelian group object in **Grd**. Conversely given an object $\partial : A_1 \to A_0$ in **pair**(**Ab**) we define the abelian group object $G(\partial)$ in **Grd** as follows. The set of objects of $G(\partial)$ is

$$G(\partial)_0 = A_0.$$

The set of morphisms of $G(\partial)$ is the product set

$$G(\partial)_1 = A_1 \times A_0$$

where $(H, x) \in A_1 \times A_0$ is a morphism

(1) $$(H, x) = H + x : \partial(H) + x \Rightarrow x.$$

The identity of x is $0_x^\square = (0, x) = 0 + x = x = \partial(0) + x \Rightarrow x$. Composition of

(2) $$x \overset{H+x}{\Longleftarrow} \partial(H) + x \overset{G+\partial(H)+x}{\Longleftarrow} \partial(G) + \partial(H) + x$$

is defined by

(3) $$(H + x)\square(G + \partial(H) + x) = (H + G) + x$$

for $H, G \in A_1$ and $x \in A_0$. Now it is readily seen that this way one gets an isomorphism of categories. The inverse $(H + x)^{\mathrm{op}}$ of the morphism $H + x$ is given by

(4) $$(H + x)^{\mathrm{op}} = (-H) + (\partial(H) + x) = -H + \partial(H) + x.$$

By (3) one readily checks that $(H + x)^{\mathrm{op}}\square(H + x) = 0^\square$.

$$\begin{aligned}(H + x)^{\mathrm{op}}\square(H + x) &= (-H + \partial(H) + x)\square(H + x)\\ &= (-H + H) + \partial(H) + x\\ &= 0 + \partial(H) + x\\ &= 0^\square_{\partial(H)+x}.\end{aligned}$$ □

We point out that an abelian group object in **Grd** is an abelian groupoid but not vice versa; that is, an abelian groupoid need not be an abelian group object in **Grd**.

For a product of pointed spaces we get the equation of mapping groupoids

(2.2.7) $$[\![X, A \times B]\!] = [\![X, A]\!] \times [\![X, B]\!].$$

This implies the following lemma.

2.2.8 Lemma. *If Z is an abelian group object in the category* $\mathbf{Top}^*$ *then* $[\![X,Z]\!]$ *is an abelian group object in the category of groupoids.*

Hence you can, for the abelian group object Z in $\mathbf{Top}^*$, apply Proposition (2.2.6) so that $[\![X,Z]\!]$ is determined by the homomorphism of abelian groups

$$\partial : [\![X,Z]\!]_1^0 \longrightarrow [\![X,Z]\!]_0.$$

Here $[\![X,Z]\!]_0$ is the abelian group of all pointed maps $X \to Z$ in $\mathbf{Top}^*$ and $[\![X,Z]\!]_1^0$ is the abelian group of all (f,H) where $H : f \Rightarrow 0$ is a track in $[\![X,Z]\!]$ from f to the trivial map $X \to * \to Z$. Moreover we obtain the exact sequence of abelian groups

$$0 \longrightarrow [X,\Omega Z] \longrightarrow [\![X,Z]\!]_1^0 \longrightarrow [\![X,Z]\!]_0 \longrightarrow [X,Z] \to 0. \tag{2.2.9}$$

Since the Eilenberg-MacLane space Z^n in (2.1) is an $\mathbb{F}$-vector space object in $\mathbf{Top}^*$ we see accordingly by (2.1.6) that we get an exact sequence of $\mathbb{F}$-vector spaces ($n \geq 1$)

$$0 \longrightarrow \tilde{H}^{n-1}(X) \longrightarrow [\![X,Z^n]\!]_1^0 \xrightarrow{\partial} [\![X,Z^n]\!]_0 \to \tilde{H}^n(X) \to 0. \tag{2.2.10}$$

Here $\tilde{H}^*(X)$ is the kernel of $H^*(X) \to H^*(*)$ induced by the inclusion $* \to X$. In fact, ∂ determines the groupoid $[\![X,Z^n]\!]$ by (2.2.6).

2.2.11 Remark. Using (21) in the appendix of (2.1) one obtains for a pointed simplicial set X the commutative diagram with exact rows and $Y = | X |$.

$$\begin{array}{ccccccccc}
\tilde{H}^{n-1}(X) & \longrightarrow & \tilde{C}^{n-1}(X,\mathbb{F})/\tilde{B}^{n-1}(X,\mathbb{F}) & \xrightarrow{d} & \tilde{Z}^n(X,\mathbb{F}) & \longrightarrow & \tilde{H}^n(X) & & \\
\| & & \downarrow{\scriptstyle j} & & \downarrow{\scriptstyle i} & & \| & & \\
\tilde{H}^{n-1}(Y) & \longrightarrow & [\![Y,Z^n]\!]_1^0 & \xrightarrow{\partial} & [\![Y,Z^n]\!]_0 & \longrightarrow & \tilde{H}^n(Y) & \longrightarrow & 0
\end{array}$$

where $\tilde{B}^{n-1}(X,\mathbb{F}) = \text{image } d : \tilde{C}^{n-2}(X,\mathbb{F}) \to \tilde{C}^{n-1}(X,\mathbb{F})$. Hence ∂ in (2.2.10) describes part of the boundary d in the cochain complex $\tilde{C}^*(X,\mathbb{F})$. In fact i and j are injective.

2.3 Track categories and track theories

A *category enriched in groupoids* $\mathcal{T}$, also termed *track category* for short, is the same as a 2-*category* all of whose 2-cells are invertible. It is thus a class of objects $ob(\mathcal{T})$, a collection of groupoids $\mathcal{T}(A,B)$ for $A,B \in Ob\mathcal{T}$ called hom-*groupoids* of $\mathcal{T}$, identities $1_A \in \mathcal{T}(A,A)_0$ and composition functors $\mathcal{T}(B,C) \times \mathcal{T}(A,B) \to \mathcal{T}(A,C)$ satisfying the usual equations of associativity and identity morphisms. For generalities on enriched categories the reader may consult Kelly [Ke]. Objects of the hom-groupoids $f \in \mathcal{T}(A,B)_0$, called *maps* in $\mathcal{T}$, constitute morphisms of an ordinary category $\mathcal{T}_0$ having the same objects as $\mathcal{T}$.

For $f, g \in \mathcal{T}(A,B)$ we shall write $f \simeq g$ (and say f is *homotopic* to g) if there exists a morphism $H : f \to g$ in $\mathcal{T}(A,B)$. Occasionally this will be also denoted as $H : f \simeq g$ or $H : f \Rightarrow g$, H sometimes called a *homotopy* or a *track* from f to g. Homotopy is a natural equivalence relation on morphisms of $\mathcal{T}_0$ and determines the *homotopy category* $\mathcal{T}_{\simeq} = \mathcal{T}_0/\simeq$. Objects of $\mathcal{T}_{\simeq}$ are once again objects in $ob(\mathcal{T})$, while morphisms of $\mathcal{T}_{\simeq}$ are homotopy classes of morphisms in $\mathcal{T}_0$. Let $q : \mathcal{T}_0 \to \mathcal{T}_{\simeq}$ be the quotient functor. Moreover let $\mathcal{T}_1$ be the disjoint union of all tracks in $\mathcal{T}$. One has the source and target functions between sets

$$\mathcal{T}_1 \overset{s}{\underset{t}{\rightrightarrows}} \mathrm{Mor}(\mathcal{T}_0) \tag{2.3.1}$$

with $qs = qt$. Here $\mathrm{Mor}(\mathcal{T}_0)$ denotes the set of morphisms in the category $\mathcal{T}_0$. We borrow from topology the following notation in a track category $\mathcal{T}$. Let

$$[A,B] = \mathcal{T}_0(A,B)/\simeq$$

be the set of homotopy classes of maps $A \to B$ and let

$$[\![A,B]\!] = \mathcal{T}(A,B)$$

be the hom-groupoid of $\mathcal{T}$ so that $[A,B]$ is the set of connected components of the groupoid $[\![A,B]\!]$.

For tracks $H : f \Rightarrow g$ in $[\![A,B]\!]$ and $H' : f \Rightarrow g$ in $[\![B,C]\!]$ we get the composed track $H' * H : f'f \Rightarrow g'g$ for the diagram

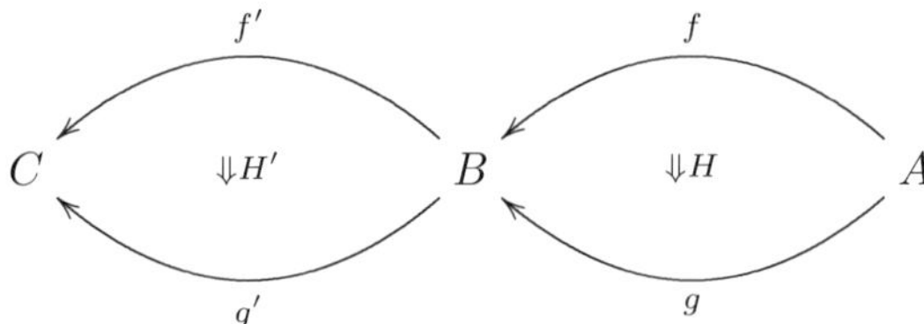

satisfying the formula

$$H' * H = (g'H')\square(H'f) = (H'g)\square(f'H).$$

We call $H' * H$ the “pasting of tracks” or the “horizontal composition” of tracks.

A map $f : A \to B$ is a *homotopy equivalence* if there exists a map $g : B \to A$ and tracks $fg \simeq 1$ and $gf \simeq 1$. This is the case if and only if the homotopy class of f is an equivalence in the homotopy category $\mathcal{T}_{\simeq}$. In this case A and B are called *homotopy equivalent* objects.

The morphisms in $\mathcal{T}_0$ are also termed 1-cells and the tracks in $\mathcal{T}_1$ are 2-cells. In particular, the category **Gpd** of groupoids is a track category. Objects are groupoids, morphisms are functors and tracks are natural transformations (since they are natural isomorphisms). Moreover any category **C** can be considered to be a track category with only identity tracks.

The leading example is the track category $[\![\mathbf{Top}^*]\!]$ of compactly generated Hausdorff spaces with basepoint $*$, given as follows. For pointed spaces A, B let $[\![A, B]\!]$ be the mapping groupoid. See (2.2.4). Hence maps are pointed maps $f, g : A \to B$ between pointed spaces and tracks $H : f \Rightarrow g$ are homotopy classes relative to $A \times \partial I$ of homotopies $H : A \times I / * \times I \to B$ with $H : f \simeq g$. In this case

$$[\![\mathbf{Top}^*]\!]_{\simeq} = \mathbf{Top}^* / \simeq \tag{2.3.2}$$

is the usual homotopy category of pointed spaces. Let $\mathbf{C} \subset \mathbf{Top}^* / \simeq$ be a full subcategory. Then $[\![\mathbf{C}]\!]$ is the track category consisting of all spaces A with $A \in Ob(\mathbf{C})$, that is $[\![\mathbf{C}]\!] \subset [\![\mathbf{Top}^*]\!]$ is a full subcategory of the track category $[\![\mathbf{Top}^*]\!]$. In particular we get the following case.

2.3.3 Definition. For a prime p the theory $\mathbf{K}_p$ of Eilenberg-MacLane spaces in (1.5.1) yields the track category

$$[\![\mathbf{K}_p]\!] \subset [\![\mathbf{Top}^*]\!].$$

Here $[\![\mathbf{K}_p]\!]$ consists of products $A = Z^{n_1} \times \cdots \times Z^{n_r}$ of Eilenberg-MacLane spaces $Z^n = K(\mathbb{Z}/p, n)$ with $n_1, \ldots, n_r \geq 1$ and $r \geq 0$. Morphisms are pointed maps between such products and tracks are homotopy classes of homotopies between such maps. We call $[\![\mathbf{K}_p]\!]$ the *track theory of Eilenberg-MacLane spaces*.

Here we use the following notion of track theory.

2.3.4 Definition. A *strong product* in a track category $\mathcal{T}$ is an object $A \times B$ equipped with maps $p_A = p_1 : A \times B \to A$, $p_B = p_2 : A \times B \to B$ in $\mathcal{T}_0$ such that the induced functor

$$[\![X, A \times B]\!] \longrightarrow [\![X, A]\!] \times [\![X, B]\!] \tag{$*$}$$

given by $f \mapsto (p_A f, p_B f)$, $(H : f \Rightarrow g) \mapsto (p_A H : p_A f \Rightarrow p_S g, p_B H : p_B f \Rightarrow p_B g)$ is an isomorphism of groupoids for all X in $\mathcal{T}$. We call $(A \times B, p_A, p_B)$ a *weak product* if $(*)$ is an equivalence of categories. Similarly a final object $*$ in $\mathcal{T}$ is *strong* if $[\![X, *]\!]$ is a groupoid with a unique morphism. Whereas a *weak final object* is an object $*$ for which $[\![X, *]\!]$ is equivalent to such a groupoid. A *track theory* is a track category $\mathcal{T}$ with a strong final object and with finite strong products. This is the analogue of a theory **T** in (1.5).

For example a product of spaces $(A \times B, p_1, p_2)$ in $\mathbf{Top}^*$ is also a strong product in the track category $[\![\mathbf{Top}^*]\!]$ by equation (2.2.7). Moreover $*$ is a strong final object in $[\![\mathbf{Top}^*]\!]$.

Hence we see

2.3.5 Lemma. *The track category* $[\![\mathbf{K}_p]\!]$ *of Eilenberg-MacLane spaces is a track theory. Products in* $\mathbf{K}_p$ *are also strong products in* $[\![\mathbf{K}_p]\!]$.

A *track functor*, or else 2-functor, between track categories is a groupoid enriched functor. For example, any object A of a track category $\mathcal{T}$ gives rise to the representable track functor

$$[\![A,-]\!] : \mathcal{T} \longrightarrow \mathbf{Gpd} \tag{2.3.6}$$

sending an object X to the groupoid $[\![A,X]\!]$. This 2-functor assigns to a map $f : X \to Y$ the functor $[\![A,f]\!] : [\![A,X]\!] \to [\![A,Y]\!]$ sending $g : A \to X$ to fg and $\gamma : g \to g'$ to γf. And this 2-functor assigns to a track $\varphi : f \Rightarrow f'$ the natural transformation $[\![A,\varphi]\!] : [\![A,f]\!] \to [\![A,f']\!]$ with components $\varphi g : fg \to fg'$.

2.3.7 Definition. A *track model* M *of a track theory* $\mathcal{T}$ is a functor $M : \mathcal{T} \to \mathbf{Grd}$ which carries strong products in $\mathcal{T}$ to products of groupoids.

As a special case of such a track model we obtain for each path-connected pointed space X in $[\![\mathbf{Top}^*]\!]$ the representable track functor

$$[\![X,-]\!] : [\![\mathbf{K}_p]\!] \longrightarrow \mathbf{Gpd} \tag{2.3.8}$$

which we call the *secondary cohomology* of X. This generalizes the cohomology of the space X since we have seen that

$$\tilde{H}^*(X) = [X,-] : \mathbf{K}_p \longrightarrow \mathbf{Set} \tag{2.3.9}$$

is a representable functor in $\mathbf{Top}^*/\simeq$. Here $[X,-]$ is a model of $\mathbf{K}_p$ which carries products in $\mathbf{K}_p$ to products of sets, see (1.5.3). Similarly the secondary cohomology $[\![X,-]\!]$ is a track model of the track theory $[\![\mathbf{K}_p]\!]$ which carries strong products in $[\![\mathbf{K}_p]\!]$ to products in $\mathbf{Gpd}$.

We have seen in (1.5.2) that models of the theory $\mathbf{K}_p$ can be identified with connected unstable algebras over the Steenrod algebra $\mathcal{A}$. We are interested in understanding a corresponding result for track models of the track theory $[\![\mathbf{K}_p]\!]$. For this reason we introduce in Section (2.5) below the secondary Steenrod algebra $[\![\mathcal{A}]\!]$.

The particular choice of Eilenberg-MacLane spaces Z^n in (2.1) yields many further properties of the track theory $[\![\mathbf{K}_p]\!]$. For example, for objects A, B in $\mathbf{K}_p$ the morphism groupoid $[\![A,B]\!]$ is an $\mathbb{F}$-vector space object in the category of groupoids since B is an $\mathbb{F}$-vector space object in $\mathbf{Top}^*$, see (2.2.7).

The track theory $[\![\mathbf{K}_p]\!]$ is very large since all maps between products of Eilenberg-MacLane spaces in $\mathbf{Top}^*$ are morphisms in $[\![\mathbf{K}_p]\!]_0$. For this reason mainly the weak equivalence type of $[\![\mathbf{K}_p]\!]$ is of interest. Here weak equivalences are defined as follows.

2.3.10 Definition. A track functor $F : \mathcal{T} \to \mathcal{T}'$ is called a *weak equivalence* between track categories if the functors $[\![A,B]\!] \to [\![F(A),F(B)]\!]$ are equivalences of groupoids for all objects A, B of $\mathcal{T}$ and each object A' of $\mathcal{T}'$ is homotopy equivalent to some object of the form $F(A)$. Such a weak equivalence induces a functor $F : \mathcal{T}_{\simeq} \to \mathcal{T}'_{\simeq}$ between homotopy categories which is an equivalence of categories.

Below we study weak equivalences between linear track extensions which are special weak equivalences as above.

2.4 Secondary cohomology operations

Let ΩY be the *loop space* of the pointed space Y. An element ΩY is a pointed map from the 1-sphere S^1 to Y. The basepoint of ΩY is the trivial map $S^1 \to * \in Y$. Let

$$\Omega_0 Y \subset \Omega Y \tag{2.4.1}$$

be the path-connected component of the basepoint in ΩY. The following lemma describes a well-known property of the track category $[\![\mathbf{Top}^*]\!]$.

2.4.2 Lemma. *Let X, Y be the path-connected pointed spaces and let $0 : X \to * \to Y$ be the trivial map. Then a track $A : 0 \Rightarrow 0$ in $[\![\mathbf{Top}^*]\!]$ can be identified with an element $A \in [X, \Omega_0 Y]$.*

Recall that $Z^n = K(\mathbb{Z}/p, n)$ is an Eilenberg-MacLane space for $n \geq 1$. Now we define

$$Z^n = * \quad \text{for } n \leq 1.$$

Then we obtain for all $n \in \mathbb{Z}$ the homotopy equivalence

$$\Omega_0 Z^n \simeq Z^{n-1} \tag{2.4.3}$$

which is well defined up to homotopy. The functor Ω_0 is compatible with products of pointed spaces; that is; $\Omega_0(X \times Y) = \Omega_0(X) \times \Omega_0(Y)$. Hence we can use (2.4.3) to define the following *loop functor*

$$L : \mathbf{K}_p \longrightarrow \mathbf{K}_p \tag{2.4.4}$$

for the theory $\mathbf{K}_p$ of Eilenberg-MacLane spaces. The functor L carries $A = Z^{n_1} \times \cdots \times Z^{n_r} \in \mathbf{K}_p$ to the object

$$L(A) = Z^{n_1 - 1} \times \cdots Z^{n_r - 1} \in \mathbf{K}_p.$$

Moreover L carries $f \in [A, B]$ with $A, B \in \mathbf{K}_p$ to the composite

$$L(f) : L(A) \simeq \Omega_0(A) \xrightarrow{\Omega_0(f)} \Omega_0(B) \simeq L(B).$$

Here the homotopy equivalence is given by (2.4.3).

We are now ready to describe classical secondary cohomology operations in the sense of Adams [A] or Kristensen [Kr1].

2.4.5 Definition. Let A, B, C be objects in $\mathbf{K}_p$. Then a *relation* (α, β) is a commutative diagram in $\mathbf{K}_p$,

(1)
$$A \xrightarrow{\alpha} B \xrightarrow{\beta} C, \qquad A \xrightarrow{0} C$$

where 0 is the trivial map, i.e., $\beta\alpha = 0$. A *secondary cohomology operation* (a, b, H) associated to the relation $\beta\alpha = 0$ in $\mathbf{K}_p$ is a diagram in $[\![\mathbf{K}_p]\!]$,

(2)
$$A \xrightarrow{a} B \xrightarrow{b} C, \qquad A \xrightarrow{0} C, \qquad \Uparrow H$$

where a (resp. b) represents α (resp. β). Here the track H exists since $\beta\alpha = 0$ in $\mathbf{K}_p$. The secondary cohomology operation (a, b, H) associated to the relation (α, β) defines a function $\theta_{(a,b,H)}$ as follows. We consider a path-connected pointed space X and the following diagram in $[\![\mathbf{Top}^*]\!]$.

(3)

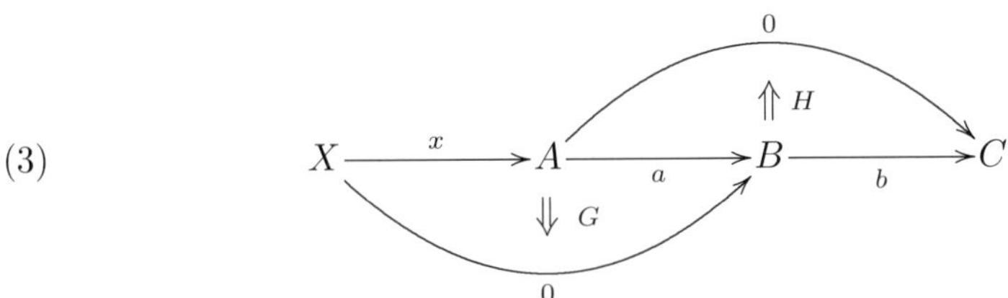

Here x represents $\xi \in [X, A]$ with $\alpha\xi = 0$. Hence tracks G exist and we get the composed track $0 \Rightarrow 0$ of the form

(4)
$$(Hx)\square(bG)^{\mathrm{op}} \in [X, \Omega_0 C] = [X, L(C)].$$

Here we use (2.4.2) and the loop functor L on $\mathbf{K}_p$ in (2.4.4). Now let $\theta_{(a,b,H)}$ be the function which carries $\xi \in [X, A]$ with $\alpha\xi = 0$ to the subset

(5)
$$\theta_{(a,b,H)}(\xi) \subset [X, L(C)]$$

consisting of all elements $(Hx)\square(bG)^{\mathrm{op}}$ with $G : ax \Rightarrow 0$ in $[\![\mathbf{Top}^*]\!]$ and x representing ξ. Of course $[X, A]$ and $[X, L(C)]$ are determined by the representable model $[X, -]$ of $\mathbf{K}_p$ which in turn can be identified with the cohomology $H^*(X)$ by (1.5.3). Therefore $\theta_{(a,b,H)}$ is a *secondary cohomology operation* in the classical sense. We shall see in Chapter 3 that $\theta_{(a,b,H)}(\xi)$ is a coset of the subgroup $\mathrm{image}((L\beta)_* : [X, L(B)] \to [X, L(C)])$. Hence $\theta_{(a,b,H)}$ is a well-defined function.

(6)
$$\begin{array}{c}\mathrm{kernel}(\alpha_* : [X, A] \to [X, B]) \\ \downarrow \theta_{(a,b,H)} \\ \mathrm{cokernel}((L\beta)_* : [X, L(B)] \to [X, L(C)])\end{array}$$

This is the typical form of a secondary cohomology operation in the literature. Two secondary cohomology operations (a, b, H) and (a', b', H') associated to the relation (α, β) are *equivalent* if there exist tracks A, B in $[\![\mathbf{K}_p]\!]$ such that the pasting of tracks in the following diagram yields the trivial track $0 \Rightarrow 0$.

(7)

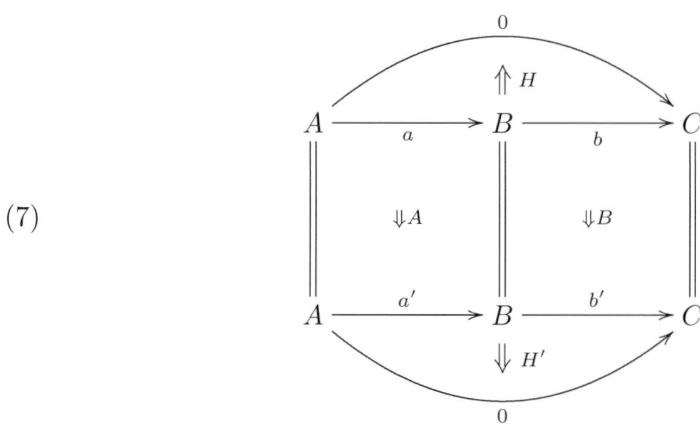

Of course the operation (6) depends only on the equivalence class (a, b, H) which in turn is well defined by the weak equivalence class of $[\![\mathbf{K}_p]\!]$. Here we use weak equivalences of linear track extensions as in Chapter 3 below which are special weak equivalences in the sense of (2.2.11).

In Section (2.6) we consider the *stable* version of secondary cohomology operations.

We describe two examples of secondary cohomology operations due to Adams [A] which are actually stable operations.

2.4.6 Example. For $p = 2$ consider the relation (α, β) in $\mathbf{K}_2$ given by $(n \geq 1)$

$$Z^n \xrightarrow{\alpha} Z^{n+1} \times Z^{n+3} \times Z^{n+4} \xrightarrow{\beta} Z^{n+5}$$

with $\alpha = (Sq^1, Sq^3, Sq^4)$ and $\beta = (Sq^4, Sq^2, Sq^1)$. Then there is a unique (stable) secondary operation (a, b, H) associated to (α, β) such that for $n = 2$ and $u \in H^2(\mathcal{C}P_\infty) = [\mathcal{C}P_\infty, Z^2]$ we have

$$\theta_{(a,b,H)}(u) = u^3 \in H^6(\mathcal{C}P_\infty) = [\mathcal{C}P_\infty, LZ^7].$$

Here $\mathcal{C}P_\infty$ is the complex projective space and $u \in H^2(\mathcal{C}P_\infty) = \mathbb{Z}/2$ is the generator. Compare the Addendum of Adams [A].

The main result of Adams [A] which implies the solution of the Hopf invariant problem is the following example. Compare also the explicit calculation in (16.6.5) below.

2.4.7 Example. Let $p = 2$. Then there are stable relations $(d(j), z_{i,j})$, $0 \leq i \leq j$, $j \neq i + 1$, in $\mathbf{K}_2$ of the form $(n \geq 1)$

$$Z^m \xrightarrow{d(j)} \times_{t=0}^{j} Z^{m+2^t} \xrightarrow{z_{i,j}} Z^{m+2^i+2^j}$$

with $d(j) = (Sq^1, Sq^2, \dots, Sq^{2^j})$ and $z_{i,j}$ chosen as in [A] page 88. Moreover let $\theta_{i,j}$ be the (stable) secondary operation associated to $(d(j), z_{i,j})$ with

$$\theta_{i,j}: \quad \begin{array}{ccc} \operatorname{kernel}(d(j)_*) & \longrightarrow & \operatorname{cokernel}(L(Z_{i,j})_*) \\ \big\downarrow & & \big\uparrow \\ H^m(X) & & H^{m+2^i+2^j-1}(X) \end{array}$$

Take $k \geq 3$. Then $u \in \operatorname{kernel}(d(k)_*) \subset H^m(X)$ satisfies

$$Sq^{2^{k+1}}(u) \in \sum_{0\leq i\leq j\leq k, j\neq i+1} a_{i,j,k}\theta_{i.j}(u) \subset H^{m+2^{k+1}}(X)$$

with appropriate $a_{i,j,k} \in \mathcal{A}_2$. See Theorem 4.6.1 [A]. We have seen in (1.1.3) that the Steenrod operations Sq^{2^i}, $i \geq 0$, are indecomposable generators of the Steenrod Algebra $\mathcal{A}_2$. The formula of Adams, however, shows that $Sq^{2^{k+1}}$ for $k \geq 3$ is decomposable with respect to secondary cohomology operations. This describes a deep and fundamental relation in the track theory $[\![\mathbf{K}_2]\!]$ of Eilenberg-MacLane spaces.

We point out that secondary cohomology operations (a, b, H) are special diagrams in the track theory $[\![\mathbf{K}_p]\!]$ of Eilenberg-MacLane spaces. Moreover the associated operations $\theta_{(a,b,H)}$ can be deduced from the track model $[\![X, -]\!]$ of $[\![\mathbf{K}_p]\!]$ in (2.2.10). In fact, any algebraic track model M of $[\![\mathbf{K}_p]\!]$ allows the definition of $\theta_{(a,b,H)}$ accordingly.

2.5 The secondary Steenrod algebra

All objects $Z^{n_1} \times \cdots \times Z^{n_r}$ in the theory $\mathbf{K}_p$ are by the construction in Section (2.1) $\mathbb{F}$-vector space objects in the category $\mathbf{Top}^*$. A morphism $\varphi \in [A, B]$ in $\mathbf{K}_p$ is *linear* in the homotopy category $\mathbf{Top}^*/\simeq$ if the diagram

(2.5.1)
$$\begin{array}{ccc} A \times A & \xrightarrow{f\times f} & B \times B \\ {\scriptstyle +}\big\downarrow & & \big\downarrow{\scriptstyle +} \\ A & \xrightarrow{f} & B \end{array}$$

homotopy commutes in $\mathbf{Top}^*$. Here f represents the homotopy class φ. If φ is linear in $\mathbf{Top}^*/\simeq$ then in general there exists no representing map $f \in \varphi$ for which diagram (2.5.1) commutes, so that $f \in \varphi$ in general cannot be chosen to be linear in $\mathbf{Top}^*$.

We use the loop functor $L : \mathbf{K}_p \to \mathbf{K}_p$ in (2.4.4) in the following definition of stable operation.

2.5.2 Definition. A *stable operation* in $\mathbf{K}_p$ of degree $k \in \mathbb{Z}$ is a sequence of maps

$$\alpha = (\alpha_n : Z^n \to Z^{n+k})_{n\in\mathbb{Z}}$$

in $\mathbf{K}_p$ with $L(\alpha_{n+1}) = \alpha_n$ for $n \in \mathbb{Z}$. Hence all α_n are linear and therefore the set $\mathcal{A}^k$ of all stable operations in $\mathbf{K}_p$ of degree k is an abelian group, in fact an $\mathbb{F}$-vector space. Moreover the composition $\beta \circ \alpha$ of stable operations given by $(\beta\circ\alpha)_n = \beta_{n+k}\circ\alpha_n$, $n \in \mathbb{Z}$, is bilinear so that composition yields the associative multiplication

$$\mathcal{A}^r \otimes \mathcal{A}^k \longrightarrow \mathcal{A}^{k+r}$$

carrying $\beta\otimes\alpha$ to $\beta\circ\alpha$. Hence $\mathcal{A} = \{\mathcal{A}^k, k \in \mathbb{Z}\}$ is a graded algebra with $\mathcal{A}^0 = \mathbb{F}$ and $\mathcal{A}^k = 0$ for $k < 0$.

The next result is a well-known consequence of (1.1.2).

2.5.3 Theorem. *The algebra $\mathcal{A}$ of stable operations in $\mathbf{K}_p$ coincides with the Steenrod algebra.*

Using power maps γ we shall define stable operations in $\mathbf{K}_p$ $(n \in \mathbb{Z})$,

$$\begin{aligned} Sq^k :\ & Z^n \longrightarrow Z^{n+k}, && \text{for } p = 2,\\ \beta :\ & Z^n \longrightarrow Z^{n+1}, && \text{for } p \text{ odd}\\ P^k :\ & Z^n \longrightarrow Z^{n+2k(p-1)}, && \text{for } p \text{ odd}\end{aligned}$$

which as well yield the isomorphism in (2.5.3), see Part II of this book.

We use the track theory $[\![\mathbf{K}_p]\!]$ to define the following secondary analogue of the Steenrod algebra in (2.5.3).

2.5.4 Definition. We fix for $n \in \mathbb{Z}$ maps in $\mathbf{Top}^*$

$$r_n : Z^n \longrightarrow \Omega_0 Z^{n+1}$$

which are homotopy equivalences defined in (2.1.7). Then let $[\![\mathcal{A}^k]\!]$ be the following groupoid, $k \geq 1$. Objects (α, H_α) in $[\![\mathcal{A}^k]\!]$ are sequences of maps in $\mathbf{Top}^*$

$$\alpha = (\alpha_n : Z^n \to Z^{n+k})_{n\in\mathbb{Z}}$$

together with sequences of tracks $H_\alpha = (H_{\alpha,n})_{n\in\mathbb{Z}}$ for the diagram

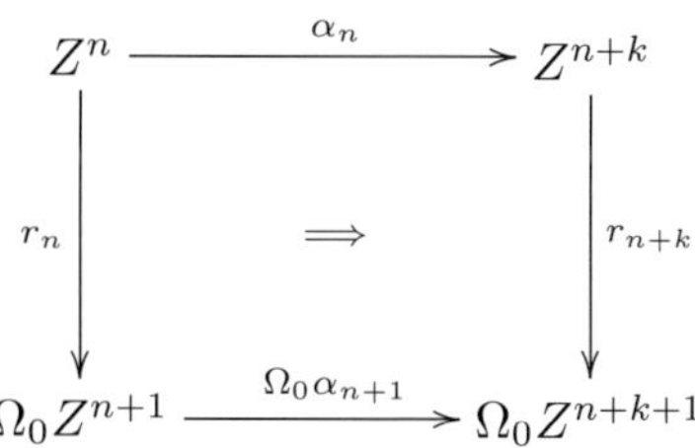

that is $H_{\alpha,n} : (\Omega_0\alpha_{n+1})r_n \Rightarrow r_{n+k}\alpha_n$. This implies that the homotopy class of α_n yields a stable operation in (2.4.2).

For $k = 0$ let $[\![\mathcal{A}^0]\!] = \mathbb{F}$ be the discrete groupoid given by $\mathcal{A}^0 = \mathbb{F}$. The elements $\alpha \in \mathbb{F}$ yield maps $\alpha_n = \alpha : Z^n \to Z^n$ given by the $\mathbb{F}$-vector space structure of Z^n. Hence α_n satisfies $r_n\alpha_n = \Omega_0(\alpha_{n+1})r_n$ so that $H_{\alpha,n}$ in this case is the trivial track.

For $k < 0$ let $[\![\mathcal{A}^k]\!] = 0$ be the trivial groupoid.

For $k > 0$ we define morphisms $H : (\alpha, H_\alpha) \Rightarrow (\beta, H_\beta)$ in the groupoid $[\![\mathcal{A}^k]\!]$ by sequences of tracks

$$H = (H_n : \alpha_n \Rightarrow \beta_n)_{n\in\mathbb{Z}}$$

in $[\![\mathbf{K}_p]\!]$ for which the pasting of tracks in the following diagram coincides with $H_{\beta,n}$.

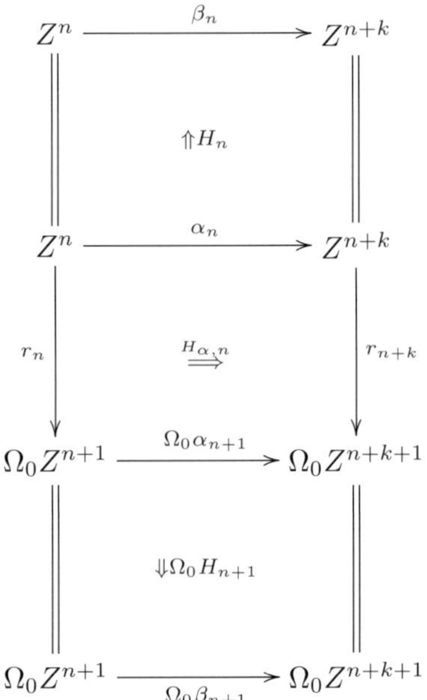

That is, the following equation holds in $[\![\mathbf{Top}^*]\!]$,

$$H_{\beta,n} = (\Omega_0 H_{n+1})r_n \square H_{\alpha,n} \square r_{n+k} H_n^{\mathrm{op}}.$$

Composition in $[\![\mathcal{A}^k]\!]$ is defined by $(H\square G)_n = H_n \square G_n$. One readily checks that $[\![\mathcal{A}^k]\!]$ is a well-defined groupoid with homotopy category $[\![\mathcal{A}^k]\!]_\simeq = \mathcal{A}^k$. Moreover one has a composition functor between groupoids

$$[\![\mathcal{A}^r]\!] \times [\![\mathcal{A}^k]\!] \xrightarrow{\circ} [\![\mathcal{A}^{k+r}]\!]$$

which is defined on objects by

$$(\alpha', H_{\alpha'}) \circ (\alpha, H_\alpha) = (\alpha'_{n+k} \circ \alpha_n, H_{\alpha',n+k} * H_{\alpha,n})_{n\in\mathbb{Z}}$$

where $*$ is the pasting operation. Moreover on morphisms $H : (\alpha, H_\alpha) \Rightarrow (\beta, H_\beta)$ and $H' : (\alpha', H_{\alpha'}) \Rightarrow (\beta', H_{\beta'})$ the composition functor is defined by

$$H \circ H' = (H'_n * H_n : \alpha'_n \circ \alpha_n \Rightarrow \beta'_n \circ \beta_n)_{n \in \mathbb{Z}}.$$

This shows that

$$[\![\mathcal{A}]\!] = ([\![\mathcal{A}^k]\!], \circ)_{k \in \mathbb{Z}}$$

is a monoid in the category of graded groupoids. We call $[\![\mathcal{A}]\!]$ the *secondary Steenrod algebra*. In fact, we shall prove that the homotopy category $[\![\mathcal{A}]\!]_\simeq = \mathcal{A}$ is the Steenrod algebra.

2.5.5 Lemma. *Each groupoid* $[\![\mathcal{A}^k]\!]$ *is an* $\mathbb{F}$*-vector space object in the category of groupoids. The composition in* $[\![\mathcal{A}]\!]$*, however, is not bilinear since maps in* $[\![\mathcal{A}^k]\!]_0$ *in general are not linear in* $\mathbf{Top}^*$*, see* (2.5.1).

Proof. We see that $[\![\mathcal{A}^k]\!]$ is an $\mathbb{F}$-vector space object since r_n in (2.1.7) is a linear map between $\mathbb{F}$-vector space objects in $\mathbf{Top}^*$. □

2.6 The stable track theory of Eilenberg-MacLane spaces

Again we use the loop functor $L : \mathbf{K}_p \to \mathbf{K}_p$ in (2.4.4) in the following definition of stable maps.

2.6.1 Definition. The *stable theory* $\mathbf{K}_p^{\text{stable}}$ of Eilenberg-MacLane spaces is defined as follows. Objects are the same as in $\mathbf{K}_p$, i.e., products

$$A = Z^{n_1} \times \cdots \times Z^{n_r}$$

with $n_1, \ldots, n_r \geq 1$, $r \geq 0$. The object A yields the sequence of spaces

$$L^{-N}(A) = Z^{N+n_1} \times \cdots \times Z^{N+n_r}$$

with $N \geq 0$. Morphisms $\alpha : A \to B$ with $B = Z^{m_1} \times \cdots \times Z^{m_k}$ are sequences of morphisms in $\mathbf{K}_p$

$$\alpha_N : L^{-N}(A) = Z^{N+n_1} \times \cdots \times Z^{N+n_r} \longrightarrow Z^{N+m_1} \times \cdots \times Z^{N+m_k}$$

with $L\alpha_N = \alpha_{N-1}$, $N \geq 1$. We call $\alpha = (\alpha_N)_{N \geq 0} : A \to B$ a *stable map* (up to homotopy) between products of Eilenberg-MacLane spaces. There is an obvious composition of such stable maps so that the category $\mathbf{K}_p^{\text{stable}}$ is well defined.

The stable theory $\mathbf{K}_p^{\text{stable}}$ can also be described in terms of the Steenrod algebra $\mathcal{A}$. Let $\mathbf{mod}_0(\mathcal{A})$ be the category of finitely generated free left $\mathcal{A}$-modules generated in degree ≥ 1. Hence an object in $\mathbf{mod}_0(\mathcal{A})$ is of the form

$$M = \mathcal{A}x_1 \oplus \cdots \oplus \mathcal{A}x_r$$

where $x_1, \ldots, x_r$ are generators of degree $\mid x_i \mid = n_i \geq 1$ for $i = 1, \ldots, r$ with $r \geq 0$. Recall that $\mathbf{C}^{\text{op}}$ denotes the *opposite category* of $\mathbf{C}$. The next result is well known.

2.6.2 Theorem. *There is an isomorphism of categories*

$$\mathbf{K}_p^{\text{stable}} = \mathbf{mod}_0(\mathcal{A})^{\text{op}}.$$

This is a consequence of the isomorphism

$$H^* : \mathbf{K}_p = \mathbf{H}_p^{\text{op}}$$

in the proof of (1.5.2). For this we need the algebraic properties of the loop functor $L : \mathbf{K}_p \to \mathbf{K}_p$ in Section (3.3) below.

We now obtain a track theory associated to $\mathbf{K}_p^{\text{stable}}$ essentially in the same way as we obtained the secondary Steenrod algebra $[\![\mathcal{A}]\!]$ in (2.5.4).

2.6.3 Definition. The track theory $[\![\mathbf{K}_p^{\text{stable}}]\!]$ is defined as follows. Objects are the same as in $\mathbf{K}_p$. Morphisms $(\alpha, H_\alpha) : A \to B$ are sequences of maps $(N \geq 0)$

$$\alpha = (\alpha_N : L^{-N}A \to L^{-N}B)$$

in $\mathbf{Top}^*$, see (2.6.1), together with tracks as in the following diagram.

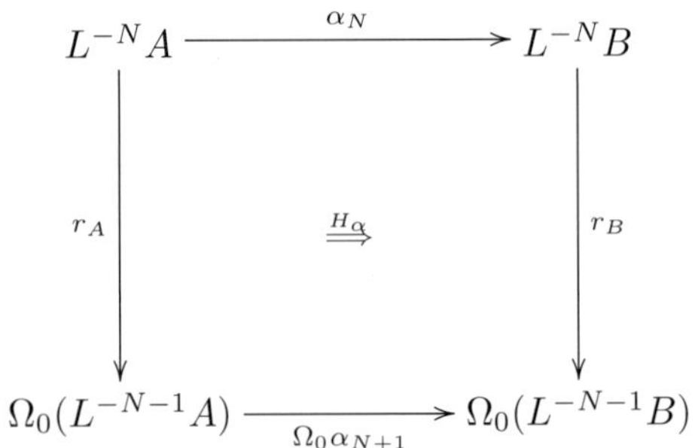

Here r_A is a product of maps r_n defined in (2.1.7). We call $(\alpha, H_\alpha) : A \to B$ a *stable map*. We say that $\alpha = (\alpha, H_\alpha)$ is *strict* if the diagram commutes in $\mathbf{Top}^*$, that is $r_B\alpha_N = (\Omega_0\alpha_{N+1})r_A$, and H_α is the trivial track. For example for a product $A \times B$ in $\mathbf{K}_p$ the projections p_A, p_B are such strict maps as follows from the definition of L and r_A above. Moreover the addition map $A \times A \xrightarrow{+} A$ is strict.

We define stable tracks $H : (\alpha, H_\alpha) \to (\beta, H_\beta)$ between stable maps in the track category $[\![\mathbf{K}_p^{\text{stable}}]\!]$ by sequences of tracks

$$H = (H_N : \alpha_N \Rightarrow \beta_N)_{N \geq 0}$$

in $[\![\mathbf{K}_p]\!]$ for which the pasting of tracks in $[\![\mathbf{Top}^*]\!]$ in the following diagram coincides with H_β.

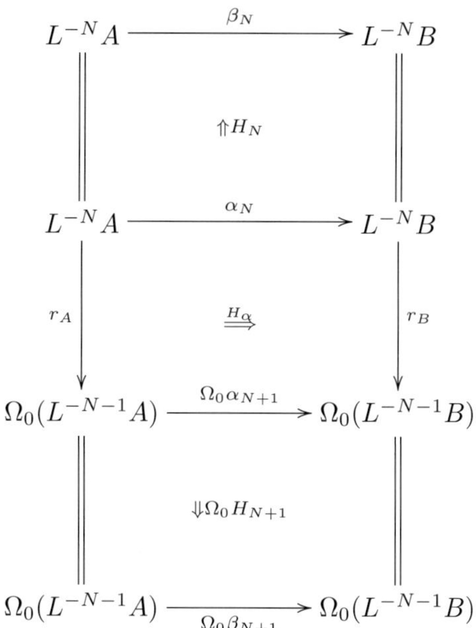

There is an obvious composition of morphisms and tracks respectively so that $[\![\mathbf{K}_p^{\mathrm{stable}}]\!]$ is a well-defined track theory with strong products $(A \times B, p_A, p_B)$. All objects are $\mathbb{F}$-vector space objects as in $\mathbf{K}_p$ since r_A above is linear.

For objects A, B in $\mathbf{K}_p$ let $[\![A, B]\!]^{\mathrm{stable}}$ be the groupoid of stable maps $A \to B$ in the track theory $[\![\mathbf{K}_p^{\mathrm{stable}}]\!]$. Then it is easy to see that for $n \geq 1$, $k \geq 1$ the forgetful functor

$$[\![Z^n, Z^{n+k}]\!]^{\mathrm{stable}} \xleftarrow{\sim} [\![\mathcal{A}^k]\!]. \tag{2.6.4}$$

is a weak equivalence of groupoids. We study further properties of the track theory $[\![\mathbf{K}_p^{\mathrm{stable}}]\!]$ in Section (3.5) and Chapter 4 below.

We have the forgetful functor

$$\phi : [\![\mathbf{K}_p^{\mathrm{stable}}]\!] \longrightarrow [\![\mathbf{K}_p]\!] \tag{2.6.5}$$

which is the identity on objects and carries $H : \alpha \Rightarrow \beta$ to $H_0 : \alpha_0 \Rightarrow \beta_0$. Accordingly a track model of $[\![\mathbf{K}_p]\!]$ is also a track model of $[\![\mathbf{K}_p^{\mathrm{stable}}]\!]$. In particular, the secondary cohomology of a pointed space $[\![X, -]\!]$ yields the track model

$$[\![\mathbf{K}_p^{\mathrm{stable}}]\!] \xrightarrow{\phi} [\![\mathbf{K}_p]\!] \longrightarrow \mathbf{Grd} \tag{2.6.6}$$

of $[\![\mathbf{K}_p^{\mathrm{stable}}]\!]$.

2.7 Stable secondary cohomology operations

We are now ready to introduce stable secondary cohomology operations.

2.7.1 Definition. Let A, B, C be objects in $\mathbf{K}_p$. Then a *stable relation* (α, β) is a commutative diagram

(1)

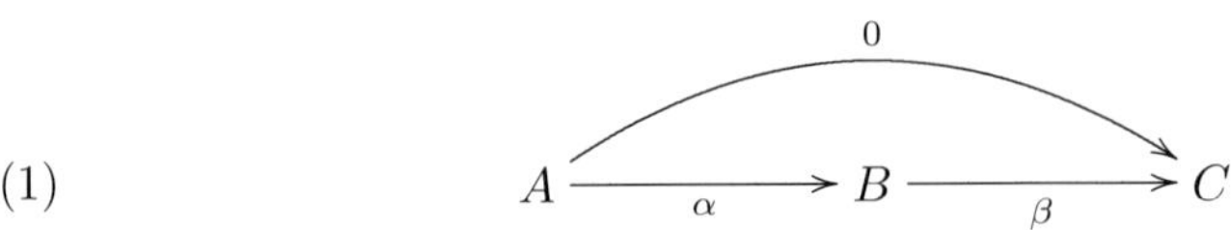

in $\mathbf{K}_p^{\text{stable}}$, i.e., $\beta\alpha = 0$. A *stable secondary cohomology operation* (a, b, H) associated to the relation $\beta\alpha = 0$ is a diagram in $[\![\mathbf{K}_p^{\text{stable}}]\!]$

(2)

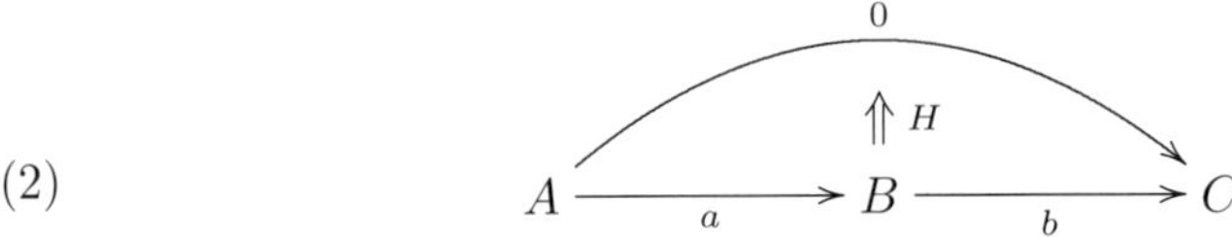

where a (resp. b) represents α (resp. β). *Equivalence* of stable secondary cohomology operations (a, b, H) and (a', b', H') associated to (α, β) is defined by a diagram in $[\![\mathbf{K}_p^{\text{stable}}]\!]$ as in (2.4.5)(7).

The example of Adams in (2.4.6) and (2.4.7) corresponds to such stable secondary cohomology operations.

One can check that stable operations in (2.6.4) correspond to "stable secondary operations" as defined by Adams 3.6 [A] in terms of the cohomology functor.

Next we describe secondary cohomology operations studied by Kristensen [Kr1].

Let $p = 2$ and let F be the free associative algebra with unit generated by symbols sq^i of degree i $(i = 1, 2, \ldots)$, that is,

$$F = T_{\mathbb{F}}(sq^1, sq^2, \ldots)$$

is the $\mathbb{F}$-tensor algebra generated by $sq^1, sq^2, \ldots$. Let R denote the kernel of the algebra map $F \to \mathcal{A}$ which carries sq^i to Sq^i. A *relation* is an element

$$r = b + \sum_{\mu=1}^{k} \alpha_\mu a_\mu \in R \tag{2.7.2}$$

with $\alpha_\mu, a_\mu, b \in F$. We choose for the stable operation Sq^i in $\mathcal{A}$ an element (sq^i, H) in $[\![\mathcal{A}]\!]_0$ so that $sq_n^i : Z^n \to Z^{n+i}$ is defined for all n. Hence any monomial α in F^k yields the corresponding composite $\alpha : Z^n \to Z^{n+k}$. Moreover an element $\beta \in F^k$ is a sum of such monomials and therefore yields a sum $\beta : Z^n \to Z^{n+k}$ of the corresponding maps. Here we use the fact that Z^n is an $\mathbb{F}$-vector space object.

Hence the relation (2.7.2) yields a diagram in $[\![\mathbf{K}_2]\!]$ with $A = Z^{n+|a_1|} \times \cdots \times Z^{n+|a_k|}$ and $N = | r |$,

$$Z^n \xrightarrow{a} A \xrightarrow{\alpha} Z^{n+N}, \quad \Downarrow H, \quad Z^n \xrightarrow{b} Z^{n+N} \tag{2.7.3}$$

where the track H exists since $r \in R$. The diagram (2.7.3) is a *secondary cohomology operation* associated to r in the sense of Kristensen [Kr1]. In fact consider the diagram in $[\![\mathbf{Top}^*]\!]$

(2.7.4)

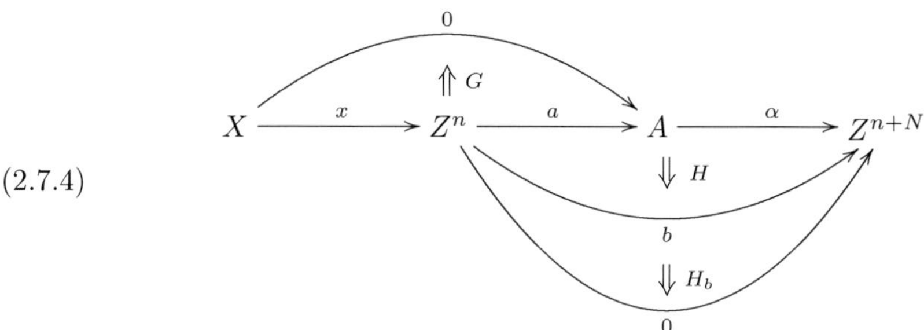

where a canonical track H_b is given. (We shall see in (5.5.1) below that such a canonical track H_b is defined if excess $(b) > n$. In fact, in this case $b : Z^n \to * \to Z^{n+N}$ can be chosen to be the trivial map and then H_b is even the identity track.) Then pasting of tracks in (2.7.4) yields as in (2.4.5) the operation $\theta_{(a,\alpha,H')}$ with $H' = H_b \square H$ of the form

$$\begin{array}{c} \mathrm{kernel}([X, Z^n] \xrightarrow{a_*} [X, A]) \subset H^n(X) \\ \Big\downarrow \theta_{(a,\alpha,H')} \\ \mathrm{cokernel}([X, A] \xrightarrow{\alpha_*} [X, Z^{n+N}]) = H^{n+N}(X)/im(\alpha_*) \end{array}$$

This operation coincides with the operation Qu^r of Kristensen [Kr1] page 74 for appropriate θ with $\Delta\theta = r$ and vice versa. Many results of Kristensen on the operations Qu^r can be derived from the track calculus in the next chapter, see also Chapter 5 where we describe a new approach concerning the Kristensen operations.

Chapter 3

Calculus of Tracks

In this chapter we describe certain basic facts concerning the calculus of tracks in topology. In particular we introduce linear track extensions and we show that the track theories $[\![\mathcal{A}]\!]$, $[\![\mathbf{K}_p^{\mathrm{stable}}]\!]$, $[\![\mathbf{K}_p]\!]$ of Chapter 2 are such linear track extensions. They represent a characteristic cohomology class $k_{\mathcal{A}}$, k_p^{st} and k_p respectively which determines the track theory up to weak track equivalence.

3.1 Maps and tracks under and over a space

Recall that we work in the category **Top** of compactly generated Hausdorff spaces. In particular the product $X \times Y$ in **Top** is compactly generated so that $X \times Y$ is a CW-complex if X and Y are CW-complexes; see Gray [G]. As usual let $IX = [0,1] \times X$ be the cylinder object in **Top**. We have the canonical maps

$$X \xrightarrow{i_t} IX \xrightarrow{q} X$$

for $t \in [0,1]$ with $i_t(x) = (t,x)$ and $q(t,x) = x$. A map $H : IX \to Y$ is a homotopy $H : f \simeq g$ with $f = Hi_0$ and $g = Hi_1$. We also denote a homotopy by $H_t : f \simeq g$ where $H_t = Hi_t$. Here $H_t : X \to Y$ is a map in **Top** for $t \in [0,1]$ with $H_0 = f$ and $H_1 = g$.

We shall use the following category $\mathbf{Top}_B^A$ of *spaces under A and over B*. Objects are diagrams $A \to X \to B$ and morphisms are commutative diagrams

(3.1.1)

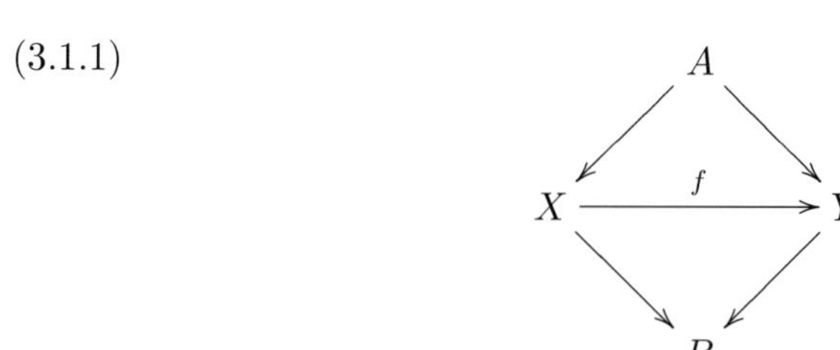

in **Top**. We define a *cylinder object* $A \to I_A X \to B$ of $A \to X \to B$ by a push out diagram.

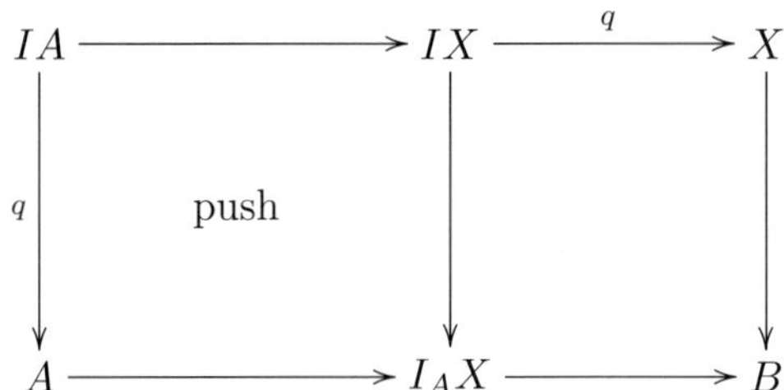

Let $\partial I_A X = i_0 X \cup i_1 X \subset I_A X$ be the *boundary* of $I_A X$. Using this cylinder in $\mathbf{Top}_B^A$ we obtain the notion of *homotopy under A and over B* and of *homotopy equivalence under A and over B*.

We also use for maps $f, g : X \to Y$ under A and over B the *tracks* $f \simeq g$ *under A and over B* which are equivalence classes of maps $H', H : I_A X \to Y$ under A and over B with $H' i_0 = H i_0 = f$ and $H' i_1 = H i_1 = g$ and $H' \sim H$ are equivalent if and only if there is a homotopy $H' \simeq H$ under the boundary $\partial I_A X$ and over B. Tracks are also denoted by $H : f \Rightarrow g$.

As a special case we may choose $A = \emptyset$ (empty set) so that we get the category $\mathbf{Top}_B^{\emptyset} = \mathbf{Top}_B$ of *spaces over* B. Or we can choose $B = *$ (point) so that $\mathbf{Top}_*^A = \mathbf{Top}^A$ is the category of *spaces under* A.

The maps f under A and over B in (3.1.1) are the objects in the groupoid

(3.1.2) $$[\![X, Y]\!]_B^A.$$

The morphisms in this groupoid denoted by $H : f \Rightarrow g$ are the tracks under A and over B. The set of components

(1) $$[X, Y]_B^A \;=\; \pi_0 [\![X, Y]\!]_B^A$$

is the set of homotopy classes of maps under A and over B. If $A = B = *$ is a point, then we write

(2) $$[\![X, Y]\!] = [\![X, Y]\!]_*^* \text{ and } [X, Y] = [X, Y]_*^*$$

for the groupoid of pointed maps $f : X \to Y$.

If $H : f \Rightarrow g$ and $G : g \Rightarrow h$ are tracks in $[\![X, Y]\!]_B^A$, then addition of homotopies yields the *composed track*

(3) $$G \square H : f \Rightarrow h.$$

This is the composition in the category $[\![X, Y]\!]_B^A$. The identity track of f is denoted by $0^\square : f \Rightarrow f$ so that $0^\square \in \mathrm{Aut}(f)$ is the neutral element. The inverse of a track $H : f \Rightarrow g$ is $H^{\mathrm{op}} : g \Rightarrow f$ so that $H \square (H^{\mathrm{op}}) = (H^{\mathrm{op}}) \square H = 0^\square$.

The groupoids $[\![X,Y]\!]^A_B$ are the morphism groupoids in the track category $[\![\mathbf{Top}^A_B]\!]$ and the sets $[X,Y]^A_B$ are the morphism sets in the homotopy category $\mathbf{Top}^A_B/\simeq$ so that

$$(4) \qquad [\![\mathbf{Top}^A_B]\!]_\simeq \;=\; \mathbf{Top}^A_B/\simeq .$$

An operation for maps (like the product $f\times g : X\times V \to Y\times W$ of maps $f : X \to Y$ and $g : V \to W$ in $\mathbf{Top}^*$) induces the corresponding operation for tracks. For example if $H : f \Rightarrow f'$ and $G : g \Rightarrow g'$ are tracks, then $H \times G : f \times g \Rightarrow f' \times g'$ is the track defined by the homotopy $H_t \times G_t$, $t \in I$.

We have the tracks $f \times G : f \times g \Rightarrow f \times g'$ and $H \times g : f \times g \Rightarrow f' \times g$ and using these tracks we obviously get the commutative diagram of tracks

$$(3.1.3) \qquad \begin{array}{ccc} f\times g & \xrightarrow{f\times G} & f\times g' \\ {\scriptstyle H\times g}\downarrow & \searrow^{H\times G} & \downarrow{\scriptstyle H\times g'} \\ f'\times g & \xrightarrow{f'\times G} & f'\times g' \end{array}$$

satisfying

$$H \times G = (f' \times G)\Box(H \times g) = (H \times g')\Box(f \times G).$$

A further operation for maps $f : X \to Y$ in $\mathbf{Top}^*$ is given by the loop space functor $\Omega : \mathbf{Top}^* \to \mathbf{Top}^*$. For a track $H : f \Rightarrow g$ in $[\![X,Y]\!]$ we thus obtain the track

$$(3.1.4) \qquad \Omega H : \Omega f \Longrightarrow \Omega g \text{ in } [\![\Omega X, \Omega Y]\!]$$

where ΩH is defined by the homotopy ΩH_t, $t \in I$. In fact

$$\Omega : [\![X,Y]\!] \longrightarrow [\![\Omega X, \Omega Y]\!]$$

is a functor between groupoids.

The following lemma is well known.

3.1.5 Lemma. *Let $E \to B$ and $E' \to B$ be fibrations in* **Top** *and let $f : E \to E'$ be a map over B which is a homotopy equivalence in* **Top**. *Then f is also a homotopy equivalence in $\mathbf{Top}_B$, i.e., there exists a map $g : E' \to E$ over B and homotopies $fg \simeq 1$ and $gf \simeq 1$ over B.*

Proof. The category **Top** is a fibration category in the sense of [BAH] and hence the result follows from the dual of II.2.12 [BAH]. □

The category $\mathbf{Top}_B$ has products defined by the pull back $E \times_B E'$. Using such products we define a *group object* $(E \to B, \mu, \nu, e)$ in $\mathbf{Top}_B$ by structure maps over B

$$
\begin{array}{rcll}
\mu & : & E\times_B E\to E & \text{(multiplication)}\\
\nu & : & E\to E & \text{(inverse)}\\
o=o_E & : & B\to E & \text{(neutral)}
\end{array}
\tag{3.1.6}
$$

satisfying the usual identities. If $B=*$ is a point, such a group object is the same as a *topological group*. A group object E in $\mathbf{Top}_B$ is via the section o_E also an object under B. Homomorphisms between group objects in $\mathbf{Top}_B$ are maps over B compatible with the structure maps. In particular such a homomorphism f is also a map under and over B, since $fo=o$.

3.1.7 Lemma. *Let $E\to B$ and $E'\to B$ be group objects in $\mathbf{Top}_B$ and fibrations in $\mathbf{Top}$. Let $f:E\to E'$ be a homomorphism between group objects in $\mathbf{Top}_B$ which is a homotopy equivalence in $\mathbf{Top}$. Then f is a homotopy equivalence under and over B.*

Proof. By Lemma (3.1.5) we know that f is a homotopy equivalence over B. Hence we have a map $g:E'\to E$ and homotopies $H_t:E\to E$, $G_t:E'\to E'$ over B with $t\in[0,1]$ and $H_t:gf\simeq 1_E$, $G_t:fg\simeq 1_{E'}$. We define the map

$$g:E'\longrightarrow E$$

by $\bar{g}(\mu(\nu(gop),g)=-gop+g$. Here we use additive notation for the group structure. Then $\bar{g}$ is again a map over B and we get

$$\bar{g}o=-gopo+go=-go+go-o$$

so that $\bar{g}$ is a map under B. Next we define accordingly $\bar{H}_t=-H_top+H_t$ and $\bar{G}_t=-G_top+G_t$ which are also maps under and over B. We have for $t=0$ the equations

$$
\begin{array}{rcl}
\bar{H}_0 &=& -gfop+gf \;=\; -gop+gf\\
&=& -gopf+gf \;=\; (-gop+g)f \;=\; \bar{g}f,\\
\bar{G}_0 &=& fgop+fg \;=\; f(-gop+g) \;=\; f\bar{g},
\end{array}
$$

and for $t=1$ we get $\bar{H}_1=-op+1_E=1_E$ and $\bar{G}_1=-op+1_{E'}=1_{E'}$. This completes the proof of the lemma. □

3.2 The partial loop operation

Let $S^1=I/\partial I$ be the 1-sphere with basepoint $*\in S^1$ given by ∂I. We consider the *free loop space* $\Omega_*X=X^{S^1}$ with basepoint $S^1\to *\to X$. We have the following maps.

(3.2.1)
$$\begin{array}{ccccc} \Omega X & \xrightarrow{\pi} & \Omega_* X & \xrightarrow{p} & X \\ & & \uparrow{\scriptstyle j} & & \\ & & X & & \end{array}$$

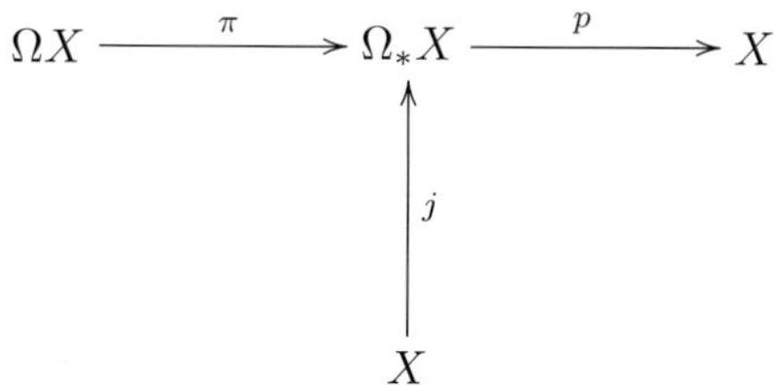

Here π is the inclusion of the loop space $\Omega X = p^{-1}(*)$ which is the fiber of the fibration p given by $p(\sigma) = \sigma(*)$ for $\sigma \in \Omega_* X$. The section j of p carries $x \in X$ to the free loop $j(x) : S^1 \to \{x\} \subset X$.

We say that a map $g : X \times Y \to B$ in **Top*** is *trivial on* Y if the composite

$$g(0,1) : Y \longrightarrow X \times Y \longrightarrow B$$

is homotopic to the zero map 0. Let

$$[X \times Y, B]_2 \subset [X \times Y, B]$$

be the subset of all homotopy classes trivial on Y.

An H-group is a group object in the homotopy category **Top**$^*/\simeq$. A map f between H-groups is H-linear if f is a homomorphism of group objects. For example the loop space is an H-group and a map Ωg is H-linear.

Now let B be an H-group. Then the *partial loop operation* L is the function

(3.2.2)
$$L : [X \times Y, B]_2 \longrightarrow [(\Omega X) \times Y, \Omega B]_2$$

defined as follows. For $\eta : X \times Y \to B$ (trivial on Y) the map $L(\eta)$ is up to homotopy the unique map (trivial on Y) for which the following diagram homotopy commutes in **Top***.

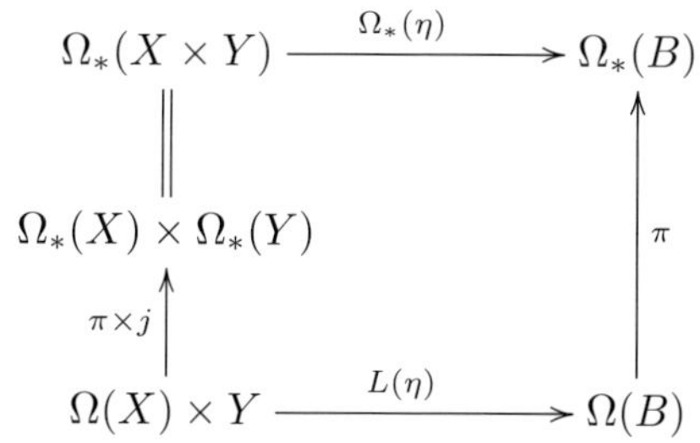

$$\begin{array}{ccc} \Omega_*(X \times Y) & \xrightarrow{\Omega_*(\eta)} & \Omega_*(B) \\ \| & & \\ \Omega_*(X) \times \Omega_*(Y) & & \uparrow{\scriptstyle \pi} \\ \uparrow{\scriptstyle \pi \times j} & & \\ \Omega(X) \times Y & \xrightarrow{L(\eta)} & \Omega(B) \end{array}$$

Here we use the free loop space in (3.2.1) above. Compare the discussion of the partial loop operation in [BAH], [BOT], [BJ3]. If $Y = *$ is a point then the partial loop operation satisfies $L(\eta) = \Omega(\eta)$ so that L generalizes the loop functor Ω. The partial loop operation is dual to the partial suspension described in [BAH].

The partial loop operation satisfies the following rules.

1. The projection p_1 satisfies $L(p_1 : X \times Y \to X) = (p_1 : (\Omega X) \times Y \to \Omega X)$.
2. The map $L(\eta)$ is linear in ΩX, that is for $\alpha, \beta \in [Z, \Omega X]$ and $f \in [Z, Y]$ we have
$$L(\eta)(\alpha + \beta, f) = L(\eta)(\alpha, f) + L(\eta)(\beta, f).$$
3. The composite
$$A \times Z \xrightarrow{(\xi, fp_2)} X \times Y \xrightarrow{\eta} B$$
with $\xi \in [A \times Z, X]_2$ satisfies
$$L(\eta(\xi, fp_2)) = L(\eta)(L(\xi), fp_2).$$

It is well known that the loop space ΩY of an H-group Y is actually an abelian H-group so that $[X, \Omega Y]$ is an abelian group for all pointed spaces X.

3.2.3 Proposition. *Let Y be an H-group and $f : X \to Y$ be a pointed map in* **Top***. *Then one has an isomorphism of groups*
$$\sigma_f : [X, \Omega Y] = \mathrm{Aut}(f)$$
where $\mathrm{Aut}(f)$ *is the automorphism group in the groupoid* $[\![X, Y]\!]$. *Hence* $\mathrm{Aut}(f)$ *is an abelian group. Moreover the loop space functor* $\Omega : [\![X, Y]\!] \to [\![\Omega X, \Omega Y]\!]$ *yields the following commutative diagram.*

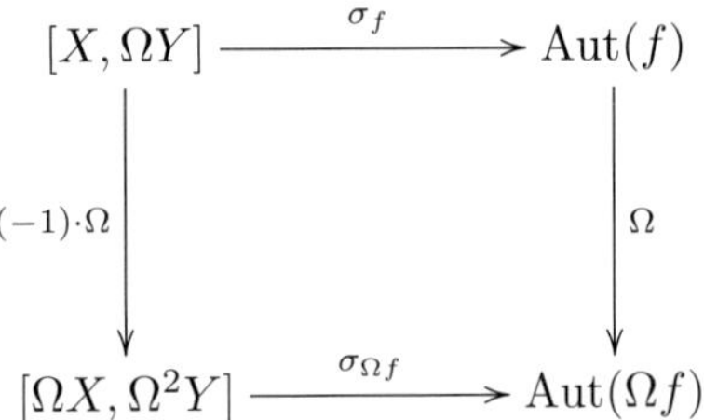

Proof. We observe that we have the canonical bijection
$$\mathrm{Aut}(f) = [X, \Omega_* Y]_Y$$
where X is a space over Y by $f : X \to Y$ and where we use the free loop space in (3.2.1). Let p_2, p_1 be the projections $Y \times Y \to Y$. We can form the composite
$$\sigma :\ \Omega(Y) \times Y \xrightarrow{\pi \times j} \Omega_* Y \times \Omega_* Y = \Omega_*(Y \times Y) \xrightarrow{\Omega_*(\bar{\mu})} \Omega_*(Y)$$
where $\bar{\mu} = p_2 + p_1 = \mu T$ is the composite of the interchange map $T : Y \times Y \to Y \times Y$ and the H-group structure map $\mu : Y \times Y \to Y$ with $\mu = p_1 + p_2$. The map σ is

homotopic to a map σ' over Y and σ' is a homotopy equivalence over Y by (3.1.5). Hence σ' induces

$$\mathrm{Aut}(f) = [X, \Omega_* Y]_Y \xrightarrow{\sigma'_*} [X, \Omega(Y) \times Y]_Y = [X, \Omega Y]$$

and this yields the isomorphism of groups in (3.2.3). Compare [BAH], [BJ2].

The loop space functor Ω with $\Omega X = (X, *)^{(S^1,*)}$ satisfies

$$\begin{aligned} \Omega\Omega_*(X) &= (X^{S^1}, 0)^{(S^1,*)} &&= (X, *)^{(S^1 \times S^1 / S^1 \times *, *)}, \\ \Omega_*\Omega(X) &= ((X, *)^{(S^1,*)})^{S^1} &&= (X, *)^{(S^1 \times S^1 / * \times S^1, *)}, \end{aligned}$$

so that the interchange map $T : S^1 \times S^1 \to S^1 \times S^1$ induces a homomorphism $\Omega\Omega_* X = \Omega_* \Omega X$. The map $T : S^1 \wedge S^1 \to S^1 \wedge S^1$ is a map of degree -1. This yields the equation $\sigma_{\Omega f}(-1)\Omega = \Omega\sigma_f$. $\square$

3.2.4 Proposition. *Let Y be an H-group and $H : f \Rightarrow g$ be a track in $[\![X, Y]\!]$. Then for $\alpha \in [X, \Omega Y]$ the following equation holds,*

$$\sigma_g(\alpha) \square H \;=\; H \square \sigma_f(\alpha).$$

We denote this track by $H \oplus \alpha : f \Rightarrow g$. Here $\oplus$ is a transitive and effective action of the abelian group $[X, \Omega Y]$ on the set $T(f, g)$ of all tracks $f \Rightarrow g$ in $[\![X, Y]\!]$. The loop space functor $\Omega : [\![X, Y]\!] \to [\![\Omega X, \Omega Y]\!]$ satisfies $\Omega(H \oplus \alpha) = (\Omega H) \oplus (-\Omega\alpha)$.

The proposition generalizes the representability of tracks $0 \Rightarrow 0$ in (2.3.2). If X is $(n-1)$-connected, then $[X, \Omega Z^n] = \tilde{H}^{n-1}(X)$ is trivial and hence we get by (3.2.4):

3.2.5 Corollary. *Let X be an $(n-1)$-connected space. Then the connected components of the groupoids $[\![X, Z^n]\!]$ are contractible.*

Let $y : Y \to Y'$ be a map between H-groups. Then we obtain the *difference element*

$$\nabla y \in [Y \times Y, Y']_2 \tag{3.2.6}$$

as follows. Let $\nabla y = -yp_2 + y(p_2 + p_1)$. We have

$$\begin{aligned} (\nabla y)(0, 1) &= -yp_2(0, 1) + y(p_2 + p_1)(0, 1) \\ &= -yp_2 + yp_2 \\ &= 0 \end{aligned}$$

so that ∇y is trivial on $(0, 1)Y \subset Y \times Y$. The difference element satisfies for a composite $gf : X \to Y \to Z$ of maps between H-groups the formula

$$\nabla(gf) = (\nabla g)(\nabla f, fp_2) : X \times X \longrightarrow Y \times Y \longrightarrow Z. \tag{1}$$

If g is H-linear then $\nabla g = gp_1$ so that in this case $\nabla(gf) = g\nabla f$. The partial loop operation

$$L(\nabla y) \in [(\Omega Y) \times Y, \Omega Y']_2 \tag{2}$$

is defined by (3.2.2).

3.2.7 Proposition. *Let $H : f \Rightarrow g$ be a track in $[\![X, Y]\!]$ where Y is an H-group. Let $x : X' \to X$ and $y : Y \to Y'$ be maps in* **Top*** *where Y' is also an H-group. Then the following formulas hold for $\alpha \in [X, \Omega Y]$,*

$$\begin{aligned} (H \oplus \alpha)x &= (Hx) \oplus (\alpha x), \\ y(H \oplus \alpha) &= (yH) \oplus L(\nabla y)(\alpha, f). \end{aligned}$$

These are formulas in $[\![X', Y]\!]$ and $[\![X, Y']\!]$ respectively. If y is linear we get $\nabla y = yp_1$ so that in this case $y(H \oplus \alpha) = (yH) \oplus (\Omega y)\alpha$ holds.

One finds proofs of (3.2.4) and (3.2.7) in [BAH] and [BUT] where we describe further properties of the partial loop operation, see also [BJ3].

For example we have for the multiplication map (2.2.1) the element $\mu_{n,m} \in [Z^n \times Z^m, Z^{n+m}]$. If we apply the loop functor Ω we get the trivial element

$$\Omega\mu_{n,m} = 0 \in [\Omega Z^n \times \Omega Z^m, \Omega Z^{n+m}]. \tag{3.2.8}$$

The partial loop operation L, however, yields the map $L\mu_{n,m}$ which via the homotopy equivalences (2.1.7) can be identified with $\mu_{n-1,m}$, that is, the diagram

$$\begin{array}{ccc} (\Omega Z^n) \times Z^m & \xrightarrow{L\mu_{n,m}} & \Omega Z^{n+m} \\ \uparrow \wr & & \uparrow \wr \\ Z^{n-1} \times Z^m & \xrightarrow{\mu_{n-1,m}} & Z^{n+m-1} \end{array} \tag{3.2.9}$$

homotopy commutes. Compare for example [BOT] 6.1.12 p. 328.

3.2.10 Lemma. *For the map $\mu = \mu_{n,m} : Z^n \times Z^m \to Z^{n+m}$ we get the element*

$$L\nabla\mu \in [\Omega(Z^n \times Z^m) \times Z^n \times Z^m, \Omega(Z^{n+m})]_2$$

which via the homotopy equivalence (2.1.7) is represented by the map

$$\bar{\mu} : Z^{n-1} \times Z^{m-1} \times Z^n \times Z^m \longrightarrow Z^{n+m-1}$$

which carries (x, y, x_2, y_2) to $x \cdot y_2 + (-1)^{nm} y \cdot x_2$. Here we use the product defined by μ in (2.1.1).

Proof. For $a \in Z^n$, $b \in Z^m$ we have $\mu(a,b) = a \cdot b$. Hence $\nabla\mu$ is defined by

$$\begin{aligned}(\nabla\mu)(x_1,y_1,x_2,y_2) &= -x_2 \cdot y_2 + (x_1+x_2)\cdot(y_1+y_2)\\ &= x_1\cdot y_1 + x_1 \cdot y_2 + x_2 \cdot y_1\\ &= x_1 \cdot y_1 + x_1\cdot y_2 + \tau(y_1 \cdot x_2)\end{aligned}$$

with $\mathrm{sign}(\tau) = (-1)^{nm}$. If we apply the partial loop operation L we get by (3.2.8) and (3.2.9) the equation $(L\nabla\mu) = \bar{\mu}$ where we use (2.1.7) as an identification. □

3.3 The partial loop functor for Eilenberg-MacLane spaces

In (2.3.4) we consider the *loop functor*

(3.3.1) $$L : \mathbf{K}_p \longrightarrow \mathbf{K}_p$$

on the theory of Eilenberg-MacLane spaces. This loop functor is a special case of the *partial loop functor*

(3.3.2) $$L_X : \mathbf{K}_p(X) \longrightarrow \mathbf{K}_p(X)$$

where X is an object in $\mathbf{K}_p$. Here $\mathbf{K}_p(X)$ is the following category with the same objects A, B as in $\mathbf{K}_p$. Morphisms $a : A \to B$ in $\mathbf{K}_p(X)$ are commutative diagrams in $\mathbf{K}_p$.

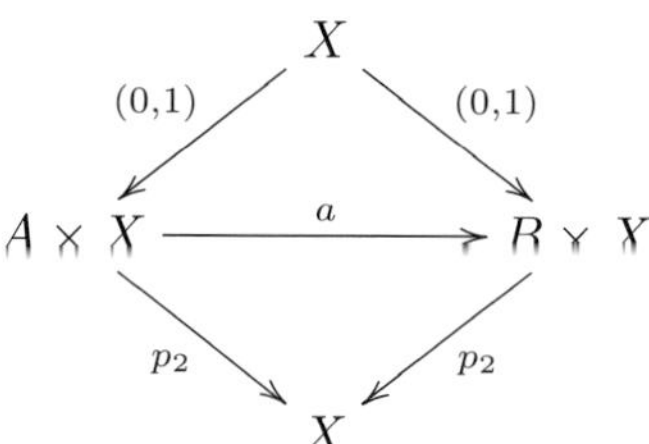

The map a is given by coordinates (α, p_2) with $\alpha \in [A \times X, B]_2$. We define L_X on objects in the same way as L; that is, for $A = Z^{n_1} \times \cdots \times Z^{n_r}$ we have $L_X A = LA = Z^{n_1-1} \times \cdots \times Z^{n_r-1}$ as in (2.3.4). On morphisms $a = (\alpha, p_2) : A \to B$ in $\mathbf{K}_p(X)$ we define $L_X(a) = (L\alpha, p_2)$ where $L\alpha$ is given by the composite

$$L : [A \times X, B]_2 \xrightarrow{L_0} [\Omega_0(A) \times X, \Omega_0 B]_2 = [L(A) \times X, LB]_2.$$

Here L_0 is defined by the partial loop operation and we use the homotopy equivalence $\Omega_0(A) \simeq L(A)$ in (2.3.3). If $X = *$ is a point then L_X coincides with L in (3.3.1).

3.3.3 Lemma. *The category $\mathbf{K}_p(X)$ is a theory with products as in $\mathbf{K}_p$ and L_X is a functor which preserves finite products. Moreover each object in $\mathbf{K}_p(X)$ is an* *abelian group object and L_X carries maps to linear maps. The functors L_X are natural in X in the sense that for all $f : X \to Y$ in $\mathbf{K}_p$ the following diagram of functors commutes.*

$$\begin{array}{ccc} \mathbf{K}_p(X) & \xrightarrow{L_X} & \mathbf{K}_p(X) \\ \downarrow{\scriptstyle f_*} & & \downarrow{\scriptstyle f_*} \\ \mathbf{K}_p(Y) & \xrightarrow{L_Y} & \mathbf{K}_p(Y) \end{array}$$

Here f_* is the identity on objects and f_* carries the morphism $a = (\alpha, p_2)$ to $f_*(a) = (\alpha(1 \times f), p_2)$.

Compare [BJ4]. We now want to compute the loop functor and the partial loop functor explicitly in terms of cohomology groups. Special cases are already considered in the examples (3.2.7)...(3.2.9) above.

3.3.4 Definition. For an unstable $\mathcal{A}$-module X let unstable $\mathcal{A}$-modules $X/\sim$ and $X/\approx$ be given by (see (1.1.12))

$$\begin{aligned} X/\sim &= \operatorname{image}(X \to \tilde{U}(X)), \\ X/\approx &= \operatorname{image}(X \to \tilde{U}(X) \to \tilde{U}(X)/\tilde{U}(X)\cdot\tilde{U}(X)) \\ &= \tilde{U}(X)/\tilde{U}(X)\cdot\tilde{U}(X). \end{aligned}$$

Then the free unstable $\mathcal{A}$-modules $F(n) = \Sigma^n(\mathcal{A}/B(n))$ admit a unique $\mathcal{A}$-linear map of degree (-1),

$$\tilde{\tilde{\Omega}} : F(n)/\approx \longrightarrow F(n-1)/\sim$$

which carries the generator $[n]$ to the generator $[n-1]$. This yields the composite map of degree (-1),

$$\tilde{\Omega} : \tilde{H}(n) \xrightarrow{q} F(n)/\approx \xrightarrow{\tilde{\tilde{\Omega}}} F(n-1)/\sim \xrightarrow{i} \tilde{H}(n-1)$$

where q is the quotient map and i is the inclusion. Compare (1.1.12)(5).

3.3.5 Lemma. *The following diagram commutes.*

$$\begin{array}{ccc} [Z^n, Z^k] & \xrightarrow{L} & [Z^{n-1}, Z^{k-1}] \\ \| & & \| \\ \tilde{H}(n)^k & \xrightarrow{\tilde{\Omega}} & \tilde{H}(n-1)^{k-1} \end{array}$$

Here L is given by the loop functor and $\tilde{\Omega}$ is defined in (3.3.4).

This lemma is a consequence of (1.1.13) and (3.2.8) and (2.4.3).

If A and B are connected algebras, then also $A \otimes B$ is a connected algebra and we have

$$(A \otimes B)^{\sim} \quad = \quad \tilde{A} \oplus \tilde{B} \oplus (\tilde{A} \otimes \tilde{B}).$$

Hence we have inclusion and projection

$$\tilde{A} \oplus \tilde{B} \xrightarrow{i} (A \otimes B)^{\sim} \xrightarrow{q} \tilde{A} \oplus \tilde{B}.$$

We define the composite Ω by the following diagram.

$$\begin{array}{ccc} (H(n_1) \otimes \cdots \otimes H(n_r))^{\sim} & \xrightarrow{\Omega} & (H(n_1-1) \otimes \cdots \otimes H(n_r-1))^{\sim} \\ \downarrow q & & \uparrow i \\ \tilde{H}(n_1) \oplus \cdots \oplus \tilde{H}(n_r) & \xrightarrow{\tilde{\Omega} \oplus \cdots \oplus \tilde{\Omega}} & \tilde{H}(n_1-1) \oplus \cdots \oplus \tilde{H}(n_r-1) \end{array}$$

3.3.6 Lemma. *For the object $X = Z^{n_1} \times \cdots \times Z^{n_r}$ in $\mathbf{K}_p$ we have the commutative diagram.*

$$\begin{array}{ccc} [X, Z^k] & \xrightarrow{L} & [LX, Z^{k-1}] \\ \| & & \| \\ (\tilde{H}(n_1) \otimes \cdots \otimes \tilde{H}(n_r))^k & \xrightarrow{\Omega} & (\tilde{H}(n_1-1) \otimes \cdots \otimes \tilde{H}(n_r-1))^{k-1} \end{array}$$

Here L is given by the loop functor and Ω is defined above.

The map $(0,1) : Y \to X \times Y$ in $\mathbf{K}_p$ induces the map $(0,1)^* : H^*(X \times Y) \to H^*(Y)$. Let

$$H^*(X \times Y)_2 = \mathrm{kernel}(0,1)^*.$$

Then the Künneth formula shows

$$H^*(X \times Y)_2 = \tilde{H}^*(X) \otimes H^*(Y).$$

For $Y = Z^{n_1} \times \cdots \times Z^{n_r}$ as in (2.3.5) we now get:

3.3.7 Lemma. *The partial loop operation L in $\mathbf{K}_p$ is determined by the commutative diagram (X, Y objects in $\mathbf{K}_p$)*

$$\begin{array}{ccc} [X \times Y, Z^k]_2 & \xrightarrow{L} & [(LX) \times Y, Z^{k-1}]_2 \\ \| & & \| \\ ((\tilde{H}^* X) \otimes H^*(Y))^k & \xrightarrow{\Omega \otimes 1} & ((\tilde{H}^*(LX)) \otimes H^*(Y))^{k-1} \end{array}$$

where Ω is defined in (3.3.6) *above.*

This is essentially a consequence of (3.2.9).

Next we consider a commutative graded algebra H and a graded module M. Then $M \otimes H$ is a right H-module by $(m \otimes x) \cdot y = m \otimes (x \cdot y)$ and a left H-module by $y \cdot (m \otimes x) = (-1)^{|y||m|}(m \otimes (y \cdot x)$. A (linear) *derivation*

$$D : H \longrightarrow M \otimes H \tag{3.3.8}$$

is a map of degree (-1) satisfying

$$D(x \cdot y) = (Dx) \cdot y + (-1)^{|x|} x \cdot (Dy).$$

Given an object Y in $\mathbf{K}_p$ we obtain the unique derivation of degree (-1),

$$\tilde{\nabla} : H^*(Y) \longrightarrow \tilde{H}^* L(Y) \otimes H^*(Y)$$

for which the following diagram commutes where $Y = Z^{n_1} \times \cdots \times Z^{n_r}$.

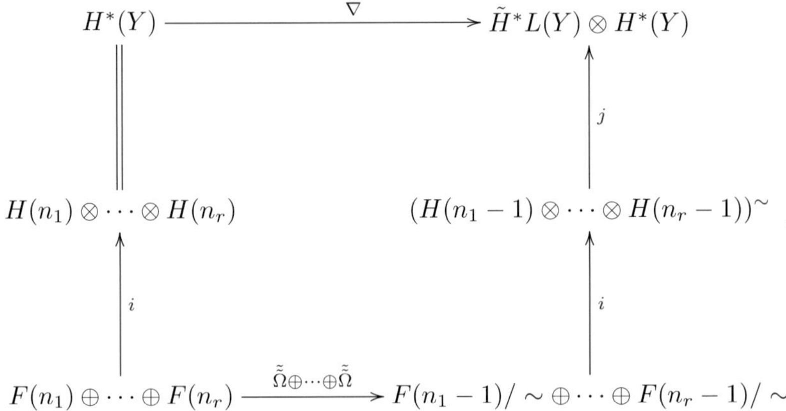

Here $\tilde{\tilde{\Omega}}$ and the morphisms i are defined in (3.3.4) and j carries x to $x \otimes 1$.

Recall that $L\nabla(y)$ is needed in (3.2.7). We now obtain the following result on $L\nabla$.

3.3.9 Lemma. *For the difference element ∇ in (3.2.6) the following diagram commutes where Y is an object in $\mathbf{K}_p$.*

$$\begin{array}{ccc}
[Y, Z^k] & \xrightarrow{\;L\nabla\;} & [L(Y) \times Y, Z^{k-1}] \\
\| & & \| \\
H^*(Y)^k & \xrightarrow{\;\tilde{\nabla}\;} & (\tilde{H}^*(LY) \otimes H^*(Y))^{k-1}
\end{array}$$

Here $\tilde{\nabla}$ is the derivation above and $L\nabla$ carries $f : Y \to Z^k$ to the partial loop operation applied to the difference element ∇f.

Proof. For a linear map $f : Y \to Z^k$ we have $\nabla f = fp_1$ and $L\nabla f = (\Omega f)p_1$. This shows that $(L\nabla)i$ is defined on $F(n_1) \oplus \cdots \oplus F(n_r)$ by $\tilde{\tilde{\Omega}} \oplus \cdots \oplus \tilde{\tilde{\Omega}}$ in the same way as $\tilde{\nabla}$ above. Hence it remains to show that $L\nabla$ is a derivation. This is a consequence of (3.2.10). In fact, we have for $a : Y \to Z^n$, $b : Y \to Z^m$ the formula $a \cdot b = \mu(a, b)$ with $\mu = \mu_{n,m}$. Hence we get

$$\begin{aligned}
(L\nabla)(a \cdot b) &= L\nabla(\mu(a, b)) \\
&= L((\nabla\mu)(\nabla(a, b), (a, b)p_2), && (3.2.6)(1) \\
&= (L\nabla\mu)(L\nabla\mu(a, b), (a, b)p_2), && (3.2.2)(3) \\
&= \bar{\mu}((L\nabla a, L\nabla b), (ap, bp_2)), && (3.2.10) \\
&= (L\nabla a) \cdot b + (-1)^{nm}(L\nabla b) \cdot a, && (3.2.10) \\
&= (L\nabla a) \cdot b + (-1)^{nm} \cdot (-1)^{n(m-1)} a \cdot (L\nabla b) \\
&= (L\nabla a) \cdot b + (-1)^n a \cdot L(\nabla b).
\end{aligned}$$

□

3.4 Natural systems

We introduce the notion of natural systems on a category **C** and we describe a particular natural system on the theory $\mathbf{K}_p$ of Eilenberg-MacLane spaces.

3.4.1 Definition. Let **C** be a category. Then the category $F\mathbf{C}$ of factorizations in **C** is defined as follows. Objects of $F\mathbf{C}$ are morphisms $f : B \to A$ and morphisms $(\alpha, \beta) : f \to g$ in $F\mathbf{C}$ are commutative diagrams

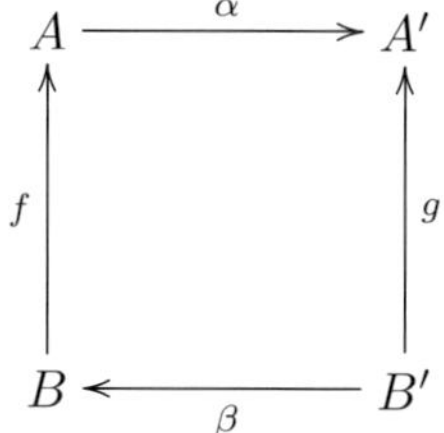

in the category **C**. A *natural system* (of abelian groups) on **C** is a functor $D : F\mathbf{C} \to \mathbf{Ab}$. Here **Ab** denotes the category of abelian groups. We write $D(f) = D_f \in \mathbf{Ab}$ and $D(\alpha, \beta) = \alpha_*\beta^*$. In the situation $\xleftarrow{f}\xleftarrow{g}\xleftarrow{h}$ the induced homomorphisms f_* and h^* will be denoted by

$$f_* : D_g \to D_{fg},\ \xi \mapsto f\xi = f_*(\xi),$$

$$h^* : D_g \to D_{gh},\ \xi \mapsto \xi h = h^*(\xi).$$

We have the forgetful functor

$$\phi : F\mathbf{C} \longrightarrow \mathbf{C} \times \mathbf{C}^{\mathrm{op}} \tag{3.4.2}$$

which carries $f : B \to A$ to (A, B) and carries (α, β) to (α, β). Hence any functor $M : \mathbf{C} \times \mathbf{C}^{\mathrm{op}} \to \mathbf{Ab}$, termed a $\mathbf{C}$-bimodule, yields the natural system $M\phi$ also denoted by M.

For example let $\mathbf{C}$ be an additive category and let $L_i : \mathbf{C} \to \mathbf{C}$, $i = 1, 2$, be additive functors. Then we obtain the $\mathbf{C}$-bimodule

$$\mathrm{Hom}(L_1, L_2) : \mathbf{C} \times \mathbf{C}^{\mathrm{op}} \longrightarrow \mathbf{Ab} \tag{3.4.3}$$

which carries (A, B) to the abelian group $\mathrm{Hom}(L_1B, L_2A)$ of morphisms $L_1B \to L_2A$ in $\mathbf{C}$. If L_2 is the identity functor we write $\mathrm{Hom}(L_1, -)$.

3.4.4 Definition. A natural system D on a category $\mathbf{C}$ is said to be *compatible with products* if for any product diagram $p_k : X_1 \times \cdots \times X_n \to X_k$, $k = 1, \ldots, n$, and any morphisms $f : Y \to X_1 \times \cdots \times X_n$ the homomorphism

$$D_f \longrightarrow D_{p_1 f} \times \cdots \times D_{p_n f}$$

defined by $\xi \mapsto (p_1\xi, \ldots, p_n\xi)$ is an isomorphism. In a dual way we define compatibility with sums (sum=coproduct).

For example $\mathrm{Hom}(L_1, L_2)$ above is compatible with products in the additive category $\mathbf{C}$ and also compatible with sums.

Recall that the category of stable operations $\mathbf{K}_p^{\mathrm{stable}}$ in (2.5) is an additive category with

$$\lambda : \mathbf{K}_p^{\mathrm{stable}} = \mathbf{mod}_0(\mathcal{A})^{\mathrm{op}}.$$

We obtain a bimodule on $\mathbf{K}_p^{\mathrm{stable}}$ by the *shift functor* $L^{-1} : \mathbf{K}_p^{\mathrm{stable}} \to \mathbf{K}_p^{\mathrm{stable}}$ which carries $A = Z^{n_1} \times \cdots \times Z^{n_r}$ to $L^{-1}A = Z^{n_1+1} \times \cdots \times Z^{n_r+1}$. Hence

$$\mathrm{Hom}(L^{-1}, -) : \mathbf{K}_p^{\mathrm{stable}} \times (\mathbf{K}_p^{\mathrm{stable}})^{\mathrm{op}} \longrightarrow \mathbf{Ab} \tag{3.4.5}$$

is well defined. This bimodule carries (A, B) to $[L^{-1}B, A]^{\mathrm{stable}}$ or, using the equivalence λ, we have

$$\begin{aligned} \mathrm{Hom}(L^{-1}B, A) &= [L^{-1}B, A]^{\mathrm{stable}} \\ &= \mathrm{Hom}_{\mathcal{A}}(\lambda(A), \lambda L^{-1}B) \\ &= \mathrm{Hom}_{\mathcal{A}}(\lambda A, \Sigma\lambda B). \end{aligned}$$

Here Σ is the *shift functor* on $\mathbf{mod}_0(\mathcal{A})$ which carries $\mathcal{A}x_1 \oplus \cdots \oplus \mathcal{A}x_r$ with $\mid x_i \mid = n_i$ to $\mathcal{A}(\Sigma x_1) \oplus \cdots \oplus \mathcal{A}(\Sigma x_r)$ with $\mid \Sigma x_i \mid = n_i + 1$. We have $\Sigma(\lambda B) = \lambda(L^{-1}B)$.

Next let A be a graded algebra over the field $\mathbb{F}$, for example the Steenrod algebra, and let M be a *graded A-bimodule*. Then we can consider A as a (graded) monoid which is a category with one object. Moreover M yields canonically a natural system on A by setting

$$M_a = M^{|a|} \text{ for } a \in \mathcal{A} \tag{3.4.6}$$

and $b_* : M_a \to M_{b \cdot a}$ carries x to $b \cdot x$ and $c^* : M_a \to M_{a \cdot c}$ carries x to $x \cdot c$.

For the algebra $\mathcal{A}$ we obtain the $\mathcal{A}$-bimodule $\Sigma\mathcal{A}$ defined by $(\Sigma\mathcal{A})^n = \mathcal{A}^{n-1}$ and $\alpha \cdot (\Sigma x) = (-1)^{|\alpha|}\Sigma(\alpha \cdot x)$ and $(\Sigma x) \cdot \beta = \Sigma(x \cdot \beta)$ for $\alpha, \beta \in \mathcal{A}$ and $x \in \mathcal{A}$. By (3.4.6) we consider $\Sigma\mathcal{A}$ as a natural system on the monoid $\mathcal{A}$.

Let $\mathbf{mod}_0(A)^{\mathrm{op}}$ be the category of finitely generated free *right* A-modules with generators in degree ≥ 1. This is the opposite of the category $\mathbf{mod}_0(A)$ of finitely generated free left A-modules in (2.5.2). Morphisms are A-linear maps of degree 0. Let $\mathrm{Hom}(V, W)$ be the $\mathbb{F}$-vector space of such morphisms $V \to W$ in $\mathbf{mod}_0(A)^{\mathrm{op}}$. Then $\mathrm{Hom}(V, W)$ is an A-bimodule with the action defined by

$$(a \cdot \alpha \cdot b)(x) = \alpha(x \cdot a)b$$

for $x \in V$, $\alpha \in \mathrm{Hom}(V, W)$, $a, b \in A$. Given an A-bimodule M as above we define the natural system $\bar{M}$ on $\mathbf{mod}_0(A)^{\mathrm{op}}$ by

(3.4.7) $$\bar{M}_\alpha = \mathrm{Hom}(V, W) \otimes_{A-A} M$$

for $\alpha : V \to W$. Here $\otimes_{A-A}$ is the bimodule tensor product, see MacLane [MLH].

For example let $\mathcal{A} = A$ be the Steenrod algebra and let $M = \Sigma\mathcal{A}$ be the $\mathcal{A}$-bimodule above. Then the natural system $\overline{\Sigma\mathcal{A}}$ is defined on $\mathbf{mod}_0(\mathcal{A})^{\mathrm{op}}$ by

(1) $$(\overline{\Sigma\mathcal{A}})_\alpha = \mathrm{Hom}(V, W) \otimes_{\mathcal{A}-\mathcal{A}} \Sigma\mathcal{A}$$

for $\alpha : V \to W$ in $\mathbf{mod}_0(\mathcal{A})^{\mathrm{op}}$. We have the isomorphism of categories

(2) $$\mathbf{mod}_0(\mathcal{A})^{\mathrm{op}} = \mathbf{K}_p^{\mathrm{stable}}$$

and using this isomorphism as an identification we get the isomorphism of natural systems

(3) $$\overline{\Sigma\mathcal{A}} = \mathrm{Hom}(L^{-1}, -)$$

where the right-hand side is defined by (3.4.5).

Finally we need the following natural system $\mathcal{L}$ on $\mathbf{K}_p$. For $f : B \to A$ in $\mathbf{K}_p$ let

(3.4.8) $$\mathcal{L}_f = [B, LA]$$

where LA is given by the loop functor L on $\mathbf{K}_p$. The natural system $\mathcal{L}$, however, does not coincide with the bimodule $[-, L]$ since induced maps f_*, h^* for $A \xleftarrow{f} B \xleftarrow{g} C \xleftarrow{h} D$ in $\mathbf{K}_p$ satisfy

(1) $$\begin{aligned} &h^* : \mathcal{L}_g = [C, LB] \to \mathcal{L}_{gh} = [D, LB], \\ &h^*(\xi) = \xi h, \end{aligned}$$

(2) $$\begin{aligned} &f_* : \mathcal{L}_g = [C, LB] \to \mathcal{L}_{fg} = [C, LA], \\ &f_*(\xi) = (L\nabla)(f)(\xi, g). \end{aligned}$$

Only in case f is linear do we get the formula $f_*(\xi) = (Lf)\xi$ which holds in general for the bimodule $[-, L]$. We have seen in (3.3.8) that $L\nabla$ can be described by the derivation $\tilde{\nabla}$. The definition of $f_*(\xi)$ in (2) corresponds to the formula in (3.2.7).

3.5 Track extensions

For a track theory $\mathcal{T}$ and for a map $f : A \to B$ in $\mathcal{T}$ the automorphism group of f in the groupoid $[\![A, B]\!] = \mathcal{T}(A, B)$ is denoted by $\mathrm{Aut}(f) = \mathrm{Hom}(f, f)$. Any track $\eta : f \Rightarrow g$ induces a group homomorphism

$$(-)^\eta : \mathrm{Aut}(g) \longrightarrow \mathrm{Aut}(f)$$

which carries $\alpha \in \mathrm{Aut}(g)$ to $\alpha^\eta = -\eta + \alpha + \eta$. On the other hand composition in $\mathcal{T}$ yields for $\xleftarrow{f}\xleftarrow{g}\xleftarrow{h}$ in $\mathcal{T}_0$ the homomorphisms

$$h^* : \mathrm{Aut}(g) \longrightarrow \mathrm{Aut}(gh), \ \xi \mapsto \xi h,$$

$$f_* : \mathrm{Aut}(g) \longrightarrow \mathrm{Aut}(fg), \ \xi \mapsto f\xi.$$

3.5.1 Definition. A *linear track extension* of a category $\mathbf{C}$ by a natural system D denoted by

$$D \xrightarrow{\sigma} \mathcal{T}_1 \rightrightarrows \mathcal{T}_0 \xrightarrow{q} \mathbf{C}$$

is a track category $\mathcal{T}$ equipped with a functor $q : \mathcal{T}_0 \to \mathbf{C}$ and a collection of isomorphisms of groups

$$\sigma_f : D_{q(f)} \longrightarrow \mathrm{Aut}(f)$$

where $f : A \to B$ is a map in $\mathcal{T}_0$. Moreover the following properties are satisfied.

(1) The functor q is full and is the identity on objects, i.e., $Ob(\mathcal{T}) = Ob(\mathbf{C})$. In addition for $f, g : A \to B$ in $\mathcal{T}_0$ we have $q(f) = q(g)$ if and only if $f \simeq g$. In other words the functor q identifies $\mathbf{C}$ with $\mathcal{T}_\simeq$. We also write $q(f) = [f]$. Hence for any $\varphi : f \Rightarrow g$ we have $[f] = [g]$.

(2) For $\varphi : f \Rightarrow g$ and $\xi \in D_{|f|} = D_{|g|}$ we have

$$\sigma_f(\xi) = \sigma_g(\xi)^\varphi.$$

Equivalently we have

$$\varphi \Box \sigma_f(\xi) = \sigma_g(\xi) \Box \varphi$$

and this element is denoted by $\varphi \oplus \xi$.

(3) For any three maps $\xleftarrow{f}\xleftarrow{g}\xleftarrow{h}$ in $\mathcal{T}_0$ and any $\xi \in D_{|g|}$ one has

$$\begin{aligned} f\sigma_g(\xi) &= \sigma_{fg}([f]\xi), \\ \sigma_g(\xi)h &= \sigma_{gh}(\xi[h]). \end{aligned}$$

We say that a track category is *linear* if it occurs as a linear track extension – of its own homotopy category, necessarily – by some natural system D. Clearly a linear track category has abelian hom-groupoids by the definition above. The result in [BJ1] shows that also the converse is true; that is:

If $\mathcal{T}$ is a track category in which all hom-groupoids are abelian groupoids, then $\mathcal{T}$ is a linear track category.

This already shows by (3.2.3) that the track categories

$$[\![\mathcal{A}]\!],\ [\![\mathbf{K}_p^{\text{stable}}]\!],\ [\![\mathbf{K}_p]\!]$$

described in Chapter 2 are linear track categories. We shall describe details as follows.

3.5.2 Theorem. *The secondary Steenrod algebra is a linear track extension*

$$\Sigma\mathcal{A} \longrightarrow [\![\mathcal{A}]\!]_1 \rightrightarrows [\![\mathcal{A}]\!]_0 \longrightarrow \mathcal{A}\,.$$

Here $\Sigma\mathcal{A}$ *is the natural system on the graded monoid* $\mathcal{A}$ *given by* (3.4.6).

Proof. If (α, H_α), (β, H_β) in $[\![\mathcal{A}]\!]_0^k$ both represent the same stable operation in $\mathbf{K}_p$ that is an element in $\mathcal{A}^k$, see (2.4.3), then for sufficiently large n we have a track $H_n : \alpha_n \Rightarrow \beta_n$ in $[\![Z^n, Z^{n+k}]\!]$, for example $n > k$. Moreover for $n > k+1$ the function Ω is a bijection on $[Z^n, Z^{n+k}]$ so that H_n determines a unique track $H : (\alpha, H_\alpha) \Rightarrow (\beta, H_\beta)$. Hence we get $[\![\mathcal{A}]\!]_{\simeq} = \mathcal{A}$. Moreover we define

$$\sigma_{(\alpha,H_\alpha)} : (\Sigma\mathcal{A})^k = \mathcal{A}^{k-1} \cong \operatorname{Aut}(\alpha, H_\alpha)$$

as follows. For the stable operation $\xi \in \mathcal{A}^{k-1}$ with $\xi = (\xi_n)_{n\in\mathbb{Z}}$ and

$$\xi_n \in [Z^n, Z^{n+k-1}] = [Z^n, \Omega_0 Z^{n+k}]$$

let $\sigma_{(\alpha,H_\alpha)}(\xi) = ((-1)^n \sigma_{\alpha_n}(\xi_n))_{n\in\mathbb{Z}}$. Here

$$\sigma_{\alpha_n} : [Z^n, \Omega_0 Z^{n+k}] \cong \operatorname{Aut}(\alpha_n)$$

is defined in (3.2.3). □

3.5.3 Theorem. *The track category of stable secondary operations is a linear track extension*

$$\operatorname{Hom}(L^{-1}, -) \longrightarrow [\![\mathbf{K}_p^{\text{stable}}]\!]_1 \rightrightarrows [\![\mathbf{K}_p^{\text{stable}}]\!]_0 \longrightarrow \mathbf{K}_p^{\text{stable}}\,.$$

Here $\operatorname{Hom}(L^{-1}, -)$ *is the natural system in* (3.4.5).

The proof of this result is similar to the proof of (3.5.2).

3.5.4 Theorem. *The track category of Eilenberg-MacLane spaces is a linear track extension*

$$\mathcal{L} \longrightarrow [\![\mathbf{K}_p]\!]_1 \rightrightarrows [\![\mathbf{K}_p]\!]_0 \longrightarrow \mathbf{K}_p\,.$$

Here $\mathcal{L}$ *is the natural system in* (3.4.7).

This is a direct consequence of the track calculus results in Section (3.2).

3.6 Cohomology of categories

We recall from [BW], [BAH] the following definition of cohomology of a category.

3.6.1 Definition. Let $\mathbf{C}$ be a small category and let D be a natural system on $\mathbf{C}$. Let $N_n(\mathbf{C})$ be the set of sequences $(\lambda_1, \dots, \lambda_n)$ of n composable morphisms in $\mathbf{C}$ (which are the n-simplices of the *nerve* of $\mathbf{C}$). For $n = 0$ let $N_0(\mathbf{C}) = Ob(\mathbf{C})$ be the set of objects in $\mathbf{C}$. The nth cochain group $F^n = F^n(\mathbf{C}, D)$ is the abelian group of all functions

(1) $$c : N_n(\mathbf{C}) \longrightarrow \bigcup_{g \in \mathrm{Mor}(\mathbf{C})} D_g = D^{\cdot}$$

with $c(\lambda_1, \dots, \lambda_n) \in D_{\lambda_1 \circ \dots \circ \lambda_n}$. Addition in F^n is given by adding pointwise in the abelian group D_g. The coboundary $\delta : F^{n-1} \to F^n$ is defined by the formula

(2) $$\begin{aligned}(\delta c)(\lambda_1, \dots, \lambda_n) &= (\lambda_1)_* c(\lambda_2, \dots, \lambda_n) \\ &\quad + \textstyle\sum_{i=1}^{n-1} (-1)^i c(\lambda_1, \dots, \lambda_i \lambda_{i+1}, \dots, \lambda_n) \\ &\quad + (-1)^n \lambda_n^* c(\lambda_1, \dots, \lambda_{n-1}).\end{aligned}$$

For $n = 1$ we have $(\delta c)(\lambda) = \lambda_* c(A) - \lambda^* c(B)$ for $\lambda : A \to B \in N_1(\mathbf{C})$. One can check that $\delta c \in F^n$ for $c \in F^{n-1}$ and that $\delta\delta = 0$. Whence the *cohomology groups*

(3) $$H^n(\mathbf{C}, D) = H^n(F^*(\mathbf{C}, D), \delta)$$

are defined, $n \geq 0$. These groups are discussed in [BW], [BAH], [JP].

A functor $\phi : \mathbf{C}' \to \mathbf{C}$ induces the homomorphism

(3.6.2) $$\phi^* : H^n(\mathbf{C}, D) \longrightarrow H^n(\mathbf{C}', \phi^* D)$$

where $\phi^* D$ is the natural system given by $(\phi^* D)_f = D_{\phi(f)}$. On cochains the map ϕ^* is given by the formula

$$(\phi^* f)(\lambda'_1, \dots, \lambda'_n) = f(\phi\lambda'_1, \dots, \phi\lambda'_n)$$

where $(\lambda'_1, \dots, \lambda'_n) \in N_n(\mathbf{C}')$. In IV.5.8 of [BAH] we show

3.6.3 Proposition. *Let $\phi : \mathbf{C} \to \mathbf{C}'$ be an equivalence of categories. Then ϕ^* is an isomorphism of groups.*

A natural transformation $\tau : D \to D'$ between natural systems induces a homomorphism

(3.6.4) $$\tau_* : H^n(\mathbf{C}, D) \longrightarrow H^n(\mathbf{C}', D)$$

by $(\tau_* f)(\lambda_1, \dots, \lambda_n) = \tau_\lambda f(\lambda_1, \dots, \lambda_n)$ where $\tau_\lambda : D_\lambda \to D'_\lambda$ with $\lambda = \lambda_1 \circ \dots \circ \lambda_n$ is given by the transformation τ. Now let

$$D'' \overset{\iota}{\rightarrowtail} D \overset{\tau}{\twoheadrightarrow} D'$$

be a short exact sequence of natural systems on $\mathbf{C}$. Then we obtain as usual the natural long exact sequence

$$\longrightarrow H^n(\mathbf{C}, D') \xrightarrow{\iota_*} H^n(\mathbf{C}, D) \xrightarrow{\tau_*} H^n(\mathbf{C}, D'') \xrightarrow{\beta} H^{n+1}(\mathbf{C}, D') \longrightarrow$$

where β is the Bockstein homomorphism. For a cocycle c'' representing a class $\{c''\}$ in $H^n(\mathbf{C}, D'')$ we obtain $\beta\{c''\}$ by choosing a cochain c as in (3.6.1)(1) with $\tau c = c''$. This is possible since τ is surjective. Then $\tau^{-1}\delta c$ is a cocycle which represents $\beta\{c''\}$.

3.6.5 Remark. The cohomology (3.6.1) generalizes the *cohomology of a group*. In fact, let G be a group and let $\mathbf{G}$ be the corresponding category with a single object and with morphisms given by the elements in G. A right G-module D yields a natural system $\bar{D} : F\mathbf{G} \to \mathbf{Ab}$ by $\bar{D}g = D$ for $g \in G$. The induced maps are given by $f^*(x) = x^f$ and $h_*(y) = y$, $f, h \in G$. Then the classical definition of the cohomology $H^n(G, D)$ coincides with the definition of

$$H^n(\mathbf{G}, \bar{D}) = H^n(G, D)$$

given by (3.6.1).

3.6.6 Definition. Let $\mathbf{C}$ be a *ringoid*, i.e., a small category in which all morphism sets $\mathbf{C}(A, B)$ are abelian groups and composition is bilinear. (A ringoid is also called a pre-additive category or a category enriched in the category $\mathbf{Ab}$ of abelian groups.) Let $D : \mathbf{C} \times \mathbf{C}^{\mathrm{op}} \to \mathbf{Ab}$ be a $\mathbf{C}$-bimodule, that is, D is additive as a functor in $\mathbf{C}$ and $\mathbf{C}^{\mathrm{op}}$. Then we call a cochain

$$c \in F^n(\mathbf{C}, D) \tag{1}$$

multilinear if for all $i = 1, \dots, n$ and $\lambda_i, \lambda'_i \in \mathbf{C}(A_i, A_{i-1})$ we have

$$c(\lambda_1, \dots, \lambda_i + \lambda'_i, \dots, \lambda_n) = c(\lambda_1, \dots, \lambda_i, \dots, \lambda_n) + c(\lambda_1, \dots, \lambda'_i, \dots, \lambda_n)$$

in $D(A_n, A_0)$. For $n \geq 1$ let

$$LF^n(\mathbf{C}, D) \subset F^n(\mathbf{C}, D) \tag{2}$$

be the subgroup of multilinear cochains. The coboundary δ in (3.6.1) restricts to $LF^n(\mathbf{C}, D) \to LF^{n+1}(\mathbf{C}, D)$ for $n \geq 0$ where we set $LF^0(\mathbf{C}, D) = F^0(\mathbf{C}, D)$. Hence the cohomology

$$HH^n(\mathbf{C}, D) = H^n LF^*(\mathbf{C}, D) \tag{3}$$

is defined which we call the *Hochschild cohomology* of $\mathbf{C}$ with coefficients in D. Moreover (2) induces the natural homomorphism

$$HH^n(\mathbf{C}, D) \longrightarrow H^n(\mathbf{C}, D). \tag{4}$$

If $\mathbf{C} = A$ is a (graded) algebra and M an A-bimodule, then M is a natural system on the monoid A and by (3) the cohomology $HH^n(A, M)$ is defined. This is the classical Hochschild cohomology of the algebra A with coefficients in the bimodule M.

Now let $\mathbf{mod}_0(A)^{\mathrm{op}}$ be the category of finitely generated free right A-modules with generators of degree ≥ 1.

Then M defines the natural system $\bar{M}$ on $\mathbf{mod}_0(A)^{\mathrm{op}}$ as in (3.4.7). In this case

$$(5) \qquad HH^n(A, M) = HH^n(\mathbf{mod}_0(A)^{\mathrm{op}}, \bar{M}).$$

This leads to the following definition of *MacLane cohomology*:

$$(6) \qquad HML^n(A, M) = H^n(\mathbf{mod}_0(A)^{\mathrm{op}}, \bar{M})$$

This is a special case of the cohomology defined in (3.6.1). MacLane cohomology is also called *topological Hochschild cohomology*. Compare Pirashvili [P] and Pirashvili-Waldhausen [PW]. By (4), (5) we have the natural transformation

$$(7) \qquad HH^n(A, M) \longrightarrow HML^n(A, M)$$

which can be studied by a spectral sequence, Pirashvili [P]. We also have the forgetful map

$$(8) \qquad \phi : HML^n(A, M) \longrightarrow H^n(A, M)$$

where M is considered as a natural system on the graded monoid A, see (3.4.6), and where $H^n(A, M)$ is defined by (3.6.1). We point out that (8) in general is not an isomorphism.

For each linear track extension $\mathcal{T}$,

$$D \longrightarrow \mathcal{T}_1 \rightrightarrows \mathcal{T}_0 \longrightarrow \mathbf{C}$$

a characteristic cohomology class in $H^3(\mathbf{C}, D)$ is defined, compare [BD] where this class is termed the 'universal Toda bracket' of $\mathcal{T}$. We recall the definition as follows.

3.6.7 Definition. The element

$$\langle c \rangle = \langle \mathcal{T} \rangle \in H^3(\mathbf{C}, D)$$

represented by the following cocycle: Choose for each morphism f in $\mathcal{T}_{\simeq} = \mathbf{C}$ a representative 1-arrow of $\mathcal{T}$ denoted $s(f) \in f$. Furthermore choose a track $\mu(f, g) : s(f)s(g) \Rightarrow s(fg)$. Then for each composable triple f, g, h the composite track in

the diagram

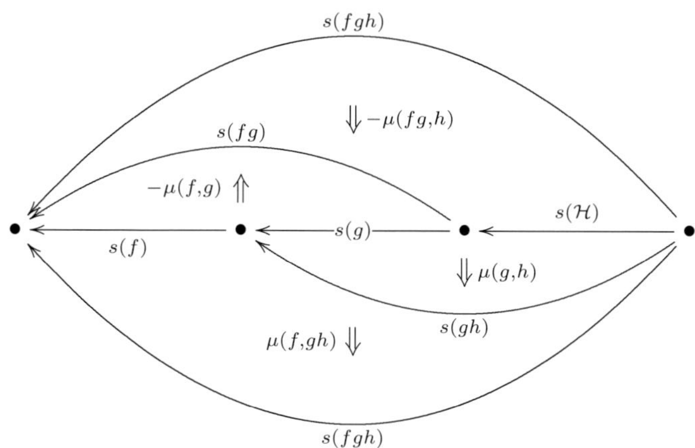

determines an element in $Aut(s(fgh))$ and hence, going back via σ, an element $c(f,g,h) \in D_{fgh}$. It can be checked that this determines a 3-cocycle of $\mathcal{T}_{\simeq}$ with coefficients in D, and that both choosing a different section s or different tracks $\mu(f,g)$ leads to a cohomologous cocycle. One can thus obtain a uniquely determined cohomology class $\langle \mathcal{T} \rangle$ represented by the cocycle c termed the *characteristic class* of $\mathcal{T}$.

3.6.8 Definition. Let D be a natural system on the small category $\mathbf{C}$. Then we define the category $\mathbf{Track}(\mathbf{C}, D)$ as follows. Objects are linear track extensions

$$D \longrightarrow \mathcal{T}_1 \rightrightarrows \mathcal{T}_0 \longrightarrow \mathbf{C}$$

and morphisms are track functors $\mathcal{T} \to \mathcal{T}'$ for which the diagram

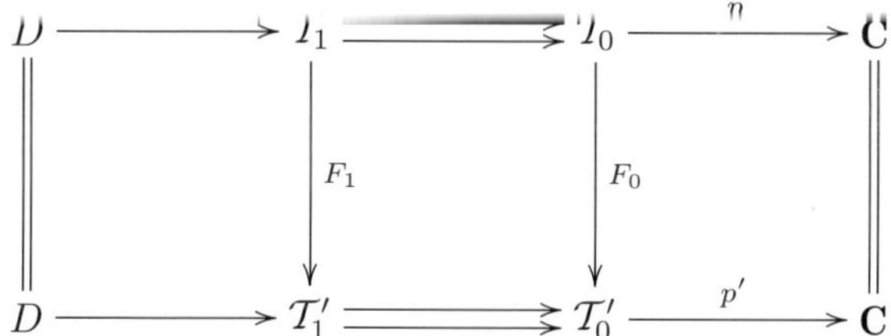

commutes, that is $F_1(\varphi \oplus \xi) = F_1(\varphi) \oplus \xi$ and $p'F_0 = p$. Let $\pi_0\mathbf{Track}(\mathbf{C}, D)$ be the set of connected components of the category $\mathbf{Track}(\mathbf{C}, D)$.

In [BD] we show the following result.

3.6.9 Theorem. *The function*

$$\pi_0\mathbf{Track}(\mathbf{C}, D) \longrightarrow H^3(\mathbf{C}, D)$$

which carries the component of $\mathcal{T}$ to the characteristic class $\langle \mathcal{T} \rangle$ is well defined. Moreover this function is a binatural bijection.

We call a morphism in $\mathbf{Track}(\mathbf{C}, D)$ a *weak track equivalence* (under D and over $\mathbf{C}$). Hence the theorem shows that a cohomology class $\xi \in H^3(\mathbf{C}, D)$ determines a linear track extension $\mathcal{T}_\xi$ up to such weak track equivalences.

Using the characteristic class in (3.6.7) one has the following well-defined cohomology classes

$$(3.6.10) \qquad \begin{cases} k_{\mathcal{A}} & = \langle [\![\mathcal{A}]\!] \rangle & \in H^3(\mathcal{A}, \Sigma\mathcal{A}), \\ k_p^{\text{stable}} & = \langle [\![\mathbf{K}_p^{\text{stable}}]\!] \rangle & \in H^3(\mathbf{K}_p^{\text{stable}}, \operatorname{Hom}(L^{-1}, -)) \\ & & = HML^3(\mathcal{A}, \Sigma\mathcal{A}), \\ k_p & = \langle [\![\mathbf{K}_p]\!] \rangle & \in H^3(\mathbf{K}_p, \mathcal{L}). \end{cases}$$

For this we use the linear track extensions in (3.5.2), (3.5.3) and (3.5.4). Here the natural map

$$HML^3(\mathcal{A}, \Sigma\mathcal{A}) \longrightarrow H^3(\mathcal{A}, \Sigma\mathcal{A})$$

carries k_p^{stable} to $k_{\mathcal{A}}$, see (3.6.6)(8). The element of interest is k_p^{stable} and not $k_{\mathcal{A}} = \phi k_p^{\text{stable}}$ with ϕ as in (3.6.6)(8).

3.7 Secondary cohomology and the obstruction of Blanc

The cohomology $H^*(X)$ of a path connected pointed space X is a connected unstable $\mathcal{A}$-algebra which by (1.5.3) equivalently can be described by the model $[X, -]$ of the theory $\mathbf{K}_p$. This model corresponds to the subcategory $[X, \mathbf{K}_p]$ with

$$(3.7.1) \qquad \mathbf{K}_p \subset [X, \mathbf{K}_p] \subset \mathbf{Top}^*/\simeq$$

defined as follows. Objects of $[X, \mathbf{K}_p]$ are X and the objects A, B in $\mathbf{K}_p$. Morphisms are the identity of X, all maps $X \to A$ in $\mathbf{Top}^*/\simeq$ with $A \in \mathbf{K}_p$ and all maps $A \to B$ in $\mathbf{K}_p$. The category $[X, \mathbf{K}_p]$ is completely determined by the model $[X, -]$ or by the unstable $\mathcal{A}$-algebra $H^*(X)$.

We define a natural system $\mathcal{L}^{H^*(X)}$ on the category $[X, \mathbf{K}_p]$ as an extension of the natural system $\mathcal{L}$ on $\mathbf{K}_p$ as follows. For the identity 1_X of X let $\mathcal{L}_{1_X}^{H^*(X)} = 0$ be the trivial group and for $f : X \to A$ with $A \in \mathbf{K}_p$ let

$$(3.7.2) \qquad \mathcal{L}_f^{H^*(X)} = [X, LA]$$

and for $g : A \to B$ with $A, B \in \mathbf{K}_p$ let $\mathcal{L}_g^{H^*(X)} = \mathcal{L}_g = [A, LB]$ be defined as in (3.4.7). Induced maps for $\mathcal{L}^{H^*(X)}$ are defined in the same way as in (3.4.7)(1),(2).

Finally let $[\![X, \mathbf{K}_p]\!]$ be the track category corresponding to $[X, \mathbf{K}_p]$ above with

$$(3.7.3) \qquad [\![\mathbf{K}_p]\!] \subset [\![X, \mathbf{K}_p]\!] \subset [\![\mathbf{Top}^*]\!].$$

Hence $[\![X, \mathbf{K}_p]\!]$ is completely determined by the track model $[\![X, -]\!] = \tilde{\mathcal{H}}^*(X)$ of $[\![\mathbf{K}_p]\!]$ which is the secondary cohomology of X; see (2.2.10).

3.7.4 Theorem. *The secondary cohomology yields a linear track extension*

$$\mathcal{L}^{H^*(X)} \longrightarrow [\![X,\mathbf{K}_p]\!]_1 \rightrightarrows [\![X,\mathbf{K}_p]\!]_0 \longrightarrow [X,\mathbf{K}_p]$$

which by (3.6.9) *is determined up to weak track equivalence by the cohomology class*

$$k_X = \langle [\![X,\mathbf{K}_p]\!] \rangle \in H^3([X,\mathbf{K}_p],\mathcal{L}^{H^*(X)}).$$

We point out that the cohomology $H^3([X,\mathbf{K}_p],\mathcal{L}^{H^*(X)})$ is algebraically determined by the connected unstable $\mathcal{A}$-algebra H^*X.

The element k_X describes the secondary cohomology of X up to weak equivalence. Moreover k_X is an invariant of the homotopy type of X. In fact, for a map $f : X \to Y$ between path connected spaces in $\mathbf{Top}^*/\simeq$ we get the induced functor α and the induced natural transformation β,

$$\begin{aligned} \alpha &= f^* : [Y,\mathbf{K}_p] \longrightarrow [X,\mathbf{K}_p], \\ \beta &= f_* : f^*\mathcal{L}^{H^*(X)} \longrightarrow \mathcal{L}^{H^*(Y)}. \end{aligned}$$

The maps α,β induce the homomorphisms

$$H^3([X,\mathbf{K}_p],\mathcal{L}^{H^*(X)}) \xrightarrow{\alpha^*} H^3([Y,\mathbf{K}_p],f^*\mathcal{L}^{H^*(X)}) \xleftarrow{\beta_*} H^3([Y,\mathbf{K}_p],\mathcal{L}^{H^*(Y)}).$$

Now the following equation holds,

$$\alpha^* k_X = \beta_* k_Y. \tag{3.7.5}$$

This is the *naturality of the invariant* k_X.

The inclusion of categories $\mathbf{K}_p \subset [X,\mathbf{K}_p]$ induces the homomorphism i^* in the exact sequence of a pair of categories:

(3.7.6)

$$\begin{array}{ccccc} H^3([X,\mathbf{K}_p],\mathcal{L}^{H^*(X)}) & \xrightarrow{i^*} & H^3(\mathbf{K}_p,\mathcal{L}) & \xrightarrow{\delta} & H^4([X,\mathbf{K}_p],\mathbf{K}_p;\mathcal{L}^{H^*(X)}). \\ k_X & \mapsto & k_p & & \end{array}$$

Here i^* carries k_X to the element $k_p = \langle [\![\mathbf{K}_p]\!] \rangle$. Hence exactness implies $\delta(k_p) = 0$. This leads to a first obstruction for the following realization problem.

Let H be a connected unstable $\mathcal{A}$-algebra. We say H is *realizable* if there exists a path connected space X with $H \cong H^*(X)$ in $\mathcal{K}_p^0$. In general H need not be realizable.

Recall that $V^\# = \mathrm{Hom}(V,\mathbb{F})$ is the dual vector space of V with $(V^\#)^\# = V$ if V is finite dimensional. Assume for a moment that H is of finite dimension and let $H^\#$ be the dual of H. Then $H^\#$ is a connected unstable $\mathcal{A}$-coalgebra.

Blanc [Bl] discovered a sequence of obstructions $(n \geq 1)$

$$\chi_n \in QH^{n+2}(H^\#,\Sigma^n H^\#)$$

where QH^* denotes the Quillen cohomology. Here χ_n is defined for $n > 1$ if $\chi_1 = \cdots = \chi_{n-1} = 0$.

3.7.7 Theorem ([Bl]). *The finite-dimensional connected unstable $\mathcal{A}$-algebra H is realizable if and only if $\chi_n = 0$ for all $n \geq 1$.*

We now define a first obstruction which plays the role of Blanc's obstruction χ_1. Let H be a connected unstable $\mathcal{A}$-algebra and let M_H be the model of $\mathbf{K}_p$ corresponding to H; see (1.5.2). Then we define the category $[M_H, \mathbf{K}_p]$ together with the inclusion

$$\mathbf{K}_p \subset [M_H, \mathbf{K}_p] \tag{3.7.8}$$

such that for $M_H = [X, -]$ and $H = H^*(X)$ this category coincides with $[X, \mathbf{K}_p]$ above. Objects in $[M_H, \mathbf{K}_p]$ are an object $*_H$, and the objects A in $\mathbf{K}_p$. Morphisms are the identity of $*_H$, and the elements of $M_H(A)$ which are morphisms $*_H \to A$ and the morphisms $A \to B$ in $\mathbf{K}_p$. The composite

$$*_H \xrightarrow{\alpha} A \xrightarrow{f} B$$

with $\alpha \in M_H(A)$ is defined by $f_*(\alpha) \in M_H(B)$. Here f_* is defined by the functor $M_H : \mathbf{K}_p \to \mathbf{Set}$. We point out that the category $[M_H, \mathbf{K}_p]$ is completely determined by $\mathbf{K}_p$ and the connected unstable $\mathcal{A}$-algebra H.

We define a natural system $\mathcal{L}^H$ on the category $[M_H, \mathbf{K}_p]$ as an extension of the natural system $\mathcal{L}$ on $\mathbf{K}_p$ as follows; compare (3.7.2). For the identity 1_* of $*_H$ let $\mathcal{L}^H_{1_*} = 0$ and for $\alpha : *_H \to A$, $\alpha \in M_H(A)$, let

$$\mathcal{L}^H_\alpha = M_H(LA). \tag{3.7.9}$$

We define induced maps for $\mathcal{L}^H$ as in (3.4.7)(1),(2). In particular $f : A \to B$ induces $\mathcal{L}^H_\alpha \to \mathcal{L}^H_{f\alpha}$ with $f_*(\xi) = (L\nabla f)_*(\xi, \alpha)$ where

$$(\xi, \alpha) \in M_H(LA) \times M_H(A) = M_A(L(A) \times A)$$

and $L\nabla f : L(A) \times A \to LB$ as in (3.3.8). Hence the natural system $\mathcal{L}^H$ also uses the derivation $\tilde{\nabla}$ in (3.3.9). Of course $\mathcal{L}^H$ is completely defined by the unstable $\mathcal{A}$-algebra H. If $H = H^*(X)$ then $\mathcal{L}^H$ coincides with the natural system $\mathcal{L}^{H^*(X)}$ in (3.7.2).

We now consider an exact sequence as in (3.7.6).

$$\begin{array}{ccccc} H^3([M_H, \mathbf{K}_p], \mathcal{L}^H) & \xrightarrow{i^*} & H^3(\mathbf{K}_p, \mathcal{L}) & \xrightarrow{\delta} & H^4([M_H, \mathbf{K}_p], \mathbf{K}_p; \mathcal{L}^H). \\ & & k_p & \mapsto & \delta(k_p) \end{array} \tag{3.7.10}$$

Here k_p is the class of the secondary Steenrod algebra. The sequence coincides with (3.7.2) if H is realizable by the space X and in this case we know $\delta(k_p) = 0$. Hence we get:

3.7.11 Theorem. *The connected unstable $\mathcal{A}$-algebra H determines the group $H^4([M_H, \mathbf{K}_p], \mathbf{K}_p; \mathcal{L}^H)$ and the element $\delta(k_p)$ is a first obstruction for the realizability of H. In particular $\delta(k_p) \neq 0$ implies that H is not realizable.*

We claim that there is a connection between Blanc's obstruction χ_1 in (3.7.7) and the element $\delta(k_p)$ but we do not work out details since this is not needed in this book.

3.8 Secondary cohomology as a stable model

We have the forgetful functor

$$\mathbf{K}_p^{\text{stable}} \xrightarrow{\phi} \mathbf{K}_p$$

which is the identity on objects and carries the stable map α to α_0. This functor preserves products. Hence a model of $\mathbf{K}_p$ is also a model of $\mathbf{K}_p^{\text{stable}}$. Using (2.5.2) we see that

3.8.1 Proposition. *There is an isomorphism of categories*

$$\mathbf{model}(\mathbf{K}_p^{\text{stable}}) \;=\; \mathbf{Mod}_0(\mathcal{A}).$$

Here $\mathbf{Mod}_0(\mathcal{A})$ is the category of graded $\mathcal{A}$-modules M with $M_i = 0$ for $i \leq 0$. Of course the isomorphism is compatible with the isomorphism in (1.5.2) in the sense that the following diagram commutes.

(3.8.2)
$$\begin{array}{ccc} \mathbf{model}(\mathbf{K}_p^{\text{stable}}) & =\!=\!=\!=\!= & \mathbf{Mod}_0(\mathcal{A}) \\ \Big\uparrow \phi^* & & \Big\uparrow \phi \\ \mathbf{model}(\mathbf{K}_p) & =\!=\!=\!=\!= & \mathcal{K}_p^0 \end{array}$$

Here ϕ on the right-hand side is the obvious forgetful functor. A pointed path connected space X yields the model $[X,-]$ of $\mathbf{K}_p$ and hence the model $[X,-]\phi$ of $\mathbf{K}_p^{\text{stable}}$. The model $[X,-]\phi$ can be identified by (3.8.1) with the $\mathcal{A}$-module $\tilde{H}^*(X)$.

We now describe the stable analogue of the constructions in (3.7) above. The space X yields the category

(3.8.3)
$$\mathbf{K}_p^{\text{stable}} \subset [X, \mathbf{K}_p^{\text{stable}}]$$

defined as follows. Objects are X and the objects A, B of $\mathbf{K}_p^{\text{stable}}$. Morphisms are the identity of X, all maps $X \to A$ in $\mathbf{Top}^*/\simeq$ and all maps $A \to B$ in $\mathbf{K}_p^{\text{stable}}$. Composition is defined by the functor ϕ above. The category $[X, \mathbf{K}_p^{\text{stable}}]$ is completely determined by the model $[X,-]\phi$ or by the $\mathcal{A}$-module $H^*(X)$. Let H be any $\mathcal{A}$-module in $\mathbf{Mod}_0(\mathcal{A})$ corresponding to the model M_H of $\mathbf{K}_p^{\text{stable}}$ by (3.8.1). Then we obtain more generally the category

(3.8.4)
$$\mathbf{K}_p^{\text{stable}} \subset [M_H, \mathbf{K}_p^{\text{stable}}]$$

which is defined similarly as in (3.7.8). If $M_H = [X,-]\phi$ then this category coincides with $[X, \mathbf{K}_p^{\text{stable}}]$ above.

We define a natural system R^H on the category $[M_H, \mathbf{K}_p^{\text{stable}}]$ as an extension of $Hom(L^{-1}, -)$ on $\mathbf{K}_p^{\text{stable}}$. Let $R^H_{1_X} = 0$ and for $f : A \to B$, $A \in \mathbf{K}_p^{\text{stable}}$, let

$$R^H_f = [X, LA]. \tag{3.8.5}$$

The induced map $\alpha_* : R^H_f \to R^H_{\alpha f}$ for $\alpha : A \to B$ in $\mathbf{K}_p^{\text{stable}}$ is defined by $\alpha_*(x) = L(\alpha_0)(x)$ for $x \in [X, LA]$. Finally let $[\![X, \mathbf{K}_p^{\text{stable}}]\!]$ be the track category corresponding to $[X, \mathbf{K}_p^{\text{stable}}]$ above with

$$[\![\mathbf{K}_p^{\text{stable}}]\!] \subset [\![X, \mathbf{K}_p^{\text{stable}}]\!]. \tag{3.8.6}$$

Here $[\![X, \mathbf{K}_p^{\text{stable}}]\!]$ is completely determined by the track model $[\![X, -]\!]\phi$ where we use the forgetful track functor

$$\phi : [\![\mathbf{K}_p^{\text{stable}}]\!] \longrightarrow [\![\mathbf{K}_p]\!]$$

see (2.6.5). Now we get the following stable analogue of (3.7.4).

3.8.7 Theorem. *The secondary cohomology $\mathcal{H}^*(X)\phi$ yields a linear track extension*

$$R^{H^*(X)} \longrightarrow [\![X, \mathbf{K}_p^{\text{stable}}]\!]_1 \rightrightarrows [\![X, \mathbf{K}_p^{\text{stable}}]\!]_0 \longrightarrow [X, \mathbf{K}_p^{\text{stable}}]$$

which by (3.6.9) *is determined up to weak equivalence by the cohomology class*

$$k_X^{\text{stable}} = < [\![X, \mathbf{K}_p^{\text{stable}}]\!] > \in H^3([X, \mathbf{K}_p^{\text{stable}}], R^{H^*(X)}).$$

Here the cohomology H^3 is completely determined by the $\mathcal{A}$-module $H^*(X)$.

We now can define an obstruction for the realizability of an $\mathcal{A}$-module H in a similar way as in Section (3.7). For this we leave it to the reader to formulate the stable analogue of (3.7.11) above.

Chapter 4

Stable Linearity Tracks

Maps in $\mathbf{K}_p^{\text{stable}}$ are $\mathbb{F}$-linear. The stable maps in $[\![\mathbf{K}_p^{\text{stable}}]\!]_0$, however, need not be linear. Therefore there arises a linearity track describing the deviation from linearity. It turns out that linearity tracks can be chosen canonically. The properties of linearity tracks serve as axioms for a Γ-track algebra. the secondary Steenrod algebra $[\![\mathcal{A}]\!]$ is a Γ-track algebra which determines the linear extension $[\![\mathbf{K}_p^{\text{stable}}]\!]$ up to weak equivalence. Moreover secondary cohomology $[\![X, \mathcal{A}]\!]$ is a Γ-track module.

4.1 Weak additive track extensions

Let $\mathbb{F}$ be a field, for example $\mathbb{F} = \mathbb{Z}/p$. An $\mathbb{F}$-*ringoid* is a category enriched in the category of $\mathbb{F}$-vector spaces, that is, morphism sets are vector spaces and composition is $\mathbb{F}$-bilinear. An $\mathbb{F}$-*additive category* $\mathbf{K}$ is an $\mathbb{F}$-ringoid in which products exist. Such products are also coproducts and are called "biproducts" or "direct sums"; see MacLane [MLC]. For example, the category $\mathbf{K} = \mathbf{K}_p^{\text{stable}}$ is an $\mathbb{F}$-additive category isomorphic to $\mathbf{mod}_0(\mathcal{A})^{\text{op}}$; see (2.5.2).

Let D be an $\mathbb{F}$-*biadditive* $\mathbf{K}$-bimodule, that is, D is a functor

(4.1.1) $$D : \mathbf{K} \times \mathbf{K}^{\text{op}} \longrightarrow \mathbf{Vec}_{\mathbb{F}},$$

where $\mathbf{Vec}_{\mathbb{F}}$ is the category of $\mathbb{F}$-vector spaces, and $D(A, B)$ with $A, B \in \mathbf{K}$ is additive in A and B.

By (3.6.6)(5) we know that the homomorphism

(4.1.2) $$HH^3(\mathbf{K}, D) \longrightarrow H^3(\mathbf{K}, D)$$

is defined.

4.1.3 Definition. Let $\mathbf{K}$ be $\mathbb{F}$-additive and let D be an $\mathbb{F}$-biadditive $\mathbf{K}$-bimodule as above. Then a linear track extension

$$D \longrightarrow \mathcal{T}_1 \rightrightarrows \mathcal{T}_0 \longrightarrow \mathbf{K}$$

is a *weak* $\mathbb{F}$-*additive track extension* if the following properties hold. The track category $\mathcal{T}$ has a strong zero object $*$ with $[\![*, X]\!]$ and $[\![X, *]\!]$ consisting only of exactly one morphism for all objects X in $\mathcal{T}$. Moreover finite strong products $A \times B$ exist in $\mathcal{T}$, see (2.3.4), and each object A is an $\mathbb{F}$-vector space object in $\mathcal{T}_0$.

For example for $\mathbf{K} = \mathbf{K}_p^{\text{stable}}$ we have the $\mathbb{F}$-biadditive $\mathbf{K}$-bimodule $D = \text{Hom}(L^{-1}, -)$ and we have seen in (2.6) and (3.5.3) that the stable track theory $\mathcal{T} = [\![\mathbf{K}_p^{\text{stable}}]\!]$ of Eilenberg-MacLane spaces is a weak $\mathbb{F}$-additive track extension with properties as in (4.1.3).

Weak coproducts in a track category are the categorical dual of weak products as defined in (2.3.4), that is:

4.1.4 Definition. A *weak coproduct* or a *weak sum* $A \vee B$ in a track category $\mathcal{T}$ is an object $A \vee B$ equipped with maps $i_A = i_1 : A \to A \vee B$ and $i_B = i_2 : B \to A \vee B$ such that the induced functor (i_1^*, i_2^*):

$$(*) \qquad [\![A \vee B, X]\!] \longrightarrow [\![A, X]\!] \times [\![B, X]\!]$$

is an equivalence of groupoids for all objects X in $\mathcal{T}$. The coproduct is *strong* if $(*)$ is an isomorphism of groupoids.

4.1.5 Proposition. *Let $\mathcal{T}$ be a weak $\mathbb{F}$-additive track extension as in* (4.1.3). *Then strong products $A \times B$ in $\mathcal{T}$ are also weak coproducts by the inclusions*

$$\begin{aligned} i_1 &= 1 \times 0 : A = A \times * \longrightarrow A \times B, \\ i_2 &= 0 \times 1 : B = * \times B \longrightarrow A \times B. \end{aligned}$$

4.1.6 Corollary. *Strong products $A \times B$ in $[\![\mathbf{K}_p^{\text{stable}}]\!]$ are weak coproducts.*

Proof of (4.1.5). We have to show that $A \times B$ is a weak coproduct. Hence we have to show that $(*)$ in (4.1.4) is an equivalence of categories; that is, a full and faithful functor which is also representative. Using the linear extension (4.2.3) we easily see that $(*)$, in fact, is full and faithful since D is a $\mathbf{K}$-bimodule which is additive in each variable. It remains to check that $(*)$ is representative, that is, for each $\alpha : A \to X$ and $\beta : B \to X$ there exist

$$\begin{cases} \xi & : \quad A \times B \longrightarrow X, \\ H & : \quad \xi i_1 \Rightarrow \alpha, \\ G & : \quad \xi i_2 \Rightarrow \beta. \end{cases}$$

This is clear since the homotopy category $\mathcal{T}_{\simeq} = \mathbf{K}$ is an $\mathbb{F}$-additive category in which products are also coproducts. □

4.1.7 Definition. A weak $\mathbb{F}$-additive track extension is *strong* $\mathbb{F}$-*additive* if all strong products $A \times B$ are also strong coproducts.

4.1.8 Proposition. *The characteristic class $\langle \mathcal{T} \rangle$ of a strong $\mathbb{F}$-additive track extension $\mathcal{T}$ is in the image of Hochschild cohomology in* (4.1.2).

We shall see that $\langle [\![\mathbf{K}_p^{\text{stable}}]\!] \rangle$ is not in the image of (4.1.2) so that $[\![\mathbf{K}_p^{\text{stable}}]\!]$ is not weakly equivalent to a strong $\mathbb{F}$-additive track extension.

4.2 Linearity tracks

Let $\mathcal{T}$ be a weak $\mathbb{F}$-additive track extension as in (4.1.3), for example $\mathcal{T} = [\![\mathbf{K}_p^{\mathrm{stable}}]\!]$. Then we have for each object A in $\mathcal{T}$ the addition map $+_A : A \times A \to A$ of the $\mathbb{F}$-vector space object A. Moreover for each morphism $a : A \to B$ in $\mathcal{T}_0$ there is a diagram in $\mathcal{T}_0$

(4.2.1)
$$\begin{array}{ccc} A \times A & \xrightarrow{a \times a} & B \times B \\ {\scriptstyle +_A}\downarrow & \overset{\Gamma_a}{\Longrightarrow} & \downarrow{\scriptstyle +_B} \\ A & \xrightarrow[a]{} & B \end{array}$$

where $\Gamma_a : a(+_A) \Rightarrow (+_B)(a \times a)$. Since

$$(i_1^*, i_2^*) : [\![A \times A, B]\!] \longrightarrow [\![A, B]\!] \times [\![A, B]\!]$$

is an equivalence of groupoids (see (4.1.5)) there is a *unique track* Γ_a with

$$i_1^* \Gamma_a = 0_a^\square = i_2^* \Gamma_a$$

where $0_a^\square : a \Rightarrow a$ is the trivial track. We call Γ_a the *linearity track* for a. We obtain for $x, y : X \to A$ the map $x + y = (+_A)(x, y) : X \to A$ and hence we get the linearity track

(4.2.2)
$$\Gamma_a^{x,y} = \Gamma_a(x, y) : a(x + y) \Rightarrow ax + ay$$

in $\mathcal{T}_1$. We also use the following diagram where $A^{\times n} = A \times \cdots \times A$ is the n-fold product.

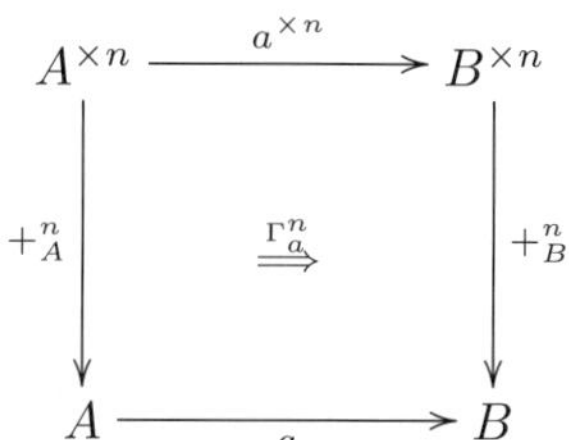

Here a *unique track* Γ_a^n is given with $\Gamma_a^n i_r = 0_a^\square$ for $r = 1, \ldots, n$. For $n = 2$ this coincides with (4.2.1). We write for $(x_1, \ldots, x_n) : X \to A^{\times n}$,

(4.2.3)
$$\Gamma_a^{x_1,\ldots,x_n} = \Gamma_a^n(x_1, \ldots, x_n) : a(x_1 + \cdots + x_n) \Rightarrow ax_1 + \cdots + ax_n$$

generalizing (4.2.2). One can check the following equation ($n \geq 3$)

$$\Gamma_a^{x_1,\ldots,x_n} = (\Gamma_a^{x_1,\ldots,x_{n-1}} + ax_n) \square \Gamma_a^{x_1,\ldots,x_{n-2},x_{n-1}+x_n}$$

so that inductively $\Gamma_a^{x_1,\dots,x_n}$ can be expressed only in terms of $\Gamma_a^{x,y}$ above. For maps $x, y : X \to A$ we obtain $x + y$ by the composite

$$x + y = (+_A)(x, y) : X \to A \times A \to A.$$

Similarly we get for tracks $H : x \Rightarrow x'$, $G : y \Rightarrow y'$ the composite

$$H + G = (+_A)(H, G) : x + y \Rightarrow x' + y' \tag{4.2.4}$$

where $(H, G) : (x, y) \Rightarrow (x', y')$ is a track in $\mathcal{T}$ since $A \times A$ is a strong product in $\mathcal{T}$. If $G = 0_y^\square : y \Rightarrow y$ is the identity track we write, compare (2.2.6),

$$H + y = H + 0_y^\square : x + y \Rightarrow x' + y$$

and accordingly if $H = 0_x^\square : x \Rightarrow x$ is the identity track we get

$$x + G = 0_x^\square + G : x + y \Rightarrow x + y'.$$

We use this notation in the following theorem.

4.2.5 Theorem. *Let $\mathcal{T}$ be a weak $\mathbb{F}$-additive track extension. Then linearity tracks in $\mathcal{T}$ satisfy the following equations* $(1), \dots, (7)$.

(1) $\Gamma_a^{xd,yd} = \Gamma_a^{x,y} d$.

(2) $\Gamma_{ba}^{x,y} = \Gamma_b^{ax,ay} \square b\Gamma_a^{x,y}$.

(3) $\Gamma_a^{x,y} = \Gamma_a^{y,x}$.

(4) $\Gamma_{a+a'}^{x,y} = \Gamma_a^{x,y} + \Gamma_{a'}^{x,y}$.

(5) $\Gamma_a^{w,x,y} = (\Gamma_a^{w,x} + ay) \square \Gamma_a^{w+x,y} = (aw + \Gamma_a^{x,y}) \square \Gamma_a^{w,x+y}$.

Equivalently the following diagram commutes.

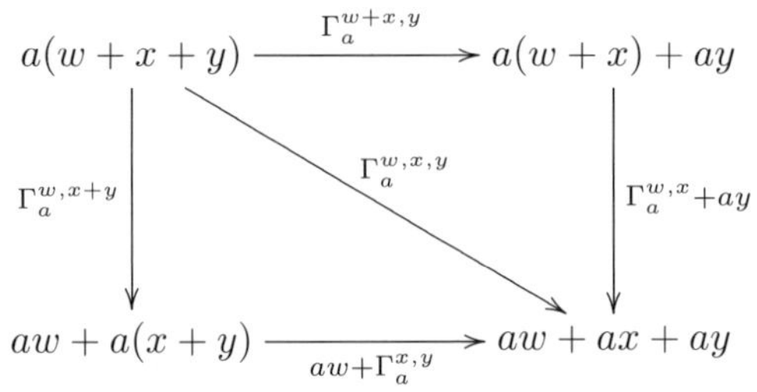

$$\begin{array}{ccc}
a(w+x+y) & \xrightarrow{\Gamma_a^{w+x,y}} & a(w+x)+ay \\
\Big\downarrow{\scriptstyle \Gamma_a^{w,x+y}} & \searrow^{\Gamma_a^{w,x,y}} & \Big\downarrow{\scriptstyle \Gamma_a^{w,x}+ay} \\
aw + a(x+y) & \xrightarrow[aw+\Gamma_a^{x,y}]{} & aw + ax + ay
\end{array}$$

This implies $\Gamma_a^{0,y} = 0^\square$ if we set $w = x = 0$ since $\Gamma_a^{0,0} = \Gamma_a^{0,0}0 = 0$ by (1).

(6) *For* $H : x \Rightarrow x'$ *and* $G : y \Rightarrow y'$ *the following diagram of tracks commutes.*

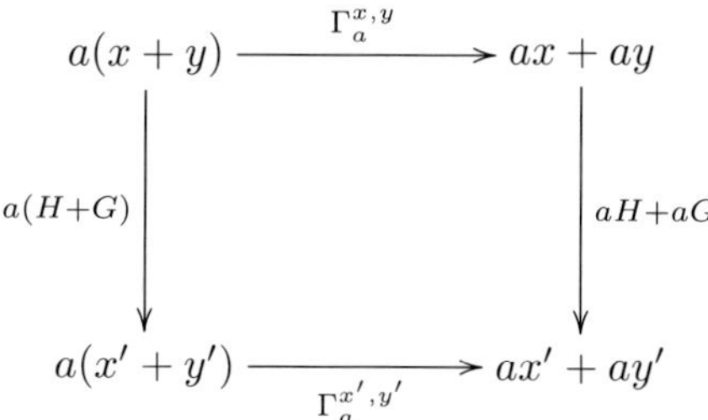

(7) *For* $A : a \Rightarrow a'$ *the following diagram of tracks commutes.*

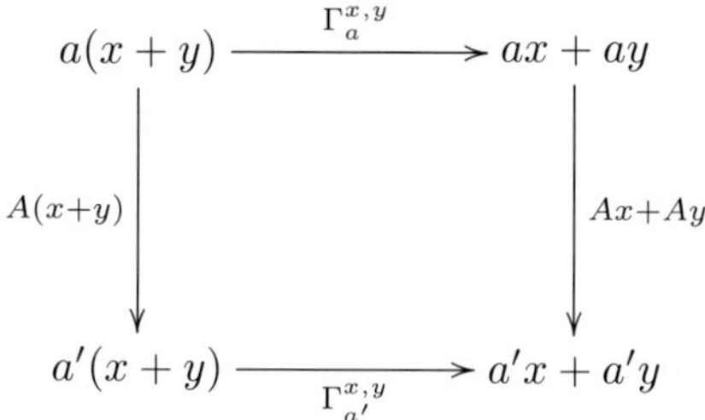

Proof of (4.2.5). Equation (1) is clear. Moreover (2) follows from

$$\begin{array}{ccccc} A \times A & \xrightarrow{a \times a} & B \times B & \xrightarrow{b \times b} & C \times C \\ \big\downarrow & \overset{\Gamma_a}{\Longrightarrow} & \big\downarrow & \overset{\Gamma_b}{\Longrightarrow} & \big\downarrow \\ A & \xrightarrow[a]{} & B & \xrightarrow[b]{} & C \end{array}$$

with $\Gamma_b * \Gamma_a = \Gamma_{ba}$ by uniqueness. Let $T = (p_2, p_1) : A \times A \to A \times A$ be the interchange map. Then $\Gamma_a T = \Gamma_a$ since $(\Gamma_a T)i_1 = \Gamma_a i_2 = 0^{\square} = \Gamma_a i_1 = (\Gamma_a T)i_2$. Hence we get (3). Again we see

$$\Gamma_{a+a'} = \Gamma_a + \Gamma_{a'}$$

since $(\Gamma_a + \Gamma_{a'})i_k = 0^{\square} + 0^{\square} = 0^{\square}$ for $k = 1, 2$. This implies (4). Again one readily checks that Γ_a^3 in (4.2.1) can be expressed by the composite C of tracks in the diagram

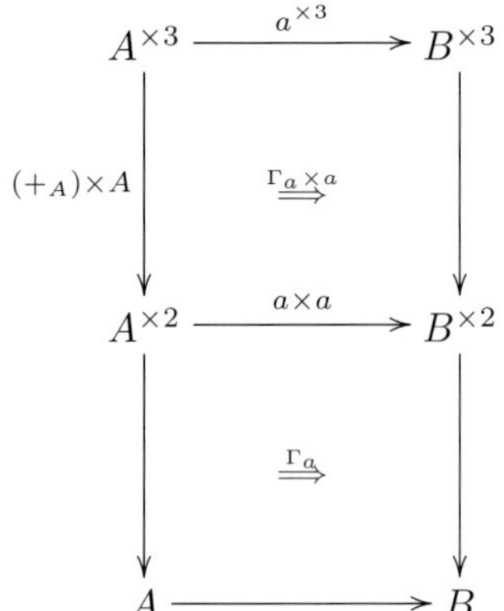

since $Ci_k = 0^\square$ for $k = 1, 2, 3$. This implies (5). Finally we get (6) by the definition $\Gamma_a^{x,y} = \Gamma_a(x, y)$ so that pasting yields

$$\begin{aligned}\Gamma_a * (H, G) &= \Gamma_a^{x',y'} \square a(H + G)\\ &= (aH + aG)\square \Gamma_a^{x,y}.\end{aligned}$$

Next we get (7) by considering the composite C of tracks in the following diagram.

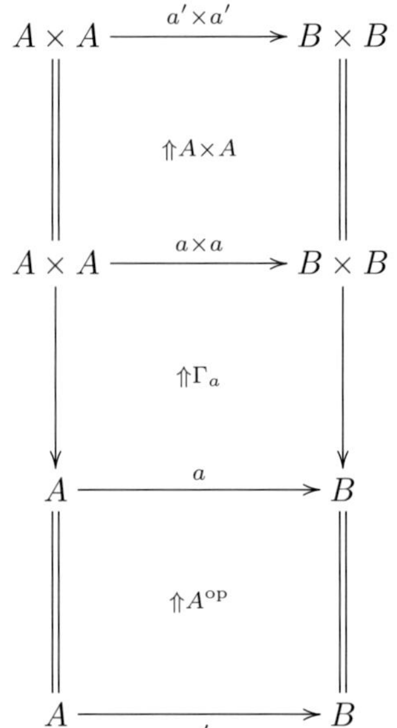

Now $Ci_k = 0^\square$ for $k = 1, 2$ implies $C = \Gamma_{a'}$. □

For $n \in \mathbb{N} = \{1, 2, \dots\}$ and an object A in $\mathcal{T}$ we obtain the map

(4.2.6) $$n \cdot 1_A : A \longrightarrow A$$

by the $\mathbb{F}$-vector space structure of $[\![A,A]\!]_0$. This map depends only on the element $\bar{n} = n \cdot 1 \in \mathbb{F}$ given by n. Moreover using (4.2.3) we have the track

$$\Gamma(n)_a = \Gamma_a^{x_1,\dots,x_n} : a(n \cdot 1_A) \Rightarrow (n \cdot 1_B)a = n \cdot a \tag{4.2.7}$$

where $x_1 = \cdots = x_n = 1_A$. This track actually depends on $n \in \mathbb{N}$ and is not well defined by $\bar{n} = n \cdot 1 \in \mathbb{F} = \mathbb{Z}/p$. (For example for $n = 2$ and $a : Z^m \to Z^{m+k}$ representing Sq^k we know by (4.5.8) that $\Gamma(n)_a : 0 \Rightarrow 0$ represents Sq^{k-1}.)

4.2.8 Lemma. *For $n, n' \in \mathbb{N}$ we have*

$$\Gamma(n' \cdot n)_a = ((n' \cdot 1_B)\Gamma(n)_a)\Box(\Gamma(n')_a(n \cdot 1_A)).$$

If $p \mid n'$ and $p \mid n$, then the lemma shows that $\Gamma(n'n)_a = 0^\Box : 0 \Rightarrow 0$ is the identity track of the zero map $0 : A \to * \to A$.

Proof. The lemma is a consequence of the diagram

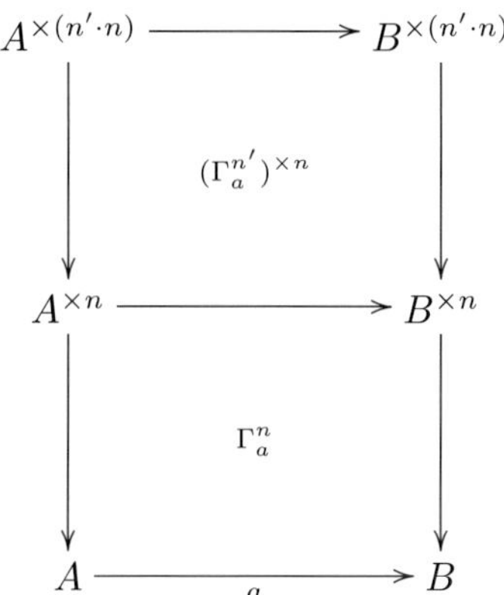

where Γ_a^n is defined in (4.2.3). Pasting of tracks in the diagram yields $\Gamma_a^{n'\cdot n}$. This is seen by the uniqueness property of $\Gamma_a^{n'\cdot n}$. □

4.2.9 Lemma. *For $n, n' \in \mathbb{N}$ the track $\Gamma(n+n')_a$ is the composite*

$$\begin{aligned} \Gamma(n+n')_a \quad &: \quad a(n \cdot 1_A + n' \cdot 1_A) \Rightarrow a(n1_A) + a(n'1_A) \Rightarrow na + n'a, \\ \Gamma(n+n')_a \quad &= \quad (\Gamma(n)_a + \Gamma(n')_a)\Box\Gamma_a^{n1_A,n'1_A}. \end{aligned}$$

This shows by (4.2.8) that

4.2.10 Proposition. $\Gamma(n)_a = \Gamma(n')_a$ *if $n \equiv n'$ mod $\mathbb{Z}/p^2$. For $n \le 0$ we choose k such that $n + kp^2 > 0$ and we define $\Gamma(n)_a = \Gamma(n + kp^2)$.*

Proof. Let $n' = n + p^2 \cdot m$. Then we have

$$\Gamma(n')_a = (\Gamma(n)_a + 0^\Box)\Box\Gamma_a^{n1_A,0} = \Gamma(n)_a.$$

Compare (4.2.5)(5). □

Since the ring $\mathbb{Z}/p^2$ in proposition (4.2.10) plays a major role we introduce the following notation:

4.2.11 Notation. Let p be a prime number. Then $\mathbb{F}$ is the field $\mathbb{F} = \mathbb{Z}/p$ and $\mathbb{G}$ is the *ring* $\mathbb{G} = \mathbb{Z}/p^2$. Clearly $\mathbb{F}$ is a $\mathbb{G}$-module by the surjection $\mathbb{G} \twoheadrightarrow \mathbb{F}$. We consider standard free $\mathbb{G}$-modules

$$V = \mathbb{G}^n = \mathbb{G} \oplus \cdots \oplus \mathbb{G}$$

with $n = \dim(V) \geq 0$. Hence $V = \mathbb{G}^n$ has the standard *inclusions* i_j and *projections* p_j,

$$\mathbb{G} \xrightarrow{i_j} V \xrightarrow{p_j} \mathbb{G}$$

for $j = 1, \dots, n$. Let

$$\nabla_V = \sum_{j=1}^{n} p_j : V \longrightarrow \mathbb{G}$$

be the *folding map* and let

$$+_V : V \oplus V \longrightarrow V$$

be the *addition map* with $+_V(x \oplus y) = x + y$ for $x, y \in V$. Of course $\nabla_{\mathbb{G}^2} = +_{\mathbb{G}}$. Each $\mathbb{G}$-linear map $\varphi : V = \mathbb{G}^n \to W = \mathbb{G}^m$ is given by a *matrix* (φ^i_j) with $\varphi^i_j \in \mathbb{G}$ defined by

$$\varphi^i_j = p_j \varphi i_i : \mathbb{G} \longrightarrow \mathbb{G}$$

for $i = 1, \dots, \dim(V)$ and $j = 1, \dots, \dim(W)$. We say that $\varphi : V \to V$ is a *permutation* (of coordinates) if $\varphi(x_1, \dots, x_n) = (x_{\sigma 1}, \dots, x_{\sigma n})$ where σ is a permutation, $x_i \in \mathbb{G}$.

Now let A be an object in $\mathcal{T}$ and let $\varphi : V = \mathbb{G}^n \to W = \mathbb{G}^m$ be a $\mathbb{G}$-linear map. Then we define

$$A \otimes \varphi : A^{\times n} \longrightarrow A^{\times m} \tag{4.2.12}$$

by $(A \otimes \varphi)(x_1, \dots, x_n) = (\sum_i \varphi^i_1 x_i, \dots, \sum_i \varphi^i_m x_i)$. More precisely, the map $A \otimes \varphi$ is defined by the projections p_i, p_j of the products via the formula:

$$p_j(A \otimes \varphi) = \sum_{i=1}^{n} \varphi^i_j p_i : A^{\times n} \longrightarrow A, \quad j = 1, \dots, m.$$

One readily checks that $A \otimes \varphi$ is functorial, that is $A \otimes 1 = 1$ and $(A \otimes \varphi)(A \otimes \psi) = A \otimes (\varphi\psi)$. Moreover since $p_i \in [\![A^{\times n}, A]\!]_0$ is an element in an $\mathbb{F}$-vector space we see that the map $A \otimes \varphi$ depends only on $\varphi \otimes \mathbb{F}$. We point out that $A \otimes \varphi$ is an $\mathbb{F}$-*linear map* between $\mathbb{F}$-vector space objects in $\mathcal{T}_0$.

Now let $a : A \to B$ be a map in $\mathcal{T}_0$ which need not be $\mathbb{F}$-linear. Then we obtain the diagram in $\mathcal{T}$.

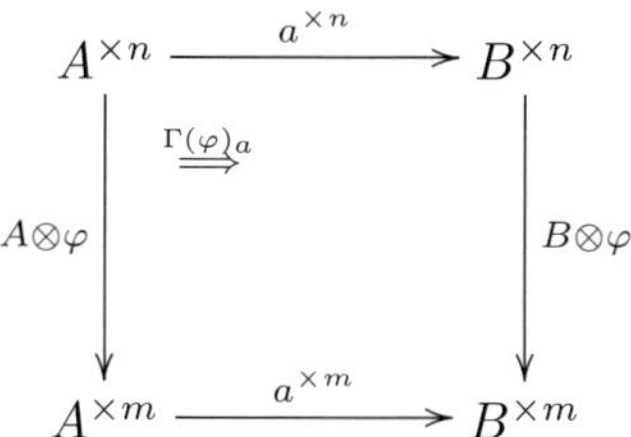

Here a track $\Gamma(\varphi)_a$ exists since a is $\mathbb{F}$-linear in the homotopy category $\mathcal{T}_{\simeq}$. The object $A^{\times n}$ is a weak coproduct with inclusions $i_i^A : A \to A^{\times n}$ for $i = 1, \dots, n$ and the object $B^{\times m}$ is a strong product with projections $p_j^B : B^{\times m} \to B$. Therefore there is a *unique track* $\Gamma(\varphi)_a$ satisfying the equation

$$p_j^B \Gamma(\varphi)_a i_i^A = \Gamma(\varphi_j^i)_a : a(\varphi_j^i 1_A) \Rightarrow (\varphi_j^i 1_B)a \tag{4.2.13}$$

where φ_j^i is the matrix of φ. Here we use the track $\Gamma(\lambda)_a$ for $\lambda \in \mathbb{G}$ defined by (4.2.10). We call $\Gamma(\varphi)_a$ the *linearity track* for φ and a. We define for $x = (x_1, \dots, x_n) : X \to A^{\times n}$ the track

$$\Gamma(\varphi)_a^x = \Gamma(\varphi)(x_1, \dots, x_n) : a^{\times m}(A \otimes \varphi)x \Rightarrow (B \otimes \varphi)a^{\times n}x.$$

For example if $\varphi = \nabla_V : V \to \mathbb{G}$ is the folding map then one readily checks that

$$\Gamma(\nabla_V)_a^x = \Gamma_a^{x_1,\dots,x_n}$$

coincides with the linearity track defined in (4.2.3). Therefore $\Gamma(\varphi)_a^x$ is a generalization of the linearity tracks considered in theorem (4.2.5).

4.2.14 Lemma. *Let $\varphi : V = \mathbb{G}^n \to V = \mathbb{G}^n$ be a permutation or let $a : A \to B$ be an $\mathbb{F}$-linear map in $\mathcal{T}_0$. Then $\Gamma(\varphi)_a = 0^\square$ is the trivial track.*

Proof. If a is linear and $\lambda \in \mathbb{Z}$, then $a(\lambda \cdot 1_A) = (\lambda \cdot 1_B)a$ and $\Gamma(\lambda)_a$ is the trivial track since Γ_a^n in (4.2.3) is the trivial track. This implies that $\Gamma(\varphi)_a$ is the trivial track. If φ is a permutation we use an argument as in (4.2.5)(3). $\square$

4.2.15 Theorem. *Let $\mathcal{T}$ be a weak $\mathbb{F}$-additive track extension, for example $\mathcal{T} = [\![\mathbf{K}_p^{\text{stable}}]\!]$. Then the linearity tracks $\Gamma(\varphi)_a^x$ in $\mathcal{T}$ above satisfy:*

(1) $\Gamma(\varphi)_a^{xd} = \Gamma(\varphi)_a^x d$.

(2) $\Gamma(\varphi)_{ba}^x = (\Gamma(\varphi)_b^{a^{\times n}x}) \square (b^{\times m}\Gamma(\varphi)_a^x)$.

(3) $\Gamma(\psi\varphi)_a^x = ((B \otimes \psi)\Gamma(\varphi)_a^x) \square (\Gamma(\psi)_a^{(A\otimes\varphi)x})$.

(4) *For* $H : a \Rightarrow a'$, $G : x \Rightarrow x'$, *with* $G = (G_1 : x_1 \Rightarrow x_1', \ldots, G_n : x_n \Rightarrow x_n')$ *the following diagram of tracks commutes.*

$$\begin{array}{ccc}
a^{\times m} \cdot ((1 \otimes \varphi)x) & \xrightarrow{\Gamma(\varphi)_a^x} & (1 \otimes \varphi)(a^{\times n} \cdot x) \\
\Big\downarrow {\scriptstyle H^{\times m} \cdot ((1 \otimes \varphi)G)} & & \Big\downarrow {\scriptstyle (1 \otimes \varphi)(H^{\times n} \cdot G)} \\
(a')^{\times m} \cdot ((1 \otimes \varphi)x') & \xrightarrow{\Gamma(\varphi)_{a'}^{x'}} & (1 \otimes \varphi)((a')^{\times n} \cdot x')
\end{array}$$

Proof. Equation (1) is obvious. Moreover (2) follows from pasting of tracks in the following diagram.

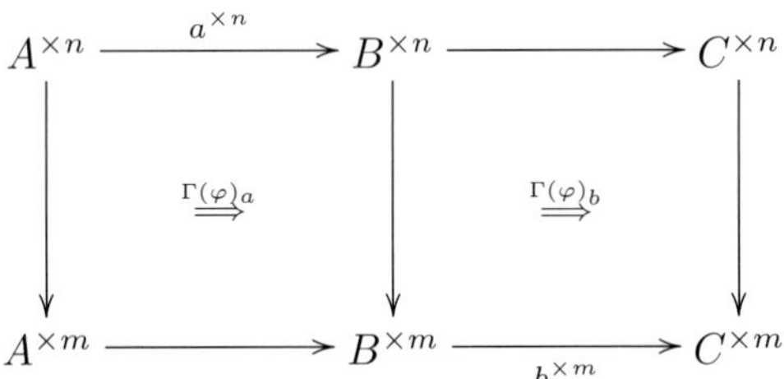

The uniqueness property for $\Gamma(\varphi)_a$ shows that pasting yields $\Gamma(\varphi)_{ba}$. Here we use (4.2.5)(2) applied to $\Gamma(\lambda)_{ba}$ in (4.2.7).

Next we obtain (3) by pasting in the following diagram.

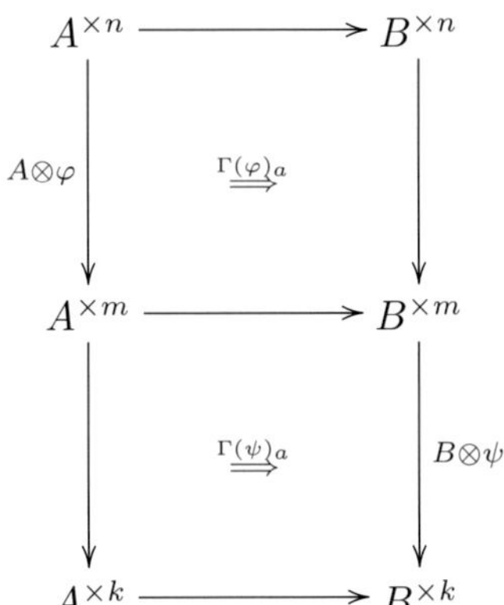

Again the uniqueness property of $\Gamma(\psi\varphi)_a$ shows that the pasting of tracks in this diagram yields $\Gamma(\psi\varphi))_a$. More precisely, we have for the track

$$G = (B \otimes \psi)\Gamma(\varphi)_a \Box \Gamma(\psi)_a(A \otimes \varphi)$$

the equation $(o \le t \le k)$

$$\begin{aligned} p_t G i_i &= p_t(B\otimes\psi)\Gamma(\varphi)_a i_i \Box p_t\Gamma(\psi)_a(A\otimes\varphi)i_i \\ &= (\sum_j \psi_t^j p_j)\Gamma(\varphi)_a i_i \Box p_t\Gamma(\psi)_a(\varphi_1^i 1_A,\dots,\varphi_m^i 1_A) \\ &= (\sum_j \psi_t^j \Gamma(\varphi_j^i)_a)\Box(p_t\Gamma(\psi)_a(\varphi_1^i 1_A\times\cdots\times\varphi_m^i 1_A)\Delta_A^m) \end{aligned}$$

with $\Delta_A^m = (1_A,\dots,1_A) : A \to A^{\times m}$ the m-fold diagonal. Here we have for the projection $p_j^B : B^{\times m} \to B$,

$$\sum_j \psi_t^j\Gamma(\varphi_j^i)_a = (\sum_j p_j^B)(\times_j \psi_t^j\Gamma(\varphi_j^i)_a)\Delta_A^m.$$

Hence we obtain G' in $[\![A^{\times m}, A]\!]$, with

$$p_t G i_i = G'\Delta_A^m,$$

defined by

$$G' = (\sum_j p_j)(\times_j\psi_t^j\Gamma(\varphi_j^i)_a)\Box(p_t\Gamma(\psi)_a(\varphi_1^i 1_A\times\cdots\times\varphi_m^i 1_A)).$$

Here G' is the unique homotopy satisfying

$$\begin{aligned} G' i_j &= \psi_t^j\Gamma(\varphi_j^i)_a\Box p_t\Gamma(\psi)_a i_j(\varphi_j^i\cdot 1_A), \\ &= \psi_t^j\Gamma(\varphi_j^i)_a\Box\Gamma(\psi_t^j)_a(\varphi_j^i\cdot 1_A), \\ &= \Gamma(\psi_t^j\cdot\varphi_j^i)_a, \text{ see (4.2.8).} \end{aligned}$$

On the other hand we have for $\lambda_j = \psi_t^j\varphi_j^i$,

$$\begin{aligned} p_t\Gamma(\psi\varphi)_a i_i &= \Gamma(\sum \psi_t^j\varphi_j^i)_a \\ &= (\sum_j \Gamma(\psi_t^j\cdot\varphi_j^i)_a)\Box\Gamma_a^{\lambda_1 1_A,\dots,\lambda_m 1_A} \\ &= (\sum_j \Gamma(\psi_t^j\cdot\varphi_j^i)_a)\Box(\Gamma_a^m((\lambda_1\cdot 1_A)\times\cdots\times(\lambda_m\cdot 1_A))\Delta_A^m) \\ &= G''\Delta_A^m \end{aligned}$$

with

$$G'' = ((\sum_j p_j)(\times_j\Gamma(\psi_t^j\cdot\varphi_j^i)_a))\Box\Gamma_a^m(\lambda_1\cdot 1_A\times\cdots\times\lambda_m 1_A).$$

Here G'' is the unique track with

$$\begin{aligned} G'' i_j &= \Gamma(\psi_t^j\cdot\varphi_j^i)_a\Box\Gamma_a^m(0\times\cdots\times\lambda_j 1_A\times\cdots\times 0) \\ &= \Gamma(\psi_t^j\cdot\varphi_j^i)_a, \text{ see (4.2.5)(5).} \end{aligned}$$

Hence we see that $G' = G''$. This proves $p_t G i_i = p_t \Gamma(\psi \cdot \varphi)_a i_i$ and hence $G = \Gamma(\psi \cdot \varphi)_a$. Hence the proof of (3) is complete.

Finally we obtain (4) similarly as in the proof of (4.2.5)(6),(7). □

Recall that $+_V : V \oplus V \to V$ is the addition map, see (4.2.11).

4.2.16 Theorem. *Let $\mathcal{T}$ be a weak $\mathbb{F}$-additive track extension like $\mathcal{T} = [\![\mathbf{K}_p^{\mathrm{stable}}]\!]$. Then the linearity tracks $\Gamma(\varphi)_a^x$ in $\mathcal{T}$ satisfy the following equations, $\varphi : V \to W$, $a : A \to B$.*

(1) $\Gamma(\varphi)_{a+a'}^x = \Gamma(\varphi)_a^x + \Gamma(\varphi)_{a'}^x$.

(2) $\Gamma(\varphi + \varphi')_a^x = (\Gamma(\varphi)_a^x + \Gamma(\varphi')_a^x) \Box \Gamma(+_W)_a^{((A\otimes\varphi)x,(A\otimes\varphi')x)}$.

(2)′ $\Gamma(\varphi \oplus \psi)_a^{(x,x')} = (\Gamma(\varphi)_a^x, \Gamma(\psi)_a^{x'})$.

(3) $\Gamma(\varphi)_a^{x+x'} = ((B\otimes\varphi)\Gamma(+_V)_a^{(x,x')})^{\mathrm{op}} \Box (\Gamma(\varphi)_a^x + \Gamma(\varphi)_a^{x'}) \Box \Gamma(+_W)_a^{((A\otimes\varphi)x,(A\otimes\varphi)x')}$.

We point out that $\Gamma(+_V)_a^{(x,x')}$ can be expressed by

$$
(4.2.17) \qquad \begin{aligned} \Gamma(+_V)_a^{(x,x')} &= \Gamma_{a^{\times n}}^{x,x'}, \text{ see (4.2.2)} \\ &= (\Gamma_a^{p_1 x, p_1 x'}, \dots, \Gamma_a^{p_n x, p_n x'}). \end{aligned}
$$

Here p_j is the jth projection of $A^{\times m}$ and $\Gamma_a = \Gamma(+_{\mathbb{G}})_a$ is defined as in (4.2.1).

Proof. We have $a + a' = +_B(a, a') : A \to B^{\times 2} \to B$. Hence we get by (4.2.15)(2),

$$
(4) \qquad \Gamma(\varphi)_{+_B(a,a')}^x = \Gamma(\varphi)_{+_B}^{(a,a')^{\times n} x} \Box (+_B)^{\times m} \Gamma(\varphi)_{(a,a')}^x.
$$

Here $+_B$ is $\mathbb{F}$-linear and we can apply (4.2.14). Hence we get the proof of (1) by

$$
(5) \qquad \begin{aligned} \Gamma(\varphi)_{a+a'}^x &= (+_B)^{\times m} \Gamma(\varphi)_{(a,a')}^x \\ &= (+_{B^{\times m}})(\Gamma(\varphi)_a^x, \Gamma(\varphi)_{a'}^x) \\ &= \Gamma(\varphi)_a^x + \Gamma(\varphi)_{a'}^x. \end{aligned}
$$

Next we consider the composite

$$
\varphi + \varphi' = \psi(\varphi, \varphi') : V \to W \oplus W \to W
$$

where $\psi = +_W$. Hence we get by (4.2.15)(3) the equation

$$
(6) \qquad \Gamma(\varphi + \varphi')_a^x = (B \otimes \psi)\Gamma(\varphi, \varphi')_a^x \Box \Gamma(\psi)_a^{(A\otimes(\varphi,\varphi'))x}.
$$

Here we have

$$
\begin{aligned} (B \otimes \psi)\Gamma(\varphi, \varphi')_a^x &= (+_{B^{\times m}})(\Gamma(\varphi)_a^x, \Gamma(\varphi')_a^x) \\ &= \Gamma(\varphi)_a^x + \Gamma(\varphi')_a^x. \end{aligned}
$$

Finally we consider the composite

(7) $$x + x' = (A \otimes \eta)(x, x') : X \to A^{\times n} \times A^{\times n} \to A^{\times n}$$

where $A \otimes \eta = +_{A^{\times n}}$ is given by $\eta = +_V$. Hence we get by (4.2.15)(3),

(8) $$\begin{aligned} \Gamma(\varphi)_a^{x+x'} &= \Gamma(\varphi)_a^{(A\otimes\eta)(x,x')} \\ &= ((B \otimes \varphi)\Gamma(\eta)_a^{(x,x')})^{\mathrm{op}} \square \Gamma(\varphi\eta)_a^{x,x'}, \end{aligned}$$

Moreover $\varphi\eta = (\varphi, \varphi) = \psi(\varphi \oplus \varphi)$ with $\psi = +_W$ so that by (4.2.15)

(9) $$\begin{aligned} \Gamma(\varphi\eta)_a^{(x,x')} &= \Gamma(\psi(\varphi \oplus \varphi))_a^{(x,x')} \\ &= ((B \otimes \psi)\Gamma(\varphi \oplus \varphi)_a^{(x,x')}) \square \Gamma(\psi)_a^{(A\otimes(\varphi\oplus\varphi))(x,x')}. \end{aligned}$$

Here we have

(10) $$\begin{aligned} (B \otimes \psi)\Gamma(\varphi \times \varphi)_a^{x,x'} &= (+_{B^{\times m}})(\Gamma(\varphi)_a^x, \Gamma(\varphi)_a^{x'}) \\ &= \Gamma(\varphi)_a^x + \Gamma(\varphi)_a^{x'}. \end{aligned}$$

This proves (3). □

4.3 The Γ-structure of the secondary Steenrod algebra

The linearity tracks $\Gamma(\varphi)_a^x$ in $[\![\mathbf{K}_p^{\mathrm{stable}}]\!]$ yield a Γ-structure of the secondary Steenrod-algebra $[\![\mathcal{A}]\!]$. The Γ-structure is part of the following notion of a Γ-track algebra. Let p be a prime and $\mathbb{F} = \mathbb{Z}/p$ and $\mathbb{G} = \mathbb{Z}/p^2$.

4.3.1 Definition. A Γ-*track algebra* $([\![A]\!], \Gamma)$ is a monoid $[\![A]\!]$ in the category of graded groupoids such that the groupoid $[\![A^k]\!]$ in degree $k \in \mathbb{Z}$ is a $\mathbb{G}$-module object in the category of groupoids. Moreover $[\![A^k]\!] = 0$ is trivial for $k < 0$ and $[\![A^0]\!]$ is a discrete groupoid with $1 \in [\![A^0]\!]$ such that $\mathbb{G} \twoheadrightarrow [\![A^0]\!]$, $x \mapsto x \cdot 1$, is surjective. The monoid structure yields *multiplication functors*

(1) $$[\![A^k]\!] \times [\![A^r]\!] \longrightarrow [\![A^{k+r}]\!]$$

carrying $(H : f \Rightarrow g, G : x \Rightarrow y)$ to $H \cdot G : f \cdot x \Rightarrow g \cdot y$. The element $1 \in [\![A^0]\!]$ is the unit of the associative multiplication (1), that is $H \cdot 1 = 1 \cdot H = H$.

Moreover the $\mathbb{G}$-module object $[\![A^r]\!]$ yields the *addition functor*

(2) $$[\![A^r]\!] \times [\![A^r]\!] \xrightarrow{+} [\![A^r]\!]$$

carrying $(G : x \Rightarrow y, G' : x' \Rightarrow y')$ to $G + G' : x + x' \Rightarrow y + y'$. Hence $[\![A^r]\!]_1$ and $[\![A^r]\!]_0$ are $\mathbb{G}$-modules, see also (2.2.6).

Multiplication *preserves zero-elements* and is *left linear*, that is

$$(3)\qquad \begin{cases} f\cdot 0 = 0\cdot x = 0, \\ H\cdot 0 = 0\cdot G = 0 = 0^{\square} : 0 \Rightarrow 0, \\ (f+f')\cdot x = f\cdot x + f'\cdot x, \\ (H+H')\cdot G = H\cdot G + H'\cdot G. \end{cases}$$

The multiplication, however, need not be right linear. But there are given *linearity tracks* in $[\![A^{k+r}]\!]$,

$$(4)\qquad \Gamma_f^{x,y} : f(x+y) \Longrightarrow fx + fy,$$

for which the following diagram of tracks commutes.

$$(5)\qquad \begin{array}{ccc} f(x+x') & \xrightarrow{\Gamma_f^{x,x'}} & fx+fx' \\ {\scriptstyle H\cdot(G+G')}\big\downarrow & & \big\downarrow{\scriptstyle H\cdot G+H\cdot G'} \\ g(y+y') & \xrightarrow[\Gamma_g^{y,y'}]{} & gy+gy' \end{array}$$

The linearity tracks are part of the following Γ-*structure* of $[\![A]\!]$. For $a \in [\![A]\!]_0$ and a $\mathbb{G}$-linear map $\varphi : V = \mathbb{G}^n \to W = \mathbb{G}^m$ the Γ-structure is a function (non-linear) between graded $\mathbb{G}$-modules

$$(6)\qquad \Gamma(\varphi)_a : [\![A]\!]_0 \otimes V \longrightarrow [\![A]\!]_1 \otimes W$$

carrying $x \in [\![A]\!]_0 \otimes V$ to $\Gamma(\varphi)_a^x \in [\![A]\!]_1 \otimes W$.

Here V is concentrated in degree 0 so that $x = (x_1, \dots, x_n)$ with $x_i \in [\![A^k]\!]_0$ and $k = |\, x \,|$. For $d \in [\![A]\!]_0$ let $x\cdot d = (x_1\cdot d, \dots, x_n\cdot d)$ and $d\cdot x = (d\cdot x_1, \dots, d\cdot x_n)$ and we use similar notation for $H = (H_1, \dots, H_m) \in [\![A]\!]_1 \otimes W$. Moreover for composable $H, G \in [\![A]\!]_1 \otimes W$ let $H\square G = (H_1\square G_1, \dots, H_m\square G_m)$.

Using this notation $\Gamma(\varphi)_a^x$ is a track

$$(7)\qquad \Gamma(\varphi)_a^x : a\cdot((1\otimes\varphi)x) \Longrightarrow (1\otimes\varphi)(a\cdot x).$$

In particular we have the linearity track

$$(8)\qquad \Gamma_a^{x_1,\dots,x_n} = \Gamma(\nabla_V)_a^x$$

which yields the track (4) as a special case $n = 2$.

The following equations hold:

(9) $\Gamma(\varphi)^x_a = 0^\square$ if φ is a permutation of coordinates,

(10) $\Gamma(\varphi)^{xd}_a = \Gamma(\varphi)^x_a \cdot d,\ d \in [\![A]\!]_0$.

(11) $\Gamma(\varphi)^x_{ba} = \Gamma(\varphi)^{ax}_b \square b \cdot \Gamma(\varphi)^x_a$.

(12) $\Gamma(\psi\varphi)^x_a = ((1 \otimes \psi)\Gamma(\varphi)^x_a)\square\Gamma(\psi)^{(1\otimes\varphi)x}_a$.

(13) $\Gamma(\varphi)^x_{a+a'} = \Gamma(\varphi)^x_a + \Gamma(\varphi)^x_{a'}$.

(14) $\Gamma(\varphi+\varphi')^x_a = (\Gamma(\varphi)^x_a + \Gamma(\varphi')^x_a)\square\Gamma(+_W)^{((1\otimes\varphi)x,(1\otimes\varphi')x)}_a$.

(15) $\Gamma(\varphi\oplus\psi)^{(x,x')}_a = (\Gamma(\varphi)^x_a, \Gamma(\psi)^{x'}_a)$.

(16) $\Gamma(\varphi)^{x+x'}_a = ((1\otimes\varphi)\Gamma(+_V)^{(x,x')}_a)^{\mathrm{op}}\square(\Gamma(\varphi)^x_a + \Gamma(\varphi)^{x'}_a)\square\Gamma(+_W)^{((1\otimes\varphi)x,(1\otimes\varphi')x)}_a$.

Moreover, for $H : a \Rightarrow a'$ and $G : x \Rightarrow x'$ with $G = (G_1 : x_1 \Rightarrow x'_1, \dots, G_n : x_n \Rightarrow x'_n)$ the following diagram of tracks commutes.

(17)
$$\begin{array}{ccc} a\cdot((1\otimes\varphi)x) & \xrightarrow{\Gamma(\varphi)^x_a} & (1\otimes\varphi)(a\cdot x) \\ \Big\downarrow{\scriptstyle H\cdot((1\otimes\varphi)G)} & & \Big\downarrow{\scriptstyle (1\otimes\varphi)(H\cdot G)} \\ a'\cdot((1\otimes\varphi)x') & \xrightarrow{\Gamma(\varphi)^{x'}_{a'}} & (1\otimes\varphi)(a'\cdot x') \end{array}$$

That is, more generally than in (5), we have

(18)
$$((1\otimes\varphi)(H\cdot G))\square\Gamma(\varphi)^x_a = \Gamma(\varphi)^{x'}_{a'}\square(H\cdot((1\otimes\varphi)G)).$$

This completes the definition of the Γ-track algebra $([\![A]\!], \Gamma)$.

4.3.2 Definition. An element $a \in [\![A]\!]_0$ in a Γ-track algebra $[\![A]\!]$ is called *linear* if for all x and φ we have $\Gamma(\varphi)^x_a = 0^\square$. Moreover $[\![A]\!]$ is a *strict* Γ-track algebra if all $a \in [\![A]\!]_0$ are linear. We shall see that a strict Γ-track algebra is a *track algebra over* $\mathbb{G}$ or a *pair algebra over* $\mathbb{G}$, see (5.1.5) below. We shall prove in the next chapter that each Γ-track algebra over $\mathbb{G}$ can be "strictified".

4.3.3 Remark. It is not clear how to define a Γ-track algebra by a minimal list of properties, so that the long list of properties described in (4.3.1) can be deduced from the minimal list. Certainly the equations in Theorem (4.2.5) should be part of such a minimal list since one readily checks:

4.3.4 Lemma. *All equations in* (4.2.5) *hold in a* Γ*-track algebra.*

Proof. (4.2.5)(5) for example is a consequence of $\nabla_V = (+_{\mathbb{G}} \oplus \mathbb{G})(+_{\mathbb{G}}) = (\mathbb{G} \oplus +_{\mathbb{G}})(+_{\mathbb{G}})$ for $V = \mathbb{G}^3$. □

4.3.5 Proposition. *Let $(\llbracket A \rrbracket, \Gamma)$ be a Γ-track algebra. Then $A = \llbracket A \rrbracket_{\simeq}$ is a graded algebra over $\mathbb{G}$ and there is an A-bimodule D such that*

$$D \longrightarrow \llbracket A \rrbracket_1 \rightrightarrows \llbracket A \rrbracket_0 \longrightarrow A$$

is a (graded) linear track extension.

Proof. The linearity tracks show that multiplication in the homotopy category $A = \llbracket A \rrbracket_{\simeq} = \llbracket A \rrbracket / \simeq$ is bilinear. Hence A is an algebra over $\mathbb{G}$ with $1 \in A^0$. The A-bimodule D with $D^n = 0$ for $n \leq 0$ is defined by

$$D = \mathrm{kernel}(\llbracket A \rrbracket_1^0 \xrightarrow{\partial} \llbracket A \rrbracket_0).$$

Here the A-bimodule structure of D is given by $\alpha \cdot H \cdot \beta = a \cdot H \cdot b$ with $a, b \in \llbracket A \rrbracket_0$ representing $\alpha, \beta \in A$, $H \in D$. We have for $H, G \in \llbracket A \rrbracket_1^0$ the equation $H \cdot G = ((\partial H) \cdot G) \square (H \cdot 0) = (H \cdot (\partial G)) \square (0 \cdot G)$; this shows $(\partial H) \cdot G = H \cdot (\partial G)$. Hence we get $a \cdot H = 0$ and $H \cdot b = 0$ if $a, b \in \mathrm{image}(\partial)$. Therefore D is a left A-module. In fact, D is also a right A-module since for $H \in D$ we get by (4.3.1)(5)

$$\begin{aligned} H \cdot (b + \partial G) &= (\Gamma_0^{b,\partial G})^{\mathrm{op}} (H \cdot b + H \cdot \partial G) \Gamma_0^{b,\partial G} \\ &= H \cdot b. \end{aligned}$$

Therefore $\alpha \cdot H \cdot \beta$ above is well defined. Moreover

$$\begin{aligned} H \cdot (b + b') &= (\Gamma_0^{b,b'})^{\mathrm{op}} (H \cdot b + H \cdot b') \Gamma_0^{b,b'} \\ &= H \cdot b + H \cdot b' \end{aligned}$$

so that D is a right A-module. Now we define the natural system D in (3.4.6) and one can check the properties in (3.5.1) with $a \in \alpha \in A$,

$$\sigma_a : D_\alpha = D^{|\alpha|} \cong \mathrm{Aut}(a)$$

carrying $H \in D^{|\alpha|}$ to $H + a$. □

4.3.6 Theorem. *The secondary Steenrod algebra $\llbracket \mathcal{A} \rrbracket$ is a Γ-track algebra. Here $\llbracket \mathcal{A} \rrbracket_1$ and $\llbracket \mathcal{A} \rrbracket_0$ are graded $\mathbb{F}$-vector spaces and $\llbracket \mathcal{A}^0 \rrbracket = \mathbb{F}$.*

Proof. According to (2.4.4) we see that $\llbracket \mathcal{A} \rrbracket$ has all the structure in (4.3.1) except linearity tracks. We choose $n \geq 1$ so that for $k \geq 1$.

(1) $$\llbracket \mathcal{A}^k \rrbracket = \llbracket Z^n, Z^{n+k} \rrbracket^{\mathrm{stable}},$$

(2) $$\mathbb{F} = \llbracket \mathcal{A}^0 \rrbracket \subset \llbracket Z^n, Z^n \rrbracket^{\mathrm{stable}},$$

by (2.5.4). Now the linearity tracks defined for $\llbracket \mathbf{K}_p^{\mathrm{stable}} \rrbracket$ in (4.2.2) yield accordingly linearity tracks Γ for $\llbracket \mathcal{A} \rrbracket$. Uniqueness of linearity tracks shows that $\Gamma(\varphi)_a^x$ in $\llbracket \mathcal{A} \rrbracket$ is independent of the choice of n. □

In addition to (4.3.2) we know that $[\![\mathcal{A}]\!]$ is a linear extension, see (3.5.2),

$$\Sigma\mathcal{A} \longrightarrow [\![\mathcal{A}]\!]_1 \rightrightarrows [\![\mathcal{A}]\!]_0 \longrightarrow \mathcal{A}$$

which coincides with the extension of the Γ-track algebra $[\![\mathcal{A}]\!]$ in (4.3.5). The secondary cohomology $[\![X,-]\!]$ of a path connected pointed space X is a model of the track theory $[\![\mathbf{K}_p^{\mathrm{stable}}]\!]$, see (2.3.8). This leads to the following notion of a Γ-track module.

4.3.7 Definition. A Γ-*track module* $([\![M]\!],\Gamma)$ over a Γ-track algebra $([\![A]\!],\Gamma)$ is defined as follows. The module $[\![M]\!]$ is a graded object $([\![M^k]\!], k\in\mathbb{Z})$ in the category of groupoids. Moreover $[\![M^k]\!]$ for $k\in\mathbb{Z}$ is a $\mathbb{G}$-module object in **Gpd** and $[\![M^k]\!]=0$ is trivial for $k\le 0$. The *monoid* $[\![A]\!]$ *acts on* $[\![M]\!]$ from the left; that is, functors

$$(1)\qquad [\![A^k]\!]\times[\![M^r]\!]\longrightarrow[\![M^{k+r}]\!]$$

are given carrying $(H: f\Rightarrow g, G: x\Rightarrow y)$ to $H\cdot G: f\cdot x\Rightarrow g\cdot y$ in $[\![M]\!]$. The element $1\in[\![A^0]\!]$ is a unit of the action with $1\cdot G=G$ and $(H\cdot H')\cdot G=H\cdot(H'\cdot G)$. The $\mathbb{G}$-module object $[\![M^r]\!]$ yields the *addition functor*

$$(2)\qquad [\![M^r]\!]\times[\![M^r]\!]\longrightarrow[\![M^r]\!]$$

carrying $(G: x\Rightarrow y, G': x'\Rightarrow y')$ to $G+G': x+x'\Rightarrow y+y'$. Hence $[\![M]\!]_1$ and $[\![M]\!]_0$ are graded $\mathbb{G}$-modules, see (2.2.6).

The action preserves zero-elements and is left linear, that is, $(x\in[\![M]\!]_0$, $G\in[\![M]\!]_1$, $f,f'\in[\![A]\!]_0$, $H,H'\in[\![A]\!]_1)$,

$$(3)\qquad \begin{cases} f\cdot 0=0\cdot x=0,\\ H\cdot 0=0\cdot G=0=0^{\square}:0\Rightarrow 0,\\ (f+f')\cdot x=f\cdot x+f'\cdot x,\\ (H+H')\cdot G=H\cdot G+H'\cdot G.\end{cases}$$

The action, however, need not be right linear. But there are given *linearity tracks* in $[\![M^{k+r}]\!]$,

$$(4)\qquad \Gamma_f^{x,y}: f(x+y)\Rightarrow fx+fy.$$

They are part of the Γ-*structure* of $[\![M]\!]$. For $a\in[\![A]\!]_0$, and a $\mathbb{G}$-linear map $\varphi: V=\mathbb{G}^n\to W=\mathbb{G}^m$, the Γ-structure is a function (non-linear) between $\mathbb{G}$-modules

$$(5)\qquad \Gamma(\varphi)_a:[\![M]\!]_0\otimes V\longrightarrow[\![M]\!]_1\otimes W$$

carrying $x\in[\![M]\!]_0\otimes V$ to $\Gamma(\varphi)_a^x\in[\![M]\!]_1\otimes W$ with

$$(6)\qquad \Gamma(\varphi)_a^x: a\cdot((1\otimes\varphi)x)\Longrightarrow(1\otimes\varphi)(a\cdot x).$$

In particular we have the linearity tracks

$$\Gamma_a^{x_1,\dots,x_n} = \Gamma(\nabla_V)_a^x \tag{7}$$

which yield (4) for $n = 2$. All equations as in (4.3.1)(5),(8)...(18) hold accordingly in $[\![M]\!]$ where we use also the Γ-structure of $[\![A]\!]$.

4.3.8 Definition. An element $a \in [\![A]\!]_0$ is called $[\![M]\!]$-*linear* if a is linear in $[\![A]\!]$ as in (4.3.2) and if for all $x \in [\![M]\!]_0 \otimes V$, $\varphi : V \to W$, the track $\Gamma(\varphi)_a^x = 0^\square$ is the trivial track. We call $[\![M]\!]$ a *strict* Γ-track module if all $a \in [\![A]\!]_0$ are $[\![M]\!]$-linear. We shall see that a strict Γ-track module is the same as a *module* over a pair algebra, see (5.1.6) below. Moreover we show in the next chapter that each Γ-track module can be "strictified".

4.3.9 Theorem. *Let X be a path connected pointed space. Then the secondary cohomology $[\![M]\!]$ with $[\![M^k]\!] = [\![X, Z^k]\!]$, $k \geq 1$ is a Γ-track module over the secondary Steenrod algebra $([\![\mathcal{A}]\!], \Gamma)$ in (4.3.6). Here $[\![M]\!]_1$ and $[\![M]\!]_0$ are graded $\mathbb{F}$-vector spaces and $[\![M^k]\!] = 0$ for $k \leq 0$.*

Proof. We define for $x = (x_1, \dots, x_r) \in [\![M]\!]_0 \otimes V$ the linearity track $\Gamma(\varphi)_a^x = \Gamma(\varphi)_a(x_1, \dots, x_r)$ where we use $\Gamma(\varphi)_a$ in (4.2.13). $\square$

We use Theorem (4.3.9) for a discussion of a formula of Kristensen, see 3.5 [Kr1]. Kristensen introduces the cochain operation $d(\alpha; x, y)$ which restricted to cocycles corresponds to the following definition:

4.3.10 Definition. Let $x, y : X \to Z^n$ be pointed maps and let $\alpha \in [\![\mathcal{A}^k]\!]_0$. Then we define

$$d(\alpha; x, y) \in [\![X, Z^{n+k}]\!]_1^0$$

by the formula

$$d(\alpha; x, y) + \alpha x + \alpha y = \Gamma_\alpha^{x,y} = \Gamma_\alpha(x, y),$$

see (4.2.2). Hence

$$d(\alpha; x, y) : \alpha(x + y) - \alpha x - \alpha y \Longrightarrow 0$$

is a *cross effect track* which plays a similar role as $\Gamma_\alpha^{x,y} : \alpha(x + y) \Rightarrow \alpha x + \alpha y$.

In Theorem (4.2.5) we describe basic properties of $\Gamma_\alpha^{x,y}$ which can be translated to achieve the corresponding properties of $d(\alpha; x, y)$. The formulas, however, are more complicated. For example the derivation formula (4.2.5)(2) corresponds to the following result:

4.3.11 Lemma. $d(\beta\alpha; x, y) = d(\beta; \alpha x, \alpha y) + \beta d(\alpha; x, y) + d(\beta, \alpha(x + y) - \alpha x - \alpha y, \alpha x + \alpha y)$.

Proof. We have:

$$\begin{aligned} d(\beta\alpha; x, y) + \beta\alpha x + \beta\alpha y &= \Gamma_{\beta\alpha}^{x,y} \\ &= \Gamma_\beta^{\alpha x, \alpha y} \square \beta \Gamma_\alpha^{x,y} \\ &= \{d(\beta; \alpha x, \alpha y) + \beta\alpha x + \beta\alpha y\} \\ &\quad \square \beta\{d(\alpha; x, y) + \alpha x + \alpha y\} \end{aligned}$$

$$\begin{aligned}
&= \{d(\beta;\alpha x,\alpha y)+\beta\alpha x+\beta\alpha y\}\\
&\quad \square\{\beta\cdot d(\alpha;x,y)+\beta(\alpha x+\alpha y)\}\\
&\quad \square\Gamma_\beta^{\alpha(x+y)-\alpha x-\alpha y,\alpha x+\alpha y}, \text{ see (4.2.5)(5)},\\
&= \{d(\beta;\alpha x,\alpha y)+\beta\alpha x+\beta\alpha y\}\\
&\quad \square\{\beta\cdot d(\alpha;x,y)+\beta(\alpha x+\alpha y)\}\\
&\quad \square\{d(\beta,\alpha(x+y)-\alpha x-\alpha y,\alpha x+\alpha y)\\
&\quad +\beta(\alpha(x+y)-\alpha x-\alpha y)+\beta(\alpha x+\alpha y)\}.
\end{aligned}$$

Now the rules in (2.2.6) yield the result. □

Formula (4.3.11) was not obtained by Kristensen, but he has in 3.5 [Kr1] the following formula for $p=2$ and cocycles x,y,

(4.3.12)
$$\begin{aligned}
&d(\beta\alpha;x,y)+d(\beta;\alpha x,\alpha y)+d(\beta;\alpha(x+y),\alpha x+\alpha y)+\beta d(\alpha;x,y)\\
&\qquad=\kappa(\beta)(\alpha x)+\kappa(\beta)(\alpha y)\\
&\qquad=\kappa(\beta)(\alpha x+\alpha y)
\end{aligned}$$

where κ is the Kristensen derivation. In order to prove (4.3.12) we have to show

$$\begin{aligned}
&d(\beta;\alpha(x+y)-\alpha x-\alpha y,\alpha x+\alpha y)\\
&\qquad=d(\beta;\alpha(x+y),\alpha x,\alpha y)+\kappa(\beta)(\alpha x+\alpha y).
\end{aligned}$$

This follows from

$$d(\beta,\xi+u+v,u+v)=d(\beta;\xi,u+v)+\kappa(\beta)(u+v)$$

which in turn is a consequence of (see (4.2.5)(5))

$$(\Gamma_\beta^{\xi,u+v}+\beta(u+v))\square\Gamma_\beta^{\xi+u+v,u+v}=(\beta\xi+\Gamma_\beta^{u+v,u+v})\square\Gamma_\beta^{\xi,0}.$$

Here we have $\Gamma_\beta^{u+v,u+v}=\kappa(\beta)(u+v)$ by (4.5.8) below.

4.4 The cocycle of $[\![\mathbf{K}_p^{\text{stable}}]\!]$

We first introduce the extended cocycle of a Γ-track algebra $([\![A]\!],\Gamma)$. For this we assume that $[\![A]\!]_1$, $[\![A]\!]_0$ are graded $\mathbb{F}$-vector spaces and $[\![A^0]\!]_0=\mathbb{F}$ as in (4.3.6).

Let D be the A-bimodule given by $([\![A]\!],\Gamma)$ as in the linear track extension (4.3.5). Let $\mathbf{mod}_0(A)^{\text{op}}$ be the category of finitely generated free right A-modules with generators in degree ≥ 1. Then we have as in (3.4.7) the natural system $\bar{D}$ on $\mathbf{mod}_0(A)^{\text{op}}$ given by

(4.4.1)
$$\bar{D}_\alpha=Hom(V,W)\otimes_{A-A}D$$

for $\alpha : V \to W$ in $\mathbf{mod}_0(A)^{\mathrm{op}}$. More explicitly let $x^1, \dots, x^{n(V)}$ be a basis in V and let $y^1, \dots, y^{n(W)}$ be a basis in W. Then α is given by a matrix (α^i_j) in A with $\alpha(x^i) = \sum_j y^j \cdot \alpha^j_i$ and one gets

$$\bar{D}_\alpha = \bigoplus_{i,j} D(\alpha^i_j).$$

This equation is needed in the next definition:

4.4.2 Definition. We choose an $\mathbb{F}$-linear section s_0 of the projection π,

$$A \xrightarrow{s_0} [\![A]\!]_0 \xrightarrow{\pi} A \tag{1}$$

with $\pi s_0 = 1$. Moreover we choose for $(\beta, \alpha) \in A^k \times A^r$ the track

$$\mu_0(\beta, \alpha) : s_0(\beta)s_0(\alpha) \Rightarrow s_0(\beta\alpha) \tag{2}$$

in $[\![A^{k+r}]\!]$ and we define for

$$Y \xleftarrow{\gamma} X \xleftarrow{\beta} W \xleftarrow{\alpha} V$$

in $\mathbf{mod}_0(A)$ the cocycle $c(\gamma, \beta, \alpha)$ depending on s_0, μ_0 as follows. Let α^i_j, β^j_k, γ^k_l be the coordinates of γ, β, α respectively. Then

$$c(\gamma, \beta, \alpha) \in \bar{D}_{\gamma\beta\alpha} = \bigoplus_{i,l} D(\gamma\beta\alpha)^i_l \tag{3}$$

is the following element where $\xi_j = s_0(\beta^j_k) \cdot s_0(\alpha^i_j)$ for $j = 1, \dots, n(W)$.

$$\begin{aligned}
c(\gamma,\beta,\alpha)^i_l \quad = \quad & \sum_k \mu_0(\gamma^k_l, (\beta\alpha)^i_k) \\
& \square \sum_k s_0(\gamma^k_l) \sum_j \mu_o(\beta^j_k, \alpha^i_j) \\
& \square \{\sum_k \Gamma^{\xi_1,\dots,\xi_{n(W)}}_{s_0(\gamma^k_l)}\}^{\mathrm{op}} \\
& \square \{\sum_j (\sum_k \mu_o(\gamma^k_l, \beta^j_k) s_0(\alpha^i_j))\}^{\mathrm{op}} \\
& \square \{\sum_j \mu_o((\gamma\beta)^j_l, \alpha^i_j)\}^{\mathrm{op}}.
\end{aligned}$$

In this formula we only use s_0 and μ_0 above and the Γ-structure of the Γ-track algebra $[\![A]\!]$. We call $c(\gamma, \beta, \alpha)$ the *extended cocycle* of $[\![A]\!]$. One can check that c represents a well-defined class $\langle c \rangle \in H^3(\mathbf{mod}_0(A), \bar{D})$.

We have seen in (2.5.2) that the Steenrod algebra $\mathcal{A}$ determines the category $\mathbf{K}_p^{\text{stable}}$ of stable homotopy classes of maps between products of Eilenberg-MacLane spaces, in fact,

$$\mathbf{K}_p^{\text{stable}} = \mathbf{mod}_0(\mathcal{A})^{\text{op}}$$

We now describe the secondary analogue of this classical result:

4.4.3 Theorem. *The secondary Steenrod algebra $[\![\mathcal{A}]\!]$ together with its Γ-structure determines the linear track extension $[\![\mathbf{K}_p^{\text{stable}}]\!]$ up to weak equivalence. More precisely the extended cocycle c of $([\![\mathcal{A}]\!], \Gamma)$ defined above represents the characteristic cohomology class $k_p^{\text{stable}} = \langle [\![\mathbf{K}_p^{\text{stable}}]\!] \rangle$ in*

$$\begin{array}{rcl} H^3(\mathbf{K}_p^{\text{stable}}, \mathrm{Hom}(L^{-1}, -)) & = & H^3(\mathbf{mod}_0(\mathcal{A})^{\text{op}}, \overline{\Sigma\mathcal{A}}). \\ k_p^{\text{stable}} & \mapsto & \langle c \rangle \end{array}$$

Here the $\mathcal{A}$-bimodule $D = \Sigma\mathcal{A}$ yields the natural system $\bar{D} = \overline{\Sigma\mathcal{A}}$ as in (4.4.1).

The following result corresponds to (4.4.3). Recall that we have the linear track extension

$$R^{H^*(X)} \longrightarrow [\![X, \mathbf{K}_p^{\text{stable}}]\!]_1 \rightrightarrows [\![X, \mathbf{K}_p^{\text{stable}}]\!]_0 \longrightarrow [X, \mathbf{K}_p^{\text{stable}}]$$

in (3.8.7).

4.4.4 Theorem. *The secondary cohomology $([\![X, -]\!], \Gamma)$ as a Γ-track module over $([\![\mathcal{A}]\!], \Gamma)$, determines the linear extension $[\![X, \mathbf{K}_p^{\text{stable}}]\!]$ up to weak equivalence over $[X, \mathbf{K}_p^{\text{stable}}]$ and under $R^{H^*(X)}$.*

The proof uses a similar computation as in the proof of (4.4.3); here we also use (2.2.10). This yields an extended cocycle for $[\![X, \mathcal{A}]\!]$.

4.4.5 Corollary. *Stable secondary cohomology operations on $H^*(X)$ are completely determined by the secondary cohomology $([\![X, -]\!], \Gamma)$ considered as a Γ-track module over the secondary Steenrod algebra $([\![\mathcal{A}]\!], \Gamma)$.*

In particular, examples of Adams in (2.3.7) yield $\theta_{i,j}$ determined by the track module $([\![X, -]\!], \Gamma)$ over $([\![\mathcal{A}]\!], \Gamma)$.

Proof of (4.4.3). We show that a cocycle (as in (3.6.7)) for the linear extension

$$(1) \qquad \mathrm{Hom}(L^{-1}, -) \longrightarrow [\![\mathbf{K}_p^{\text{stable}}]\!]_1 \rightrightarrows [\![\mathbf{K}_p^{\text{stable}}]\!]_0 \longrightarrow \mathbf{K}_p^{\text{stable}}$$

can be expressed completely in terms of $[\![\mathcal{A}]\!]$.

We first choose an $\mathbb{F}$-homomorphism s_0

$$(2) \qquad \mathcal{A}^k \xrightarrow{s_0} [\![\mathcal{A}^k]\!]_0 \xrightarrow{\pi} \mathcal{A}^k$$

which splits the projection π carrying a to the homotopy class α of a. Moreover we choose for $(\beta, \alpha) \in \mathcal{A}^k \times \mathcal{A}^r$ a track

$$\mu_0(\beta, \alpha) : s_0(\beta)s_0(\alpha) \Rightarrow s_0(\beta\alpha) \text{ in } [\![\mathcal{A}^{k+r}]\!]. \tag{3}$$

Now we consider a morphism

$$\alpha : A = Z^{a_1} \times \cdots \times Z^{a_{n(A)}} \longrightarrow Z^{b_1} \times \cdots \times Z^{b_{n(B)}} = B$$

in the $\mathbb{F}$-additive category $\mathbf{K}_p^{\text{stable}}$. Then α is given by a matrix

$$\alpha = (\alpha_j^i \in \mathcal{A}^{b_j - a_i}) \tag{4}$$

with $i = 1, \dots, n(A)$ and $j = 1, \dots, n(B)$.

Here we use the equation $[Z^n, Z^{n+k}]^{\text{stable}} = \mathcal{A}^k$ for $k \in \mathbb{Z}$. We now define the section

$$s : [A, B]^{\text{stable}} \longrightarrow [\![A, B]\!]_0^{\text{stable}}$$

by setting

$$p_j^B s(\alpha) = \sum_{i=1}^{n(A)} s_0(\alpha_j^i) p_i^A. \tag{5}$$

Here we use the projections p_j^B, p_i^A of the strong products A and B above. For objects D, C, B, A we have the indices l, k, j, resp. i with

$$\begin{aligned} 1 \le l \le n(D) \quad &, \\ 1 \le k \le n(C) \quad &, \\ 1 \le j \le n(B) \quad &, \\ 1 \le i \le n(C) \quad &. \end{aligned}$$

Now consider a composite

$$C \xleftarrow{\beta} B \xleftarrow{\alpha} A \tag{6}$$

in $\mathbf{K}_p^{\text{stable}}$ with α given by (α_j^i) and β given by (β_k^j) accordingly so that $s\alpha$ and $s\beta$ are defined with

$$(\beta\alpha)_k^i = \sum_{j=1}^{n(B)} \beta_k^j \alpha_j^i. \tag{7}$$

Here $\beta_k^j \alpha_j^i$ is the product in $\mathcal{A}$. We now define a track

$$\mu(\alpha, \beta) : (s\beta)(s\alpha) \Rightarrow s(\beta\alpha) \tag{8}$$

in $[\![\mathbf{K}_p^{\text{stable}}]\!]$ in terms of μ_0 in (3) and the linearity structure Γ on $[\![\mathbf{K}_p^{\text{stable}}]\!]$ in (4.2.5). For each index k we have the projection p_k^C of the product C. Since C is a strong product it suffices to define $p_k^C\mu(\beta,\alpha)$ so that we have

$$p_k^C\mu(\beta,\alpha) : p_k^C(s\beta)(s\alpha) \Rightarrow p_k^C s(\beta\alpha) \tag{9}$$

with

$$\begin{aligned} p_k^C(s\beta)(s\alpha) &= (\textstyle\sum_j s_0(\beta_k^j)p_j^B)s\alpha \\ &= \textstyle\sum_j (s_0(\beta_k^j)p_j^B(s\alpha)) \\ &= \textstyle\sum_j s_0(\beta_k^j)(\sum_i s_0(\alpha_j^i)p_i^A) \\ &= R_1, \end{aligned} \tag{10}$$

$$\begin{aligned} p_k^C s(\beta\alpha) &= \textstyle\sum_i s_0((\beta\alpha)_k^i)p_i^A \\ &= \textstyle\sum_i s_0(\sum_j \beta_k^j\alpha_j^i)p_i^A \\ &= \textstyle\sum_i(\sum_j s_0(\beta_k^j\alpha_j^i))p_i^A \\ &\overset{\mu_0^k(\beta,\alpha)}{\Longleftarrow} \textstyle\sum_i(\sum_j s_0(\beta_k^j)s_0(\alpha_j^i))p_i^A \\ &= R_2. \end{aligned} \tag{11}$$

Here the track $\mu_0^k(\beta,\alpha)$ is defined by μ_0 in (3), that is

$$\mu_0^k(\beta,\alpha) = \sum_i(\sum_j \mu_0(\beta_k^j,\alpha_j^i))p_i^A. \tag{12}$$

Since p_i^A is a linear map between $\mathbb{F}$ vector space objects in $[\![\mathbf{K}_p^{\text{stable}}]\!]_0$ we see that

$$\begin{aligned} R_2 &= \textstyle\sum_i\sum_j(s_0(\beta_k^j)s_0(\alpha_j^i)p_i^A) \\ &= \textstyle\sum_j\sum_i(s_0(\beta_k^j)s_0(\alpha_j^i)p_i^A). \end{aligned} \tag{13}$$

Here we have the linearity track in $[\![\mathbf{K}_p^{\text{stable}}]\!]$,

$$\Gamma_k^j(\beta,\alpha) = \Gamma_{s_0(\beta_k^j)}^{s_0(\alpha_j^1)p_1^A,\dots,s_0(\alpha_j^{n(A)})p_{n(A)}^A} \tag{14}$$

with

$$\Gamma_k^j(\beta,\alpha) : s_0(\beta_k^j)(\sum_i s_0(\alpha_j^i)p_i^A) \Rightarrow \sum_i s_0(\beta_k^j)s_0(\alpha_j^i)p_i^A.$$

Hence we get

$$\Gamma_k^A(\beta,\alpha) = \sum_j \Gamma_k^j(\beta,\alpha) : R_1 \Longrightarrow R_2 \tag{15}$$

and

(16)
$$p_k^C \mu(\beta,\alpha) = \mu_0^k(\beta,\alpha)\Box\Gamma_k^A(\beta,\alpha)$$

defines $\mu(\beta,\alpha)$ in (9). Next we consider the composite

(17)
$$D \xleftarrow{\gamma} C \xleftarrow{\beta} B \xleftarrow{\alpha} A.$$

Using $\mu(\beta,\alpha)$ above the cocycle $c(\gamma,\beta,\alpha)$ is defined as in (3.6.7). In order to compute this cocycle we use the projection $p_l^D : D \to Z^{d_l}$ and the inclusion $i_i^A : Z^{a_i} \to A$ which are linear in $\mathbf{Top}^*$. We have to compute the composite

(18)
$$c_l^i = p_l^D c(\gamma,\beta,\alpha) i_i^A \in \mathcal{A}^{d_l-1-a_i}.$$

We obtain c_l^i by the following composite of tracks.

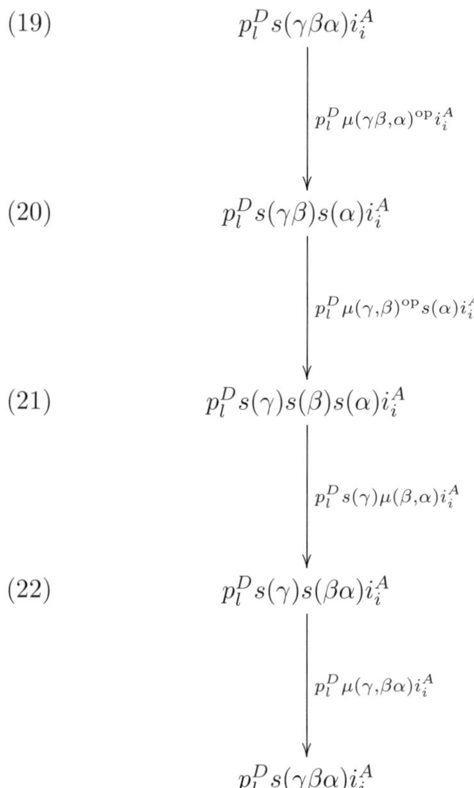

Here each track μ is given by a composite of tracks as in (15). We describe the tracks in (19),...,(21) more explicitly as follows. First we get $(19) = ((23)\Box(24))^{\mathrm{op}}$ as follows where $(24) = 0^{\Box}$ so that $(19) = (23)^{\mathrm{op}}$.

(23)
$$\mu_0^l(\gamma\beta,\alpha) i_i^A = \sum_j \mu_0((\gamma\beta)_l^j, \alpha_j^i),$$

$$
\begin{aligned}
\Gamma_l^A(\gamma\beta,\alpha)i_i^A &= \textstyle\sum_j \Gamma_l^j(\gamma\beta,\alpha)i_i^A[2mm] \\
&= \textstyle\sum_j \Gamma_{s_0(\gamma\beta)_l^j}^{0,\ldots,s_0(\alpha_j^i),\ldots,0} \\
&= 0^{\square}.
\end{aligned} \tag{24}
$$

Here we use $p_{i_1}^A i_i^A = 1$ if $i_1 = i$ and $= 0$ if $i_1 \neq 0$. Next we see that $(20) = ((25)\square(26))^{\mathrm{op}}$ with

$$
\begin{aligned}
\mu_0^l(\gamma,\beta)s(\alpha)i_i^A &= (\textstyle\sum_j(\sum_k \mu_0(\gamma_l^k,\beta_k^j))p_j^B)s(\alpha)i_i^A \\
&= (\textstyle\sum_j(\sum_k \mu_0(\gamma_l^k,\beta_k^j))p_j^B s(\alpha)i_i^A) \\
&= (\textstyle\sum_j(\sum_k \mu_0(\gamma_l^k,\beta_k^j)s_0(\alpha_j^i)),
\end{aligned} \tag{25}
$$

$$
\begin{aligned}
\Gamma_l^B(\gamma,\beta)s(\alpha)i_i^A &= \textstyle\sum_k \Gamma_l^k(\gamma,\beta)s(\alpha)i_i^A \\
&= \textstyle\sum_k \Gamma_{s_0(\gamma_l^k)}^{s_0(\beta_k^1)s_0(\alpha_1^i),\ldots,s_0(\beta_k^{n(B)})s_0(\alpha_{n(B)}^i)}.
\end{aligned} \tag{26}
$$

Next we obtain $(21) = (27)\square(28)$ as follows. Here we have $(28) = 0^{\square}$ so that $(21) = (27)$,

$$
\begin{aligned}
&p_l^D s(\gamma)(\mu_0^k(\beta,\alpha)i_i^A)_{k=1,\ldots,n(C)} \\
&\quad = (\textstyle\sum_k^{n(C)} s_0(\gamma_l^k)p_k^C)(\sum_j \mu_0(\beta_k^j,\alpha_j^i))_{k=1,\ldots,n(C)} \\
&\quad = \textstyle\sum_k^{n(C)} s_0(\gamma_l^k)\sum_j \mu_o(\beta_k^j,\alpha_j^i),
\end{aligned} \tag{27}
$$

$$
\begin{aligned}
&p_l^D s(\gamma)(\Gamma_k^A(\beta,\alpha)i_i^A)_{k=1,\ldots,n(C)} \\
&\quad = (\textstyle\sum_{k=1}^{n(C)} s_0(\gamma_l^k)p_k^C)(\sum_j \Gamma_k^j(\beta,\alpha)i_i^A)_{k=1,\ldots,n(C)} \\
&\quad = \textstyle\sum_{k=1}^{n(C)} s_0(\gamma_l^k)(\sum_j \Gamma_k^j(\beta,\alpha)i_i^A) \\
&\quad = \textstyle\sum_{k=1}^{n(C)} s_0(\gamma_l^k)(\sum_j \Gamma_{s_0(\beta_k^j)}^{0,\ldots,s_0(\alpha_j^i),\ldots,0}) \\
&\quad = 0^{\square}.
\end{aligned} \tag{28}
$$

Finally we get $(22) = (29)\square(30)$ as follows. Here we have $(30) = 0^{\square}$ so that $(22) = (29)$,

$$
\mu_0(\gamma,\beta\alpha)i_i^A = \sum_k \mu_0(\gamma_l^k,(\beta\alpha)_k^i), \tag{29}
$$

$$
\begin{aligned}
\Gamma_l^A(\gamma,\beta\alpha)i_i^A &= \textstyle\sum_k \Gamma_l^k(\gamma,\beta\alpha)i_i^A \\
&= \textstyle\sum_k \Gamma_{s_0(\gamma_l^k)}^{0,\ldots,s_0(\beta\alpha)_k^i,\ldots,0} \\
&= 0^{\square}.
\end{aligned} \tag{30}
$$

Hence we see that c_l^i in (18) is given by

$$(22)\square(21)\square(20)\square(19) = (29)\square(27)\square(26)^{\mathrm{op}}\square(25)^{\mathrm{op}}\square(23)^{\mathrm{op}}$$

where the right-hand side is well defined only by μ_0 above and the linearity tracks Γ in $[\![\mathcal{A}]\!]$. This completes the proof of (4.4.3). □

4.5 The Kristensen derivation

Let A be a (connected) graded algebra over $\mathbb{F} = \mathbb{Z}/p$ and let D be an A-bimodule. A *linear derivation* κ *of degree* r from A to D is an $\mathbb{F}$-linear map

$$\kappa : A^n \longrightarrow D^{n+r},\ n \in \mathbb{Z}, \tag{4.5.1}$$

with $\kappa(\alpha \cdot \beta) = (\kappa\alpha)\beta + (-1)^r \alpha(\kappa\beta)$.

Now let $([\![A]\!], \Gamma)$ be a Γ-track algebra for which $[\![A]\!]_1$ and $[\![A]\!]_0$ are graded $\mathbb{F}$-vector spaces as in the case of the secondary Steenrod algebra $[\![\mathcal{A}]\!]$. We have the linear track extension

$$D \longrightarrow [\![A]\!]_1 \rightrightarrows [\![A]\!]_0 \longrightarrow A \tag{4.5.2}$$

as in (4.3.5). In particular the secondary Steenrod algebra is such an extension with $D = \Sigma\mathcal{A}$. The linearity tracks $\Gamma_a^{x,y}$ in $[\![A]\!]$ define as in (4.2.3) the track

$$\Gamma_a^{x_1,\dots,x_n} : a(x_1 + \cdots + x_n) \Longrightarrow ax_1 + \cdots + ax_n$$

If $x_1 = \cdots = x_n = 1 \in \mathbb{F} = [\![A]\!]^0$ we thus get as in (4.2.7),

$$\Gamma(n)_a = \Gamma_a^{1,\dots,1} : a(n \cdot 1) \Longrightarrow n \cdot a. \tag{4.5.3}$$

Hence if p divides n we have $n \cdot 1 = 0$ and $n \cdot a = 0$ so that $\Gamma(n)_a : 0 \Rightarrow 0$ represents an element in D by the linear extension (4.5.2). Here we use the assumption that $[\![A]\!]_0$ is a graded $\mathbb{F}$-vector space.

4.5.4 Definition. Let $([\![A]\!], \Gamma)$ be a Γ-track algebra as above. Then for $p\backslash n$ a *linear derivation* of degree 0,

$$\Gamma[n] : A \longrightarrow D,$$

is defined as follows. We choose for $\alpha \in A^r$, $r \geq 1$ an element $a \in [\![A^r]\!]_0$ representing α and we define

$$\Gamma[n]_\alpha = \Gamma(n)_a$$

by the linearity track (4.5.3).

4.5.5 Lemma. *The derivation* $\Gamma[n]$ *is well defined for* $p\backslash n$.

Proof. If $H : a \Rightarrow a'$ is a track, then by (4.2.5)(6) we see that

$$\begin{array}{ccc} a(n\cdot 1) & \xrightarrow{\Gamma(n)_a} & n\cdot a \\ {\scriptstyle H(n\cdot 1)}\downarrow & & \downarrow{\scriptstyle n\cdot H} \\ a'(n\cdot 1) & \xrightarrow{\Gamma(n)_{a'}} & n\cdot a' \end{array}$$

commutes. Here for $p\backslash n$ we have $H(n\cdot 1) = 0^\square : 0 \Rightarrow 0$ and $n\cdot H = 0^\square : 0 \Rightarrow 0$. Hence we get $\Gamma(n)_a = \Gamma(n)_{a'}$. Moreover (4.2.5)(4) shows that $\Gamma[n]$ is $\mathbb{F}$-linear and (4.2.5)(2) yields the derivation property above. □

The following proposition shows that only $\Gamma[n]$ for $n = p$ is of interest.

4.5.6 Proposition. $\Gamma[k\cdot p] = k\cdot \Gamma[p]$.

Compare (4.2.8).
As a crucial example we get the following result.

4.5.7 Proposition. *The secondary Steenrod algebra* $(\llbracket \mathcal{A} \rrbracket, \Gamma)$ *is a* Γ*-track algebra which yields the derivation of the Steenrod algebra*

$$\Gamma[p] : \mathcal{A} \longrightarrow \Sigma\mathcal{A}$$

of degree 0*. This is the same as a derivation* $\Gamma[p] : \mathcal{A} \to \mathcal{A}$ *of degree* -1.

For $p = 2$ one has the *Kristensen derivation* [Kr1] of degree -1,

$$\kappa : \mathcal{A} \longrightarrow \mathcal{A}$$

which carries Sq^n to Sq^{n-1}, $n \geq 1$, and Sq^0 to 0. Using a result of [Kr1] we show:

4.5.8 Theorem. *For* $p = 2$ *the derivation* $\Gamma[p] : \mathcal{A} \to \mathcal{A}$ *in* (4.5.7) *coincides with the Kristensen derivation* κ.

Proof. Here we use the connection between algebraic cocycles (used by Kristensen) and topological cocycles (used in this book) discussed in the Appendix of 2.1. We leave the straightforward details to the reader. Recall definition (4.3.10). Kristensen [Kr1] proves

$$d(\alpha, x, x) = \kappa(\alpha)(x).$$

Now it is clear that for $p = 2$ we have

$$\Gamma[p](\alpha) = d(\alpha, x, x).$$

□

Next we obtain the computation of $\Gamma[p]$ for p odd as follows.

4.5.9 Theorem. *For p odd the derivation $\Gamma[p] : \mathcal{A} \to \mathcal{A}$ in (4.5.7) is the unique derivation which satisfies $\Gamma[p]_\beta = 1$ for the Bockstein operation $\beta \in \mathcal{A}$ and $\Gamma[p]_{P^i} = 0$ for the reduced powers P^i, $i \geq 0$.*

We prove this result in the Appendix of this section.

We generalize the definition of the derivation (4.5.4) as follows.

4.5.10 Definition. Let $\mathcal{T}$ be a weak $\mathbb{F}$-additive track extension as in (4.1.3), for example $\mathcal{T} = [\![\mathbf{K}_p^{\mathrm{stable}}]\!]$. Hence for $\alpha : A \to B$ in $\mathcal{T}_0$ the track $\Gamma(p)_\alpha : 0 \Rightarrow 0$ is defined as in (4.2.7). We define the homomorphism

$$\Gamma[p] : [A,B] \longrightarrow D(A,B)$$

as follows. Let $\alpha \in [A,B]$ and let $a : A \to B$ be a map in $\mathcal{T}_0$ which represents the homotopy class α. Then we set

$$\Gamma[p]_\alpha = \sigma^{-1}(\Gamma(p)_a : 0 \Rightarrow 0)$$

where $\sigma : D(A,B) \cong \mathrm{Aut}(0 : A \to B)$ is given by the linear extension $\mathcal{T}$, see (3.5.1).

4.5.11 Proposition. *$\Gamma[p]$ in (4.5.10) is a well-defined $\mathbb{F}$-linear map satisfying*

$$\Gamma[p]_{\beta\alpha} = \alpha^*(\Gamma[p]_\beta) + \beta_*(\Gamma[p]_\alpha).$$

This is the derivation property of $\Gamma[p]$.

This is a consequence of (4.2.15).

The track theory $[\![\mathbf{K}_p^{\mathrm{stable}}]\!]$ is a weak $\mathbb{F}$-additive track extension which by (4.5.10) yields the $\mathbb{F}$-linear map

$$\Gamma[p] : [A,B]^{\mathrm{stable}} \longrightarrow [L^{-1}A, B]^{\mathrm{stable}} \tag{4.5.12}$$

for products of Eilenberg-MacLane spaces A and B. In fact $\Gamma[p]$ is totally determined by $\Gamma[p]$ in (4.5.7) as follows. Let

$$A = Z^{a_1} \times \cdots \times Z^{a_{n(A)}}, \; B = Z^{b_1} \times \cdots \times Z^{b_{n(B)}}$$

and let $n_{i,j} = b_j - a_i$. Then we get the commutative diagram

$$\begin{array}{ccc} [A,B]^{\mathrm{stable}} & \xrightarrow{\Gamma[p]} & [L^{-1}A,B]^{\mathrm{stable}} \\ \| & & \| \\ \bigoplus_{i,j} \mathcal{A}^{n_{i,j}} & \xrightarrow{\bigoplus_{i,j}\Gamma[p]} & \bigoplus_{i,j} \mathcal{A}^{n_{i,j}-1} \end{array} \tag{4.5.13}$$

which follows easily from the additivity rule (4.2.5)(4). Hence (4.5.8) shows that for $p = 2$ the derivation $\Gamma[p]$ in (4.5.12) is determined by the Kristensen derivation κ.

Appendix to Section 4.5: Computation of $\Gamma[p]$ for p odd

We first show

4.5.14 Proposition. *$\Gamma[p]_U = 0$ for $U : Z^q \to Z^{pq}$ in* **Top***.

Proof. In this proof we use notation as in section (8.2) and (9.3) below. We consider the track $\tilde{\Gamma}$ in the diagram $(r = q - k,\ Z^r = Z^r_{\mathbb{F}})$

$$\begin{array}{ccccc} Z^r & \xrightarrow{\Delta} & (Z^r)^p & \xrightarrow{U^p} & (Z^{pr})^p \\ & & \downarrow + & \overset{\tilde{\Gamma}}{\Longrightarrow} & \downarrow + \\ & & Z^r & \xrightarrow{U} & Z^{pr} \end{array} \tag{1}$$

where $\tilde{\Gamma}$ is a track under $Z^r \vee \cdots \vee Z^r \subset (Z^r)^p$. The track $\tilde{\Gamma}$ satisfies

$$\Gamma[p]_{P_i} = \Delta^* \tilde{\Gamma} : 0 \Longrightarrow 0 \tag{2}$$

where Δ is the p-fold diagonal. According to the formula

$$U(x + y) = N\bar{U}(x, y) + U(x) + U(y) \tag{3}$$

in Section (8.2) below we get

$$U(x_1 + \cdots + x_p) = N(\sum_{i=1}^{p-1} \bar{U}(x_1 + \cdots + x_i, x_{i+1})) + U(x_1) + \cdots + U(x_p). \tag{4}$$

Hence the track $\Gamma : N \Rightarrow 0$ in Section (8.2) below yields the track $\tilde{\Gamma}$ in (7), that is

$$\tilde{\Gamma}(x_1, \ldots, x_p) = \Gamma(\sum_{i=1}^{p-1} \bar{U}(x_1 + \cdots + x_i, x_{i+1})) + U(x_1) + \cdots + U(x_p). \tag{5}$$

Hence we get

$$(\Delta^* \tilde{\Gamma})(x) = \tilde{\Gamma}(x, \ldots, x) = \Gamma(\sum_{i=1}^{p-1} \bar{U}(ix, x)) + pUx \tag{6}$$

where $pUx = 0$. Here we have $\bar{U}(ix, x) = \bar{\alpha}(i, 1)\bar{U}(x, x)$ with $\bar{\alpha}$ in Section (9.3) below. Moreover $\Gamma(a + b) = \Gamma a + \Gamma b$ so that

$$\begin{aligned} (\Delta^* \tilde{\Gamma})(x) &= (\textstyle\sum_{i=1}^{p-1} \bar{\alpha}(i, 1))\Gamma\bar{U}(x, x) \\ &= -\Gamma\bar{U}(x, x). \end{aligned} \tag{7}$$

According to the definition of $\bar{U}$ we have

$$\bar{U}(x,x) = \sum_{b\in B} b(x,x) = \sum_{b\in B} x^p = \sum_{b\in B} U(x). \tag{8}$$

Hence the proposition $\Gamma[p]_{P_i} = 0^\square$ follows from

$$\Gamma U(x) = 0^\square. \tag{9}$$

Since $\Gamma U(x) = \sum_{\alpha\in\pi} \Gamma_\alpha U(x)$ by definition of Γ we study the function

$$\chi : \mathbb{Z}/p = \pi \longrightarrow Aut(U(y)) = [X, Z^{pq-1}] \tag{10}$$

which carries α to $\Gamma_\alpha U(y)$. We get

$$\begin{aligned} \Gamma_{\alpha\beta}U(y) &= \Gamma_\beta U(y)\square\Gamma_\alpha(\beta U y) \\ &= \Gamma_\beta U y\square\Gamma_\alpha(Uy) \end{aligned} \tag{11}$$

since $\beta U y = U y$ for $\beta \in \pi$. Hence χ is a homomorphism. This shows that

$$\begin{aligned} \Gamma U(x) &= \textstyle\sum_{\alpha\in\pi} \Gamma_\alpha U(y) \\ &= \textstyle\sum_{\alpha\in\pi} \chi(\alpha) \\ &= \textstyle\sum_{r=1}^{p-1} r = \chi(1) \\ &= \tfrac{p(p-1)}{2}\chi(1) \\ &= 0 \end{aligned} \tag{12}$$

since p is odd and $p\chi(1) = 0$. □

4.5.15 Proposition. $\Gamma[p]_{P^i} = 0$ *for* $i \geq 1$.

Proof. In Section (10.8) we show that there is a stable map sP^i in $[\![Z^q, Z^{pq}]\!]^{\text{stable}}$ such that $U : Z^q \to Z^{pq}$ in $\mathbf{Top}^*$ coincides with $(sP^i)_0$, q even. Therefore the forgetful map

$$\phi : [Z^q, Z^{pq-1}]^{\text{stable}} \longrightarrow [Z^q, Z^{pq-1}]$$

carries $\Gamma[p]_{P^i}$ to $\Gamma[p]_U$. By the result in (1.1.13) we see that ϕ is injective. Therefore the result follows from (4.5.14). □

4.5.16 Proposition. $\Gamma[p]_\beta = 1$.

Proof. The inclusion $\mathbb{F} = \mathbb{Z}/p \to \mathbb{G} = \mathbb{Z}/p^2$ induces the map

$$i : Z^n_{\mathbb{F}} \longrightarrow Z^n_{\mathbb{G}} \tag{1}$$

between Eilenberg-MacLane spaces defined for $R = \mathbb{F}$ and $R = \mathbb{G}$ respectively, see Section (2.1). Moreover the addition maps $+_{\mathbb{F}} : \mathbb{F}^p \to \mathbb{F}$ and $+_{\mathbb{G}} : \mathbb{G}^p \to \mathbb{G}$ yield the following *commutative* diagram where $X^p = X \times \cdots \times X$ is the p-fold product.

(2)
$$\begin{array}{ccc} (Z^n_{\mathbb{F}})^p & \xrightarrow{i^p} & (Z^n_{\mathbb{G}})^p \\ \downarrow + & & \downarrow + \\ Z^n_{\mathbb{F}} & \xrightarrow{i} & Z^n_{\mathbb{G}} \end{array}$$

Now we apply the natural fiber sequence to the map i^p and to i and we get the following commutative diagram.

(3)
$$\begin{array}{ccc} (F)^p & \xrightarrow{\beta^p} & (Z^n_{\mathbb{F}})^p \\ \downarrow + & & \downarrow + \\ F & \xrightarrow{\beta} & Z^n_{\mathbb{F}} \end{array}$$

Here $F \simeq \Omega Z^n_{\mathbb{F}} \simeq Z^{n-1}_{\mathbb{F}}$ is given by the fiber sequence

(4)
$$F \longrightarrow Z^n_{\mathbb{F}} \longrightarrow Z^n_{\mathbb{G}} \longrightarrow Z^n_{\mathbb{F}}.$$

It is well known that the boundary map $Z^{n-1}_{\mathbb{F}} \simeq F \xrightarrow{\beta} Z^n_{\mathbb{F}}$ represents the Bockstein map, see (2.1.9). In order to compute $\Gamma[p]_\beta$ we have to be careful with respect to the homotopy equivalence

(5)
$$Z^{n-1}_{\mathbb{F}} \underset{r_{n-1}}{\xrightarrow{\sim}} \Omega Z^n_{\mathbb{F}} \underset{\bar{\pi}}{\xrightarrow{\sim}} F$$

defined in (2.1.9). Here r_{n-1} is $\mathbb{F}$-linear but $\bar{\pi}$ is not $\mathbb{F}$-linear. Therefore $\Gamma[p]_\beta$ is represented by the following diagram, $Z^n = Z^n_{\mathbb{F}}$.

(6)
$$\begin{array}{ccccccccc} Z^{n-1} & \xrightarrow{r_{n-1}} & \Omega Z^n & \xrightarrow{\Delta} & (\Omega Z^n)^p & \xrightarrow{\bar{\pi}^p} & F^p & & \\ & & & \searrow^{0} & +\downarrow & \overset{H}{\Longrightarrow} & \downarrow + & & \\ & & & & \Omega Z^n & \xrightarrow[\bar{\pi}]{} & F & \xrightarrow{\partial} & Z^n \end{array}$$
(with an arrow labelled 0 from ΩZ^n to Z^n)

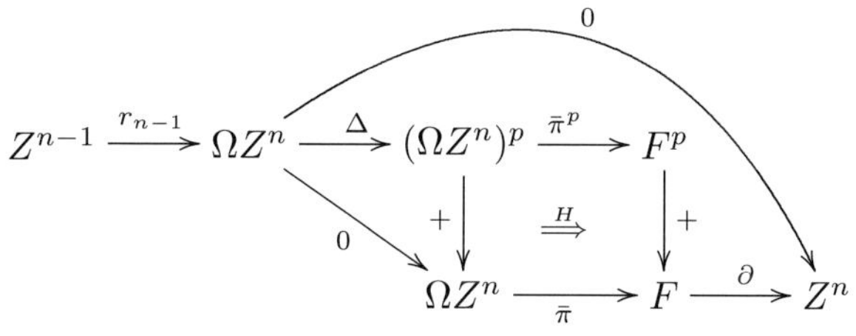

Here the track H is unique by (3.2.5). The composite $+(\bar{\pi}^p)\Delta$ satisfies

$$(+(\bar{\pi}^p)\Delta)(x) = p \cdot x$$

so that $\partial(p \cdot x) = 0$ since Z^n is a $\mathbb{F}$-vector space. The space F, however, is a $\mathbb{G}$-module so that $p \cdot x \in F$ need not be trivial. We now observe that $\pi : F \to \Omega Z^n$ yields the track

$$\pi H : 0^{\square} : 0 \Longrightarrow 0 \tag{7}$$

which is the identity track of 0. In fact by (3.2.5) the track (7) is unique. Moreover by (3.2.5) there is a unique track U in the following diagram.

$$F \xrightarrow{\pi} \Omega Z^n \xrightarrow{\bar{\pi}} F, \quad U : 1 \Longrightarrow \bar{\pi}\pi \tag{8}$$

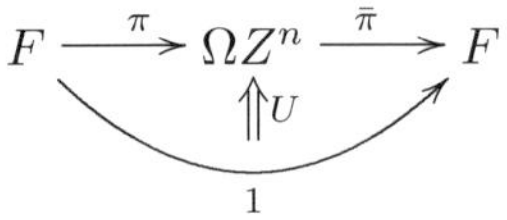

Now pasting of H and U yields

$$U * H = UH_1 \square U_0 H = U_1 H \square U H_0 \tag{9}$$

with $U_1 H = \bar{\pi}\pi H = \bar{\pi} 0^{\square} = 0^{\square}$ and $U_0 H = H$. Hence we get $H = (UH_1)^{\mathrm{op}} \square U H_0$ with $H_0 \Delta = 0$. Therefore we get

$$\begin{aligned} \Gamma[p]_\beta &= \partial H \Delta r_{n-1} \\ &= \partial U^{\mathrm{op}} H_1 \Delta r_{n-1} \\ &= \partial U^{\mathrm{op}} (\cdot p) \bar{\pi} r_{n-1}. \end{aligned} \tag{10}$$

Here the map $\cdot p : F \to F$ (carrying y to $y \cdot p \in F$) admits a factorization

$$\begin{array}{ccc} F & \xrightarrow{\cdot p} & F \\ {\scriptstyle \pi} \downarrow \wr & & \uparrow {\scriptstyle j} \\ \Omega Z^n & \xrightarrow[\Omega(i)]{} & \Omega Z^n_{\mathbb{G}} \end{array} \tag{11}$$

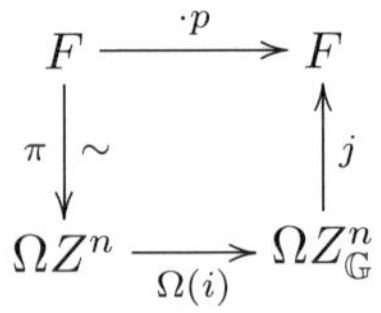

where j is the inclusion. In fact, by (2.1.3)(3) we have for $(x, \sigma) \in F$ the equation $p(x, \sigma) = (px, p\sigma) = (0, p\sigma) = (0, i\pi\sigma)$ so that $p\sigma$ is a loop. The inclusion j satisfies $\partial j = 0$ and $\pi j = \Omega(\pi : Z^n_{\mathbb{G}} \to Z^n)$ so that $\pi j(\Omega i) = 0$. Therefore we get a well-defined track

$$V = \partial U^{\mathrm{op}} j \Omega(i) : 0 \Longrightarrow 0 \tag{12}$$

representing an element in $[\Omega Z^n, \Omega Z^n]$ such that

$$\Gamma[p]_\beta = V \pi \bar{\pi} r_{n-1} = V r_{n-1}. \tag{13}$$

The following lemma shows that V is the identity of ΩZ^n in $[\Omega Z^n, \Omega Z^n]$ so that $\Gamma[p]_\beta = r_{n-1}$ represents 1. For $n \geq 1$ we see that $\Gamma[p]_\beta$ defined by the stable map β in $[\![\mathcal{A}]\!]$ coincides with $\Gamma[p]_\beta$ above. $\square$

Let $\pi : E \to B$ be a fibration in **Top*** with fiber A. Consider the fiber sequence

$$\begin{array}{c} \Omega A \xrightarrow{\Omega\pi} \Omega E \xrightarrow{j} F \xrightarrow{\partial} A \xrightarrow{i} E \xrightarrow{\pi} B \\ \Omega E \xrightarrow{\Omega\pi} \Omega B, \quad F \xrightarrow[\sim]{\pi} \Omega B \end{array} \tag{4.5.17}$$

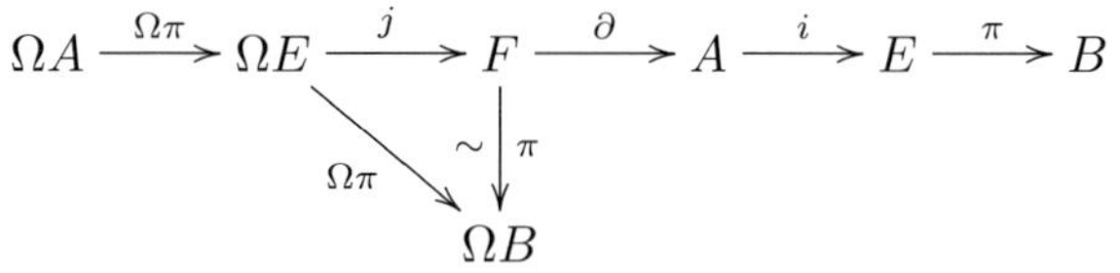

with $F = \{(x,\sigma) \in A \times (E,*)^{(I,0)}, ix = \sigma(1)\}$ and $\pi(x,\sigma) = \pi\sigma \in \Omega(B)$. Since π is a homotopy equivalence we can choose a track $\bar{U}$ as in the diagram

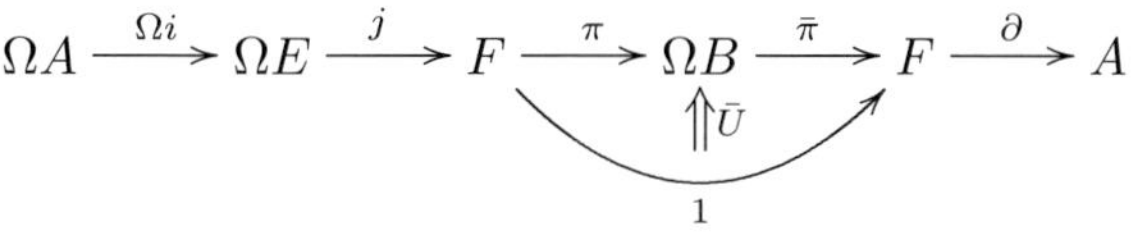

$$\Omega A \xrightarrow{\Omega i} \Omega E \xrightarrow{j} F \xrightarrow{\pi} \Omega B \xrightarrow{\bar{\pi}} F \xrightarrow{\partial} A, \qquad \bar{U} : 1 \Rightarrow \bar{\pi}\pi \ (F \to F)$$

where $\bar{\pi}$ is a homotopy inverse of π. Since $\partial j = 0$ and $\pi j \Omega(i) = \Omega(\pi)\Omega(i) = \Omega(\pi i) = 0$ this diagram represents a track $0 \Rightarrow 0$ in $[\![\Omega A, A]\!]$ and hence an element in $[\Omega A, \Omega A]$.

4.5.18 Lemma. *There exists a track $\bar{U}$ such that the track $\partial \bar{U} j \Omega(\pi)$ in the diagram above represents the identity element in $[\Omega A, \Omega A]$.*

Proof. The proof of the lemma is not so obvious though the lemma holds in any fibration category with zero object $*$, see [BAH]. Therefore it suffices to prove the dual lemma in a cofibration category with zero object $*$. We may assume that all objects are fibrant and cofibrant. We consider for a cofibration π the following diagram.

$$\begin{array}{ccccccccccc} \Sigma A & \longleftarrow & \Sigma E & \longleftarrow & F & \xleftarrow{\partial} & A & \xleftarrow{i} & E & \xleftarrow{\pi} & B \\ & & & & \| & & \| & & & & \\ \bar{F} & = & CE \cup_B CB & \xrightarrow[\sim]{} & (CE)/B & & E/B & & & & \\ & & & & \uparrow{\scriptstyle \pi\,\sim} & & \uparrow{\scriptstyle \sim} & & & & \\ & & & & \Sigma B & \xleftarrow[\bar{\partial}]{} & E \cup_B CB & & & & \end{array} \tag{1}$$

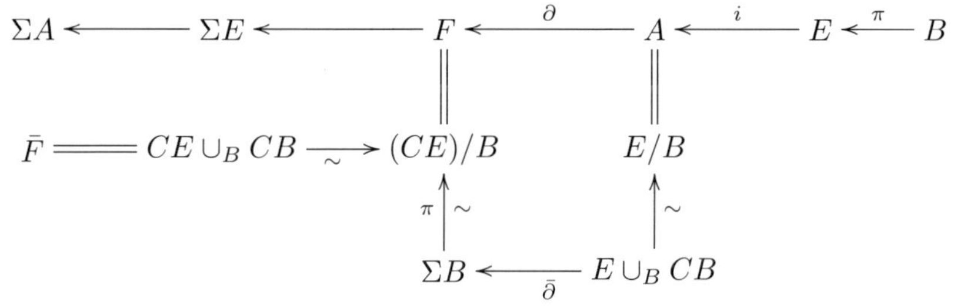

We replace the cofiber $A = E/B$ by $E \cup_B CB$. The map $\bar{\partial}$ is given by the composite

$$\bar{\partial} : E \cup_B CB \xrightarrow{q} CB/B = \Sigma B \xrightarrow{-1} \Sigma B$$

where q is the quotient map. We replace F by $\bar{F}$ in the diagram. We have to show that there exists a homotopy

$$\begin{aligned} H : I(\bar{F}) &\longrightarrow (CE)/B, \\ H_0 : \bar{F} = CE \cup_B CB &\xrightarrow{q} (CE)/B, \\ H_1 : \bar{F} = CE \cup_B CB &\xrightarrow{q} \Sigma B \xrightarrow{-1} \Sigma B, \end{aligned} \tag{2}$$

where q denotes the quotient maps. The cylinder of $\bar{F}$ satisfies

$$I(\bar{F}) = I(CB) \cup_{IB} I(CE) \tag{3}$$

and we define $H_0' = H_0 \mid I(CE)$ by the composite

$$H_0' : \quad \begin{array}{ccccccc} ICE & \xrightarrow{q} & CCE & \xrightarrow{\varphi} & CE & \longrightarrow & (CE)/B \\ & & \cup & \nearrow_{(1,1)} & & & \\ & & CE \cup_E CE & & & & \end{array} \tag{4}$$

where the extension φ of $(1,1)$ exists since CE is contractible. The restriction $H_0' \mid IB$ admits a factorization

$$IB \longrightarrow \Sigma B \xrightarrow{\pi} (CE)/B.$$

We now define $H_0'' = H_0 \mid ICB$ by an extension in the following diagram.

$$\begin{array}{ccc} ICB & \xrightarrow{H_0''} & (CE)/B \\ \uparrow & & \uparrow{\scriptstyle (\pi,-\pi)} \\ i_0CB \cup IB \cup i_1CB & \xrightarrow[(0,q,q)]{} & \Sigma B \vee \Sigma B \end{array} \tag{5}$$

The extension H_0'' exists since the obstruction $\pi + (-\pi) = 0$ vanishes. Now one can check that $H = H_0'' \cup H_0'$ is a well-defined homotopy as above. Moreover the following diagram commutes where q are quotient maps and r is the inclusion.

$$\begin{array}{ccccc} I(CB \cup_B E) & \xrightarrow{q} & \Sigma(E/B) & & \\ & & & \nwarrow^{\Sigma(i)} & \\ \downarrow{\scriptstyle r} & & & & \Sigma(E) \\ & & & \nearrow_{q} & \\ I(CB \cup_B CE) & \xrightarrow[H]{} & (CE)/B & & \end{array} \tag{6}$$

This proves (4.5.18). □

4.6 Obstruction to linearity of cocycles

In this section we show that the Kristensen derivation is actually an obstruction to the $\mathbb{F}$-linearity of cocycles representing the characteristic class of $[\![\mathbf{K}_p^{\text{stable}}]\!]$. For this we introduce the following natural map between cohomology groups of categories.

4.6.1 Definition. Let $\mathbb{F} = \mathbb{Z}/p$. Let $\mathbf{A}$ be an $\mathbb{F}$-additive category and let D be an $\mathbb{F}$-additive $\mathbf{A}$-bimodule. Then we define the linear map

$$\Gamma_p : H^3(\mathbf{A}, D) \longrightarrow H^1(\mathbf{A}, D)$$

as follows. Let $\langle c \rangle \in H^3(\mathbf{A}, D)$ be represented by the cocycle c. We may assume that the cocycle c is *normalized* with respect to 0-maps, sums and products. For this compare the Appendix of Baues-Tonks [BT]. Then $\Gamma_p(\langle c \rangle)$ is represented by the 1-cocycle

$$d^c : \mathbf{A} \longrightarrow D$$

which carries $\alpha : A \to B$ to $d^c(\alpha) \in D(A, B)$ given by the following formula. Let

$$A^p : A \longrightarrow A^{\times p} = A \times \cdots \times A$$

be the p-fold diagonal in $\mathbf{A}$ and let

$$A_p : A^{\times p} \longrightarrow A$$

be the p-fold codiagonal in $\mathbf{A}$. We have $A_p A^p = 0$ since multiplication by p is trivial. We now set

$$d^c(\alpha) = c(\alpha, A_p, A^p) - c(B_p, \alpha^{\times p}, A^p) + c(B_p, B^p, \alpha).$$

4.6.2 Proposition. *The linear map* Γ_p *in* (4.6.1) *is well defined.*

Proof. We first check that d^c is a cocycle, that is, $\delta d^c = 0$, or

$$0 = (\delta d^c)(\beta, \alpha) = \beta d^c_p(\alpha) - d^c_p(\beta\alpha) + d^c_p(\beta)\alpha$$

for $E \xleftarrow{\beta} B \xleftarrow{\alpha} A$ in $\mathbf{A}$. We know that $\delta c = 0$ since c is a cocycle. Hence we have the following formulas:

$$\begin{aligned} 0 &= (\delta c)(\beta, \alpha, A_p, A^p) \\ &= \beta c(\alpha, A_p, A^p) - c(\beta\alpha, A_p, A^p) + c(\beta, \alpha A_p, A^p) \\ &\quad - \underline{c(\beta, \alpha, 0)} + \underline{c(\beta, \alpha, A_p) A^p}. \end{aligned}$$

$$\begin{aligned} 0 &= (\delta c)(\beta, B_p, \alpha^{\times p}, A^p) \\ &= \beta c(B_p, \alpha^{\times p}, A^p) - c(\beta B_p, \alpha^{\times p}, A^p) + c(\beta, B_p \alpha^{\times p}, A^p) \\ &\quad - c(\beta, B_p, \alpha^{\times p} A^p) + \underline{c(\beta, B_p, \alpha^{\times p}) A^p}. \end{aligned}$$

$$\begin{aligned} 0 &= (\delta c)(\beta, B_p, B^p, \alpha) \\ &= \beta c(B_p, B^p, \alpha) - c(\beta B_p, B^p, \alpha) + \underline{c(\beta, 0, \alpha)} \\ &\quad - c(\beta, B_p, B^p \alpha) + c(\beta, B_p, B^p)\alpha. \end{aligned}$$

$$\begin{aligned} 0 &= (\delta c)(E_p, \beta^{\times p}, B^p, \alpha) \\ &= \underline{E_p c(\beta^{\times p}, B^p, \alpha)} - c(E_p \beta^{\times p}, B^p, \alpha) + c(E_p, \beta^{\times p} B^p, \alpha) \\ &\quad - c(E_p, \beta^{\times p}, B^p \alpha) + c(E_p, \beta^{\times p}, B^p)\alpha. \end{aligned}$$

Since c is normalized with respect to zero maps, sums and products the underlined terms vanish. We write $d^c(\alpha) = \xi(\alpha) + \eta(\alpha)$ where $\xi(\alpha) = c(\alpha, A_p, A^p) - c(B_p, \alpha^{\times p}, A^p)$ and $\eta(\alpha) = c(B_p, B^p, \alpha)$. By definition of ξ we have

$$\begin{aligned}(\delta\xi)(\beta,\alpha) &= -c(\beta\alpha, A_p, A^p) + c(E_p, (\beta\alpha)^{\times p}, A^p) \\ &\quad +\beta c(\alpha, A_p, A^p) + c(\beta, B_p, A^p)\alpha \\ &\quad -\beta c(B_p, \alpha^{\times p}, A^p) - c(E_p, \beta^{\times p}, B^p)\alpha.\end{aligned}$$

Hence the equations above show that $(\delta\xi)(\beta,\alpha)$ is the sum

$$\begin{array}{llllll} - & c(\beta\alpha, A_p, A^p) & + & c(E_p, (\beta\alpha)^{\times p}, A^p) & + & c(\beta\alpha, A_p, A^p) \\ - & c(\beta, \alpha A_p, A^p) & - & \beta c(B_p, B^p, \alpha) & + & c(\beta B_p, B^p, \alpha) \\ + & c(\beta, B_p, B^p\alpha) & - & c(\beta B_p, \alpha^{\times p}, A^p) & + & c(\beta, B_p\alpha^{\times p}, A^p) \\ - & c(\beta, B_p, \alpha^{\times p}A^p) & - & c(E_p\beta^{\times p}, B^p, \alpha) & + & c(E_p, \beta^{\times p}B^p, \alpha) \\ - & c(E_p, \beta^{\times p}, B^p\alpha). & & & & \end{array}$$

Here we have $\alpha^{\times p}A^p = B^p\alpha$, $\beta^{\times p}B^p = E^p\beta$, $\alpha A_p = B_p\alpha^{\times p}$ and $\beta B_p = E_p\beta^{\times p}$ so that $\delta(\xi)(\beta,\alpha)$ is the sum

$$\begin{aligned}\delta(\xi)(\beta,\alpha) &= c(E_p, (\beta\alpha)^{\times p}, A^p) - c(\beta B_p, \alpha^{\times p}, A^p) - c(E_p, \beta^{\times p}, B^p\alpha) \\ &\quad -\beta c(B_p, B^p, \alpha) + c(E_p, E^p\beta, \alpha).\end{aligned}$$

On the other hand we have

$$\begin{aligned} 0 &= (\delta c)(E_p, E^p, \beta, \alpha) \\ &= \underline{E_p c(E^p, \beta, \alpha)} - \underline{c(0, \beta, \alpha)} + c(E_p, E^p\beta, \alpha) \\ &\quad -c(E_p, E^p, \beta\alpha) + c(E_p, E^p, \beta)\alpha \end{aligned}$$

$$\begin{aligned} 0 &= (\delta c)(E_p, \beta^{\times p}, \alpha^{\times p}, A^p) \\ &= \underline{E_p c(\beta^{\times p}, \alpha^{\times p}, A^p)} - c(E_p\beta^{\times p}, \alpha^{\times p}, A^p) + c(E_p, (\beta\alpha)^{\times p}, A^p) \\ &\quad -c(E_p, \beta^{\times p}, \alpha^{\times p}A^p) + \underline{c(E_p, \beta^{\times p}, \alpha^{\times p})A^p}. \end{aligned}$$

Therefore we get

$$\begin{aligned}(\delta\xi)(\beta,\alpha) &= -\beta c(B_p, B^p, \alpha) + c(E_p, E^p\beta, \alpha) \\ &= -\beta c(B_p, B^p, \alpha) + c(E_p, E^p, \beta, \alpha) - c(E_p, E^p, \beta)\alpha \\ &= -(\delta\eta)(\beta,\alpha)\end{aligned}$$

where η is defined above. Hence we see $\delta(d^c) = \delta\xi + \delta\eta = 0$. This completes the proof that d^c is a cocycle.

Next let $c = \delta f$ be a coboundary where f is a normalized cochain. Then we have

$$c(\alpha,\beta,\gamma) = (\delta f)(\alpha,\beta,\gamma) = \alpha f(\beta,\gamma) - f(\alpha\beta,\gamma) + f(\alpha,\beta\gamma) - f(\alpha,\beta)\gamma.$$

Hence the definition of d^c yields

$$\begin{aligned} d^c(\alpha) &= \alpha f(A_p, A^p) - \underline{B_p f(\alpha^{\times p}, A^p)} + \underline{B_p f(B^p, \alpha)} \\ &\quad -(f(\alpha A_p, A^p) - f(B_p \alpha^{\times p}, A^p) + \underline{f(0,\alpha)}) \\ &\quad +(\underline{f(\alpha,0)} - f(B_p, \alpha^{\times p} A^p) + f(B_p, B^p \alpha)) \\ &\quad -(f(\alpha, A_p)A^p - f(B_p, \alpha^{\times p})A^p + f(B_p, B^p)\alpha). \end{aligned}$$

This shows that

$$d^c(\alpha) = \alpha f(A_p, A^p) - f(B_p, B^p)\alpha$$

and hence d^c is a coboundary. This completes the proof that Γ_p is well defined. □

4.6.3 Proposition. *The composition*

$$HH^3(\mathbf{A}, D) \longrightarrow H^3(\mathbf{A}, D) \xrightarrow{\Gamma_p} H^1(\mathbf{A}, D)$$

is trivial. Here we use the natural map (3.6.6)(4).

Proof. Assume that c is normalized and multilinear. We have inclusions $i_r^A : A \to A^{\times p}$ and projections $p_r^A : A^{\times p} \to A$ for $1 \le r \le p$ and the equations

$$A_p = \sum_r p_r^A, \; A^p = \sum_r i_r^A,$$

$$\alpha^{\times p} = \sum_r i_r^B \alpha p_r^A$$

hold. Hence multilinearity of c shows

$$\begin{aligned} d^c(\alpha) &= \sum_{t,s} c(\alpha, p_t^A, i_s^A) + \sum_{r,t} c(p_r^B, i_t^B, \alpha) \\ &\quad - \sum_{r,s,t} c(p_r^B, i_t^B \alpha p_t^A, i_s^A). \end{aligned}$$

Since c is normalized we see that

$$c(\alpha, p_t^A, i_s^A) = \begin{cases} 0 & \text{for } t \neq s \\ c(\alpha, 1_A, 1_A) & \text{for } t = s, \end{cases}$$

$$c(p_r^B, i_t^B, \alpha) = \begin{cases} 0 & \text{for } t \neq r \\ c(1_B, 1_B, \alpha) & \text{for } t = r, \end{cases}$$

$$c(p_r^B, i_t^B, \alpha p_t^A, i_s^A) = \begin{cases} c(1_A, \alpha, 1_A) & \text{for } r = t = s \\ 0 & \text{otherwise.} \end{cases}$$

Since multiplication by p is trivial we hence get $d^c(\alpha) = 0$. □

Proposition (4.6.3) shows that Γ_p is an obstruction to the linearity of cocycles, that is: Let $x \in H^3(\mathbf{A}, D)$ and let $\Gamma_p(x) \neq 0$. Then x is not in the image of $HH^3(\mathbf{A}, D)$ and hence x cannot be represented by a trilinear cocycle. We apply this in Theorem (4.6.5) below.

4.6.4 Proposition. *Let $\mathcal{T}$ be a weak $\mathbb{F}$-additive track extension as in* (4.1.3)*, for example $\mathcal{T} = [\![\mathbf{K}_p^{\text{stable}}]\!]$. Then the characteristic class $\langle\mathcal{T}\rangle$ yields the element*

$$\Gamma_p(\langle\mathcal{T}\rangle) \in H^1(\mathbf{K}, D)$$

which is represented by the derivation $\Gamma[p]$ in (4.5.10)*, that is $\Gamma[p] \in \Gamma_p(\langle\mathcal{T}\rangle)$.*

Proof. We have to compare the definitions of Γ_p and $\Gamma[p]$ and we have to use the definition of the characteristic class $\langle\mathcal{T}\rangle$. In fact, the comparison gave us the intuition to define the operator Γ_p by the somewhat obscure formula for $d^c(\alpha)$ in (4.6.1). We choose first a cocycle c representing $\langle\mathcal{T}\rangle$ as follows, see (3.6.7). Let $s : \mathbf{K}(A, B) \to \mathcal{T}_0(A, B)$ be compatible with products and with the $\mathbb{F}$-vector space structures of A, B in $\mathbf{K}$ and $\mathcal{T}_0$ respectively. This implies that $c(B_p, B^p, \alpha)$ represented by the track diagram (3.6.7) is trivial, since all tracks in this track diagram are trivial tracks. Moreover we see that the sum $c(\alpha, A_p, A^p) - c(B_p, \alpha^{\times p}, A^p)$ is represented by the tracks in the following diagram.

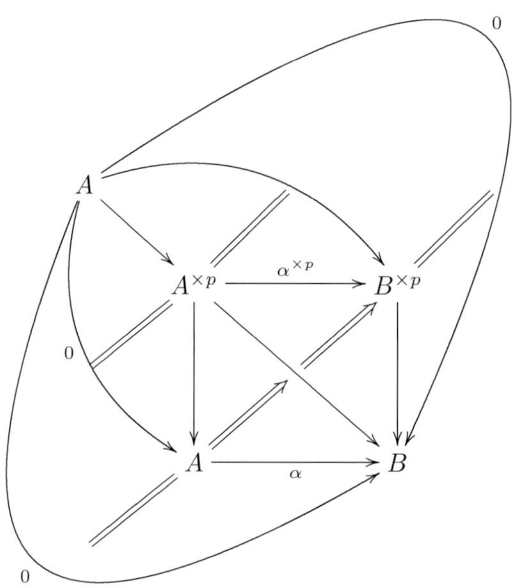

Here $=$ denotes trivial tracks. This proves $\Gamma[p] \in \Gamma_p(\langle\mathcal{T}\rangle)$. □

4.6.5 Theorem. *For the characteristic class k_p^{stable} of $[[\mathbf{K}_p^{\text{stable}}]]$ the element*

$$\Gamma_p(k_p^{\text{stable}}) \in H^1(\mathbf{K}_p^{\text{stable}}, \text{Hom}(L^{-1}, -))$$

is represented by the Kristensen derivation $\Gamma[p]$ and hence $\Gamma_p(k_p^{\text{stable}}) \neq 0$, see (4.5.13)*. This shows that k_p^{stable} cannot be represented by a linear cocycle, $p = 2$.*

This is a consequence of (4.6.4) and (4.6.3).

Chapter 5

The Algebra of Secondary Cohomology Operations

In this chapter we show that the secondary Steenrod algebra which is a Γ-track algebra $(\llbracket\mathcal{A}\rrbracket, \Gamma)$ can be canonically "strictified". This yields a new secondary algebra $\mathcal{B}$ in which multiplication is bilinear. The secondary algebra $\mathcal{B}$ is well defined up to isomorphism; so that $\mathcal{B}$ is the true algebra of (stable) secondary cohomology operations generalizing the Steenrod algebra $\mathcal{A}$.

5.1 Track algebras, pair algebras and crossed algebras

Let R be a ring and let $\mathbf{Mod}(R)$ be the category of (left) R-modules and R-linear maps.

An R-module object in the category of groupoids $\mathbf{Grd}$ is an abelian group object in $\mathbf{Grd}$ together with a left action of R. Let $\mathbf{pair}(\mathbf{Mod}(R))$ be the category of pairs in $\mathbf{Mod}(R)$, then we obtain as in (2.2.6):

5.1.1 Proposition. *The category of R-module objects in* $\mathbf{Grd}$ *and R-linear maps is isomorphic to the category* $\mathbf{pair}(\mathbf{Mod}(R))$.

Let M be an R-module object in $\mathbf{Grd}$ with $\partial_0, \partial_1 : M_1 \to M_0$ given by source and target. Then we define the pair

$$(1) \qquad \partial_0^0 : M_1^0 \to M_0 \text{ in } \mathbf{pair}(\mathbf{Mod}(R))$$

where $M_1^0 = \{H : a \Rightarrow 0 \in M_1\} = \text{kernel}(\partial_1)$ and $\partial_0^0(H : a \Rightarrow 0) = a$. That is ∂_0^0 is the restriction of ∂_0. Conversely let

$$(2) \qquad \partial : M_1^0 \longrightarrow M_0$$

be an object in $\mathbf{pair}(\mathbf{Mod}(R))$. Then we define $M_1 = M_1^0 \oplus M_0$ and for $H \in M_1^0$, $x \in M_0$ we write $H + x = (H, x) \in M_1$. Then $\partial_0(H + x) = \partial(H) + x$ and $\partial_1(H + x) = x$ so that $H + x : \partial(H) + x \Rightarrow x$.

Composition of tracks is defined by

(3) $$(H+x)\Box(G+\partial(H)+x) = H+G+x.$$

This yields the R-module object M in **Grd** associated to ∂, see (2.2.6).

We consider a pair $\partial : M_1^0 \to M_0$ in **pair**(**Mod**(R)) as a chain complex concentrated in degree 0 and 1. Now let R be a commutative ring. Then we get for pairs $X = (\partial_X : X_1 \to X_0)$, $Y = (\partial_Y : Y_1 \to Y_0)$ in **Mod**(R) the tensor product of chain complexes $X \otimes Y$ (with $\otimes = \otimes_R$) defined by

$$X_1 \otimes Y_1 \xrightarrow{d_2} X_1 \otimes Y_0 \oplus X_0 \otimes Y_1 \xrightarrow{d_1} X_0 \otimes Y_0,$$
$$d_2(a \otimes b) = (\partial a) \otimes b - a \otimes (\partial b),$$
$$d_1(a \otimes y) = (\partial a) \otimes y,$$
$$d_1(x \otimes b) = x \otimes (\partial b),$$

with $x \in X_0$, $y \in Y_0$, $a \in X_1$, $b \in Y_1$. Hence d_1 induces the boundary map

$$\partial_\otimes : (X_1 \otimes Y_0 \oplus X_0 \otimes Y_1)/im(d_2) \longrightarrow X_0 \otimes Y_0$$

which again is a pair in **Mod**(R). This shows that (**pair**(**Mod**(R)), $\bar{\otimes}$) is a monoidal category with the product

(5.1.2) $$X\bar{\otimes}Y = \partial_\otimes$$

defined above.

A (non-negatively) *graded pair* X in **Mod**(R) is a sequence of pairs X_i, $i \in \mathbb{Z}$, in **Mod**(R) (with $X_i = 0$ for $i < 0$). Then X is the same as an R-linear map of degree 0,

$$X = (\partial : X_1 \longrightarrow X_0)$$

where X_1, X_0 are (non-negatively) graded objects in **Mod**(R). For such graded pairs X, Y we get the $X\bar{\otimes}Y$ satisfying

(5.1.3) $$(X\bar{\otimes}Y)^k = \bigoplus_{n+m=k} X^n\bar{\otimes}Y^m.$$

This is a monoidal structure of the *category of graded pairs* in **Mod**(R). Morphisms are R-linear maps of degree 0. We now describe the concept of algebra in the category of graded groupoids. Such algebras can be introduced in three different ways, as 'track algebras', 'pair algebras' or 'crossed algebras'.

5.1.4 Definition. A (graded) *track algebra* over R is a monoid A in the category of graded groupoids such that A^k, $k \in \mathbb{Z}$, is an R-module object in **Grd** with $A^k = 0$ for $k < 0$. Moreover the monoid multiplication is a functor

$$A^n \times A^m \longrightarrow A^{n+m}$$

which is R-bilinear ($n, m \in \mathbb{Z}$).

5.1.5 Definition. A (graded) *pair algebra* over R is a monoid A in the monoidal category of graded pairs in $\mathbf{Mod}(R)$ with multiplication (see (5.1.3))

$$\mu : A\bar{\otimes}A \longrightarrow A.$$

Moreover $A_i = 0$ for $i < 0$.

5.1.6 Definition. A (graded) *crossed algebra* A over R is a graded pair

$$\partial : A_1 \longrightarrow A_0 \tag{1}$$

in $\mathbf{Mod}(R)$ with $A_1^n = A_0^n = 0$ for $n < 0$ such that A_0 is a graded algebra in $\mathbf{Mod}(R)$ and A_1 is an A_0-bimodule and ∂ is an A_0-bimodule map. Moreover for $a, b \in A_1$ the formula

$$\partial(a) \cdot b = a \cdot \partial(b) \tag{2}$$

holds in A_1. We write

$$\pi_0(A) = \text{cokernel}(\partial), \tag{3}$$

$$\pi_1(A) = \text{kernel}(\partial). \tag{4}$$

Then it is easily seen that $\pi_0(A)$ is an algebra over R and that $\pi_1(A)$ is a $\pi_0(A)$-bimodule. Hence we have the exact sequence of A_0-bimodules

$$0 \longrightarrow \pi_1(A) \longrightarrow A_1 \xrightarrow{\partial} A_0 \longrightarrow \pi_0(A) \longrightarrow 0.$$

Remark. A crossed algebra is the same as a "crossed module" considered in Baues-Minian [BM]. Such crossed modules correspond to classical crossed modules for groups considered by J.H.C. Whitehead. We here prefer the notion "crossed algebra" since we will also consider "modules over a crossed algebra".

5.1.7 Proposition. *The categories of track algebras, pair algebras, and crossed algebras respectively are equivalent to each other.*

In fact, using (5.1.1) we see that a track algebra yields a pair algebra and vice versa. Moreover the definition of $\bar{\otimes}$ in (5.1.2) shows that a pair algebra yields a crossed algebra and vice versa. Given a track algebra A we obtain the associated crossed algebra by

$$\partial : A_1^0 \longrightarrow A_0$$

as in (5.1.1)(1). The A_0-bimodule structure of A_1^0 is defined by

$$f \cdot H \cdot g = 0_f^\square \cdot H \cdot 0_g^\square \tag{1}$$

for $f, g \in A_0$, $H \in A^0$. Here $o_f^\square : f \Rightarrow f$ is the trivial track. Moreover for $H : f \Rightarrow g$, $G : x \Rightarrow y$ in A^1 we have the formula

$$\begin{aligned} H \cdot G &= (g \cdot G)\square(H \cdot x) \\ &= (H \cdot y)\square(f \cdot G) \end{aligned} \tag{2}$$

which for $g = 0$ and $y = 0 =$ implies

$$H \cdot G = H \cdot x = f \cdot G \tag{3}$$

and this corresponds to the formula $H \cdot (\partial G) = (\partial H) \cdot G$ in a crossed algebra.

According to the equivalent concepts

$$\text{track algebra} \;=\; \text{pair algebra} \;=\; \text{crossed algebra} \tag{5.1.8}$$

we can define a module over a track algebra also in three different but equivalent ways. We do this first for the case of track algebras.

5.1.9 Definition. A (left) module M over a track algebra A is a graded object M in the category of groupoids **Grd** such that M^k, $k \in \mathbb{Z}$, is an R-module object in **Grd**. Moreover the monoid A acts from the left on M and the action is a functor

$$A^n \times M^k \longrightarrow M^{n+k}$$

which is R-linear $(n, k \in \mathbb{Z})$. Compare (5.1.4).

Next we consider modules over a pair algebra.

5.1.10 Definition. A (left) *module* M over a pair algebra A is a graded pair M in **Mod**(R) together with an R-linear map of degree 0,

$$\mu : A \bar{\otimes} M \longrightarrow M$$

which is an action of the monoid A on M.

5.1.11 Definition. A (left) *module* M over a crossed algebra A is a graded pair $M = (\partial : M_1 \to M_0)$ in **Mod**(R) such that M_1 and M_0 are left A_0-modules and ∂ is A_0-linear. Moreover a commutative diagram of A_0-linear maps

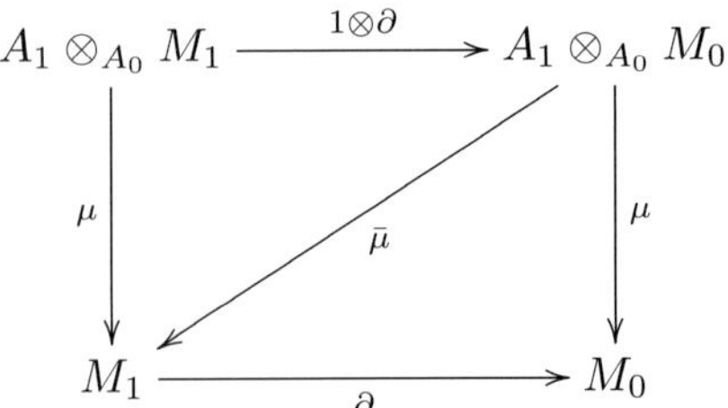

is given where $\mu(a \otimes x) = (\partial a) \cdot x$ for $a \in A_1$ and $x \in M_1$ or $x \in M_0$.

According to the equivalent concepts in (5.1.8) we also see that the concepts

$$\begin{aligned} \text{module over a track algebra} &= \text{module over a pair algebra} \\ &= \text{module over a crossed algebra} \end{aligned} \tag{5.1.12}$$

are equivalent. In fact, in addition to (5.1.7) we get:

5.1.13 Proposition. *Let A be a track algebra corresponding to the pair algebra A' and to the crossed algebra A''. Then the categories of (left) modules over A, or A', or A'' are equivalent to each other.*

The proof uses similar arguments as in (5.1.7).

5.2 The Γ-pseudo functor

Let $(\llbracket A \rrbracket, \Gamma)$ be a Γ-track algebra and let $i_E : E \subset A$ be a graded set of generators of the graded algebra $A = \llbracket A \rrbracket_{\simeq}$. We can choose a lift s'' of the inclusion i_E as in the following commutative diagram.

(5.2.1)

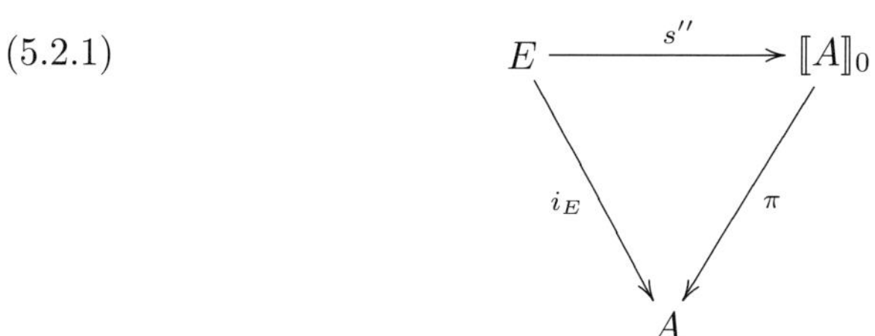

Let $\mathrm{Mon}(E)$ be the free monoid generated by E and $T_{\mathbb{G}}(E) = \mathbb{G}\,\mathrm{Mon}(E)$ be the free $\mathbb{G}$-module generated by $\mathrm{Mon}(E)$. Then

$$T_{\mathbb{G}}(E) = \bigoplus_{n \geq 0} (\mathbb{G}E)^{\otimes n} \tag{1}$$

is the $\mathbb{G}$-tensor algebra generated by $\mathbb{G}E$ where $\mathbb{G}E$ is the free $\mathbb{G}$-module generated by E. The function s'' above yields a commutative diagram.

$$\begin{array}{ccc} T_{\mathbb{G}}(E) = \mathbb{G}\,\mathrm{Mon}(E) & \xrightarrow{\ s\ } & \llbracket A \rrbracket_0 \\ \uparrow & & \| \\ \mathrm{Mon}(E) & \xrightarrow{\ s'\ } & \llbracket A \rrbracket_0 \\ \uparrow & & \| \\ E & \xrightarrow[\ s''\ ]{} & \llbracket A \rrbracket_0 \end{array} \tag{2}$$

Here the vertical arrows are the inclusions. Since $\llbracket A \rrbracket_0$ is a graded monoid we obtain the unique monoid homomorphism s' of degree 0 on $\mathrm{Mon}(E)$ extending s''. Since $\llbracket A \rrbracket_0$ is a graded $\mathbb{G}$-module we obtain the unique $\mathbb{G}$-*linear map* s extending s'. The map s, however, is not multiplicative for the multiplication in the tensor algebra $T_{\mathbb{G}}(E)$, that is, for $a, b \in T_{\mathbb{G}}(E)$ the element $s(a \cdot b)$ does not coincide with the element $s(a) \cdot s(b)$. If $a, b \in \mathrm{Mon}(E)$ we have $s(a \cdot b) = s'(a \cdot b) = (s'a) \cdot (s'b) = (sa) \cdot (sb)$. Moreover we get

$$s(a) \cdot s(b) = s(a \cdot b) \tag{3}$$

for $a \in T_{\mathbb{G}}(E)$ and $b \in \mathrm{Mon}(E)$ since multiplication in $\llbracket A \rrbracket_0$ is left linear.

5.2.2 Definition. We write $a = \sum_{i=1}^{n(a)} n_a^i a_i$ with $n_a^i \in \mathbb{G}$, $n_a^i \neq 0$, and $a_i \in \mathrm{Mon}(E)$ pairwise distinct for $i = 1, \dots, n(a)$. Let

$$(1) \qquad \begin{cases} \varphi_a & : \quad V_a = \mathbb{G}^{n(a)} \longrightarrow \mathbb{G}, \\ \varphi_a & = \quad \sum_{i=1}^{n(a)} n_a^i p_i \end{cases}$$

be given by a and let $\hat{a} = (a_1, \dots, a_{n(a)})$ be the tuple associated to a. Then we have $s\hat{a} = (sa_1, \dots, sa_{n(a)}) \in [\![A]\!]_0 \otimes V_a$ and the equation

$$(2) \qquad s(a) = \sum_{i=1}^{n(a)} n_a^i s(a_i) = (1 \otimes \varphi_a)(s\hat{a})$$

holds. Hence we get for $x \in T_{\mathbb{G}}(E)$ the track

$$(3) \qquad \Gamma(x, a) = \Gamma(\varphi_a)_{sx}^{s\hat{a}} : (sx) \cdot (sa) \Rightarrow s(x \cdot a)$$

since $(sx) \cdot (sa) = (sx) \cdot (1 \otimes \varphi_a)(s\hat{a})$ and $s(x \cdot a) = (1 \otimes \varphi_a)(sx) \cdot (s\hat{a})$. Here we use (5.2.1)(3). Now (3) is the trivial track

$$(4) \qquad \Gamma(x, a) = 0_{s(x \cdot a)}^{\square} \text{ if } a \in \mathrm{Mon}(E).$$

5.2.3 Theorem. *For $a, b, c \in T_{\mathbb{G}}(E)$ we have the formula*

$$\Gamma(ab, c)\square\Gamma(a, b)(sc) = \Gamma(a, bc)\square(sa)\Gamma(b, c).$$

Both sides are tracks $(sa)(sb)(sc) \Rightarrow s(abc)$.

The formula in Theorem (5.2.3) shows that pasting of Γ-tracks in the following diagram yields the identity track. This exactly is the property of a *pseudo functor* $(s, \Gamma) : T_{\mathbb{G}}(E) \to [\![A]\!]$, compare for example Fantham-Moore [FM].

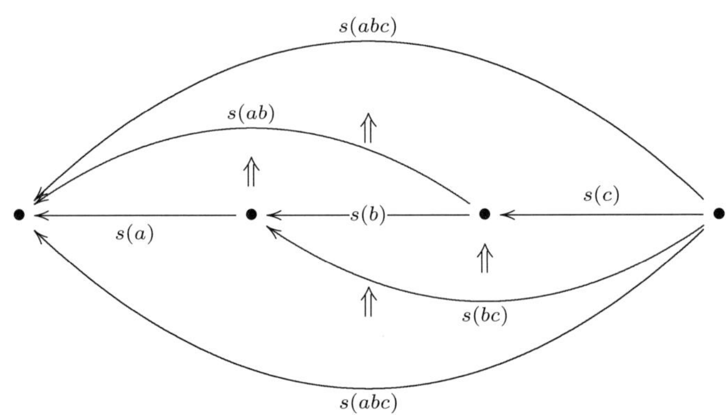

Proof. Let

$$\Gamma(a,b,c) = \Gamma(ab,c)\square\Gamma(a,b)(sc) \tag{1}$$

and let

$$\Gamma'(a,b,c) = \Gamma(a,bc)\square(sa)\Gamma(b,c). \tag{2}$$

We have to show (1) = (2) for $a,b,c \in T_{\mathbb{G}}(E)$.

We now consider the case that $a,c \in T_{\mathbb{G}}(E)$ and $b \in \mathrm{Mon}(E)$. Then we get $\Gamma(a,b) = 0^{\square}$ so that

$$\begin{aligned} \Gamma(a,b,c) &= \Gamma(\varphi_c)^{s\hat{c}}_{s(ab)} \\ &= \Gamma(\varphi_c)^{s\hat{c}}_{s(a)(sb)}, \qquad \text{since } b \in \mathrm{Mon}(E) \\ &= \Gamma(\varphi_c)^{(sb)(s\hat{c})}_{sa}\square(sa)\Gamma(\varphi_c)^{s\hat{c}}_{sb}, \qquad \text{see (4.3.1)(11)}, \end{aligned} \tag{3}$$

$$\Gamma'(a,b,c) = \Gamma(\varphi_{bc})^{s\widehat{(bc)}}_{sa}\square(sa)\Gamma(\varphi_c)^{s\hat{c}}_{sb}. \tag{4}$$

Here we have $\varphi_{bc} = \varphi_c$ and $s\widehat{(bc)} = (sb)(s\hat{c})$ since $b \in \mathrm{Mon}(E)$. This shows that (1) = (2) if $a,c \in T_{\mathbb{G}}(E)$ and $b \in \mathrm{Mon}(E)$.

Now we consider for fixed $a,c \in T_{\mathbb{G}}(E)$ the functions d and d' with

$$\begin{aligned} d(x) &= \Gamma(a,x,c), \\ d'(x) &= \Gamma'(a,x,c). \end{aligned} \tag{5}$$

We know $d(x) = d'(x)$ for $x \in \mathrm{Mon}(E)$. Assume now that for $x \in T_{\mathbb{G}}(E)$ we have $d(x) = d'(x)$, then we show for $y \in \mathrm{Mon}(E)$ that

$$d(x+y) = d'(x+y).$$

This proves that $d = d'$. In fact, we only need to consider the following two cases with $n(x) \geq 1$, see (5.2.2).

$$y \neq x_i \text{ for all } i = 1,\dots,n(x). \tag{I}$$

$$y = x_1 \text{ and } n_1^x \neq -1. \tag{II}$$

In case (I) we have

$$\varphi_{x+y} = (\varphi_x, 1) : V_{x+y} = V_x \oplus \mathbb{G} \longrightarrow \mathbb{G}, \tag{6}$$

and in case (II) we have

$$\varphi_{x+y} = \varphi_x + p_1^x : V_{x+y} = V_x \longrightarrow \mathbb{G}. \tag{7}$$

By definition of d we get

(8) $$d(x+y) = \Gamma(\varphi_c)^{s\hat{c}}_{s(ax+ay)}$$

(9) $$\square(\varphi_{x+y})^{s(x+y)\hat{}}_{sa}(sc),$$

(10) $$d(x) = \Gamma(\varphi_c)^{s\hat{c}}_{s(ax)}\square\Gamma(\varphi_x)^{s\hat{x}}_{sa}(sc),$$

(11) $$(8) = \Gamma(\varphi_c)^{s\hat{c}}_{s(ax)} + \Gamma(\varphi_c)^{s\hat{c}}_{s(ay)}, \quad (4.3.1)(13).$$

Now we compute (9). We get in case (I) and also in case (II) the formula

(12) $$(9) = \Gamma(\varphi_x)^{s\hat{x}}_{sa}(sc) + (sa)(sy)(sc)$$

(13) $$\square\Gamma^{sx,sy}_{sa}(sc).$$

We prove (9) = (12)□(13) first in case (I). Then we get:

(14) $$(9) = \Gamma(+_{\mathbb{G}}(\varphi_x \oplus 1))^{s\hat{x},sy}_{sa}(sc) \quad \text{case (I)},$$

(15) $$(9) = (1 \otimes +_{\mathbb{G}})\Gamma(\varphi_x \oplus 1)^{s\hat{x},sy}_{sa}(sc), \quad (4.3.1)(12),$$

(16) $$\square\Gamma^{(1\otimes(\varphi_x\oplus 1))(s\hat{x},sy)}_{sa}(sc),$$

(17) $$(15) = (\Gamma(\varphi_x)^{s\hat{x}}_{sa} + \Gamma(1)^{sy}_{sa})(sc), \quad (4.3.1)(13).$$

Here (15) = (12) by (4.3.1)(3) and (16) = (13) by (4.3.1)(10). This completes the proof of (9) = (12)□(13) in case (I). Now we prove this in case (II). Then we have

(a) $$(9) = \Gamma(\varphi_x + p_1^x)^{s\hat{x}}_{sa}(sc),$$

(b) $$= (\Gamma(\varphi_x)^{s\hat{x}}_{sa} + \Gamma(p_1^x)^{s\hat{x}}_{sa})(sc)$$

(c) $$\square\Gamma(+_{\mathbb{G}})^{(1\otimes\varphi_x)s\hat{x},(1\otimes p_1^x)s\hat{x}}_{sa}(sc).$$

Here $\Gamma(p_1^x)$ is the trivial track of $(sa)(sy)$ since $x_1 = y$. This shows $(b) = (12)$. Moreover we have $(c) = (13)$. This completes the formula $(9) = (12)\square(13)$ in case (II).

Since (9) = (12)□(13) in case (I) and case (II) we get

$$d(x+y) = (8)\square(9) = (11)\square(12)\square(13).$$

Here we have by (10)

(18) $$(12) = \{\Gamma(\varphi_c)^{s\hat{c}}_{s(ax)}\}^{\mathrm{op}} + s(ay)(sc)$$

(19) $$\square\{d(x) + s(ay)(sc)\},$$

so that $d(x+y) = (11)\square(18)\square(19)\square(13)$, that is:

(20) $$d(x+y) = \{\Gamma(\varphi_c)^{s\hat{c}}_{s(ax)} + \Gamma(\varphi_c)^{s\hat{c}}_{s(ay)}\}$$

(21) $$\square\{\Gamma(\varphi_c)^{s\hat{c}}_{s(ax)}\}^{\mathrm{op}} + s(ay)sc$$

(22) $$\square d(x) + s(ay)(sc)$$

(23) $$\square\Gamma^{sx,sy}_{sa}(sc).$$

Here we get

$$(24)\qquad (20)\square(21) = s(axc) + \Gamma(\varphi_c)^{s\hat{c}}_{s(ay)},$$

$$(25)\qquad (20)\square(21) = s(axc) + \{\Gamma(\varphi_c)^{(sy)(s\hat{c})}_{sa}\}$$

$$(26)\qquad \square\{s(axc) + (sa)\Gamma(\varphi_c)^{s\hat{c}}_{sy}\}, \quad (4.3.1)(11).$$

Now we use the assumption

$$(27)\qquad d(x) = d'(x) = \Gamma(\varphi_{xc})^{s\widehat{(xc)}}_{sa}\square(sa)\Gamma(\varphi_c)^{s\hat{c}}_{sx}.$$

Hence we get

$$(28)\qquad d(x+y) = s(axc) + \Gamma(\varphi_{yc})^{s\widehat{(yc)}}_{sa}, \text{ since } \varphi_{yc} = \varphi_c,$$

$$(29)\qquad \square s(axc) + (sa)\Gamma(\varphi_c)^{s\hat{c}}_{sy}$$

$$(30)\qquad \square\Gamma(\varphi_{xc})^{s\widehat{(xc)}}_{sa} + s(ay)(sc)$$

$$(31)\qquad \square(sa)\Gamma(\varphi_c)^{s\hat{c}}_{sx} + s(ay)(sc)$$

$$(32)\qquad \square\Gamma^{(sx)(sc),(sy)(sc)}_{sa}.$$

Equivalently we get

$$(33)\qquad d(x+y) = \{\Gamma(\varphi_{xc})^{s\widehat{(xc)}}_{sa} + \Gamma(\varphi_{yc})^{s\widehat{(yc)}}_{sa}\}$$

$$(34)\qquad \square\{(sa)\Gamma(\varphi_c)^{s\hat{c}}_{sx} + (sa)\Gamma(\varphi_c)^{s\hat{c}}_{sy}\}$$

$$(35)\qquad \square\Gamma^{(sx)(sc),(sy)(sc)}_{sa}.$$

On the other hand we get by (2)

$$(36)\qquad d'(x+y) = \Gamma(\varphi_{xc+yc})^{s\widehat{(xc+yc)}}_{sa}$$

$$(37)\qquad \square(sa)\Gamma(\varphi_c)^{s\hat{c}}_{s(x+y)},$$

$$(38)\qquad (37) = (sa)(\Gamma(\varphi_c)^{s\hat{c}}_{sx} + \Gamma(\varphi_c)^{s\hat{c}}_{sy}), \quad (4.3.1)(13),$$

$$(39)\qquad (37) = \{\Gamma^{s(xc),s(yc)}_{sa}\}^{\mathrm{op}}, \quad (4.3.1)(5),$$

$$(40)\qquad \square\{(sa)\Gamma(\varphi_c)^{s\hat{c}}_{sx} + (sa)\Gamma(\varphi_c)^{s\hat{c}}_{sy}\}$$

$$(41)\qquad \square\Gamma^{(sx)(sc),(sy)(sc)}_{sa}.$$

Here (40) = (34) and (41) = (35). Hence $d(x+y) = d'(x+y)$ follows from the equation (33) = (36)□(39) or equivalently (36) = (33)□(39)^{op}, that is

$$(42)\qquad \Gamma(\varphi_{xc+yc})^{s\widehat{(xc+yc)}}_{sx} = \{\Gamma(\varphi_{xc})^{s\widehat{(xc)}}_{sa} + \Gamma(\varphi_{yc})^{s\widehat{(yc)}}_{sa}\}\square\Gamma^{s(xc),s(yc)}_{sa}.$$

This formula is a consequence of the following lemma which we also need in the next section. Hence the proof of (5.2.3) is complete. □

5.2.4 Lemma. *For $a, x, y \in T_{\mathbb{G}}(E)$ we have*

$$\Gamma(\varphi_{x+y})_{sa}^{s(x+y)\hat{}} = (\Gamma(\varphi_x)_{sa}^{s\hat{x}} + \Gamma(\varphi_y)_{sa}^{s\hat{y}})\Box\Gamma_{sa}^{sx,sy}.$$

Proof. We first prove the formula for $y \in \mathrm{Mon}(E)$. Then we have case (I) and case (II) as in the proof of (5.2.3).

In case (I) we know that $y \neq x_i$ for all i. This shows $\varphi_{x+y} = +_{\mathbb{G}}(\varphi_x \oplus \varphi_y)$ and also $s(x+y)\hat{} = (s\hat{x}, s\hat{y})$. Therefore (5.2.4) is a consequence of (4.3.1)(12), that is:

$$\begin{aligned}
&(1) & \Gamma(\varphi_{x+y})_{sa}^{s(x+y)\hat{}} &= \Gamma(+_{\mathbb{G}}(\varphi_x \oplus \varphi_y))_{sa}^{s\hat{x},s\hat{y}} \\
&(2) & &= (1 \otimes +_{\mathbb{G}})(\Gamma(\varphi_x)_{sa}^{s\hat{x}}, \Gamma(\varphi_y)_{sa}^{s\hat{y}}) \\
&(3) & &\Box\Gamma_{sa}^{(1\otimes(\varphi_x\oplus\varphi_y))(s\hat{x},s\hat{y})}.
\end{aligned}$$

This yields the proof of the lemma for $y \in \mathrm{Mon}(E)$ and case (I).

In case (II) we have $x_1 = y$ and $n = n_1^x \neq -1$. Let $z = x - nx_1$. Then we have $x = z + ny$ and $x + y = z + my$ where $m = n + 1 \neq 0$. Moreover $\hat{z}$ does not contain y so that

$$\begin{aligned}
&(4) & \varphi_x &= +_{\mathbb{G}}(\varphi_z \oplus \varphi_{ny}), \\
&(5) & \varphi_{x+y} &= +_{\mathbb{G}}(\varphi_z \oplus \varphi_{my}).
\end{aligned}$$

Here we have $\varphi_{ny} = n\varphi_y$, $\varphi_{my} = m\varphi_y$. Moreover we get

$$(6) \qquad s(\hat{x}) = (s(\hat{z}), s(\hat{y})) = s(x+y)\hat{}.$$

Using (4.3.1)(12) we get:

$$\begin{aligned}
&(7) & \Gamma(\varphi_x)_{sa}^{s(x)\hat{}} &= \Gamma(+_{\mathbb{G}}(\varphi_z \oplus \varphi_{ny}))_{sa}^{s(z)\hat{},s(y)\hat{}} \\
&(8) & &= \{\Gamma(\varphi_{zc})_{sa}^{s(z)\hat{}} + \Gamma(n\varphi_y)_{sa}^{s(y)\hat{}}\} \\
&(9) & &\Box\Gamma_{sa}^{s(z),ns(y)}, \\
&(10) & \Gamma(\varphi_{x+y})_{sa}^{s(x+y)\hat{}} &= \Gamma(+_{\mathbb{G}}(\varphi_z \oplus \varphi_{my}))_{sa}^{s(z)\hat{},s(y)\hat{}} \\
&(11) & &= \{\Gamma(\varphi_z)_{sa}^{s(z)\hat{}} + \Gamma(m\varphi_{yc})_{sa}^{s(y)\hat{}}\} \\
&(12) & &\Box\Gamma_{sa}^{s(z),ms(y)}.
\end{aligned}$$

Here we have $m = 1 + n$ so that by (4.3.1)(14)

$$\begin{aligned}
&(13) & \Gamma(m\varphi_y)_{sa}^{s(y)\hat{}} &= \{\Gamma(\varphi_y)_{sa}^{s(y)\hat{}} + \Gamma(n\varphi_y)_{sa}^{s(y)\hat{}}\} \\
&(14) & &\Box\Gamma_{sa}^{s(y),ns(y)}.
\end{aligned}$$

Hence we get

$$\begin{aligned}
&(15) & (10) = &\{\Gamma(\varphi_z)_{sa}^{s(z)\hat{}} + \Gamma(n\varphi_y)_{sa}^{s(y)\hat{}} + \Gamma(\varphi_y)_{sa}^{s(y)\hat{}}\} \\
&(16) & &\Box\{s(a)s(z) + \Gamma_{sa}^{s(y),ns(y)}\} \\
&(17) & &\Box\Gamma_{sa}^{s(z),(n+1)s(y)}.
\end{aligned}$$

By (4.2.5)(5) we see

$$(17) = \{s(a)s(z) + \Gamma_{sa}^{ns(y),s(y)}\}^{\mathrm{op}} \tag{18}$$

$$\square\{\Gamma_{sa}^{s(z),ns(y)} + s(a)s(y)\} \tag{19}$$

$$\square\Gamma_{sa}^{s(z)+ns(y),s(y)}. \tag{20}$$

Here (20) is part of (5.2.4). Moreover $(18)^{\mathrm{op}} = (16)$ by (4.2.5)(3). This shows $(10) = (15)\square(19)\square(20)$, that is:

$$(10) = \{\Gamma(\varphi_z)_{sa}^{\widehat{s(z)}} + \Gamma(n\varphi_y)_{sa}^{\widehat{s(y)}} + \Gamma(\varphi_y)_{sa}^{\widehat{s(y)}}\} \tag{21}$$

$$\square\{\Gamma_{sa}^{s(z),ns(y)} + (sa)s(y)\} \tag{22}$$

$$\square\Gamma_{sa}^{s(x),s(y)}. \tag{23}$$

Now $(7) = (8)\square(9)$ and (8) is part of (21) and (9) is part of (22). This shows

$$(10) = \{(7) + \Gamma(\varphi_y)_{sa}^{\widehat{s(y)}}\}\square(23)$$

and this proves the lemma in case (II), $y \in \mathrm{Mon}(E)$. Now the proof of the lemma is complete for $y \in \mathrm{Mon}(E)$.

Now assume the formula in (5.2.4) holds for (x, y) with $x, y \in T_{\mathbb{G}}(E)$ and let $v \in \mathrm{Mon}(E)$. Then we show that the formula holds for $(x, y + v)$.

In fact, since $v \in \mathrm{Mon}(E)$ we have shown that for $w = x + y$ the formula holds for (w, v) so that

$$\Gamma(\varphi_{w+v})_{sa}^{\widehat{s(w+v)}} = \{\Gamma(\varphi_w)_{sa}^{s\hat{w}} + \Gamma(\varphi_v)_{sa}^{s\hat{v}}\}\square\Gamma_{sa}^{s(w),sv}. \tag{10}$$

Here the assumption on (x, y) yields a formula for $\Gamma(\varphi_w)_{sa}^{s\hat{w}}$ so that we get

$$\Gamma(\varphi_{w+v})_{sa}^{\widehat{s(w+v)}} = \{\Gamma(\varphi_x)_{sa}^{s\hat{x}} + \overbrace{\Gamma(\varphi_y)_{sa}^{s\hat{y}} + \Gamma(\varphi_v)_{sa}^{s\hat{v}}}^{A}\} \tag{11}$$

$$\square\{\Gamma_{sa}^{sx,sy} + (sa)(sv)\} \tag{12}$$

$$\square\Gamma_{sa}^{s(x+y),v}. \tag{13}$$

Here $(12)\square(13)$ coincide by (4.2.5)(5) with $(14)\square(15)$,

$$\{(sa)(sx) + \Gamma_{sa}^{sy,sv}\} \tag{14}$$

$$\square\Gamma_{sa}^{sy,sy+sv}. \tag{15}$$

Since Lemma (5.2.4) holds for (y, v) we get $A\square(14) = \Gamma(\varphi_{y+v})_{sa}^{\widehat{s(y+v)}}$ and this yields the formula:

$$\begin{aligned}\Gamma(\varphi_{w+v})_{sa}^{\widehat{s(w+v)}} &= \Gamma(\varphi_x)_{sa}^{s\hat{x}} + \Gamma(\varphi_{y+v})_{sa}^{\widehat{s(y+v)}} \\ &\quad \square\Gamma_{sa}^{sy,s(y+v)}.\end{aligned}$$

Hence the lemma also holds for $(x, y + v)$ and the proof of the lemma is complete.

□

5.3 The strictification of a Γ-track algebra

Let p be a prime and $\mathbb{F} = \mathbb{Z}/p$ and $\mathbb{G} = \mathbb{Z}/p^2$. We show that each Γ-track algebra $[\![A]\!]$ over $\mathbb{F}$ as considered in (4.3.1) is weakly equivalent to a track algebra $[\![A, E, s]\!]$ over $\mathbb{G}$ termed the strictification of $[\![A]\!]$. Here it is crucial that we alter the ground ring from $\mathbb{F}$ to $\mathbb{G}$. In fact, we have seen in (4.6) that in general there is no track algebra over $\mathbb{F}$ which is weakly equivalent to $[\![A]\!]$.

Let $([\![A]\!], \Gamma)$ be a Γ-track algebra as in (4.3.1). We choose a graded set E of generators of the graded algebra A. Since $A_0 = \mathbb{F}$ we choose only generators in degree ≥ 1. We choose a lift s'' as in the diagram

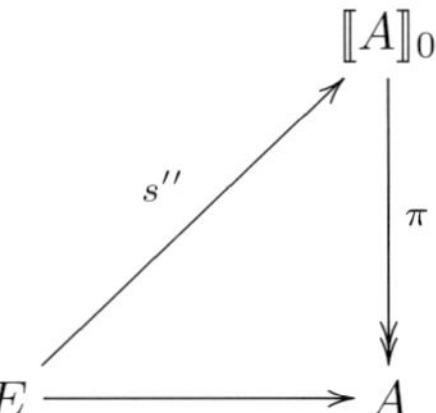

so that $s''(e)$ represents the homotopy class $e \in E \subset A$. Then the $\mathbb{G}$-linear map

$$s : T_{\mathbb{G}}(E) \longrightarrow [\![A]\!]_0 \tag{5.3.1}$$

is defined as in (5.2.1) and we have the Γ-tracks $\Gamma(a, b) : s(a) \cdot s(b) \Rightarrow s(a \cdot b)$ in (5.2.2) so that

$$(s, \Gamma) : T_{\mathbb{G}}(E) \longrightarrow [\![A]\!]$$

is a pseudo functor as proved in (5.2.3).

5.3.2 Definition. Using Γ-tracks we define the Γ-product $H \bullet G$ of tracks as follows. Let $f, g, x, y \in T_{\mathbb{G}}(E)$ and let

$$\begin{aligned} H :\ sf &\Longrightarrow sg, \\ G :\ sx &\Longrightarrow sy \end{aligned}$$

be tracks in $[\![A]\!]$. Then the Γ-*product* is the track

$$\begin{aligned} H \bullet G &: s(f \cdot x) \Longrightarrow s(g \cdot y), \\ H \bullet G &= \Gamma(g, y)\square(H \cdot G)\square\Gamma(f, x)^{\mathrm{op}} \end{aligned}$$

where $H \cdot G$ is defined by multiplication in $[\![A]\!]$, see (4.3.1)(1). Hence the Γ-product corresponds to pasting in the diagram.

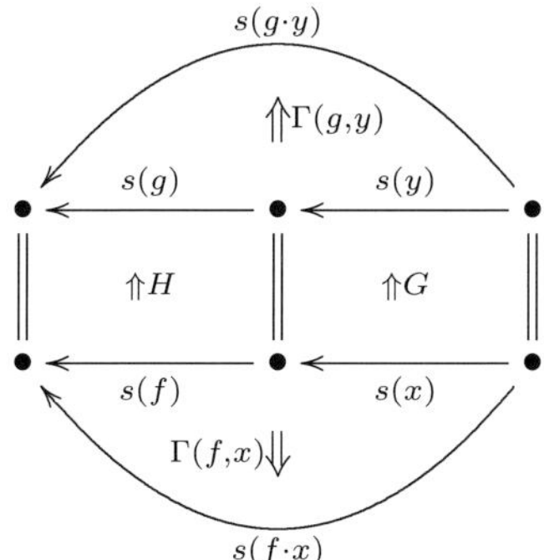

The pseudo functor property of (s,Γ) shows immediately:

5.3.3 Proposition. *The* Γ*-product is associative, that is*

$$(H \bullet G) \bullet F = H \bullet (G \bullet F),$$

and the unit $1 \in T_{\mathbb{G}}(E)$ *is a unit for the* Γ*-product, that is* $1 \bullet H = H \bullet 1 = H$.

Here we use the notation

$$\begin{aligned} f \bullet G &= 0^{\square}_{sf} \bullet G, \\ H \bullet x &= H \bullet 0^{\square}_{sx} \end{aligned}$$

where $0^{\square}_{sf} : sf \Rightarrow sf$ and $0^{\square}_{sx} : sx \Rightarrow sx$ are the identity tracks. One readily checks the formula

$$\begin{aligned} H \bullet G &= (g \bullet G)\square(H \bullet x), \\ &= (H \bullet y)\square(f \bullet G). \end{aligned} \tag{5.3.4}$$

5.3.5 Theorem. *The* Γ*-product is bilinear.*

Proof. We have for $H' : s(f') \Rightarrow s(g')$ the equations

$$(H + H') \bullet G = \Gamma(g + g', y)\square((H + H') \cdot G)\square\Gamma(f + f', x)^{\mathrm{op}}. \tag{1}$$

Here we have by (4.3.1)(13) the equation $\Gamma(g + g', y) = \Gamma(g, y) + \Gamma(g', y)$ so that by (4.3.1)(3) we get

$$\begin{aligned} (H + H') \bullet G = &\{\Gamma(g, y) + \Gamma(g', y)\}, \\ &\square\{H \cdot G + H' \cdot G\}, \\ &\square\{\Gamma(f, x) + \Gamma(f', x)\}^{\mathrm{op}}. \end{aligned} \tag{2}$$

This shows $(H + H') \bullet G = (H \bullet G) + (H' \bullet G)$.

Next we consider $G' : s(x') \Rightarrow s(y')$ and

$$H \bullet (G + G') = \Gamma(g, y + y')\Box(H \cdot (G + G'))\Box\Gamma(f, x + x')^{\mathrm{op}}. \tag{3}$$

Here we can apply (4.3.1)(5) so that

$$H \cdot (G + G') = (\Gamma_{sy}^{sy+sy'})^{\mathrm{op}}\Box(H \cdot G + H \cdot G')\Box\Gamma_{sf}^{sx,sx'}. \tag{4}$$

On the other hand

$$H \bullet G = \Gamma(g, y)\Box(H \cdot G)\Box\Gamma(f, x)^{\mathrm{op}}, \tag{5}$$

$$H \bullet G' = \Gamma(g, y')\Box(H \cdot G')\Box\Gamma(f, x')^{\mathrm{op}}. \tag{6}$$

Hence we have

$$\begin{aligned} H \bullet G + H \bullet G' = {} & \{\Gamma(g, y) + \Gamma(g, y')\}, \\ & \Box(H \cdot G + H \cdot G'), \\ & \Box\{\Gamma(f, x) + \Gamma(f, x')\}^{\mathrm{op}}. \end{aligned} \tag{7}$$

Therefore $H \bullet (G + G') = H \bullet G + H \bullet G'$ is a consequence of

$$\Gamma(g, y + y')\Box\{\Gamma_{sg}^{sy,sy'}\}^{\mathrm{op}} = \Gamma(g, y) + \Gamma(g, y'), \tag{8}$$

$$\Gamma(\varphi_{y+y'})_{sg}^{s(y+y')^{\frown}} = (\Gamma(\varphi_y)_{sg}^{s(y)^{\frown}} + \Gamma(\varphi_{y'})_{sg}^{s(y')^{\frown}})\Box\Gamma_{sg}^{sy,sy'}. \tag{9}$$

But this formula is proved in (5.2.4). □

5.3.6 Definition. Let $(\llbracket A \rrbracket, \Gamma)$ be a Γ-track algebra and let E be a set of generators of the algebra A and let $s : T_{\mathbb{G}}(E) \to \llbracket A \rrbracket_0$ be defined as in (5.2.1), (5.3.1). Then we obtain the *track algebra* $\llbracket A, E, s \rrbracket$ over A as follows, see (5.1.4). Let

$$\llbracket A, E, s \rrbracket_0 = T_{\mathbb{G}}(E). \tag{1}$$

For $x, y \in T_{\mathbb{G}}(E)$ a track $G : x \Rightarrow y$ in $\llbracket A, E, s \rrbracket$ is a triple

$$G = (y, \hat{G}, x) : x \Longrightarrow y \tag{2}$$

where $\hat{G} : sx \Rightarrow sy$ is a track in $\llbracket A \rrbracket$. *Composition* of such tracks is defined by composition in $\llbracket A \rrbracket$

$$(y, \hat{G}, x)\Box(x, \hat{H}, z) = (y, \hat{G}\Box\hat{H}, z) \tag{3}$$

and $(y, 0_{sy}^{\Box}, y)$ is the trivial track of y. The *product* of tracks in $\llbracket A, E, s \rrbracket$ is defined by the Γ-product in $\llbracket A \rrbracket$, that is

$$(y, \hat{G}, x) \cdot (f, \hat{H}, g) = (y \cdot f, \hat{G} \bullet \hat{H}, x \cdot g). \tag{4}$$

By (5.3.2) this product is associative with unit $(1, 0^{\square}_{s1}, 1) = 1$. The set $[\![A, E, s]\!]_1$ of tracks is also given by the following pull back diagram.

(5)
$$\begin{array}{ccc} [\![A, E, s]\!]_1 & \xrightarrow{\bar{s}} & [\![A]\!]_1 \\ \downarrow{\scriptstyle(\partial_1,\partial_2)} & & \downarrow{\scriptstyle(\partial_1,\partial_0)} \\ T_{\mathbb{G}}(E) \oplus T_{\mathbb{G}}(E) & \xrightarrow{s \oplus s} & [\![A]\!]_0 \oplus [\![A]\!]_0 \end{array}$$

Since $s, \partial_0, \partial_1$ are $\mathbb{G}$-linear we see that $[\![A, E, s]\!]_1$ is a graded $\mathbb{G}$-module. By (5.3.5) the product (4) is bilinear. This shows that $[\![A, E, s]\!]$ is a well-defined track algebra over $\mathbb{G}$ termed the *strictification* of the Γ-track algebra $([\![A]\!], \Gamma)$.

We have seen in (5.1.5) that the track algebra $[\![A, E, s]\!]$ can be equivalently described as a pair algebra or as a crossed algebra over $\mathbb{G}$. According to (4.3.5) we have the (graded) linear track extension

(5.3.7)
$$D \longrightarrow [\![A]\!]_1 \rightrightarrows [\![A]\!]_0 \longrightarrow A$$

of a Γ-track algebra $([\![A]\!], \Gamma)$. According to the definition of D and A we see that

$$\begin{aligned} A &= \pi_0([\![A, E, s]\!]), \\ D &= \pi_1([\![A, E, s]\!]) \end{aligned}$$

where we use the crossed algebra associated to $[\![A, E, s]\!]$, see (5.1.6).

5.3.8 Theorem. *The graded linear track extension*

$$D \longrightarrow [\![A, E, s]\!]_1 \rightrightarrows [\![A, E, s]\!]_0 \longrightarrow A$$

is weakly equivalent to (5.1.6).

Proof. A cocycle for $[\![A, E, s]\!]$ is easily seen to coincide with the corresponding cocycle for $[\![A]\!]$, see (5.3.7). Hence the result follows from (3.6.9). □

The strictification $[\![A, E, s]\!]$ depends on the choice of generators E and the choice of s'' in (5.2.1). Let s''_0 be a further lift as in (5.2.1) with $\pi s''_0 = i_E$. Then there exists a track

$$S'' : s'' \Longrightarrow s''_0,$$

that is for $e \in E$ we have $S''_e : s''(e) \Rightarrow s''_0(e)$ in $[\![A]\!]$.

5.3.9 Theorem. *The track $S'' : s'' \Rightarrow s''_0$ induces an isomorphism of track algebras*

$$\bar{S} : [\![A, E, s]\!] \cong [\![A, E, s_0]\!]$$

for which $\bar{S}_0$ is the identity of $T_{\mathbb{G}}(E)$.

Proof. Let s_0 be defined by s_0'' in the same way as in (5.2.1)(2). Then we obtain by S'' in the same way the track $S : s \Rightarrow s_0$ in $[\![A]\!]$. We define the isomorphism $\bar{S}_1$ on $(y, \hat{G}, x) \in [\![A, E, s]\!]_1$ by

$$\bar{S}(y, \hat{G}, x) = (y, S_y \square \hat{G} \square S_x^{\mathrm{op}}, x).$$

One readily checks that $\bar{S}$ is a well-defined isomorphism of track algebras. Here we need (4.3.1)(16). □

5.4 The strictification of a Γ-track module

In a similar way as in (5.3) we can strictify a *module* over a Γ-track algebra $[\![A]\!]$. Let E and s be given as in (5.2.1).

Let $[\![M]\!]$ be a module over the Γ-track algebra $([\![A]\!], \Gamma)$, see (4.3.7). Then $M = [\![M]\!]_{\simeq}$ is a left A-module and we can choose a set E_M of generators of degree ≥ 1 of the A-module M. Moreover we choose a lift s_M'',

(5.4.1)
$$\begin{array}{ccc} & & [\![M]\!]_0 \\ & \overset{s_M''}{\nearrow} & \downarrow \pi \\ E_M & \longrightarrow & M \end{array}$$

of the inclusion $E_M \subset M$ so that $s_M''(e)$ represents the homotopy class $e \in E_M \subset M$. Let $\mathbb{G}E_M$ be the free $\mathbb{G}$-module generated by E_M. Then

(1)
$$T_{\mathbb{G}}(E) \otimes \mathbb{G}E_M = \mathbb{G}(\mathrm{Mon}(E) \times E_M)$$

is the free $T_{\mathbb{G}}(E)$-module generated by E_M, see (5.2.1)(1). Similarly as in (5.2.1)(2) we get the following commutative diagram.

(2)
$$\begin{array}{ccc} \mathbb{G}(\mathrm{Mon}(E) \times E_M) & \xrightarrow{s_M} & [\![M]\!]_0 \\ \cup & & \| \\ \mathrm{Mon}(E) \times E_M & \xrightarrow{s_M'} & [\![M]\!]_0 \\ \cup & & \| \\ E_M & \xrightarrow{s_M''} & [\![M]\!]_0 \end{array}$$

Here s_M' is the s'-equivariant map extending s_M'' and s_M is the $\mathbb{G}$-linear map extending s_M'. Hence we get the $\mathbb{G}$-linear map

(3)
$$s_M : T_{\mathbb{G}}(E) \otimes \mathbb{G}E_M \longrightarrow [\![M]\!]_0.$$

We can define for $b \in T_{\mathbb{G}}(E)$ and $y \in T_{\mathbb{G}}(E) \otimes \mathbb{G}E_M$ the Γ-*track*

(5.4.2)
$$\Gamma_M(b, y) : s(b) \cdot s_M(y) \Longrightarrow s_M(b \cdot y)$$

in the same way as in (5.2.2). Moreover (5.2.3) holds accordingly so that the Γ-action $H \bullet G$ for $H : sf \Rightarrow sg$ in $[\![A]\!]$ and $G : s_M(x) \Rightarrow s_M(y)$ is defined by

$$H \bullet G = \Gamma_M(g \cdot y)\square(H \cdot G)\square\Gamma(f, x)^{\mathrm{op}} \tag{5.4.3}$$

as in (5.3.2). This action satisfies (5.3.3) accordingly and also satisfies $\mathbb{G}$-bilinearity as in (5.3.5). The corresponding proofs are easily generalized to the case of actions. Hence we obtain the following definition corresponding to (5.3.6).

5.4.4 Definition. Let $([\![A]\!], \Gamma)$ be a Γ-track algebra with strictification $[\![A, E, s]\!]$. Let M be a module over $([\![A]\!], \Gamma)$ as in (4.3.1) and let (E_M, s_M) be chosen as in (5.4.1). Then we obtain the $[\![A, E, s]\!]$-module $[\![M, E_M, s_M]\!]$ as follows. Here we use the notation in (5.1.6). Let

$$[\![M, E_M, s_M]\!]_0 = T_{\mathbb{G}}(E) \otimes \mathbb{G}E_M \tag{1}$$

be the free $[\![A, E, s]\!]_0 = T_{\mathbb{G}}(E)$-module generated by E_M. For $x, y \in [\![M, E_M, s_M]\!]_0$ a track $G : x \Rightarrow y$ in $[\![M, E_M, s_M]\!]$ is a triple

$$G = (y, \hat{G}, x) : x \Longrightarrow y \tag{2}$$

where $\hat{G} : s_M x \Rightarrow s_M y$ is a track in $[\![M]\!]$. We define composition and action in the same way as in (5.3.6)(3),(4) and we obtain the $\mathbb{G}$-module structure as in (5.3.6)(5). Then it is easily seen that $[\![M, E_M, s_M]\!]$ is a well-defined left $[\![A, E, s]\!]$-module which is termed the *strictification* of $[\![M]\!]$.

The strictification $[\![M, E_M, s_M]\!]$ satisfies a result similar to (5.3.8). Moreover the strictification is well defined up to isomorphism by E_M since a result similar as in (5.3.9) holds.

5.5 The strictification of the secondary Steenrod algebra

For a prime p we have the Steenrod algebra $\mathcal{A}$ over $\mathbb{F} = \mathbb{Z}/p$ together with the canonical set of algebra generators

$$E_{\mathcal{A}} = \begin{cases} \{Sq^i \mid i \geq 1\} & \text{for } p = 2, \\ \{\beta\} \cup \{P^i \text{ and } P^i_\beta \mid i \geq 1\} & \text{for } p \text{ odd.} \end{cases} \tag{5.5.1}$$

Here the generator $P^i_\beta \in E_{\mathcal{A}}$ is mapped by the inclusion $E_{\mathcal{A}} \to \mathcal{A}$ to the composite element βP^i. We need these extra generators P^i_β for the "instability condition" defined below. Let $([\![A]\!], \Gamma)$ be the secondary Steenrod algebra which is a Γ-track algebra. Hence we can apply the strictification in (5.3). For this we choose a lift s

as in the diagram

(1)
$$\begin{array}{ccc} & & [\![\mathcal{A}]\!]_0 \\ & \overset{s}{\nearrow} & \downarrow \pi \\ E_{\mathcal{A}} & \xrightarrow[i_E]{} & \mathcal{A} \end{array}$$

where i_E is the inclusion and $\pi s = i_E$. Hence the lift s chooses for each element $e \in E_{\mathcal{A}}$ a stable map $s(e)$ in $[\![\mathcal{A}]\!]_0$ representing the homotopy class e. The stable map $s(e)$ is given by a sequence of maps $s(e)_n$ and a sequence of tracks as in the following diagram, $n \in \mathbb{Z}$, $k =| e |$.

(2)
$$\begin{array}{ccc} Z^n & \xrightarrow{s(e)_n} & Z^{n+k} \\ \downarrow & \overset{H_{e,n}}{\Longrightarrow} & \downarrow \\ \Omega Z^{n+1} & \xrightarrow[\Omega(s(e)_{n+1})]{} & \Omega(Z^{n+1+k}) \end{array}$$

Compare (2.4.4). In Section (10.8) below we choose the stable map $s(e)$ for $e \in E_{\mathcal{A}}$ such that the following *instability condition* (2a), (2b), (2c) is satisfied. For $e = Sq^k$ $(p = 2)$ we choose $s(e)$ in such a way that

(2a) $$s(Sq^k)_n = 0 : Z^n \longrightarrow * \longrightarrow Z^{n+k}$$

is the trivial map for $k > n$ and also $H_{e,n} = 0^\square$ is the trivial track for $k > n+1$, see (1.1.6). Similarly for $e = P^k$ (p odd) we choose

(2b) $$s(P^k)_n = 0 : Z^n \longrightarrow * \longrightarrow Z^{n+2k(p-1)}$$

for $2k > n$ and also $H_{e,n} = 0^\square$ for $2k > n+1$. Compare (1.1.6). Moreover for $e = P^k_\beta$ (p odd) we choose

(2c) $$s(P^k_\beta)_n = 0 : Z^n \longrightarrow * \longrightarrow Z^{n+2k(p-1)+1}$$

for $2k+1 > n$ and also $H_{e,n} = 0^\square$ for $2k+1 > n+1$. Compare (1.1.6). For $e = \beta$ there is no condition of instability since we assume $Z^0 = *$ is a point.

We have for $a = s(Sq^k)$ and $x, y : X \longrightarrow Z^n$ the linearity track $\Gamma^{x,y}_a : a(x,y) \Longrightarrow ax + ay$ in (4.2.2). By the instability condition (2a) the track $\Gamma^{x,y}_a : 0 \Longrightarrow 0$ represents an element in $H^{n+k-1}(X)$ for $k > n$. In fact, we shall show in (10.8) the *delicate linearity track formula*

(2d) $$\Gamma^{x,y}_{s(Sq^k)} = \begin{cases} x \cdot y & \text{for } k = n+1, \\ 0 & \text{for } k > n+1. \end{cases}$$

For $x = y$ we know that $\Gamma^{x,x}_{s(Sq^k)} : 0 \Longrightarrow 0$ is the element $\kappa(Sq^k)(x) = Sq^{k-1}(x)$ in $H^{n+k-1}(x)$, see (4.5.8). Here κ is the Kristensen derivation. The delicate formula above is compatible with this result since $\Gamma^{x,x}_{s(Sq^k)} = Sq^{k-1}(x) = x \cdot x$ for $|x| = k-1$. If the prime p is odd a delicate formula such as (2d) does not arise since we get

$$\text{(2e)} \qquad \begin{aligned} \Gamma^{x,y}_{s(P^k)} &= 0 \quad \text{for } 2k > n, \qquad \text{and} \\ \Gamma^{x,y}_{s(P^k_\beta)} &= 0 \quad \text{for } 2k+1 > n. \end{aligned}$$

The lift s in (1) defines as in (5.2.1) the $\mathbb{G}$-linear map

$$\text{(3)} \qquad s : T_{\mathbb{G}}(E_{\mathcal{A}}) \longrightarrow T_{\mathbb{F}}(E_{\mathcal{A}}) \longrightarrow [\![\mathcal{A}]\!]_0$$

which together with the Γ-tracks $\Gamma(a,b) : s(a)s(b) \Rightarrow s(a \cdot b)$ in $[\![\mathcal{A}]\!]_1$ is a pseudo functor, see (5.2.3).

We now can define the *excess* $e(\alpha)$ of an element $\alpha \in T_{\mathbb{F}}(E_{\mathcal{A}})$ in such a way that the map

$$\text{(4)} \qquad s(\alpha)_n : Z^n \longrightarrow * \longrightarrow Z^{n+|\alpha|}$$

is the trivial map for $e(\alpha) > n$ and $H_{\alpha,n} = 0^\square$ is the trivial track for $e(\alpha) > n+1$.

If $a_1, \ldots, a_k \in \mathrm{Mon}(E_{\mathcal{A}})$ are pairwise distinct and $\alpha = n_1 a_1 + \cdots + n_k a_k$ with $n_i \in \mathbb{F} - \{0\}$, then $e(\alpha) = \mathrm{Min}(e(a_1), \ldots, e(a_k))$. Moreover for a monomial $a = e_1 \cdot \cdots \cdot e_r \in \mathrm{Mon}(E_{\mathcal{A}})$ with $e_1, \ldots, e_r \in E_{\mathcal{A}}$ put for $p = 2$,

$$\text{(5)} \qquad e(a) = \mathrm{Max}_j(| e_j | - | e_{j+1} \cdot \cdots \cdot e_r |).$$

Moreover for p odd put

$$\text{(6)} \qquad e(a) = \mathrm{Max}_j \begin{cases} 2 | e_j | - | e_{j+1} \cdot \cdots \cdot e_r | & \text{for} \quad e_j \in \{P^1, P^2, \ldots\}, \\ 2 | e_j | + 1 - | e_{j+1} \cdot \cdots \cdot e_r | & \text{for} \quad e_j \in \{P^1_\beta, P^2_\beta, \ldots\}, \\ 1 & \text{for} \quad e_j = \beta. \end{cases}$$

Now one readily checks that (4) holds by use of (2a), (2b), (2c).

5.5.2 Definition. The *strictification of the secondary Steenrod algebra* is the track algebra $[\![\mathcal{A}, E_{\mathcal{A}}, s]\!]$ defined by s in (5.5.1), see (5.3.6). This track algebra can be equivalently described as a pair algebra or a crossed algebra as in (5.1). The crossed algebra $\mathcal{B} = \mathcal{B}(s)$ corresponding to $[\![\mathcal{A}, E_{\mathcal{A}}, s]\!]$ is given by

$$\text{(1)} \qquad \partial : \mathcal{B}_1 \longrightarrow \mathcal{B}_0$$

where $\mathcal{B}_0 = T_{\mathbb{G}}(E_{\mathcal{A}})$ is the $\mathbb{G}$-tensor algebra generated by $E_{\mathcal{A}}$. Moreover $\mathcal{B}_1$ is the $\mathbb{G}$-module consisting of pairs (H, x) where $x \in \mathcal{B}_0$ and $H : sx \Rightarrow 0$ is a track in $[\![\mathcal{A}]\!]$. In degree 0 the crossed algebra $\mathcal{B}$ coincides with the following diagram.

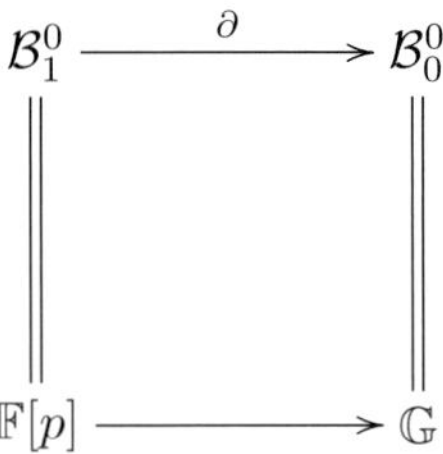

Here $\mathbb{F}[p]$ is the $\mathbb{F}$-vector space generated by the element $[p] = (0^{\square}, p\cdot 1) \in \mathcal{B}_1$ where $p\cdot 1 \in T_{\mathbb{G}}(E_{\mathcal{A}})$ and $0^{\square}$ is the trivial track of the 0-element in the discrete groupoid $[\![\mathcal{A}^0]\!] = \mathbb{F}$, see (2.4.4). The boundary map ∂ in (1) is defined by $\partial(H,x) = x$. The $\mathcal{B}_0$-bimodule structure of $\mathcal{B}_1$ is defined by

(2)
$$\begin{aligned} (H,x)\cdot y &= (H \bullet y, x\cdot y), \\ y\cdot(H,x) &= (y\bullet H, y\cdot x). \end{aligned}$$

Here $H{\bullet}y : s(x{\cdot}y) \Rightarrow 0$ is the track $H{\bullet}y = (H{\cdot}sy)\square\Gamma(x,y)^{\mathrm{op}}$ and $y{\bullet}H : s(y{\cdot}x) \Rightarrow 0$ is the track $y\bullet H = (sy\cdot H)\square\Gamma(y,x)^{\mathrm{op}}$ where we use the Γ-tracks of the pseudo functor (s,Γ), see (5.5.1)(3) and (5.2.3). In Section (5.3) we have shown that $\mathcal{B}$ is a well-defined crossed algebra, see (5.1.6), with

(3)
$$\begin{aligned} \pi_o\mathcal{B} &= \mathrm{cokernel}(\partial) &&= \mathcal{A}, \\ \pi_1\mathcal{B} &= \mathrm{kernel}(\partial) &&= \Sigma\mathcal{A}. \end{aligned}$$

Moreover two lifts s, s_0 as in (5.5.1) together with a track $S : s \Rightarrow s_0$ in $[\![\mathcal{A}]\!]$ yield the isomorphism

(4)
$$\bar{S} : \mathcal{B}(s) \cong \mathcal{B}(s_0)$$

which is the identity on $T_{\mathbb{G}}(E_{\mathcal{A}})$ and on $\pi_1(\mathcal{B}(s)) = \Sigma\mathcal{A} = \pi_1\mathcal{B}(s_0)$, see (5.3.9). Therefore $\mathcal{B}$ is well defined up to such isomorphisms of a crossed algebra. We call $\mathcal{B}$ the *crossed algebra of secondary cohomology operations*.

As the main goal of the book we will discuss properties of the crossed algebra $\mathcal{B}$ which hopefully will lead to a computation of $\mathcal{B}$. In Chapter 11 we shall see that $\mathcal{B}$ has the additional structure of a secondary Hopf algebra.

In order to compute the crossed algebra $\mathcal{B}$ we choose the following set of generators $E^1_{\mathcal{A}}$ of the ideal $I_{\mathbb{G}}(\mathcal{A}) = \mathit{kernel}(T_{\mathbb{G}}(\mathcal{A}) \to \mathcal{A})$. Let

(5.5.3)
$$E^1_{\mathcal{A}} \subset I_{\mathbb{G}}(E_{\mathcal{A}}) \subset T_{\mathbb{G}}(E_{\mathcal{A}})$$

be the subset consisting of $p = p\cdot 1$, where 1 is the unit of the algebra $T_{\mathbb{G}}(E_{\mathcal{A}})$, and of the *Adem relations* (1.1) considered as elements of $T_{\mathbb{G}}(E_{\mathcal{A}})$. Moreover $\beta^2 \in E^1_{\mathcal{A}}$

and $P^n_\beta - \beta P^n \in E^1_{\mathcal{A}}$ $(n \geq 1)$ if p is odd. Then it is clear that $E^1_{\mathcal{A}}$ generates the ideal $I_{\mathbb{G}}(E_{\mathcal{A}})$. We now choose a lift t as in the following diagram where j_E is the inclusion.

(1)
$$\begin{array}{ccc} & & \mathcal{B}_1 \\ & \overset{t}{\nearrow} & \downarrow \partial \\ E^1_{\mathcal{A}} & \xrightarrow[j_E]{} & I_{\mathbb{G}}(E_{\mathcal{A}}) \end{array}$$

That is, t carries a relation $r \in E^1_{\mathcal{A}}$ to a pair $t(r) = (H(r), r) \in \mathcal{B}_1$ where $H(r) : s(r) \Rightarrow 0$ is a track in $[\![\mathcal{A}]\!]$. For $r = p \in E^1_{\mathcal{A}}$ we have $s(r) = 0$ and $H(r) = 0^\square$ is the trivial track of 0. The map t in (1) induces the following commutative diagram of crossed algebras with exact rows.

(2)
$$\begin{array}{ccccccccc} 0 & \longrightarrow & \Sigma\mathcal{A} & \xrightarrow{i} & \mathcal{B}_1 & \xrightarrow{\partial} & \mathcal{B}_0 & \xrightarrow{\pi} & \mathcal{A} & \longrightarrow & 0 \\ & & \uparrow \Gamma^t_{\mathcal{B}} & & \uparrow t & & \| & & \| & & \\ 0 & \longrightarrow & K_{\mathcal{A}} & \xrightarrow{j} & [E^1_{\mathcal{A}}] & \xrightarrow{d} & \mathcal{B}_0 & \longrightarrow & \mathcal{A} & \longrightarrow & 0 \end{array}$$

Here d is the *free crossed algebra* generated by elements $[r]$ with $r \in E^1_{\mathcal{A}}$ and $d[r] = r$, that is,

$$[E^1_{\mathcal{A}}] = (\mathcal{B}_0 \otimes (\mathbb{G}E^1_{\mathcal{A}}) \otimes \mathcal{B}_0)/U$$

where $\mathbb{G}E^1_{\mathcal{A}}$ is the free $\mathbb{G}$-module generated by $E^1_{\mathcal{A}}$ and U is the $\mathcal{B}_0$-submodule of $V = \mathcal{B}_0 \otimes (\mathbb{G}E^1_{\mathcal{A}}) \otimes \mathcal{B}_0$ generated by the elements $\bar{d}(a) \cdot b - a(\bar{d}b)$ for $a, b \in V$. Here $\bar{d} : V \to \mathcal{B}_0$ is the unique $\mathcal{B}_0$-bimodule map with $\bar{d}[r] = r$ for $r \in E^1_{\mathcal{A}}$. Since $\bar{d}U = 0$ we get the induced map d in the diagram. Moreover the map t in the diagram is the algebra map between crossed algebras which is the identity on $\mathcal{B}_0$ and satisfies $t[r] = t(r)$ with t as in (5.5.3)(1). Then

(3)
$$K_{\mathcal{A}} = kernel(d : [E^1_{\mathcal{A}}] \longrightarrow \mathcal{B}_0)$$

is a well-defined $\mathcal{A}$-bimodule termed the bimodule of *relations among relations* and the induced map

(4)
$$\Gamma^t_{\mathcal{B}} : K_{\mathcal{A}} \longrightarrow \Sigma\mathcal{A}$$

is a map between $\mathcal{A}$-bimodules depending on the choice of t in (5.5.3)(1). Kristensen [Kr4] studies a "Massey product operator M" which corresponds to $\Gamma^t_{\mathcal{B}}$ and claims that a formula for M can be found. The computation of $\Gamma^t_{\mathcal{B}}$ is equivalent to the computation of the crossed algebra $\mathcal{B}$ since

(5)
$$\begin{array}{ccc} \Sigma\mathcal{A} & \longrightarrow & \mathcal{B}_1 \\ \uparrow \Gamma^t_{\mathcal{B}} & & \uparrow t \\ K_{\mathcal{A}} & \longrightarrow & [E^1_{\mathcal{A}}] \end{array}$$

is a *push out diagram* in the category of $\mathcal{B}_0$-bimodules. Here the $\mathcal{A}$-bimodule $K_{\mathcal{A}}$ is completely determined by generators $E_{\mathcal{A}}$ and relations $E^1_{\mathcal{A}}$ in the Steenrod algebra. The $\mathcal{A}$-bimodule map $\Gamma^t_{\mathcal{B}}$, however, depends on the crossed algebra $\mathcal{B}$ and can be considered as an additional structure of the Steenrod algebra $\mathcal{A}$. Kristensen [Kr4] and Kristensen-Madsen [KrM1] compute certain elements $[a,b,c]$ in $K_{\mathcal{A}}$ and [Kr4] indicates a method how to determine the map $\Gamma^t_{\mathcal{B}}$ though there is not a definition of the bimodule $K_{\mathcal{A}}$ of relations among relations in [Kr4].

Now assume that $t_0 : E^1_{\mathcal{A}} \to \mathcal{B}_1$ is a further lift as in (1). Then there exists a map $\Delta : E^1_{\mathcal{A}} \to \Sigma\mathcal{A}$ with

$$t_0(e) = t(e) + i\Delta(e)$$

for $e \in E^1_{\mathcal{A}}$. The map Δ induces a $\mathcal{B}_0$-bimodule map $\Delta : [E^1_{\mathcal{A}}] \to \Sigma\mathcal{A}$ such that

$$\Gamma^{t_0}_{\mathcal{B}} = \Gamma^t_{\mathcal{B}} + \Delta j. \tag{6}$$

Hence the class

$$\Gamma_{\mathcal{B}} = \{\Gamma^t_{\mathcal{B}}\} \in Hom_{\mathcal{A}-\mathcal{A}}(K_{\mathcal{A}}, \Sigma\mathcal{A})/j^* Hom_{\mathcal{B}_0-\mathcal{B}_0}([E^1_{\mathcal{A}}], \Sigma\mathcal{A}) \tag{7}$$

is independent of the choice of t and of the choice of s defining $\mathcal{B} = \mathcal{B}(s)$. Each element in the class $\Gamma_{\mathcal{B}}$ can serve as a map $\Gamma^t_{\mathcal{B}}$ in (5) which defines $\mathcal{B}_1$ as a push out. Hence the computation of the class $\Gamma_{\mathcal{B}}$ is equivalent to the computation of the isomorphism type of $\mathcal{B}$. In Baues-Pirashvili [BP] we show that there is an isomorphism

$$HML^3(\mathcal{A}, \Sigma\mathcal{A}) \quad\cong\quad \mathrm{Hom}_{\mathcal{A}-\mathcal{A}}(K_{\mathcal{A}}, \Sigma\mathcal{A})/j^* \,\mathrm{Hom}_{\mathcal{B}_0-\mathcal{B}_0}([E^1_{\mathcal{A}}], \Sigma\mathcal{A}) \tag{8}$$

carrying the class k_p^{stable} to $\Gamma_{\mathcal{B}}$.

Recall that we obtained in (4.5.7) the degree 0 derivation $\Gamma[p] : \mathcal{A} \to \Sigma\mathcal{A}$ which for $p = 2$ coincides with the Kristensen derivation χ in (4.5.8). The map $\Gamma^t_{\mathcal{B}}$ extends $\Gamma[p]$ since we prove:

5.5.4 Theorem. *For $x \in \mathcal{B}_0$ with $\pi(x) = \xi \in \mathcal{A}$ we get the element*

$$[p]\cdot x - x\cdot [p] \in K_{\mathcal{A}}$$

and the map $\Gamma^t_{\mathcal{B}}$ satisfies the formula

$$\Gamma^t_{\mathcal{B}}([p]\cdot x - x\cdot[p]) = \Gamma[p](\xi).$$

Proof. We have $d([p]\cdot x) = p\cdot x = d(x\cdot[p])$ so that $[p]\cdot x - x\cdot[p] \in K_{\mathcal{A}}$. Moreover we get by definition of $t([p]) = (0,p)$ the formula

$$\begin{aligned}
t([p]\cdot x - x\cdot[p]) &= (0,p)\cdot x - x\cdot(0,p)\\
&= (0\bullet x, p\cdot x) - (x\bullet 0, x\cdot p)\\
&= ((0\cdot x)\square\Gamma(0,x)^{\mathrm{op}}, p\cdot x) - ((x\cdot 0)\square\Gamma(x,0)^{\mathrm{op}}, p\cdot x).
\end{aligned}$$

Here $0 \cdot x = 0^{\square}$ and $x \cdot 0 = 0^{\square}$ are the identity tracks of 0. Moreover for $x \in \mathrm{Mon}(E_{\mathcal{A}})$ the track $\Gamma(0,x) = 0^{\square}$ is the identity track of 0. Therefore we get

$$t([p] \cdot x - x \cdot [p]) = (\Gamma(x,0)^{\mathrm{op}}, 0)$$

for $x \in \mathrm{Mon}(E_{\mathcal{A}})$. Here

$$\Gamma(x,0) : sx \cdot sp \Longrightarrow s(x \cdot p) = p \cdot s(x)$$

is the opposite of the track

$$\Gamma(p)_{sx} : (sx)(p \cdot 1) \Longrightarrow p \cdot sx$$

which represents $\Gamma[p](\xi)$. Hence

$$\Gamma^t_{\mathcal{B}}([p] \cdot x - x \cdot [p]) = \Gamma[p](\xi)$$

for $x \in \mathrm{Mon}(E)$. Since $\Gamma^t_{\mathcal{B}}$ and $\Gamma[p]$ are $\mathbb{F}$-linear the result (5.5.4) follows. $\square$

Let x be an element of degree $| x | \geq 1$. Then we obtain for the crossed algebra $\mathcal{B}$ the *free right $\mathcal{B}$-module* $x \cdot \mathcal{B}$ generated by x, see (5.1.7). Let $\mathbf{mod}_0(\mathcal{B})^{\mathrm{op}}$ be the track category of finitely generated free right $\mathcal{B}$-modules

(5.5.5) $$x_1 \cdot \mathcal{B} \oplus \cdots \oplus x_r \cdot \mathcal{B}$$

with generators of degree $| x_i | \geq 1$ for $i = 1, \dots, r$. Morphisms (0-cells) are $\mathcal{B}$-linear maps and tracks (1-cells) are natural transformations between such maps (considered as functors between graded groupoids). Then one gets the linear track extension

$$\overline{\Sigma\mathcal{A}} \longrightarrow \mathbf{mod}_0(\mathcal{B})_1^{\mathrm{op}} \rightrightarrows \mathbf{mod}_0(\mathcal{B})_0^{\mathrm{op}} \longrightarrow \mathbf{mod}_0(\mathcal{A})^{\mathrm{op}}$$

where $\mathbf{mod}_0(\mathcal{A})^{\mathrm{op}}$ is defined as in (2.5.2) and $\overline{\Sigma\mathcal{A}}$ is the natural system given by the $\mathcal{A}$-bimodule $\Sigma\mathcal{A}$, see (4.4.1). We have seen in (2.5.2) that

$$\mathbf{K}_p^{\mathrm{stable}} = \mathbf{mod}_0(\mathcal{A})^{\mathrm{op}}.$$

This result has the following secondary analogue

5.5.6 Theorem. *The linear track extension given by* $\mathbf{mod}_0(\mathcal{B})^{\mathrm{op}}$ *is weakly equivalent to the linear track extension given by* $[\![\mathbf{K}_p^{\mathrm{stable}}]\!]$ *in* (2.5.3)

Proof. The extended cocycle $\langle c \rangle$ in (4.4.3) is exactly a cocycle for the linear track extension $\mathbf{mod}_0(\mathcal{B})^{\mathrm{op}}$. Hence the result follows from (3.6.9). $\square$

Kristensen introduced Massey products for the Steenrod algebra $\mathcal{A}$. They can be easily derived from the crossed algebra $\mathcal{B}$ as follows.

5.5.7 Definition. Let $A = (a^j)$, $B = (b^i_j)$ and $C = (c_i)$ be matrices with $i = 1, \dots, s$ and $j = 1, \dots, t$ and entries

$$a^j, b^i_j, c_i \in \mathcal{B}_0 = T_{\mathbb{G}}(E_{\mathcal{A}}). \tag{1}$$

Moreover assume that products AB and BC have entries in $I_{\mathbb{G}}(E_{\mathcal{A}})$. Then we can choose matrices X, Y with entries in $\mathcal{B}_1$ such that

$$X = (x^i) \quad \text{satisfies} \quad \partial X = AB, \tag{2}$$

$$Y = (y_j) \quad \text{satisfies} \quad \partial Y = BC. \tag{3}$$

Then one readily checks that

$$\begin{aligned} \partial(X \cdot C - A \cdot Y) &= \partial(X) \cdot C - A \cdot \partial(Y) \\ &= ABC - ABC \\ &= 0 \end{aligned}$$

so that $XC - AY$ represents an element in $\Sigma\mathcal{A}$. The *Massey product*

$$\langle A, B, C \rangle \subset \Sigma\mathcal{A} \tag{4}$$

is the set of all elements $X \cdot C - A \cdot Y$ with X and Y satisfying (2) and (3). This set is a coset of the subgroup

$$\sum_{i=1}^{t} (\Sigma\mathcal{A}) \cdot \pi(c_i) + \sum_{j=1}^{s} \pi(a^j)(\Sigma\mathcal{A}) \subset \Sigma\mathcal{A}$$

where $\pi(c_i), \pi(a_j) \in \mathcal{A}$ are given by the quotient map $\pi : T_{\mathbb{G}}(E_{\mathcal{A}}) \to \mathcal{A}$.

We point out that the crossed algebra structure of $\mathcal{B}$ yields obvious properties of the triple Massey product $\langle A, B, C \rangle$ which also can be understood as a Massey product in the linear track extension $\mathbf{mod}_0(\mathcal{B})^{\mathrm{op}}$ in (5.5.6).

Kristensen-Pedersen [KrP] and Kristensen [Kr4] define the Massey product in terms of the secondary Steenrod algebra $[\![\mathcal{A}]\!]$ and therefore Γ-tracks are involved in their definition. Since we know that $([\![\mathcal{A}]\!], \Gamma)$ has the strictification $\mathcal{B}$, the definition in (5.5.7) corresponds directly to the classical definition of a *matrix Massey product*, see Massey-Petersen [MaP].

5.6 The strictification of secondary cohomology and Kristensen operations

Let X be a path-connected pointed space and let $E_X \subset \tilde{H}^*(X)$ be a set of generators of the $\mathcal{A}$-module $\tilde{H}^*(X)$. We choose a lift s as in the diagram

(5.6.1)
$$\begin{array}{ccc} & & [\![X, Z^*]\!]_0 \\ & \overset{s_X}{\nearrow} & \downarrow \pi \\ E_X & \xrightarrow[i_X]{} & \tilde{H}^*(X) \end{array}$$

where i_X is the inclusion with $\pi s_X = i_X$. Hence the lift s_X chooses for each element $e \in E_X$ a continuous map $s_X(e) : X \to Z^{|e|}$ representing the homotopy class $e \in [X, Z^{|e|}] = H^{|e|}(X)$. For example we can choose $E_X = [\![X, Z^*]\!]_0$ and s_X the identity. This is the *natural choice* of E_X which is very large but functorial in X. As in (5.4.1) the lift s_X and s in (5.5.1)(3) determine the $\mathbb{G}$-linear map

(1)
$$s_X : T_{\mathbb{G}}(E_{\mathcal{A}}) \otimes \mathbb{G}E_X \longrightarrow [\![X, Z^*]\!]_0.$$

Moreover for $b \in T_{\mathbb{G}}(E_{\mathcal{A}})$ and $y \in T_{\mathbb{G}}(E_{\mathcal{A}}) \otimes \mathbb{G}E_X$ we have the Γ-track

(2)
$$\Gamma_X(b, y) : s(b) \cdot s_X(y) \Longrightarrow s_X(b \cdot y)$$

in $[\![X, Z^*]\!]_1$. Recall that $\mathcal{B}$ is the crossed algebra in (5.5.2) which is the strictification of the secondary Steenrod algebra.

5.6.2 Definition. The *strictified secondary cohomology* $\mathcal{H}^*(X, E_X, s_X)$ is the $\mathcal{B}$-module, see (5.1.7), defined as follows: Let

(1)
$$\mathcal{H}^*(X, E_X, s_X)_0 = T_{\mathbb{G}}(E_{\mathcal{A}}) \otimes \mathbb{G}E_X$$

and let $\mathcal{H}^*(X, E_X, s_X)_1$ be given by the following pull back diagram.

(2)
$$\begin{array}{ccc} \mathcal{H}^*(X, E_X, s_X)_1 & \longrightarrow & [\![X, Z^*]\!]_1^0 \\ \partial \downarrow & \text{pull} & \downarrow \partial \\ \mathcal{H}^*(X, E_X, s_X)_0 & \xrightarrow[s_X]{} & [\![X, Z^*]\!]_0 \end{array}$$

Hence an element in $\mathcal{H}^*(X, E_X, s_X)_1$ is a pair (H, y) with $y \in T_{\mathbb{G}}(E_{\mathcal{A}}) \otimes \mathbb{G}E_X$ and $H : s_X(y) \Rightarrow 0$ in $[\![X, Z^*]\!]_1^0$. Moreover $\partial(H, y) = 0$. Now ∂ is a well-defined $\mathcal{B}$-module by setting

(3)
$$\begin{aligned} a \cdot (H, y) &= (a \bullet H, a \cdot y) \quad \text{for } a \in \mathcal{B}_0 = T_{\mathbb{G}}(E_{\mathcal{A}}), \\ (G, a) \cdot z &= (G \bullet z, a \cdot z) \end{aligned}$$

for $a \in \mathcal{B}_0 = T_{\mathbb{G}}(E_{\mathcal{A}})$, $(G, a) \in \mathcal{B}_1$, and $z \in \mathcal{H}^*(X, E_X, s_X)_0$. The Γ-products $a \bullet H$ and $G \bullet z$ are defined by Γ-tracks, that is

(4)
$$\begin{aligned} a \bullet H &= ((sa)H) \square \Gamma(a, y)^{\mathrm{op}}, \\ G \bullet z &= (G(s_X z)) \square \Gamma(a, z)^{\mathrm{op}}. \end{aligned}$$

According to (2.2.10) we get

(5)
$$\begin{aligned} \pi_0 \mathcal{H}^*(X, E_X, s_X) &= \text{cokernel}(\partial) &&= \tilde{H}^*(X), \\ \pi_1 \mathcal{H}^*(X, E_X, s_X) &= \text{kernel}(\partial) &&\cong \Sigma \tilde{H}^*(X). \end{aligned}$$

Moreover two lifts s_X, s_X^0 as in (5.6.1) together with a track $S : s_X \Rightarrow s_X^0$ in $[\![X, Z^*]\!]_1$ yield the isomorphism of $\mathcal{B}$-modules

(6)
$$\bar{S} : \mathcal{H}^*(X, E_X, s_X) \cong \mathcal{H}^*(X, E_X, s_X^0)$$

which is the identity on $T_{\mathbb{G}}(E_{\mathcal{A}}) \otimes \mathbb{G}E_X$ and on $\pi_1 \mathcal{H}^*(X, E_X, s_X) = \Sigma \tilde{H}^*(X) = \pi_1 \mathcal{H}^*(X, E_X, s_X^0)$. Therefore the strictified cohomology is well defined up to such isomorphism.

We now can introduce secondary cohomology operations as follows.

5.6.3 Definition. Recall that

(1)
$$I_{\mathbb{G}}(E_{\mathcal{A}}) = \text{kernel}(\pi : T_{\mathbb{G}}(E_{\mathcal{A}}) \longrightarrow \mathcal{A})$$

and let

(2)
$$I_{\mathbb{G}}(E_X) = \text{kernel}(\pi : T_{\mathbb{G}}(E_{\mathcal{A}}) \otimes \mathbb{G}E_X \longrightarrow \tilde{H}^*(X)).$$

A *relation* is an element

(3)
$$r = b + \sum_{\nu=1}^{k} \alpha_\nu a_\nu \in I_{\mathbb{G}}(E_{\mathcal{A}})$$

with $\alpha_\nu, b_\nu, b \in T_{\mathbb{G}}(E_{\mathcal{A}})$. A *secondary cohomology operation associated* to r is an element

(4)
$$H \in \mathcal{B}_1 \text{ with } \partial H = r.$$

If the element b has *excess* $e(b) > n$, then the stable map $s(b)_n = 0 : Z^m \to * \to Z^{m+|b|}$ for $m \leq n$, compare (5.5.1). If $e(b) > n$, then each element $x \in T_{\mathbb{G}}(E_{\mathcal{A}}) \otimes \mathbb{G}E_X$ with $| x | \leq n$ yields the Γ-track

$$\Gamma(b, x)^{\mathrm{op}} : s(b \cdot x) \Longrightarrow s(b) \cdot s(x) = 0$$

so that

(5)
$$\bar{\Gamma}(b, x) = (\Gamma(b, x)^{\mathrm{op}}, b \cdot x) \in \mathcal{H}^*(X, E_X, s_X)_1$$

satisfies $\partial\bar{\Gamma}(b,x) = b \cdot x$. Now we assume that

$$a_\nu \cdot x \in I_{\mathbb{G}}(E_{\mathcal{A}}) \text{ for } \nu = 1\dots,k \tag{6}$$

so that there are elements

$$H_\nu^x \in \mathcal{H}^*(X, E_X, s_X)_1 \text{ with } \partial H_\nu^x = a_\nu \cdot x. \tag{7}$$

Then an easy computation shows that

$$\begin{aligned}
&\partial(H \cdot x - \bar{\Gamma}(b,x) - \sum_{\nu=1}^{k} \alpha_\nu \cdot H_\nu^x) \\
&\quad = (b + \sum_\nu \alpha_\nu \cdot r_\nu) \cdot x - b \cdot x - \sum_\nu \alpha_\nu \cdot a_\nu \cdot x = 0.
\end{aligned}$$

Hence we get the coset of elements

$$\begin{aligned}
\theta_H(x) &= \{H \cdot x - \bar{\Gamma}(b,x) - \textstyle\sum_{\nu=1}^k \alpha_\nu \cdot H_\nu^x \mid \partial H_\nu^x = \alpha_\nu \cdot x\} \\
&\in (\Sigma\tilde{H}^*(X))/(\textstyle\sum_{\nu=1}^k \pi(\alpha_\nu) \cdot \Sigma\tilde{H}^*(X)).
\end{aligned} \tag{8}$$

This is the element defined by the Kristensen operation in (2.7.4). Now it is easy to develop the properties of the operation θ_H by use of the $\mathcal{B}$-module structure of $\mathcal{H}^*(X, E_X, s_X)$.

In particular we get the following results where equality holds modulo the total indeterminancy.

5.6.4 Theorem. *Assume $\theta_H(x)$ and $\theta_H(y)$ are defined. Then*

$$\begin{aligned}
\theta_H(x+y) &= \theta_H(x) + \theta_H(y) && \text{if } |x| = |y| < e(b) - 1, \\
\theta_H(x+y) &= \theta_H(x) + \theta_H(y) - d(sb; sx, sy) && \text{if } |x| = |y| = e(b) - 1.
\end{aligned}$$

Compare Theorem 4.3 [Kr1] and see (4.3.10) above.

Proof. By (5.6.3)(8) we see that

$$\theta_H(x+y) = \theta_H(x) + \theta_H(y) + \Delta(b; x, y)$$

where $\Delta(b;x,y) \in \Sigma\tilde{H}^*(X)$ is given by $\bar{\Gamma}(b;x,y) - \bar{\Gamma}(b,x) - \bar{\Gamma}(b,y)$. According to (4.3.10) we get

$$\begin{aligned}
\Gamma(b, x+y) - \Gamma(b,x) - \Gamma(b,y) &= \Gamma_{sb}^{sx,sy} - sb \cdot sx - sb \cdot sy \\
&= d(sb; sx, sy).
\end{aligned}$$

Compare (4.3.10). Now one can check that $(\Gamma_{sb}^{sx,sy})_n$ is the trivial track for $n < e(b) - 1$. See (5.5.1)(2a)(2b). □

5.6.5 Theorem. *For $c \in T_{\mathbb{G}}(E_{\mathcal{A}})$ we have*

$$\theta_{H\cdot c}(x) = \theta_H(c\cdot x) \qquad \text{and} \qquad \theta_{c\cdot H}(x) = c\theta_X(x).$$

Compare Theorems 5.2 and 5.3 in [Kr1]. Also the following result corresponds to 5.3 [Kr1].

5.6.6 Theorem. *Assume $\theta_H(x)$ and $\theta_G(x)$ are defined. Then*

$$\theta_{H+G}(x) = \theta_H(x) + \theta_G(x).$$

Proof. Here we use the fact that

$$\bar{\Gamma}(b_H + b_G, x) = \bar{\Gamma}(b_H, x) + \bar{\Gamma}(b_G, x)$$

as follows from (4.3.1)(13). Here $b_H = b$ is given by the relation $r = \partial H$ and similarly b_G is given by the relations ∂G. □

5.7 Two-stage operation algebras

An Ω-spectrum X is a sequence of pointed CW-spaces $X_n, n \geq 1$, together with homotopy equivalences

$$X_n \xrightarrow{\simeq} \Omega X_{n+1}.$$

Given Ω-spectra X, Y let

(5.7.1) $$[X,Y]_k^{\text{stable}},\ k \geq 0,$$

be the set of all sequences $\alpha = (\alpha_n, n \geq 1)$ with $\alpha_n \in [X_n, Y_{n+k}]$ such that

$$\begin{array}{ccc} X_n & \xrightarrow{\alpha_n} & Y_{n+k} \\ \downarrow{\simeq} & & \downarrow{\simeq} \\ \Omega X_{n+1} & \xrightarrow[\Omega\alpha_{n+1}]{} & \Omega Y_{n+k+1} \end{array}$$

commutes in $\mathbf{Top}^*/\simeq$. It is easy to see that $[X,Y]_*^{\text{stable}}$ is a non-negatively graded abelian group. Moreover for Ω-spectra X, Y, Z we have the bilinear composition law

(5.7.2) $$[Y,Z]_*^{\text{stable}} \otimes [X,Y]_*^{\text{stable}} \longrightarrow [X,Z]_*^{\text{stable}}.$$

In particular $[X,X]_*^{\text{stable}}$ is a graded algebra termed the *operation algebra* of X.

For example we have the Eilenberg-MacLane spectrum $K = \{K(\mathbb{F}, n), n \geq 1\}$ and the operation algebra

$$\mathcal{A} \ = \ [K, K]_*^{\text{stable}}$$

is the Steenrod algebra, see (2.5.2).

We now associate with an element $k \in \mathcal{A}$ a 2-*stage* Ω-*spectrum* $P(k) = \{P_n(k), n \geq 1\}$ where $P_n(k)$ is the homotopy fiber of

$$K(\mathbb{F}, n) \xrightarrow{k} K(\mathbb{F}, n+ \mid k \mid).$$

Then we get the 2-*stage operation algebra*

$$\mathcal{A}(k) \ = \ [P(k), P(k)]_*^{\text{stable}} \tag{5.7.3}$$

which is considered in Kristensen-Madsen [KrM2]. We now describe $\mathcal{A}(k)$ in terms of the crossed algebra $\mathcal{B} = (\partial : \mathcal{B}_1 \to \mathcal{B}_0)$.

5.7.4 Definition. Let $\hat{k}_1$, $\hat{k}_2$ be elements in $\mathcal{B}_0 = T_{\mathbb{G}}(E_{\mathcal{A}})$. Then we define the graded $\mathbb{G}$-module

$$\mathcal{B}(\hat{k}_1, \hat{k}_2) = \{(\alpha, \beta, G) \in \mathcal{B}_0 \times \mathcal{B}_0 \times \mathcal{B}_1;\ \alpha \cdot \hat{k}_1 - \hat{k}_2 \cdot \beta = \partial G\}/\sim$$

with $\mid (\alpha, \beta, G) \mid = \mid \alpha \mid$. The relation $\sim$ is defined as follows. Let $(\alpha, \beta, G) \sim (\alpha', \beta', G')$ if and only if there exist $A, B \in \mathcal{B}_1$ with

$$\begin{cases} \partial A &= \alpha - \alpha', \\ \partial B &= \beta - \beta', \\ G &= G' + A \cdot \hat{k}_1 - \hat{k}_2 \cdot B. \end{cases}$$

We define for $\hat{k}_1, \hat{k}_2, \hat{k}_3 \in \mathcal{B}_0$ the composition law

$$\mathcal{B}(\hat{k}_2, \hat{k}_3) \otimes \mathcal{B}(\hat{k}_1, \hat{k}_2) \xrightarrow{\circ} \mathcal{B}(\hat{k}_1, \hat{k}_3)$$

by

$$(\alpha', \beta', G) \circ (\alpha, \beta, G) \ = \ (\alpha'\alpha, \beta'\beta, G' \cdot \beta + \alpha' \cdot G).$$

One can check that $\circ$ is compatible with the relation $\sim$ above.

5.7.5 Theorem. *Let $k_1, k_2 \in \mathcal{A}$ and let $\hat{k}_1, \hat{k}_2 \in \mathcal{B}_0$ be elements which represent k_1 and k_2 respectively. Then there is a canonical map of graded abelian groups*

$$\mathcal{B}(\hat{k}_1, \hat{k}_2) \ \to \ [P(k_1), P(k_2)]_*^{\text{stable}}$$

which is compatible with the composition law. In particular if $\hat{k} \in \mathcal{B}_0$ represents $k \in \mathcal{A}$ then

$$\mathcal{B}(\hat{k}, \hat{k}) \ \to \ \mathcal{A}(k)$$

is a map between algebras which is a surjection in degree $\leq |k| - 2$ and an isomorphism in degree $< |k| - 2$.

Proof. The definition of $\mathcal{B}(\hat{k}_1, \hat{k}_2)$ corresponds to the category of homotopy pairs in [BUT]. □

5.7.6 Remark. For $k = Sq^{(0.1)} = Sq^3 + Sq^2Sq^1$ Kristensen-Madsen [KrM2] compute the operation algebra $\mathcal{A}(k)$ by the formula

$$\mathcal{A}(k) = \Lambda(T) \odot X(k)$$

where $\Lambda(T)$ is the exterior algebra over $\mathbb{F}$ generated by an element T of degree 5 and $X(k)$ is the $\mathbb{F}$-algebra defined in [KrM2]. Moreover $\odot$ denotes the semi-tensor product of Massey-Peterson.

Part II

Products and Power Maps in Secondary Cohomology

The Eilenberg-MacLane spaces Z^n have an "additive structure" since they are $\mathbb{F}$-vector space objects. They also have a "multiplicative structure" by the multiplication maps $\mu : Z^n \times Z^n \to Z^{n+m}$. The theory in Part I is based on the additive structure of Z^n which is also defined in the category of stable maps between products of Eilenberg-MacLane spaces. In the following Part II we consider the multiplicative structure of the spaces Z^n and we study unstable maps between products of Eilenberg-MacLane spaces. In particular we construct power maps and power tracks which correspond to the power algebra structure of the cohomology $H^*(X)$ in Chapter 1.

Chapter 6

The Algebra Structure of Secondary Cohomology

It is a fundamental result of algebraic topology that the cohomology $H^*(X)$ of a space X is a (commutative graded) algebra. In this chapter we consider the secondary analogue of this result. The multiplicative structure of the Eilenberg-MacLane spaces Z^n constructed in Section (2.1) and the action of the permutation group σ_n on Z^n lead canonically to the algebraic concept of "secondary permutation algebra". We show that the secondary cohomology of a pointed space is naturally a secondary permutation algebra.

6.1 Permutation algebras

Let k be a commutative ring and let R be a (non-graded) k-*algebra* with *unit* i and *augmentation* ϵ,

(6.1.1) $$k \xrightarrow{i} R \xrightarrow{\epsilon} k.$$

Here i and ϵ are algebra maps with $\epsilon i = 1$. We assume that R is free as a module over k. For example, let k be a field and G be a group together with a homomorphism $\epsilon : G \to k^*$ where k^* is the group of units in the field k. Then ϵ induces an augmentation

(1) $$\epsilon : k[G] \to k$$

where $k[G]$ is the *group algebra* of G. Here $k[G]$ is the free k-module with basis G and ϵ carries the basis element $g \in G$ to $\epsilon(g)$. In particular we have for the *permutation group* σ_n (which is the group of bijections of the set $\{1, \ldots, n\}$) the sign-homomorphism

(2) $$\text{sign} : \sigma_n \to \{1, -1\} \to k^*$$

which induces the *sign-augmentation*

(3) $$\epsilon = \epsilon_{\text{sign}} : k[\sigma_n] \to k,$$

or we can use the *trivial augmentation*

$$\epsilon \;=\; \epsilon_{\text{trivial}} : k[\sigma_n] \longrightarrow k$$

with $\epsilon(\alpha) = 1$ for $\alpha \in \sigma_n$. Later we shall consider the ring $k = \mathbb{G} = \mathbb{Z}/p^2\mathbb{Z}$ where p is a prime. In this case $(G[\sigma_n], \epsilon)$ is given by the sign-augmentation if p is odd and the trivial augmentation if p is even, that is $\varepsilon(\alpha) = \text{sign}(\alpha)^p$ for $\alpha \in \sigma_n$. For k-modules A, B we use the *tensor product*

(4) $$A \otimes B = A \otimes_k B.$$

A homomorphism $f : A \to B$ is termed a k-*linear* map. If A and B are R-modules, then the map f is R-*linear* if in addition $f(r \cdot x) = r \cdot f(x)$ for $r \in R, x \in A$. If R and K are k-algebras, then also $R \otimes K$ is a k-algebra with augmentation

(5) $$\epsilon : R \otimes K \xrightarrow{\epsilon \otimes \epsilon} k \otimes k = k.$$

The multiplication in $R \otimes K$ is defined as usual by $(\alpha \otimes \beta) \cdot (\alpha' \otimes \beta') = (\alpha\alpha') \otimes (\beta\beta')$. Moreover, if X is an R-module and Y is a K-module, then $X \otimes Y$ is an $R \otimes K$-module by $(\alpha \otimes \beta) \cdot (x \otimes y) = (\alpha x) \otimes (\beta y)$.

We now consider the sequence $R_* = \{R_n, n \geq 0\}$ of augmented k-algebras

(6.1.2) $$R_n = k[\sigma_n]$$

where σ_n is the permutation group. We have the algebra maps

(1) $$i_{n,m} = \odot : R_n \otimes R_m \longrightarrow R_{n+m}$$

induced by the inclusion $\sigma_n \times \sigma_m \subset \sigma_{n+m}$. The algebra map $i_{n,m}$ carries $\alpha \otimes \beta$ to $\alpha \odot \beta$. For $\gamma \in R_k$ we get

(2) $$(\alpha \odot \beta) \odot \gamma = \alpha \odot (\beta \odot \gamma)$$

in R_{n+m+k}. Since $\odot$ is an algebra map we have

(3) $$(\alpha \cdot \alpha') \odot (\beta \cdot \beta') = (\alpha \odot \beta) \cdot (\alpha' \odot \beta')$$

where $\alpha \cdot \alpha'$ denotes the product in R_n. Let $1_n \in R_n$ be the unit element of R_n with $1_n \odot 1_m = 1_{n+m}$. For $n = 0$ we have $R_0 = k$ and $1_0 \in R_0$ satisfies $1_0 \odot \alpha = \alpha \odot 1_0 = \alpha$. As in (2.1.1) let

(4) $$\tau_{n,m} \in \sigma_{n+m} \subset R_{n+m}$$

be the permutation exchanging the block $\{1, \ldots, n\}$ and the block $\{m+1, \ldots, m+n\}$ for $n, m \geq 1$ with $\tau_{n,m}(1) = m + 1$. Then the following properties hold.

$$
(5) \qquad \begin{aligned}
&\tau_{m,n}\tau_{n,m} = 1_{n+m},\\
&\tau_{m,0} = \tau_{0,m} = 1_m,\\
&\tau_{n,m}(\alpha \odot \beta) = (\beta \odot \alpha)\tau_{n,m} \text{ for } \alpha \in R_n,\ \beta \in R_m.\\
&\tau_{m+n,k} = (\tau_{m,k} \odot 1_n)(1_m \odot \tau_{n,k}).
\end{aligned}
$$

We call R_* with this structure a *coefficient algebra.*

6.1.3 Definition. A *permutation algebra* V is a sequence of R_n-modules V^n, $n \in \mathbb{Z}$, with $V^0 = k$ and $V^i = 0$ for $i < 0$ together with k-linear maps

$$(1) \qquad V^n \otimes V^m \longrightarrow V^{n+m}$$

carrying $x \otimes y$ to $x \cdot y$. For $z \in V^k$ we have in V^{n+m+k}

$$(2) \qquad (x \cdot y) \cdot z = x \cdot (y \cdot z),$$

and for $\alpha \in R_n,\ \beta \in R_m$ we have

$$(3) \qquad (\alpha x) \cdot (\beta y) = (\alpha \odot \beta)(x \cdot y).$$

Moreover, $1 \in k = V^0$ is a unit of the multiplication (1) with $1 \cdot x = x \cdot 1 = x$. In addition, the multiplication (1) satisfies in V^{n+m} the equation

$$(4) \qquad \tau_{x,y}(x \cdot y) = y \cdot x.$$

Here $\tau_{x,y} = \tau_{n,m} \in \sigma_{n+m}$ is the interchange element for $x \in V^n$ and $y \in V^m$.

A map $f : V \to W$ between permutation algebras is given by R_n-linear maps $f : V^n \to W^n$ with $f(1) = 1$ and $f(x \cdot y) = f(x) \cdot f(y)$. This defines the category **Perm**(k) of permutation algebras over k.

6.1.3 (a) Example. The coefficient algebra R_* is a permutation algebra. In fact

$$V^n = R_n = k[\sigma_n] \text{ for } n \geq 0$$

is an R_n-module by the left action

$$R_n \otimes V^n \longrightarrow V^n$$

which carries $\alpha \otimes x$ to $\alpha \bullet x = \alpha \cdot x \cdot \bar{\alpha}$. Here the involution $\alpha \mapsto \bar{\alpha}$ of R_n is given by the inverse $\bar{\alpha} = \alpha^{-1}$ for $\alpha \in \sigma_n$ so that for $\alpha, x \in \sigma_n$ we have $\alpha \bullet x = \alpha x \alpha^{-1}$. Moreover, we have the multiplication

$$V^n \otimes V^m \longrightarrow V^{n+m}$$

which carries $x \otimes y$ to $x \odot y$. Now associativity of the multiplication holds by (6.1.2)(2). Moreover, we have

$$\begin{aligned}(\alpha \bullet x) \odot (\beta \bullet y) &= (\alpha x \bar{\alpha}) \odot (\beta y \bar{\beta}) \\ &= (\alpha \odot \beta)(x \odot y)(\bar{\alpha} \odot \bar{\beta}) \\ &= (\alpha \odot \beta) \bullet (x \odot y)\end{aligned}$$

so that (6.1.2)(3) holds. Finally (6.1.3)(4) is satisfied since

$$\begin{aligned}\tau_{x,y} \bullet (x \odot y) &= \tau_{x,y}(x \odot y)\bar{\tau}_{x,y} \\ &= (y \odot x)\tau_{x,y}\bar{\tau}_{x,y} \\ &= y \odot x.\end{aligned}$$

Here we use (6.1.2)(5).

6.1.3 (b) Example. Let $V = \hat{k}$ be defined by $V^n = k$ for $n \geq 0$. Then V is a permutation algebra. Here V^n is the R_n-module with the action $\alpha \bullet x$ given by $\alpha \bullet x = x$ for $\alpha \in \sigma_n$. Moreover, the multiplication $V^n \otimes V^m \to V^{n+m}$ is multiplication in k which is commutative. Using the augmentation $\epsilon : R_n \to k$ for $n \geq 0$ in (6.1.1)(3) we get the map

$$\epsilon : R_* \longrightarrow \hat{k}$$

between permutation algebras in **Perm**(k).

Permutation algebras were also considered in Stover [St]. As in [St] we obtain the category of R_*-modules as follows.

Let R_* be a coefficient algebra with interchange elements $\tau_{m,n} \in R_{m+n}$ as in (6.2). An *R_*-module* V is a sequence of (left) R_n-modules $V^n, n \geq 0$. A *map* or an *R_*-linear map* $f : V \to W$ between R_*-modules is given by a sequence of R_n-linear maps $f^n : V^n \to W^n$ for $n \geq 0$. The commutative ring k (concentrated in degree 0) is an R_*-module. Moreover, using the augmentation ϵ of $R_n, n \geq 0$, we see that each graded k-module M with $M^n = 0$ for $n < 0$ is an R_*-module which we call an *ϵ-module*. For $x \in M^m$ we write $|x| = m$ where $|x|$ is the *degree* of x.

Given R_*-modules $V_1, \dots, V_k$ we define the *R_*-tensor product* $V_1 \bar{\otimes} \cdots \bar{\otimes} V_k$ by

$$(V_1 \bar{\otimes} \cdots \bar{\otimes} V_k)^n = \bigoplus_{n_1 + \cdots + n_k = n} R_n \otimes_{R_{n_1} \otimes \cdots \otimes R_{n_k}} V_1^{n_1} \otimes \cdots \otimes V_k^{n_k} \tag{6.1.4}$$

where we use the algebra map $\odot : R_{n_1} \otimes \ldots \otimes R_{n_k} \to R_n$ given by the structure of the coefficient algebra R_* in (6.1.2). One readily checks associativity

$$\begin{aligned}&(V_{1,1} \bar{\otimes} \cdots \bar{\otimes} V_{1,k_1}) \bar{\otimes} \cdots \bar{\otimes} (V_{s,1} \bar{\otimes} \cdots \bar{\otimes} V_{s,k_s}) \\ &= V_{1,1} \bar{\otimes} \cdots \bar{\otimes} V_{1,k_1} \bar{\otimes} \cdots \bar{\otimes} V_{s,1} \bar{\otimes} \cdots \bar{\otimes} V_{s,k_s}.\end{aligned}$$

Compare Stover [St] 2.9. Moreover, the interchange element τ in R_* yields the isomorphism

$$T : V\bar{\otimes}W \cong W\bar{\otimes}V$$

which carries $v \otimes w$ to $\tau_{w,v} w \otimes v$ where $\tau_{w,v} = \tau_{m,n} \in R_{m+n}$ for $w \in W^m, v \in V^n$. Of course we have $k\bar{\otimes}V = V = V\bar{\otimes}k$.

6.1.5 Definition. An *algebra A over R_** is given by a R_*-module A with $A^0 = k$ and $A^i = 0$ for $i < 0$ and a R_*-linear map $\mu : A\bar{\otimes}A \to A$, $\mu(a \otimes b) = a \cdot b$, which is associative in the sense that the diagram

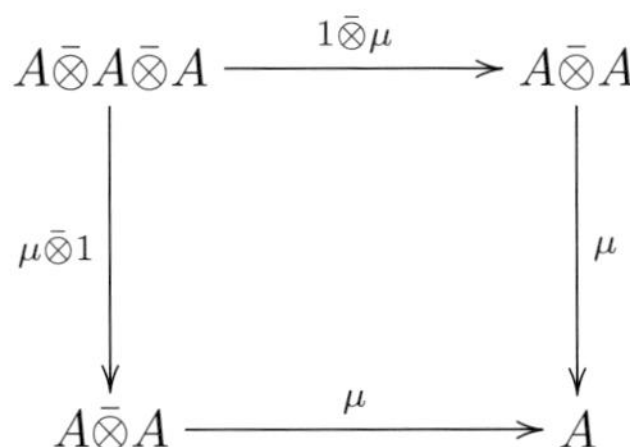

commutes and has a unit $1 \in k = A^0$. Moreover, A is *τ-commutative* if

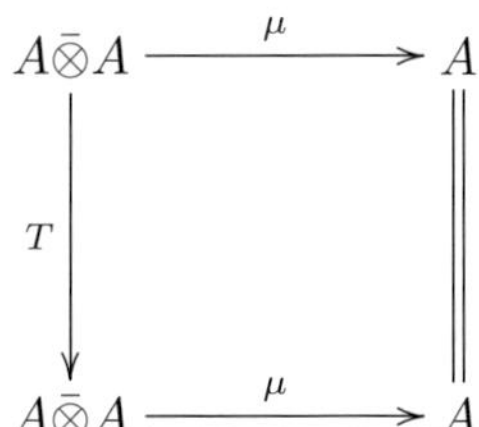

commutes. Then one readily checks that a *permutation algebra* in (6.1.3) is the same as a τ-commutative algebra A over R_*, see [BSC].

6.1.6 Definition. Given an algebra A over R_* we say that an R_*-module V is an *A-module* if a map $m : A\bar{\otimes}V \to V$ is given such that

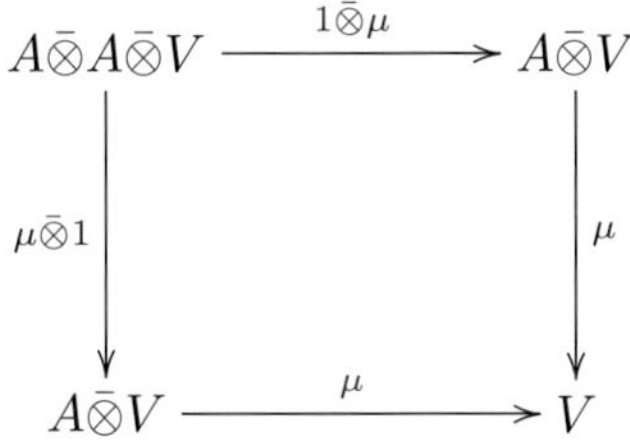

commutes. Hence for $a \cdot x = \mu(a \otimes x)$ with $a \in A, x \in V$ we have $(\alpha a) \cdot (\beta x) = (\alpha \odot \beta)(a \cdot x)$ and $(a \cdot b) \cdot x = a \cdot (b \cdot x)$. Moreover, $1 \cdot x = x$ is satisfied for the unit $1 \in k = A^0$.

For an R_*-module V let $V_{(k)}$ be the underlying graded k-module. If A is a k-algebra over R_*, then $A_{(k)}$ is a graded k-algebra in the usual sense with $A^0_{(k)} = k$.

6.1.7 Lemma. *Let A be a permutation algebra and let V be an A-module. Then $V_{(k)}$ is an $A_{(k)}$-bimodule by defining*

$$a \cdot x \cdot b = a \cdot \tau_{b,x}(b \cdot x)$$

for $a, b \in A, x \in V$.

Proof. We write $1_x = 1_n \in R_n$ for $x \in V^n$. Now we have for $a, b \in A$,

$$\begin{aligned}(a \cdot x) \cdot b &= \tau_{b,a\cdot x} b \cdot (a \cdot x) = \tau_{b,a\cdot x}(b \cdot a) \cdot x \\ &= \tau_{b,a\cdot x}(\tau_{a,b} a \cdot b) \cdot x \\ &= \tau_{b,a\cdot x}(\tau_{a,b} \odot 1_x)(a \cdot b \cdot x),\end{aligned}$$

$$a \cdot (x \cdot b) = a \cdot \tau_{b,x}(b \cdot x) = (1_a \odot \tau_{b,x})(a \cdot b \cdot x).$$

Here we have $\tau_{b,a\cdot x}(\tau_{a,b} \odot 1_x) = 1_a \odot \tau_{b,x}$ by one of the equations in (6.1.2)(5). □

Recall that each non-negatively graded k-module M is an R_*-module by use of $\epsilon : R_* \to k$. Such an R_*-module is termed an ϵ-module, see (6.1.4). A permutation algebra A for which A is an *ϵ-module* is the same as a *commutative graded algebra* over k with $A^0 = k$ since we have, by (6.1.3)(4),

$$\begin{aligned} y \cdot x &= \epsilon(\tau_{n,m})(x \cdot y) \\ &= (-1)^{|x||y|} x \cdot y. \end{aligned} \tag{6.1.8}$$

Moreover, we obtain as a special case of (6.1.7) the well-known lemma:

6.1.9 Lemma. *Let H be a commutative graded k-algebra and let M be an H-module. Then M is an H-bimodule by defining*

$$a \cdot x \cdot b = a \cdot (-1)^{|b||x|} b \cdot x$$

for $a, b \in H$ and $x \in M$.

6.1.10 Definition. Let V be a graded k-module concentrated in degree ≥ 1. Then the free R_*-module $R_* \odot V$ generated by V is given by

$$(R_* \odot V)^n = R_n \otimes V^n.$$

For an R_*-module W we obtain the *tensor algebra over R_** by

$$\overline{T}(W) = \bigoplus_{n \geq 0} W^{\bar{\otimes} n}$$

where $W^{\bar{\otimes} 0} = k$ and $W^{\bar{\otimes} n}$ is the n-fold $\bar{\otimes}$-product $W \bar{\otimes} \cdots \bar{\otimes} W$ defined in (6.1.4). For the usual tensor algebra $T(V)$ over k we get

$$\overline{T}(R_* \odot V) = R_* \odot T(V)$$

so that $R_* \odot T(V)$ is the free k-algebra over R_* generated by V with the multiplication

$$(\alpha \otimes x) \cdot (\beta \otimes y) = \alpha \odot \beta \otimes x \cdot y$$

for $\alpha, \beta \in R_*$ and $x, y \in T(V)$. Let

$$K_\tau \subset R_* \odot T(V) = A$$

be the R_*-submodule generated by elements $1 \otimes y \cdot x - \tau_{x,y} \otimes x \cdot y$ for $x, y \in T(V)$. Then K_τ generates the ideal $A \cdot K_\tau \cdot A$ and the R_*-quotient module

$$\mathrm{Perm}(V) = A/A \cdot K_\tau \cdot A$$

is the *free permutation algebra* generated by the graded k-module V. That is Perm is a functor

$$\mathrm{Perm} : \mathbf{Mod}(k)^{\geq 1} \longrightarrow \mathbf{Perm}(k) \tag{6.1.11}$$

where $\mathbf{Mod}(k)^{\geq 1}$ is the category of graded k-modules concentrated in degree ≥ 1 and $\mathbf{Perm}(k)$ is the category of permutation algebras in (6.1.3). Moreover, the functor Perm is left adjoint to the forgetful functor which carries a permutation algebra A to $\tilde{A} = A/A^0$.

Let W be a k-module (non-graded). Then the permutation group σ_n acts on the n-fold tensor product $W^{\otimes n}$ by permuting the factors. This action is used in the following result. Moreover, we have for $n, m \geq 1$ the inclusion

$$\sigma_n \subset \sigma_{n \cdot m}$$

which carries a permutation in σ_n to the corresponding permutation of the blocks $\{1, \dots, m\}$, $\{m+1, \dots, 2m\}$, …, $\{(n-1)m+1, \dots, n \cdot m\}$ in $\sigma_{n \cdot m}$. Hence for $n_1 \cdot m_1 + \cdots + n_k \cdot m_k = r$ we get the inclusion

$$\sigma_{n_1} \times \cdots \times \sigma_{n_k} \subset \sigma_{n_1 \cdot m_1} \times \cdots \times \sigma_{n_k \cdot m_k} \subset \sigma_r$$

which yields the ring homomorphism

$$R_{n_1} \otimes \cdots \otimes R_{n_k} \longrightarrow R_r$$

needed in the following formula.

6.1.12 Proposition. *Let $V = (V^m, m \in \mathbb{Z})$ be a graded k-module concentrated in degree ≥ 1. Then we have for $r \geq 0$,*

$$\mathrm{Perm}(V)^r = \bigoplus R_r \otimes_{R_{n_1} \otimes \cdots \otimes R_{n_k}} (V^{m_1})^{\otimes n_1} \otimes \cdots \otimes (V^{m_k})^{\otimes n_k}.$$

Here the direct sum is taken over the index set:

$$\begin{gathered} n_1 \cdot m_1 + \cdots + n_k \cdot m_k = r, \\ 1 \leq m_1 < m_2 < \cdots < m_k, \\ n_1, \dots, n_k \geq 0, \\ k \geq 0. \end{gathered}$$

For example, if V is concentrated in degree 1, then

$$\mathrm{Perm}(V) = T(V) = \bigoplus_{n\geq 0} V^{\otimes n} \tag{6.1.13}$$

is the tensor algebra with the action of σ_n on $V^{\otimes n}$ by permutation of factors. We have the natural transformation

$$\epsilon : \mathrm{Perm}(V) \longrightarrow \Lambda(V) \tag{6.1.14}$$

where $\Lambda(V)$ is the free commutative graded k-algebra generated by V. The transformation is induced by ϵ. We have

$$\Lambda(V) = E(V^{\mathrm{odd}}) \otimes S(V^{\mathrm{even}}) \tag{6.1.15}$$

where V^{odd} (V^{even}) is the part of V concentrated in odd (even) degrees. Moreover, $E(V^{\mathrm{odd}})$ denotes the exterior algebra and $S(V^{\mathrm{even}})$ is the symmetric algebra or polynomial algebra.

Now let A and B be permutation algebras. Then the R_*-tensor product $A\bar{\otimes}B$ is also a permutation algebra with the multiplication

$$(a \otimes b) \cdot (x \otimes y) = (1 \odot \tau_{x,b} \odot 1)(a \cdot x) \otimes (b \cdot y). \tag{6.1.16}$$

We have inclusions

$$\begin{aligned} i_1 : \quad & A = A\bar{\otimes}k && \longrightarrow A\bar{\otimes}B, \\ i_2 : \quad & B = k\bar{\otimes}B && \longrightarrow A\bar{\otimes}B. \end{aligned}$$

and one can check:

6.1.17 Lemma. $(A\bar{\otimes}B, i_1, i_2)$ *is a coproduct in the category* **Perm**(k) *of permutation algebras.*

6.2 Secondary permutation algebras

In (5.1.6) we introduced the notion of a crossed algebra which is the notion "crossed module" in the context of algebras. We now modify this concept for permutation algebras as follows. Let k be a commutative ring.

6.2.1 Definition. A *crossed permutation algebra* (A, ∂) over k is a permutation algebra A_0 as in (6.1.3) together with an A_0-module A_1 as in (6.1.6) and an A_0-linear map (of degree 0)

$$\partial : A_1 \longrightarrow A_0$$

satisfying for $x, y \in A_1$ the equation

$$(\partial x) \cdot y = \tau_{\partial y,x}(\partial y) \cdot x.$$

Here we use the notation in (6.1.4) so that $\tau_{\partial y,x} = \tau_{n,m}$ with $\mid y \mid = \mid \partial y \mid = n$ and $\mid x \mid = \mid \partial x \mid = m$.

6.2.2 Lemma. *A crossed permutation algebra* (A, ∂) *yields for the underlying* k*-modules a crossed algebra*

$$\partial : (A_1)_{(k)} \longrightarrow (A_0)_{(k)}$$

in the sense of (5.1.6).

Proof. In fact $(A_0)_{(k)}$ is a graded algebra over k and $(A_1)_{(k)}$ is an $(A_0)_{(k)}$-bimodule by using the definition in (6.1.7). Hence the equation in (6.2.1) is equivalent to $(\partial x) \cdot y = x \cdot (\partial y)$ in a crossed algebra. □

6.2.3 Example. Let $f : A_0 \to B$ be a map between permutation algebras in **Perm**(k) and let $A_1 = \text{kernel}(f)$. Then the inclusion

$$\partial : A_1 \longrightarrow A_0$$

is a crossed permutation algebra. Here A_1 is an A_0-module by multiplication in A_0. Moreover, we have for $x, y \in A_1$

$$(\partial x) \cdot y = \tau_{\partial y, x}(\partial y) \cdot x$$

since this equation holds in A_0, see (6.1.3)(4).

For an R_*-module V let $I(R_*) \odot_{R_*} V$ be the R_*-module defined in degree n by

(6.2.4) $$(I(R_*) \odot_{R_*} V)^n = I(R_n) \otimes_{R_n} V^n$$

Here we use the R_n-bimodule $I(R_n) = \text{kernel}(\epsilon : R_n \to k)$. We have the R_*-linear map

$$\mu : I(R_*) \odot_{R_*} V \longrightarrow V$$

which carries $a \otimes x$ to $a \cdot x$ for $a \in I(R_n) \subset R_n$ and $x \in V^n$.

In addition to the notion of a crossed permutation algebra in (6.2.1) we need the following concept which is motivated by the properties of secondary cohomology in Section (6.3).

6.2.5 Definition. A *secondary permutation algebra* is defined by a commutative diagram of R_*-linear maps

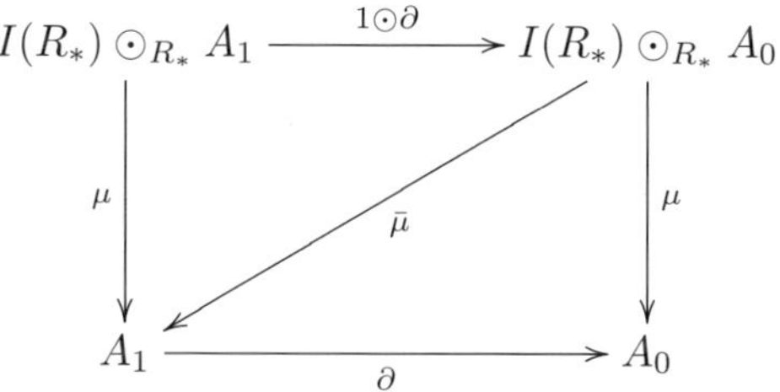

where ∂ is a crossed permutation algebra as in (6.2.1) and for $a, b \in A_0$, $\beta \in I(R_*)$ the equation

$$a \cdot \bar{\mu}(\beta \otimes b) = \bar{\mu}((1 \odot \beta) \otimes (a \cdot b))$$

holds. A map between such secondary permutation algebras is a map as in (6.2.1) which is compatible with $\bar{\mu}$. Let **secalg**(k) be the category of secondary permutation algebras over k.

6.2.6 Remark. Recall the concept of a module over a crossed algebra in (5.1.11) which has similarities but does not agree with the concept of a secondary permutation algebra above. However, in each degree n a secondary permutation algebra is a module over the crossed algebra $I(R_n) \to R_n$ (concentrated in degree 0). This generalizes the fact that a permutation algebra in degree n is a module over R_n (where R_n is also concentrated in degree 0). Compare also the discussion of secondary modules in [BSC].

Given a permutation algebra A_0 we define the A_0-bimodule structure of $I(R_*) \odot_{R_*} A_0$ by

$$\left\{ \begin{array}{lll} a \cdot (\beta \otimes b) & = & (1 \odot \beta) \otimes (a \cdot b), \\ (\alpha \otimes a) \cdot b & = & (\alpha \odot 1) \otimes (a \cdot b). \end{array} \right. \tag{6.2.7}$$

We have seen in (6.1.7) that an A_0-module A_1 is also an A_0-bimodule.

6.2.8 Lemma. *The equation in* (6.2.6) *is equivalent to the condition that*

$$\bar{\mu} : I(R_*) \odot_{R_*} A_0 \longrightarrow A_1$$

is a map of A_0*-bimodules.*

Proof. The equation in (6.2.6) shows that $\bar{\mu}$ is an A_0-linear map of left A_0-modules. Moreover, we get for $x = \bar{\mu}(\alpha \otimes a)$ with $| x |=| a |$ and hence $\tau_{b,x} = \tau_{b,a}$ the equations

$$\begin{aligned} \bar{\mu}(\alpha \otimes a) \cdot b &= \tau_{b,x} b \cdot \bar{\mu}(\alpha \otimes a) \\ &= \tau_{b,x} \bar{\mu}((1 \odot \alpha) \otimes (b \cdot a)) \\ &= \bar{\mu}(\tau_{b,a}(1 \odot \alpha) \otimes b \cdot a) \\ &= \bar{\mu}((\alpha \odot 1)\tau_{b,a} \otimes b \cdot a) \\ &= \bar{\mu}((\alpha \odot 1) \otimes \tau_{b,a} b \cdot a) \\ &= \bar{\mu}((\alpha \odot 1 \otimes a \cdot b) \\ &= \bar{\mu}((\alpha \otimes a) \cdot b). \end{aligned}$$

□

Similarly as in (6.2.2) we now get

6.2.9 Lemma. *A secondary permutation algebra A yields for the underlying k-modules a crossed algebra* $A_{(k)}$*,*

$$\partial_{(k)} : (A_1)_{(k)} \longrightarrow (A_0)_{(k)}$$

in the sense of (5.1.6)*, such that*

$$H = \pi_0 A_{(k)} = cokernel(\partial_{(k)})$$

is a commutative graded k-algebra and

$$D = \pi_1 A_{(k)} = kernel(\partial_{(k)})$$

is an H-bimodule with the H-bimodule structure in (6.1.8).

Hence secondary permutation algebras are "appropriate resolutions" of *commutative* graded algebras like the cohomology of a space, while crossed algebras as in (5.1.6) are "resolutions" for graded algebras in general.

6.2.10 Definition. Let A be a permutation algebra and let V be a k-module concentrated in degree ≥ 1 and let

$$d : V \longrightarrow A$$

be a k-linear map of degree 0. Then the *free* secondary permutation algebra $A(d)$ generated by d is defined by the following universal property. We have $A(d)_0 = A$ and $V \xrightarrow{i} A(d)_1$ with $\partial i = d$ and for each commutative diagram of k-linear maps

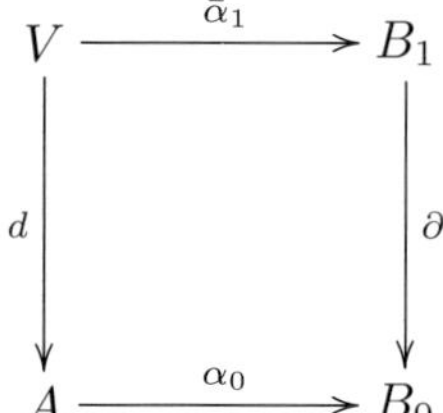

where $B = (\partial : B_1 \to B_0) \in$ **secalg** and $\alpha_0 \in$ **Perm**, there is a unique map $\alpha = (\alpha_0, \alpha_1) : A(d) \to B$ in **secalg** for which α_1 extends $\bar{\alpha}_1$, that is, the following diagram commutes.

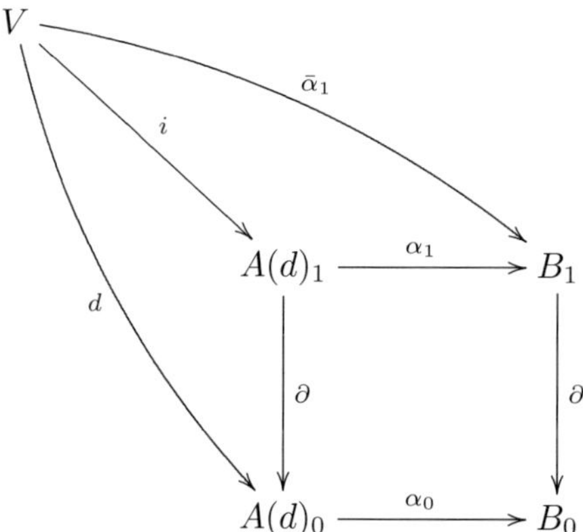

6.2.11 Proposition. *The free secondary permutation algebra $A(d)$ exists.*

Proof. We construct $A(d)_1$ and ∂ as follows. Recall the definition of $R_* \odot V$ in (6.1.10). Then we first obtain the following push out diagram in the category of R_*-modules.

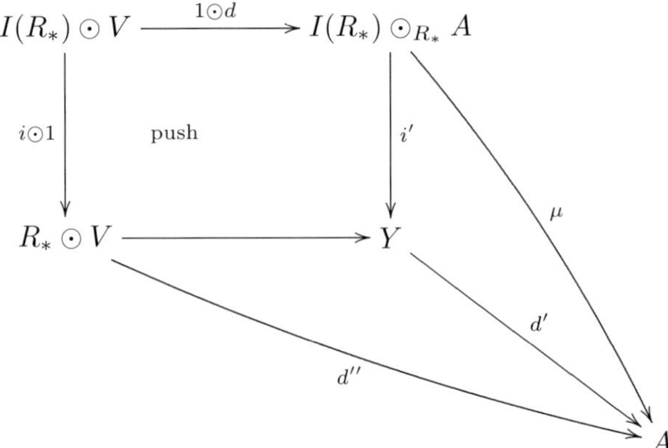

Here d'' is defined by $d''(\alpha \otimes x) = \alpha \cdot d(x)$. The pair (μ, d'') induces the R_*-linear map d' which thus determines the map of A-modules ∂' and ∂ in the following commutative diagram.

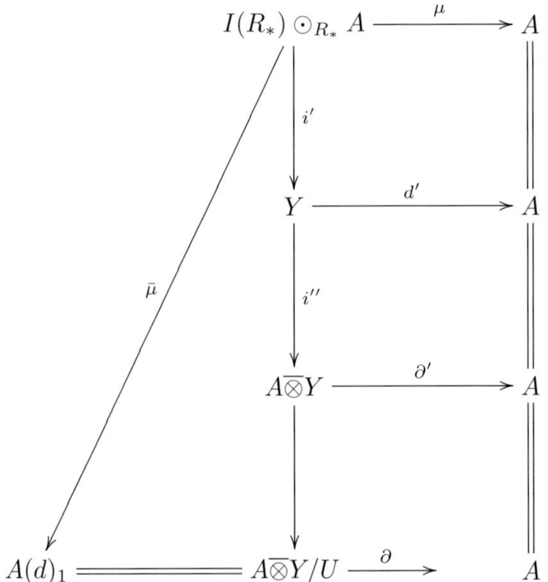

Here i'' is defined by $i''(y) = 1 \otimes y$ and ∂' is defined by $\partial'(a \otimes y) = a \cdot d'(y)$. Let U be the A-submodule of $A \overline{\otimes} Y$ generated by the elements

$$\begin{array}{rcl} (\partial' x) \cdot y & - & \tau_{\partial' y, x}(\partial' y) \cdot x, \\ a \cdot j(\beta \otimes b) & - & j(1 \odot \beta \otimes a \cdot b), \text{with } j = i'' i \end{array}$$

with $x, y \in A\overline{\otimes}Y$ and $a, b \in A$, $\beta \in I(R_*)$. One readily checks that U is in the kernel of ∂' so that ∂' induces the A-module map ∂ on the quotient $A(d)_1 = A\overline{\otimes}Y/U$.

Now one can check that $(A(d), \partial, \bar{\mu})$ is a well-defined secondary permutation algebra with the universal property in (6.2.10). □

6.3 Secondary cohomology as a secondary permutation algebra

Let k be a commutative ring and let $\mathbf{Top}_0^*$ be the category of path-connected pointed spaces and pointed maps. We define the *secondary cohomology functor*

$$\mathcal{H}^* : (\mathbf{Top}_0^*)^{\mathrm{op}} \longrightarrow \mathbf{secalg}(k). \tag{6.3.1}$$

Here the right-hand side is the category of secondary permutation algebras over k in (6.2.5). For the ring k we have Eilenberg-MacLane spaces

$$Z^n = K(k, n)$$

defined in Section (2.1). Then a space X in $\mathbf{Top}_0^*$ yields $\mathcal{H}^n(X) = \mathcal{H}^n(X, k)$ as in (2.2.10) by

$$\mathcal{H}^n(X)_1 = [\![X, Z^n]\!]_1^0 \xrightarrow{\partial} [\![X, Z^n]\!]_0 = \mathcal{H}^n(X)_0 \tag{1}$$

with $n \geq 1$. Moreover, let $\mathcal{H}^i(X)_1 = 0$ for $i \leq 0$ and $\mathcal{H}^0(X)_0 = k$ and $\mathcal{H}^1(X)_0 = 0$ for $i < 0$. By construction of Z^n in (2.1.4) the permutation group σ_n acts via k-linear maps on Z^n. Therefore $\mathcal{H}^n(X)_1$ and $\mathcal{H}^n(X)_0$ are R_n-modules and ∂ is R_n-linear for $n \geq 0$. Here $R_n = k[\sigma_n]$ is the group algebra. The multiplication map

$$\mu = \mu_{m,n} : Z^m \times Z^n \longrightarrow Z^{m+n}$$

in (2.1.1) is k-bilinear and $(\sigma_m \times \sigma_n \subset \sigma_{m+n})$-equivariant and satisfies $\mu_{n,m}T = \tau\mu_{m,n}$ by (2.1.2). Therefore $\mathcal{H}^*(X)$ is a *crossed permutation algebra* as in (6.2.1) with *multiplication* induced by $\mu_{n,m}$. That is, for $x \in \mathcal{H}^m(X)_0$, $y \in \mathcal{H}^n(X)_0$ we define

$$x \cdot y = \mu_{m,n}(x, y) : X \longrightarrow Z^m \times Z^n \longrightarrow Z^{m+n}. \tag{2}$$

Moreover, for $a \in \mathcal{H}^m(X)_1$, $b \in \mathcal{H}^n(X)_1$ we define similarly $a \cdot y = \mu_{m,n}(a, y)$ and $x \cdot b = \mu_{m,n}(x, b)$. One readily checks that $(\mathcal{H}^*(X), \partial)$ is a well-defined crossed permutation algebra. Moreover, $\mathcal{H}^*(X)$ is a *secondary permutation algebra*, as in (6.2.8), by the map

$$\bar{\mu} : I(R_n) \odot_R \mathcal{H}^n(X)_0 \longrightarrow \mathcal{H}^n(X)_1 \tag{3}$$

defined as follows. We know that the mapping groupoid $[\![Z^n, Z^n]\!]$ has contractible connected components, see (3.2.5). Moreover, the action of σ_n on Z^n yields the

k-linear map $R_n \to [\![Z^n, Z^n]\!]_0$ which carries $\sigma \in \sigma_n$ to $\sigma \cdot 1_{Z^n}$. Here the homotopy class of σ is given by $\epsilon(\sigma) = sign(\sigma) \in \{1, -1\}$, see (2.1.3). Therefore there is a unique track

(4) $$\Gamma_\sigma : \sigma \cdot 1_{Z^n} \Longrightarrow \epsilon(\sigma) \cdot 1_{Z^n}$$

where 1_{Z^n} is the identity on Z^n. We call Γ_σ the *permutation track* of $\sigma \in R_n$. The k-module structure of Z^n yields

(5) $$\Gamma_{\sigma-\epsilon\sigma} = \Gamma_\sigma - \epsilon(\sigma) \cdot 1_{Z^n} : (\sigma - \epsilon\sigma) \cdot 1_{Z^n} \Longrightarrow 0$$

and we define $\bar{\mu}$ in (3) by composition of x and $\Gamma_{\sigma-\epsilon\sigma}$, that is

(6) $$\bar{\mu}((\sigma - \epsilon\sigma) \otimes x) = \Gamma_{\sigma-\epsilon\sigma} \circ x$$

for $(x : X \to Z^n) \in \mathcal{H}^n(X)_0$. For elements x, y as in (2) we obtain the *interchange track* by use of (4), that is,

(7) $$\begin{aligned} &T(x,y) : x \cdot y \Longrightarrow (-1)^{|x||y|} y \cdot x, \\ &T(x,y) = \Gamma_{\tau(y,x)} \circ (y \cdot x) \in [\![X, Z^{|x|+|y|}]\!]. \end{aligned}$$

Here $\tau(y,x)$ is the interchange permutation with $\tau(y,x) \cdot y \cdot x = x \cdot y$.

6.3.2 Lemma. *$(\mathcal{H}^*(X), \partial, \bar{\mu})$ is a well-defined secondary permutation algebra.*

This shows that we have a well-defined functor $\mathcal{H}^*$ as in (6.3.1).

Proof. The diagram in (6.2.8) commutes since

(1) $$\partial\bar{\mu}((\sigma - \epsilon\sigma) \otimes x) = (\partial\Gamma_{\sigma-\epsilon\sigma}) \circ x = (\sigma - \epsilon\sigma) \cdot x$$

and since for $a \in \mathcal{H}^*(X)_1$ we have:

(2) $$\begin{aligned} \bar{\mu}(1 \odot \partial)((\sigma - \epsilon\sigma) \otimes a) &= \Gamma_{\sigma-\epsilon\sigma} \circ \partial a \\ &= (\partial\Gamma_{\sigma-\epsilon\sigma}) \circ a, \quad \text{see (5.1.5)(3)}, \\ &= (\sigma - \epsilon\sigma) \cdot a. \end{aligned}$$

Moreover, the equation in (6.2.8) holds since

(3) $$a \cdot \bar{\mu}((\sigma - \epsilon\sigma) \otimes b) = a \cdot (\Gamma_{\sigma-\epsilon\sigma} \circ b),$$

(4) $$\bar{\mu}((1 \odot (\sigma - \epsilon\sigma)) \otimes (a \cdot b)) = \Gamma_{1\odot(\sigma-\epsilon\sigma)} \circ (a \cdot b).$$

Here $\circ$ is composition and $\cdot$ is multiplication defined in (6.3.1)(2). Now (3) coincides with (4) since for $\mu : Z^m \wedge Z^n \to Z^{m+n}$ in (2.1.5) we have:

(5) $$\mu \circ (1_{Z^m} \wedge \Gamma_{\sigma-\epsilon\sigma}) = \Gamma_{1\odot(\sigma-\epsilon\sigma)} \circ \mu.$$

In fact, both sides are tracks

$$(1 \odot (\sigma - \epsilon\sigma))\mu \Longrightarrow 0$$

in $[\![Z^m \wedge Z^n, Z^{m+n}]\!]$ and this track is unique by (3.2.5). □

6.4 Induced homotopies

Let $f, g : X \to Y$ be maps in $\mathbf{Top}_0^*$ and let $H : f \Rightarrow g$ be a track. Then we obtain the diagram

$$\begin{array}{ccc} \mathcal{H}^*(Y)_1 & \xrightarrow{f_1^*, g_1^*} & \mathcal{H}^*(X)_1 \\ \Big\downarrow \partial & & \Big\downarrow \partial \\ \mathcal{H}^*(Y)_0 & \xrightarrow{f_0^*, g_0^*} & \mathcal{H}^*(X)_0 \end{array}$$

where $f^* = (f_0^*, f_1^*)$ and $g^* = (g_0^*, g_1^*)$ are maps $\mathcal{H}^*(Y) \to \mathcal{H}^*(X)$ in the category $\mathbf{secalg}(k)$.

Now the track $H : f \Rightarrow g$ defines the induced R_*-linear map

(6.4.1)
$$\begin{aligned} &H^* : \mathcal{H}^*(Y)_0 \longrightarrow \mathcal{H}^*(X)_1, \\ &H^*(x) = x \circ H - x \circ g. \end{aligned}$$

We have the following formulas:

(1)
$$\begin{aligned} \partial H^*(x) &= \partial(x \circ H - x \circ g) \\ &= x \circ f - x \circ g \\ &= f_0^*(x) - g_0^*(x) \\ &= (f_0^* - g_0^*)(x). \end{aligned}$$

Using (2.3.1) we get for $a \in \mathcal{H}^*(X)_1$,

(2)
$$\begin{aligned} a * H &= (0H)\square(af) \\ &= af \\ &= (ag)\square(\partial a)H. \end{aligned}$$

Therefore H^* in (6.4.1) satisfies

$$\begin{aligned} H^*(\partial a) &= (\partial a)H - (\partial a)g \\ &= ((ag)^{\mathrm{op}}\square af) - (\partial a)g \\ &= ((ag)^{\mathrm{op}} - (\partial a)g)\square(af - (\partial a)g) \\ &= (-ag)\square(af - (\partial a)g) && (2.2.6) \\ &= -ag + af && (2.2.6) \\ &= f_1^*(a) - g_1^*(a). \end{aligned}$$

Next we consider for the diagonal map Δ the commutative diagram

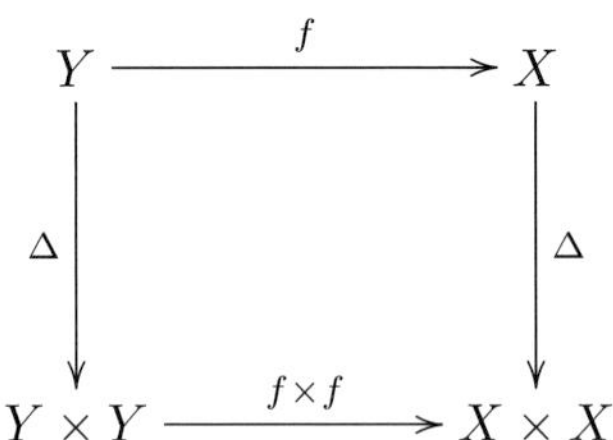

and the track $H \times H : f \times f \Rightarrow g \times g$ with $\Delta H = (H \times H)\Delta$. Here we have

$$\begin{aligned} H \times H &= (g \times H)\square(H \times f) \\ &= (H \times g)\square(f \times H). \end{aligned}$$

For $x, y \in \mathcal{H}^*(Y)_0$ w get the product $x \cdot y = \mu(x \times y)\Delta$. Hence we have

$$\begin{aligned} H^*(x \cdot y) &= (x \cdot y)H - (x \cdot y)g \\ &= \mu(x \times y)\Delta H - (x \cdot y)g \\ &= \mu(x \times y)((H \times g)\square(f \times H))\Delta - (xg) \cdot (yg) \\ &= (\mu(x \times y)(H \times g)\Delta)\square(\mu(x \times y)(f \times H)\Delta) - (xg) \cdot (yg) \\ &= (xH \cdot yg)\square(xf \cdot yH) - xg \cdot yg \\ &= ((H^*x + xg) \cdot yg)\square(xf \cdot (H^*(y) + yg)) - xg \cdot yg \\ &= \underbrace{H^*x \cdot yg + xg \cdot yg)}_{A} \square \underbrace{(xf \cdot H^*(y) + xf \cdot yg)}_{B} - xg \cdot yg. \end{aligned}$$

Thus we get

$$(3) \qquad \begin{aligned} H^*(x \cdot y) &= (A - xg \cdot yg)\square(B - xg \cdot yg) \\ &= (H^*x \cdot yg)\square(xf \cdot H^*y + xf \cdot yg - xg \cdot yg) \\ &= (H^*x \cdot yg) + (xf \cdot H^*y), \quad \text{see (2.2.6)(3).} \end{aligned}$$

Finally, we consider the connection of H^* and $\bar{\mu}$ in (6.3.1)(3). For the tracks

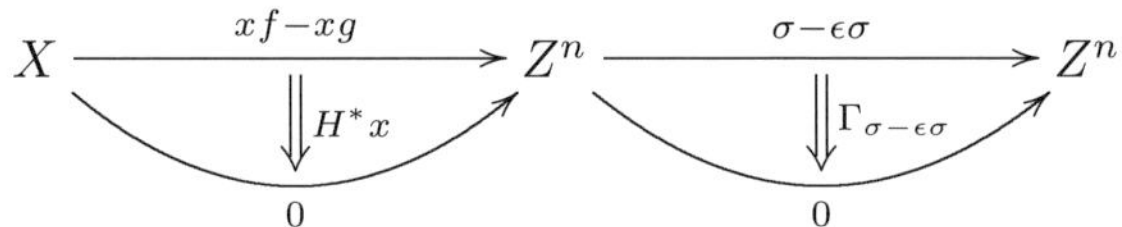

we get the formula

$$(4) \qquad \begin{aligned} (\sigma - \epsilon\sigma)H^*x &= (\Gamma_{\sigma-\epsilon\sigma})(xf - xg) \\ &= \bar{\mu}((\sigma - \epsilon\sigma) \otimes (xf - xg)). \end{aligned}$$

Hence the following diagram commutes.

$$\begin{array}{ccc} I(R_*)\odot_{R_*}\mathcal{H}^*(Y)_0 & \xrightarrow{1\odot H^*} & I(R_*)\odot_{R_*}\mathcal{H}^*(X)_1 \\ \Big\downarrow{\scriptstyle 1\odot(f_0^*-g_0^*)} & & \Big\downarrow{\scriptstyle \mu} \\ I(R_*)\odot_{R_*}\mathcal{H}^*(X)_0 & \xrightarrow{\bar{\mu}} & \mathcal{H}^*(X)_1 \end{array}$$

The properties of H^* above lead to the following definition:

6.4.2 Definition. Let $f = (f_0, f_1)$ and $g = (g_0, g_1)$ be maps $(A, \partial) \to (B, \partial)$ between secondary permutation algebras in the category **secalg**(k) in (6.2.8). A *homotopy* or *track* $H : f \Rightarrow g$ is an R_*-linear map

$$H : A_0 \longrightarrow B_1$$

with the following properties ($x, y \in A_0$):

(1) $$\partial H = f_0 - g_0,$$

(2) $$H\partial = f_1 - g_1,$$

(3) $$H(x \cdot y) = (Hx) \cdot (g_0 y) + (f_0 x) \cdot (Hy),$$

(4) and the following diagram commutes.

$$\begin{array}{ccc} I(R_*)\odot_{R_*}A_0 & \xrightarrow{1\odot H} & I(R_*)\odot_{R_*}B_1 \\ \Big\downarrow{\scriptstyle 1\odot(f_0-g_0)} & & \Big\downarrow{\scriptstyle \mu} \\ I(R_*)\odot_{R_*}B_0 & \xrightarrow{\bar{\mu}} & B_1 \end{array}$$

Here property (4) is redundant so that a homotopy is an R_*-linear map satisfying (1), (2) and (3). In fact, since B is a secondary permutation algebra we have by (6.2.5) the equation ($\xi \in I(R_q)$, $x \in A_0, |x| = q$),

$$\xi \cdot H(x) = \bar{\mu}(\xi \odot \partial H(x)) = \bar{\mu}(\xi \odot (f_0 - g_0)(x))$$

and hence diagram (4) commutes.

6.4.3 Proposition. *The category* **secalg**(k) *with tracks as in* (6.4.2) *is a track category* $[\![\mathbf{secalg}(k)]\!]$ *as in* (2.3.1).

Proof. The composition of tracks $H : f \Rightarrow g$ and $G : g \Rightarrow h$ is given by the sum

$$G\Box H = G + H$$

and the trivial track $0^{\Box}$ of f is the trivial map $0 : A_0 \to B_1$. Composition is defined by

$$\begin{aligned} H \circ a &= H \circ a_0, \\ b \circ H &= b_1 \circ H. \end{aligned}$$

Now one readily checks the properties in (2.3.1). □

6.4.4 Theorem. *Secondary cohomology* $\mathcal{H}^*$ *is a track functor* (2.3.6) *from the track category* $[\![\mathbf{Top}_0^*]\!]^{\mathrm{op}}$ *to the track category* $[\![\mathbf{secalg}(k)]\!]$ *above. This functor carries a track* $H : f \Rightarrow g$ *in* $\mathbf{Top}^*$ *to the track* $H^* : f^* \Rightarrow g^*$ *in* (6.4.1).

Hence we have the induced functor between homotopy categories

(6.4.5) $$\mathcal{H}^* : (\mathbf{Top}_0^*/\simeq)^{\mathrm{op}} \longrightarrow \mathbf{secalg}(k)/\simeq .$$

Therefore the homotopy type of $\mathcal{H}^*(X) = \mathcal{H}^*(X, k)$ is an invariant of the homotopy type of X. This invariant also carries some information on Steenrod squares for $k = \mathbb{F} = \mathbb{Z}/2$ as we see in the next section.

6.5 Squaring maps

Let $A = (\partial : A_1 \to A_0)$ be a secondary permutation algebra over the commutative ring k as in (6.2.5) with

$$\bar{\mu} : I(R_*) \odot_{R_*} A_0 \longrightarrow A_1.$$

Then we have the exact sequence

$$0 \longrightarrow \pi_1(A) \longrightarrow A_1 \xrightarrow{\partial} A_0 \longrightarrow \pi_0(A) \longrightarrow 0.$$

Here $\pi_1(A) = \mathrm{kernel}(\partial)$ and $\pi_0(A) = \mathrm{cokernel}(\partial)$ are graded k-modules. The following result defines the *squaring map* Sq for A. Let $\tau_{n,n} \in \sigma_{2n}$ be defined as in (6.1.2)(4) with $\mathrm{sign}(\tau_{n,n}) \in k$.

6.5.1 Proposition. *Let* $n \geq 1$ *with* $\mathrm{sign}(\tau_{n,n}) = 1$. *Then there is a well-defined* k*-linear map*

$$Sq : \pi_0(A)^n \longrightarrow \pi_1(A)^{2n}$$

which carries the element $\{x\} \in \pi_0 A$ *represented by* $x \in A_0^n$ *to*

$$\bar{\mu}((\tau_{n,n} - 1) \otimes (x \cdot x)) \in \pi_1 A.$$

Moreover, we get $2Sq = 0$.

For example if n is even we have for $\tau = \tau_{n,n}$ the equation $\epsilon(\tau) = \text{sign}(\tau) = (-1)^{n\cdot n} = 1$.

Proof. We have $\tau - 1 \in I(R_*)$ and we have

$$\text{(1)}\quad \begin{aligned} \partial\bar{\mu}((\tau-1)\otimes(x\cdot x)) &= (\tau-1)(x\cdot x) \\ &= \tau(x\cdot x) - x\cdot x \\ &= x\cdot x - x\cdot x \\ &= 0. \end{aligned}$$

since $\tau(x\cdot x) = x\cdot x$ by (6.1.3)(4). Therefore $Sq(x) = \bar{\mu}((\tau-1)\otimes(x\cdot x)) \in \pi_1(A)$. Now let $y = x + \partial a$ for $a \in A_1$. We have to show $Sq(x) = Sq(y)$. This is equivalent to $\Delta = 0$ where

$$\text{(2)}\quad \Delta = \bar{\mu}((\tau-1)\otimes(x\cdot b + b\cdot x + b\cdot b))$$

with $b = \partial a$. We have $\tau\tau = 1$ so that

$$0 = \tau\tau - 1 = (\tau-1)\tau + (\tau-1).$$

Hence we get

$$\text{(3)}\quad \begin{aligned} 0 &= \bar{\mu}(((\tau-1)\tau + (\tau-1))\otimes(b\cdot x)) \\ &= \bar{\mu}(\tau-1)\otimes\tau(b\cdot x)) + \bar{\mu}((\tau-1)\otimes b\cdot x) \\ &= \bar{\mu}(\tau-1)\otimes x\cdot b) + \bar{\mu}((\tau-1)\otimes b\cdot x) \\ &= \bar{\mu}((\tau-1)\otimes(x\cdot b + b\cdot x)). \end{aligned}$$

Moreover, we get for $b = \partial a$ the formula

$$\partial(b\cdot a) = b\cdot(\partial a) = b\cdot b$$

so that

$$\text{(4)}\quad \begin{aligned} \bar{\mu}((\tau-1)\otimes(b\cdot b)) &= \bar{\mu}((\tau-1)\otimes\partial(b\cdot a)) \\ &= (\tau-1)(b\cdot a) \\ &= \tau(\partial a)\cdot a - (\partial a)\cdot a \\ &= 0, \qquad \text{see (6.2.1).} \end{aligned}$$

By (3) and (4) we see that Δ in (2) is trivial so that Sq in (6.5.1) is a well-defined function.

For $r \in k$ we get

$$\text{(5)}\quad \begin{aligned} Sq(r\cdot x) &= \mu((1-\tau)\otimes(rx\cdot rx)) \\ &= r^2\mu((1-\tau)\otimes(x\cdot x)) \\ &= r^2 Sq(x). \end{aligned}$$

Moreover, for $x, y \in A_0^n$ we get

$$\begin{aligned} Sq(x+y) &= \mu((1-\tau)\otimes(x+y)\cdot(x+y)) \\ &= Sq(x)+Sq(y)+\mu((1-\tau)\otimes(x\cdot y+y\cdot x)) \\ &= Sq(x)+Sq(y). \end{aligned} \tag{6}$$

Here we use the argument in (3) where we replace b by y. Now we have $Sq(0)=0$ and $0=Sq(a-a)=Sq(a)+Sq(-a)$ where $Sq(-a)=(-1)^2Sq(a)=Sq(a)$ by (5). Therefore $2Sq=0$. □

For $A=\mathcal{H}^*(X)$ in (6.3.1) we have

$$\begin{aligned} \pi_0\mathcal{H}^*(X) &= H^*(X), \\ \pi_1\mathcal{H}^*(X) &= \Sigma\tilde{H}^*(X), \end{aligned}$$

where $H^*(X)$ is the cohomology ring of X. In this case we get:

6.5.2 Theorem. *For $\mathbb{F}=\mathbb{Z}/2$ the ϵ-crossed permutation algebra $\mathcal{H}^*(X)$ yields by* (6.5.1) *the squaring map*

$$Sq : H^n(X) \longrightarrow (\Sigma\tilde{H}^*X)^{2n} = H^{2n-1}(X)$$

which coincides with the Steenrod square Sq^{n-1} for $n \geq 1$.

Proof. This is a consequence of (4.5.8). In fact the map

$$\alpha : Z^n \xrightarrow{\Delta} Zn \times Z^n \xrightarrow{\mu_{n,n}} Z^{2n}$$

which carries $z \in Z^n$ to $z\cdot z$ represents the Steenrod square Sq^n by ($\mathcal{K}$2) in (1.1.7). For the stable map α we have the track Γ_α in (4.2.1) as in the diagram

$$\begin{array}{ccc} Z^n\times Z^n & \xrightarrow{\alpha\times\alpha} & Z^{2n}\times Z^{2n} \\ {\scriptstyle +}\downarrow & \overset{\Gamma_\alpha}{\Longrightarrow} & \downarrow{\scriptstyle +} \\ Z^n & \xrightarrow[\alpha]{} & Z^{2n} \end{array}$$

which for $x : X \to Z^n$ yields $\Gamma_\alpha^{x,x} = \Gamma_\alpha(x,x) : \alpha(2x)=0 \Rightarrow 2\alpha(x)=0$ and we know by (4.5.8) that $\Gamma_\alpha^{x,x}=\Gamma[2](x)$ is given by $Sq^{n-1}(x)$. On the other hand we get for $y, z \in Z^n$ the formulas

$$\begin{aligned} (\alpha+)(y,z) &= \alpha(y+z) \\ &= (y+z)\cdot(y+z) \\ &= y^2+z^2+yz+zy, \end{aligned}$$

$$\begin{aligned} +(\alpha\times\alpha)(y,z) &= \alpha y+\alpha z \\ &= y^2+z^2. \end{aligned}$$

Here we have $zy = \tau_{n,n} yz$ so that for $\tau = \tau_{n,n}$ the track $\Gamma_\tau : \tau \Rightarrow \epsilon(\tau) = 1$ in (6.3.1)(4) yields the track

$$\Gamma_\tau yz : zy = \tau yz \Longrightarrow yz$$

and hence the track

$$yz + \Gamma_\tau yz : yz + zy \Longrightarrow yz + yz = 0.$$

Therefore the track Γ_α above coincides with the track $\bar{\Gamma}_\alpha$ where

$$\bar{\Gamma}_\alpha(y,z) = y^2 + z^2 + yz + \Gamma_\tau yz : (\alpha+)(y,z) \Longrightarrow +(\alpha \times \alpha)(y,z)$$

since $i_1^* \bar{\Gamma}_\alpha = 0_\alpha^\square$ and $i_2^* \bar{\Gamma}_\alpha = 0_\alpha^\square$. Here we use the uniqueness in (4.2.1) and $[Z^n \wedge Z^n, \Omega Z^{2n}] = 0$. On the other hand we have

$$\begin{aligned}
\bar{\Gamma}_\alpha(x,x) &= x^2 + x^2 + x^2 + \Gamma_\tau x^2 \\
&= -x^2 + \Gamma_\tau x^2 \\
&= (\Gamma_\tau - 1) \circ x^2 \\
&= (\Gamma_{\tau - 1}) \circ x^2, \qquad \text{since } \Gamma_\tau - 1 = \Gamma_{\tau-1}, \\
&= \bar{\mu}((\tau - 1) \otimes x^2) \\
&= Sq(x).
\end{aligned}$$

This proves the result. □

6.6 Secondary cohomology of a product space

We have seen in (6.1.17) that $A_0 \bar{\otimes} B_0$ is the coproduct in the category **Perm**(k). If U is an A_0-module and W is a B_0-module as in (6.1.6), then $U \bar{\otimes} W$ is an $A_0 \bar{\otimes} B_0$-module by

(6.6.1) $$(a \otimes b) \cdot (u \otimes w) = (1 \odot \tau_{u,b} \odot 1)(a \cdot u \otimes b \cdot w).$$

Now let A and B be crossed permutation algebras as in (6.2.1). Then $X_0 = A_0 \bar{\otimes} B_0$ is a permutation algebra and by (6.6.1) $A_1 \bar{\otimes} B_1$, $A_0 \bar{\otimes} B_1$ and $B_0 \bar{\otimes} A_1$ are X_0-modules. We obtain the chain complex of X_0-modules and X_0-linear maps:

(6.6.2) $$A_1 \bar{\otimes} B_1 \xrightarrow{d_2} A_0 \bar{\otimes} B_1 \oplus B_0 \bar{\otimes} A_1 \xrightarrow{d_1} A_0 \bar{\otimes} B_0,$$

$$\begin{aligned}
d_2(a \otimes b) &= (\partial a) \otimes b - \tau_{b,a}(\partial b) \otimes a, \\
d_1(x \otimes b) &= x \otimes (\partial b), \\
d_1(y \otimes a) &= \tau_{a,y}((\partial a) \bar{\otimes} y),
\end{aligned}$$

for $a \in A_1$, $x \in A_0$ and $b \in B_1$, $y \in B_0$. We have $d_1 d_2 = 0$ since

$$\begin{aligned} d_1 d_2(a \otimes b) &= d_1((\partial a) \otimes b - \tau_{b,a}(\partial b) \otimes a) \\ &= (\partial a) \otimes (\partial b) - \tau_{b,a}\tau_{\partial a,\partial b}(\partial a) \otimes (\partial b) \\ &= 0. \end{aligned}$$

Here we see $\tau_{\partial a,\partial b} = \tau_{a,b}$ and $\tau_{b,a}\tau_{a,b} = 1$. The chain complex above yields the X_0-module

(1) $$X_1 = (A \bar{\otimes} B)_1 = \operatorname{cokernel}(d_2)$$

and d_1 induces the X_0-linear map

(2) $$A \bar{\otimes} B = (\partial : (A \bar{\otimes} B)_1 \longrightarrow (A \bar{\otimes} B)_0 = A_0 \bar{\otimes} B_0).$$

6.6.3 Proposition. *$A \bar{\otimes} B$ is a well-defined crossed permutation algebra which is the coproduct of A and B in the category of crossed permutation algebras.*

Proof. We have to check that $\partial : X_1 \to X_0$ satisfies the formula in (6.2.1), that is, for $u, v \in X_1$ we have

(1) $$(\partial u) \cdot v = \tau_{v,u}(\partial v) \cdot u.$$

In fact for $u = x \otimes b$, $v = x' \otimes b'$ we prove (1) as follows.

(2) $$\begin{aligned} (\partial u) \cdot v &= (x \otimes \partial b) \cdot (x' \otimes b') \\ &= (1 \odot \tau_{x',b} \odot 1)(x \cdot x' \otimes (\partial b) \cdot b') \\ &= (1 \odot \tau_{x',b} \odot 1)(\tau_{x',x} \odot \tau_{b',b})(x' \cdot x \otimes (\partial b') \cdot b) \\ &= (1 \odot \tau_{x',b} \odot 1)(\tau_{x',x} \odot \tau_{b',b})(1 \odot \tau_{\partial b',x} \odot 1)(x' \otimes \partial b')(x \otimes b) \\ &= \tau_{v,u}(\partial v) \cdot u. \end{aligned}$$

In a similar way we prove (1) for $u = y \otimes a$ and $v = y' \otimes a'$. Moreover for $u = x \otimes b$ and $v = y \otimes a$ we get

(3) $$\begin{aligned} (\partial u) \cdot v &= (x \otimes \partial b) \cdot (y \otimes a) \\ &= (1 \odot \tau_{y,b} \odot 1)(x \cdot y \otimes (\partial b) \cdot a) \\ &= (1 \odot \tau_{y,b} \odot 1)(\tau_{y,x} \odot \tau_{a,b})(y \cdot x \otimes (\partial a) \cdot b) \\ &= (1 \odot \tau_{y,b} \odot 1)(\tau_{y,x} \odot \tau_{a,b})(1 \odot \tau_{a,x} \odot 1)(y \otimes \partial a) \cdot (x \otimes b) \\ &= \tau_{v,u}(\partial v) \cdot u. \end{aligned}$$

This completes the proof of (1) and hence $A \bar{\otimes} B$ is a well-defined crossed permutation algebra. The inclusions

$$\begin{aligned} i_1 : \; A = A \bar{\otimes} k &\longrightarrow A \bar{\otimes} B, \\ i_2 : \; B = k \bar{\otimes} B &\longrightarrow A \bar{\otimes} B \end{aligned}$$

are defined by $k \subset A$ and $k \subset B$ respectively. Moreover maps $f : A \to D$ and $g : B \to D$ between crossed permutation algebras yield a unique map

$$(f, g) : A \bar{\otimes} B \longrightarrow D \tag{4}$$

with $(f, g)i_1 = f$ and $(f, g)i_2 = g$. We obtain

$$(f, g)_0 = (f_0, g_0) : A_0 \bar{\otimes} B_0 \longrightarrow D_0 \tag{5}$$

by (6.1.17). Moreover we define $(f, g)_1 : (A \bar{\otimes} B)_1 \to D_1$ by

$$\begin{array}{rcl} (f, g)_1(x \otimes b) & = & (f_0 x) \cdot (g_1 b), \\ (f, g)_1(y \otimes a) & = & (g_0 y) \cdot (f_1 a). \end{array} \tag{6}$$

One can check that (f, g) is $(f, g)_0$-equivariant and that $\partial(f, g)_1 = (f, g)_0\partial$ so that the map (f, g) is well defined. □

Now let A and B be secondary permutation algebras with

$$\begin{array}{rcl} \bar{\mu}_A : I(R_*) \odot_{R_*} A_0 & \longrightarrow & A_1, \\ \bar{\mu}_B : I(R_*) \odot_{R_*} B_0 & \longrightarrow & B_1 \end{array}$$

as in (6.2.5). Then we obtain the following diagram where $A \bar{\otimes} B$ is the coproduct of the underlying crossed permutation algebras, see (6.6.3).

(6.6.4)

$$\begin{array}{ccc} A_0 \bar{\otimes}(I(R_*) \odot_{R_*} B_0) \oplus B_0 \bar{\otimes}(I(R_*) \odot_{R_*} A_0) & \xrightarrow{\tilde{\mu}} & (A \bar{\otimes} B)_1 \\ \Big\downarrow \tilde{q} & \text{push} & \Big\downarrow q \\ I(R_*) \odot_{R_*} (A_0 \bar{\otimes} B_0) & \xrightarrow{\bar{\mu}} & (A \bar{\bar{\otimes}} B)_1 \end{array}$$

This is a push out diagram of $(A_0 \bar{\otimes} B_0)$-modules and $A_0 \bar{\otimes} B_0$-linear maps. The map $\tilde{\mu}$ is defined by $A_0 \bar{\otimes} \mu_B \oplus B_0 \bar{\otimes} \mu_A$, see (6.6.2), and $\tilde{q}$ is defined by

$$\begin{array}{rcl} \tilde{q}(a \otimes (\lambda \otimes b)) & = & (1 \odot \lambda) \otimes (a \otimes b), \\ \tilde{q}(b' \otimes (\lambda' \otimes a')) & = & (\lambda' \odot 1) \otimes (a' \otimes b'), \end{array} \tag{1}$$

with $a \in A_0$, $\lambda \otimes b \in I(R_*) \odot_{R_*} B_0$ and $b' \in B_0$, $\lambda' \otimes a' \in I(R_*) \otimes_{R_*} A_0$. The maps

$$\left\{ \begin{array}{lcl} \partial : (A \bar{\otimes} B)_1 & \longrightarrow & (A \bar{\otimes} B)_0 = A_0 \bar{\otimes} B_0, \\ \mu : I(R_*) \otimes_{R_*} (A_0 \bar{\otimes} B_0) & \longrightarrow & A_0 \bar{\otimes} B_0, \end{array} \right. \tag{2}$$

satisfy $\partial \tilde{\mu} = \mu \tilde{q}$ so that one obtains by (∂, μ) the induced map

$$\partial : (A \bar{\otimes} B)_1 \longrightarrow (A \bar{\bar{\otimes}} B)_0 = A_0 \bar{\otimes} B_0.$$

6.6.5 Proposition. *$(A \bar{\bar{\otimes}} B, \partial, \bar{\mu})$ is a well-defined secondary permutation algebra which is the coproduct of A and B in the category* **secalg**(k)*. Moreover we have the natural isomorphism of commutative graded algebras*

$$\pi_0(A \bar{\bar{\otimes}} B) = \pi_0 A \otimes \pi_0 B$$

and there is a natural map of $\pi_0 A \otimes \pi_0 B$-modules

$$i_1 : \pi_o A \otimes \pi_1 B \oplus \pi_1 A \otimes \pi_0 B \longrightarrow \pi_1(A \bar{\bar{\otimes}} B).$$

Let X and Y be path-connected pointed spaces and let $X \times Y$ be the product space. The projections $p_1 : X \times Y \to X$ and $p_2 : X \times Y \to Y$ induce maps

$$\begin{array}{rcl} p_1^* : \mathcal{H}^*(X) & \longrightarrow & \mathcal{H}^*(X \times Y), \\ p_2^* : \mathcal{H}^*(Y) & \longrightarrow & \mathcal{H}^*(X \times Y), \end{array}$$

which in turn yield the binatural map in **secalg**(k)

$$j = (p_1^*, p_2^*) : \mathcal{H}^*(X) \bar{\bar{\otimes}} \mathcal{H}^*(Y) \longrightarrow \mathcal{H}^*(X \times Y) \tag{6.6.6}$$

where the left-hand side is the coproduct in (6.6.5). The Künneth theorem shows for a *field* k that the map j induces an isomorphism $j_0 = \pi_0(j)$.

(1)

$$\begin{array}{ccc} \pi_0(\mathcal{H}^*(X) \bar{\bar{\otimes}} \mathcal{H}^*(Y)) & \xrightarrow{\quad j_0 \quad} & \pi_0\mathcal{H}^*(X \times Y) \\ \| & & \| \\ (\pi_0\mathcal{H}^*(X)) \otimes (\pi_0\mathcal{H}^*(Y)) & & \| \\ \| & & \| \\ H^*(X) \otimes H^*(Y) & =\!=\!=\!=\!= & H^*(X \times Y) \end{array}$$

Moreover the map j induces the map $j_1 = \pi_1(j)$ for which the following diagram commutes,
(2)

$$\begin{array}{ccc} \pi_1(\mathcal{H}^*(X) \bar{\bar{\otimes}} \mathcal{H}^*(Y)) & \xrightarrow{\quad j_1 \quad} & \pi_1\mathcal{H}^*(X \times Y) \\ \uparrow{\scriptstyle i_1} & & \| \\ & & \Sigma\tilde{H}^*(X \times Y) \\ & & \| \\ (\Sigma\tilde{H}^*(X)) \otimes H^*(Y) \oplus H^*(X) \otimes (\Sigma\tilde{H}^*(Y)) & \xrightarrow[\quad \tilde{j}_1 \quad]{} & \Sigma(H^*(X) \otimes H^*(Y))\tilde{} \end{array}$$

Here $\tilde{j}_1$ is defined by $\tilde{j}_1(\Sigma x \otimes y) = \Sigma(x \otimes y)$ and $\tilde{j}_1(x \otimes \Sigma y) = (-1)^{|x|}\Sigma(x \otimes y)$ and i_1 is the map in (6.6.5). For the algebra $H = H^*X \otimes H^*Y$ the map $j_1 = \pi_1(j)$ in

(2) is an H-linear map between H-modules which is natural in X and Y. Diagram (2) shows that $j_1 = \pi_1(j)$ is surjective.

Using the push forward induced by $j_1 = \pi_1(j)$ we obtain the map in **secalg**(k)

$$(j_1)_*(\mathcal{H}^*(X) \bar{\bar{\otimes}} \mathcal{H}^*(Y)) \xrightarrow{\sim} \mathcal{H}^*(X \times Y) \tag{3}$$

which for a field k is a weak equivalence (i.e., induces isomorphisms in π_0 and π_1). Therefore up to weak equivalence the secondary permutation algebra $\mathcal{H}^*(X \times Y)$ is determined by $\mathcal{H}^*(X)$, $\mathcal{H}^*(Y)$ and the map j_1 in (2). This is a kind of *secondary Künneth theorem*. The computation of j_1 remains unclear.

Chapter 7

The Borel Construction and Comparison Maps

We first describe properties of the Borel construction on the classifying space of a group G. Then we introduce comparison maps between Borel constructions with fiber an Eilenberg-MacLane space. In the next chapter we deduce from comparison maps the power maps between Eilenberg-MacLane spaces.

7.1 The Borel construction

For a discrete group G the universal covering

$$p : EG \longrightarrow BG$$

for the classifying space BG can be chosen to be a functor in G, that is, a homomorphism $a : G \to \pi$ between groups induces a commutative diagram in $\mathbf{Top}^*$.

(7.1.1)
$$\begin{array}{ccc} EG & \xrightarrow{Ea} & E\pi \\ \downarrow & & \downarrow \\ BG & \xrightarrow{Ba} & B\pi \end{array}$$

Moreover $EG \to BG$ is compatible with products of groups $G = G_1 \times G_2$, $BG = BG_1 \times BG_2$, $EG = EG_1 \times EG_2$. For a G-space X we obtain the *Borel construction*

(7.1.2)
$$p : E = EG \times_G X \to BG$$

which is a fibration in **Top** with fibre $X = p^{-1}(*)$. We obtain E by the quotient space

(1)
$$EG \times_G X = EG \times X / \sim$$

with $(\tilde{x}, \alpha w) \sim (\tilde{x}\alpha, w)$ for $\alpha \in G, w \in X$ and $\tilde{x} \in EG$. Given $x \in BG$ let $\tilde{x} \in EG$ be a point with $p(\tilde{x}) = x$. If the G-space X has a fixpoint $*$ we obtain a section

(2) $$o : BG \to EG \times_G X$$

with $o(x) = (\tilde{x}, *)$. We say that $EG \times_G X$ is *good* if the inclusion

(3) $$i : X \vee BG \to EG \times_G X$$

is a closed cofibration in **Top**. Here $X \vee BG$ is the one-point union and $i|X$ is the inclusion of the fiber and $i|BG = o$. For example, $EG \times_G X$ is good if X is a CW-complex with a cellular action of G and zero cell $*$. In fact, then $EG \times_G X$ is a CW-complex and i in (3) is the inclusion of a subcomplex.

The Borel construction is functorial in the following sense. Let $a : G \to \pi$ be a homomorphism between groups and let X be a G-space and let Y be a π-space and let $f : X \to Y$ be an a-equivariant map, that is, $f(\alpha \cdot w) = a(\alpha) \cdot f(w)$ for $\alpha \in G$, $w \in X$. Then we obtain the induced map $f_\#$ for which the following diagram commutes.

(7.1.3) $$\begin{array}{ccc} EG \times_G X & \xrightarrow{f_\#} & E\pi \times_\pi Y \\ \downarrow & & \downarrow \\ BG & \xrightarrow{Ba} & B\pi \end{array}$$

Here $f_\#$ carries $(\tilde{x}, w)$ to $((Ea)\tilde{x}, fw)$. If f carries a fixpoint $* \in X$ to a fixpoint $* \in Y$, then $f_\# o = o(Ba)$ for the section o above.

The Borel construction is compatible with products in the following sense. Let X, Y be G-spaces so that the product $X \times Y$ is a G-space with the *diagonal action* $\alpha(w, v) = (\alpha w, \alpha v)$ for $\alpha \in G$, $w \in X$, $v \in Y$. Then the Borel constructions $E_G^X = EG \times_G X$, $E_G^Y = EG \times_G Y$ and $E_G^{X \times Y} = EG \times_G (X \times Y)$ are defined and we get

(7.1.4) $$E_G^{X \times Y} = E_G^X \times_{BG} E_G^Y$$

so that $E_G^{X \times Y}$ is a *product* in $\mathbf{Top}_{BG}$. If X and Y have G-fix points $*$, then $X \vee Y$ is a G-subspace of $X \times Y$ and the smash product $X \wedge Y$ is a G-space. Moreover the Borel construction $E_G^{X \wedge Y} = EG \times_G (X \wedge Y)$ is a *smash product* in $\mathbf{Top}_{BG}^{BG}$, that is

(7.1.5) $$E_G^{X \wedge Y} = E_G^X \wedge_{BG} E_G^Y.$$

Here the right-hand side is defined by the push out in **Top**

$$\begin{array}{ccc} E_G^X \cup_{BG} E_G^Y & \longrightarrow & E_G^X \times_{BG} E_G^Y \\ \downarrow & \text{push} & \downarrow \\ BG & \longrightarrow & E_G^X \wedge_{BG} E_G^Y \end{array}$$

with $E_G^X \cup_{BG} E_G^Y$ being the push out of $E_G^X \xleftarrow{o} BG \xrightarrow{o} E_G^Y$.

Using the compatibility of the Borel construction with products in (7.1.4) we get

7.1.6 Lemma. *If X is a topological group and G acts via automorphisms of the topological group X then $EG \times_G X \to BG$ is a group object in $\mathbf{Top}_{BG}$, see* (3.1.6).

The multiplication $\mu : X \times X \to X$ induces the multiplication $\mu_\# : E_G^X \times_{BG} E_G^X = E_G^{X \times X} \to E_G^X$.

7.1.7 Proposition. *Let X and Y be topological groups and assume the discrete group G acts on X (resp. Y) via automorphisms of the topological group. Let $f : X \to Y$ be a G-equivariant map and a homomorphism of topological groups. If X and Y are CW-spaces and f is a weak homotopy equivalence, then*

$$f_\sharp = EG \times_G f : EG \times_G X \to EG \times_G Y$$

is a homotopy equivalence under and over BG.

Proof. Also $EG \times_G X$ and $EG \times_G Y$ are CW-spaces and $f_\sharp$ is a weak homotopy equivalence which thus is a homotopy equivalence. Hence we can apply Lemma (3.1.7) since $f_\sharp$ is also a homomorphism of group objects in $\mathbf{Top}_{BG}$. □

We recall from 5.2.4 Baues [BOT] the following result on cohomology groups with local coefficients. Let A be an abelian group. Then a pointed CW-space X is an Eilenberg-MacLane space of type $K(A,n)$ or a *$K(A,n)$-space* if $\pi_n X = A$ and $\pi_j X = 0$ for $j \neq n$. We also write in this case $X = K(A,n)$.

Assume now that $K(A,n)$ is a G-space with fixpoint $*$. Then A is a G-module denoted by $\widetilde{A}$ and we have the Borel construction $L(\widetilde{A},n) = EG \times_G K(A,n)$ yielding the fibration

$$K(A,n) \longrightarrow L(\widetilde{A},n) \xrightarrow{p} BG \tag{7.1.8}$$

as in Baues [BOT] p. 300. We consider a closed cofibration $V \subset W$ between pointed connected CW-spaces and a commutative diagram.

$$\begin{array}{ccc} V & \xrightarrow{o(\varphi|V)} & L(\widetilde{A},n) \\ \downarrow & & \downarrow \\ W & \xrightarrow{\varphi} & BG \end{array}$$

Recall that $[W, L(\widetilde{A},n)]^V_{BG}$ is the set of homotopy classes of maps $W \to L(\widetilde{A},n)$ under V and over BG. According to 5.2.4 Baues [BOT] we have the isomorphism

$$[W, L(\widetilde{A},n)]^V_{BG} \cong H^n(W,V;\varphi^*\widetilde{A}). \tag{7.1.9}$$

Here $\varphi^*\widetilde{A}$ is the $\pi_1(W)$-module induced by $\pi_1(\varphi) : \pi_1 W \to \pi_1 BG = G$ and the right-hand side is the cohomology with local coefficients in $\varphi^*\widetilde{A}$.

The pair (W, V) is *n-connected* if $\pi_j(W, V) = 0$ for $j \le n$. This is the case if and only if there is a homotopy equivalence of pairs $(W, V) \simeq (\overline{W}, \overline{V})$ where $\overline{W}$ is a CW-complex with subcomplex $\overline{V}$ such that $\overline{V}$ contains the n-skeleton $\overline{W}^n$ so that the complement $\overline{W} - \overline{V}$ has only cells in dimension $> n$.

Let π be a group and let Y be a π-space with fixpoint $*$. We consider for $i = 0, 1$ a commutative diagram of the form

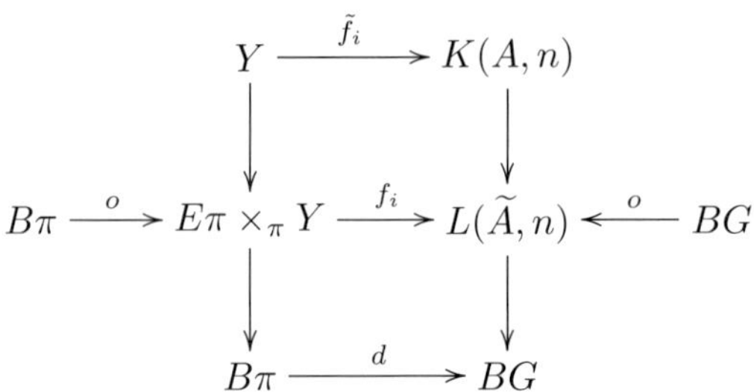

with $f_i o = do$ so that f_i is a map under $B\pi$ and over BG. The next result generalizes Corollary (3.2.5); many constructions of tracks in this book rely on this result.

7.1.10 Proposition. *Let $E\pi \times_\pi Y$ be good as in (7.1.2)(3) and let Y be an $(n-1)$-connected CW-space, $n \ge 1$. If there is a homotopy in* $\mathbf{Top}^*$

$$\widetilde{H} : \tilde{f}_0 \simeq \tilde{f}_1, \qquad \widetilde{H} : IY \to K(A, n),$$

then there exists a homotopy $H : f_0 \simeq f_1$ under $B\pi$ and over BG and the track $f_0 \simeq f_1$ under $B\pi$ and over BG is unique. Moreover H can be chosen to be an extension of $\widetilde{H}$. If $\tilde{f}_0 = \tilde{f}_1$ there is a unique track $f_0 \simeq f_1$ under $B\pi \vee Y$ and over BG. This shows that the groupoid

$$[\![E\pi \times_\pi Y, L(\tilde{A}, n)]\!]_{BG}^{B\pi \vee Y}$$

is contractible.

7.1.11 Addendum. *Assume only f_0 and a homotopy $\widetilde{H} : \tilde{f}_0 \simeq \tilde{f}_1$ in* $\mathbf{Top}^*$ *with $\widetilde{H}_t : Y \to K(A, n)$ are given as above. Then there exists f_1 with $\tilde{f}_1 = f_1 | Y$ and an extension $H : f_0 \simeq f_1$ of $\widetilde{H}$ where H_t is a map under $B\pi$ and over BG.*

Proof of (7.1.10). Let $E = E\pi \times_\pi Y$ and $L = L(\widetilde{A}, n)$. Then we have the following commutative diagram.

$$\begin{array}{ccc} E_0 = I(Y \vee B\pi) \cup E \cup E & \xrightarrow{\;k\;} & L \\ \downarrow{\scriptstyle j} & & \downarrow{\scriptstyle p} \\ IE & \xrightarrow{\;dpq\;} & BG \end{array}$$

Here j is the inclusion $j = (I(i), i_0, i_1)$ (see (1.2.4)) and k is the map given by (f_0, f_1) on $E \cup E$ and $odq : IB\pi \to L$ and by the homotopy $\widetilde{H} : IY \to K(A,n) \subset L$.

The map p is a fibration and the map j is a cofibration. Hence we can apply obstruction theory as in Theorem 5.4.3 in Baues [BOT].

The pair (IE, E_0) is readily seen to be $(n+1)$-connected and the obstructions are in $H^{m+1}(IE, E_0, \tilde{\pi}_m K(A,n))$ which is the zero group for all m. This shows that there is a map $H : IE \to L$ with $pH = dpq$ and $Hj = k$ as required in the lemma. In a similar way one checks uniqueness of H. For this we consider two such homotopies $H, H' : f_0 \simeq f_1$ under $B\pi$ and over BG which yield the commutative diagram

$$\begin{array}{ccc} I(IB\pi \cup E \cup E) \cup IE \cup IE & \xrightarrow{k'} & L \\ \downarrow{\scriptstyle j'} & & \downarrow{\scriptstyle p} \\ IIE & \xrightarrow{dpqq} & BG \end{array}$$

with $j' = (Ij'', i_0, i_1)$ and $j'' = (Io, i_0, i_1)$ and $k' = (k'', H, H')$ and $k'' = (odq, f_0, f_1)q$. Here j' is again $(n+1)$-connected and obstruction theory yields a map $F : IIE \to L$ with $pF = dpqq$ and $Fj' = k'$. This implies the uniqueness of H up to homotopy.

Finally we prove the addendum by the commutative diagram

$$\begin{array}{ccc} I(Y \vee B\pi) \cup E & \xrightarrow{k''} & L \\ \downarrow{\scriptstyle j''} & & \downarrow{\scriptstyle p} \\ IE & \xrightarrow{dpq} & B\pi \end{array}$$

with $j'' = (I(i), i_0)$ and k'' being restrictions of j and k respectively. Since j'' is a homotopy equivalence we again obtain by obstruction theory a map $H : IE \to L$ with $Hj'' = k''$ and $pH = dpq$. □

7.2 Comparison maps

We have seen in Section (2.1) that the Eilenberg-MacLane space Z^n is a topological R-module and that the symmetric group σ_n acts on Z^n via R-linear automorphisms. Given a homomorphism $i : G \to \sigma_n$ the space Z^n is thus a G-space. On the other hand we have the composite homomorphism

(7.2.1) $$G \xrightarrow{i} \sigma_n \xrightarrow{sign} \{-1, 1\}.$$

Here $\{-1, 1\}$ acts on Z^n by the automorphism $-1 : Z^n \to Z^n$ which carries $w \in Z^n$ to $-w$ with $-w$ defined by the R-module structure of Z^n. Hence G acts via (sign)i on Z^n and this action is termed the sign-*action* $Z^n_\pm$ of G on Z^n. In this section we compare the G-space Z^n and the G-space $Z^n_\pm$.

We observe that for $\sigma \in G$ the maps

$$\begin{cases} \sigma & : \quad Z^n \to Z^n \quad \text{with } \sigma(w) = \sigma \cdot w, \\ \mathrm{sign}(\sigma) & : \quad Z^n \to Z^n \quad \text{with } \mathrm{sign}(\sigma)(w) = \mathrm{sign}(i\sigma) \cdot w \end{cases}$$

are homotopic. Moreover (6.3.1)(4) shows that there is a unique track

(7.2.2) $$\Gamma_\sigma : \sigma \Rightarrow \mathrm{sign}(\sigma).$$

This is a first way of connecting the G-space Z^n with the G-space $Z^n_\pm$ studied in Chapter 6.

Using the Borel construction we have a further more subtle comparison between the G-spaces Z^n and $Z^n_\pm$ as follows.

7.2.3 Definition. For the G-spaces Z^n and $Z^n_\pm$ the Borel constructions $EG \times_G Z^n$ and $EG \times_G Z^n_\pm$ are defined. A *comparison map* λ_G is a map for which the following diagram commutes in **Top**.

$$\begin{array}{ccc} BG \vee Z^n & = & BG \vee Z^n_\pm \\ \downarrow j & & \downarrow j \\ EG \times_G Z^n & \xrightarrow{\lambda_G} & EG \times_G Z^n_\pm \\ \downarrow p & & \downarrow \\ BG & = & BG \end{array}$$

Here j is given by the section o of p and by the inclusion of the fiber. Hence λ_G is a map under and over BG which is the identity on fibers.

7.2.4 Theorem. *Comparison maps λ_G in* (7.2.3) *exist and for two such comparison maps λ_G, λ'_G there exists a unique track $o : \lambda_G \Rightarrow \lambda'_G$ under $BG \vee Z^n$ and over BG.*

The theorem states that the groupoid

(7.2.5) $$[\![EG \times_G Z^n, EG \times_G Z^n_\pm]\!]^{BG \vee Z^n}_{BG}$$

is contractible. The tracks in this groupoid are termed *canonical tracks*. Theorem (7.2.4) is the crucial connection between the G-spaces Z^n and $Z^n_\pm$ used in this paper. The uniqueness of tracks in (7.2.4) is a consequence of (7.1.10). For the construction of comparison maps we need the following lemma.

7.2.6 Lemma. *There exist topological R-modules with an action of σ_n via linear automorphisms together with σ_n-equivariant R-linear maps*

$$Z^n \xrightarrow{f_1} Y' \xleftarrow{g_1} Y'' \xrightarrow{f_2} K^n_\pm$$

which are homotopy equivalences in **Top***. *Here $K^n_\pm$ has the* sign-*action of σ_n.*

Proof. For $S(S^n)$ in $\Delta\,\mathbf{Mod}$ we obtain the chain complex $C = NS(S^n)$ in $\mathbf{Ch}_+$ with $H_*C = H_nS^n = R$ concentrated in degree n. Hence we obtain the following diagram of chain complexes in $\mathbf{Ch}_+$.

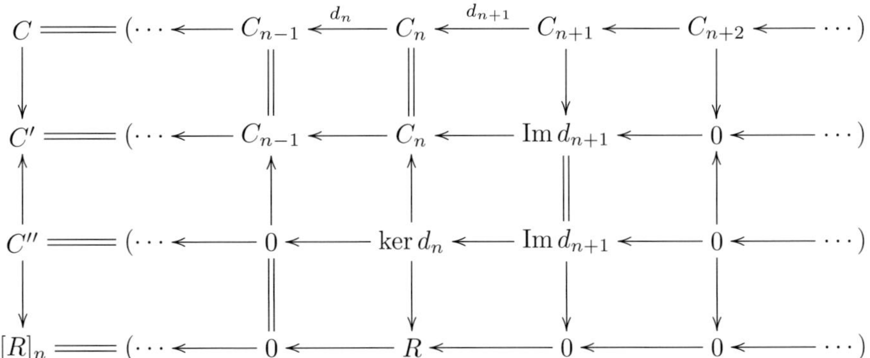

Here $[R]_n$ is the chain complex which is R concentrated in degree n. All chain maps induce isomorphisms in homology and are equivariant with respect to the action of σ_n. Hence we get

$$S(S^n) \cong \Gamma NS(S^n) = \Gamma C \to \Gamma C' \leftarrow \Gamma C'' \to \Gamma[R]_n$$

and therefore we get the following diagram.

$$\begin{array}{ccccccc} |\Phi S(S^n)| & \longrightarrow & |\Phi\Gamma C'| & \longleftarrow & |\Phi\Gamma C''| & \longrightarrow & |\Phi\Gamma[R]_n| \\ \| & & \| & & \| & & \| \\ Z^n & & Y' & & Y'' & & K^n_\pm \end{array}$$

Here $K^n_\pm$ is the small model of the Eilenberg-MacLane space $K(R,n)$ for example used by Kristensen [Kr1]. □

7.2.7 (Construction of the comparison map). Let $Y'_\pm, Y''_\pm$ and $Z^n_\pm$ be the topological R-modules in (7.2.6) with the sign-action. Then we get σ_n-equivariant R-linear maps

(1) $$Z^n \xrightarrow{f_1} Y' \xleftarrow{g_1} Y'' \xrightarrow{f_2} K^n_\pm \xleftarrow{g_2} Y''_\pm \xrightarrow{f_3} Y'_\pm \xleftarrow{g_3} Z^n_\pm$$

which are homotopy equivalences in $\mathbf{Top}^*$. Here $g_2 = f_2$ and $f_3 = g_1$ and $g_3 = f_1$ are equations of R-linear maps. Of course the induced map

(2) $$(g_3)_*^{-1}(f_3)_*(g_2)_*^{-1}(f_2)_*(g_1)_*^{-1}(f_1)_* = \mathrm{id}$$

is the identity on $\pi_n(Z^n)$. Let Y_i be the ith space in the sequence (1). Since all Y_i are realizations of simplicial groups, the Borel constructions $EG \times_G Y_i$ are good for all subgroups G of σ_n; see (7.1.2).

For a subgroup $G \subset \sigma_n$ the maps g_i in (1) induce maps

(3) $$(g_i)_\sharp : EG \times_G Y_{2i+1} \to EG \times_G Y_{2i}$$

which by (7.1.7) are homotopy equivalences under and over BG. We choose a homotopy inverse $\bar{g}_i$ under and over BG of $(g_i)_\sharp$. We now get a *comparison map* λ_G under $BG \vee Z^n$ and over BG

(4) $$\lambda_G : EG \times_G Z^n \to EG \times_G Z^n_\pm$$

by the composite $\lambda'_G = \bar{g}_3(f_3)_\sharp\bar{g}_2(f_2)_\sharp\bar{g}_1(f_1)_\sharp$ of maps $(f_i)_\sharp$ and $\bar{g}_i$ as follows. Using (2) we see that λ'_G induces on fibers a map $\lambda' : Z^n \to Z^n_\pm$ which is homotopic in **Top*** to the identity of the space Z^n. Moreover λ'_G is a map under and over BG. Now we can use the addendum (7.1.11) which shows that the homotopy $\lambda' \simeq 1$ has an extension $H : \lambda'_G \simeq \lambda_G$ under and over BG which defines the map λ_G in (4).

7.2.8 Remark. Karoubi in 2.5[Ka1] indicates the construction of Z^n in (2.1.4) though he does not give details. Our proof of (7.2.6) is more direct than an argument used by Karoubi in 2.12[Ka1].

7.3 Comparison tracks

The comparison maps are endowed with additional structure given by comparison tracks. We here describe three types of comparison tracks termed linear tracks, smash tracks and diagonal tracks respectively.

Since Z^n is a σ_n-space, also the r-fold product

(7.3.1) $$(Z^n)^{\times r} = Z^n \times \cdots \times Z^n$$

is a σ_n-space with the diagonal action. Moreover this product is a topological R-module since Z^n is one. Let homomorphisms

(1) $$G \xrightarrow[a]{} \pi \xrightarrow[i]{} \sigma_n$$

between groups be given and let

(2) $$f : (Z^n)^{\times r} \longrightarrow (Z^n)^{\times k}$$

with $r, k \geq 1$ be an *R-linear* continuous map which is a-equivariant. Here $(Z^n)^{\times r}$ is a G-space by $ia : G \to \sigma_n$ and $(Z^n)^{\times k}$ is a G-space by $i : \pi \to \sigma_n$. Hence we obtain the following diagram of Borel constructions.

(7.3.2) $$\begin{array}{ccc} EG \times_G (Z^n)^{\times r} & \xrightarrow{\lambda^r_G} & EG \times_G (Z^n_\pm)^{\times r} \\ \downarrow f_\# & \Rightarrow & \downarrow f^\pm_\# \\ E\pi \times_\pi (Z^n)^{\times k} & \xrightarrow{\lambda^k_G} & E\pi \times_\pi (Z^n_\pm)^{\times k} \end{array}$$

Here λ^r_G (and λ^k_π) are r-fold (resp. k-fold) products of comparison maps using (7.1.4) and $f_\#$ is defined as in (7.1.3).

7.3.3 Proposition. *There is a unique track*

$$L^{a,f} : \lambda_\pi^k f_\# \Rightarrow f_\#^\pm \lambda_G^r$$

under $BG \vee (Z^n)^{\times r}$ *and over* $B\pi$. *This track is termed the* linear track *for diagram* (7.3.2). *If* $\pi = G$ *and* $a = 1$ *is the identity of* G, *we write* $L^f = L^{1,f}$.

If $r = k = 1$ and $\pi = G$, then λ_π and λ_G are two different comparison maps for $G = \pi$. In this case the linear track L^1 of the identity 1 of Z^n is the same as a canonical track in (7.2.4).

The proposition is an easy consequence of (7.1.10). Uniqueness of the linear comparison tracks implies the following formula: Let $b : H \to G$ be a homomorphism between groups and let $g : (Z^n)^{\times t} \to (Z^n)^{\times r}$ be a b-equivariant R-linear map so that $L^{b,g}$ is defined as in (7.3.3). Then the following *composition formula for linear tracks* holds where $\square$ is the composition of tracks as in (2.2.1).

(7.3.4)
$$L^{ab,fg} = f_\#^\pm L^{b,g} \square L^{a,f} g_\#.$$

This formula also shows that linear tracks are *compatible* with canonical tracks in (7.2.4) in the following way. Let $L_G^1 : \lambda_G' \Rightarrow \lambda_G$ and $L_\pi^1 : \lambda_\pi \Rightarrow \lambda_\pi'$ be canonical tracks, then the track addition

$$f_\#^\pm (L_G^1)^r \square L^{a,f} \square (L_\pi^1)^k f_\#$$

is the linear track $(\lambda_\pi')^k f_\# \Rightarrow f_\#^\pm (\lambda_G')^r$.

For a tuple of numbers $n = (n_1, \dots, n_r)$ with $n_i \geq 1$ for $i = 1, \dots, r$ let

(7.3.5)
$$\sigma_{(n)} = \sigma_{n_1} \times \cdots \times \sigma_{n_r}$$

be the product of permutation groups. Let $|n| = n_1 + \cdots + n_r$ so that $\sigma_{(n)} \subset \sigma_{|n|}$. The group $\sigma_{(n)}$ acts on the product

(1)
$$Z^{\times(n)} = Z^{n_1} \times \cdots \times Z^{n_r}$$

of topological R-modules. Let

(2)
$$f : Z^{\times(n)} \longrightarrow Z^{|n|}$$

be a *multilinear map* (linear in each variable $x_i \in Z^{n_i}$) and let f be G-equivariant where G acts by a given homomorphism $G \to \sigma_{(n)} \to \sigma_{|n|}$. Then we obtain the following diagram of Borel constructions.

(7.3.6)
$$\begin{array}{ccc} EG \times_G Z^{\times(n)} & \xrightarrow{\lambda_G^r} & EG \times_G Z_\pm^{\times(n)} \\ \downarrow{\scriptstyle f_\#} & \Rightarrow & \downarrow{\scriptstyle f_\#} \\ EG \times_G Z^{|n|} & \xrightarrow{\lambda_G} & EG \times_G Z_\pm^{|n|} \end{array}$$

Here λ_G^r is again an r-fold product of comparison maps using (7.1.4) and $f_\#$ is defined as in (7.1.3). We have the smash product map

$$\hat{p} : Z^{\times(n)} = Z^{n_1} \times \cdots \times Z^{n_r} \longrightarrow Z^{\wedge(n)} = Z^{n_1} \wedge \cdots \wedge Z^{n_r}$$

where the right-hand side is the r-fold smash product. Since f is multilinear there is a unique factorization

$$f : \; Z^{\times(n)} \xrightarrow{\hat{p}} Z^{\wedge(n)} \xrightarrow{\hat{f}} Z^{|n|}$$

of f in (7.3.5). Here $\hat{f}$ is again G-equivariant and by (7.1.5) we obtain the diagram

(7.3.7)
$$\begin{array}{ccc} EG \times_G Z^{\wedge(n)} & \xrightarrow{\lambda_G^{\wedge r}} & EG \times_G Z_{\pm}^{\wedge(n)} \\ {\scriptstyle \hat{f}_\#}\downarrow & \Rightarrow & \downarrow{\scriptstyle \hat{f}_\#} \\ EG \times_G Z^{|n|} & \xrightarrow{\lambda_G} & EG \times_G Z_{\pm}^{|n|} \end{array}$$

where $\lambda_G^{\wedge r}$ is the r-fold smash product of comparison maps over BG. We have

$$\hat{p}_\# \lambda_G^r = \lambda_G^{\wedge r} \hat{p}_\#.$$

Now we obtain for the diagrams (7.3.6) and (7.3.7) the following tracks.

7.3.8 Proposition. *There is a unique track*

$$S^{\hat{f}} : \; \lambda_G \hat{f}_\# \Rightarrow \hat{f}_\# \lambda_G^{\wedge r}$$

under $BG \vee Z^{\wedge(n)}$ and over BG for (7.3.7) which defines the track

$$S^f = S^{\hat{f}} (\hat{p})_\# : \; \lambda_G f_\# \Rightarrow f_\# \lambda_G^r.$$

The track S^f is termed the smash track *for* (7.3.6).

The proposition again is a consequence of (7.1.10). Uniqueness of $S^{\hat{f}}$ implies the following compatibility with composition of multilinear maps. Let $n^i = (n_1^i, \ldots, n_{r_i}^i)$ be a tuple of numbers ≥ 1 for $i = 1, \ldots, r$ such that $|n^i| = n_i$. Moreover let $n^* = (n^1, \ldots, n^r)$ be the composed tuple so that $\sigma_{(n^*)} \subset \sigma_{(n)} \subset \sigma_{|n|}$. Let $G \to \sigma_{(n^*)}$ be a given homomorphism and let $f^i : Z^{\times(n^i)} \to Z^{n_i}$ be a G-equivariant multilinear map. Then the composition

$$f(f^1 \times \cdots \times f^r) : \; Z^{\times(n^*)} \longrightarrow Z^{|n|}$$

is again multilinear and G-equivariant so that $S^{f(f^1 \times \cdots \times f^r)}$ is defined. Now we have the *composition formula for smash tracks*

(7.3.9)
$$S^{f(f^1 \times \cdots \times f^r)} = f_\#(S^{f^1} \times \cdots \times S^{f^r}) \square S^f (f^1 \times \cdots \times f^r)_\#.$$

Here $S^{f^1} \times \cdots \times S^{f^r}$ is the product of smash tracks defined by use of (7.1.4). This formula again shows the compatibility of smash tracks and canonical tracks similarly as in (7.3.4).

Finally we consider comparison tracks which are deduced from a diagonal map Δ. For this let $n = (n_1, \dots, n_r)$ and $\rho \to \sigma_{(n)} \subset \sigma_{|n|}$ be given such that the composite

$$\rho \longrightarrow \sigma_{(n)} \subset \sigma_{|n|} \xrightarrow{\text{sign}} \{-1, 1\}$$

is trivial; that is, the sign-action of ρ on $Z_{\pm}^{|n|}$ is trivial. Moreover let

(7.3.10) $$g\colon\ Z^{\times(n)} \longrightarrow Z^{|n|}$$

be a ρ-equivariant multilinear map. Then we get the composite map

(1) $$C(g) : E\rho \times_{\rho} Z^{\times(n)} \xrightarrow{g\#} E\rho \times_{\rho} Z^{|n|} \xrightarrow{\lambda_\rho} B\rho \times Z^{|n|} \xrightarrow{p_2} Z^{|n|}$$

where p_2 is the projection and λ_ρ is a comparison map.

Let $k \geq 1$ and let $\pi \to \sigma_k$ be a homomorphism between groups for which again the composite

$$\pi \longrightarrow \sigma_k \xrightarrow{\text{sign}} \{-1, 1\}$$

is trivial. Then the composite

$$\pi \times \rho \longrightarrow \sigma_k \times \sigma_{|n|} \subset \sigma_k \int \sigma_{|n|} \subset \sigma_{k|n|} \xrightarrow{\text{sign}} \{1, -1\}$$

is also trivial, so that the sign-action of $\pi \times \rho$ on $Z_{\pm}^{k|n|}$ is trivial. Let

(2) $$f : (Z^{|n|})^{\times k} \longrightarrow Z^{k|n|}$$

be a $\pi \times \rho$-equivariant multilinear map. Here ρ acts diagonally on $(Z^{|n|})^{\times k}$ via $\rho \to \sigma_{(n)} \subset \sigma_{|n|}$ and π acts via $\pi \to \sigma_k$ by *permuting* the factors of the k-fold product $(Z^{|n|})^{\times k}$. Then we consider the composite

(3) $$(Z^{\times(n)})^{\times k} \xrightarrow{g^{\times k}} (Z^{|n|})^{\times k} \xrightarrow{f} Z^{k|n|}$$

which is again a $\pi \times \rho$-equivariant multilinear map. Here π acts on the k-fold product on the left-hand side by permuting the factors of the product. The group π acts also on the k-fold product $(EG \times_G Z^{\times(n)})^{\times k}$ by permuting factors and there is a canonical *diagonal map* with $X = Z^{\times k}$,

(4) $$D\colon\ E(\pi \times \rho) \times_{\pi \times \rho} X^{\times k} \longrightarrow E\pi \times_{\pi} (E\rho \times_{\rho} X)^{\times k}.$$

Here $E(\pi \times \rho) = E(\pi) \times E(\rho)$ and D carries $(\tilde{x}, \tilde{y}, x_1, \dots, x_k)$ with $\tilde{x} \in E(\pi)$, $\tilde{y} \in E(\rho)$ and $x_1, \dots, x_k \in X$ to the element $(\tilde{x}, (\tilde{y}, x_1), \dots, (\tilde{y}, x_k))$. One readily checks by the definition of the Borel construction in (7.1.2) that D is a well-defined map.

The diagonal map leads to the following diagram.

(7.3.11)

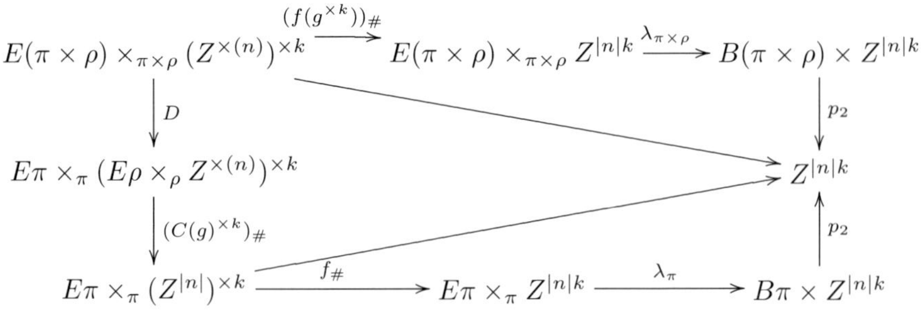

$$
\begin{array}{ccccc}
E(\pi\times\rho)\times_{\pi\times\rho}(Z^{\times(n)})^{\times k} & \xrightarrow{(f(g^{\times k}))_\#} & E(\pi\times\rho)\times_{\pi\times\rho}Z^{|n|k} & \xrightarrow{\lambda_{\pi\times\rho}} & B(\pi\times\rho)\times Z^{|n|k} \\
\downarrow{\scriptstyle D} & & & & \downarrow{\scriptstyle p_2} \\
E\pi\times_\pi(E\rho\times_\rho Z^{\times(n)})^{\times k} & & & & Z^{|n|k} \\
\downarrow{\scriptstyle (C(g)^{\times k})_\#} & & & & \uparrow{\scriptstyle p_2} \\
E\pi\times_\pi(Z^{|n|})^{\times k} & \xrightarrow{f_\#} & E\pi\times_\pi Z^{|n|k} & \xrightarrow{\lambda_\pi} & B\pi\times Z^{|n|k}
\end{array}
$$

Here λ_π and $\lambda_{\pi\times\rho}$ are comparison maps. The map $C(g)^{\times k}$ is π-equivariant since π acts by permuting factors. According to the notation in (7.3.10)(1) the top triangle of the diagram yields

$$C(f(g^{\times k})) = p_2\lambda_{\pi\times\rho}(fg^{\times k})_\#$$

and the bottom triangle of the diagram yields

$$C(f) = p_2\lambda_\pi f_\#.$$

The left-hand column in diagram (7.3.11) induces the following commutative diagram.

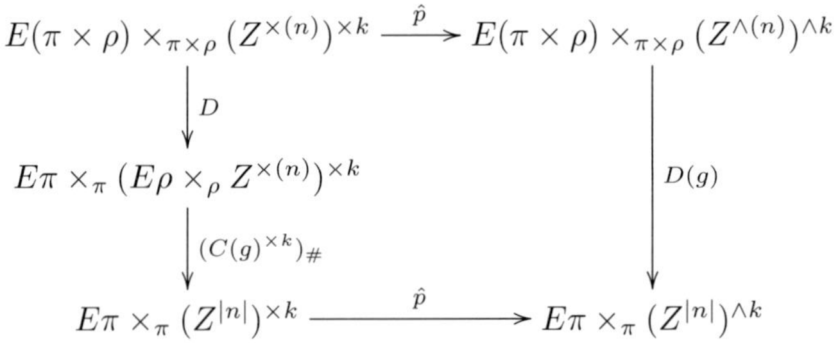

$$
\begin{array}{ccc}
E(\pi\times\rho)\times_{\pi\times\rho}(Z^{\times(n)})^{\times k} & \xrightarrow{\hat p} & E(\pi\times\rho)\times_{\pi\times\rho}(Z^{\wedge(n)})^{\wedge k} \\
\downarrow{\scriptstyle D} & & \\
E\pi\times_\pi(E\rho\times_\rho Z^{\times(n)})^{\times k} & & \downarrow{\scriptstyle D(g)} \\
\downarrow{\scriptstyle (C(g)^{\times k})_\#} & & \\
E\pi\times_\pi(Z^{|n|})^{\times k} & \xrightarrow{\hat p} & E\pi\times_\pi(Z^{|n|})^{\wedge k}
\end{array}
$$

Here $\hat p$ are the quotient maps. The induced map $D(g)$ is well defined since $C(g)$ in (7.3.10) carries section points $o(x) = (\tilde x, *) \in E\rho\times_\rho Z^{\times(n)}$ to the basepoint $* = 0 \in Z^{|n|}$. Multilinearity of f and $f(g^{\times k})$ yield the following diagram.

(7.3.12)

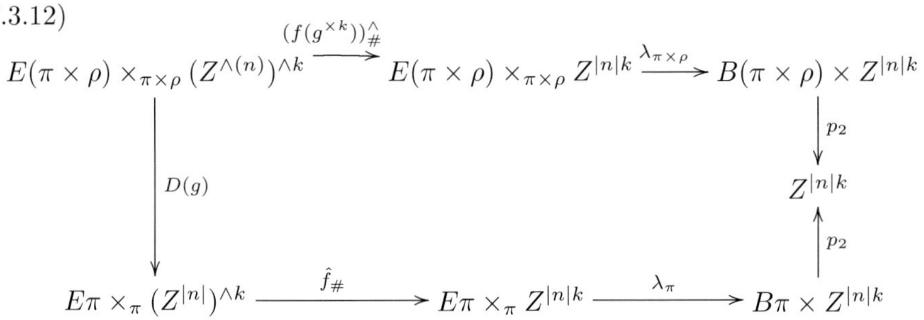

$$
\begin{array}{ccccc}
E(\pi\times\rho)\times_{\pi\times\rho}(Z^{\wedge(n)})^{\wedge k} & \xrightarrow{(f(g^{\times k}))^\wedge_\#} & E(\pi\times\rho)\times_{\pi\times\rho}Z^{|n|k} & \xrightarrow{\lambda_{\pi\times\rho}} & B(\pi\times\rho)\times Z^{|n|k} \\
 & & & & \downarrow{\scriptstyle p_2} \\
\downarrow{\scriptstyle D(g)} & & & & Z^{|n|k} \\
 & & & & \uparrow{\scriptstyle p_2} \\
E\pi\times_\pi(Z^{|n|})^{\wedge k} & \xrightarrow{\hat f_\#} & E\pi\times_\pi Z^{|n|k} & \xrightarrow{\lambda_\pi} & B\pi\times Z^{|n|k}
\end{array}
$$

Now we obtain for the diagram (7.3.11) and (7.3.12) the following tracks.

7.3.13 Proposition. *There is a unique track*

$$D^{\hat{f},\hat{g}} : p_2\lambda_\pi \hat{f}_\# D(g) \Rightarrow p_2\lambda_{\pi\times\rho}(f(g^{\times k}))^\wedge_\#$$

under $B(\pi \times \rho) \vee (Z^{\wedge(n)})^{\wedge k}$ *and over* $B\pi$ *for* (7.3.12) *which defines the track*

$$D^{f,g} = D^{\hat{f},\hat{g}}(\hat{p})_\# : C(f)(C(g)^{\times k})_\# D \Rightarrow C(f(g^{\times k}))$$

for diagram (7.3.11). *The track* $D^{f,g}$ *is termed the* diagonal track *for* (7.3.11).

The proposition is a consequence of (7.1.10). If we alter the comparison maps λ_ρ, λ_π and $\lambda_{\pi\times\rho}$ by canonical tracks (7.2.4), then the diagonal track $D^{f,g}$ is compatible with this alteration analogously as in (7.3.4).

Diagonal tracks are compatible with composition as follows. For this let $t \geq 1$ and let $\tau \to \sigma_t$ be a homomorphism between groups such that the composite

$$\tau \longrightarrow \sigma_t \xrightarrow{\text{sign}} \{-1, 1\}$$

is trivial. Then τ acts on a t-fold product $Y^{\times t}$ by permutation of factors. According to the definition of diagonal maps D in (7.3.10)(4) we obtain the following diagram.
(7.3.14)

$$\begin{array}{ccc}
E(\tau \times \pi \times \rho) \times_{\tau\times\pi\times\rho} (X^{\times k})^{\times t} & \xrightarrow{D} & E(\tau \times \pi) \times_{\tau\times\pi} ((E\rho \times_\rho X)^{\times k})^{\times t} \\
\Big\downarrow D & & \Big\downarrow D \\
E\tau \times_\tau (E(\pi \times \rho) \times_{\pi\times\rho} (X^{\times k}))^{\times t} & \xrightarrow{(D^{\times t})_\#} & E\tau \times_\tau (E\pi \times_\pi (E\rho \times_\rho X)^{\times k})^{\times t}
\end{array}$$

Here $D^{\times t}$ is τ-equivariant since τ acts by permuting the factors of the t-fold product. Using the definition of D one readily checks:

7.3.15 Lemma. *Diagram* (7.3.14) *commutes.*

Now we consider the composition of maps:

$$(Z^{\times(n)})^{\times kt} \xrightarrow{g^{\times kt}} (Z^{|n|})^{\times kt} \xrightarrow{f^{\times t}} (Z^{k|n|})^{\times t} \xrightarrow{h} Z^{kt|n|}. \tag{7.3.16}$$

Here h is multilinear and h is $\tau \times \rho \times \pi$-equivariant. Since the corresponding sign-actions are trivial we can define

$$C(g),\ C(f) \quad \text{and} \quad C(h),\ C(f(g^{\times k})), C(h(f^{\times t})),\ C(h(f^{\times t})(g^{\times kt}))$$

as in (7.3.10)(1). Moreover the diagonal tracks

$$D^{f,g},\ D^{h,f},\ D^{h,f(g^{\times k})} \quad \text{and} \quad D^{h(f^{\times t}),g}$$

are defined as in (7.3.15).

Now we get the following diagram of tracks.

(7.3.17)
$$\begin{array}{ccc}
C(h)(C(f)C(g)^{\times k}_{\#}D)^{\times t}_{\#}D & = & C(h)C(f)^{\times t}_{\#}C(g)^{\times tk}_{\#}DD \\
\Big\Downarrow {\scriptstyle C(h)(D^{f,g})^{\times t}_{\#}D} & & \Big\| \\
C(h)C(fg^{\times k})^{\times t}_{\#}D & & \\
\Big\Downarrow {\scriptstyle D^{h,fg^{\times k}}} & & \\
C(h(f(g^{\times k}))^{\times t}) & & \\
\Big\Uparrow {\scriptstyle D^{hf^{\times t},g}} & & \\
C(hf^{\times t})C(g)^{\times (tk)}_{\#}D & & \\
\Big\Uparrow {\scriptstyle D^{k,t}C(g)^{\times (tk)}_{\#}D} & & \Big\| \\
C(h)C(f)^{\times t}_{\#}DC(g)^{\times (tk)}_{\#}D & = & C(h)C(f)^{\times t}_{\#}C(g)^{\times (tk)}_{\#}DD
\end{array}$$

In the top row we use (7.3.15) and in the bottom row we use the naturality of D.

7.3.18 Proposition. *The diagram of tracks* (7.3.17) *commutes.*

This follows from the uniqueness of tracks in (7.3.13) if we consider diagram (7.3.17) on the level of smash products similarly as in (7.3.12).

Chapter 8

Power Maps and Power Tracks

The power maps introduced in this chapter yield the crucial ingredient for the definition of Steenrod operations in the next chapter. In the literature the power map was only considered as a homotopy class of maps. We here observe that the power map as a map in $\mathbf{Top}^*$ is well defined up to a canonical track. Moreover we describe certain homotopy commutative diagrams associated with power maps. These diagrams are used to prove

- the linearity of Steenrod operations,
- the Cartan formula, and
- the Adem relation respectively.

We show that there are in fact well-defined tracks for these diagrams which we call the linearity track, the Cartan track and the Adem track. These power-tracks are defined by the comparison tracks in Section (7.3). The power tracks correspond exactly to the relations in a power algebra; see (1.2.6) and (8.5.4).

8.1 Power maps

We define the *power map*

(8.1.1) $$U : Z^q \xrightarrow{\Delta} (Z^q)^{\times p} \xrightarrow{\mu} Z^{qp}$$

which carries $x \in Z^q$ to the p-fold product $U(x) = x^p = x \cdot \dots \cdot x$. This map has the factorization $U = \mu\Delta$ where Δ is the diagonal and μ is given by the multiplication map μ in (2.1.1) with $\mu(x_1, \dots, x_p) = x_1 \cdot \dots \cdot x_p$. Moreover U is a pointed map, that is $U(0) = 0$.

The symmetric group σ_p acts on $(Z^q)^{\wedge p}$ by permuting the factors Z^q and acts on Z^q trivially. Moreover $\sigma_p \subset \sigma_{pq}$ carries a permutation $\alpha \in \sigma_p$ to the corresponding permutation of q-blocks in σ_{pq}. Then it is clear that U is σ_p-equivariant and hence U induces for each subgroup $G \subset \sigma_p$ the following map.

(1)
$$\begin{array}{ccc} BG \times Z^q & \xrightarrow{U_\sharp} & EG \times_G Z^{pq} \\ \| & & \uparrow \mu_\sharp \\ EG \times_G Z^q & \xrightarrow{\Delta_\sharp} & EG \times_G (Z^q)^{\wedge p} \end{array}$$

Hence via the comparison map λ_G we get the composite

(2)
$$BG \times Z^q \xrightarrow{U_\sharp} EG \times_G Z^{pq} \xrightarrow{\lambda_G} EG \times_G Z^{pq}_{\pm}.$$

If q is even or if $-1 = 1$ in R or if q is odd and G is contained in the alternating group, then the sign-action of G of $Z^{pq}_{\pm}$ is trivial so that in this case we get

(3)
$$EG \times_G Z^{pq}_{\pm} = BG \times Z^{pq}.$$

Moreover the following diagram commutes.

(4)
$$\begin{array}{ccc} BG \vee Z^q & \xrightarrow{1 \vee U} & BG \vee Z^{pq} \\ \downarrow & & \downarrow \\ BG \times Z^q & \xrightarrow{\lambda_G U_\sharp} & EG \times_G Z^{pq}_{\pm} \\ \downarrow & & \downarrow \\ BG & \xrightarrow{1} & BG \end{array}$$

This is readily seen since λ_G is a map under $BG \vee Z^{pq}$ and over BG.

Let p be a prime and let $R = \mathbb{Z}/p\mathbb{Z}$ be the field of p elements and let $\pi = \mathbb{Z}/p\mathbb{Z}$ be the cyclic group of order p. Then $B\pi$ is a $K(R,1)$-space and we fix a homotopy equivalence in $\mathbf{Top}^*$

(8.1.2)
$$h_\pi : Z^1 \xrightarrow{\simeq} B\pi$$

where Z^1 is defined as in (2.1.4). Here h_π induces the identity in homology $(h_\pi)_* = 1 : R = H_1 Z^1 \to H_1 B\pi = \pi$. By (3.2.5) the component

$$[\![Z^1, B\pi]\!]_1 \subset [\![Z^1, B\pi]\!]$$

of such maps in the groupoid $[\![Z^1, B\pi]\!]$ is contractible.

8.1.3 Definition. The group $\pi = \mathbb{Z}/p$ is a subgroup of σ_p by using cyclic permutations. For this subgroup the condition used in (8.1.1)(3) holds. In fact the subgroup $\pi = \mathbb{Z}/p \subset \sigma_p$ is generated by the permutation T_π which sends i to $(i+1)$ *mod* p. The sign of this permutation is

$$\mathrm{sign}(T_\pi) = (-1)^{p-1}.$$

Since $(-1)^{p-1} = 1(\mathrm{mod}\ p)$ the sign action of π on Z^{pq} is trivial. Hence we get the composite

$$\gamma : Z^1 \times Z^q \xrightarrow{h_\pi \times 1} B\pi \times Z^q \xrightarrow{\lambda_\pi U_\#} B\pi \times Z^{pq} \xrightarrow{p_2} Z^{pq}$$

where h_π is defined as in (8.1.2) and $\lambda_\pi U_\#$ is the map in (8.1.1)(2) and p_2 is the projection. We call such a composite also a *power map*. Formally such a power map γ is a triple $(\gamma, h_\pi, \lambda_\pi)$ so that h_π and λ_π are part of the definition of a power map.

Using (8.1.1)(4) we see that the following diagram commutes for each power map γ.

(8.1.4)
$$\begin{array}{ccc} Z^1 \vee Z^q & \xrightarrow{(0,U)} & Z^{pq} \\ \downarrow & & \| \\ Z^1 \times Z^q & \xrightarrow{\gamma} & Z^{pq} \end{array}$$

Power maps depend on the choice of λ_π and h_π but we have the following crucial observation.

8.1.5 Proposition. *There is a well-defined contractible subgroupoid* $\underline{\underline{\gamma}}$ *with*

$$\underline{\underline{\gamma}} \subset [\![Z^1 \times Z^q, Z^{pq}]\!]^{Z^1 \vee Z^q}.$$

The objects of $\underline{\underline{\gamma}}$ *are the power maps.*

Proof. The subgroupoid $\underline{\underline{\gamma}}$ is the image of the functor

$$\begin{array}{c} [\![Z^1, B\pi]\!]_1 \times [\![E\pi \times_\pi Z^{pq}, E\pi \times_\pi Z^{pq}_\pm]\!]_{B\pi}^{B\pi \vee Z^{pq}} \\ \downarrow \\ [\![Z^1 \times Z^q, Z^{pq}]\!]^{Z^1 \vee Z^q} \end{array}$$

which carries $H : h_\pi \Rightarrow h'_\pi$ and $G : \lambda_\pi \Rightarrow \lambda'_\pi$ to the composite

$$Z^1 \times Z^q \xrightarrow[H \times 1]{} B\pi \times Z^q \xrightarrow[GU_\#]{} B\pi \times Z^{pq} \xrightarrow[p_2]{} Z^{pq}$$

where we use the composition in a 2-category. Since a product of contractible groupoids is contractible, we see that the image of the functor is a contractible groupoid. Hence the track $(\gamma, h_{pi}, \lambda_\pi) \Rightarrow (\gamma', h'_{pi}, \lambda'_\pi)$ in $\underline{\underline{\gamma}}$ is the composite $p_2(GU_\#) * (H \times 1)$ where G and H are the unique tracks above. $\square$

We call the tracks in the contractible subgroupoid $\underline{\underline{\gamma}}$ the *canonical tracks* for power maps. We also say that a power map is well defined up to a canonical track. Indeed the subgroupoid $\underline{\underline{\gamma}}$ is well defined since it only depends on the structure of the Eilenberg-MacLane spaces Z^n in (2.1).

Remark. The homotopy class of the power map γ was considered by Karoubi [Ka2] in order to define Steenrod operations. Also Milgram used the homotopy class of the power map; compare 27.11 and 27.13 in Gray [G]. We are interested in the secondary structure of cohomology operations. Therefore we think of γ as a map in **Top*** and not as a homotopy class of maps in **Top**$^*/\simeq$. For this it is a crucial observation that power maps form a contractible groupoid as defined in (8.1.5).

8.1.6 Definition. For maps $v : X \to Z^1$ and $x : X \to Z^q$ let $\gamma_v(x)$ be the composite

$$\gamma_v(x) : X \xrightarrow{(v,x)} Z^1 \times Z^q \xrightarrow{\gamma} Z^{pq}$$

where γ is a power map in $\underline{\underline{\gamma}}$. We consider γ_v also as a functor

$$\gamma_v : [\![X, Z^q]\!] \longrightarrow [\![X, Z^{pq}]\!].$$

We prove below that this functor induces on π_0 the following commutative diagram.

$$\begin{array}{ccc} \pi_0[\![X, Z^q]\!] & \xrightarrow{\pi_0\gamma_v} & \pi_0[\![X, Z^{pq}]\!] \\ \| & & \| \\ H^qX & \xrightarrow[\gamma_{\bar{v}}]{} & H^{pq}X \end{array}$$

Here $\bar{v} \in H^1X = \pi_0[\![X, Z^1]\!]$ is represented by v and $\gamma_{\bar{v}}$ is the function in (1.2.7) defining the power algebra structure of H^*X.

8.2 Linearity tracks for power maps

We now consider linearity properties of the power map γ. Using the R-module structure of Z^n we define the *cross effect map*

(8.2.1)
$$\begin{aligned} U^{cr} &: Z^q \times Z^q \to Z^{pq} \\ U^{cr}(x,y) &= U(x+y) - U(x) - U(y) \\ &= (x+y)^p - x^p - y^p, \end{aligned}$$

and the cross effect map

(8.2.2)
$$\begin{aligned} \gamma^{cr} &: Z^1 \times Z^q \times Z^q \to Z^{pq} \\ \gamma^{cr}(\alpha, x, y) &= \gamma(\alpha, x+y) - \gamma(\alpha, x) - \gamma(\alpha, y). \end{aligned}$$

Here U^{cr} carries $Z^q \vee Z^q$ to 0 and γ^{cr} carries $Z^1 \times (Z^q \vee Z^q)$ to 0. Then (8.1.4) shows that the diagram

$$\begin{array}{ccc} Z^1 \times (Z^q \vee Z^q) \cup \{0\} \times Z^q \times Z^q & \xrightarrow{0 \cup U^{cr}} & Z^{pq} \\ \downarrow & & \| \\ Z^1 \times Z^q \times Z^q & \xrightarrow{\gamma^{cr}} & Z^{pq} \end{array}$$

commutes. Moreover we consider the following diagram in which p_{23} is the following projection.

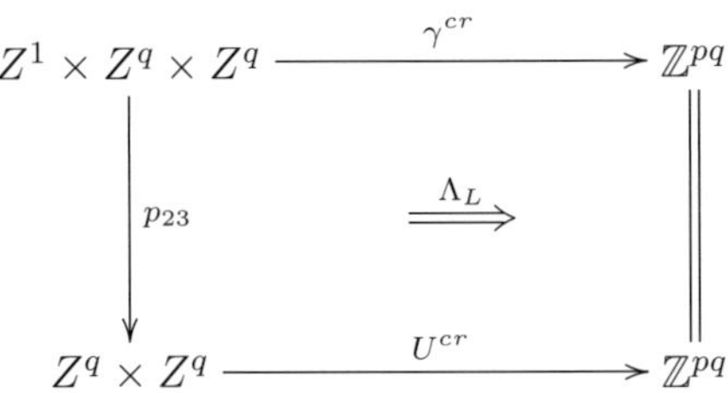

8.2.3 Theorem. *Linear comparison tracks in* (7.3.3) *induce the track*

$$\Lambda_L : U^{cr} p_{23} \Rightarrow \gamma^{cr}$$

under $Z^1 \times (Z^q \vee Z^q) \cup \{0\} \times Z^q \times Z^q$ *which we call the* linearity track *for* γ. *Moreover* Λ_L *is compatible with canonical tracks in* $\underline{\underline{\gamma}}$.

Here "compatibility" means that a canonical track $H : \gamma \Rightarrow \gamma'$ in $\underline{\underline{\gamma}}$ yields the track $H^{cr} : \gamma^{cr} \Rightarrow (\gamma')^{cr}$ in the obvious way and the linearity track Λ'_L for γ' satisfies $H^{cr} \square \Lambda_L = \Lambda'_L$.

Proof. Let $N : Z^{pq} \to Z^{pq}$ be the π-*norm map* defined by

(1) $$N(x) = \sum_{\alpha \in \pi} \alpha \cdot x$$

where we use the action of π on Z^{pq} given by $\pi \subset \sigma_p \subset \sigma_{pq}$. Then N is a π-equivariant linear map where π acts trivially on the source space Z^{pq} and via $\pi \to \sigma_{pq}$ on the target space Z^{pq}. There is a map

(2) $$\bar{U} : Z^q \times Z^q \longrightarrow Z^{pq}$$

with

$$N\bar{U} = U^{cr}$$

and $\bar{U}(x, 0) = \bar{U}(0, x) = 0$ for $x \in Z^q$. In fact $U^{cr}(x, y) = (x + y)^p - x^p - y^p$ is the sum of all monomials that contain k factors x and $(p - k)$ factors y, where $1 \leq k \leq p - 1$. The cyclic group $\pi = \mathbb{Z}/p$ permutes such factors freely. We choose a basis B consisting of monomials $b(x,y)$ whose permutations under π give each monomial exactly once. Then

$$\bar{U}(x,y) = \sum_{b\in B} b(x,y)$$

is defined by this basis B. The map $\bar{U}$ yields the commutative diagram

(3)
$$\begin{array}{ccc} B\pi \times Z^q \times Z^q & \xrightarrow{1\times\overline{U}} & B\pi \times Z^{pq} \\ \downarrow{\scriptstyle 1} & & \downarrow{\scriptstyle N_\sharp} \\ B\pi \times Z^q \times Z^q & \xrightarrow{U^{cr}_\sharp} & E\pi \times_\pi Z^{pq} \end{array}$$

where U^{cr} is π-equivariant in the same way as the map U in (8.1.1).

Using (7.3.4) we obtain the linear track $L^N : \lambda_\pi N_\# \Rightarrow 1\times N$ for the following diagram.

(4)
$$\begin{array}{ccc} B\pi \times Z^{pq} & \xrightarrow{1\times 1} & B\pi \times Z^{pq} \\ \downarrow{\scriptstyle N_\sharp} & & \downarrow{\scriptstyle 1\times N} \\ E\pi \times_\pi Z^{pq} & \xrightarrow{\lambda_\pi} & B\pi \times Z^{pq} \end{array}$$

Next let $\nu : Z^{pq} \times Z^{pq} \times Z^{pq} \to Z^{pq}$ be the π-equivariant linear map defined by $\nu(x,y,z) = x - y - z$. Then (7.3.4) yields the linear track $L^\nu : \lambda_\pi \nu_\# \Rightarrow \nu_\# \lambda^3_\pi$ for the following diagram.

(5)
$$\begin{array}{ccc} E\pi \times_\pi (Z^{pq} \times Z^{pq} \times Z^{pq}) & \xrightarrow{\lambda^3_\pi} & B\pi \times (Z^{pq} \times Z^{pq} \times Z^{pq}) \\ \downarrow{\scriptstyle \nu_\sharp} & & \downarrow{\scriptstyle 1\times\nu} \\ E\pi \times_\pi Z^{pq} & \xrightarrow{\lambda_\pi} & B\pi \times Z^{pq} \end{array}$$

Now we have for $\alpha \in Z_1$ and $\beta = h_\pi \alpha \in B\pi$ the equations:

$$\begin{aligned} \gamma^{cr}(\alpha,x,y) &= \gamma(\alpha,x+y) - \gamma(\alpha,x) - \gamma(\alpha,y) \\ &= p_2\lambda_\pi U_\sharp(\beta,x+y) - p_2\lambda_\pi U_\sharp(\beta,x) - p_2\lambda_\pi U_\sharp(\beta,y) \\ &= p_2(1\times\nu)\lambda^3_\pi U^+_\sharp(\beta,x,y) \end{aligned}$$

where $U^+ : Z^q \times Z^q \to Z^{pq} \times Z^{pq} \times Z^{pq}$ carries (x,y) to $(U(x+y), Ux, Uy)$. Hence we have

$$\begin{aligned} \gamma^{cr} &= p_2(1\times\nu)\lambda^3_\pi U^+_\sharp(h_\pi \times 1) \simeq p_2\lambda_\pi\nu_\sharp U^+_\sharp(h_\pi\times 1) \\ &= p_2\lambda_\pi U^{cr}_\sharp(h_\pi\times 1) = p_2\lambda_\pi N_\sharp(h_\pi \times \overline{U}) \simeq p_2(1\times N)(h_\pi\times\overline{U}) \\ &= p_2(h_\pi \times N\overline{U}) = (N\overline{U})p_{23} = U^{cr}p_{23}. \end{aligned}$$

This defines the track

$$\begin{aligned}\Lambda_L &= (p_2 L^\nu U^+_\#(h_\pi \times 1))\Box(p_2 L^N(h_\pi \times \bar{U}))^{\mathrm{op}}\\ \Lambda_L &: U^{cr} p_{23} \Rightarrow \gamma^{cr}.\end{aligned}$$

Hence the proof of (8.2.3) is complete. □

For maps $v : X \to Z^1$, $x : X \to Z^q$, $y : X \to Z^q$ in **Top*** the linearity track Λ_L yields the track

(8.2.4)
$$\begin{aligned} L_v^{x,y} &: \quad U^{cr}(x,y) \Longrightarrow \gamma_v^{cr}(x,y),\\ L_v^{x,y} &= \quad \Lambda_L(v,x,y) \in [\![X, Z^{pq}]\!]\end{aligned}$$

with $U^{cr}(x,y) = U(x+y) - Ux - Uy$ and $\gamma_v^{cr}(x,y) = \gamma_v(x+y) - \gamma_v(x) - \gamma_v(y)$; see (8.1.6). According to the construction of Λ_L in the proof of (8.2.3) we can describe the track $L_v^{x,y}$ by the following diagram which is based on the equations

(1) $$\nu U^+ = U^{cr} = N\bar{U}.$$

We indicate in the diagram only the arrows; the objects • are appropriate Borel constructions. Subdiagrams with a number 1 or 2 are homotopy commutative with a fixed track; all other subdiagrams are commutative. Let $w = h_\pi v : X \to B\pi$.

(2)

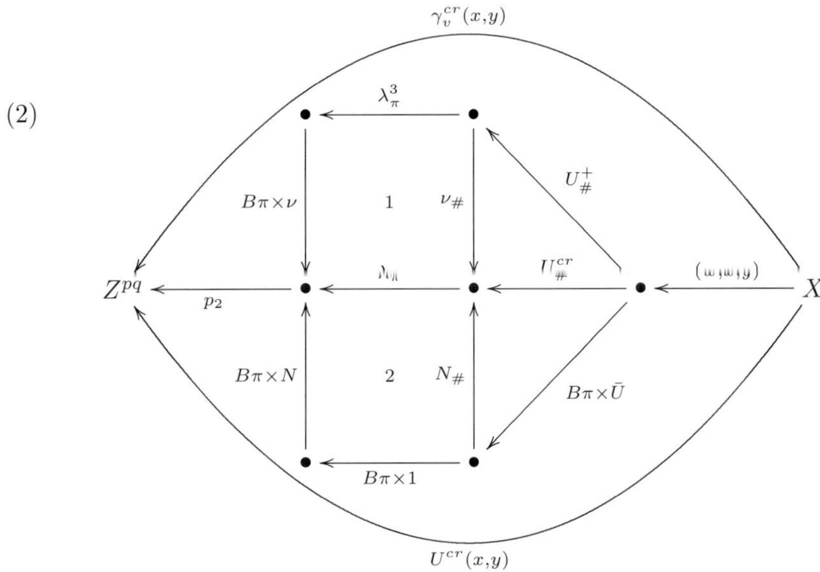

Here the subdiagram 1 is given by (8.2.3)(5) and subdiagram 2 is given by (8.2.3)(4). The map $\lambda_\pi^3 = \lambda_\pi \bar{\times} \lambda_\pi \bar{\times} \lambda_\pi$ is a product over $B\pi$ according to (7.1.4) while $B\pi \times X$ denotes the product in **Top**. The π-norm map $N : Z^{pq} \to Z^{pq}$ in (8.2.3)(1) admits by (7.2.2) the well-defined track

(8.2.5) $$\Gamma = \sum_{\alpha \in \pi} \Gamma_\alpha : N \Longrightarrow N^\pm = 0$$

where

$$N^{\pm}(x) = \sum_{\alpha\in\pi} \mathrm{sign}(\alpha)\cdot x = p\cdot x = 0 \in Z^{pq}.$$

Here we use the fact that for $\alpha \in \pi \subset \sigma_{pq}$ we have $\mathrm{sign}(\alpha) = 1$. According to the definition of $\bar{U}$ in (8.2.3)(2) we have

$$U(x+y) - Ux - Uy = U^{cr}(x,y) = N\bar{U}(x,y)$$

so that $\Gamma\bar{U}(x,y) : U^{cr}(x,y) \Rightarrow 0$. Therefore the track

$$(8.2.6)\qquad \begin{aligned} \Gamma_0^{x,y} &: \quad U(x+y) \Longrightarrow Ux + Uy \\ \Gamma_0^{x,y} &= \quad \Gamma\bar{U}(x,y) + Ux + Uy \end{aligned}$$

is defined. This yields by $L_v^{x,y}$ in (8.2.4) the track

$$(8.2.7)\qquad \begin{aligned} \Gamma_v^{x,y} &: \quad \gamma_v(x,y) \Longrightarrow \gamma_v(x) + \gamma_v(y) \\ \Gamma_v^{x,y} &= \quad (\Gamma\bar{U}(x,y)\Box(L_v^{x,y})^{\mathrm{op}}) + \gamma_v(x) + \gamma_v(y). \end{aligned}$$

For $v = 0$ the track $L_v^{x,y}$ is the identity track so that $\Gamma_v^{x,y}$ for $v = 0$ coincides with (8.2.6).

Moreover we define for $r \in \mathbb{F}$ the *linearity track*

$$(8.2.8)\qquad L(r)_v^x : \gamma_v(rx) \Longrightarrow r\cdot\gamma_v(x)$$

as follows. We have $U(rx) = r^pU(x) = rU(x)$ since $r^p = r$ in $\mathbb{F} = \mathbb{Z}/p$. Therefore we get the following diagram where $r : Z^{pq} \to Z^{pq}$ carries x to $r\cdot x$.

$$\begin{array}{ccccccccc}
Z^{pq} & \xleftarrow{p_2} & B\pi\times Z_{\pm}^{pq} & \xleftarrow{\lambda_\pi} & E\pi\times_\pi Z^{pq} & \xleftarrow{U_\#} & Z^q & \xleftarrow{x} & X \\
\downarrow r & & \downarrow r_\# & \overset{L^r}{\Leftarrow} & \downarrow r_\# & & \downarrow r & & \\
Z^{pq} & \xleftarrow{p_2} & B\pi\times Z_{\pm}^{pq} & \xleftarrow[\lambda_\pi]{} & E\pi\times_\pi Z^{pq} & \xleftarrow{U_\#} & Z^q & &
\end{array}$$

Hence we can define

$$L(r)_v^x = p_2L^rU_\# x.$$

Here L^r is the linear track in (7.3.4). We point out that, for example, $\Gamma_v^{x,x}$ defined in (8.2.7) is a track $\Gamma_v^{x,x} : \gamma_v(2x) \Rightarrow 2\gamma_v(x)$ which, however, in general does not coincide with $L(r)_v^x : \gamma_v(2x) \Rightarrow 2\gamma_v(x)$.

In a similar way we get for a permutation $\sigma \in \sigma_q$ inducing $\sigma : Z^q \to Z^q$ the permutation track

$$(8.2.9)\qquad P(\sigma)_v = P(\sigma)_v^x : \gamma_v(\sigma x) \Longrightarrow \sigma^p\cdot\gamma_v(x)$$

as follows. We have $U(\sigma x) = \sigma^p U(x)$ where $\sigma^p \in \sigma_{pq}$ is the permutation for which $(\sigma x)^p = \sigma^p x^p$ where $x^p = x \cdot \dots \cdot x$ is the p-fold product. Hence we get the diagram

$$\begin{array}{ccccccccc}
Z^{pq} & \xleftarrow{p_2} & B\pi \times Z^{pq}_{\pm} & \xleftarrow{\lambda_\pi} & E\pi \times_\pi Z^{pq} & \xleftarrow{U_\#} & Z^q & \xleftarrow{x} & X \\
\downarrow{\scriptstyle \sigma^p} & & \downarrow{\scriptstyle \sigma^p_\#} & \overset{L^{\sigma^p}}{\Leftarrow} & \downarrow{\scriptstyle (\sigma^p)_\#} & & \downarrow{\scriptstyle \sigma} & & \\
Z^{pq} & \xleftarrow{p_2} & B\pi \times Z^{pq}_{\pm} & \xleftarrow[\lambda_\pi]{} & E\pi \times_\pi Z^{pq} & \xleftarrow{U_\#} & Z^q & &
\end{array}$$

which yields the definition

$$P(\sigma)^x_v = p_2 L^{\sigma^p} U_\# x.$$

Here L^{σ^p} is the linear track in (7.3.4).

We point out that the linearity track $\Gamma^{x,y}_v$ is also defined if we replace x, y by tracks $\bar{x} : x \Rightarrow x'$ and $\bar{y} : y \Rightarrow y'$ in $[\![X, Z^q]\!]$. In fact, such tracks are represented by homotopies

$$\bar{\bar{x}}, \bar{\bar{y}} : IX \longrightarrow Z^q \text{ with } IX = [0,1] \times X / [0,1] \times *$$

so that $\Gamma^{\bar{\bar{x}},\bar{\bar{y}}}_v$ is defined in $[\![IX, Z^q]\!]$. Using the diagonal of $\Gamma^{\bar{\bar{x}},\bar{\bar{y}}}_v$ we get the track

(8.2.10) $$\Gamma^{\bar{x},\bar{y}}_v : \gamma_v(x+y) \Rightarrow \gamma_v(x') + \gamma_v(y')$$

and the following diagram of tracks in $[\![X, Z^q]\!]$ commutes.

$$\begin{array}{ccc}
\gamma_v(x+y) & \xrightarrow{\Gamma^{x,y}_v} & \gamma_v(x) + \gamma_v(y) \\
\downarrow{\scriptstyle \gamma_v(\bar{x}+\bar{y})} & \searrow^{\Gamma^{\bar{x},\bar{y}}_v} & \downarrow{\scriptstyle \gamma_v(\bar{x}) + \gamma_v(\bar{y})} \\
\gamma_v(x'+y') & \xrightarrow[\Gamma^{x',y'}_v]{} & \gamma_v(x') + \gamma_v(y')
\end{array}$$

8.3 Cartan tracks for power maps

Next we consider the following diagram with $\Delta_{13}(\alpha, x, y) = (\alpha, x, \alpha, y)$.

$$\begin{array}{ccc}
Z^1 \times Z^q \times Z^{q'} & \xrightarrow{1 \times \mu} & Z^1 \times Z^{q+q'} \\
\downarrow{\scriptstyle \Delta_{13}} & & \downarrow{\scriptstyle \gamma} \\
Z^1 \times Z^q \times Z^1 \times Z^{q'} & \overset{\Lambda_C}{\Longleftarrow} & Z^{p(q+q')} \\
\downarrow{\scriptstyle \gamma \times \gamma} & & \downarrow{\scriptstyle \sigma} \\
Z^{pq} \times Z^{pq'} & \xrightarrow{\mu} & Z^{pq+pq'}
\end{array}$$

Here $\sigma \in \sigma_{pq+pq'}$ is the permutation for which

$$\sigma(x \cdot y)^p = x^p \cdot y^p \tag{8.3.1}$$

with $x \in Z^q$ and $y \in Z^{q'}$. One has $\text{sign}(\sigma) = (-1)^{\frac{qq'p(p-1)}{2}}$. We observe that the diagram restricted to $Z^1 \vee Z^q \times Z^{q'}$ commutes.

8.3.2 Theorem. *Linear comparison tracks* (7.3.3) *and smash tracks* (7.3.8) *induce the track*

$$\Lambda_C : \sigma\gamma(1 \times \mu) \Rightarrow \mu(\gamma \times \gamma)\Delta_{13}$$

under $Z^1 \vee Z^q \times Z^q$ *which we call the* Cartan track *for* γ. *Moreover* Λ_C *is compatible with canonical tracks in* $\underline{\underline{\gamma}}$.

Proof. Let $n = pq + pq' = p(q + q')$. We have the following commutative diagram.

(1)
$$\begin{array}{ccc} Z^q \times Z^{q'} & \xrightarrow{\mu} & Z^{q+q'} \\ & & \downarrow U \\ \downarrow U\times U & & Z^n \\ & & \downarrow \sigma \\ Z^{pq} \times Z^{pq'} & \xrightarrow{\mu} & Z^n \end{array}$$

The group π acts trivially on $Z^q \times Z^{q'}$ and acts via cyclic permutation of the $(q + q')$-blocks on Z^n. This shows that all maps in the diagram are actually π-equivariant. Hence we get for induced maps on Borel constructions:

$$\sigma_\sharp U_\sharp(1 \times \mu) = \mu_\sharp(U \times U)_\sharp. \tag{2}$$

Moreover using (7.3.3) we see that the diagram

(3)
$$\begin{array}{ccc} E\pi \times_\pi Z^n & \xrightarrow{\lambda_\pi} & B\pi \times Z^n \\ \downarrow \sigma_\sharp & & \downarrow 1\times\sigma \\ E\pi \times_\pi Z^n & \xrightarrow{\lambda_\pi} & B\pi \times Z^n \end{array}$$

homotopy commutes under $B\pi \vee Z^n$ and over $B\pi$ and the corresponding linear track $L^\sigma : \lambda_\pi\sigma_\# \Rightarrow (1 \times \sigma)\lambda_\pi$ is well defined.

Moreover we obtain for the π-equivariant bilinear map $\mu : Z^{pq} \times Z^{pq'} \to Z^n$ defined by the multiplication (2.1.1) the smash track

(4) $$S^{\mu} : \lambda_{\pi}\mu_{\#} \Rightarrow (1 \times \mu)\lambda_{\pi}^2$$

for the following diagram.

(5) $$\begin{array}{ccc} E\pi \times_{\pi} (Z^{pq} \times Z^{pq'}) & \xrightarrow{\lambda_{\pi}^2} & B\pi \times Z^{pq} \times Z^{pq'} \\ \downarrow{\scriptstyle \mu_{\#}} & & \downarrow{\scriptstyle 1\times\mu} \\ E\pi \times Z^n & \xrightarrow{\lambda_{\pi}} & B\pi \times Z^n \end{array}$$

This is a track over and under $B\pi \vee Z^{pq} \times Z^{pq'}$. Hence we get the track

$$\Lambda_C = (p_2 S^{\mu}(U \times U)_{\#}(h_{\pi} \times 1))\square(p_2 L^{\sigma} U_{\#}(h_{\pi} \times \mu))^{\mathrm{op}}$$

given by the composite

$$\begin{aligned} \sigma\gamma(1 \times \mu) &= p_2(1 \times \sigma)\lambda_{\pi} U_{\sharp}(h_{\pi} \times \mu) \\ &\simeq p_2 \lambda_{\pi} \sigma_{\sharp} U_{\sharp}(h_{\pi} \times \mu) \\ &= p_2 \lambda_{\pi} \mu_{\sharp}(U \times U)_{\sharp}(h_{\pi} \times 1) \\ &\simeq p_2(1 \times \mu)\lambda_{\pi}^2 (U \times U)_{\sharp}(h_{\pi} \times 1) \\ &= \mu(\gamma \times \gamma)\Delta_{13}. \end{aligned}$$

This is the Cartan track. □

For maps $v : X \to Z^1$, $x : X \to Z^q$, $y : X \to Z^{q'}$ in **Top*** the Cartan track Λ_C induces the track

(8.3.3) $$\begin{aligned} C_v^{x,y} &: \quad \sigma(x,y)\gamma_v(x \cdot y) \Longrightarrow \gamma_v(x) \cdot \gamma_v(y) \\ C_v^{x,y} &= \quad \Lambda_C(v,x,y) \in [\![X, Z^{p(q+q')}]\!]. \end{aligned}$$

Here the permutation $\sigma(x,y) = \sigma \in \sigma_{pq+pq'}$ is defined as in (8.3.1) for $q = \mid x \mid$ and $q' = \mid y \mid$. Moreover the product $\cdot$ is defined by multiplication maps μ in (2.1.2). According to the definition of Λ_C in the proof of (8.3.1) we can describe $C_v^{x,y}$ by the following diagram which is based on the equation

(1) $$\sigma U \mu = \mu(U \times U)$$

corresponding to $\sigma U(x \cdot y) = (Ux) \cdot (Uy)$.

We use the same convention as in (8.2.4).

(2)

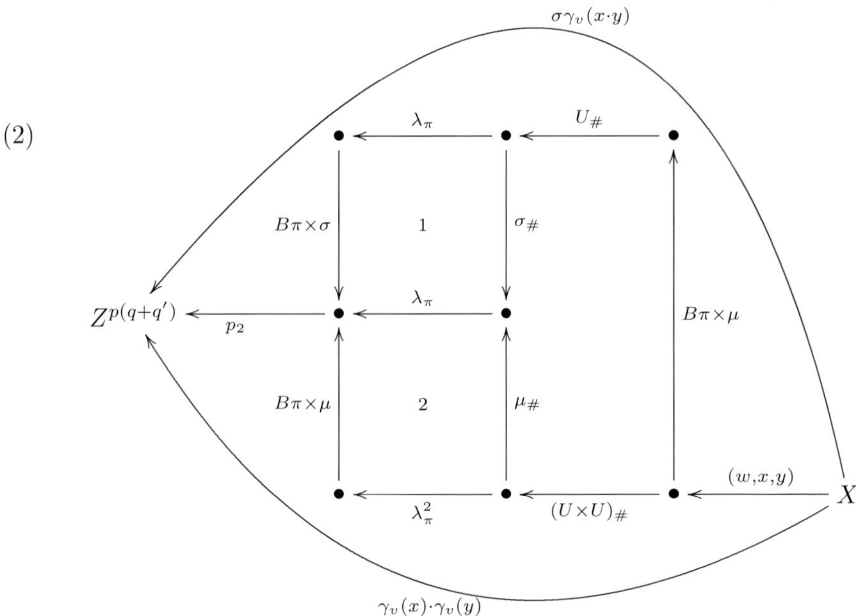

Subdiagram 1 is given by (8.3.2)(3) and subdiagram 2 is given by (8.3.2)(5). All other subdiagrams are commutative. The map $\lambda_\pi^2 = \lambda_\pi \bar{\times} \lambda_\pi$ is a product over $B\pi$ according to (7.1.4) and $B\pi \times X$ is the product in **Top**.

8.4 Adem tracks for power maps

Now let $\pi = \rho = \mathbb{Z}/p$. Then the product group $\pi \times \rho$ is contained in σ_{p^2} and we have the following commutative diagram of groups.

$$\begin{array}{ccccccccccc}
\pi \times \rho & \subset & \sigma_p \times \sigma_p & \subset & \sigma_p \int \sigma_p & \subset & \sigma_{p^2} & \subset & \sigma_{p^2 q} \\
\downarrow{\scriptstyle T} & & & & & & & & \downarrow{\scriptstyle (_)^\tau} \\
\rho \times \pi & \subset & \sigma_p \times \sigma_p & \subset & \sigma_p \int \sigma_p & \subset & \sigma_{p^2} & \subset & \sigma_{p^2 q}
\end{array}$$

The rows are the canonical inclusions as in McClure [MC], p. 254. The map T is the interchange map with $T(\alpha, \beta) = (\beta, \alpha)$ and $(\_)^\tau$ carries ξ to $\tau\xi\tau^{-1}$. If we consider σ_{p^2} as the permutation group of the product set $\pi \times \rho$, then T is such a permutation which yields $\tau \in \sigma_{p^2} \subset \sigma_{p^2 q}$. Hence for $(\alpha, \beta) \in \pi \times \rho$ we have the equation in $\sigma_{p^2 q}$,

(8.4.1) $$\tau(\alpha, \beta) = (\beta, \alpha)\tau.$$

Therefore the map

$$\tau_\# : E(\pi \times \rho) \times_{\pi\times\rho} Z^{p^2q} \to E(\rho \times \pi) \times_{\rho\times\pi} Z^{p^2q} \tag{1}$$

which carries $((\tilde{v}, \tilde{w}), x)$ to $(\tilde{w}, \tilde{v}, x)$ is well defined with $\tilde{v} \in E\pi$, $\tilde{w} \in E\rho$ and $x \in Z^{p^2q}$. In fact (8.4.1) shows that $\tau : Z^{p^2q} \to Z^{p^2q}$ is a T-equivariant linear map since for $\tau(x) = \tau \cdot x$ we have $\tau((\alpha, \beta) \cdot x) = T(\alpha, \beta) \cdot \tau(x)$. Hence τ induces the map $\tau_\#$ as in (7.3.2). We point out that the sign of $\tau \in \sigma_{p^2q}$ is

$$\mathrm{sign}(\tau) = (-1)^{\frac{q(p-1)p}{2}}. \tag{2}$$

Next we consider the following diagram obtained by the interchange map T and by the power map γ in (8.1.3).

$$\begin{array}{ccccc} Z^1 \times Z^1 \times Z^q & \xrightarrow{1\times\gamma} & Z^1 \times Z^{pq} & \xrightarrow{\gamma} & Z^{p^2q} \\ \Big\downarrow{\scriptstyle T\times 1} & & \xRightarrow{\Lambda_A} & & \Big\downarrow{\scriptstyle \tau} \\ Z^1 \times Z^1 \times Z^q & \xrightarrow{1\times\gamma} & Z^1 \times Z^{pq} & \xrightarrow{\gamma} & Z^{p^2q} \end{array}$$

We observe that $\gamma(1 \times \gamma)(T \times 1)$ and $\tau\gamma(1 \times \gamma)$ both restricted to $Z^1 \times Z^1 \vee Z^q$ coincide with (o, U^2) where $U^2 : Z^q \to Z^{p^2q}$ carries x to $U^2(x) = UU(x) = x^{p^2}$. In addition we show:

8.4.2 Theorem. *Linear comparison tracks* (7.3.8) *and diagonal tracks* (7.3.13) *induce the track*

$$\Lambda_A : \gamma(1 \times \gamma)(T \times 1) \Rightarrow \tau\gamma(1 \times \gamma)$$

under $Z^1 \times Z^1 \vee Z^q$ *which we call the* Adem track for γ. *Moreover* Λ_A *is compatible with canonical tracks in* $\underline{\underline{\gamma}}$.

Proof. We have the multilinear maps

$$\begin{aligned} f = \mu : (Z^{pq})^{\times p} &\longrightarrow Z^{p^2q}, \\ g = \mu : (Z^q)^{\times p} &\longrightarrow Z^{pq}, \end{aligned}$$

which yield the composite

$$f(g^{\times p}) : ((Z^q)^{\times p})^{\times p} \longrightarrow Z^{p^2q}. \tag{1}$$

Here g is $\rho = \mathbb{Z}/p$-equivariant and f is $\pi \times \rho = \mathbb{Z}/p \times \mathbb{Z}/p$-equivariant. Moreover the sign-actions are trivial so that we can apply (7.3.13). Hence we obtain the diagonal track

$$D^{f,g} : C(f)(C(g)^{\times p})_\# D \Rightarrow C(f(g^{\times p})) \tag{2}$$

for diagram (7.3.11) with f and g above.

Next one readily checks that the following diagram commutes.

(3)
$$\begin{array}{ccc} B(\pi\times\rho)\times Z^q & \xrightarrow{U^2_\sharp} & E(\pi\times\rho)\times_{\pi\times\rho} Z^{p^2q} \\ \uparrow{\scriptstyle T\times 1} & & \uparrow{\scriptstyle \tau\#} \\ B(\rho\times\pi)\times Z^q & \xrightarrow{U^2_\sharp} & E(\rho\times\pi)\times_{\rho\times\pi} Z^{p^2q} \end{array}$$

Here $U^2 = U\circ U$ carries x to $x^{p\cdot p}$, see (8.1.1)(1). Moreover $\tau_\#$ is induced by the T-equivariant linear map τ in (8.4.1)(1). For this map we have by (7.3.3) the linear comparison track

(4)
$$L^{T,\tau} : \lambda_{\pi\times\rho}\tau_\# \Rightarrow (T\times\tau)\lambda_{\rho\times\pi}$$

for the following diagram.

(5)
$$\begin{array}{ccc} E(\pi\times\rho)\times_{\pi\times\rho} Z^{p^2q} & \xrightarrow{\lambda_{\pi\times\rho}} & B(\pi\times\rho)\times Z^{p^2q} \\ \uparrow{\scriptstyle \tau\#} & \Longrightarrow & \uparrow{\scriptstyle T\times\tau} \\ E(\rho\times\pi)\times_{\rho\times\pi} Z^{p^2q} & \xrightarrow{\lambda_{\rho\times\pi}} & B(\rho\times\pi)\times Z^{p^2q} \end{array}$$

Next we have the equations:

(6)
$$\begin{aligned} \gamma(1\times\gamma) &= (p_2\lambda_\pi U_\#(h_\pi\times 1))(Z^1\times(p_2\lambda_\rho U_\#(h_\rho\times 1))) \\ &= (p_2\lambda_\pi U_\#)(B\pi\times p_2\lambda_\rho U_\#)(h_\pi\times h_\rho\times 1) \\ &= C(f)\Delta^{(p)}_\#(B\pi\times C(g)\Delta^{(p)}_\#)(h_\pi\times h_\rho\times 1). \end{aligned}$$

Here the diagonal maps $\Delta = \Delta^{(n)} : X \to X^{\times n}$ are defined in (8.1.1). Moreover the following diagram commutes.

(7)
$$\begin{array}{ccc} B\pi\times B\rho\times (Z^q)^{\times p} & \xrightarrow{\Delta^{(p)}_\#} & E(\pi\times\rho)\times_{\pi\times\rho}((Z^q)^{\times p})^{\times p} \\ \downarrow{\scriptstyle B\pi\times C(g)} & & \downarrow{\scriptstyle C(g)^{\times p}_\# D} \\ B\pi\times Z^{pq} & \xrightarrow{\Delta^{(p)}_\#} & E\pi\times_\pi (Z^{pq})^{\times p} \end{array}$$

Here the right-hand side is defined as in diagram (7.3.11). Commutativity is easily seen by the definition of the diagonal maps. Now (6) and (7) yield the first equation of the following composition of tracks.

$$
(8)\qquad
\begin{aligned}
\gamma(1\times\gamma) &= C(f)(C(g)^{\times p}_{\#}D)\Delta^{(p)}_{\#}(B\pi\times\Delta^{(p)}_{\#})(h_\pi\times h_\rho\times 1)\\
&= C(f)(C(g)^{\times p}_{\#}D)\Delta^{(p^2)}_{\#}(h_\pi\times h_\rho\times 1)\\
&\simeq C(f(g^{\times p}))\Delta^{(p^2)}_{\#}(h_\pi\times h_\rho\times 1)\ \text{ see (2)},\\
&= p_2\lambda_{\pi\times\rho}(f(g^{\times p}))_{\#}\Delta^{(p^2)}_{\#}(h_\pi\times h_\rho\times 1)\\
&= p_2\lambda_{\pi\times\rho}U^2_{\#}(h_\pi\times h_\rho\times 1)\\
&= p_2\lambda_{\rho\times\pi}U^2_{\#}(h_\rho\times h_\pi\times 1).
\end{aligned}
$$

Here the last equation holds since $\rho=\pi=\mathbb{Z}/p$. Now we can apply (8) and (5) in order to get

$$
(8.4.3)\qquad
\begin{aligned}
\gamma(1\times\gamma)(T\times 1) &\simeq p_2\lambda_{\pi\times\rho}U^2_{\#}(h_\pi\times h_\rho\times 1)(T\times 1), && \text{see (8)},\\
&= p_2\lambda_{\pi\times\rho}U^2_{\#}(T\times 1)(h_\rho\times h_\pi\times 1)\\
&= p_2\lambda_{\pi\times\rho}\tau_{\#}U^2_{\#}(h_\rho\times h_\pi\times 1), && \text{see (3)},\\
&\simeq p_2(T\times\tau)\lambda_{\rho\times\pi}U^2_{\#}(h_\rho\times h_\pi\times 1), && \text{see (5)},\\
&= \tau p_2\lambda_{\rho\times\pi}U^2_{\#}(h_\rho\times h_\pi\times 1)\\
&\simeq \tau\gamma(1\times\gamma), && \text{see (8)}.
\end{aligned}
$$

This is the Adem track which uses the linear comparison track (5) and uses twice the diagonal track in (8). □

Remark. The proof of the Adem track above is actually less complicated than the proof 2.7 of Karoubi [Ka2] who follows the line of proof in Steenrod-Epstein [SE] p. 117. In fact, our argument does not need the cohomology of the symmetric group σ_{p^2} (with local coefficients if q is odd).

Let $v,w:X\to Z^1$ and $x:X\to Z^q$ be maps in **Top***. Then the Cartan track Λ_A induces the track

$$
(8.4.4)\qquad
\begin{aligned}
A^x_{v,w} &: \gamma_v\gamma_w(x)\Longrightarrow\tau_x\cdot\gamma_w\gamma_v(x)\\
A^x_{v,w} &= \Lambda_A(w,v,x)\in[\![X,Z^{p^2q}]\!].
\end{aligned}
$$

Here $\tau_x=\tau\in\sigma_{p^2}\subset\sigma_{p^2q}$ is defined as in (8.4.1) for $q=\mid x\mid$.

8.5 Cohomology as a power algebra

For a prime p and $\mathbb{F}=\mathbb{Z}/p$ the Eilenberg-MacLane space $Z^n=K(\mathbb{F},n)$ yields the cohomology groups of a pointed space X by the well-known equation

$$
(8.5.1)\qquad \tilde H^n(X)=\tilde H^n(X,\mathbb{Z}/p)=[X,Z^n].
$$

Hence the power map $\gamma : Z^1 \times Z^n \to Z^{pn}$ induces the operation $(n \geq 1)$

$$\gamma_x : H^n(X) \longrightarrow H^{pn}(X) \tag{8.5.2}$$

which for $x \in H^1$ carries $y \in H^n(X)$ to the composite

$$\gamma_x(y) : X \xrightarrow{(x,y)} Z^1 \times Z^n \xrightarrow{\gamma} Z^{pn}.$$

We have seen in Section (1.5) that the cohomology $H^*(X)$ corresponds to a model of the theory $\mathbf{K}_p$ of Eilenberg-MacLane spaces. We now define for each such model

$$M \in \mathbf{model}(\mathbf{K}_p),$$

the algebra M^* with $M^n = M(Z^n)$, $n \geq 1$, and multiplication

$$M^n \times M^m = M(Z^n \times Z^m) \xrightarrow{\mu_*} M(Z^{n+m}) = M^{n+m}$$

induced by μ in (2.1.1). Moreover we define (M^*, γ) by

$$\gamma_x : M^n \longrightarrow M^{pn} \tag{8.5.3}$$

for $x \in M^1$, $n \geq 1$. Here γ_x carries $y \in M^n$, $n \geq 1$, to $\gamma_*(x, y)$ where γ_* is the composite

$$M^1 \times M^n = M(Z^1 \times Z^n) \xrightarrow{\gamma_*} M(Z^{pn}) = M^{pn}$$

induced by the power map γ.

8.5.4 Theorem. *For a path-connected space X the cohomology $(H^*(X), \gamma)$ defined by (8.5.2) is a power algebra. More generally for each model M of $\mathbf{K}_p$ the algebra (M^*, γ) defined by (8.5.3) is a power algebra.*

Proof. If $x = 0$ is represented by $x : X \to * \in Z^1$ we see that $\gamma(x, y)$ is represented by the composite $X \xrightarrow{y} Z^n \subset Z^1 \times Z^n \xrightarrow{\gamma} Z^{pn}$ and we can use (8.1.4). Hence $\gamma_0(y) = y^p$. Next we see by (8.2.4) that γ_x is a homomorphism of R-vector spaces. Moreover (8.3.3) shows

$$\gamma_x(y \cdot z) = \operatorname{sign}(\sigma)\gamma_x(y) \cdot \gamma_x(z).$$

This is equation (1.2.6)(ii). Finally (8.4.4) shows

$$\gamma_x\gamma_y(z) = \operatorname{sign}(\tau)\gamma_y\gamma_x(z).$$

This is equation (1.2.6)(iii). The Bockstein map $\beta : Z^1 \to Z^2$ defined as in (8.5.13) below shows that $H^*(X)$ and M^* are also β-algebras in β-$\mathbf{Alg}_0$. □

Using (8.5.1) the power map γ defines an element

$$\begin{aligned} \gamma \in H^{pq}(Z^1 \times Z^q) &= (H^*(Z^1) \otimes H^*(Z^q))^{pq} \\ &= \bigoplus_{i=0}^{pq} H^i(Z^1) \otimes H^{pq-i}(Z^q) \\ &= \bigoplus_{i=0}^{pq} w_i \otimes H^{pq-i}(Z^q). \end{aligned}$$

Here $w_i = \omega_i(x)$ where $x = w_1 \in H^1(Z^1) = \mathbb{Z}/p$ is a generator, see (1.2.3)(4). Hence one obtains well-defined elements $D_i \in H^{pq-i}(Z^q)$ with

(8.5.5)
$$\gamma = \sum_{i=0}^{pq} w_i \otimes D_i .$$

This is an equation for the homotopy class of γ. The elements D_i essentially coincide with those in Steenrod-Epstein [SE] and are used for the following definition of *Steenrod operations*:

8.5.6 Definition. For $p = 2$ let

$$\mathrm{Sq}^j \in H^{q+j}(Z^q) = \left[Z^q, Z^{q+j}\right]$$

be defined by $\mathrm{Sq}^j = D_{q-j}$ and for p odd let

$$P^j \in H^{q+2j(p-1)}(Z^q) = \left[Z^q, Z^{q+2j(p-1)}\right]$$

be defined by $(-1)^j \vartheta_q P^j = D_{(q-2j)(p-1)}$. Here $\vartheta_q = (-1)^{\frac{m(q^2+q)}{2}} \cdot (m!)^q$ where $m = \frac{p-1}{2}$ is an element in the field $R = \mathbb{Z}/p$.

8.5.7 Remark. We point out that the sign for P^j above coincides with the sign of Steenrod-Epstein [SE] p. 112 and not with the sign of Karoubi [Ka2] p. 705. In fact, Karoubi takes the sign from formula (vi) of McClure [MC] p. 259; though formula (4) p. 260 is the appropriate formula. Then the sign of McClure and Steenrod-Epstein coincide provided we identify w_1 with the generator b used by McClure. Karoubi also uses a choice of generators x_i in $H^*(B\mathbb{Z}/p)$ which do not coincide with the generators w_i used by Steenrod-Epstein, see (1.2.3)(4).

The elements Sq^j and P^j in (8.5.6) are directly deduced from the power map γ via formula (8.5.5). Moreover the power map γ was obtained easily by the power function U and the comparison map λ_π. Therefore Definition (8.5.6) is a best possible direct way to introduce the elements Sq^j and P^j. These elements coincide with the classical elements since we have the following result.

8.5.8 Theorem. *The elements* Sq^j *and* P^j *defined in* (8.5.6) *by use of the power map* γ *coincide with the corresponding Steenrod operations.*

Proof. Karoubi [Ka2] shows that Sq^0 and P^0 are represented by the identity of Z^q. Therefore the linearity track and the Cartan track imply that Sq^j, P^j satisfy the axioms for Steenrod operations in Steenrod-Epstein [SE]. The uniqueness theorem chapter VIII [SE] thus implies that the elements (8.5.6) coincide with the corresponding classical elements. □

It is possible to describe all elements D_i in (8.5.5) in terms of Steenrod operations. If $p = 2$ we have for the homotopy class of γ the equation

(8.5.9)
$$\gamma = \sum_j w_{q-j} \otimes \mathrm{Sq}^j .$$

If p is odd we have accordingly

$$(8.5.10)\quad \gamma = \vartheta_q \sum_j (-1)^j w_{(q-2j)(p-1)} \otimes P^j + \vartheta_q \sum_j (-1)^j w_{(q-2j)(p-1)-1} \otimes \beta P^j$$

where β is the Bockstein operator. The sum is taken over $j \in \mathbb{Z}$ with $w_n = 0$ for $n < 0$ and $P^j = 0$ and $\mathrm{Sq}^j = 0$ for $j < 0$. Comparing (8.5.10) with (8.5.5) we see that for p odd many D_i are actually trivial. Formula (8.5.10) corresponds to McClure [MC] (4) p. 260 or to Steenrod-Epstein [SE] p. 119. See also 1.10 [Ka2].

8.5.11 Corollary. *Let V be a finitely generated $\mathbb{Z}/p$-vector space and $V^\# = \mathrm{Hom}(V, \mathbb{Z}/p)$. Then we have the isomorphism of power algebras natural in V,*

$$(H^*(B(V^\#), \gamma) \;=\; (E_\beta(V), \gamma).$$

Here the right-hand side is defined by (1.2.8).

Proof. We use (1.1.8), (1.1.9) and (1.2) and (8.5.8). □

The Bockstein homomorphism ($q \geq 1$)

$$(8.5.12)\qquad \beta : H^q(X) \longrightarrow H^{q+1}(X)$$

associated with the short exact sequence $0 \to \mathbb{Z}/p \to \mathbb{Z}/p^2 \to \mathbb{Z}/p \to 0$ is induced by a map

$$\beta : Z^q \longrightarrow Z^{q+1}$$

which is well defined up to a canonical track. That is by (8.5.1) the Bockstein homomorphism is the composite

$$H^q(X) = [X, Z^q] \xrightarrow{\beta_*} [X, Z^{q+1}] = H^{q+1}(X).$$

In the next section (8.6) we show:

8.5.13 Theorem. *For p odd the composite of the Bockstein map β and the power map γ,*

$$Z^1 \times Z^q \xrightarrow{\gamma} Z^{pq} \xrightarrow{\beta} Z^{pq+1},$$

is null homotopic.

As an application of (8.5.13) and (8.5.4) we get

8.5.14 Theorem. *Let p be odd and let X be a path connected space and M be a model of $\mathbf{K}_p$. Then $(H^*(X), \gamma)$ and (M^*, γ) in (8.5.4) are Bockstein power algebras.*

Proof. We define β on $H^*(X)$ by (8.5.12). Moreover we define β on M^* by the induced map

$$M^q = M(Z^q) \xrightarrow{\beta_*} M(Z^{q+1}) = M^{q+1}$$

induced by the Bockstein map $\beta : Z^q \to Z^{q+1}$. □

8.5.15 Theorem. *Let X be a path connected space and let M be a model of the theory $\mathbf{K}_p$ of Eilenberg-MacLane spaces. Then the power algebras $(H^*(X),\gamma)$ and (M^*,γ) in (8.5.4) are unitary extended power algebras for $p=2$ and are unitary extended Bockstein power algebras for p odd.*

Proof. We define the extended structure of $(H^*(X),\gamma)$ by

$$E_\beta(V)\otimes H^*(X) \;=\; H^*(B(V^\#)\times X)$$

where we use the Künneth formula and the power algebra structure of the right-hand side given by (8.5.4). We use for the model M of the theory $\mathbf{K}_p$ the isomorphism of categories

$$\mathcal{K}_p^0 \underset{b}{\overset{a}{\rightleftarrows}} \mathbf{model}(\mathbf{K}_p)$$

in (1.5.2). Then bM in $\mathcal{K}$ and $H^*B(V^\#)$ in $\mathcal{K}$ have a tensor product $H^*B(V^\#)\otimes bM$ in $\mathcal{K}$ and we define the power algebra

$$E_\beta(V)\otimes M^* \;=\; (a(H^*B(V^\#)\otimes bM),\gamma)$$

where γ is defined by (8.5.4). This is the extended structure of (M^*,γ). These algebras are unitary by (8.5.8) and (8.5.10). Moreover for p odd these power algebras are Bockstein algebras by (8.5.14). □

8.5.16 Corollary. *The functor Ψ in (1.5.4) is well defined for $p=2$ and p odd.*

8.6 Bockstein tracks for power maps

Let $Z^n = Z^n_{\mathbb{F}}$ with $\mathbb{F}=\mathbb{Z}/p$. We have seen in (2.1.11) that there is a well-defined contractible subgroupoid

$$\underline{\underline{\beta}} \subset \lambda Z^n, Z^{n+1}\rrbracket.$$

The objects of $\underline{\underline{\beta}}$ are the Bockstein maps β. Moreover we have the well-defined contractible subgroupoid

$$\underline{\underline{\gamma}} \subset \llbracket Z^1\times Z^q, Z^{pq}\rrbracket.$$

The objects of $\underline{\underline{\gamma}}$ are the power maps γ. The morphisms in $\underline{\underline{\beta}}$ and $\underline{\underline{\gamma}}$ are termed *canonical tracks*. We now show:

8.6.1 Theorem. *Let p be odd. Then there is a well-defined track*

$$\Lambda_B : \beta\gamma \Longrightarrow 0 \;\; in \;\; \llbracket Z^1\times Z^q, Z^{pq+1}\rrbracket$$

under $Z^1\times *$ which we call the Bockstein track *for γ. Moreover Λ_B is compatible with canonical tracks in $\underline{\underline{\gamma}}$ and $\underline{\underline{\beta}}$.*

The following facts are needed in the proof of (8.6.1). Recall that for $\mathbb{G} = \mathbb{Z}/p^2$, the short exact sequence

$$0 \longrightarrow \mathbb{F} \xrightarrow{i} \mathbb{G} \xrightarrow{\pi} \mathbb{F} \longrightarrow 0$$

induces the fibration

$$Z^n \xrightarrow{i} Z^n_{\mathbb{G}} \xrightarrow{\pi} Z^n.$$

Since the diagram

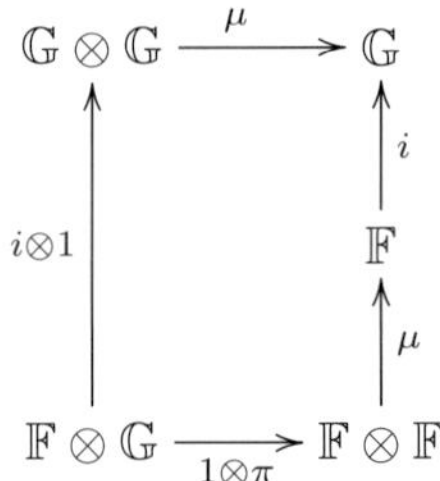

commutes, the following diagram is also commutative $(i + j = n)$.

(8.6.2)
$$\begin{array}{ccc}
Z^i_{\mathbb{G}} \times Z^j_{\mathbb{G}} & \xrightarrow{\mu} & Z^n_{\mathbb{G}} \\
\uparrow{\scriptstyle i\times 1} & & \uparrow{\scriptstyle i} \\
 & & Z^n \\
 & & \uparrow{\scriptstyle \mu} \\
Z^i \times Z^j_{\mathbb{G}} & \xrightarrow[1\times\pi]{} & Z^i \times Z^j
\end{array}$$

Here μ is the multiplication map.

Moreover we need the following notation on cubes. Let $I^k = I \times \cdots \times I$ be the k-dimensional cube with $I = [0,1]$ the unit interval and $S^1 = I/\{0,1\}$. We have the quotient map

$$p_0 : (I^k, \partial I^k) \longrightarrow (S^1 \wedge \cdots \wedge S^1, *) = (S^k, *).$$

Here the boundary ∂I^k of I^k is given by the union of *faces*

$$i^\epsilon_j : I^{k-1} \longrightarrow \partial I^k$$

with $\epsilon = \in \{0,1\}$ and $j = 1, \dots, k$ and

$$i^\epsilon_j(t_1, \dots, t_{k-1}) = (t_1, \dots, t_{j-1}, \epsilon, t_{j+1}, \dots, t_{k-1}).$$

We need the map

(8.6.3)
$$\rho : \partial I^k \longrightarrow \bigvee_{j=1}^{k} S^{k-1}$$

defined by

$$\text{(1)} \qquad \begin{cases} \rho(i_j^0) &= *, \\ \rho(i_j^1) &= i_j p_0. \end{cases}$$

Here $i_j : S^{k-1} \subset \bigvee_{j=1}^k S^{k-1}$ is the inclusion of index j. Hence the map ρ carries faces of level 0 to the basepoint and carries faces of level 1 of the form $I^{k-1} = i_j^1 I^{k-1}$ via the quotient map p_0 to the sphere $S^{k-1} = i_j S^{k-1}$.

We now choose a homeomorphism of pairs (χ, χ_0),

$$\text{(2)} \qquad \begin{array}{ccc} (I^k, \partial I^k) & \xleftarrow{(\chi,\chi_0)} & (S^{k-1} \wedge (I,0), S^{k-1}) \\ \downarrow{\scriptstyle p_0} & & \downarrow{\scriptstyle S^{k-1}\wedge p_0} \\ (S^k, *) & \xleftarrow{\bar{\chi}} & (S^{k-1} \wedge S^1, *) \end{array}$$

such that the map $\bar{\chi} : S^k \to S^k$ induced by χ is homotopic to the identity of S^k. Then the well-known *homotopy addition lemma* implies that the composite $\partial\chi_0$ admits a homotopy

$$\text{(3)} \qquad \rho\chi_0 \simeq \sum_{j=1}^k (-1)^{j-1} i_j.$$

We introduce the mapping space

$$\text{(4)} \qquad \Omega^{(k-1)} Z^{n+k} = (Z^{n+k}, 0)^{(\partial I^k, *)}$$

with $* = (0, \dots, 0) \in I^k$. Then ρ induces the map

$$\text{(5)} \qquad \rho^* : \times_{j=1}^k \Omega^{k-1} Z^{n+k} \longrightarrow \Omega^{(k-1)} Z^{n+k}.$$

Moreover we introduce the following pull back of spaces.

$$\text{(6)} \qquad \begin{array}{ccc} F^n_{(k)} & \longrightarrow & (Z^{n+k}_{\mathbb{G}}, 0)^{(I^k,*)} \\ \downarrow{\scriptstyle \partial^k} & & \downarrow{\scriptstyle \partial_0} \\ \Omega^{(k-1)} Z^{n+k} & \xrightarrow[i]{} & \Omega^{(k-1)} Z^{n+k}_{\mathbb{G}} \end{array}$$

Here i is induced by the inclusion $Z^{n+k} \to Z^{n+k}_{\mathbb{G}}$ and ∂_0 is induced by the inclusion $\partial I^k \subset I^k$. We get the map

$$\text{(7)} \qquad \pi^k : F^n_{(k)} \longrightarrow \Omega^k Z^{n+k}$$

which carries $(a,b) \in F^n_{(k)}$ to the unique map $\pi^k(a,b) : S^k \to Z^{n+k}$ for which the following diagram commutes.

$$\begin{array}{ccc} I^k & \xrightarrow{\ b\ } & Z^{n+k}_{\mathbb{G}} \\ {\scriptstyle p_0}\big\downarrow & & \big\downarrow{\scriptstyle \pi} \\ S^k & \xrightarrow[\pi^k(a,b)]{} & Z^{n+k} \end{array}$$

Recall that the Bockstein map β is defined by (see (2.1.9)

(8.6.4)
$$Z^n \underset{s_n}{\xrightarrow{\ \sim\ }} \Omega Z^{n+1} \underset{\pi}{\xleftarrow{\ \sim\ }} F^n \underset{\partial}{\longrightarrow} Z^{n+1}$$

where F^n is the fiber of $Z^{n+1} \to Z^{n+1}_{\mathbb{G}}$, that is

(1)
$$F^n = \{(x,\sigma) \in Z^{n+1} \times (Z^{n+1}_{\mathbb{G}}, 0)^{(I,0)},\ \sigma(1) = x\}.$$

In fact F^n can be identified with $F^n_{(1)}$ in (8.6.3)(6). Using π^k and ∂^k in (8.6.3) we get the following commutative diagram in which the horizontal homeomorphisms are induced by χ in (8.6.3)(2).

(2)
$$\begin{array}{ccc} \Omega^k Z^{n+k} & \underset{\approx}{\xrightarrow{\ \bar{\chi}^*\ }} & \Omega^{k-1}\Omega Z_{n+k} \\ {\scriptstyle \pi^k}\big\uparrow & a & \big\uparrow{\scriptstyle \Omega^{k-1}\pi} \\ F^n_{(k)} & \underset{\approx}{\xrightarrow{\ \chi^*\ }} & \Omega^{k-1}F^{n+k-1} \\ {\scriptstyle \partial^k}\big\downarrow & b & \big\downarrow{\scriptstyle \Omega^{k-1}\partial} \\ \Omega^{(k-1)}Z^{n+k} & \underset{\approx}{\xrightarrow{\ \chi_0^*\ }} & \Omega^{k-1}Z^{n+k} \end{array}$$

Here the right-hand side is defined by π and ∂ in (8.6.4). For $k \geq 1$ we need the maps, see (2.1.7),

(8.6.5)
$$r^k_n,\ s^k_n :\ Z^n \longrightarrow \Omega^k Z^{n+k}.$$

Using the inclusion $i_{\mathbb{F}} : S^1 \to Z^1$ in (2.1.7) we write $\hat{t} = i_{\mathbb{F}}(t)$. Then

(1)
$$r^k_n(x)(t_1 \wedge \cdots \wedge t_k) = x \cdot \hat{t}_1 \cdot \cdots \cdot \hat{t}_k, \qquad \text{and}$$

(2)
$$s^k_n(x)(t_1 \wedge \cdots \wedge t_k) = \hat{t}_1 \cdot \cdots \cdot \hat{t}_k \cdot x.$$

Hence we have

$$(3) \qquad \tau_{k,n} \cdot s_n^k(x) = r_n^k(x)$$

with

$$\operatorname{sign}(\tau_{k,n}) = (-1)^{k \cdot n},$$

where $\tau_{k,n}$ is the interchange permutation.

One readily checks that

$$(4) \qquad \begin{aligned} r_n^k &= (\Omega^{k-1} r_{n+k-1}) r_n^{k-1}, \\ s_n^k &= (\Omega^{k-1} s_{n+k-1}) s_n^{k-1}. \end{aligned}$$

Let $n = i_1 + \cdots + i_k$ with $i_1, \ldots, i_k \geq 1$, $k \geq 1$. Then we have the following commutative diagram where μ is the multiplication map.

$$(8.6.6) \qquad \begin{array}{ccc} Z^{i_1} \times \cdots \times Z^{i_k} & \xrightarrow{\quad \mu \quad} & Z^n \\ \downarrow{\scriptstyle s_{i_1} \times \cdots \times s_{i_k}} & 1 & \downarrow{\scriptstyle s_n^k} \\ \Omega Z^{i_1+1} \times \cdots \times \Omega Z^{i_k+1} & \xrightarrow[\tau_k \bar{\mu}]{} & \Omega^k Z^{n+k} \end{array}$$

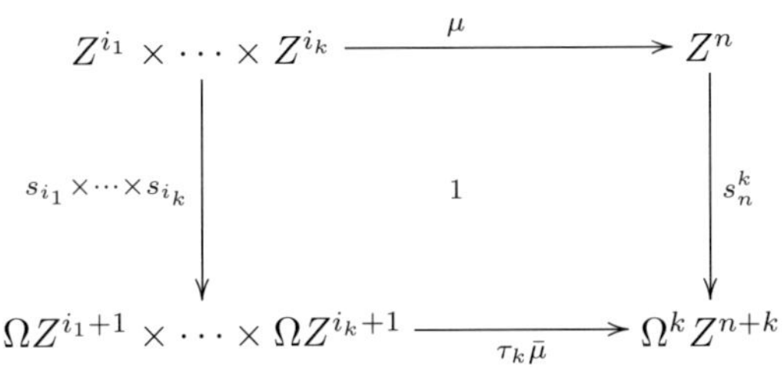

For $t_1, \ldots, t_k \in S^1$ and for $y_1 \in Z^{i_1}, \ldots, y_k \in Z^{i_k}$ let $\tau_{(k)} = \tau_k^{y_1, \ldots, y_k}$ be the interchange permutation for which

$$(1) \qquad \tau_{(k)}(\hat{t}_1 \cdot y_1 \cdot \hat{t}_2 \cdot y_2 \cdot \cdots \cdot \hat{t}_k \cdot y_k) = \hat{t}_1 \cdot \cdots \cdot \hat{t}_k \cdot y_1 \cdot \cdots \cdot y_k$$

with $\operatorname{sign}(\tau_{(k)}) = (-1)^{i_1 + (i_1+i_2) + \cdots + (i_1+i_2+\cdots+i_{k-1})}$. The map $\bar{\mu}$ in (8.6.4) is the *multiplication of loops*, that is

$$(2) \qquad \bar{\mu}(\sigma_1, \ldots, \sigma_k) = \sigma_1 \boxtimes \cdots \boxtimes \sigma_k$$

where the exterior product (see (2.1.5)) on the right-hand side is considered as a map $S^1 \wedge \cdots \wedge S^1 \longrightarrow Z^{n+k}$ carrying $t_1 \wedge \cdots \wedge t_k$ to $\sigma_1(t_1) \cdot \cdots \cdot \sigma_k(t_k)$. Diagram (8.6.4) commutes since

$$\begin{aligned} s_n^k \mu(y_1, \ldots, y_k)(t_1 \wedge \cdots \wedge t_k) &= \hat{t}_1 \cdot \cdots \cdot \hat{t}_k \cdot y_1 \cdot \cdots \cdot y_k \\ &= \tau_{(k)}(\hat{t}_1 \cdot y_1 \cdot \hat{t}_2 \cdot y_2 \cdot \cdots \cdot \hat{t}_k \cdot y_k) \\ &= \tau_{(k)}(\bar{\mu}(s_{i_1} y_1, \ldots, s_{i_k} y_k))(t_1 \wedge \cdots \wedge t_k). \end{aligned}$$

Next we embed diagram (8.6.6) into the following commutative diagram.

(8.6.7)
$$\begin{array}{ccc}
Z^{i_1}\times\cdots\times Z^{i_k} & \xrightarrow{\quad\mu\quad} & Z^n \\
{\scriptstyle s_{i_1}\times\cdots\times s_{i_k}}\downarrow\wr & 1 & \wr\downarrow {\scriptstyle s_n^k} \\
\Omega Z^{i_1+1}\times\cdots\times\Omega Z^{i_k+1} & \xrightarrow[\tau_{(k)}\bar{\mu}]{} & \Omega^k Z^{n+k}\\
{\scriptstyle \pi\times\cdots\times\pi}\uparrow\wr & 2 & \wr\uparrow{\scriptstyle\pi^k}\\
F^{i_1}\times\cdots\times F^{i_k} & \xrightarrow[\tau_{(k)}\bar{\bar{\mu}}]{} & F^n_{(k)}\\
\downarrow\bar{\partial} & 3 & \downarrow\partial^k\\
\times_{j=1}^k\Omega^{k-1}Z^{n+k} & \xrightarrow[\tau_{(k)}\rho^*]{} & \Omega^{(k-1)}Z^{n+k}
\end{array}$$

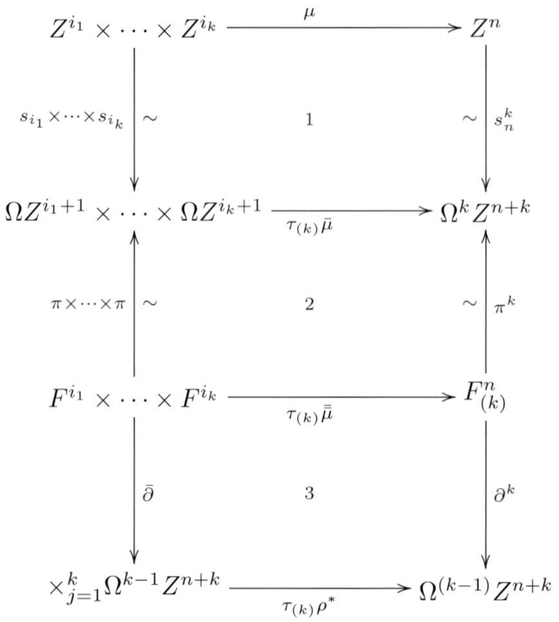

Subdiagram 1 is given by (8.6.4) and subdiagram 2 is given by (8.6.3)(7) with

(1) $$\bar{\bar{\mu}}((x_1,\sigma_1),\dots,(x_k,\sigma_k)) = \sigma_1\boxtimes\cdots\boxtimes\sigma_k.$$

Here the exterior product on the right-hand side is considered as a map

(2) $$\sigma_1\boxtimes\cdots\boxtimes\sigma_k : I^k \longrightarrow Z_{\mathbb{G}}^{n+k}$$

for $\sigma_1 : I \longrightarrow Z_{\mathbb{G}}^{i_1+1},\dots,\sigma_k : I \longrightarrow Z_{\mathbb{G}}^{i_k+1}$. Now we use (8.6.2) to see that $\sigma_1\boxtimes\cdots\boxtimes\sigma_k\,|_{\partial I^k}$ maps to $Z^{n+k}\subset Z_{\mathbb{G}}^{n+k}$. Therefore $\bar{\bar{\mu}}$ is well defined.

In fact, we have by (8.6.2) the formula

(3) $$\begin{aligned}&(\sigma_1\boxtimes\cdots\boxtimes\sigma_{j-1})\cdot i(x_j)\cdot(\sigma_{j+1}\boxtimes\cdots\boxtimes\sigma_k)\\ &\quad= i[(\pi\sigma_1)\boxtimes\cdots\boxtimes(\pi\sigma_{j-1})\cdot x_j\cdot(\pi\sigma_{j+1}\boxtimes\cdots\boxtimes(\pi\sigma_k)].\end{aligned}$$

Hence we define $\bar{\partial}=(\bar{\partial}_1,\dots,\bar{\partial}_k)$ by

(4) $$\begin{aligned}&\bar{\partial}_j((x_1,\sigma_1),\dots,(x_k,\sigma_k))\\ &\quad= (\pi\sigma_1)\boxtimes\cdots\boxtimes(\pi\sigma_{j-1})\cdot x_j\cdot(\pi\sigma_{j+1}\boxtimes\cdots\boxtimes(\pi\sigma_k).\end{aligned}$$

Here the right-hand side is a map $I^{k-1}\to S^{k-1}\to Z^{n+k}$ which defines an element in $\Omega^{k-1}Z^{n+k}$. Now (3) shows that subdiagram 3 commutes. Let $\tau_k^{(j)}$ be the permutation of Z^{n+k} which satisfies

(5) $$\begin{aligned}&\tau_k^{(j)}(\hat{t}_1\cdot y_1\cdot\dots\cdot\hat{t}_{j-1}\cdot y_{j-1}\cdot x_j\cdot\hat{t}_{j+1}\cdot y_{j+1}\cdot\dots\cdot\hat{t}_k\cdot y_k)\\ &\quad= y_1\cdot\dots\cdot y_j\cdot x_j\cdot y_{j+1}\cdot\dots\cdot y_k\cdot\hat{t}_1\cdot\dots\cdot\hat{t}_{j-1}\cdot\hat{t}_{j+1}\cdot\dots\cdot\hat{t}_k.\end{aligned}$$

Here we have $| y_j | = i_j$ and $| x_j | = i_j + 1$. Now $\bar{\partial}_j$ yields the following commutative diagram with $j = 1, \dots, k$.

(6)

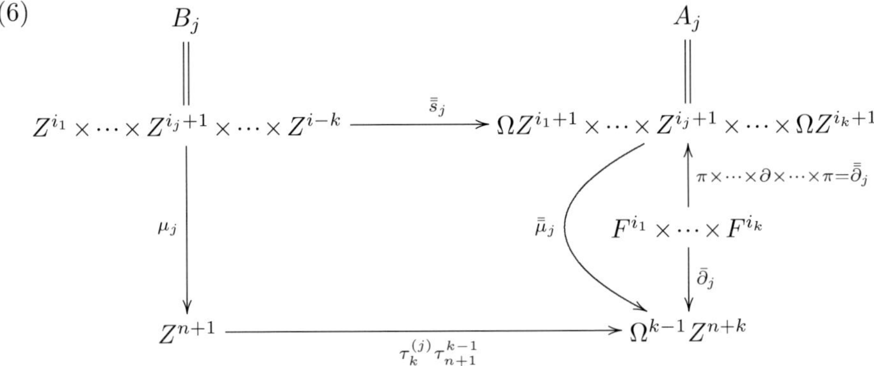

Here μ_j is the multiplication map and $\bar{\bar{s}}_j = s_{i_1} \times \cdots \times 1 \times \cdots \times s_{i_k}$ and $\bar{\bar{\mu}}_j$ carries $(\sigma_1, \dots, x_j, \dots, \sigma_k)$ to $(\sigma_1 \boxtimes \cdots \boxtimes \sigma_{j-1}) \cdot x_j \cdot (\sigma_{j+1} \boxtimes \cdots \boxtimes \sigma_k)$. Equations (4) and (5) show that diagram (6) commutes.

The right-hand side of (8.6.7) is "equivalent" to the Bockstein map $\beta(-1)^{n(k-1)}$. For this we use the following in which the indicated tracks are unique.

(7)

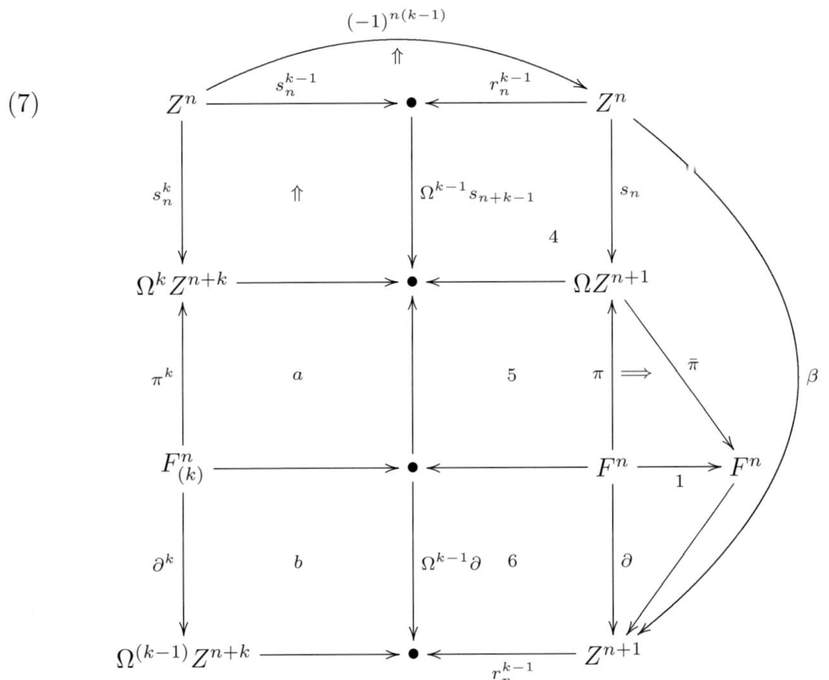

Here 4, 5 and 6 are defined as in (2.1.9)(6). Subdiagrams without tracks are commutative. Using (6) and (7) above we see that (8.6.7) implies the *derivation property* of the Bockstein map β, see $(\mathcal{K}1)$ in (1.1.7).

We are now ready to introduce the following diagram for the power map $U : Z^q \to Z^{pq}$ with $U(x) = x^p$. Recall that the group $G = \mathbb{Z}/p \subset \sigma_{pq}$ acts on Z^{pq} by cyclic permutation of q-blocks and G acts trivially on Z^q and U is a G-equivariant map.

8.6.8 Lemma. *For all primes p there is a commutative diagram of G-equivariant maps as follows.*

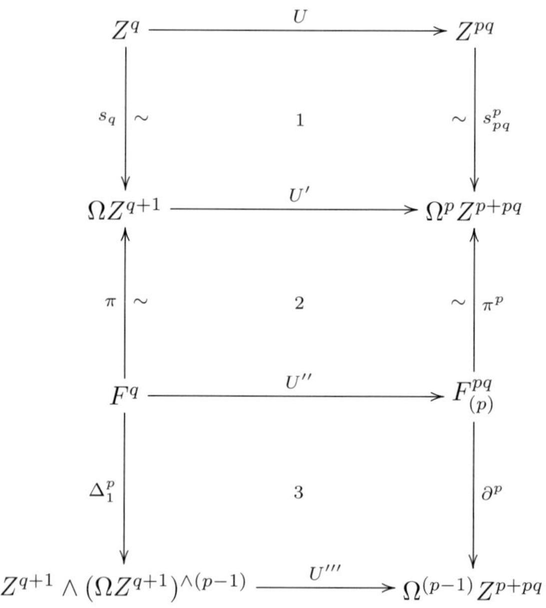

The maps s_q and π are defined as in (8.6.4) and the maps s^p_{pq}, π^p, ∂^p are defined in (8.6.7). Moreover we put

(1) $$U'(a) = \tau_{(p)}\bar{\mu}(a, \dots, a) \text{ for } a \in \Omega Z^{q+1}, \text{ and}$$

(2) $$U''(b) = \tau_{(p)}\bar{\bar{\mu}}(b, \dots, b) \text{ for } b \in F^q.$$

Here $\tau_{(k)}\bar{\mu}$ and $\tau_{(k)}\bar{\bar{\mu}}$ for $k = p$ and $i_1 = \cdots = i_p = q$ are defined in (8.6.7). Next let Δ^p_1 be given by

(3) $$\Delta^p_1(b) = (\partial b) \wedge (\pi a) \wedge \cdots \wedge (\pi a).$$

Moreover we define for $x \in Z^{q+1}$, $a_1, \dots, a_{p-1} \in \Omega Z^{q+1}$,

(4) $$U'''(x \wedge a_1 \wedge \cdots \wedge a_{p-1}) = \rho^*(A)$$

where A is the p-tuple

(5) $$A = (\tau_p a_1 \boxtimes \cdots \boxtimes a_{j-1} \cdot x \cdot a_j \boxtimes \cdots \boxtimes a_{p-1} \mid j = 1, \dots, p).$$

Now (8.6.7) shows that the diagram is commutative. The crucial observation is that all maps in the diagram are G-equivariant. The group G acts trivially on all spaces at the left-hand side of the diagram. Moreover $\alpha \in G$ yields the element $\epsilon_\alpha = \alpha \in G \subset \sigma_p$ and the element $\alpha \in G \subset \sigma_{pq}$ so that $\epsilon_\alpha \odot \alpha \in \sigma_{p+pq}$ is defined. We define the action of $\alpha \in G$ on $(y : S^p \to Z^{p+pq}) \in \Omega^p Z^{p+pq}$ by

(6) $$\alpha \cdot y = (\epsilon_\alpha \odot \alpha) y(\epsilon_\alpha^{-1}).$$

Here ϵ_α^{-1} acts on $S^p = S^1 \wedge \cdots \wedge S^1$ by permuting the S^1-coordinates. The map s_{pq}^p in the diagram is G-equivariant since

$$\begin{aligned} s_{pq}^p(\alpha x)(t) &= i(t) \cdot \alpha x \text{ for } t \in S^p,\ x \in Z^{pq} \\ &= (1 \odot \alpha) i(t) \cdot x \\ &= (\epsilon_\alpha \odot \alpha)(\epsilon_\alpha^{-1} \odot 1) i(t) \cdot x \\ &= (\epsilon_\alpha \odot \alpha)(i(\epsilon_\alpha^{-1} t) \cdot x) \\ &= (\alpha s_{pq}^p(x))(t). \end{aligned}$$

Similarly as in (6) we define the action of α on $(a, b) \in F_{(p)}^{pq}$ and $a \in \Omega^{(p-1)} Z^{p+pq}$ by

(7) $$\begin{cases} \alpha(a, b) &= (\alpha a, \alpha b), \\ \alpha a &= (\epsilon_\alpha \odot \alpha) a(\epsilon_\alpha^{-1}), \\ \alpha b &= (\epsilon_\alpha \odot \alpha) b(\epsilon_\alpha^{-1}). \end{cases}$$

Here ϵ_α^{-1} acts on the cube $(I^p, \partial I^p)$ by permuting coordinates. This shows that π^p and ∂^p are G-equivariant. Moreover U' and U'' are G-equivariant. In fact for U' we have

(8) $$\begin{aligned} \alpha U'(a) &= \alpha \tau_{(p)} \bar{\mu}(a, \dots, a) \\ &= (\epsilon_\alpha \odot \alpha) \tau_{(p)} (a \boxtimes \cdots \boxtimes a) \epsilon_\alpha^{-1} \\ &= \tau_{(p)} (a \boxtimes \cdots \boxtimes a) \\ &= U'(a). \end{aligned}$$

Compare the definition of the permutation $\tau_{(p)}$ in (8.6.6)(1). Similar equations show that U'' is G-equivariant.

We now show the fact that also U''' in (4) is G-equivariant, that is

(9) $$\alpha U''' = U''' \text{ for } \alpha \in G.$$

Proof of (9). We have

$$(\alpha U''')(x \wedge a_1 \wedge \cdots \wedge a_{p-1}) = \alpha \rho^*(A)$$

where A is the p-tuple in (5). According to the definition of ρ we obtain for $\beta = (\epsilon_\alpha)^{-1}$ the commutative diagram

$$\text{(10)}\quad \begin{array}{ccccccc} I^{p-1} & \xrightarrow{i_j^1} & \partial I^p & \xrightarrow{\rho} & \bigvee_{i=1}^p S^{p-1} & \xleftarrow{i_j} & S^{p-1} \\ \downarrow{\scriptstyle\beta} & & \downarrow{\scriptstyle\beta} & & \downarrow{\scriptstyle\beta} & & \downarrow{\scriptstyle\beta^j} \\ I^{p-1} & \xrightarrow[i_{\beta j}^1]{} & \partial I^p & \xrightarrow{\rho} & \bigvee_{i=1}^p S^{p-1} & \xleftarrow[i_{\beta j}]{} & S^{p-1} \end{array}$$

with $\beta^j \in \sigma_{p-1}$ given by

$$\beta^j : \{1,\dots,p-1\} = \{1,\dots,p\} - \{j\} \xrightarrow{\beta} \{1,\dots,p\} - \{\beta j\} = \{1,\dots,p-1\}.$$

Here the equations are the monotone bijections. Hence we get for $A = (A_1,\dots,A_p)$

$$\begin{aligned} \alpha\rho''(x \wedge a_1 \wedge \cdots \wedge a_{p-1}) &= \alpha\rho^*(A) \\ &= (\epsilon_\alpha \odot \alpha)(\beta\rho)^*(A) \text{ with } \beta = \epsilon_\alpha^{-1} \\ &= \rho^*(\epsilon_\alpha \odot \alpha)(A)\beta \\ &= \rho^*(B_1,\dots,B_p). \end{aligned}$$

Let $\bar{j} = \beta j$ and let $\hat{a}_t = a_{t-1}$ for $t \in \mathbb{Z}/p$.

$$\text{(11)}\quad \begin{aligned} B_j &= (\epsilon_\alpha \odot \alpha)\tau_p(a_1 \boxtimes \cdots \boxtimes a_{\bar{j}-1} \cdot x \cdot \hat{a}_{\bar{j}+1} \boxtimes \cdots \boxtimes \hat{a}_p)\beta^j \\ &= (\epsilon_\alpha \odot \alpha)\tau_p(a_{\beta 1} \boxtimes \cdots \boxtimes a_{\beta(j-1)} \cdot x \cdot \hat{a}_{\beta(j+1)} \boxtimes \cdots \boxtimes \hat{a}_{\beta p}) \\ &= \tau_p(a_1 \boxtimes \cdots \boxtimes a_{j-1} \cdot x \cdot a_j \boxtimes \cdots \boxtimes a_{p-1}) \\ &= A_j. \end{aligned}$$

Compare the definition of τ_k in (8.6.6)(1) for $p = k$. By (11) the proof of (9) is complete. This also completes the proof of (8.6.8) □

8.6.9 Lemma. *Let p be odd. The composite $U'''\Delta_1^p$ in (8.6.8) is null homotopic and there is a well-defined track $U'''\Delta_1^p \Rightarrow 0$.*

Proof. If q is odd then the diagonal

$$\Delta : Z^q \longrightarrow Z^q \wedge Z^q$$

is null homotopic with a well-defined track $\Delta \Rightarrow 0$, since $\tau_{q,q}\Delta = \Delta$ with $\mathrm{sign}(\tau_{q,q}) = (-1)^q = -1$. This shows that $\Delta_1^p \Rightarrow 0$ if q is odd. If q is even we see that $U''' \Rightarrow 0$ as follows. Here the track $U''' \Rightarrow 0$ is unique by (3.2.5). According to (8.6.3)(3) the map U''' has the degree

$$d = \sum_{j=1}^p \epsilon_j \cdot (-1)^{j-1}$$

where ϵ_j is the sign of the interchange of x and A_j in $x \cdot A_j$ with $A_j = a_1 \boxtimes \cdots \boxtimes a_{j-1}$. Hence

$$\epsilon_j = (-1)^{((j-1)(q+1))\cdot(q+1)} = (-1)^{j-1}$$

since q is even. Therefore we get $d = p = 0$. □

8.6.10 Definition. Let $\partial : F \to X$ and $\partial' : F' \to X'$ be G-equivariant maps between G-spaces in **Top***. We say that ∂ and ∂' are *weakly G-equivalent* if there exists a commutative diagram of G-spaces in **Top*** and G-equivariant maps

$$\begin{array}{ccccccccc} F & \longleftarrow & F_1 & \longrightarrow & F_2 & \longleftarrow \cdots \longrightarrow & F_n & \longleftarrow & F' \\ \downarrow{\scriptstyle \partial} & & \downarrow & & \downarrow & & \downarrow & & \downarrow{\scriptstyle \partial'} \\ X & \longleftarrow & X_1 & \longrightarrow & X_2 & \longleftarrow \cdots \longrightarrow & X_n & \longleftarrow & X' \end{array}$$

in which all horizontal arrows are homotopy equivalences in **Top***. The homotopy inverses need not be G-equivariant.

8.6.11 Lemma. *The G-equivariant map $\partial^p : F^{pq}_{(p)} \to \Omega^{(p-1)} Z^{p+pq}$ with the G-action in* (8.6.8)(7) *is weakly G-equivalent to the same map ∂^p with trivial G-action.*

Proof. Recall that $F^{pq}_{(p)}$ is the space of pair maps

(1) $$(I^p, \partial I^p) \longrightarrow (Z^{p+pq}_{\mathbb{G}}, Z^{p+pq}_{\mathbb{F}})$$

with G-action on $F^{pq}_{(p)}$ given by a G-action on $Z^{p+pq}_{\mathbb{G}}$ and on I^p. There is a G-equivariant homeomorphism

(2) $$(C\partial I^p, \partial I^p) \approx (I^p, \partial I^p)$$

where C is the reduced cone of a G-space in **Top***. We now use first the method in (7.2.6) to show that the pair of G-spaces $(Z^{p+pq}_{\mathbb{G}}, Z^{p+pq}_{\mathbb{F}})$ is weakly G-equivalent to the same pair with trivial G-action. This shows that ∂^p in (8.6.11) is weakly G-equivalent to the same map ∂^p with the G-action given only by the G-action on ∂I^p via (2). Now we use the space of maps (1) considered as a space of maps in the category of simplicial groups where the simplicial groups $Z^{p+pq}_{\mathbb{G}}, Z^{p+pq}_{\mathbb{F}}$ are abelian. Hence this is the space of maps in the category of abelian simplicial groups and since by the Dold-Kan equivalence abelian simplicial groups are equivalent to chain complexes we can apply the method in (7.2.6) to the singular chain complex given by ∂I^p with the induced G-action. This shows by (2) the result in (8.6.11). □

8.6.12 Corollary. *The weak G-equivalence* (8.6.11) *yields comparison maps λ_G together with an induced track in the following diagram.*

$$\begin{array}{ccc} EG \times_G F^{pq}_{(p)} & \xrightarrow[\sim]{\lambda'_G} & BG \times F^{pq}_{(p)} \\ \downarrow{\scriptstyle \partial^p_\#} & \overset{H_G}{\Longrightarrow} & \downarrow{\scriptstyle 1 \times \partial^p} \\ EG \times_G \Omega^{(p-1)} Z^{p+pq} & \xrightarrow[\lambda''_G]{\sim} & BG \times \Omega^{(p-1)} Z^{p+pq} \end{array}$$

Here λ'_G, λ''_G are homotopy equivalences under and over BG which are the identity on fibers, see (7.2.3). *Moreover H_G is a track under and over BG.*

The track in (8.6.12) induces the track in (8.6.1).

Proof of (8.6.1). We obtain the following diagram of Borel constructions with $G = \mathbb{Z}/p$.

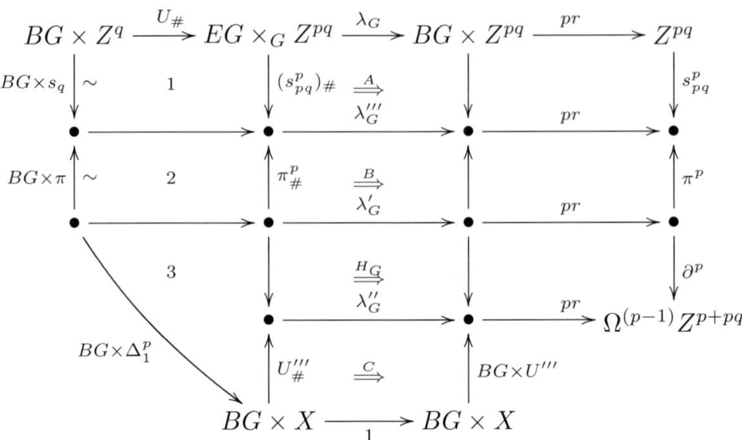

Here we set $X = Z^{q+1} \wedge (\Omega Z^{q+1})^{\wedge(p-1)}$ and the map λ_G is the comparison map (7.2.3) and the track H_G is defined in (8.6.12). We define λ'''_G by the composite

$$\lambda'''_G = (BG \times \pi^p)\lambda'_G h$$

where h is a homotopy inverse of $\pi^p_\#$ under and over BG. Now the tracks A, B and C in the diagram are the unique tracks under and over BG given by (7.1.10). The commutative subdiagrams 1, 2 and 3 are obtained by applying the Borel construction to the corresponding diagrams in (8.6.8). The top row of the diagram corresponds to the power map γ. Therefore the diagram together with (8.6.9) and (8.6.7)(7) yields the track $\gamma\beta \Rightarrow 0$. □

Chapter 9

Secondary Relations for Power Maps

In Chapter 8 we have defined the

$$\begin{aligned} \text{linearity tracks} \quad & \Gamma_v^{x,y}, \quad L(r)_v^x, \ P(\sigma)_v^x, \\ \text{Cartan tracks} \quad & C_v^{x,y}, \\ \text{Adem tracks} \quad & A_v^{x,y}. \end{aligned}$$

Moreover we have by (6.3.1) the permutation tracks $\Gamma_\sigma(x)$. All these tracks are well defined and natural in X. In this chapter we describe relations for these tracks. We do not yet consider relations for the Bockstein track in (8.6).

9.1 A list of secondary relations

Consider a diagram of tracks in $[\![X, Z^n]\!]$.

(9.1.1)
$$\begin{array}{ccc} a & \xrightarrow{A} & b \\ {\scriptstyle G}\downarrow & & \downarrow{\scriptstyle B} \\ f & \xrightarrow[H]{} & g \end{array}$$

We choose the orientation of this diagram compatible with the arrow A. According to the orientation we obtain the automorphisms $d_1, \ldots, d_4$ as follows:

$$\begin{aligned} d_1 &= G^{\mathrm{op}} H^{\mathrm{op}} BA \in \mathrm{Aut}(a) \overset{\sigma_a}{=} [X, Z^{n-1}], \\ d_2 &= AG^{\mathrm{op}} H^{\mathrm{op}} B \in \mathrm{Aut}(b) \overset{\sigma_b}{=} [X, Z^{n-1}], \\ d_3 &= BAG^{\mathrm{op}} H^{\mathrm{op}} \in \mathrm{Aut}(g) \overset{\sigma_g}{=} [X, Z^{n-1}], \\ d_4 &= H^{\mathrm{op}} BAG^{\mathrm{op}} \in \mathrm{Aut}(f) \overset{\sigma_f}{=} [X, Z^{n-1}]. \end{aligned}$$

Here the isomorphisms $\sigma_a, \sigma_b, \sigma_g, \sigma_f$ are defined in (3.2.3). By (3.2.4) we know that the equation

(1) $$d = \sigma_a d_1 = \sigma_b d_2 = \sigma_g d_3 = \sigma_f d_4 \in [X, Z^{n-1}]$$

holds. We call this element the *primary element* represented by the oriented diagram (9.1.1). By (3.2.4) we also have the equation

(2) $$(H\square G) \oplus d = B\square A.$$

This primary element is trivial if and only if the diagram commutes. If we change the orientation of (9.1.1) then we alter the primary element by the sign -1.

In Chapter 8 we have seen that there are the following well-defined tracks in $[\![X, Z^*]\!]$ which are natural in X.

(9.1.2) $$\begin{array}{llll} L(r)_v^X & : & \gamma_v(rx) \Longrightarrow r\gamma_v(x) \text{ for } r \in \mathbb{F}, & \text{see (8.2.8)} \\ P(\sigma)_v^x & : & \gamma_v(\sigma x) \Longrightarrow \sigma^p\gamma_v(x) \text{ for } \sigma \in \sigma_{|x|}, & \text{see (8.2.9)} \\ \Gamma_v^{x,y} & : & \gamma_v(x+y) \Longrightarrow \gamma_v(x) + \gamma_v(y), & \text{see (8.2.7)} \\ C_v^{x,y} & : & \sigma(x,y)\gamma_v(x\cdot y) \Longrightarrow \gamma_v(x)\cdot\gamma_v(y), & \text{see (8.3.3)} \\ A_{v,w}^x & : & \gamma_v\gamma_w(x) \Longrightarrow \tau_x\gamma_w\gamma_v(x). & \text{see (8.4.4)} \end{array}$$

Moreover we have for $\sigma \in \sigma_n$ the permutation track (6.3.1)(4)

$$\Gamma_\sigma : \sigma \Longrightarrow \text{sign}(\sigma) : Z^n \longrightarrow Z^n.$$

Now we describe relations between tracks which are diagrams as in (9.1.1) defining primary elements. These primary elements are also natural in X and hence can be expressed in terms of primary cohomology operations.

First the equation $x + y = y + x$ yields the following relation.

(9.1.3) $$\begin{array}{ccc} \gamma_v(x+y) & \xrightarrow{\Gamma_v^{x,y}} & \gamma_v(x) + \gamma_v(y) \\ \| & & \| \\ \gamma_v(y+x) & \xrightarrow{\Gamma_v^{y,x}} & \gamma_v(y) + \gamma_v(x) \end{array}$$

We prove in Section (9.2) that this diagram commutes.

Next the associativity $(x+y)+z = x+(y+z)$ yields the following relation.

(9.1.4) $$\begin{array}{ccc} \gamma_v(x+y+z) & \xrightarrow{\Gamma_v^{x+y,z}} & \gamma_v(x+y) + \gamma_v(z) \\ \downarrow{\scriptstyle \Gamma_v^{x,y+z}} & & \downarrow{\scriptstyle \Gamma_v^{x,y}+\gamma_v(z)} \\ \gamma_v(x) + \gamma_v(y+z) & \xrightarrow{\gamma_v(x)+\Gamma_v^{y,z}} & \gamma_v(x) + \gamma_v(y) + \gamma_v(z) \end{array}$$

We prove in Section (9.2) that this diagram commutes.

Moreover $x + 0 = x$ yields the following relation.

(9.1.5)
$$\begin{array}{ccc} \gamma_v(x+0) & \xrightarrow{\Gamma_v^{x,0}} & \gamma_v(x) + \gamma_v(0) \\ \| & & \| \\ \gamma_v(x) & \xrightarrow[0^\square]{} & \gamma_v(x) \end{array}$$

Also this diagram commutes, see Section (9.2).

We now define inductively

(9.1.6)
$$\begin{array}{rcl} \Gamma_v^{x_1,\dots,x_r} & = & (\Gamma_v^{x_1,\dots,x_{r-1}} + \gamma_v(x_r))\square\Gamma_v^{x_1+\dots+x_{r-1},x_r}, \\ \Gamma_v^{x_1,\dots,x_r} & : & \gamma_v(x_1+\dots+x_r) \Longrightarrow \gamma_v(x_1) + \dots + \gamma_v(x_r), \end{array}$$

and for $x_1 = \dots = x_r$ we set

(9.1.7)
$$\begin{array}{rcl} \Gamma(r)_v^x & = & \Gamma_v^{x,\dots,x}, \\ \Gamma(r)_v^x & : & \gamma_v(rx) \Longrightarrow r\gamma_v(x). \end{array}$$

Hence we get for $r \in \mathbb{N}$ the following relation.

(9.1.8)
$$\begin{array}{ccc} \gamma_v(rx) & \xrightarrow{\Gamma(r)_v^x} & r\gamma_v(x) \\ \| & & \| \\ \gamma_v(rx) & \xrightarrow{L(r)_v^x} & r\gamma_v(x) \end{array}$$

Here we use $\mathbb{N} \to \mathbb{F}$ mapping r to $r \cdot 1 = r \in \mathbb{F}$. Relation (9.1.8) describes $L(r)_v^x$ in terms of linearity tracks $\Gamma_v^{x,x'}$ and vice versa. The primary element of (9.1.8) in general is non-trivial and is computed in section (9.3.6). For $r = p^2 - 1$ and p odd, diagram (9.1.8) commutes so that

(1)
$$L(-1)_v^x = L(p^2-1)_v^x = \Gamma(p^2-1)_v^x.$$

This *sign track* is needed in the next relation.

Using permutation tracks Γ_σ and $P(\sigma)_v^x$ we get the following relation which can be used to replace $P(\sigma)_v^x$.

(9.1.9)
$$\begin{array}{ccc} \gamma_v(\sigma x) & \xrightarrow{P(\sigma)_v^x} & \sigma^p\gamma_v(x) \\ \Big\downarrow{\scriptstyle \gamma_v(\Gamma_\sigma)} & & \Big\downarrow{\scriptstyle \Gamma_{\sigma^p}} \\ \gamma_v(\mathrm{sign}(\sigma)\cdot x) & \xrightarrow{L(\mathrm{sign}(\sigma))_v^x} & \mathrm{sign}(\sigma)\cdot\gamma_v(x) \end{array}$$

We show in Section (9.4) that this diagram commutes.

The equation $\tau(x,y)\cdot x\cdot y = y\cdot x$ yields the following relation with $P = P(\tau(x,y))_v^{x\cdot y}$.

(9.1.10)
$$\begin{array}{ccc}
\sigma(y,x)\gamma_v(\tau(x,y)\cdot x\cdot y) & \xrightarrow{\sigma(y,x)\cdot P} & \sigma(y,x)\tau(x,y)^p\gamma_v(x\cdot y) \\
\| & & \| \\
\sigma(y,x)\gamma_v(y\cdot x) & & \tau(x^p,y^p)\sigma(x,y)\gamma_v(x\cdot y) \\
\Big\downarrow{\scriptstyle C_v^{y,x}} & & \Big\downarrow{\scriptstyle \tau(x^p,y^p)C_v^{x,y}} \\
\gamma_v(y)\cdot\gamma_v(x) & = & \tau(x^p,y^p)\gamma_v(x)\cdot\gamma_v(y)
\end{array}$$

In Section (9.5) we prove that this diagram commutes.

Next the associativity $x\cdot(y\cdot z) = (x\cdot y)\cdot z)$ yields the following relation.

(9.1.11)
$$\begin{array}{ccc}
\sigma(x,y,z)\gamma_v(x\cdot y\cdot x) & \xrightarrow{(\sigma(x,y)\times 1)C_v^{xy,z}} & (\sigma(x,y)\cdot\gamma_v(x\cdot y))\cdot\gamma_v(z) \\
\Big\downarrow{\scriptstyle (1\times\sigma(y,z))C_v^{x,yz}} & & \Big\downarrow{\scriptstyle C_v^{x,y}\cdot\gamma_v(z)} \\
\gamma_v(x)\cdot(\sigma(y,z)\gamma_v(y\cdot z)) & \xrightarrow{\gamma_v(x)C_v^{y,z}} & \gamma_v(x)\cdot\gamma_v(y)\cdot\gamma_v(z)
\end{array}$$

Also this diagram commutes, see section (9.5). Here we have $\sigma(x,y,z) = (\sigma(x,y)\times 1)\sigma(x\cdot y,z) = (1\times\sigma(y,z))\sigma(x,y\cdot z)$.

The distributivity $(x+x')\cdot y = x\cdot y + x'\cdot y$ yields the following relation.

(9.1.12)
$$\begin{array}{ccc}
\sigma(x,y)\gamma_v((x+x')\cdot y) & \xrightarrow{C_v^{x+x',y}} & \gamma_v(x+x')\cdot\gamma_v(y) \\
\Big\downarrow{\scriptstyle \sigma(x,y)\cdot\Gamma_v^{xy,x'y}} & & \Big\downarrow{\scriptstyle \Gamma_v^{x,x'}\gamma_v(y)} \\
\sigma(x,y)\gamma_v(x\cdot y)+\sigma(x,y)\gamma_v(x'\cdot y) & \xrightarrow{C_v^{x,y}+C_v^{x',y}} & \gamma_v(x)\cdot\gamma_v(y)+\gamma_v(x')\cdot\gamma_v(y)
\end{array}$$

We compute this relation in Section (9.6). If p is odd this diagram commutes.

Next we consider relations for the Adem track $A_{v,w}^x$. For the sum $x+x'$ we obtain the following relation.

(9.1.13)
$$\begin{array}{ccc}
\gamma_v\gamma_w(x+x') & \xrightarrow{A_{v,w}^{x+x'}} & \tau_x\gamma_w\gamma_v(x+x') \\
\Big\downarrow{\scriptstyle \gamma_v\Gamma_w^{x,x'}} & & \Big\downarrow{\scriptstyle \tau_x\gamma_w\Gamma_v^{x,x'}} \\
\gamma_v(\gamma_w x+\gamma_w x') & & \tau_x\gamma_w(\gamma_v x+\gamma_v x') \\
\Big\downarrow{\scriptstyle \Gamma_v^{\gamma_w x,\gamma_w x'}} & & \Big\downarrow{\scriptstyle \tau_x\Gamma_w^{\gamma_v x,\gamma_v x'}} \\
\gamma_v\gamma_w(x)+\gamma_v\gamma_w(x') & \xrightarrow[A_{v,w}^x+A_{v,w}^{x'}]{} & \tau_x(\gamma_w\gamma_v(x)+\gamma_w\gamma_v(x'))
\end{array}$$

For the product $x \cdot y$ we obtain the following relation with $\tilde{\tau} = \tau_x \odot \tau_y$.

(9.1.14)

$$
\begin{array}{ccc}
\sigma(x^p, y^p)\sigma(x,y)^p\gamma_v\gamma_w(x \cdot y) & \xrightarrow{\sigma(x^p,y^p)\sigma(x,y)^p A^{x \cdot y}_{v,w}} & \sigma(x^p, y^p)\sigma(x,y)^p\tau_{x \cdot y}\gamma_w\gamma_v(x \cdot y) \\
\Big\uparrow {\scriptstyle \sigma(x^p,y^p)P(\sigma(x,y))_v} & & \Big\uparrow {\scriptstyle \tilde{\tau}\sigma(x^p,y^p)P(\sigma(x,y))_w} \\
\sigma(x^p, y^p)\gamma_v(\sigma(x,y)\gamma_w(x \cdot y)) & & \tilde{\tau}\sigma(x^p, y^p)\gamma_w(\sigma(x,y)\gamma_v(x \cdot y)) \\
\Big\downarrow {\scriptstyle \sigma(x^p,y^p)\gamma_v(C^{x,y}_w)} & & \Big\downarrow {\scriptstyle \tilde{\tau}\sigma(x^p,y^p)\gamma_w(C^{x,y}_v)} \\
\sigma(x^p, y^p)\gamma_v(\gamma_w(x) \cdot \gamma_w(y)) & & \tilde{\tau}\sigma(x^p, y^p)\gamma_w(\gamma_v(x) \cdot \gamma_v(y)) \\
\Big\downarrow {\scriptstyle C_v^{\gamma_w(x),\gamma_w(y)}} & & \Big\downarrow {\scriptstyle \tilde{\tau}C_w^{\gamma_v(x),\gamma_v(y)}} \\
\gamma_v\gamma_w(x) \cdot \gamma_v\gamma_w(y) & \xrightarrow[A^x_{v,w} \cdot A^y_{v,w}]{} & (\tau_x \odot \tau_y)(\gamma_w\gamma_v(x) \cdot \gamma_w\gamma_v(y))
\end{array}
$$

Here we use

$$\sigma(x^p, y^p)\sigma(x,y)^p\tau_{xy} = (\tau_x \odot \tau_y)\sigma(x^p, y^p)\sigma(x,y)^p.$$

We now define inductively:

(9.1.15) $\qquad C_v^{x_1,\ldots,x_r} : \sigma(x_1, \ldots, x_r)\gamma_v(x_1 \cdot \cdots \cdot x_r) \Longrightarrow \gamma_v(x_1) \cdot \cdots \gamma_v(x_r).$

For $r = 2$ this track is given by (8.3.3). For $r = 3$ this is the composite in (9.1.11). Moreover for $r \geq 3$ we set

$$C_v^{x_1,\ldots,x_r} = (C_v^{x_1,\ldots,x_{r-1}} \cdot \gamma_v(x_r))\square(\sigma(x_1, \ldots, x_{r-1}) \times 1)C_v^{x_1 \cdot \cdots \cdot x_{r-1}, x_r}$$

and

$$\sigma(x_1, \ldots, x_r) = (\sigma(x_1, \ldots, x_{r-1}) \times 1)\sigma(x_1 \cdot \cdots \cdot x_{r-1}, x_r).$$

For $x_1 = \cdots = x_r$ we get as a special case

$$
\begin{array}{rcl}
C(r)^x_v & : & \sigma(r,x)\gamma_v(x^r) \Longrightarrow \gamma_v(x)^r \text{ where} \\
C(r)^x_v & = & C_v^{x,\ldots,x} \text{ and } \sigma(r,x) = \sigma(x, \ldots, x) \text{ with } r\text{-times } x.
\end{array}
$$

In particular since $U(x) = x^p$ we have

$$C(p)^x_v : \sigma(p,x)\gamma_v U(x) \Longrightarrow U(\gamma_v(x)).$$

This track is used in the next relation. Since for $v = 0$ we have $\gamma_v(x) = U(x)$ we get the following relation.

(9.1.16)

$$\begin{array}{ccc} \sigma(p,x)\gamma_v U(x) & \xrightarrow{\sigma(p,x)A^x_{v,o}} & \sigma(p,x)\tau_x U\gamma_v(x) \\ \| & & \downarrow \Gamma_{\sigma(p,x)\tau_x} \\ \sigma(p,x)\gamma_v U(x) & \xrightarrow[C(p)^x_v]{} & U\gamma_v(x) \end{array}$$

Here we assume that $p = 2$ or $|\, x \,|$ even so that

$$\mathrm{sign}(\sigma(p,x)\tau_x) = (-1)^{p|x|\bar{p}} = 1$$

with $\bar{p} = p(p-1)/2$. If p is odd and $|\, x \,|$ odd, then there is a canonical track $U\gamma_v x \Rightarrow 0$.

Interchanging v and w yields the following relation.

(9.1.17)

$$\begin{array}{ccc} \gamma_v\gamma_w(x) & \xrightarrow{A^x_{v,w}} & \tau_x\gamma_w\gamma_v(x) \\ \| & & \| \\ \tau_x\tau_x\gamma_v\gamma_w(x) & \xleftarrow[\tau_x A^x_{w,v}]{} & \tau_x\gamma_w\gamma_v(x) \end{array}$$

Here we use the fact that $\tau_x\tau_x = 1$.

Next we get the following hexagon relation for $u, v, w : X \to Z^1$.

(9.1.18)

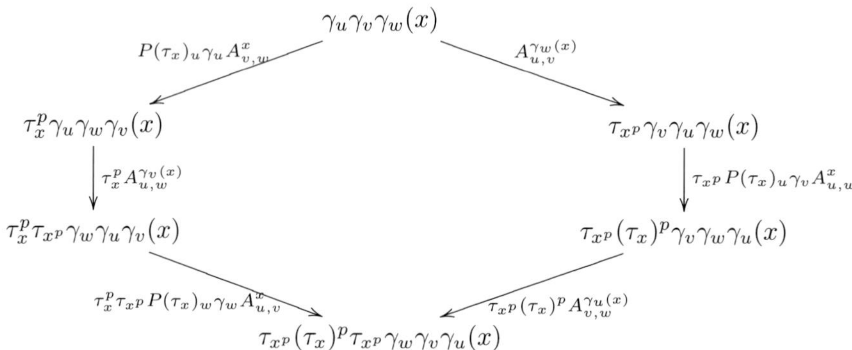

Here we use the fact that

$$(\tau_x)^p\tau_{x^p}(\tau_x)^p = \tau_{x^p}(\tau_x)^p\tau_{x^p}.$$

9.2 Secondary linearity relations

In this section we compute the relations (9.1.3), (9.1.4), (9.1.5) concerning the linearity track

$$\Gamma_v^{x,y} : \gamma_v(x+y) \Longrightarrow \gamma_v(x) + \gamma_v(y)$$

defined in (8.2.7). For $v = 0$ this is the track

$$\Gamma_0^{x,y} : U(x+y) \Longrightarrow U(x) + U(y)$$

defined in (8.2.6) where U is the power map with $U(x) = x^p$. Let X be a pointed space and let $x, y, z : X \to Z^q$ be pointed maps. Then $U(x)$, $U(x+y)$, $U(x+y+z)$ are objects in the groupoid $[\![X, Z^{pq}]\!]$.

9.2.1 Proposition. *The track*

$$\Gamma_0^{x,y} : U(x+y) \Rightarrow Ux + Uy$$

in $[\![X, Z^{pq}]\!]$ *is natural in* X *with the following properties:*

(i) $\Gamma_0^{x,y} = \Gamma_0^{y,x}$,

(ii) $(Ux + \Gamma_0^{y,z}) \Box \Gamma_0^{x,y+z} = (\Gamma_0^{x,y} + Uz) \Box \Gamma_0^{x+y,z}$,

(iii) $\Gamma_0^{0,0}$ *is the identity track of* 0.

By (i) and (ii) the following diagrams in the groupoid $[\![X, Z^{pq}]\!]$ commute.

$$\begin{array}{ccc} U(x+y) & \xrightarrow{\Gamma_0^{x,y}} & Ux + Uy \\ \| & & \| \\ U(y+x) & \xrightarrow{\Gamma_0^{y,x}} & Uy + Ux \end{array}$$

$$\begin{array}{ccc} U(x+y+z) & \xrightarrow{\Gamma_0^{x+y,z}} & U(x+y) + U(z) \\ {\scriptstyle \Gamma_0^{x,y+z}} \downarrow & & \downarrow {\scriptstyle \Gamma_0^{x,y} + U(z)} \\ U(x) + U(y+z) & \xrightarrow{U(x) + \Gamma_0^{y,z}} & U(x) + U(y) + U(z) \end{array}$$

Proof of (9.2.1). The π-norm map $N : Z^{pq} \to Z^{pq}$ with $N(x) = \sum_{\alpha \in \pi} \alpha \cdot x$ for $\pi = \mathbb{Z}/p$ admits by (7.2.2) a track

$$\Gamma = \sum_{\alpha \in \pi} \Gamma_\alpha : N \Rightarrow N^{\pm} = 0$$

with $N^{\pm}(x) = \sum_{\alpha \in \pi} \mathrm{sign}(\alpha) \cdot x = p \cdot x = 0 \in Z^{pq}$. Compare (8.2.5). Here we use the fact that the composite

$$\pi \subset \sigma_{pq} \xrightarrow{\mathrm{sign}} \{1, -1\}$$

is the trivial homomorphism and that Z^{pq} is a $\mathbb{Z}/p$-vector space object. We know that $\Gamma : N \Rightarrow 0$ is unique. This shows that for $\alpha_0 \in \pi$ we have

$$(2) \qquad \Gamma\alpha_0 = \Gamma$$

with $\alpha_0 : Z^{pq} \to Z^{pq}$ carrying x to $\alpha_0 \cdot x$. In fact (2) is true since $N\alpha_0 = N$ and $0\alpha_0 = 0$. Moreover uniqueness shows that Γ is additive, that is $\Gamma(a+b) = \Gamma a + \Gamma b$ for maps $a, b : Y \to Z^{pq}$. According to the proof of (8.2.3) we have

$$(3) \qquad U(x+y) - Ux - Uy = N\bar{U}(x,y)$$

so that

$$(4) \qquad \Gamma_0^{x,y} = \Gamma\bar{U}(x,y) + Ux + Uy : U(x+y) \Rightarrow Ux + Uy$$

is well defined. We have by use of the basis B in the proof of (8.1.4) the formula

$$\bar{U}(x,y) = \sum_{b \in B} b(x,y).$$

In fact $\Gamma_0^{x,y}$ does not depend on the choice of the basis B defining $\bar{U} = \bar{U}_B$. A different basis B_0 yields elements $\beta_b \in \pi$ with $\beta_b \cdot b(x,y) \in B_0$, $b \in B$, so that by (2) we have

$$\begin{aligned} \Gamma\bar{U}_{B_0} &= \Gamma \sum_{b \in B} \beta_b \cdot b(x,y) \\ &= \sum_{b \in B} \Gamma\beta_b \cdot b(x,y) = \sum_{b \in B} \Gamma b(x,y) \\ &= \Gamma \sum_{b \in B} b(x,y) = \Gamma\bar{U}_B. \end{aligned}$$

If $p = 2$ then B contains exactly one element b, for example $b(x,y) = x \cdot y$. If $p \geq 3$ then let $B' \subset B$ be the subset of monomials $b(x,y)$ for which the number of factors x in $b(x,y)$ is even. Then we get for $b \in B'$ an element $\alpha_b \in \pi$ with $\alpha_b \cdot b(x,y) \in B$ so that for $p \geq 3$,

$$(5) \qquad \bar{U}(x,y) = \sum_{b \in B'} (b(x,y) + \alpha_b \cdot b(y,x)).$$

Hence by (2) we get for $p \geq 3$,

$$\begin{aligned} \Gamma\bar{U}(x,y) &= \Gamma(\sum_{b \in B'} (b(x,y) + \alpha_b \cdot b(y,x)) \\ &= \sum_{b \in B'} \Gamma b(x,y) + \Gamma\alpha_b b(y,x) \\ &= \sum_{b \in B'} \Gamma b(x,y) + \Gamma b(y,x), \end{aligned}$$

so that $\Gamma\bar{U}(x,y) = \Gamma\bar{U}(y,x)$. This proves (i) for $p \geq 3$. In fact, since (4) does not depend on B we may assume $\alpha_b = 1$ for $b \in B'$. For $p = 2$ we have $\Gamma\bar{U}(x,y) = \Gamma(x \cdot y) = \Gamma\alpha(y \cdot x) = \Gamma(y \cdot x) = \Gamma\bar{U}(y,x)$ where α is the generator of $\pi = \mathbb{Z}/2$. Hence (i) also holds for $p = 2$.

Next we consider (ii). According to (4) we have to show

$$\begin{aligned} &(Ux + \Gamma\bar{U}(y,z) + Uy + Uz)\square(\Gamma\bar{U}(x, y+z) + Ux + U(y+z)) \\ &\quad = (\Gamma\bar{U}(x,y) + Ux + Uy + Uz)\square(\Gamma\bar{U}(x+y,z) + U(x+y) + Uz). \end{aligned} \tag{6}$$

Using (3) this is equivalent to

$$\begin{aligned} \Gamma\bar{U}(y,z)\square(\Gamma\bar{U}(x,y+z) + N\bar{U}(y,z)) &= \Gamma\bar{U}(y,z) + \Gamma\bar{U}(x,y+z) \\ = \Gamma\bar{U}(x,y)\square(\Gamma\bar{U}(x+y,z) + N\bar{U}(x,y)) &= \Gamma\bar{U}(x,y) + \Gamma\bar{U}(x+y,z). \end{aligned} \tag{7}$$

These are tracks from $U(x+y+z) - Ux - Uy - Uz$ to 0 in $[\![X, Z^{pq}]\!]$. Hence it remains to show

$$\Gamma(\bar{U}(y,z) + \bar{U}(x,y+z)) = \Gamma(\bar{U}(x,y) + \bar{U}(x+y,z)). \tag{8}$$

This is obviously true for $p = 2$. For $p \geq 3$ we consider

$$U(x+y+z) - Ux - Uy - Uz = N(\bar{U}(x,y) + \bar{U}(x,z) + \bar{U}(y,z) + \bar{\bar{U}}(x,y,z)). \tag{9}$$

Here $\bar{\bar{U}}$ is a sum of monomials of length p containing x, y and z at least as one factor. The group $\pi = \mathbb{Z}/p$ acts on such monomials freely. We choose a basis B'' consisting of such monomials $b(x,y,z)$ whose permutations under π give each such monomial exactly once so that $\bar{\bar{U}}(x,y,z) = \sum_{b \in B''} b(x,y,z)$. Now we get

$$\bar{U}(x, y+z) = \bar{U}(x,y) + \bar{U}(x,z) + \bar{\bar{U}}_R(x,y,z), \tag{10}$$

$$\bar{U}(x+y,z) = \bar{U}(x,z) + \bar{U}(y,z) + \bar{\bar{U}}_L(x,y,z). \tag{11}$$

Here R and L are a basis of monomials $b_R(x,y,z)$ and $b_L(x,y,z)$ respectively with

$$\begin{cases} b_R(x,y,z) = \alpha_b b(x,y,z), & b \in B'', \\ b_L(x,y,z) = \beta_b b(x,y,z), & b \in B'', \end{cases} \tag{12}$$

where $\alpha_b, \beta_b \in \pi$. Hence we get by (2)

$$\Gamma\bar{\bar{U}}_R(x,y,z) = \Gamma\bar{\bar{U}}(x,y,z) = \Gamma\bar{\bar{U}}_L(x,y,z)$$

and this implies (8) and equivalently (ii). □

Next we consider for $v : X \to Z^1$, $x : X \to Z^q$ the composite of maps

$$\gamma_v(x) = \gamma(v,x) : X \longrightarrow Z^1 \times Z^q \xrightarrow{\gamma} Z^{pq}$$

where $\gamma \in \underline{\underline{\gamma}}$ is a power map. Compare (8.1.6). We have by (8.1.1)(4) and (8.1.3) the equations

$$\text{(9.2.2)} \qquad \begin{cases} \gamma_0(x) = U(x) = x^p & \text{for} \quad v = 0, \\ \gamma_v(0) = 0 & \text{for} \quad x = 0. \end{cases}$$

Here $\gamma_0(x) = U(x)$ satisfies the relations in (9.2.1). More generally such relations hold for $\gamma_v(x)$ as follows.

9.2.3 Theorem. *The linearity track*

$$\Gamma_v^{x,y} : \gamma_v(x+y) \Rightarrow \gamma_v(x) + \gamma_v(y)$$

in $[\![X, Z^{pq}]\!]$ *is natural in* X *with the following properties*

(i) $\Gamma_v^{x,y} = \Gamma_v^{y,x}$,

(ii) $(\gamma_v(x) + \Gamma_v^{y,z})\Box\Gamma_v^{x,y+z} = (\Gamma_v^{x,y} + \gamma_v(z))\Box\Gamma_v^{x+y,z}$,

(iii) $\Gamma_v^{x,0}$ *is the identity track of* $\gamma_v(x)$.

These are the *secondary linearity relations* for the power map $\gamma \in \underline{\underline{\gamma}}$. Compare (9.1.3), (9.1.4) and (9.1.5).

Naturality means that a map $f : Y \to X$ induces a functor

$$f^* : [\![X, Z^{pq}]\!] \longrightarrow [\![Y, Z^{pq}]\!]$$

between groupoids which satisfies

$$\text{(9.2.4)} \qquad f^*\Gamma_v^{x,y} = \Gamma_{vf}^{xf,yf}.$$

For $v = 0$ Theorem (9.2.3) corresponds to (9.2.1). By (i) and (ii) in (9.2.3) the following diagrams commute in the groupoid $[\![X, Z^{pq}]\!]$.

$$\text{(i)} \qquad \begin{array}{ccc} \gamma_v(x+y) & \xrightarrow{\Gamma_v^{x,y}} & \gamma_v(x) + \gamma_v(y) \\ \| & & \| \\ \gamma_v(y+x) & \xrightarrow{\Gamma_v^{y,x}} & \gamma_v(y) + \gamma_v(x) \end{array}$$

Here we use the commutativity of the vector space addition $+$.

$$\text{(ii)} \qquad \begin{array}{ccc} \gamma_v(x+y+z) & \xrightarrow{\Gamma_v^{x+y,z}} & \gamma_v(x+y) + \gamma_v(z) \\ \downarrow{\scriptstyle \Gamma_v^{x,y+z}} & & \downarrow{\scriptstyle \Gamma_v^{x,y}+\gamma_v(z)} \\ \gamma_v(x) + \gamma_v(y+z) & \xrightarrow{\gamma_v(x)+\Gamma_v^{y,z}} & \gamma_v(x) + \gamma_v(y) + \gamma_v(z) \end{array}$$

Poof of (9.2.3). By (8.2.3) we have the track

$$L_v^{x,y} = \Lambda_L(v,x,y) : U(x+y) - Ux - Uy \longrightarrow \gamma_v(x+y) - \gamma_v(x) - \gamma_v(y)$$

and by (8.2.6) we have the track

$$\Gamma_0^{x,y} - Ux - Uy : U(x+y) - Ux - Uy \longrightarrow 0.$$

Recall that H^{op} denotes the inverse of the track H. Then we define

(1) $$\Gamma_v^{x,y} = ((\Gamma_0^{x,y} - Ux - Uy)\square(L_v^{x,y})^{\mathrm{op}}) + \gamma_v(x) + \gamma_v(y).$$

Using (9.2.1) we see that equation (i) in (9.2.3) is equivalent to

(2) $$\Lambda_L^{\mathrm{op}}(v,x,y) = \Lambda_L^{\mathrm{op}}(v,y,x).$$

According to the proof of (8.2.3) we have for $w = h_\pi v$,

(3) $$\Lambda_L^{\mathrm{op}}(v,x,y) = (p_2 L^N(w, \bar{U}(x,y))\square(p_2 L^\nu U_\#^+(w,x,y))^{\mathrm{op}}.$$

Here L^N and L^ν are the linearity tracks in the proof of (8.2.3).

If $p \geq 3$ we can assume that $\bar{U}(x,y) = \bar{U}(y,x)$.

For $p = 2$ we have $\bar{U}(x,y) = x{\cdot}y$. In this case we get $L^N(w, x{\cdot}y) = L^N(w, y{\cdot}x)$ since for the generator $\alpha \in \pi = \mathbb{Z}/2$ we have $N\alpha = N$ so that

(4) $$L^N = L^{N\alpha} = (N^\pm)_\# L^\alpha \square L^N \alpha_\# = L^N \alpha_\#, \text{ by (7.3.4)}.$$

Here we use the fact that $L^\alpha = B\pi \times \alpha = 0^\square$ is the trivial track.

Hence we have for $p \geq 2$,

$$L^N(w, \bar{U}(x,y)) = L^N(w, \bar{U}(y,x)).$$

For the proof of (2) we still have to check that

$$L^\nu U_\#^+(w,x,y) = L^\nu U_\#^+(w,y,x)$$

holds. This follows from the fact that ν in the proof of (8.2.3) satisfies $\nu(1 \times T) = \nu$ where $T : Z^{pq} \times Z^{pq} \to Z^{pq} \times Z^{pq}$ is the interchange map. In fact, we have by (7.3.4)

$$L^\nu = L^{\nu(1\times T)} = (\nu^\pm)_\# L^{1\times T} \square L^\nu (1 \times T)_\# = L^\nu (1 \times T)_\#,$$

since $L^{1\times T}$ by definition of λ_π^3 in the proof of (8.2.3) is the trivial track $0^\square$. This completes the proof of (2) and hence (9.2.3)(i) is true.

For the proof of (9.2.3)(ii) we use the notation

(5) $$\begin{cases} H_v^{x,y} = \Gamma_v^{x,y} - \gamma_v x - \gamma_v y : \gamma_v^{cr}(x,y) \to 0, \\ \gamma_v^{cr}(x,y) = \gamma_v(x+y) - \gamma_v(x) - \gamma_v(y). \end{cases}$$

Then (9.2.3)(ii) is equivalent to the following diagram.

$$(6)\qquad \begin{array}{ccc} H_v^{y,z}\square(H_v^{x,y+z}+\gamma_v^{cr}(y,z)) & =\!= & H_v^{x,y}\square(H_v^{x+y,z}+\gamma_v^{cr}(x,y)) \\ \| & & \| \\ H_v^{y,z}+H_v^{x,y+z} & & H_v^{x,y}+H_v^{x+y,z} \end{array}$$

Here we have by (1) above the equation

$$(7)\qquad H_v^{x,y} = H_0^{x,y}\square\Lambda_L(v,x,y)^{\mathrm{op}},$$

hence (6) is equivalent to

$$(8)\qquad \begin{aligned} &(H_0^{y,z}\square\Lambda_L(v,y,z)^{\mathrm{op}}) + (H_0^{x,y+z}\square\Lambda_L(v,x,y+z)^{\mathrm{op}}) \\ &\quad = (H_0^{x,y}\square\Lambda_L(v,x,y)^{\mathrm{op}}) + (H_0^{x+y,z}\square\Lambda_L(v,x+y,z)^{\mathrm{op}}). \end{aligned}$$

Since (6) holds for $v=0$ by (9.2.1) we see that (8) is equivalent to

$$(9)\qquad \Lambda_L(v,y,z)^{\mathrm{op}}+\Lambda_L(v,x,y+z)^{\mathrm{op}}=\Lambda_L(v,x,y)^{\mathrm{op}}+\Lambda_L(v,x+y,z)^{\mathrm{op}}.$$

According to (3) the tracks in (9) are given by the tracks

$$\begin{array}{lcllcl} T_1 &=& p_2L^N(w,\bar U(y,z)) & \text{and} & T_2 &=& p_2L^\nu U_\#^+(w,y,z)^{\mathrm{op}}, \\ R_1 &=& p_2L^N(w,\bar U(x,y+z)) & \text{and} & R_2 &=& p_2L^\nu U_\#^+(w,x,y+z)^{\mathrm{op}}, \\ T_1' &=& p_2L^N(w,\bar U(x,y)) & \text{and} & T_2' &=& p_2L^\nu U_\#^+(w,x,y)^{\mathrm{op}}, \\ R_1' &=& p_2L^N(w,\bar U(x+y,z)) & \text{and} & R_2' &=& p_2L^\nu U_\#^+(w,x+y,z)^{\mathrm{op}}. \end{array}$$

In fact (9) is equivalent to the equation

$$(10)\qquad T_1\square T_2+R_1\square R_2 \;=\; T_1'\square T_2'+R_1'\square R_2'.$$

This again is equivalent to

$$(11)\qquad (T_1+R_1)\square(T_2+R_2) \;=\; (T_1'+R_1')\square(T_2'+R_2').$$

Let $a=\gamma_w(x+y+z)-\gamma_w x-\gamma_w y-\gamma_w z$ and let $b=U(x+y+z)-Ux-Uy-Uz$ and let

$$(12)\qquad \begin{cases} c &=& p_2\lambda_\pi\nu_\# U_\#^+(w,y,z)+p_2\lambda_\pi\nu_\# U_\#^+(w,x,y+z), \\ d &=& p_2\lambda_\pi\nu_\# U_\#^+(w,x,y)+p_2\lambda_\pi\nu_\# U_\#^+(w,x+y,z). \end{cases}$$

Then (11) yields the composite of tracks

$$(13)\qquad \begin{cases} a \xrightarrow{T_2+R_2} c \xrightarrow{T_1+R_1} b, \\ a \xrightarrow{T_2'+R_2'} c \xrightarrow{T_1'+R_1'} b. \end{cases}$$

For the proof of (11) we consider the following commutative diagram with $Z = Z^{pq}$ and $A : Z \times Z \to Z$, $A(x,y) = x+y$ and $\nu : Z \times Z \times Z \to Z$, $\nu(x,y,z) = x-y-z$.

(14)
$$\begin{array}{ccccc} Z \times Z & \xrightarrow{A} & Z & \xleftarrow{A} & Z \times Z \\ \uparrow{\scriptstyle \nu\times\nu} & & & & \uparrow{\scriptstyle \nu\times\nu} \\ Z^{\times 3} \times Z^{\times 3} & \xleftarrow{\nu'} & Z^{\times 6} & \xrightarrow{\nu''} & Z^{\times 3} \times Z^{\times 3} \end{array}$$

Here we set for $x = (x_1, \dots, x_6) \in Z^{\times 6}$

$$\begin{aligned} \nu'(x) &= (x_5, x_2, x_3, x_6, x_1, x_5), \\ \nu''(x) &= (x_4, x_1, x_2, x_6, x_4, x_3), \end{aligned}$$

so that $A(\nu \times \nu)\nu'(x) = x_5 - x_2 - x_3 + x_6 - x_1 - x_5 = x_6 - x_1 - x_2 - x_3 = x_4 - x_1 - x_2 + x_6 - x_4 - x_3 = A(\nu \times \nu)\nu''(x)$. For the element

(15) $$U^{++}(x,y,z) = (U(x), U(y), U(z), U(x+y), U(y+z), U(x+y+z))$$

we thus get

(16) $$\begin{cases} \nu' U^{++}(x,y,z) &= (U^+(y,z), U^+(x, y+z)), \\ \nu'' U^{++}(x,y,z) &= (U^+(x,y), U^+(x+y, z)), \end{cases}$$

where $U^+(x,y) = U(x+y) - Ux - Uy$ as in the proof of (8.2.3). According to (7.3.4) we get

(17) $$\begin{aligned} L^{A(\nu\times\nu)\nu'} &= (A(\nu \times \nu))_\# L^{\nu'} \square L^{A(\nu\times\nu)}(\nu')_\# \\ &= L^{A(\nu\times\nu)}(\nu')_\# \end{aligned}$$

since $L^{\nu'}$ is the trivial track by definition of $\lambda^6_\#$. Here we use the fact that ν' is given by permutation and diagonal. Similarly we get

(18) $$L^{A(\nu\times\nu)\nu''} = L^{A(\nu\times\nu)}(\nu'')_\#$$

so that by (14)

(19) $$L^{A(\nu\times\nu)}(\nu')_\# = L^{A(\nu\times\nu)}(\nu'')_\#.$$

Moreover by (7.3.4) we have

(20) $$L^{A(\nu\times\nu)} = (A_\# L^{\nu\times\nu}) \square (L^A(\nu \times \nu)_\#)$$

with $L^{\nu\times\nu} = L^\nu \times_{B\pi} L^\nu$. One can check that

(21) $$p_2 A_\# (L^\nu \times_{B\pi} L^\nu) \nu'_\# U^{++}_\#(w,x,y,z) = (T_2 + R_2)^{\mathrm{op}},$$

(22) $$p_2 A_\# (L^\nu \times_{B\pi} L^\nu) \nu''_\# U^{++}_\#(w,x,y,z) = (T'_2 + R'_2)^{\mathrm{op}},$$

so that by (20) we get for $V = U_{\#}^{++}(w, x, y, z)$,

$$p_2(L^{A(\nu\times\nu)}\square L^A(\nu\times\nu)_{\#}^{\mathrm{op}})\nu'_{\#}V = (T_2 + R_2)^{\mathrm{op}},$$
$$p_2(L^{A(\nu\times\nu)}\square L^A(\nu\times\nu)_{\#}^{\mathrm{op}})\nu''_{\#}V = (T'_2 + R'_2)^{\mathrm{op}}.$$

This implies by (19) the following equation with

$$\begin{cases} K' &= p_2 L^A(\nu\times\nu)_{\#}\nu'_{\#}V, \\ K'' &= p_2 L^A(\nu\times\nu)_{\#}\nu''_{\#}V, \end{cases}$$

(23) $$(K')^{\mathrm{op}}\square(T_2 + R_2) = (K'')^{\mathrm{op}}\square(T'_2 + R'_2).$$

On the other hand one can check that for the norm map $N : Z^{pq} \to Z^{pq}$ we have

(24) $$T_1 + R_1 = p_2 A_{\#} L^{N\times N}(w, \bar{U}(y,z), \bar{U}(x, y+z)),$$

(25) $$T'_1 + R'_1 = p_2 A_{\#} L^{N\times N}(w, \bar{U}(x,y), \bar{U}(x+y, z)).$$

Here we use $L^{N\times N} = L^N \times_{B\pi} L^N$. We have for the addition map $A : Z^{pq}\times Z^{pq} \to Z^{pq}$ the equation $A(N\times N) = NA$ so that by (7.3.4)

(26) $$\begin{aligned} L^N A_{\#} &= N_{\#}L^A\square L^N A_{\#} = L^{NA} = L^{A(N\times N)} \\ &= (A_{\#}L^{N\times N})\square(L^A(N\times N)_{\#}). \end{aligned}$$

Here $N_{\#}L^A$ is the trivial track since L^A is defined on the trivial fibration with $\lambda_\pi = 1$ the identity. Hence we get

$$A_{\#}L^{N\times N} = (L^N A_{\#})\square(L^A(N\times N)_{\#})^{\mathrm{op}}.$$

This implies by (24) and (25) that

(27) $$T_1 + R_1 = (p_2 L^N A_{\#})(w, \bar{U}(y,z), \bar{U}(x, y+z))\square(K')^{\mathrm{op}}$$

with

(28) $$\begin{aligned} K' &= p_2 L^A(N\times N)_{\#}(w, \bar{U}(y,z), \bar{U}(x, y+z)) \\ &= p_2 L^A(\nu\times\nu)_{\#}\nu'_{\#}V \end{aligned}$$

as in (23). Similarly we get

(29) $$T'_1 + R'_1 = (p_2 L^N A_{\#})(w, \bar{U}(x,y), \bar{U}(x+y, z))\square(K'')^{\mathrm{op}}$$

with

$$\begin{aligned} K'' &= p_2 (L^A)^{\mathrm{op}}(N\times N)_{\#}(w, \bar{U}(x,y), \bar{U}(x+y, z)) \\ &= p_2 (L^A)^{\mathrm{op}}(\nu\times\nu)_{\#}\nu''_{\#}V \end{aligned}$$

as in (23).

We now show

$$(T_1 + R_1)\square(K')^{\mathrm{op}} = (T_1' + R_1')\square(K'')^{\mathrm{op}}. \tag{30}$$

Then (30) and (23) imply (11) and hence the proof of (9.2.3) is complete. By (27), (29) equation (30) is equivalent to

$$p_2 L^N A_\#(w, \bar{U}(y,z), \bar{U}(x, y+z)) = p_2 L^N A_\#(w, \bar{U}(x,y), \bar{U}(x+y, z)) \tag{31}$$

In fact by (9.2.1)(10), (11) equation (31) is equivalent to

$$p_2 L^N(w, U^0 + \bar{\bar{U}}_R(x,y,z)) = p_2 L^N(w, U^0 + \bar{\bar{U}}_L(x,y,z)) \tag{32}$$

with $U^0 = \bar{U}(x,y) + \bar{U}(x,z) + \bar{U}(y,z)$. Using (26) we see that (32) is equivalent to

$$p_2 L^N(w, \bar{\bar{U}}_R(x,y,z)) = p_2 L^N(w, \bar{\bar{U}}_L(x,y,z)). \tag{33}$$

This formula can be proved inductively by (9.2.1)(12) and (26) since

$$p_2 L^N(w, \alpha\cdot x) = p_2 L^N(w,x) \quad \text{by (4)} \quad \text{for} \quad \alpha \in \pi.$$

Now the proof of (9.2.3)(ii) is complete.

Finally we consider the proof of (9.2.3)(iii). For this we first observe that $\Gamma_v^{0,0}$ is the identity track of $0 = \gamma_v(0)$. This can be derived directly from definition (1) above. Next we set $x = y = 0$ in formula (9.2.3)(ii) and we get

$$(\gamma_v(0) + \Gamma_v^{0,z})\square\Gamma_v^{0,z} \;=\; (\Gamma_v^{0,0} + \gamma_v(z))\Gamma_v^{0,z}.$$

This implies

$$\Gamma_v^{0,z} \;=\; \Gamma_v^{0,0} + \gamma_v(z)$$

where the right-hand side is the identity track of $\gamma_v(z)$. This proves (9.2.3)(iii) by (i) so that the proof of (9.2.3) is complete. □

9.3 Relations for iterated linearity tracks

Given maps $x_i : X \to Z^q$ with $i = 1, 2, \ldots$ we define inductively as in (9.1.7) for $r \geq 2$,

$$\Gamma_v^{x_1,\ldots,x_r} : \gamma_v\Big(\sum_{i=1}^{r} x_i\Big) \Rightarrow \sum_{i=1}^{r} \gamma_v(x_i). \tag{9.3.1}$$

For $r = 2$ this is the linearity track in (8.2.7) and for $r > 2$ we set

$$\Gamma_v^{x_1,\ldots,x_r} \;=\; (\Gamma_v^{x_1,\ldots,x_{r-1}} + \gamma_v(x_r))\square\Gamma_v^{x_1+\cdots+x_{r-1},x_r}. \tag{1}$$

This definition corresponds to the bracket of length r of the form $(\dots((1,2),3)\dots,r)$. But by (9.2.3)(ii) any other bracket of length r can be used to define $\Gamma_v^{x_1,\dots,x_r}$ so that the iterated linearity track (9.3.1) is independent of this bracket. Moreover (9.2.3)(i) shows that for any permutation σ of $(1,\dots,r)$ we have

$$\Gamma_v^{x_1,\dots,x_r} = \Gamma_v^{x_{\sigma 1},\dots,x_{\sigma r}}. \tag{2}$$

Also the iterated linearity track is natural in X by (9.2.4); that is, for a map $f : Y \to X$ we have

$$f^*\Gamma_v^{x_1,\dots,x_r} = \Gamma_{vf}^{x_1 f,\dots,x_r f}. \tag{3}$$

The track $\Gamma_v^{x,y,z}$ coincides with the composition of tracks in (9.2.3)(ii).

If $x_1 = \dots = x_r = x$ we get as in (9.1.7) the track ($r \geq 1$)

$$\Gamma(r)_v^x = \Gamma_v^{x,\dots,x} : \gamma_v(r \cdot x) \Rightarrow r\gamma_v(x). \tag{9.3.2}$$

This is the identity track for $r = 1$. For $r = p$ we have $p \cdot x = 0$ and $p \cdot \gamma_v(x) = 0$ so that

$$\Gamma(p)_v^x : 0 \Longrightarrow 0$$

represents an element in $[X, Z^{pq-1}]$.

9.3.3 Definition. There is a well-defined element

$$\bar{\gamma} \in [Z^1 \times Z^q, Z^{pq-1}] \qquad \text{with} \qquad \Gamma(p)_v^x = \bar{\gamma}(v,x).$$

This follows from naturality. In fact, for $X = Z^1 \times Z^q$ we have the projections $p_1 = v : X \to Z^1$ and $p_2 = x : X \to Z^q$ so that in this case $\bar{\gamma} = \Gamma(p)_{p_1}^{p_2}$. The element $\bar{\gamma}$ is computed in the next result.

Recall that we defined in (4.5.7) the linear derivation

$$\Gamma[p] : \mathcal{A} \longrightarrow \Sigma\mathcal{A}$$

on the Steenrod algebra $\mathcal{A}$. For $p = 2$ this is the Kristensen derivation and for p odd recall Theorem (4.5.9). Moreover we have the formulas (8.5.10) and (8.5.11) expressing $\gamma \in [Z^1 \times Z^q, Z^{pq}]$ in terms of elements in $\mathcal{A}$. If we apply $\Gamma[p]$ to these elements we get the element $\bar{\gamma} \in [Z^1 \times Z^q, Z^{pq-1}]$, that is:

9.3.4 Theorem. *For $p = 2$ we have*

$$\bar{\gamma}(v,x) = \sum_j v^{q-j} \cdot Sq^{j-1}(x).$$

Moreover if p is odd we have

$$\bar{\gamma}(v,x) = \vartheta_q \sum_j (-1)^j w_{(q-2j)(p-1)-1}(v) \cdot P^j(x).$$

9.3.5 Definition. Let p be a prime and $r \in \mathbb{N} = \{1, 2, \ldots\}$. It is well known that $r^p - r$ is divisible by p so that the function, termed a *Fermat quotient*

$$\begin{aligned} &\alpha_0 : \mathbb{N} \longrightarrow \mathbb{N}, \\ &a_0(r) = (r^p - r)/p, \end{aligned} \tag{1}$$

is well defined. For $\mathbb{F} = \mathbb{Z}/p$ and $\mathbb{G} = \mathbb{Z}/p^2$ the function α_0 induces

$$\alpha : \mathbb{G} \to \mathbb{F} \tag{2}$$

with $\alpha(r \cdot 1) = \alpha_0(r) \cdot 1$. Here 1 denotes the unit in $\mathbb{F}$ and $\mathbb{G}$. Let $\bar{\alpha}_0(x, y)$ be the universal polynomial over $\mathbb{Z}$ satisfying

$$p\bar{\alpha}_0(x, y) = (x + y)^p - x^p - y^p. \tag{3}$$

For example $\bar{\alpha}_0(x, y) = x \cdot y$ for $p = 2$ and $\bar{\alpha}_0(x, y) = x^2 y + xy^2$ for $p = 3$. Then $\bar{\alpha}_0$ induces a function

$$\bar{\alpha} : \mathbb{F} \times \mathbb{F} \longrightarrow \mathbb{F} \tag{4}$$

with $\bar{\alpha}(r \cdot 1, t \cdot 1) = \bar{\alpha}_0(r, t) \cdot 1$. Now α in (2) satisfies

$$\alpha(r) = \sum_{j=1}^{r-1} \bar{\alpha}(j, 1). \tag{5}$$

Moreover the function $\bar{U}$ in (8.2.3)(2) with $N\bar{U} = U^{cr}$ satisfies

$$\bar{U}(rx, tx) = \bar{\alpha}(r, t) U(x). \tag{6}$$

One readily checks that $\alpha(i) = 0$, $\alpha(p) = -1$ and if p is odd $\alpha(p^2 - 1) = 0$. Moreover one proves (5) by the equation

$$\begin{aligned} p\alpha_0(r) &= \sum_{j=1}^{r-1} p\bar{\alpha}_o(j, 1) \\ &= \sum_{j=1}^{r-1} ((j + 1)^p - j^p - 1), \text{ see (3)} \\ &= r^p - 1 - (r - 1) = r^p - r. \end{aligned}$$

Using the track $\Gamma(r)^x_v$ in (9.3.2) and the track $L(r)^x_v$ in (8.2.8) we obtain the following result which computes the relation (9.1.7).

9.3.6 Theorem. *The tracks* ($r \in \mathbb{N}$)

$$\Gamma(r)^x_v,\ L(r)^x_v : \gamma_v(rx) \Longrightarrow r\gamma_v(x)$$

satisfy the equation (*see* (3.2.4))

$$\Gamma(r)^x_v = L(r)^x_v \oplus (\frac{r - r^p}{p} \bar{\gamma}(v, x))$$

with $\bar{\gamma}(v,x) \in [X, Z^{pq-1}]$, $q = \mid x \mid$, *given by* (9.3.4). *In particular we get for* p *odd*

$$\Gamma(p^2-1)^x_v = L(p^2-1)^x_v = L(-1)^x_v.$$

We prove the theorem in (9.3) below.

We point out that for $r = p$ we have $rx = 0$ and $r\gamma_v(x) = 0$ and $L(p)^x_v : 0 \Rightarrow 0$ is the identity track. Hence for $r = p$ the theorem shows that $\Gamma(p)^x_v : 0 \Rightarrow 0$ represents $\bar{\gamma}(v,x)$. This, in fact, holds by definition in (9.3.3). Moreover, for $r = 1$ the tracks $\Gamma(1)^x_v$ and $L(1)^x_v$ are both identity tracks.

9.3.7 Proposition. *The track* $\Gamma(r)^x_v$ *satisfies for* $r, t \in \mathbb{N}$ *the equations*

$$\begin{aligned} \Gamma(r+t)^x_v &= (\Gamma(r)^x_v + \Gamma(t)^x_v)\Box\Gamma^{rx,tx}_v, \\ \Gamma(r\cdot t)^x_v &= (r\cdot\Gamma(t)^x_v)\Box\Gamma(r)^{tx}_v \\ &= (t\cdot\Gamma(r)^x_v)\Box\Gamma(t)^{rx}_v. \end{aligned}$$

Moreover, if $r \equiv t$ *modulo* p^2, *then* $\Gamma(r)^x_v = \Gamma(t)^x_v$.

Compare (4.2.8), (4.2.9) and (4.2.10).

Proof. For $r, t > 0$ the equations hold since they correspond to certain brackets of length $r+t$ or $r\cdot t$, and $\Gamma(r)^x_v$ is independent of the choice of bracket, see (9.3.1)(1). □

9.3.8 Corollary. *For* $r = p^2$ *the track*

$$\Gamma(p^2)^x_v = 0^\Box : \gamma_v(p^2x) = 0 \Longrightarrow p^2\gamma_v(x)$$

is the identity track of the trivial map. For $r = p^2-1$ *the track*

$$\Gamma(p^2-1)^x_v : \gamma_v(-x) \Longrightarrow -\gamma_v(x)$$

satisfies the equation

$$\Gamma(p^2-1)^x_v = (\Gamma^{-x,x}_v)^{\mathrm{op}} - \gamma_v(x).$$

Proof. We have for $r = p^2-1$ and $r' = 1$ the following equation by (9.3.7),

$$\begin{aligned} \Gamma(p^2)^x_v &= (\Gamma(p^2-1)^x_v + \Gamma(1)^x_v)\Box\Gamma^{(p^2-1)x,x}_v \\ &= (\Gamma(p^2-1)^x_v + \gamma_v(x))\Box\Gamma^{-x,x}_v \end{aligned}$$

and $\Gamma(p^2)^x_v = 0^\Box$ is the trivial track by the second equation in (9.3.7). □

We now consider for $r, t \in \mathbb{N}$ the following diagram of tracks in $[\![X, Z^{pq}]\!]$.

(9.3.9)
$$\begin{array}{ccc} \gamma_v(rx+tx) & \xrightarrow{\Gamma^{rx,tx}_v} & \gamma_v(rx)+\gamma_v(tx) \\ {\scriptstyle L(r+t)^x_v}\downarrow & & \downarrow{\scriptstyle L(r)^x_v + L(t)^x_v} \\ (r+t)\gamma_v(x) & = & r\gamma_v(x)+t\gamma_v(x) \end{array}$$

9.3.10 Proposition. *There is a natural element* $\bar{\gamma}_v^x \in [X, Z^{pq-1}]$ *so that the primary element of* (9.3.9) *is given by* $\bar{\alpha}(r,t)\bar{\gamma}_v^x$ *with* $\bar{\alpha}$ *defined in* (9.3.5).

Proof. We subtract on both sides of (9.3.4) the track $L(r)_v^x + L(t)_r^v$ so that we get the following equivalent relation.

(1)

$$\begin{array}{ccc} \gamma_v^{cr}(rx,tx) & \xleftarrow{\;L_v^{rx,tx}\;} & U^{cr}(rx,tx) = 0 \\ L_0 \downarrow & \searrow \Gamma_0 & \swarrow \Gamma\bar{U}(rx,tx) \\ 0 & = & 0 \end{array}$$

Here we set

$$\begin{aligned} L_0 &= L(r+t)_v^x - L(v)_v^x - L(t)_v^x, \\ \Gamma_0 &= \Gamma_v^{rx,tx} - \gamma_v(rx) - \gamma_v(tx) \\ &= \Gamma\bar{U}(rx,tx)\Box(L_v^{rx,tx})^{\mathrm{op}}, \text{ see (8.2.7).} \end{aligned}$$

We have the equation

$$\begin{aligned} U^{cr}(rx,tx) &= U(rx+tx) - U(rx) - U(tx) \\ &= (r+t)^p Ux - r^p Ux - t^p Ux \\ &= 0 \end{aligned}$$

since $(r+t)^p = r^p + t^p$ in $\mathbb{F} = \mathbb{Z}/p$. According to (8.2.4) the composite track $L_0 \Box L_v^{rx,tx} : 0 \Rightarrow 0$ is represented by the following diagram with $f = (r+t)_\# \bar{\times} r_\# \bar{\times} t_\#$ is a product over $B\pi$ and $w = h_\pi v$.

(2)

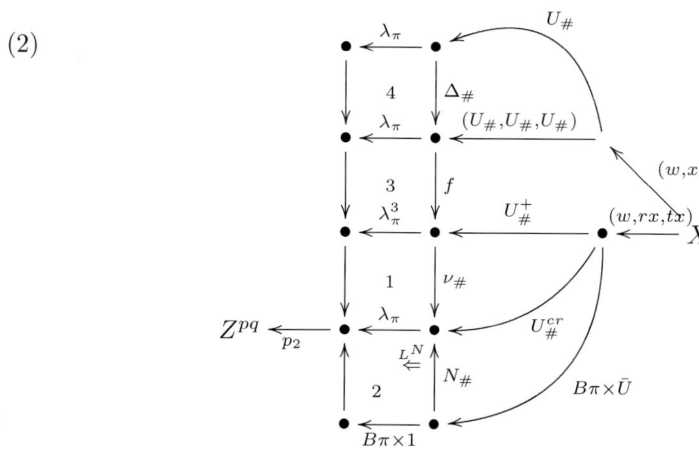

Here 3 is given by $L^{r+t} \bar{\times} L^r \bar{\times} L^t$ and 4 is the commutative diagram defined by the diagonal map Δ. Since $\nu f \Delta = 0$ we see that pasting of 1,3,4 yields the identity track, see (7.3.9). This shows

(3) $$L_0 \Box L_v^{rx,tx} = p_2(L^N)^{\mathrm{op}}(B\pi \times \bar{U})(w, rx, tx).$$

We now use the equation

$$\bar{U}(rx, tx) = \bar{\alpha}(r,t)U(x),$$

see (9.3.5). Hence we get for $a = \bar{\alpha}(r,t) \in \mathbb{F}$ the following diagram representing 3.

(4)

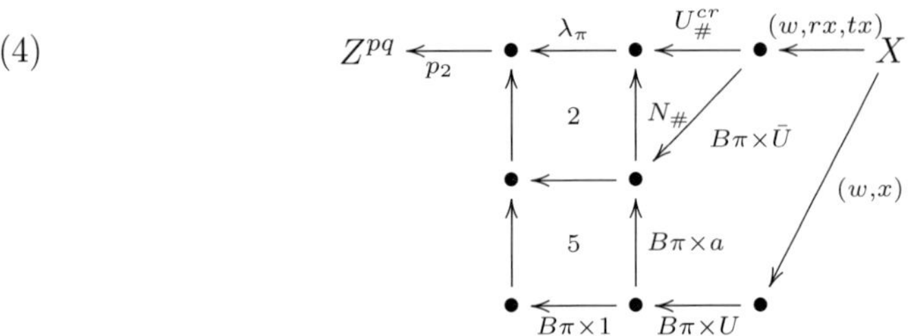

Here 5 is a commutative diagram. Since N is linear we have $Na = aN$ so that the pasting of 2 and 5 is given by the pasting of 6 and 2 in the following diagram, see (7.3.9).

(5)

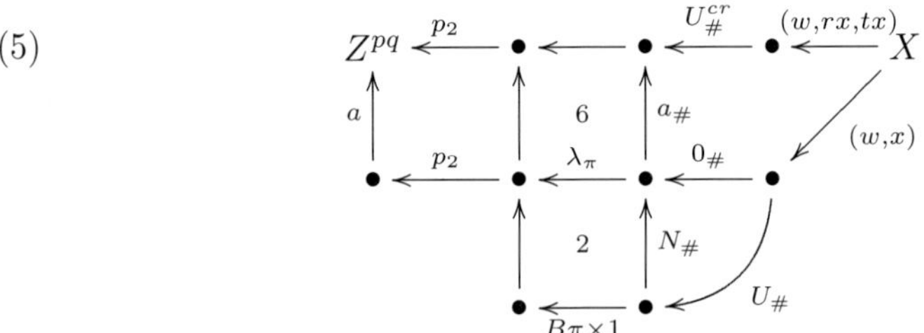

Since 6 is composed with $0_\#$ with $NU = 0$ we see that 3 coincides with

(6) $$L_0 \square L_v^{rx,tx} = a \cdot p_2(L^N)^{\mathrm{op}} U_\#(w,x).$$

On the other hand we have for $\Gamma : N \Rightarrow 0$,

(7) $$\begin{aligned} \Gamma\bar{U}(rx,tx) &= \Gamma a U x \\ &= a\Gamma U x. \end{aligned}$$

Uniqueness shows that $\Gamma a = a\Gamma$. This proves that the primary element of (9.3.9) is given by

(8) $$(6)^{\mathrm{op}} \square (7) = a \cdot \bar{\gamma}_v^x \text{ with}$$

(9) $$\bar{\gamma}_v^x = p_2(L^N)U_\#(w,x)\square \Gamma U x.$$

This completes the proof of (9.3.10). □

9.3.11 Corollary. *Let* Δ_r *be the primary element of* (9.3.6), *that is*

$$\Gamma(r)^x_v \oplus \Delta_r = L(r)^x_v.$$

Then Δ_r *satisfies the formula* $\Delta_1 = 0$ *and*

$$\Delta_{r+t} = \Delta_r + \Delta_t - \bar{\alpha}(r,t)\bar{\gamma}^x_v.$$

Proof. According to (9.3.10) we have for $a = \bar{\alpha}(r,t)$ and $L_r = L(r)^x_v$ the formula

$$L_{r+t} \oplus (a\bar{\gamma}^x_v) = (L_r + L_t)\Box\Gamma^{rx,tx}_v$$

so that for $\Gamma_r = \Gamma(r)^x_v$,

$$\Gamma_{r+t} \oplus (\Delta_{r+t} + a\bar{\gamma}^x_v) = ((\Gamma_r \oplus \Delta_r) + (\Gamma_t \oplus \Delta_t))\Box\Gamma^{rx,tx}_v.$$

Here we have by (9.3.7)

$$\Gamma_{r+t} = (\Gamma_r + \Gamma_t)\Gamma^{rx,tx}_v.$$

Therefore we get the formula in (9.3.11). □

Proof of (9.3.6). We have by (9.3.11) the formula

$$\begin{aligned} \Delta_1 &= 0, \\ \Delta_{r+1} &= \Delta_r - \bar{\alpha}(r,1)\bar{\gamma}^x_v. \end{aligned}$$

This shows inductively by (9.3.5)(5) that

$$\Delta_r = -\alpha(r)\bar{\gamma}^x_v.$$

We know that $-\Delta_p = \bar{\gamma}(v,x)$, see the remark following (9.3.6). Since $\alpha(p) = -1$ we get

$$\bar{\gamma}^x_v = -\bar{\gamma}(v,x).$$

Therefore we get

$$\Delta_r = \alpha(r)\cdot\bar{\gamma}(v,x)$$

and the proof of (9.3.6) is complete. □

9.4 Permutation relations

According to (7.2.2) we have for $\sigma \in \sigma_q$ the track in $[\![Z^q, Z^q]\!]$,

$$\Gamma_\sigma : \sigma \Rightarrow \mathrm{sign}(\sigma) : Z^q \to Z^q.$$

Here we have $\mathrm{sign}(\sigma) = \mathrm{sign}(\sigma)^p$ since Z^q is an $\mathbb{F}$-vector space with $\mathbb{F} = \mathbb{Z}/p\mathbb{Z}$. For a pointed map $x : X \to Z^q$ we obtain therefore the track in $[\![X, Z^q]\!]$,

(9.4.1) $$\Gamma_\sigma = \Gamma_\sigma(x) : \ \sigma x \Rightarrow \mathrm{sign}(\sigma)^p x$$

which we call the *permutation track*. One readily checks the relations

(1) $\Gamma_{\sigma\tau}(x) = (\mathrm{sign}(\sigma)\Gamma_\tau(x))\square\Gamma_\sigma(\tau x)$ for $\sigma, \tau \in \sigma_q$,

(2) $\Gamma_\sigma(x+y) = \Gamma_\sigma(x) + \Gamma_\sigma(y)$ for $x, y : X \to Z^q$.

Moreover for the product $x \cdot y : X \to Z^{q+q'}$ of $x : X \to Z^q$, $y : X \to Z^{q'}$ and $\sigma_1 \in \sigma_q, \sigma_2 \in \sigma_{q'}$ we get

(3) $$\Gamma_{\sigma_1\times\sigma_2}(x \cdot y) = \Gamma_{\sigma_1}(x) \cdot \Gamma_{\sigma_2}(y).$$

This readily follows from (6.3.2)(5).

Now let $v : X \to Z^1$ be a pointed map and let

$$\gamma_v(x) : X \xrightarrow{(v,x)} Z^1 \times Z^q \xrightarrow{\gamma} Z^{pq}$$

as in (8.1.6). Moreover let $\sigma^p \in \sigma_{pq}$ be the permutation for which $(\sigma x)^p = \sigma^p x^p$ where $x^p = x \cdot \dots \cdot x$ is the p-fold product. We define the permutation track

(9.4.2) $$P(\sigma)^x_v : \gamma_v(\sigma x) \Rightarrow \sigma^p \gamma_v(x)$$

as in (8.2.9), namely we have for $w = h_\pi v$ the equation

$$\begin{aligned}
\gamma_v(\sigma x) &= p_2\lambda_\pi U_\#(w, \sigma x)\\
&= p_2\lambda_\pi U_\#\sigma_\#(w, x)\\
&= p_2\lambda_\pi(\sigma^p)_\# U_\#(w, x)\\
&\Rightarrow p_2(\sigma^p)_\#\lambda_\pi U_\#(w, x)\\
&= \sigma^p p_2\lambda_\pi U_\#(w, x)\\
&= \sigma^p\gamma_v(x),
\end{aligned}$$

so that $P(\sigma)^x_v = p_2(L^{\sigma^p})U_\#(w, x)$.

9.4.3 Theorem. *The permutation track $P(\sigma)^x_v$ can be described in terms of the permutation tracks Γ_σ in (9.2.1), since the following diagram commutes.*

$$\begin{array}{ccc}
\gamma_v(\sigma x) & \xrightarrow{P(\sigma)^x_v} & \sigma^p\gamma_v(x)\\
{\scriptstyle \gamma_v(\Gamma_\sigma)}\big\downarrow & & \big\downarrow{\scriptstyle \Gamma_{\sigma^p}}\\
\gamma_v(\mathrm{sign}(\sigma)^p \cdot x) & \xrightarrow{L(\mathrm{sign}(\sigma)^p)^x_v} & \mathrm{sign}(\sigma)^p \cdot \gamma_v(x)
\end{array}$$

We shall need this result in the proof of (9.3.3)(i).

Proof. We recall that for $r \in \mathbb{Z}$ the track $L(r)^x_v : \gamma_v(r \cdot x) \Rightarrow r \cdot \gamma_v(x)$ defined in (8.2.8) is given by

(1) $$L(r)^x_v : p_2(L^{r\cdot})U_\#(w, x) : \gamma_v(r \cdot x) \Longrightarrow r \cdot \gamma_v(x)$$

with $w = h_\pi v$. Here $r\cdot : Z^q \to Z^q$ is multiplication by r, $r \geq 1$. This is a linear map so that the linear track $L^{r\cdot}$ is defined. We have to compare (1) for $r = \mathrm{sign}(\sigma)^p$ with

$$(2) \qquad P(\sigma)^x_v = p_2(L^{\sigma^p})U_\#(w,x) : \gamma_v(\sigma x) \Longrightarrow \sigma^p \gamma_v(x).$$

For this we consider the cylinder $IZ^q = Z^q \times [0,1]/* \times [0,1]$. We choose maps

$$(3) \qquad \begin{array}{rcl} \Gamma_\sigma : IZ^q & \longrightarrow & Z^q \\ \Gamma_{\sigma^p} : IZ^{pq} & \longrightarrow & Z^{pq} \end{array}$$

representing the permutation tracks Γ_σ and Γ_{σ^p}. We consider IZ^q as a triple

$$IZ^q = (Z^q, IZ^q, Z^q)$$

with inclusions i_0 and i_1 respectively of the boundary. We also have the triple

$$(IZ^q)^{\wedge p} = ((Z^q)^{\wedge p}, (IZ^q)^{\wedge p}, (Z^q)^{\wedge p})$$

with inclusions $i_0^{\wedge p}$ and $i_1^{\wedge p}$. Moreover, we have the π-invariant inclusion

$$j : I(Z^q)^{\wedge p} \longrightarrow (IZ^q)^{\wedge p}$$

which carries $(t, x_1, \ldots, x_p)$ to $((t,x_1), \ldots, (t,x_p))$. The group $\pi = \mathbb{Z}/p$ acts by permuting coordinates in $X^{\wedge p}$. The map j is a homotopy equivalence in **Top** and a map between triples which is the identity on the boundary. Therefore

$$(4) \qquad j_\# : B\pi \times_\pi I(Z^q)^{\wedge p} \longrightarrow B\pi \times_\pi (IZ^q)^{\wedge p}$$

has a homotopy inverse $\bar{j}$ over $B\pi$ which is also a map of triples and is the identity on the boundary.

Now consider the following diagram corresponding to (1) and (2) respectively with $i = 0, 1$.

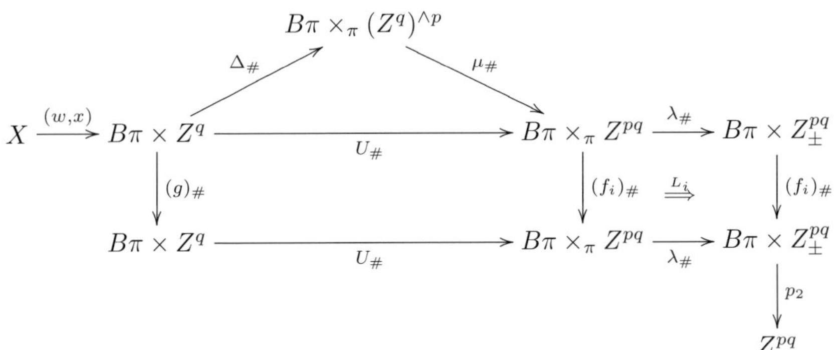

Here we set $g_1 = r$ and $f_1 = r^p = r$ and $g_0 = \sigma$ and $f_0 = \sigma^p$ and $L_1 = L^{r\cdot}$ and $L_0 = L^{\sigma^p}$ are linear tracks defining (1) and (2) respectively. We consider the following diagram of maps between triples.

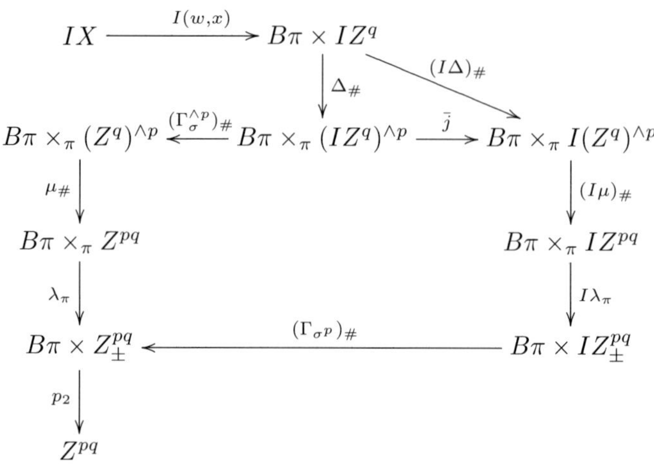

Let $H = \lambda_\# \mu_\# (\Gamma_\sigma^{\wedge p})_\#$ and $G = (\Gamma_{\sigma^p})_\# (I\lambda_\pi)(I\mu)_\# \bar{j}$. Then we obtain homotopies $H' = p_2 H \Delta_\# I(w,x)$ and $G' = p_2 G \Delta_\# I(w,x)$ such that the corresponding tracks satisfy

$$\gamma_v(\Gamma_\sigma) = H' : \gamma_v(\sigma x) \Longrightarrow \gamma_v(rx), \tag{5}$$

$$\Gamma_{\sigma^p} = G' : \sigma^p \gamma_v(x) \Longrightarrow r\gamma_v(x). \tag{6}$$

Here we have (6) since there is a homotopy over $B\pi$ and under the boundary

$$\bar{j}\Delta_\# \simeq (I\Delta)_\#. \tag{7}$$

This is a consequence of the following commutative diagram.

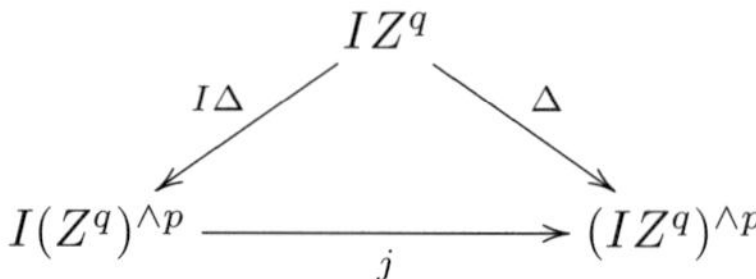

Given a triple $X = (A \subset X \supset B)$ the boundary ∂IX of the cylinder IX is defined by

$$\begin{aligned} i \;:\; & \partial IX \subset IX, \\ & \partial IX = IA \cup IB \cup i_0 X \cup i_1 X. \end{aligned} \tag{8}$$

Hence we get the following diagram.

(9)
$$\begin{array}{ccc} B\pi \times_\pi \partial(I(IZ^q)^{\wedge p}) & \xrightarrow{i_\#} & B\pi \times_\pi I(IZ^q)^{\wedge p} \\ {\scriptstyle F}\downarrow & \swarrow {\scriptstyle \bar{F}} & \\ B\pi \times Z^{pq}_\pm & & \end{array}$$

Here the map F is given by

(10)
$$F = L_0\mu_\# \cup L_1\mu_\# \cup H \cup G.$$

Now obstruction theory as in (7.1.10) shows that there is a map $\bar{F}$ over $B\pi$ extending F. The existence of $\bar{F}$ shows by (1) and (2) and (4) and (5) that the diagram in (9.4.3) commutes. □

9.5 Secondary Cartan relations

We consider pointed maps

$$x : X \to Z^q, \; y : X \to Z^{q'}, \; z : X \to Z^{q''}$$

so that products $x \cdot y$, $x \cdot y \cdot z$ are defined as in (2.1.5). We have $\tau(x, y) \in \sigma_{q+q'}$ with

(9.5.1)
$$\tau(x, y)x \cdot y = y \cdot x.$$

Moreover let $\sigma(x, y) \in \sigma_{pq+pq'}$ be the permutation with

(9.5.2)
$$\sigma(x, y)(x \cdot y)^p = x^p \cdot y^p.$$

Hence $\sigma(x, y)$ coincides with σ in (8.2.3). We have the following rules:

(1)
$$\sigma(y, x)\tau(x, y)^p = \tau(x^p, y^p)\sigma(x, y),$$

(2)
$$\begin{aligned}\sigma(x, y, z) &= (\sigma(x, y) \times 1)\sigma(x \cdot y, z) \\ &= (1 \times \sigma(y, z))\sigma(x, y \cdot z).\end{aligned}$$

Here $\sigma(x, y, z)$ is the permutation with $\sigma(x, y, z)(xyz)^p = x^p y^p z^p$. Moreover let

$$\gamma_v(x) : X \xrightarrow{(v,x)} Z^1 \times Z^q \xrightarrow{\gamma} Z^{pq}$$

be defined as in (8.1.6).

9.5.3 Theorem. *The Cartan track Λ_C in (8.2.3) induces the track*

$$C = C_v^{x,y} : \sigma(x, y)\gamma_v(x \cdot y) \Rightarrow \gamma_v(x) \cdot \gamma_v(y)$$

in $[\![X, Z^{p(q+q')}]\!]$ which is natural in X and for which the following diagrams (i), (ii) *commute. These diagrams are the* secondary Cartan relations.

(i)
$$\begin{array}{ccc}
\sigma(y,x)\gamma_v(\tau(x,y)\cdot x\cdot y) & \xrightarrow{\sigma(y,x)\cdot P} & \sigma(y,x)\tau(x,y)^p\gamma_v(x\cdot y) \\
\| & & \| \\
\sigma(y,x)\gamma_v(y\cdot x) & & \tau(x^p,y^p)\sigma(x,y)\gamma_v(x\cdot y) \\
\downarrow C_v^{y,x} & & \downarrow \tau(x^p,y^p)C_v^{x,y} \\
\gamma_v(y)\cdot\gamma_v(x) & = & \tau(x^p,y^p)\gamma_v(x)\cdot\gamma_v(y)
\end{array}$$

Here P is the track in (9.4.3), $P = P(\tau(x,y))_v^{x\cdot y}$.

(ii)
$$\begin{array}{ccc}
\sigma(x,y,z)\gamma_v(x\cdot y\cdot x) & \xrightarrow{(\sigma(x,y)\times 1)C_v^{xy,z}} & (\sigma(x,y)\cdot\gamma_v(x\cdot y))\cdot\gamma_v(z) \\
\downarrow (1\times\sigma(y,z))C_v^{x,yz} & & \downarrow C_v^{x,y}\cdot\gamma_v(z) \\
\gamma_v(x)\cdot(\sigma(y,z)\gamma_v(y\cdot z)) & \xrightarrow{\gamma_v(x)C_v^{y,z}} & \gamma_v(x)\cdot\gamma_v(y)\cdot\gamma_v(z)
\end{array}$$

Proof of (9.5.3). We define for $w = h_\pi v$ and $\sigma = \sigma(x,y)$,

(1)
$$\begin{aligned} C = C_v^{x,y} &= \Lambda_C(v,x,y) \\ &= (p_2 S^\mu(U\times U)_\#(w,x,y))\square(p_2 L^\sigma U_\#(w,x\cdot y))^{\mathrm{op}}. \end{aligned}$$

Compare the proof of (8.3.2). Now (i) is equivalent to the following equation (see (9.4.3)).

(2)
$$C_v^{y,x} = (\bar\tau\cdot C_v^{x,y})\square p_2\bar\sigma_\#(L^{\tau^p})U_\#(w,x,y),$$

$$\begin{cases} \tau = \tau(x,y) & , \quad \sigma = \sigma(x,y), \\ \bar\tau = \tau(x^p,y^p) & , \quad \bar\sigma = \sigma(y,x). \end{cases}$$

By (9.5.2)(1) we have the equation $\bar\sigma\tau^p = \bar\tau\sigma$. Hence by (7.3.4) we get

(3)
$$\bar\sigma_\# L^{\tau^p}\square L^{\bar\sigma}(\tau^p)_\# = \bar\tau L^\sigma\square L^{\bar\tau}\sigma_\#.$$

Hence we get

(4)
$$(\bar\tau_\# L^\sigma)^{\mathrm{op}}\square(\bar\sigma_\# L^{\tau^p}) = L^{\bar\tau}\sigma_\#\square(L^{\bar\sigma}\tau^p_\#)^{\mathrm{op}}.$$

Therefore (2) is equivalent to

(5)
$$C_v^{y,x} = \bar\tau p_2 S^\mu(U\times U)_\#(w,x,y)\square(L^{\bar\tau}\sigma_\#\square(L^{\bar\sigma}\tau^p_\#)^{\mathrm{op}})U_\#(w,xy).$$

Since $\tau^p U = U\tau$ we see that (2) is equivalent to

(6)
$$p_2 S^{\bar\mu}(U\times U)_\#(w,y,x) = p_2\bar\tau_\# S^\mu(U\times U)_\#(w,x,y)\square L^{\bar\tau}\sigma_\# U_\#(w,xy).$$

We have the equation $\bar{\tau}\mu = \bar{\mu}T$. Hence we know by (7.3.9) that

$$S^{\bar{\mu}T} = \bar{\mu}_\# S^T \Box S^\mu T_\# = S^\mu T_\# = S^{\bar{\tau}\mu} = \bar{\tau}_\# S^\mu \Box S^{\bar{\tau}} \mu_\# \tag{7}$$

where $S^{\bar{\tau}} = L^{\bar{\tau}}$ since $\bar{\tau}$ is linear and where $S^T = L^T = 0^\Box$ is the trivial track. Moreover $\sigma U(x \cdot y) = \sigma(x \cdot y)^p = x^p \cdot y^p = \mu(U \times U)(x, y)$. This shows by (7) that (6) holds. Hence the proof of (i) is complete.

For the proof of (ii) we have to consider:

$$C_1 = C_v^{x,y} = (p_2 S^{\mu_1}(U \times U)_\#(w, x, y))\Box(p_2 L^{\sigma_1} U_\#(w, xy))^{\mathrm{op}}$$

with $\mu_1(x, y) = x \cdot y$, $\sigma_1 = \sigma(x, y)$,

$$C_2 = C_v^{y,z} = (p_2 S^{\mu_2}(U \times U)_\#(w, y, z))\Box(p_2 L^{\sigma_2} U_\#(w, yz))^{\mathrm{op}}$$

with $\mu_2(y, z) = y \cdot z$, $\sigma_2 = \sigma(y, z)$,

$$C_3 = C_v^{xy,z} = (p_2 S^{\mu_3}(U \times U)_\#(w, xy, z))\Box(p_2 L^{\sigma_3} U_\#(w, xyz))^{\mathrm{op}}$$

with $\mu_3(x \cdot y, z) = x \cdot y \cdot z$, $\sigma_3 = \sigma(x \cdot y, z)$,

$$C_4 = C_v^{x,yz} = (p_2 S^{\mu_4}(U \times U)_\#(w, x, yz))\Box(p_2 L^{\sigma_4} U_\#(w, xyz))^{\mathrm{op}}$$

with $\mu_4(x, y \cdot z) = x \cdot y \cdot z$, $\sigma_4 = \sigma(x, y \cdot z)$. We have to show

$$(C_1 \cdot \gamma_v(z))\Box(\sigma_L \cdot C_3) = (\gamma_v(x) \cdot C_2)\Box(\sigma_R \cdot C_4) \tag{8}$$

with $\sigma_R = 1 \times \sigma(y, z)$ and $\sigma_L = \sigma(x, y) \times 1$. We have by (9.5.2) the equation

$$\sigma_L \sigma_3 = \sigma(x, y, z) = \sigma_R \sigma_4 \tag{9}$$

so that

$$(\sigma_L)_\# L^{\sigma_3} \Box L^{\sigma_L}(\sigma_3)_\# = L^{\sigma(x,y,z)} = (\sigma_R)_\# L^{\sigma_4} \Box L^{\sigma_R}(\sigma_4)_\#.$$

Hence (8) is equivalent to

$$\begin{aligned} &(C_1 \gamma_v(z))\Box \sigma_L p_2 S^{\mu_3}(U \times U)_\#(w, xy, z)\Box p_2 L^{\sigma_L}(\sigma_3)_\# U_\#(w, xyz) \\ &= (\gamma_v(x) C_2)\Box \sigma_R p_2 S^{\mu_4}(U \times U)_\#(w, x, yz)\Box p_2 L^{\sigma_R}(\sigma_4)_\# U_\#(w, xyz). \end{aligned} \tag{10}$$

Here we have

$$\begin{aligned} (\mu_3)_\#(U \times U)_\#(w, xy, z) &= (\sigma_3)_\# U_\#(w, xyz), \\ (\mu_4)_\#(U \times U)_\#(w, x, yz) &= (\sigma_4)_\# U_\#(w, xyz), \end{aligned}$$

since $\sigma_3(xyz)^p = (xy)^p z^p$ and $\sigma_4(xyz)^p = x^p(yz)^p$. Hence (10) is equivalent to

$$\begin{aligned} &(C_1 \gamma_v(z))\Box p_2 \left[(\sigma_L)_\# S^{\mu_3} \Box L^{\sigma_L}(\mu_3)_\#\right](U \times U)_\#(w, xy, z) \\ &= (\gamma_v(x) C_2)\Box p_2 \left[(\sigma_R)_\# S^{\mu_4} \Box L^{\sigma_R}(\mu_4)_\#\right](U \times U)_\#(w, x, yz). \end{aligned} \tag{11}$$

Using (7.3.9) we see that (11) is equivalent to:

$$(12)\qquad \begin{aligned} &(C_1\gamma_v(z))\square \overbrace{p_2S^{\sigma_L\mu_3}(U\times U)_\#(w,xy,z)}^{A}\\ &=(\gamma_v(x)C_2)\square \underbrace{p_2S^{\sigma_R\mu_4}(U\times U)_\#(w,x,yz)}_{B}.\end{aligned}$$

We have the equation $\mu_3(\mu_1\times 1)=\mu_4(1\times\mu_2)$ so that by (7.3.9)

$$(13)\qquad \begin{aligned} &(\mu_3)_\#(S^{\mu_1}\times\lambda_\#)\square S^{\mu_3}(\mu_1\times 1)_\#\\ &=(\mu_4)_\#(\lambda_\pi\times S^{\mu_2})\square S^{\mu_4}(1\times\mu_2)_\#.\end{aligned}$$

Moreover $\gamma_v(z)=p_2\lambda_\pi U_\#(w,z)$ so that

$$\begin{aligned} A&=(p_2S^{\mu_1}(U\times U)_\#(w,x,y))\gamma_v(z)\\ &=p_2(\mu_3)_\#(S^{\mu_1}\times\lambda_\pi)(U\times U\times U)_\#(w,x,y,z),\end{aligned}$$

$$\begin{aligned} B&=\gamma_v(x)\cdot(p_2S^{\mu_2}(U\times U)_\#(w,y,z))\\ &=p_2(\mu_4)_\#(\lambda_\pi\times S^{\mu_2})(U\times U\times U)_\#(w,x,y,z).\end{aligned}$$

Using the definition of C_1 and C_2 we see that (12) is equivalent to

$$(14)\qquad \begin{aligned} &A\square\overbrace{((p_2L^{\sigma_1}U_\#(w,xy)^{\mathrm{op}}\gamma_v(z))}^{A''}\square A'\\ &=B\square\underbrace{(\gamma_v(x)(p_2L^{\sigma_2}U_\#(w,yz))^{\mathrm{op}})}_{B''}\square B'.\end{aligned}$$

Hence (13) implies that (14) is equivalent to

$$(15)\qquad \begin{aligned} &p_2(S^{\mu_3})^{\mathrm{op}}(\mu_1\times 1)_\#(U\times U\times U)_\#(w,x,y,z)\square A''\square A'\\ &=p_2(S^{\mu_4})^{\mathrm{op}}(1\times\mu_2)_\#(U\times U\times U)_\#(w,x,y,z)\square B''\square B'.\end{aligned}$$

Since $U(x)U(y)=\sigma(x,y)U(xy)$ and $U(y)U(z)=\sigma(y,z)U(yz)$ we see that

$$(9.5.4)\qquad \begin{aligned} (\mu_1\times 1)_\#(U\times U\times U)_\#(w,x,y,z)&=(\sigma_L)_\#(U\times U)_\#(w,xy,z),\\ (1\times\mu_2)_\#(U\times U\times U)_\#(w,x,y,z)&=(\sigma_R)_\#(U\times U)_\#(w,x,yz).\end{aligned}$$

On the other hand we get for A'',B'' in (14)

$$\begin{aligned} A''&=p_2(\mu_3)_\#(L^{\sigma_1}\times\lambda_\pi)^{\mathrm{op}}(U\times U)_\#(w,xy,z),\\ B''&=p_2(\mu_4)_\#(\lambda_\pi\times L^{\sigma_2})^{\mathrm{op}}(U\times U)_\#(w,x,yz).\end{aligned}$$

This implies that (15) is equivalent to

$$(16)\qquad \begin{aligned} &p_2((S^{\mu_3})^{\mathrm{op}}(\sigma_L)_\#\square(\mu_3)_\#(L^{\sigma_1}\times\lambda_\pi)^{\mathrm{op}})(U\times U)_\#(w,xy,z)\square A'\\ &=p_2((S^{\mu_4})^{\mathrm{op}}(\sigma_R)_\#\square(\mu_4)_\#(\lambda_\pi\times L^{\sigma_2})^{\mathrm{op}})(U\times U)_\#(w,x,yz)\square A''.\end{aligned}$$

Now we have by definition of $\sigma_L = \sigma_1 \times 1$ and $\sigma_R = 1 \times \sigma_2$ the equation $\mu_3\sigma_L = \sigma_3(\sigma_1 \times 1)$ and $\mu_4\sigma_R = \mu_4(1 \times \sigma_2)$. This shows that the left-hand side of (16) is equal to

$$p_2(S^{\mu_3\sigma_L})^{\mathrm{op}}(U \times U)_{\#}(w, xy, z)\square A' = 0^{\square} \tag{17}$$

and the right-hand side of (16) is equal to

$$p_2(S^{\mu_4\sigma_R})^{\mathrm{op}}(U \times U)_{\#}(w, x, yz)\square B' = 0^{\square}. \tag{18}$$

Since both tracks (17) and (18) are the trivial track we see that (16) holds. This completes the proof of (ii). $\square$

9.6 Cartan linearity relation

We consider pointed maps $v : X \to Z^1$ and

$$x, x' : X \longrightarrow Z^q, \; y : X \longrightarrow Z^{q'}$$

so that $x + x'$ and $x \cdot y$, $x' \cdot y$ are defined. We have the linearity track

$$\Gamma_v^{x,x'} : \gamma_v(x + x') \Longrightarrow \gamma_v(x) + \gamma_v(x')$$

in (9.1.4) and the Cartan track

$$C_v^{x,y} : \sigma(x,y)\gamma_v(x \cdot y) \Longrightarrow \gamma_v(x) \cdot \gamma_v(y)$$

in (9.3.3). These tracks yield following the diagram.

(9.6.1)

$$\begin{array}{ccc} \sigma(x,y)\gamma_v((x+x')\cdot y) & \xrightarrow{\quad C_v^{x+x',y} \quad} & \gamma_v(x+x')\cdot\gamma_v(y) \\ \Big\downarrow{\scriptstyle \sigma(x,y)\Gamma_v^{xy,x'y}} & & \Big\downarrow{\scriptstyle \Gamma_v^{x,x'}\gamma_v(y)} \\ \sigma(x,y)\gamma_v(x\cdot y) + \sigma(x,y)\gamma_v(x'\cdot y) & \xrightarrow{\; C_v^{x,y} + C_v^{x',y} \;} & \gamma_v(x)\cdot\gamma_v(y) + \gamma_v(x')\cdot\gamma_v(y) \end{array}$$

For $v = 0$ we have $\gamma_0 x = Ux = x^p$ and $C_0^{x,y}$ is the identity track. Therefore we obtain the following as a special case $v = 0$ of diagram (9.6.1).

(9.6.2)

$$\begin{array}{ccc} \sigma(x,y)U((x+x')\cdot y) & = & U(x+x')\cdot U(y) \\ \Big\downarrow{\scriptstyle \sigma(x,y)\Gamma_0^{xy,x'y}} & & \Big\downarrow{\scriptstyle \Gamma_0^{x,x'}U(y)} \\ \sigma(x,y)(U(x\cdot y) + U(x'\cdot y)) & = & (U(x)+U(x'))\cdot U(y) \end{array}$$

9.6.3 Theorem. *For the tracks in* (9.6.2) *we have*

$$\sigma(x,y)\Gamma_0^{xy,x'y} \oplus \Delta(x,x',y) = \Gamma_0^{x,x'}U(y)$$

where $\Delta(x,x',y) \in [X, Z^{pq-1}]$ *is given by the formula*

$$\Delta(x,x',y) = \begin{cases} x \cdot x' \cdot Sq^{|y|-1}(y) & \text{for } p = 2, \\ 0 & \text{for } p \text{ odd.} \end{cases}$$

9.6.4 Theorem. *The primary element of* (9.6.1) *does not depend on* v *and hence is given by* $\Delta(x,x',y)$ *in* (9.6.3).

Proof of (9.6.4). For the inverse σ^{-1} of $\sigma = \sigma(x,y)$ we obtain the track

$$C_1 = \sigma^{-1}C_v^{x,y} : \gamma_v(x \cdot y) \Longrightarrow \sigma^{-1}\gamma_v(x) \cdot \gamma_v(y).$$

According to the diagram in (8.2.4) we obtain C_1 by diagram 2 below where $w = h_\pi v$ and

$$\bar{\mu} = \sigma^{-1}\mu_{pq,pq'} : Z^{pq} \times Z^{pq'} \longrightarrow Z^{p(q+q')}.$$

We point out that $\bar{\mu}$ is π-equivariant since for $\alpha \in \pi$,

$$\begin{aligned} \bar{\mu}(\alpha x, \alpha y) &= \sigma^{-1}(\alpha x \cdot \alpha y) = \sigma^{-1}(\alpha \odot \alpha)(x \cdot y) \\ &= \alpha\sigma^{-1}(x \cdot y) = \alpha\bar{\mu}(x,y). \end{aligned}$$

Hence the map $\bar{\mu}_\#$ between Borel constructions in the following diagram is defined.

(2)

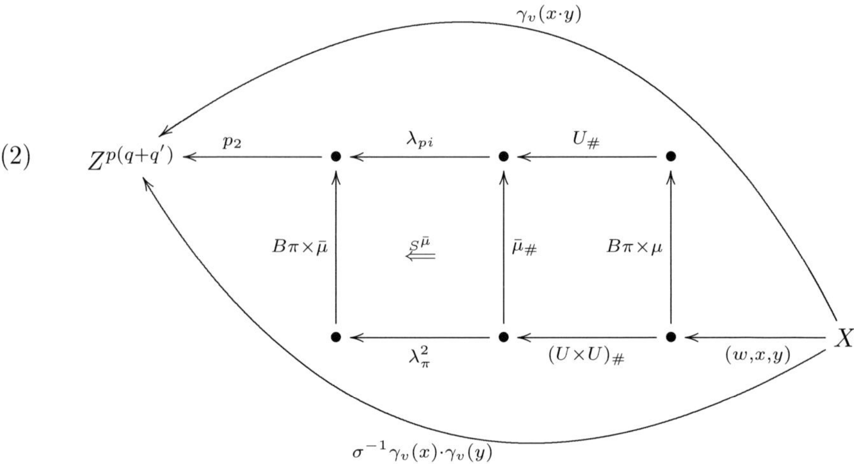

Similar diagrams are obtained for the tracks

(3) $$C_2 = \sigma^{-1}C_v^{x',y} : \gamma_v(x' \cdot y) \Longrightarrow \sigma^{-1}\gamma_v(x) \cdot \gamma_v(y),$$

(4) $$C_3 = \sigma^{-1}C_v^{x+x',y} : \gamma_v((x+x') \cdot y) \Longrightarrow \sigma^{-1}\gamma_v(x+x') \cdot \gamma_v(y).$$

Using the definition of $\Gamma_v^{x,y}$ in (9.1.4)(1) we see that commutativity of the diagram in (9.4.1) is equivalent to the commutativity of the following diagram where we use $L_v^{x,y}$ in (8.2.4).

(5)
$$\begin{array}{ccc}
\gamma_v^{cr}(xy,x'y) & \xrightarrow{C_3-C_1-C_2} & \sigma^{-1}(\gamma_v^{cr}(x,x')\cdot\gamma_v(y)) \\
\uparrow{\scriptstyle L_v^{xy,x'y}} & & \uparrow{\scriptstyle \sigma^{-1}L_v^{x,x'}\cdot\gamma_v(y)} \\
U^{cr}(xy,x'y) & & \sigma^{-1}U^{cr}(x,x')\cdot\gamma_v(y) \\
{\scriptstyle \Gamma\bar{U}(xy,x'y)}\searrow & & \swarrow{\scriptstyle \sigma^{-1}(\Gamma\bar{U}(x,x'))\cdot\gamma_v(y)} \\
 & 0 &
\end{array}$$

Here $\Gamma : N \Rightarrow 0$ is the track for the norm map $N : Z^{pq} \to Z^{pq}$ in the proof of (9.1.1). As in (8.2.3) let $\nu : Z^{pq} \times Z^{pq} \times Z^{pq} \to Z^{pq}$ be defined by $\nu(x,y,z) = x-y-z$. Then diagram (2) shows that the track $C_3 - C_1 - C_2$ is given by diagram (7) below with

(6)
$$\begin{aligned}
w_1 &= (w, xy, x'y) : X \longrightarrow Z^1 \times Z^{pq} \times Z^{pq'}, \\
w_2 &= (w, (x+x', y), (x,y), (x',y)) : X \longrightarrow Z^1 \times (Z^q \times Z^{q'})^3.
\end{aligned}$$

Moreover let U^+ be defined as in the proof of (8.2.3) and recall that products like $\lambda_\pi^3 = \lambda_\pi \bar{\times} \lambda_\pi \bar{\times} \lambda_\pi$ are products over $B\pi$, where we use the symbol $\bar{\times}$ to denote the *product over* $B\pi$.

(7)
$$\begin{array}{ccccccc}
\bullet & \xleftarrow{p_2} & \bullet & & & & \\
 & & \uparrow{\scriptstyle B\pi\times\nu} & & & & \\
 & & \bullet & \xleftarrow{\lambda_\pi^3} & \bullet & \xleftarrow{U_\#^+} & \bullet & \\
 & & \uparrow{\scriptstyle B\pi\times\bar{\mu}^3} & \overset{(S^{\bar{\mu}})^3}{\Longleftarrow} & \uparrow{\scriptstyle \bar{\mu}_\#^3} & & \nwarrow{\scriptstyle w_1} & \\
 & & \bullet & \xleftarrow{(\lambda_\pi^2)^3} & \bullet & \xleftarrow{(U\times U)_\#^3} & \bullet & \xleftarrow{w_2} & X
\end{array}$$

Now diagram (7) is embedded into the large diagram (11) below which represents the composite of tracks

(8)
$$(\sigma^{-1}L_v^{x,x'}\cdot\gamma_v(y))^{\mathrm{op}}\square(C_3-C_1-C_2)\square L_v^{xy,x'y}$$

in diagram (5). Let

(9)
$$w_3 = (w,(x,x'),y) : X \longrightarrow Z^1 \times Z^q \times Z^q \times Z^{q'}$$

be similarly defined as w_1 and w_2 in (6) and let $\Delta_\#$ be given by the diagonal map with

(10)
$$\begin{cases}
\Delta : (Z^{pq})^4 \longrightarrow (Z^{pq})^6, \\
\Delta(x,y,z,u) = ((x,u),(y,u),(z,u)).
\end{cases}$$

(11)

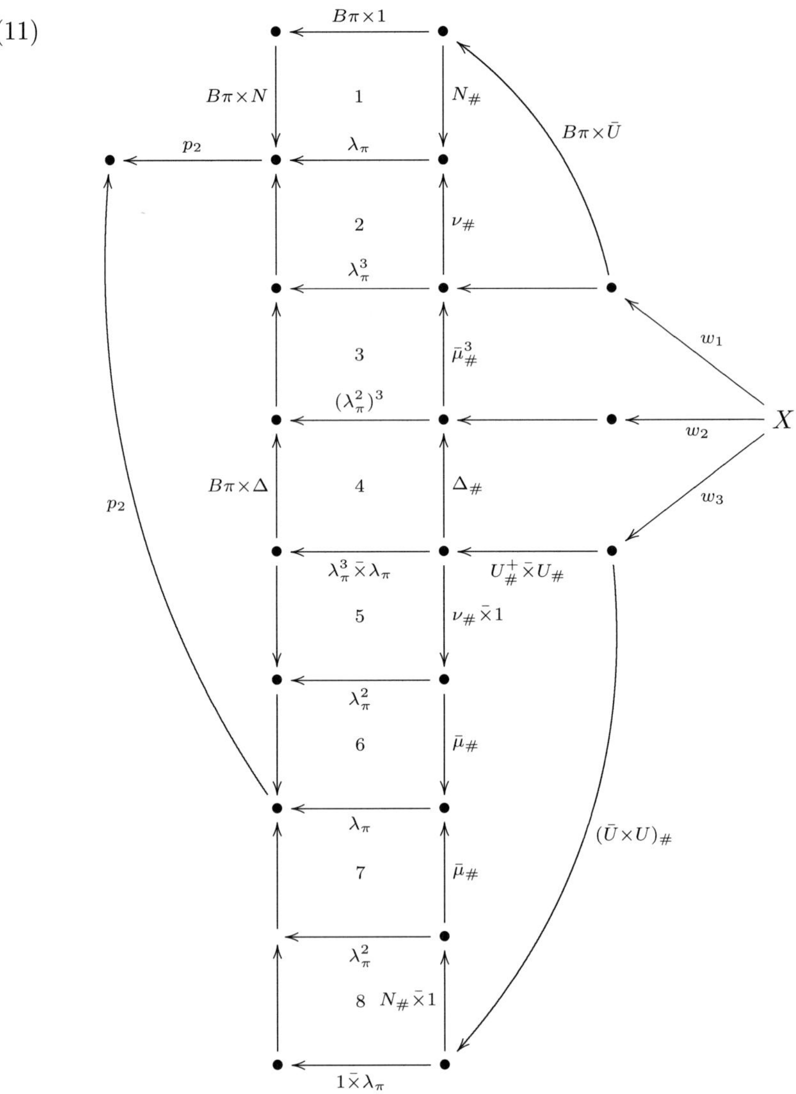

All subdiagrams numbered $1, \dots, 8$ in diagram (11) are diagrams together with linear tracks or smash tracks. The other subdiagrams of (11) commute. Since Δ is a diagonal map also subdiagram 4 commutes.

Subdiagrams 1, 2 correspond to the defining diagram in (8.2.4) of $L_v^{xy,x'y}$.

Subdiagram 3 is given by diagram (7) and yields the track $C_3 - C_1 - C_2$.

Subdiagrams 6 and 7 are opposite to each other and therefore cancel as a composite of tracks. This shows that subdiagrams 5, 6, 7, 8 yield by (8.2.4) the track $\sigma^{-1} L_v^{x,x'} \cdot \gamma_v(y)$. Here 8 is the track $L^N \bar{\times} \lambda_\pi$ and 5 is the track $L^\nu \bar{\times} \lambda_\pi$.

Hence we proved that diagram (11) represents the track (8).

We now observe that

$$\nu\bar{\mu}^3\Delta = \bar{\mu}(\nu \times 1) \tag{12}$$

since we have

$$\begin{aligned}
\nu\bar{\mu}^3\Delta(x,y,z,u) &= \nu\bar{\mu}^3((x,u),(y,u),(z,u)) \\
&= \sigma^{-1}x \cdot u - \sigma^{-1}y \cdot u - \sigma^{-1}z \cdot u \\
&= \sigma^{-1}(x-y-z)\cdot u \\
&= \bar{\mu}(\nu \times 1)(x,y,z,u).
\end{aligned}$$

Equation (12) shows by (7.3.9) that the tracks 2, 3, 4, 5, 6 cancel each other. Therefore the composite track (8) represented by diagram (11) is also represented by the following diagram.

(13)

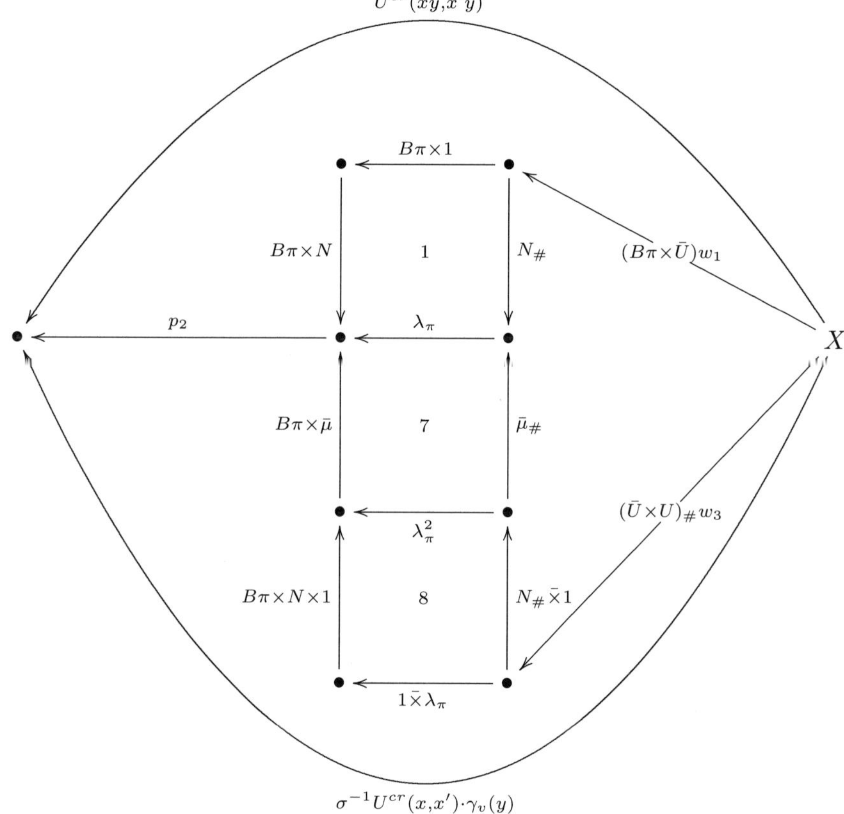

We describe further details of diagram (13) as follows.

Let $BZ^{pq} = B\pi \times Z^{pq}$ be the trivial fibration and let $EZ^{pq} = E\pi \times_\pi Z^{pq}$ be the Borel construction for the π-space Z^{pq}. Then subdiagram 8 together with objects is explicitly given by the following.

(14)
$$\begin{array}{ccc} B\pi \times Z^{pq} \times Z^{pq'} & \xleftarrow{\lambda_\pi^2} & E(Z^{pq} \times Z^{pq'}) \\ {\scriptstyle B\pi\times N\times 1}\uparrow & 8 & \uparrow{\scriptstyle N_\# \bar{\times} 1} \\ B\pi \times Z^{pq} \times Z^{pq'} & \xleftarrow[1\bar{\times}\lambda_\pi]{} & EZ^{pq}\bar{\times}EZ^{pq'} \end{array}$$

Here we use the notation $\bar{\times}$ for the product over $B\pi$, see (7.1.4). Using the definition of $\bar{U}$ in (8.2.3)(2) we see that

(15)
$$\begin{aligned} \bar{U}(x \cdot y, x' \cdot y) &= \sigma^{-1}\bar{U}(x, x') \cdot Uy \\ &= \bar{\mu}(\bar{U} \times U)(x, x', y). \end{aligned}$$

Hence we can replace $(B\pi \times \bar{U})w_1$ in (13) by

(16)
$$(B\pi \times \bar{U})w_1 = (B\pi \times \bar{\mu}(\bar{U} \times U))w_3.$$

The triangle in (13) commutes since

(17)
$$N\bar{\mu}(\bar{U} \times U) = \bar{\mu}(N \times 1)(\bar{U} \times U).$$

In fact (17) holds by the following computation.

$$\begin{aligned} N\bar{\mu}(\bar{U} \times U)(x, x', y) &= N\sigma^{-1}\bar{U}(x, x') \cdot Uy \\ &= N\bar{U}(x \cdot y, x' \cdot y) \\ &= U^{cr}(x \cdot y, x' \cdot y) \\ &= \sigma^{-1}U^{cr}(x, x') \cdot U(y) \\ &= \bar{\mu}(N \times 1)(\bar{U} \times U)(x, x', y). \end{aligned}$$

We observe that (17) admits a refinement since

(18)
$$N\bar{\mu}(1 \times U) = \bar{\mu}(N \times U).$$

In fact, we prove (18) by the equations

$$\begin{aligned} N\bar{\mu}(1 \times U)(z, y) = N\bar{\mu}(z, Uy) &= \sum_{\alpha\in\pi} \alpha\bar{\mu}(z, Uy) = \sum_{\alpha\in\pi} \bar{\mu}(\alpha z, \alpha Uy) \\ &= \sum_{\alpha\in\pi} \bar{\mu}(\alpha z, Uy) \quad \text{since} \quad \alpha Uy = Uy \\ &= \bar{\mu}(\sum_{\alpha\in\pi} \alpha z, Uy) = \bar{\mu}(Nz, Uy) = \bar{\mu}(N \times U)(z, y). \end{aligned}$$

We now embed diagram (13) into the following slightly larger diagram obtained from (13) by adding subdiagrams 9, 10. We also use (18) and

$$w_4 = (B\pi \times \bar{U} \times 1)w_3 = (w, \bar{U}(x, x'), y).$$

(19)

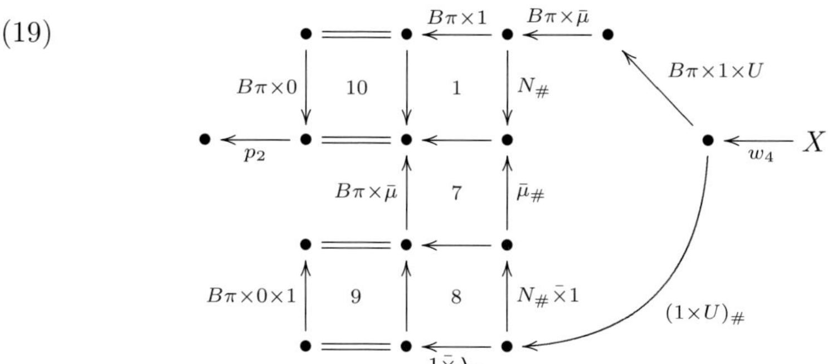

Moreover the subdiagrams (10) and (9) are tracks given by $\Gamma : N \Rightarrow 0$ in (5), namely $10 = B\pi \times \Gamma$ and $9 = B\pi \times \Gamma \times Z^{pq'}$. We used Γ to define tracks in (5) namely we have

$$\text{(20)} \quad \begin{aligned} \Gamma\bar{U}(xy, x'y) &= p_2 10(B\pi \times \bar{\mu})(B\pi \times \bar{U} \times U)w_3, \\ \sigma^{-1}(\Gamma\bar{U}(x, x')) \cdot \gamma_v(y) &= p_2(B\pi \times \bar{\mu})9(1\bar{\times}\lambda_\pi)(\bar{U} \times U)_\# w_3. \end{aligned}$$

This shows that diagram (19) describes the composite of tracks $0 \Rightarrow 0$ in diagram (5). Therefore diagram (5) commutes provided we can show that diagram (19) describes the identity track $0 \Rightarrow 0$. For this we embed (19) into the following diagram.

(21)

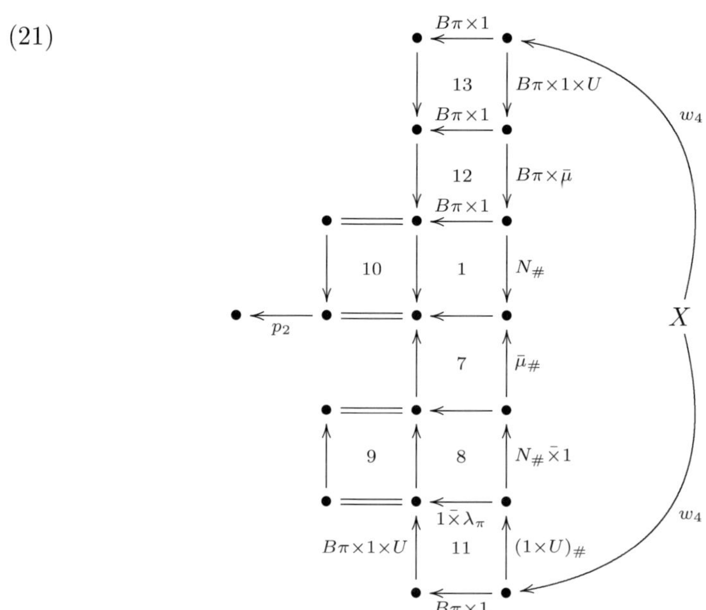

Diagram 11 is again a linear track which we are allowed to add to diagram (19) since 11 is composed with the zero map of 9. Moreover 12 and 13 are commutative diagrams. By equation (18) and by use of (7.3.9) we see that the pasting $P = 11 * 8 * 7$ coincides with the pasting $P = 1 * 12 * 13$. Let

(22)
$$\begin{aligned} Q &= p_2 10(B\pi \times \bar{\mu})(B\pi \times 1 \times U)w_4, \\ R &= p_2(B\pi \times \bar{\mu})9(B\pi \times 1 \times U)w_4. \end{aligned}$$

Then (21) describes the following composite of track $0 \Rightarrow 0$,

$$(21) = R\square(p_2 P w_4)\square(p_2 P w_4)^{\mathrm{op}}\square Q^{\mathrm{op}} = R\square Q^{\mathrm{op}}.$$

This shows that (21) does not depend on v or $w = h_\pi v$ since

$$R\square Q^{\mathrm{op}} = (\Gamma\bar{\mu}(1 \times U)^{\mathrm{op}}\square\bar{\mu}(\Gamma \times Z^{pq'})(1 \times U))(\bar{U}(x, x'), y).$$

This completes the proof of (9.6.4). $\square$

Proof of (9.6.3). According to the definition in (8.2.6) we have $\Gamma : N \Rightarrow 0$ and

(1)
$$\Gamma_0^{x,y} = \Gamma\bar{U}(x, y) + U(x) + U(y).$$

Hence for $\Delta = \Delta(x, x', y)$ and $\sigma = \sigma(x, y)$ we get

$$\begin{aligned} &\sigma(\Gamma\bar{U}(xy, x'y) + U(xy) + U(x'y)) \oplus \Delta \\ &\quad = (\sigma\Gamma\bar{U}(xy, x'y) \oplus \Delta) + \sigma(U(xy) + U(x'y)) \\ &\quad = (\Gamma\bar{U}(x, x')) \cdot U(y) + U(x)U(y) + U(x')U(y). \end{aligned}$$

Therefore Δ can be computed by

(2)
$$(\sigma\Gamma\bar{U}(xy, x'y)) \oplus \Delta = (\Gamma\bar{U}(x, x')) \cdot U(y)$$

or equivalently by

(3)
$$\begin{aligned} \Delta &= \sigma\Gamma^{\mathrm{op}}\bar{U}(xy, x'y)\square(\Gamma\bar{U}(x, x')) \cdot U(y) \\ &= \sigma\Gamma^{\mathrm{op}}\sigma^{-1}\sigma\bar{U}(xy, x'y)\square(\Gamma\bar{U}(x, x')) \cdot U(y) \\ &= \sigma\Gamma^{\mathrm{op}}\sigma^{-1}(\bar{U}(x, x') \cdot U(y))\square(\Gamma\bar{U}(x, x')) \cdot U(y). \end{aligned}$$

Here we have

(4)
$$\sigma\Gamma^{\mathrm{op}}\sigma^{-1} = \sigma(\sum_{\alpha\in\pi}\Gamma_\alpha)\sigma^{-1} = \sum_{a\in\pi}\Gamma_{\alpha\odot\alpha}$$

since $\sigma\alpha\sigma^{-1} = \alpha \odot \alpha$. Moreover we have

(5)
$$\begin{aligned} (\Gamma\bar{U}(x, x')) \cdot U(y) &= ((\textstyle\sum_{\alpha\in\pi}\Gamma_\alpha)\bar{U}(x, x') \cdot U(y) \\ &= \textstyle\sum_{\alpha\in\pi}(\Gamma_\alpha\bar{U}(x, x')) \cdot U(y) \\ &= \textstyle\sum_{\alpha\in\pi}\Gamma_{\alpha\odot 1}(\bar{U}(x, x') \cdot U(y)), \text{ see (9.4.1).} \end{aligned}$$

Therefore (3) shows for $z = \bar{U}(x, x')$,

$$\Delta = \sum_{\alpha\in\pi} (\Gamma^{\mathrm{op}}_{\alpha\odot\alpha}(z \cdot U(y))\Box\Gamma_{\alpha\odot 1}(z \cdot U(y)). \tag{6}$$

Here the right-hand side is well defined since for $\alpha \in \pi$ we have $\alpha U(y) = U(y)$ and

$$\begin{aligned} (\alpha \odot \alpha)(z \cdot U(y)) &= \alpha z \cdot \alpha U(y) \\ &= \alpha z \cdot U(y) \\ &= (\alpha \odot 1)(z \cdot U(y)). \end{aligned}$$

Since $\alpha \odot \alpha = (1 \odot \alpha)(\alpha \odot 1)$ and $\mathrm{sign}(\alpha) = 1$ we get by (9.4.1)(1),

$$\Gamma_{\alpha\odot\alpha} = \Gamma_{\alpha\odot 1}\Box\Gamma_{1\odot\alpha}.$$

Therefore we have

$$\begin{aligned} &\Gamma^{\mathrm{op}}_{\alpha\odot\alpha}(z \cdot U(y))\Box\Gamma_{\alpha\odot 1}(z \cdot U(y)) \\ &\quad = \Gamma^{\mathrm{op}}_{1\odot\alpha}(z \cdot U(y))\Box\Gamma^{\mathrm{op}}_{\alpha\odot 1}(z \cdot U(y))\Box\Gamma_{\alpha\odot 1}(z \cdot U(y)), \\ &\quad = \Gamma^{\mathrm{op}}_{1\odot\alpha}(z \cdot U(y)), \\ &\quad = z \cdot (\Gamma^{\mathrm{op}}_{\alpha} U(y)), \text{ see (9.4.1).} \end{aligned}$$

This shows that Δ in (6) is given by

$$-\Delta = z \cdot \sum_{\alpha\in\pi} \Gamma_\alpha U(y). \tag{7}$$

For $p = 2$ we have $z = \bar{U}(x, x') = x \cdot x'$ and $\pi = \mathbb{Z}/2 = \{1, \tau\}$ so that

$$\begin{aligned} \textstyle\sum_{\alpha\in\pi} \Gamma_\alpha U(y) &= (\Gamma_1 + \Gamma_\tau)y^2 \\ &= y^2 + \Gamma_\tau y^2 \\ &= Sq^{|y|-1}(y) \end{aligned} \tag{8}$$

by (6.5.1). This proves that for $p = 2$ we have $\Delta(x, x', y) = x \cdot x' \cdot Sq^{|y|-1}y$. Now we use an argument as in the proof of (4.5.9) above. For $\alpha, \beta \in \pi$ we have

$$\begin{aligned} \Gamma_{\alpha\beta} U(y) &= \Gamma_\beta U(y)\Box\Gamma_\alpha(\beta U(y)), \text{ see (9.4.1)(1)} \\ &= \Gamma_\beta U(y)\Box\Gamma_\alpha U(y). \end{aligned} \tag{9}$$

Thus the function

$$\chi : \mathbb{Z}/p = \pi \longrightarrow Aut(U(y)) \cong [X, Z^{p|y|-1}]$$

which carries α to $\Gamma_\alpha U(y)$ is a homomorphism. This shows that

$$\begin{aligned}\sum_{\alpha\in\pi}\Gamma_\alpha U(y) &= \sum_{\alpha\in\pi}\chi(\alpha)\\ &= \sum_{r=1}^{p-1} r\cdot\chi(1)\\ &= p(p-1)/2\chi(1).\end{aligned} \tag{10}$$

Since $p\chi(1) = 0$ we see that the element (10) is trivial if p is odd. This shows by (7) that $\Delta(x, x', y) = 0$ if p is odd. □

Chapter 10

Künneth Tracks and Künneth-Steenrod Operations

10.1 Künneth tracks

For cohomology with coefficients in the field k we have the *Künneth formula*

(10.1.1) $$H^*(Z \times Y) = H^*(Z) \otimes H^*(Y).$$

Here Z and Y are finite type path-connected pointed spaces and $Z \times Y$ is the product space. We now describe properties of the Künneth formula on the level of tracks.

Recall that we defined the Eilenberg-MacLane spaces $Z^n = K(k,n)$ for $n \geq 1$ as in (2.1.4). We have for $n, m \geq 1$ the multiplication map

(1) $$\mu_{m,n} : Z^m \times Z^n \longrightarrow Z^{m+n}$$

in (2.1.1). For maps $f : Z \to Z^m$ and $g : Y \to Z^n$ we get the composite map

(2) $$f \boxtimes g = \mu_{m,n}(f \times g) : Z \times Y \longrightarrow Z^{m+n}.$$

Moreover if $m = 0$ and $\lambda \in k$ we set $\lambda \boxtimes g = \lambda \cdot g$ and if $n = 0$ and $\lambda \in k$ we set $f \boxtimes \lambda = \lambda \cdot f$.

We consider a map $(n \geq 1)$

(10.1.2) $$f : Z \times Y \longrightarrow Z^n \text{ in } \mathbf{Top}^*$$

which represents an element $\varphi \in H^n(Z \times Y)$. Let $\mathcal{B}$ be a basis of $H^*(Z)$. Since Z is path connected and pointed we have $H^0 Z = k$ and $1 \in k = H^0 Z$ is assumed to be the basis element $1 \in B$. By (10.1.1) we get

(1) $$H^*(Z \times Y) = \bigoplus_{b \in \mathcal{B}} b \otimes H^*(Y).$$

This shows that there are unique elements $\varphi_b \in H^{n-|b|}(Y)$ for $b \in \mathcal{B}$ (with $\varphi_b = 0$ for $n- \mid b \mid < 0$) such that

(2) $$\varphi = \sum_{b \in \mathcal{B}_n} b \otimes \varphi_b = \sum_{b \in \mathcal{B}} b \otimes \varphi_b.$$

Here $\mathcal{B}_n = \{b \in B \mid\mid b \mid \leq n\}$ is a finite set.

Now we fix maps $s(b) : Z \to Z^{|b|}$ in $\mathbf{Top}^*$ representing $b \in \mathcal{B}$ with $\mid b \mid \geq 1$. For $\mid b \mid = 0$ let $s(b) = 1 \in k$. For $n- \mid b \mid \geq 1$ we choose a map

(3) $$s(\varphi_b) : Y \longrightarrow Z^{n-|b|}$$

in $\mathbf{Top}^*$ representing φ_b. Here we set $s(\varphi_b) = f \mid_{(*\times Y)}$ if $\mid b \mid = 0$. Moreover for $n- \mid b \mid = 0$ we set $s(\varphi_b) = \varphi_b \in k = H^0(Y)$ and for $n- \mid b \mid < 0$ we set $s(\varphi_b) = 0$. Then (10.1.1)(2) yields the map

(4) $$\sum_{b \in \mathcal{B}} s(b) \boxtimes s(\varphi_b) : Z \times Y \longrightarrow Z^n$$

representing the sum in (2). Therefore there exists a track

(5) $$K : \sum_{b \in \mathcal{B}} s(b) \boxtimes s(\varphi_b) \Longrightarrow f$$

termed a *Künneth track* for f. This track can be chosen to be a track under $* \times Y$ if $* \to Z$ is a cofibration.

10.1.3 Proposition. *Let $s'(\varphi_b)$ be a further representation of φ_b for $b \in \mathcal{B}$ as above and let*

$$T : \sum_{b \in \mathcal{B}} s(b) \boxtimes s(\varphi_b) \Rightarrow \sum_{b \in \mathcal{B}} s(b) \boxtimes s'(\varphi_b)$$

be a track. Then there exists for $b \in \mathcal{B}$ with $n- \mid b \mid \geq 1$ a unique track

$$T_b : s(\varphi_b) \Longrightarrow s'(\varphi_b) \text{ in } [\![Y, Z^{n-|b|}]\!]$$

such that $T = \sum_{b \in B} s(b) \boxtimes T_b$. Here $s(b) \boxtimes T_b$ is the trivial track for $n- \mid b \mid < 0$.

We call T_b the *coordinate* of the track T associated to the element $b \in \mathcal{B}$. We can alter the track T in (10.1.3) by an element

$$\begin{cases} \alpha & \in & [Z \times Y, \Omega Z^n] = H^{n-1}(Z \times Y), \\ \alpha & = & \sum_{b \in \mathcal{B}} b \otimes \alpha_b \text{ with } \alpha_b \in H^{n-1-|b|}(Y). \end{cases}$$

Then $T \oplus \alpha$ is again a track as in (10.1.3), compare (3.2.4).

10.1.4 Proposition. *The coordinate of $T \oplus \alpha$ satisfies the formula*

$$(T \oplus \alpha)_b = T_b \oplus ((-1)^{|b|} \alpha_b).$$

Proof of (10.1.3) *and* (10.1.4). Since $f_b = s(\varphi_b)$ and $f'_b = s'(\varphi_b)$ represent φ_b, we can choose a track

(1) $$H_b : f_b \Longrightarrow f'_b.$$

Then using (3.2.4) the track T and the track

(2) $$H = \sum_{b \in \mathcal{B}_n} s(b) \boxtimes H_b$$

yield an element α with $H \oplus \alpha = T$. We claim that there is $\epsilon_b \in \{-1, 1\}$ for $b \in \mathcal{B}_n$ such that

(3) $$\sum_{b \in \mathcal{B}_n} b \boxtimes (H_b \oplus \epsilon_b \alpha_b) = (\sum_{b \in \mathcal{B}_n} b \boxtimes H_b) \oplus \alpha.$$

Hence $T_b = H_b \oplus \epsilon_b \alpha_b$ satisfies the formula in (10.1.3). For the proof of (3) we need (3.2.7) and (3.2.10). In fact for elements β_b we get

(4) $$\sum_{b \in \mathcal{B}_n} b \boxtimes (H_b \oplus \beta_b) = (A_+ \tilde{\mu})_* ((b \times H_b) \oplus (0 \times \beta_b))$$

(5) $$= (A_+ \tilde{\mu})_* (b \times H_b)_{\mathcal{B}} \oplus L\nabla (A_+ \tilde{\mu})((0 \times \beta_b); (b \times f_b)).$$

Here we have $L\nabla(A_+\tilde{\mu}) = (\Omega A_+) L\nabla \tilde{\mu}$ since A_+ is linear. Moreover by (3.2.7) we get

(6) $$L\nabla(A_+\tilde{\mu})((0 \times \beta_b); (b \times f_b)) = \sum_{b \in \mathcal{B}_n} \delta_b \beta_b \cdot b$$

with $\delta_b = (-1)^{|b|(n-|b|)}$ and $\beta_b \cdot b = (-1)^{(n-|b|-1)\cdot|b|} b \otimes \beta_b$. Hence $\epsilon_b = (-1)^{|b|}$ satisfies (3).

A similar computation yields a proof of the formula in (10.1.4). □

10.1.5 Corollary. *Let $s'(\varphi_b)$ be a further representative of φ_b for $b \in \mathcal{B}$ and let*

$$K' : \sum_{b \in \mathcal{B}} s(b) \boxtimes s'(\varphi_b) \Longrightarrow f$$

be a further Künneth track for f as in (10.1.2). *Then there exists a unique track*

$$T_b : s(\varphi_b) \Longrightarrow s'(\varphi_b) \text{ for } b \in \mathcal{B}_n$$

such that

$$\sum_{b \in \mathcal{B}} s(b) \boxtimes T_b = (K')^{\mathrm{op}} \square K.$$

Of course the track T_b depends on the choice of the Künneth tracks K and K' for f, the track T_b is the coordinate of $(K')^{\mathrm{op}} \square K$.

10.2 Künneth-Steenrod operations

Let $k = \mathbb{F} = \mathbb{Z}/p$ where p is a prime. We apply the Künneth tracks in Section (10.1) to the power maps

$$\gamma : Z^1 \times Z^q \longrightarrow Z^{pq}, \; q \geq 1.$$

In fact we have a basis

(10.2.1) $$\mathcal{B} = \{w_0, w_1, w_2, \ldots\} \subset H^*(Z^1) = E_\beta(\mathbb{F}x)$$

where $w_0 = 1$ and $w_1 = x \in H^1(Z^1) = [Z^1, Z^1]$ is represented by the identity of Z^1. According to (1.2.3)(4) and (8.5.5) we have

(1) $$w_i = x^i \text{ for } p = 2, \text{ and}$$

(2) $$w_i = \begin{cases} (-\beta x)^j & \text{for } i = 2j, \quad p \text{ odd}, \\ x \cdot (-\beta x)^j & \text{for } i = 2j+1, \quad p \text{ odd}. \end{cases}$$

Here $\beta : H^1(Z^1) \to H^2(Z^1)$ is the Bockstein homomorphism. We choose a map in **Top***

(3) $$w_i : Z^1 \longrightarrow Z^i$$

representing (1) and (2) as follows. For $p = 2$ the map w_i is the power map which carries $x \in Z^1$ to the i-fold product $x^i = x \cdot \cdots \cdot x \in Z^i$ with the product defined by (2.1.2). For p odd we choose a map

(4) $$\beta : Z^1 \to Z^2$$

representing the Bockstein operator (see (2.1.11)) and we define the map w_i in (3) by use of β. That is, w_i carries $x \in Z^1$ to the j-fold product $(-\beta x)^j \in Z^i$ for $i = 2j$ and to the product $x \cdot (-\beta x)^j \in Z^i$ for $i = 2j+1$.

Now the basis (10.2.1) yields as in Section (10.1) a *Künneth track* for γ

(10.2.2) $$K_q : \sum_{i=0}^{pq} w_i \boxtimes s(D_i) \Longrightarrow \gamma.$$

Here $s(D_i) : Z^q \to Z^{pq-i}$ is a map in **Top*** representing the class $D_i \in H^{pq-i}(Z^q)$ in (8.5.5). Moreover $s(D_i)$ satisfies further conditions described in (10.2.4), (10.2.5) and (10.2.6) below. Now let

(1) $$(v, x) : X \longrightarrow Z^1 \times Z^q$$

be a map in **Top*** with $\gamma_v(x) = \gamma \circ (v, x)$. We use the composites

(2) $$w_i(v) = w_i \circ v : X \longrightarrow Z^1 \longrightarrow Z^i,$$

(3) $$D_i(x) = (sD_i) \circ x : X \longrightarrow Z^q \longrightarrow Z^{pq-i}.$$

Here the notation $D_i(x)$ should not be confusing. In fact, if ξ is a cohomology class then $D_i(\xi)$ is the composite of ξ and the homotopy class $D_i \in [Z^q, Z^{pq-1}]$. But in (3) the element $x : X \to Z^q$ is a map in **Top*** and hence $D_i(x)$ is given by the composite $s(D_i) \circ x$ with $s(D_i)$ chosen in (10.2.2).

With this notation K_q above induces the *Künneth track*

$$K_v(x) = K_q(v,x) : \sum_{i \geq 0} w_i(v) \cdot D_i(x) \Longrightarrow \gamma_v(x). \tag{10.2.3}$$

This track is *natural* in X. That is, for a map $f : Y \to X$ we get

$$f^* K_v(x) = K_{vf}(xf).$$

In computations below we shall use such natural tracks. We obtain similar results as in (10.1.3), (10.1.4), (10.1.5) for such natural tracks. Recall that we have the set of generators (see (5.5.1))

$$E_{\mathcal{A}} \subset \mathcal{A}$$

in the Steenrod algebra $\mathcal{A}$ with

$$\begin{aligned} E_{\mathcal{A}} &= \{Sq^1, Sq^2, \dots\} && \text{for } p = 2, \\ E_{\mathcal{A}} &= \{\beta, P^1, P^2, \dots, P^1_\beta, P^2_\beta, \dots\} && \text{for } p \text{ odd.} \end{aligned}$$

For $\alpha \in E_{\mathcal{A}}$ we obtain representing maps $s(\alpha)_q : Z^q \to Z^{q+|\alpha|}$ as follows.

According to (8.5.10) with $p = 2$ the map

$$s(D_{q-i}) = s(Sq^i)_q : Z^q \longrightarrow Z^{q+i} \text{ in } \mathbf{Top}^* \tag{10.2.4}$$

represents the Steenrod operation Sq^i for $i \geq 0$. We may assume that $s(Sq^0)_q = id$ is the identity of Z^q and $s(Sq^q)_q = U$ is the power map $U : Z^q \to Z^{2q}$ with $U(x) = x \cdot x$. In this case K_q is a track under $Z^1 \vee Z^q$. Moreover we set $s(Sq^i)_q = 0$, the trivial map for $i > q$, see (5.5.1).

We call this sequence of maps $s(Sq^i)_q$ with $i \in \mathbb{Z}$ the *Künneth-Steenrod operations* (associated to K_q). Moreover we write for a map $x : X \to Z^q$ in **Top*** with $| x | = q$,

$$Sq^i(x) = s(Sq^i)_q \circ x : X \longrightarrow Z^q \longrightarrow Z^{q+i}.$$

This composite denotes a map in **Top***. When we write $Sq^j(x)$ it is understood that the map $Sq^j(x)$ is given by a Künneth-Steenrod operation.

In case p is odd various elements D_i are trivial and we choose $s(D_i) = 0$ to be the trivial map if $D_i = 0$. According to (8.5.6) the map

$$(-1)^j (\vartheta_q)^{-1} s(D_{(q-2j)(p-1)}) = (sP^j)_q : Z^q \longrightarrow Z^{q+2j(p-1)} \tag{10.2.5}$$

represents the Steenrod operation P^j for $j \geq 0$.

Here we may assume that $(sP^0)_q = id$, the identity of Z^q if $j = 0$ and if q is even, and if $j = q/2$ then $(sP^j)_q = U : Z^q \to Z^{pq}$ is the power map with $U(x) = x^p$ for $x \in Z^q$. In this case K_q is a track under $Z^1 \vee Z^q$. Again we set, see (5.5.1) and (1.1.6),

(1) $$(sP^j)_q = 0 \text{ for } 2j > q.$$

We call the sequence of maps $(sP^j)_q$ with $j \in \mathbb{Z}$ the *Künneth-Steenrod operations* (associated to K_q). We write for a map $x : X \to Z^q$ in **Top*** with $| x |= q$,

(2) $$P^j(x) = s(P^j)_q \circ x : X \longrightarrow Z^q \longrightarrow Z^{q+2j(p-1)}.$$

This composite denotes a map in **Top***. Also for p odd the map

(10.2.6) $$(-1)^j(\vartheta_q)^{-1}s(D_{(q-2j)(p-1)-1}) = s(P^j_\beta)_q : Z^q \longrightarrow Z^{q+2j(p-1)+1}$$

is part of the Künneth track. For $j = 0$ we may assume that this map represents the Bockstein operation map β, that is

(1) $$(\vartheta_q)^{-1}s(D_{q(p-1)-1}) = (s\beta)_q : Z^q \longrightarrow Z^{q+1}$$

is an element in the contractible groupoid $\underline{\underline{\beta}}$ in (2.1.11). There is a *Bockstein track*

(2) $$s(P^j_\beta)_q \Longrightarrow (s\beta)_{q+2j(p-1)}(sP^j)_q$$

where the right-hand side is given by (10.2.5). According to (10.2.6) we set

(3) $$s(P^j_\beta)_q = 0 \text{ for } 1 + 2j > q.$$

Compare the condition of *instability* in (1.1.6). We call the maps $s(P^j_\beta)$ also a *Künneth-Steenrod operation* (associated to K_q). Again we write for a map $x : X \to Z^q$ in **Top*** with $| x |= q$,

(4) $$\beta(x) = (s\beta)_q \circ x : X \longrightarrow Z^q \longrightarrow Z^{q+1},$$

(5) $$P^j_\beta(x) = s(P^j_\beta) \circ x : X \longrightarrow Z^q \longrightarrow Z^{q+2j(p-1)+1}, \ j > 0.$$

Here $\beta(x)$ and $P^j_\beta(x)$ are again maps in **Top***. By (10.2.6) we see that $P^j_\beta(x)$ plays a similar role as $P^j(x)$. The Bockstein track (2) induces the track

(6) $$P^j_\beta(x) \Longrightarrow \beta P^j(x).$$

At this point we do not understand the basic properties of the Bockstein track.

The Künneth tracks K_q are kind of "strings" connecting power maps and maps representing Steenrod operations. We shall use these strings to transform the secondary relations for power maps in Chapter 9. For $\alpha \in E_{\mathcal{A}}$ and $q \geq 1$ we have chosen above maps $s(\alpha)_q$ associated to a Künneth track K_q. In fact, we denote the pair $(s(\alpha)_q, K_q)$ by $s(\alpha)_q$ so that K_q is part of the definition of $s(\alpha)_q$. Therefore we call $s(\alpha)_q$ a Künneth-Steenrod operation.

10.2.7 Proposition. *For $\alpha \in E_{\mathcal{A}}$ let $(s(\alpha)_q, K_q)$ and $(s'(\alpha)_q, K'_q)$ be two different Künneth-Steenrod operations. Then one has a well-defined track*

$$\Gamma_\alpha : s(\alpha)_q \Longrightarrow s'(\alpha)_q.$$

Proof. We have the Künneth tracks

$$\begin{array}{rlcl} K_q : & \sum_{i\geq 0} w_i \boxtimes s(D_i) & \Longrightarrow & \gamma, \\ K'_q : & \sum_{i\geq 0} w_i \boxtimes s'(D_i) & \Longrightarrow & \gamma. \end{array}$$

Hence by (10.1.5) there is a unique track

$$T_i : s(D_i) \Longrightarrow s'(D_i)$$

such that

$$(\sum_i w_i \boxtimes T_i) = (K'_q)^{\mathrm{op}} \square K_q.$$

Now T_i yields the tracks T_α according to (10.2.4) and (10.2.5). □

10.3 Linearity tracks for Künneth-Steenrod operations

For a map $x : X \to Z^q$ we have defined in (10.2.4), (10.2.5), (10.2.6) the Künneth-Steenrod operations

$$\begin{cases} Sq^i(x) \text{ for } p = 2, \\ P^i(x) \text{ and } \beta(x) \text{ and } P^i_\beta(x) \text{ for } p \text{ odd.} \end{cases}$$

These are again maps in $\mathbf{Top}^*$ which are natural in X. Let $x, y : X \to Z^q$ be maps in $\mathbf{Top}^*$.

10.3.1 Theorem. *Künneth tracks induce well-defined tracks*

$$\begin{array}{rlcll} \Gamma^{x,y} & : & Sq^i(x+y) & \Longrightarrow & Sq^i(x) + Sq^i(y) & \text{for } i \leq |x|, \\ \Gamma^{x,y} & : & P^i(x+y) & \Longrightarrow & P^i(x) + P^i(y) & \text{for } 2i \leq |x|, \\ \Gamma^{x,y} & : & \beta(x+y) & \Longrightarrow & \beta(x) + \beta(y), & \\ \Gamma^{x,y} & : & P^i_\beta(x+y) & \Longrightarrow & P^i_\beta(x) + P^i_\beta(y) & \text{for } 2i+1 \leq |x|\,. \end{array}$$

These tracks in $[\![X, Z^*]\!]$ are natural in X. If $Sq^i(x) = U(x)$ or $P^i(x) = U(x)$ is the power map, then $\Gamma^{x,y}$ coincides with $\Gamma_0^{x,y}$ in (8.2.6).

Proof. For maps $v : X \to Z^1$ and $x, y : X \to Z^q$ in $\mathbf{Top}^*$ we have the following composite of tracks which are natural in X, see (8.2.7) and (10.2.3).

(1)
$$\begin{array}{ccc} \gamma_v(x+y) & \xrightarrow{\Gamma_v^{x,y}} & \gamma_v(x) + \gamma_v(y) \\ \uparrow{\scriptstyle K_v(x+y)} & & \uparrow{\scriptstyle K_v(x)+K_v(y)} \\ \sum_i w_i(v) \cdot D_i(x+y) & & \sum_i w_i(v) \cdot (D_i(x) + D_i(y)) \end{array}$$

According to (10.1.3) there is a unique natural track

(2) $$\Gamma^{x,y} : D_i(x+y) \Longrightarrow D_i(x) + D_i(y)$$

which is the coordinate of the composite (1). Using (10.2.4), (10.2.5), (10.2.6) we get the result. □

10.3.2 Definition. *The Künneth linearity track.*

$$\Gamma^{x,y} = \Gamma_i^{x,y} : Sq^i(x+y) \Longrightarrow Sq^i(x) + Sq^i(y)$$

is defined for all $i \geq 0$ by (10.3.1) for $i \leq \mid x \mid$ and by the *delicate linearity track formula*

$$\Gamma_i^{x,y} = \begin{cases} x \cdot y & \text{for } i = \mid x \mid +1, \\ 0 & \text{for } i > \mid x \mid +1. \end{cases}$$

Here $x \cdot y$ is a cup product in $H^*(X)$ representing a track $0 \Longrightarrow 0$, see (10.2.4). Moreover we define the *Künneth linearity tracks*

$$\begin{aligned} \Gamma^{x,y} &= \Gamma_i^{x,y} : P^i(x+y) \Longrightarrow P^i(x) + P^i(y), \\ \Gamma^{x,y} &= \Gamma_{(i)}^{x,y} : P_\beta^i(x+y) \Longrightarrow P_\beta^i(x) + P_\beta^i(y), \end{aligned}$$

for all $i \geq 0$ by (10.3.1) and by $\Gamma_i^{x,y} = 0$ for $2i > \mid x \mid$ and $\Gamma_{(i)}^{x,y} = 0$ for $2i+1 > \mid x \mid$.

We now can transform the relations (9.1.3), (9.1.4) and (9.1.5) and we get:

10.3.3 Theorem. *For $\alpha \in E_\mathcal{A}$ the track*

$$\Gamma^{x,y} : \alpha(x+y) \Longrightarrow \alpha(x) + \alpha(y)$$

in (10.3.1) *satisfies*

(i) $\Gamma^{x,y} = \Gamma^{y,x}$,

(ii) $(\alpha(x) + \Gamma^{y,z}) \square \Gamma^{x,y+z} = (\Gamma^{x,y} + \alpha(z)) \square \Gamma^{x+y,z}$,

(iii) $\Gamma^{x,0} =$ *identity track of* $\alpha(x)$.

This result is similar to properties of linearity tracks $\Gamma_\alpha^{x,y}$ in the secondary Steenrod algebra, see (4.2.5). The proof of (10.3.3), however, relies on (9.2) and the definition (10.3.1). Below we shall compare Künneth linearity tracks $\Gamma^{x,y}$ and stable linearity tracks $\Gamma_\alpha^{x,y}$ in the secondary Steenrod algebra, see section (10.8).

Proof of (10.3.3). Proposition (i) is clear since $\Gamma_v^{x,y} = \Gamma_v^{y,x}$ in (9.1.3). Also (iii) is obvious. Moreover we get (ii) as follows. Consider the commutative diagram where

K denotes appropriate Künneth tracks.

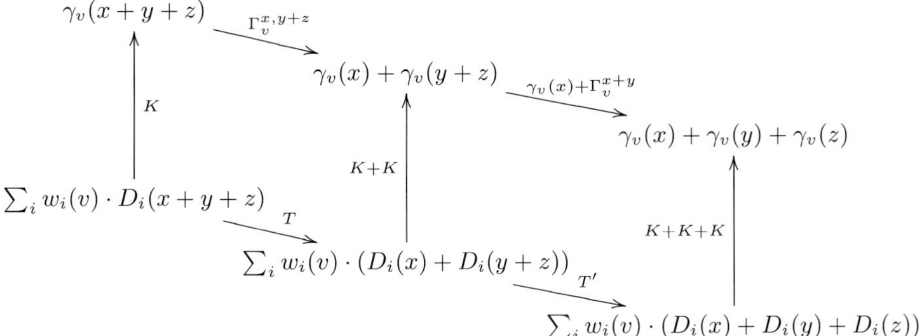

Here T has coordinates $\Gamma^{x,y+z}$ and T' has coordinates $D_i(x)+\Gamma^{y,z}$. The diagram corresponds to the left-hand side of (ii). The right-hand side of (ii) yields a similar diagram and we can apply (9.1.4). This yields (ii) by comparing coordinates of tracks using (10.1.3). $\square$

As in the theorem let $\alpha = Sq^i, P^i, \beta, P^i_\beta$ be a Künneth-Steenrod operation. Given maps $x_i : X \to Z^q$ for $i = 1, 2, \ldots, r$ we define inductively for $r \geq 2$ the natural track

(10.3.4)
$$\begin{aligned} \Gamma^{x_1,\ldots,x_r} &: \alpha(\textstyle\sum_{j=1}^r x_j) \Longrightarrow \sum_{j=1}^r \alpha(x_j), \\ \Gamma^{x_1,\ldots,x_r} &= (\Gamma^{x_1,\ldots,x_{r-1}} + \alpha(x_r))\square\Gamma^{x_1+\cdots+x_{r-1},x_r}. \end{aligned}$$

For $r = 2$ this track coincides with the track in (10.3.1). If $x_1 = \cdots = x_r = x$ we get

(1) $$\Gamma(r)^x = \Gamma^{x\,\ldots,x} : \alpha(rx) \Longrightarrow r\alpha(x).$$

Here $\Gamma(r)^x$ is the identity track and we define

(2) $$\Gamma(-1)^x = \Gamma(p^2-1)^x : \alpha(-x) \Longrightarrow -\alpha(x).$$

10.3.5 Lemma. $\Gamma^{x_1,\ldots,x_r}$ *is the Künneth coordinate of* $\Gamma_v^{x_1,\ldots,x_r}$ *in* (9.1.6). *Also* $\Gamma(r)^x$ *is the Künneth coordinate of* $\Gamma(r)^x_v$ *in* (9.1.7). *Moreover* $\Gamma(-1)^x$ *is the Künneth coordinate of* $L(-1)^x_v$ *in (9.1.8).*

Proof. We only consider $\Gamma(r)^x$. For $r \in \mathbb{Z}/p^2$ the track

$$\Gamma(r) : D_i(r\cdot x) \Longrightarrow r\cdot D_i(x)$$

is the coordinate of the composite in the following diagram.

$$\begin{array}{ccc} \gamma_v(rx) & \xrightarrow{\Gamma(r)^x_v} & r\gamma_v(x) \\ \uparrow{\scriptstyle K_v(rx)} & & \uparrow{\scriptstyle rK_v(x)} \\ \sum_i w_i(v)\cdot D_i(rx) & & \sum_i w_i(v)\cdot(rD_i(x)) \end{array}$$

Here we use (10.2.4), (10.2.5), (10.2.6) to define $\Gamma(r)^x$ for α. By (10.3.3) we see that the coordinate $\Gamma(r)^x$ coincides with (10.3.4)(1). Moreover by (9.1.8) we have $\Gamma(p^2-1)^x_v = L(-1)^x_v$ so that (10.3.4)(2) is the Künneth coordinate of $L(-1)^x_v$. □

10.3.6 Definition. Let $x : X \to Z^q$ in $\mathbf{Top}^*$ and a permutation $\sigma \in \sigma_q$ be given. For a Künneth-Steenrod operation $\alpha = Sq^i, P^i, \beta, P^i_\beta$ we define

$$P(\sigma)^x : \alpha(\sigma x) \Longrightarrow \mathrm{sign}(\sigma)\alpha(x)$$

by the composite ($\epsilon = \mathrm{sign}(\sigma)$)

$$\alpha(\sigma x) \xrightarrow{\alpha(\Gamma_\sigma)} \alpha(\epsilon x) \xrightarrow{\Gamma(\epsilon)^x} \epsilon\alpha(x).$$

Here $\Gamma_\sigma : \sigma \Rightarrow \mathrm{sign}(\sigma)$ is the track in (7.2.2) and $\Gamma(\epsilon)^x$ is defined in (10.3.4)(2). We call $P(\sigma)^x$ the *Künneth permutation track.*

10.3.7 Lemma. *$P(\sigma)^x$ is the Künneth coordinate of $\Gamma_{\sigma^p}\square P(\sigma)^x_v$ in* (9.1.9).

Proof. By (9.1.9) we know

$$\Gamma_{\sigma^p}\square P(\sigma)^x_v = L(\epsilon)^x_v\square\gamma_v(\Gamma_\sigma)$$

where $L = L(\epsilon)^x_v$. The Künneth coordinate of this track is the coordinate of the following composite.

$$\begin{array}{ccc}
\gamma_v(\sigma x) & \xrightarrow{L\square\gamma_v(\Gamma_\sigma)} & \epsilon\gamma_v(x) \\
{\scriptstyle K_v(\sigma x)}\uparrow & & \uparrow{\scriptstyle \epsilon K_v(x)} \\
\sum_i w_i(v)\cdot D_i(\sigma x) & & \sum_i w_i(v)\cdot(\epsilon D_i(x))
\end{array}$$

Now one readily checks that the coordinate of this composite is $P(\sigma)^x$ defined in (10.3.6). For this we use the diagonal of $K_v(\Gamma_\sigma)$. □

10.4 Cartan tracks for Künneth-Steenrod operations

For a map $x : X \to Z^q$ in $\mathbf{Top}^*$ we have defined in (10.2.4), (10.2.5), (10.2.6) the Künneth-Steenrod operations $\alpha(x)$ for $\alpha \in E_{\mathcal{A}}$, that is,

$$\begin{cases} Sq^i(x) & \text{for} \quad p=2, \\ P^i(x),\ \beta(x) \text{ and } P^i_\beta(x) & \text{for} \quad p \text{ odd}. \end{cases}$$

These again are maps in $\mathbf{Top}^*$ which are natural in X. Let $x : X \to Z^q$ and $y : X \to Z^q$ be maps in $\mathbf{Top}^*$.

10.4.1 Theorem. *Künneth tracks induce well-defined tracks* ($n \geq 1$)

$$\begin{aligned}
C^{x,y} : Sq^n(x \cdot y) &\Longrightarrow \sum_{i+j=n} Sq^i(x) \cdot Sq^j(y),\\
C^{x,y} : P^n(x \cdot y) &\Longrightarrow \sum_{i+j=n} P^i(x) \cdot P^j(y),\\
C^{x,y} : \beta(x \cdot y) &\Longrightarrow \beta(x) \cdot y + (-1)^{|x|} x \cdot \beta(y),\\
C^{x,y} : P^n_\beta(x \cdot y) &\Longrightarrow \sum_{i+j=n, i,j \geq 0} (P^i_\beta(x) \cdot P^j(y) + P^i(x) \cdot P^j_\beta(y)).
\end{aligned}$$

Here we have $P^0(x) = x$ *and* $P^0_\beta(x) = \beta(x)$. *These tracks in* $[\![X, Z^*]\!]$ *are natural in* X.

We call $C^{x,y}$ the *Künneth-Cartan track*. For the maps $w_i : Z^1 \to Z^i$ in (10.2.1) we need the following result. Let

$$\epsilon_{i,j} = \begin{cases} 0 & \text{if } p, i, j \text{ are odd,} \\ 1 & \text{otherwise.} \end{cases}$$

10.4.2 Proposition. *There is a well-defined track* ($i, j \geq 0$)

$$W_{i,j} : \epsilon_{i,j} w_{i+j} \Longrightarrow w_i \cdot w_j$$

for which the following diagrams (*i*), (*ii*) *commute.*

(i)

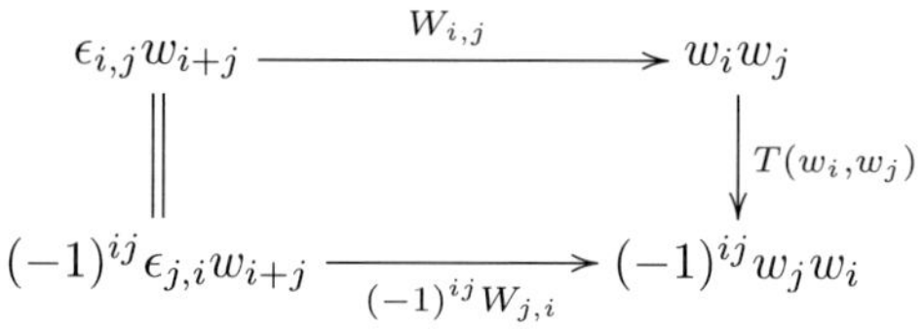

$$\begin{array}{ccc}
\epsilon_{i,j} w_{i+j} & \xrightarrow{W_{i,j}} & w_i w_j \\
\Big\| & & \Big\downarrow {\scriptstyle T(w_i, w_j)} \\
(-1)^{ij} \epsilon_{j,i} w_{i+j} & \xrightarrow[(-1)^{ij} W_{j,i}]{} & (-1)^{ij} w_j w_i
\end{array}$$

(ii)

$$\begin{array}{ccccc}
 & & \epsilon_{j,k} w_i w_{j+k} & & \\
 & \swarrow {\scriptstyle w_i W_{j,k}} & & \nwarrow {\scriptstyle \epsilon_{j,k} W_{i,j+k}} & \\
w_i w_j w_k & & & & \epsilon_{i,j,k} w_{i+j+k} \\
 & \nwarrow {\scriptstyle W_{i,j} w_k} & & \swarrow {\scriptstyle \epsilon_{i,j} W_{i+j,k}} & \\
 & & \epsilon_{i,j} w_{i+j} w_k & &
\end{array}$$

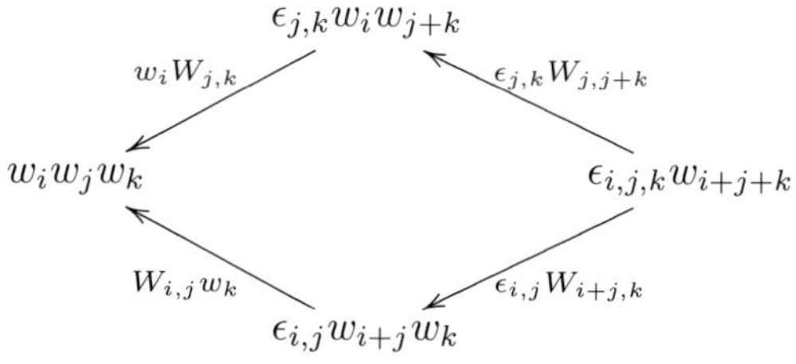

Here we use the equation

(iii) $$\epsilon_{i,j,k} = \epsilon_{i+j,k} \cdot \epsilon_{i,j} = \epsilon_{i,j+k} \cdot \epsilon_{j,k}.$$

Proof. By definition in (10.2.1)(3) we have

(1) $$w_i \cdot w_j = w_{i+j}$$

if $p = 2$ or if p odd and j even or $i = 0$ or $j = 0$. In this case $W_{i,j}$ is the identity track. If p is odd, i even, j odd we get

(2) $$\begin{aligned} w_i \cdot w_j &= (-\beta x)^{i/2} \cdot x \cdot (-\beta x)^{(j-1)/2}, \\ w_{i+j} &= x \cdot (-\beta x)^{(i+j-1)/2}, \end{aligned}$$

with $| (-\beta x)^{i/2} |= i$ even. Hence in this case the interchange track (6.3.1)(7) yields

(3) $$W_{i,j} = T(w_i, w_j)^{\mathrm{op}}.$$

Finally if p, i, j are odd we get the track

(4) $$W_{i,j} : 0 \Longrightarrow w_i \cdot w_j$$

as follows. For x in (10.2.1)(3) we have $| x |= 1$. Hence we get the interchange track

$$T(x, x) : x \cdot x \Longrightarrow -x \cdot x$$

which yields the track

$$T = (T(x, x) + x \cdot x)^{\mathrm{op}} : 0 \Longrightarrow 2x \cdot x.$$

This yields for p odd the track

$$T' = \frac{p+1}{2} T : 0 \Longrightarrow (p + 1)(x \cdot x) = x \cdot x$$

since $p(x \cdot x) = 0$. Moreover we have the following diagram.

$$\begin{array}{c} w_i \cdot w_j \;=\; x \cdot (-\beta x)^{(i-1)/2} \cdot x \cdot (-\beta x)^{(j-1)/2} \\ \Big\Uparrow {\scriptstyle x \cdot T(x, -(\beta x)^{(i-1)/2}) \cdot (-\beta x)^{(j-1)/2}} \\ x \cdot x \cdot (-\beta x)^{(i-1)/2} \cdot (-\beta x)^{(j-1)/2} \\ \Big\Uparrow {\scriptstyle T' \cdot (-\beta x)^{(i+j-2)/2}} \\ 0 \end{array}$$

The composite of these tracks is $W_{i,j}$ in (4). Now one can check that the diagrams commute. □

10.4.3 *Proof of* (10.4.1). Let $v : X \to Z^1$, $x : X \to Z^q$ and $y : X \to Z^{q'}$ be maps in **Top*** with $| x |= q$ and $| y |= q'$. The permutation $\sigma(x, y) = \sigma$ is defined as in (8.3.1) with

(1) $$\epsilon(x, y) = \text{sign } \sigma(x, y) = (-1)^{|x| \cdot |y| (p-1)p/2}.$$

With the notation in (10.4.1) let

(2) $$\bar{\epsilon}_{i,j} = (-1)^{j(p|x|-i)}\epsilon_{i,j}.$$

Then the *Künneth-Cartan track*

(3) $$C^{x,y}_{(n)} : \epsilon(x,y)D_n(x\cdot y) \Longrightarrow \sum_{i+j=n,i,j\geq 0} \bar{\epsilon}_{i,j} D_i(x)\cdot D_j(y)$$

is defined as the coordinate of the composite tracks $R^{x,y}$ in the following diagram. Here $C^{x,y}_v$ is the Cartan track (8.3.3) and $K_v(x\cdot y)$, $K_v(x)$, $K_v(y)$ are Künneth tracks.

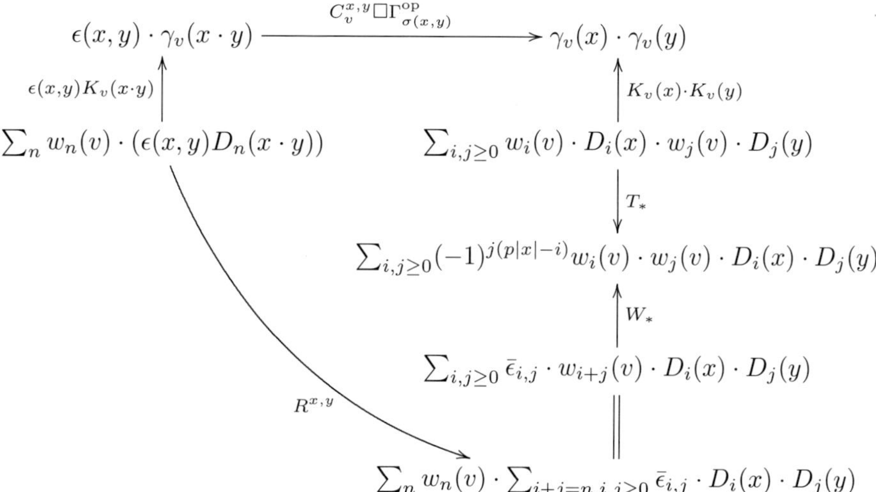

Here $\Gamma_{\sigma(x,y)}$ is given by (7.2.2) and we define T_* by use of the interchange tracks (6.3.1)(7), that is

(4) $$T_* = \sum_{i,j\geq 0} w_i(v)\cdot T(D_i(x), w_j(v))\cdot D_j(y).$$

Moreover we define W_* by the tracks in (10.4.2)

(5) $$W_* = \sum (-1)^{j(p|x|-i)} W_{i,j}\cdot D_i(x)\cdot D_j(y).$$

This completes the definition of the diagram.

Recall that by (10.2.4) we have for $p=2$ the Künneth-Steenrod operation

$$s(D_{2q-i}) = s(Sq^i)_q : Z^q \longrightarrow Z^{q+i} \text{ in } \mathbf{Top}^*$$

and for $x : X \to Z^q$ with $|\, x\,| = q$ we write

$$Sq^i x = (sSq^i)_q \circ x : X \longrightarrow Z^{q+i} \text{ in } \mathbf{Top}^*.$$

We now obtain the *Künneth-Cartan track*

$$C^{x,y} : Sq^n(x \cdot y) \Longrightarrow \sum_{i+j=n} (Sq^i x) \cdot (Sq^j y) \tag{10.4.4}$$

by the composite in the following diagram where $m = 2 \mid x \mid +2 \mid y \mid -n$, $r = 2 \mid x \mid -i$ and $s = 2 \mid y \mid -j$.

$$\begin{array}{ccc}
Sq^n(x \cdot y) & & \sum_{i+j=n}(Sq^i x) \cdot (Sq^j y) \\
\| & & \| \\
D_{2|x \cdot y|-n}(x \cdot y) & & \sum_{i+j=n}(D_{2|x|-i}x) \cdot (D_{2|y|-j}y) \\
\| & & \| \\
D_m(x \cdot y) & \xrightarrow{C^{x,y}_{(m)}} & \sum_{r+s=m}(D_r x) \cdot (D_s y)
\end{array}$$

Here $C^{x,y}_{(m)}$ is defined by (10.4)(3) above and we use the convention that $Sq^i x = 0$ for $i > 2 \mid x \mid$.

Now let p be odd. Then we have by (10.2.5) the Künneth-Steenrod operation

$$(-1)^j (\vartheta_q)^{-1} s(D_{(q-2j)(p-1)}) = (sP^j)_q : Z^q \longrightarrow Z^{q+2j(p-1)}.$$

For $x : X \to Z^q$ with $\mid x \mid = q$ we write

$$\vartheta^j_x = (-1)^j (\vartheta_q)^{-1} \text{ and } \vartheta^x_j = (\vartheta^j_x)^{-1}.$$

Recall that we write

$$P^j x = (sP^j)_q \circ x : X \longrightarrow Z^{q+2j(p-1)} \text{ in } \mathbf{Top}^*.$$

Then we define the *Künneth-Cartan track*

$$C^{x,y} : P^n(x \cdot y) \Longrightarrow \sum_{i+j=n} (P^i x) \cdot P^j y) \tag{10.4.5}$$

as follows. For $m = (|x \cdot y| - 2n)(p-1)$, $r = (\mid x \mid -2i)(p-1)$, $s = (\mid y \mid -2j)(p-1)$ we get $C^{x,y}$ by the following composite.

$$\begin{array}{c}
P^n(x \cdot y) = \vartheta^n_{xy} D_m(x \cdot y) \\
\Big\downarrow {\scriptstyle \epsilon(x,y) \cdot \vartheta^n_{xy} \cdot C^{x,y}_{(m)}} \\
\epsilon(x,y)\vartheta^n_{xy} \sum_{r+s=m, r,s \text{ even}} (-1)^{s(p|x|-r)} D_r(x) \cdot D_s(y) \\
\| \\
\sum_{i+j=n} \epsilon(x,y)\vartheta^n_{xy}\vartheta^x_i(P^i x) \cdot \vartheta^y_j(P^j y) \\
\| \\
\sum_{i+j=n}(P^i x) \cdot (P^j y)
\end{array}$$

Here we use the fact that

$$\epsilon(x,y)\vartheta^n_{xy}\vartheta^x_i \cdot \vartheta^y_j = 1$$

for $n = i + j$, compare (1.2.11)(4). We use the convention that $sD_i = 0$ if $D_i = 0$. For this reason many summands $D_r(x) \cdot D_s(y)$ are trivial.

Moreover for p odd we have by (10.2.6) the Künneth-Bockstein operation

$$(\vartheta_q)^{-1}s(D_{q\cdot(p-1)-1}) = (s\beta)_q : Z^q \longrightarrow Z^{q+1}.$$

For $x : X \to Z^q$ with $| x |= q$ we set

$$\beta x = (s\beta)_q \circ x : X \longrightarrow Z^q \longrightarrow Z^{q+1} \text{ in } \mathbf{Top}^*.$$

Then we define the *Künneth-Cartan track*

$$C^{x,y} : \beta(xy) \Longrightarrow (\beta x)\cdot y + (-1)^{|x|}x\cdot(\beta y) \tag{10.4.6}$$

as follows. Let $m =| xy | (p-1) - 1$ and $\vartheta_x = \vartheta_q$ for $| x |= q$. Moreover let

$$\begin{aligned} r_1 &=| x | (p-1) - 1 \quad , \quad s_1 =| y | (p-1),\\ r_2 &=| x | (p-1) \quad , \quad s_2 =| y | (p-1) - 1. \end{aligned}$$

Then we get $C^{x,y}$ by the following composite.

$$\begin{array}{c}
\beta(xy) = \vartheta^{-1}_{xy}D_m(xy)\\
\Big\downarrow {\scriptstyle \epsilon(x,y)\cdot\vartheta^{-1}_{xy}\cdot C^{x,y}_{(m)}}\\
\epsilon(x,y)\vartheta^{-1}_{xy}\sum_{r+s=m, r \text{ or } s \text{ even}}(-1)^{s(p|x|-r)}D_r(x)\cdot D_s(y)\\
\Big\|\\
\epsilon(x,y)\vartheta^{-1}_{xy}(-1)^{s_1(p|x|-r_1)}D_{r_1}(x)\cdot D_{s_1}(y)\\
+\epsilon(x,y)\vartheta^{-1}_{xy}(-1)^{s_2(p|x|-r_2)}D_{r_2}(x)\cdot D_{s_2}(y)\\
\Big\|\\
\epsilon(x,y)\vartheta^{-1}_{xy}\vartheta_x\cdot\beta(x)\cdot\vartheta_y\cdot y\\
+\epsilon(x,y)\vartheta^{-1}_{xy}(-1)^{s_2(p|x|-r_2)}\vartheta_x\cdot x\cdot\vartheta_y\cdot\beta(y)\\
\Big\|\\
\beta(x)\cdot y + (-1)^{|x|}x\cdot\beta(y)
\end{array}$$

By (10.4.4), (10.4.5) and (10.4.6) the Cartan tracks $C^{x,y}$ in (10.4.1) are well defined. □

10.5 The interchange relation for Cartan tracks

For the product of maps $x : X \to Z^q$, $y : X \to Z^{q'}$ in **Top*** we have the *interchange track*

(10.5.1)
$$\begin{array}{rcl} T(x,y) & : & x \cdot y \longrightarrow (-1)^{|x||y|} y \cdot x = (-1)^{p|x||y|} y \cdot x, \\ T(x,y) & = & \Gamma_{\tau(y,x)}(y \cdot x) \in [\![X, Z^{q+q'}]\!]. \end{array}$$

Here $\tau(y,x)$ is the permutation with $\tau(y,x) y \cdot x = x \cdot y$. Compare (6.3.1)(7). We now describe the connection between Künneth-Cartan tracks $C^{x,y}$ and $C^{y,x}$ in (10.4.1). For this we consider the following diagrams for $p = 2$ and p odd respectively.

(10.5.2)
$$\begin{array}{ccc} Sq^n(x \cdot y) & \xrightarrow{C^{x,y}} & \sum_{i+j=n} Sq^i(x) \cdot Sq^j(y) \\ {\scriptstyle Sq^n T(x,y)} \downarrow & & \downarrow {\scriptstyle \sum_{i+j=n} T(Sq^i(x), Sq^j(y))} \\ Sq^n(y \cdot x) & \xrightarrow{C^{y,x}} & \sum_{i+j=n} Sq^j(y) \cdot Sq^i(x) \end{array}$$

(10.5.3)
$$\begin{array}{ccc} P^n(x \cdot y) & \xrightarrow{C^{x,y}} & \sum_{i+j=n} P^i(x) \cdot P^j(y) \\ {\scriptstyle P^n T(x,y)} \downarrow & & \\ P^n(\epsilon y \cdot x) & & \downarrow {\scriptstyle \sum_{i+j=n} T(P^i x, P^j y)} \\ {\scriptstyle \Gamma(\epsilon)^{y \cdot x}} \downarrow & & \\ \epsilon P^n(y \cdot x) & \xrightarrow[\epsilon C^{y,x}]{} & \sum_{j+i=n} \epsilon P^j(y) \cdot P^i(x) \end{array}$$

Here we set $\epsilon = (-1)^{|x||y|} = (-1)^{|P^i x||P^j y|}$ since p is odd.

(10.5.4)
$$\begin{array}{ccc} \beta(x \cdot y) & \xrightarrow{C^{x,y}} & (\beta x) \cdot y + (-1)^q x \cdot (\beta y) \\ {\scriptstyle \beta T(x,y)} \downarrow & & \downarrow {\scriptstyle T(\beta x, y) + (-1)^q T(x, \beta y)} \\ \beta(\epsilon y \cdot x) & & (-1)^{(q+1)q'} y \cdot (\beta x) + (-1)^{q+q(q'+1)} (\beta y) \cdot x \\ {\scriptstyle \Gamma(\epsilon)^{y \cdot x}} \downarrow & & \| \\ \epsilon \beta(y \cdot x) & \xrightarrow[\epsilon C^{y,x}]{} & \epsilon(\beta y) \cdot x + \epsilon(-1)^{q'} y \cdot (\beta x) \end{array}$$

There is a similar diagram for $P^n_\beta(x \cdot y)$.

10.5.5 Theorem. *The interchange relations* (10.5.2), (10.5.3) *and* (10.5.4) *above are commutative diagrams of tracks in* $[\![X, Z^*]\!]$.

By definition in (10.4) the Künneth-Cartan track $C^{x,y}$ is a Künneth coordinate of the composite

(10.5.6) $$\bar{C}^{x,y}_v : \epsilon(x,y)\gamma_v(xy) \xrightarrow{\Gamma^{\mathrm{op}}_{\sigma(x,y)}} \sigma(x,y)\gamma_v(xy) \xrightarrow{C^{x,y}_v} \gamma_v(x)\cdot\gamma_v(y).$$

We now use the relation (9.5.3)(i) for studying the following diagram where $\epsilon(x,y) = \epsilon(y,x)$ and $\epsilon = (-1)^{p|x|\cdot|y|} = (-1)^{|x^p|\cdot|y^p|}$.

$$\begin{array}{ccc}
\epsilon(x,y)\gamma_v(xy) & \xrightarrow{\bar{C}^{x,y}_v} & \gamma_v(x)\cdot\gamma_v(y) \\
\downarrow{\scriptstyle \epsilon(x,y)\gamma_v(T(x,y))} & & \\
\epsilon(x,y)\gamma_v(\epsilon yx) & & \downarrow{\scriptstyle T(\gamma_v x,\gamma_v y)} \\
\downarrow{\scriptstyle \epsilon(x,y)\Gamma(\epsilon)^{yx}_v} & & \\
\epsilon(x,y)\epsilon\gamma_v(yx) & \xrightarrow[\epsilon\bar{C}^{y,x}_v]{} & \epsilon\gamma_v(y)\gamma_v(x)
\end{array}$$

10.5.7 Lemma. *This diagram is commutative.*

Proof. The commutative diagram (9.5.3)(i) is given as follows.

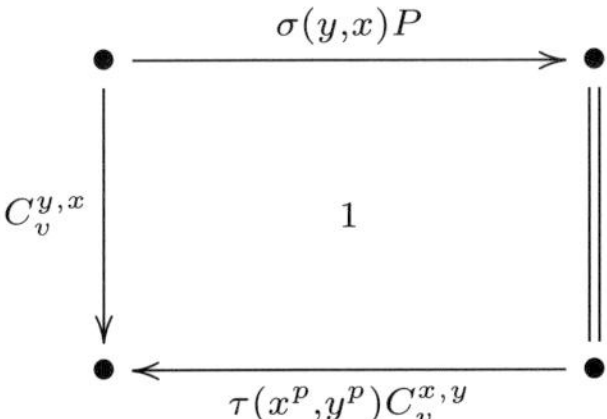

This diagram is embedded into the following commutative diagram with $\tau' = \tau(x^p, y^p)$, $\tau'' = \tau(x,y)^p$ and $\bar{\epsilon} = \epsilon(x,y) = \epsilon(y,x)$ and $\bar{\sigma} = \sigma(y,x)$ and $\sigma = \sigma(x,y)$.

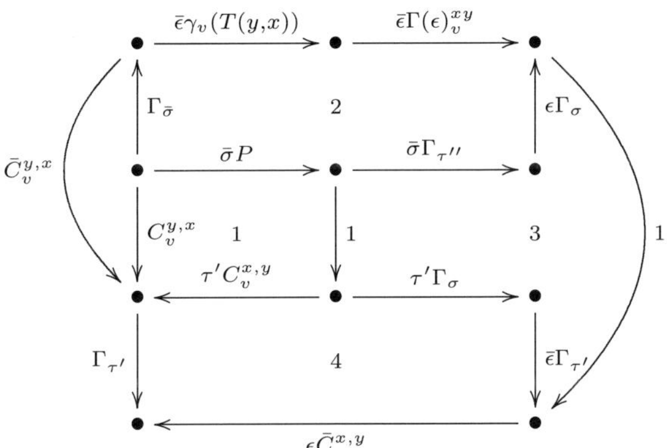

Here 2 commutes by (9.4.3) and 3 commutes by (9.4.1)(1) and 4 commutes since 4 corresponds to the pasting $\Gamma_{\tau'} * \bar{C}^{x,y}_v$. □

Proof of (10.5.5). We show that the following diagram commutes, compare (10.4)(3).

$$(1)\qquad \begin{array}{ccc} \epsilon(x,y)D_n(xy) & \xrightarrow{C^{x,y}_{(n)}} & \sum_{i,j}\bar\epsilon_{i,j}D_i(x)\cdot D_j(y) \\ \downarrow{\scriptstyle \epsilon(x,y)D_n(T(x,y))} & & \\ \epsilon(x,y)D_n(\epsilon yx) & & \downarrow{\scriptstyle \sum_{i,j}\bar\epsilon_{i,j}T(D_ix,D_jy)} \\ \downarrow{\scriptstyle \epsilon(x,y)\Gamma(\epsilon)^{yx}} & & \\ \epsilon(x,y)\epsilon D_n(yx) & \xrightarrow[\epsilon C^{y,x}_{(n)}]{} & \sum_{i,j}\bar\epsilon_{j,i}\epsilon D_j(y)\cdot D_i(x) \end{array}$$

In fact, the left-hand side of this diagram is the Künneth coordinate of the left-hand side of diagram (10.5.7). Therefore it remains to check that the right-hand side of the diagram above is the Künneth coordinate of $T(\gamma_v x,\gamma_v y)$ in (10.5.7). But this is a consequence of the following commutative diagram with $a = w_i(v)$, $b = D_i(x)$, $c = w_j(v)$, $d = D_j(y)$ and $\pm = (-1)^{|ab|\cdot|cd|}$.

$$(2)\qquad \begin{array}{ccc} abcd & \xrightarrow{T(ab,cd)} & \pm cdab \\ \downarrow{\scriptstyle aT(b,c)d} & & \downarrow{\scriptstyle \pm cT(d,a)b} \\ (-1)^{|b|\cdot|c|}acbd & \xrightarrow[T(a,c)T(b,d)]{} & \pm cadb \end{array}$$

For $T(a,c)$ we need the commutative diagram in (10.4.2)(i). This completes the proof that (1) is commutative. From (1) we deduce the result in (10.5.5) by definition of $C^{x,y}$. □

10.6 The associativity relation for Cartan tracks

For the Künneth-Cartan tracks $C^{x,y}$ we obtain the following diagrams. Let $x : X \to Z^q$, $y : X \to Z^{q'}$, $z : X \to Z^{q''}$ be maps in **Top***.

$$(10.6.1)\qquad \begin{array}{ccc} Sq^m(xyz) & \xrightarrow{C^{xy,z}_m} & \sum_{n+k=m} Sq^n(xy)Sq^k(z) \\ \downarrow{\scriptstyle C^{x,yz}_m} & & \downarrow{\scriptstyle \sum_{n,k} C^{x,y}_n\cdot Sq^k(z)} \\ \sum_{i+r=m} Sq^i(x)Sq^r(yz) & \xrightarrow[\sum_{i,r} Sq^i(x)\cdot C^{y,z}_r]{} & \sum_{i+j+k=m} Sq^i(x)Sq^j(y)Sq^k(z) \end{array}$$

$$(10.6.2)\qquad \begin{array}{ccc} P^m(xyz) & \xrightarrow{C^{xy,z}_m} & \sum_{n+k=m} P^n(xy)P^k(z) \\ \downarrow{\scriptstyle C^{x,yz}_m} & & \downarrow{\scriptstyle \sum_{n,k} C^{x,y}_n\cdot P^k(z)} \\ \sum_{i+r=m} P^i(x)P^r(yz) & \xrightarrow[\sum_{i,r} P^i(x)\cdot C^{y,z}_r]{} & \sum_{i+j+k=m} P^i(x)P^j(y)P^k(z) \end{array}$$

(10.6.3)
$$\begin{array}{ccc}
\beta^m(xyz) & \xrightarrow{C^{xy,z}} & \beta(xy)\cdot z+(-1)^{|xy|}xy\cdot\beta(z) \\
\Big\downarrow{\scriptstyle C^{x,yz}} & & \Big\downarrow{\scriptstyle C^{x,y}\cdot z+(-1)^{|xy|}xy\beta(z)} \\
\beta(x)\cdot yz(-1)^{|x|}x\beta(yz) & \xrightarrow[\beta(x)\cdot yz+(-1)^{|x|}x\cdot C^{y,z}]{} & \begin{array}{c}\beta(x)\cdot yz+(-1)^{|x|}x\cdot\beta(y)z \\ +(-1)^{|xy|}xy\cdot\beta(z)\end{array}
\end{array}$$

There is a similar diagram for $P^m_\beta(xyz)$.

10.6.4 Theorem. *The associativity relations* (10.6.1), (10.6.2) *and* (10.6.3) *are commutative diagrams of tracks in* $[\![X,Z^*]\!]$.

We use the notation $\bar{C}^{x,y}_v$ in (10.5.6) so that we derive from (9.5.3)(ii) the following diagram with $\epsilon(x,y,z)=\operatorname{sign}\sigma(x,y,z)$ and $\epsilon(x,y)=\operatorname{sign}\sigma(x,y)$; see (9.5.2).

$$\begin{array}{ccc}
\epsilon(x,y,z)\cdot\gamma_v(xyz) & \xrightarrow{\epsilon(x,y)\bar{C}^{xy,z}_v} & \epsilon(x,y)\cdot\gamma_v(xy)\cdot\gamma_v(z) \\
\Big\downarrow{\scriptstyle \epsilon(y,z)\bar{C}^{x,yz}_v} & & \Big\downarrow{\scriptstyle \bar{C}^{x,y}_v\cdot\gamma_v(z)} \\
\epsilon(y,z)\cdot\gamma_v(x)\cdot\gamma_v(yz) & \xrightarrow[\gamma_v(x)\cdot\bar{C}^{y,z}_v]{} & \gamma_v(x)\cdot\gamma_v(y)\cdot\gamma_v(z)
\end{array}$$

10.6.5 Lemma. *This diagram is commutative.*

Proof. We use (9.5.3)(ii) and (10.5.6) and (9.4.1)(1). □

Proof of (10.6.4). Recall the definition of $C^{x,y}_{(n)}$ in (10.4)(3). We show that the following diagram commutes.
(1)

$$\begin{array}{ccc}
\epsilon(x,y,z)D_m(xyz) & \xrightarrow{\epsilon(x,y)C^{xy,z}_{(m)}} & \epsilon(x,y)\sum_{n+k=m}\epsilon_{n,k}D_n(xy)D_k(z) \\
\Big\downarrow{\scriptstyle \epsilon(y,z)C^{x,yz}_{(m)}} & & \Big\downarrow{\scriptstyle \sum_{n,k}\bar{\epsilon}_{n,k}C^{x,y}_{(n)}D_k(z)} \\
\epsilon(y,z)\sum\limits_{i+r=m}\bar{\epsilon}_{i,r}D_i(x)D_r(yz) & \xrightarrow[\sum_{i,r}\bar{\epsilon}_{i,r}D_i(x)C^{y,z}_{(r)}]{} & \sum\limits_{i+j+k=m}\bar{\epsilon}_{i,j,k}D_i(x)D_j(y)D_k(z)
\end{array}$$

Here we set

(2)
$$\bar{\epsilon}_{i,j,k}=\bar{\epsilon}_{i+j,k}\cdot\bar{\epsilon}_{i,j}=\bar{\epsilon}_{i,j+k}\cdot\bar{\epsilon}_{j,k}.$$

This equation readily can be checked by (10.4)(2) and the definition of $\epsilon_{i,j}$. According to the definition of $C^{x,y}$ we see that commutativity of diagram (1) above implies the proposition in (10.6.4).

We obtain diagram (1) as the Künneth coordinate of the tracks in diagram (10.6.5), compare the definition of $C^{x,y}_{(n)}$ in (10.4). For this we multiply $K_v(x)$ with the diagram in (10.4) defining $R^{y,z}$.

This yields the following commutative diagram.

(3)

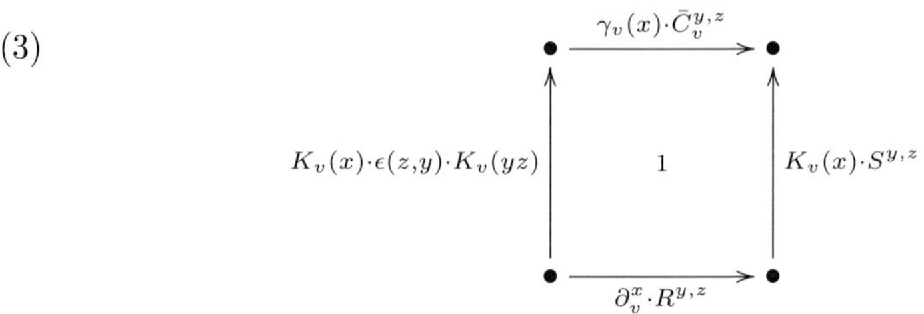

Here ∂_v^x is the source of $K_v(x)$ and $S^{y,z}$ is the composite

(4) $$S^{y,z} = (K_v(y)\cdot K_v(z))\square(T_*^{y,z})^{\mathrm{op}}\square W_*^{y,z}$$

given by the right-hand side of the diagram in (10.4) where we replace (x,y) by (y,z). We embed diagram 1 into the following commutative diagram of tracks.

(5)

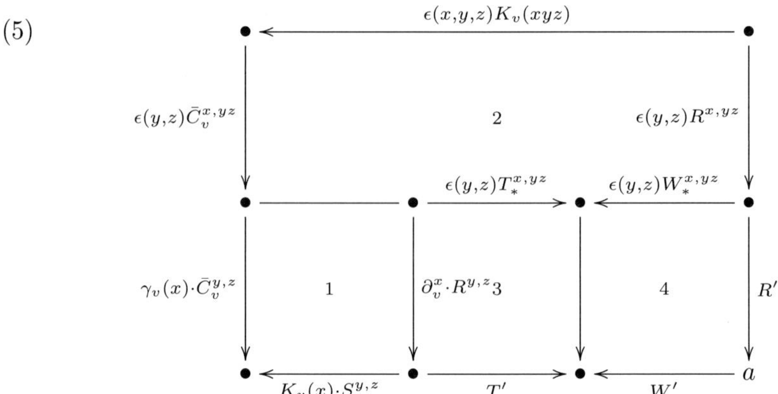

Diagram 2 is given by the diagram in (10.4), where we replace (x,y) by (x,yz), multiplied by $\epsilon(y,z)$. The object a in diagram (5) is

(6) $$a = \sum_m w_m(v)\cdot\left(\sum_{i+j+k=m}\bar{\epsilon}_{i,j,k}D_i(x)\cdot D_j(y)\cdot D_k(z)\right)$$

and the track R' has coordinates given by the bottom arrow in (1).

Hence the diagram (1) is commutative if and only if the following diagram is commutative.

(7)

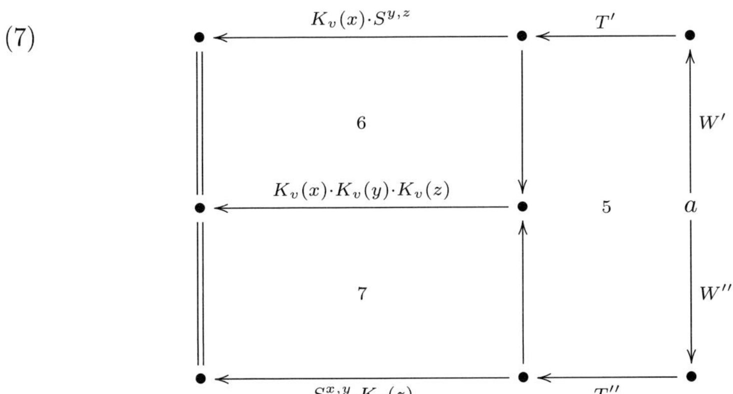

Here we define W'', T'' by the diagram similar to (5) which corresponds to the top arrow and the right-hand side of (10.6.5). Subdiagrams 6 and 7 are the obvious commutative diagrams. Moreover 5 commutes if and only if for

$$a = D_i x,\ b = D_i y,\ c = D_k z$$

the following diagram (8) commutes. We set $w_i = w_i(v)$ and we use the interchange tracks and the tracks $W_{i,j}$ in (10.4.2). Moreover we indicate signs of the coefficients $\bar{\epsilon}_{i,j}$ by $\pm$.

(8)

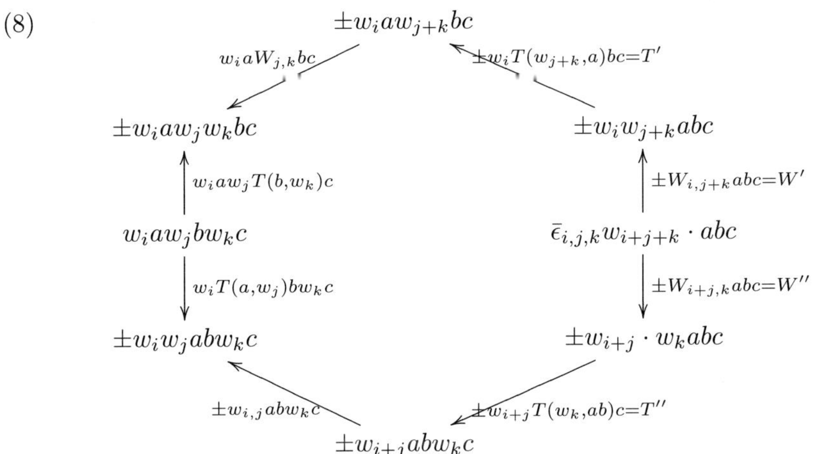

If $p = 2$ then $W_{i,j}$ is the identity track and in this case it is easy to see that (8) commutes. If p is odd one can check the commutativity of (3) by the definition of $W_{i,j}$ in (10.4.2) and by diagram (10.4.2)(ii). □

10.7 The linearity relation for Cartan tracks

For the Künneth-Cartan tracks $C^{x,y}$ in (10.4) and for the Künneth linearity tracks $\Gamma^{x,y}$ in (10.3) we obtain the following diagrams. Let $x, x' : X \to Z^q$ and $y : X \to Z^{q'}$ be maps in **Top***.

(10.7.1)
$$\begin{array}{ccc}
Sq^n((x+x')y) & \xrightarrow{\;C^{x+x',y}\;} & \sum_{i+j=n} Sq^i(x+x')\cdot Sq^j(y) \\
\Big\downarrow{\scriptstyle \Gamma_n^{xy,x'y}} & & \Big\downarrow{\scriptstyle \sum_{i,j}\Gamma_i^{x,x'}\cdot Sq^j(y)} \\
Sq^n(xy)+Sq^n(x'y) & \xrightarrow{\;C^{x,y}+C^{x',y}\;} & \sum_{i=j=n}(Sq^i(x)+Sq^i(x'))\cdot Sq^j(y)
\end{array}$$

(10.7.2)
$$\begin{array}{ccc}
P^n((x+x')y) & \xrightarrow{\;C^{x+x',y}\;} & \sum_{i+j=n} P^i(x+x')\cdot P^j(y) \\
\Big\downarrow{\scriptstyle \Gamma_n^{xy,x'y}} & & \Big\downarrow{\scriptstyle \sum_{i,j}\Gamma_i^{x,x'}\cdot P^j(y)} \\
P^n(xy)+P^n(x'y) & \xrightarrow{\;C^{x,y}+C^{x',y}\;} & \sum_{i+j=n}(P^i(x)+P^i(x'))\cdot P^j(y)
\end{array}$$

(10.7.3)
$$\begin{array}{ccc}
\beta((x+x')y) & \xrightarrow{\;C^{x+x',y}\;} & \beta(x+x')\cdot y+(-1)^{|x|}(x+x')\beta(y) \\
\Big\downarrow{\scriptstyle \Gamma^{xy,x'y}} & & \Big\downarrow{\scriptstyle \Gamma^{x,x'}\cdot y+(-1)^{|x|}(x+x')\beta(y)} \\
\beta(xy)+\beta(x'y) & \xrightarrow{\;C^{x,y}+C^{x',y}\;} & (\beta(x)+\beta(x'))\cdot y+(-1)^{|x|}(x+x')\beta(y)
\end{array}$$

There is a similar diagram for $P^n_\beta((x+x')y)$.

10.7.4 Theorem. *The relation* (10.7.1) *is a commutative diagram of tracks. The relations* (10.7.2) *and* (10.7.3) *are commutative diagrams of tracks in* $[\![X, Z^*]\!]$.

We again use the notation $\bar{C}_v^{x,y}$ in (10.5.6) so that we derive from (9.6.3) the following diagram.

$$\begin{array}{ccc}
\epsilon(x,y)\gamma_v((x+x')y) & \xrightarrow{\;\bar{C}_v^{x,y}\;} & \gamma_v(x+x')\gamma_v(y) \\
\Big\downarrow{\scriptstyle \epsilon(x,y)\Gamma_v^{xy,x'y}} & & \Big\downarrow{\scriptstyle \Gamma_v^{x,x'}\cdot\gamma_v(y)} \\
\epsilon(x,y)(\gamma_v(xy)+\gamma_v(x'y)) & \xrightarrow{\;\bar{C}_v^{x,y}+\bar{C}_v^{x',y}\;} & (\gamma_v(x)+\gamma_v(x'))\gamma_v(y)
\end{array}$$

10.7.5 Lemma. *This diagram commutes for p odd and for $p=2$ the primary element of the diagram is $x\cdot x'\cdot Sq^{|y|-1}(y)$.*

Proof. We use (9.6.3) and the track $\Gamma_{\sigma(x,y)}$ and (9.4.1)(2). □

Proof of (10.7.4). It suffices to consider the following diagram, see (10.4).

(1)
$$\begin{array}{ccc} \epsilon(x,y)D_n((x+x')y) & \xrightarrow{C^{x+x',y}_{(n)}} & \sum_{i+j=n}\bar{\epsilon}_{i,j}D_i(x+x')\cdot D_j(y) \\ \Big\downarrow{\scriptstyle \epsilon(x,y)\Gamma^{xy,x'y}_{(n)}} & & \Big\downarrow{\scriptstyle \sum_{i,j}\bar{\epsilon}_{i,j}\Gamma^{x,x'}_{(i)}\cdot D_j(y)} \\ \epsilon(x,y)(D_n(xy)+D_n(x'y)) & \xrightarrow{C^{x,y}_{(n)}+C^{x',y}_{(n)}} & \sum_{i+j=n}\bar{\epsilon}_{i,j}(D_i(x)+D_i(x'))\cdot D_j(y) \end{array}$$

The tracks in this diagram are the Künneth coordinates of the corresponding tracks in diagram (10.7.5). This is seen by (10.4) and by the following commutative diagram.

$$\begin{array}{ccc} (a+a')c & = & ac+a'c \\ \Big\downarrow{\scriptstyle T(a+a',c)} & & \Big\downarrow{\scriptstyle T(a,c)+T(a',c)} \\ \pm c(a+a') & = & \pm ca+\pm ca' \end{array}$$

For this we use (9.4.1)(2). The primary element of diagram (1) is trivial if p is odd or if $p=2$ and $n\neq|xy|$ and hence we get the result in these cases. For $p=2$ and $n=|xy|$, however, diagram (1) yields the following diagram with $|x|=q$, $|y|=q'$, $\alpha=Sq^n$, $\beta=Sq^q$.

(2)
$$\begin{array}{ccc} Sq^n((x+x')\cdot y) & \xrightarrow{C^{x+x',y}} & Sq^q(x+x')\cdot Sq^{q'}(y) \\ \Big\downarrow{\scriptstyle \Gamma^{xy,x'y}_{s\alpha}} & & \Big\downarrow{\scriptstyle \Gamma^{x,x'}_{s\beta}\cdot Sq^{q'}(y)} \\ Sq^n(xy)+Sq^n(x'y) & \xrightarrow{C^{x,y}+C^{x',y}} & (Sq^q(x)+Sq^q(x'))\cdot Sq^{q'}(y) \end{array}$$

This diagram has primary element $x\cdot x'\cdot Sq^{q'-1}(y)$. The morphism on the right-hand side of (10.7.1) does not coincide with the right-hand side of diagram (2) but is

$$\Gamma^{x,x'}_{q}\cdot Sq^{q'}(y)+\Gamma^{x,x'}_{q+1}\cdot Sq^{q'-1}(y).$$

By the delicate linearity track formula (10.3.2) we have $\Gamma^{x,x'}_{q+1}=x\cdot x'$. This shows that diagram (2) yields the commutativity of diagram (10.3.2) in case $n=|xy|$ □

10.8 Stable Künneth-Steenrod operations

Recall that the secondary Steenrod algebra $[\![\mathcal{A}]\!]$ in (2.5.4) is defined by groupoids $[\![\mathcal{A}^k]\!]$. Objects in $[\![\mathcal{A}^k]\!]_0$ are *stable maps* (α,H_α) given by a sequence of maps

(10.8.1) $$\alpha=(\alpha_q:Z^q\longrightarrow Z^{q+k})_{q\in\mathbb{Z}}$$

in **Top*** together with a sequence of tracks $H_\alpha = (H_{\alpha,q})_{q\in\mathbb{Z}}$ for the following diagram.

$$\begin{array}{ccc} Z^q & \xrightarrow{\alpha_q} & Z^{q+k} \\ {\scriptstyle r_q}\downarrow & \overset{H_{\alpha,q}}{\Longrightarrow} & \downarrow{\scriptstyle r_{q+k}} \\ \Omega_0 Z^{q+1} & \xrightarrow[\Omega_o\alpha_{q+1}]{} & \Omega_0 Z^{q+k+1} \end{array} \tag{1}$$

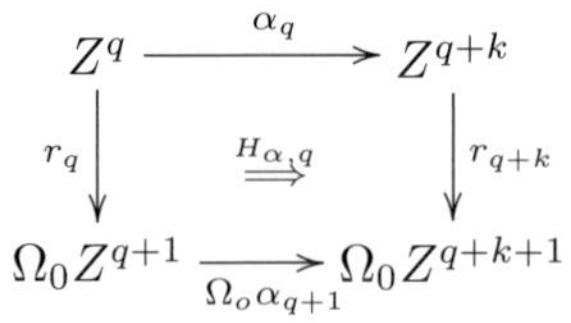

Here r_q is the homotopy equivalence in (2.1.7). Moreover a *stable track*

$$H : (\alpha, H_\alpha) \Longrightarrow (\beta, H_\beta) \tag{2}$$

in $[\![\mathcal{A}^k]\!]_1$ is a sequence of tracks

$$H = (H_q : \alpha_q \Longrightarrow \beta_q)_{q\in\mathbb{Z}}$$

in **Top*** such that

$$H_{\beta,q} = (r_{q+k}H_q)\square H_{\alpha,q}\square((\Omega_0 H_{q+1})^{\mathrm{op}} r_q). \tag{3}$$

Compare the diagram in (2.5.4). Each stable map α represents an element $\{\alpha\} \in \mathcal{A}^k$ in the Steenrod algebra $\mathcal{A}$.

10.8.2 Theorem. *A sequence* $(K_q, q \geq 1)$ *of Künneth tracks as in* (10.2.2) *induces for a Steenrod operation*

$$\alpha \in E_{\mathcal{A}} = \{Sq^1, Sq^2, \ldots\} \quad \textit{for} \quad p = 2 \textit{ and}$$
$$\alpha \in E_{\mathcal{A}} = \{\beta, P^1, P^2, \ldots, P^1_\beta, P^2_\beta, \ldots\} \quad \textit{for} \quad p \textit{ odd}$$

a well-defined stable map

$$s\alpha \in [\![\mathcal{A}^{|\alpha|}]\!]_0.$$

We call $s\alpha$ the *stable Künneth-Steenrod operation* (associated to $\alpha \in E_{\mathcal{A}}$ via Künneth tracks).

Proof of (10.8.2). As in (2.1.7) we choose a map

$$i_{\mathbb{F}} : S^1 \longrightarrow Z^1$$

which in homology induces the ring homomorphism $\mathbb{Z} \to \mathbb{F} = \mathbb{Z}/p$. Moreover we choose a track $\mathcal{B}_0$ in the following diagram.

$$S^1 \xrightarrow{i_{\mathbb{F}}} Z^1 \xrightarrow{s\beta} Z^2, \qquad \mathcal{B}_0 : s\beta\, i_{\mathbb{F}} \Longrightarrow 0 \tag{10.8.3}$$

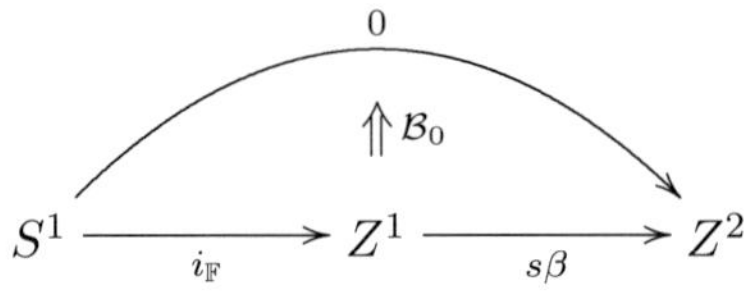

Here β is the Bockstein operator and $s\beta$ is the corresponding Künneth-Steenrod operation. For $p = 2$ we have $\beta = Sq^1$. We now define the following diagram with $\alpha \in E_{\mathcal{A}}$ and $|\alpha| = k$.

(10.8.4)
$$\begin{array}{ccc} Z^q \times S^1 & \xrightarrow{(s\alpha)_q \times 1} & Z^{q+k} \times S^1 \\ {\scriptstyle t_q}\downarrow & \overset{G_{\alpha,q}}{\Longrightarrow} & \downarrow{\scriptstyle t_{q+k}} \\ Z^{q+1} & \xrightarrow{(s\alpha)_{q+1}} & Z^{q+1+k} \end{array}$$

Here $(s\alpha)_q$ is the Künneth-Steenrod operation associated to the Künneth track K_q. Moreover we set

(1) $$t_q = \mu_{q,1}(1 \times i_{\mathbb{F}}) : Z^q \times S^1 \longrightarrow Z^q \times Z^1 \longrightarrow Z^{q+1}.$$

Hence t_q induces the map

(2) $$\hat{t}_q : Z^q \wedge Z^1 \longrightarrow Z^{q+1}$$

with $\hat{t}_q = \mu(1 \wedge i_{\mathbb{F}})$ and the adjoint of $\hat{t}_q$ is the homotopy equivalence

(3) $$r_q : Z^q \longrightarrow \Omega_0 Z^{q+1}, \; q \geq 1.$$

Compare the definition of r_q in (2.1.7). Let $X = Z^q \times S^1$ and let $x = p_1 : X \to Z^q$ and $y = i_{\mathbb{F}} p_2 : X \to S^1 \to Z^1$ be given by the projections p_1 and p_2. Then we have

(4) $$t_q = x \cdot y : X \longrightarrow Z^{q+1}.$$

If we apply $(s\alpha)_{q+1}$ to $x \cdot y$ we obtain the following Cartan tracks:

For $\alpha = Sq^n \in E_{\mathcal{A}}$ we get the composite of tracks

(5)
$$\begin{array}{ccc} (sSq^n)_{q+1} t_q & \overline{\overline{\qquad\qquad}} & Sq^n(x \cdot y) \\ \Big\downarrow & & \downarrow{\scriptstyle C^{x,y}} \\ {\scriptstyle G_{\alpha,q}} & & (Sq^n x) \cdot y + (Sq^{n-1}x) \cdot (Sq^1 y) \\ \Big\downarrow & & \downarrow{\scriptstyle (Sq^n x)\cdot y + (Sq^{n-1}x)\cdot \mathcal{B}_0} \\ t_{q+k}((sSq^n)_q \times 1) & \overline{\overline{\qquad\qquad}} & (Sq^n x) \cdot y \end{array}$$

which defines the track $G_{\alpha,q}$ in (10.8.4). Here $C^{x,y}$ is the Künneth-Cartan track in (10.4) and $\mathcal{B}_0$ is given by the track in (10.8.3).

For $\alpha = P^n \in E_{\mathcal{A}}$ we get the following track.

(6)
$$\begin{array}{ccc} (sP^n)_{q+1}t_q & \Longleftrightarrow & P^n(x\cdot y) \\ \Big\downarrow{\scriptstyle G_{\alpha,q}} & & \Big\downarrow{\scriptstyle C^{x,y}} \\ t_{q+k}((sP^n)_q \times 1) & \Longleftrightarrow & (P^n x)\cdot y \end{array}$$

For $\alpha = \beta = \mathit{Bockstein} \in E_{\mathcal{A}}$ we get the composite of following tracks.

(7)
$$\begin{array}{ccc} (s\beta)_{q+1}t_q & \Longleftrightarrow & \beta(x\cdot y) \\ & & \Big\downarrow{\scriptstyle C^{x,y}} \\ \Big\downarrow{\scriptstyle G_{\alpha,q}} & & (\beta x)\cdot y + (-1)^q x\cdot(\beta y) \\ & & \Big\downarrow{\scriptstyle (\beta x)\cdot y+(-1)^q x\cdot \mathcal{B}_0} \\ t_{q+1}((s\beta)_q \times 1) & \Longleftrightarrow & (\beta x)\cdot y \end{array}$$

Here $\mathcal{B}_0$ is given by (10.8.3).Finally for $\alpha = P^n_\beta \in E_{\mathcal{A}}$ we get the following track.

$$\begin{array}{ccc} (sP^n_\beta)_{q+1}t_q & \Longleftrightarrow & P^n_\beta(x\cdot y) \\ & & \Big\downarrow{\scriptstyle C^{x,y}} \\ \Big\downarrow{\scriptstyle G_{\alpha,q}} & & P^n_\beta(x)\cdot y + P^n(x)\cdot\beta(y) \\ & & \Big\downarrow{\scriptstyle P^n_\beta(x)\cdot y+P^n(x)\cdot \mathcal{B}_0} \\ t_{q+k}((sP^n_\beta)_q \times 1) & \Longleftrightarrow & (P^n_\beta x)\cdot y \end{array}$$

Now we consider the cofiber sequence

$$Z^q \vee S^1 \xrightarrow{j} Z^q \times S^1 \xrightarrow{\pi} Z^q \wedge S^1$$

where j is the inclusion and π is the quotient map. One readily checks that

$$G_{\alpha,q}j : 0 \Longrightarrow 0$$

is the identity track of the trivial map. This implies that in the diagram

(10.8.5)
$$\begin{array}{ccc} Z^q \wedge S^1 & \xrightarrow{(s\alpha)_q\wedge 1} & Z^{q+k}\wedge S^1 \\ \Big\downarrow{\scriptstyle \hat{t}_q} & \overset{\hat{G}_{\alpha,q}}{\Longrightarrow} & \Big\downarrow{\scriptstyle \hat{t}_{q+k}} \\ Z^{q+1} & \xrightarrow[(s\alpha)_{q+1}]{} & Z^{q+1+k} \end{array}$$

there is a unique track $\hat{G}_{\alpha,q}$ with

$$G_{\alpha,q} = \hat{G}_{\alpha,q}\pi.$$

Now let $H_{\alpha,q}$ in the following diagram be the adjoint of $\hat{G}_{\alpha,q}$.

(10.8.6)
$$\begin{array}{ccc} Z^q & \xrightarrow{(s\alpha)_q} & Z^{q+k} \\ {\scriptstyle r_q}\downarrow & \overset{H_{\alpha,q}}{\Longrightarrow} & \downarrow{\scriptstyle r_{q+k}} \\ \Omega_0 Z^{q+1} & \xrightarrow[\Omega_0(s\alpha)_{q+1}]{} & \Omega_0 Z^{q+1+k} \end{array}$$

Then the stable map $s\alpha$ for $\alpha \in E_{\mathcal{A}}$ in Theorem (10.8.2) is defined by

(10.8.7)
$$s\alpha = ((s\alpha)_q, H_{\alpha,q})$$

with $(s\alpha)_q$ the Künneth-Steenrod operation associated to K_q and $H_{\alpha,q}$ in (10.8.6). This completes the proof of (10.8.2). □

Let X be a space and let $x, x' : X \to Z^{|x|}$ be maps in **Top***. In (4.2.2) we define for the stable map $s\alpha$ with $\alpha \in E_{\mathcal{A}}$ the *stable linearity track*

$$\Gamma_{s\alpha}^{x,x'} : \alpha(x + x') \Longrightarrow \alpha(x) + \alpha(x').$$

Moreover in (10.3.2) we define the *Künneth linearity track*

$$\Gamma^{x,x'} : \alpha(x + x') \Longrightarrow \alpha(x) + \alpha(x'),$$

which for $p = 2$ satisfies the delicate linearity track formula in (10.3.2). We now show that these linearity tracks coincide.

10.8.8 Theorem. *For the stable Künneth-Steenrod operation $s\alpha$, $\alpha \in E_{\mathcal{A}}$, in* (10.8.7) *the linearity tracks satisfy*

$$\Gamma_{s\alpha}^{x,x'} = \Gamma^{x,x'} : \alpha(x + x') \Longrightarrow \alpha(x) + \alpha(x').$$

The theorem shows that all results on stable linearity tracks in Chapter 2 also hold for Künneth linearity tracks. Compare (10.3.2).

Proof. Recall from (2.6.4) that for $q_0 =| x |$ and $k =| \alpha |$ we have

(1)
$$s\alpha \in [\![\mathcal{A}^k]\!]_0 \xrightarrow{\sim} [\![Z^{q_0}, Z^{q_0+k}]\!]_0^{\text{stable}}.$$

The linearity track $\Gamma^{x,x'}_{s\alpha}$ is defined by $\Gamma_{s\alpha}$ in the following diagram in $[\![\mathbf{K}_p^{\text{stable}}]\!]$ where A is the addition map.

(2)
$$\begin{array}{ccc} Z^{q_0} \times Z^{q_0} & \xrightarrow{s\alpha \times s\alpha} & Z^{q_0+k} \times Z^{q_0+k} \\ {\scriptstyle A=A_{q_0}}\downarrow & \overset{\Gamma_{s\alpha}}{\Longrightarrow} & \downarrow{\scriptstyle A} \\ Z^{q_0} & \xrightarrow[s\alpha]{} & Z^{q_0+k} \end{array}$$

See (2.6.3) and (4.2.1). We claim that for $s\alpha$ with $\alpha \in E_{\mathcal{A}}$ as defined in (10.8.7) the component of $\Gamma_{s\alpha}$ in degree $q \geq q_0$ coincides with the Künneth linearity track Γ_q defined as follows. Let $X = \mathbb{Z}^q \times Z^q$ and let $a = p_1$, $b = p_2$ be the projections $X \to Z^q$. Then $\Gamma_q = \Gamma^{a,b}$ is a track for the following diagram in $\mathbf{Top}^*$.

(3)
$$\begin{array}{ccc} Z^q \times Z^q & \xrightarrow{(s\alpha)_q \times (s\alpha)_q} & Z^{q+k} \times Z^{q+k} \\ {\scriptstyle A_q=A}\downarrow & \overset{\Gamma_q}{\Longrightarrow} & \downarrow{\scriptstyle A=A_{q+k}} \\ Z^q & \xrightarrow[(s\alpha)_q]{} & Z^{q+k} \end{array}$$

We claim that

(4)
$$(\Gamma_{s\alpha})_q = \Gamma_q \text{ for } q \geq q_0.$$

This is clear if q is sufficiently large since both tracks in (4) coincide on $Z^q \vee Z^q$. Hence (4) is true if $\Gamma = (\Gamma_q, q \geq q_0)$ is a well-defined stable track as $\Gamma_{s\alpha}$ in (2). For this we consider the following diagram where we use the linearity of r_q so that the track $H_{\alpha,q}A_q$ is well defined. We write $a_q = (s\alpha)_q$.

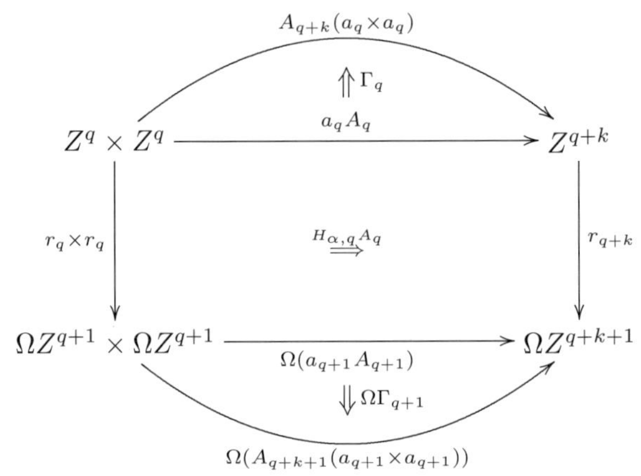

(5)

Now (4) holds if and only if pasting of tracks in (5) yields the track

$$(\Omega A_{q+k+1})(H_{\alpha,q} \times H_{\alpha,q}) = (5). \tag{6}$$

We prove (6) by considering the following adjoint diagram of (5) where

$$\Delta_{24}(a, b, c) = (a, c, b, c).$$

(7)

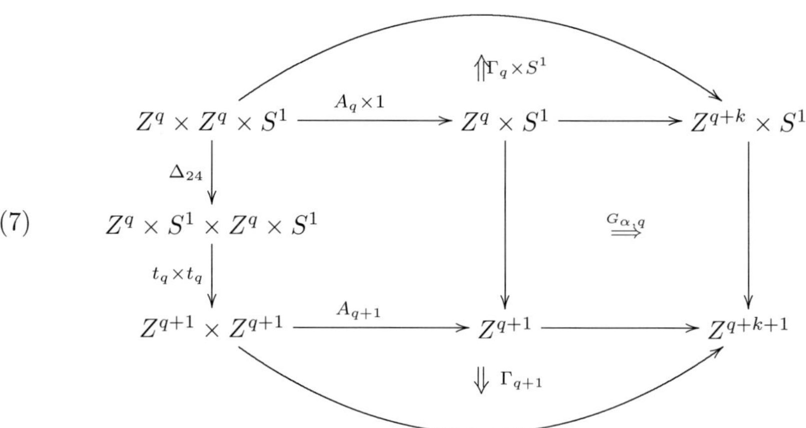

Now (6) holds if and only if pasting of tracks in (2) yields

$$A_{q+k+1}(G_{\alpha,q} \times G_{\alpha,q})\Delta_{24} = (7). \tag{8}$$

We prove (8) by use of (10.7.4). Let $X = Z^q \times Z^q \times S^1$ and let $a = p_1, a' = p_2$ and let $b = i_{\mathbb{F}} p_3$ be given by the projections of X.

We now consider the case $p = 2$ and $\alpha = Sq^k$. Then (10.7.4) shows that the following diagram of tracks commutes.

$$\begin{array}{ccc}
Sq^k((a+a')b) & \xrightarrow{\Gamma_k^{ab,a'b}} & Sq^k(ab) + Sq^k(a'b) \\
\Big\downarrow C^{a+a',b} & 1 & \Big\downarrow C^{a,b}+C^{a',b} \\
\begin{array}{c} Sq^k(a+a')b \\ +Sq^{k-1}(a+a')\cdot Sq^1 b \end{array} & \xrightarrow{\Gamma_a} & \begin{array}{c} Sq^k(a)b+Sq^k(a')b \\ +Sq^{k-1}(a)Sq^1b+Sq^{k-1}(a')Sq^1 b \end{array} \\
\Big\downarrow 0^{\square}+Sq^{k-1}(a+a')\mathcal{B}_0 & 2 & \Big\downarrow \begin{array}{c} 0^{\square}+Sq^{k-1}(a)\mathcal{B}_0 \\ +Sq^{k-1}(a')\mathcal{B}_0 \end{array} \\
Sq^k(a+a')\cdot b & \xrightarrow[\Gamma_k^{a,a'}\cdot b]{} & Sq^k(a)b + Sq^k(a')b
\end{array} \tag{9}$$

Here subdiagram 1 with

$$\Gamma_a = \Gamma_k^{a,a'} \cdot b + \Gamma_{k-1}^{a,a'} \cdot Sq^1(b)$$

commutes by (10.7.4). Moreover subdiagram 2 commutes by use of the product of tracks $\Gamma_{k-1}^{a,a'} \cdot \mathcal{B}_0$.

The column on the left-hand side of (9) is $G_{\alpha,q}(A_q \times 1)$, the column on the right-hand side of (9) is $A_{q+k+1}(G_{\alpha,q} \times G_{\alpha,q})\Delta_{24}$. Moreover

(10)
$$\begin{aligned} \Gamma_k^{ab,a'b} &= \Gamma_{q+1}(t_q \times t_q)\Delta_{24}, \\ \Gamma_k^{a,a'} \cdot b &= t_{q+k}(\Gamma_q \times S^1). \end{aligned}$$

This shows that (8) is equivalent to the commutativity of diagram (9). In a similar way we prove (8) for $\alpha \in E_{\mathcal{A}}$ and for p odd. □

Chapter 11

The Algebra of Δ-tracks

In this chapter we introduce *generalized Cartan tracks* C_α defined for each α in the algebra $T_{\mathbb{G}}(E_{\mathcal{A}})$. We show that relations of the Cartan tracks in Chapter 10 yield corresponding relations of generalized Cartan tracks. The notion of "secondary Hopf algebra" in the next chapter relies only on the relations for generalized Cartan tracks. The diagonal of the secondary Hopf algebra is deduced from the relation diagonal in this chapter.

11.1 The Hopf-algebra $T_{\mathbb{G}}(E_{\mathcal{A}})$

Let R be a commutative ring with unit and let M, N be non-negatively graded R-modules. For a prime p the tensor product $M \otimes N$ is "p-symmetric" by the *interchange isomorphism*

$$(11.1.1)\qquad \begin{aligned} T : M \otimes N &\cong N \otimes M, \\ T(x \otimes y) &= (-1)^{p|x||y|} y \otimes x. \end{aligned}$$

We shall use this sign convention for T in the presence of a prime p. We call (11.1.1) the *even sign convention* since for $p = 2$ we have $T(x \otimes y) = y \otimes x$. For odd primes p we get the usual sign rule $T(x \otimes y) = (-1)^{|x||y|}$ which we call the *odd sign convention.*

Given a graded R-algebra (A, μ_A) the tensor product $A \otimes A$ is an R-algebra via the multiplication

$$\mu_{A\otimes A} = (\mu_A \otimes \mu_A)(A \otimes T \otimes A).$$

Here T is the interchange with the even sign convention (11.1.1). That is, for $x \otimes y$, $x' \otimes y' \in A \otimes A$ the multiplication in $A \otimes A$ is defined by

$$(x \otimes y) \cdot (x' \otimes y') = (-1)^{p|y||x'|}(x \cdot x') \otimes (y \cdot y').$$

Of course for $R = \mathbb{Z}/2$ the even and the odd sign convention coincide.

11.1.2 Definition. Let A be a non-negatively graded R-algebra with unit. Then A is a *Hopf algebra* if A is augmented by

$$\epsilon : A \longrightarrow R$$

and if an algebra map

$$\Delta : A \longrightarrow A \otimes A$$

is given so that the following diagrams commute.

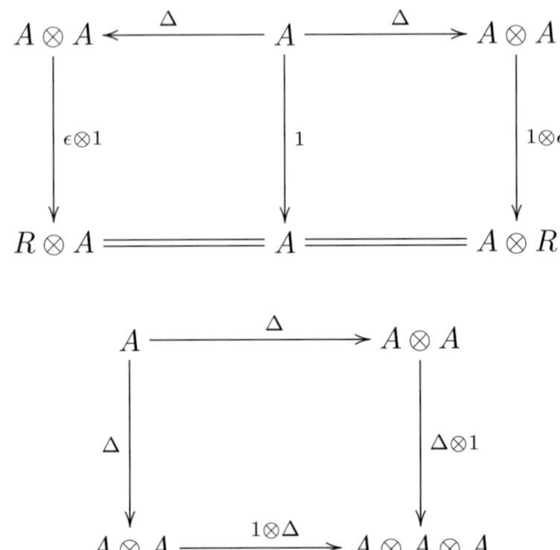

The Hopf algebra is *co-commutative* if in addition the following diagram commutes.

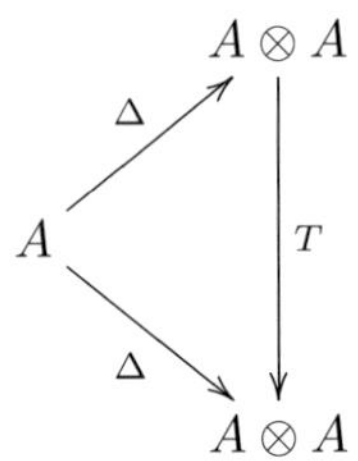

Here the algebra structure of $A \otimes A$ and the interchange T are defined by the even sign convention as in (11.1.1) above. Diagrams as in (11.1.2) are used to define a *coalgebra* in any monoidal category. Hence a Hopf algebra as above is the same as a coalgebra in the monoidal category of (non-negatively graded) algebras over R.

The Steenrod algebra $\mathcal{A}$ is a co-commutative Hopf algebra over $\mathbb{F} = \mathbb{Z}/p$ with the diagonal

$$\delta : \mathcal{A} \longrightarrow \mathcal{A} \otimes \mathcal{A} \qquad (11.1.3)$$

defined by

$$\delta(Sq^i) = \textstyle\sum_{k+l=i} Sq^k \otimes Sq^l \text{ for } p = 2$$

$$\begin{cases} \delta(\beta) = \beta \otimes 1 + 1 \otimes \beta, \\ \delta(P^i) = \sum\limits_{k+l=i} P^k \otimes P^l \text{ for } p \text{ odd.} \end{cases}$$

Compare (1.2.7). For $P^i_\beta = \beta P^i$ we have the formula

$$\begin{aligned} \delta P^i_\beta &= \delta(\beta) \cdot \delta(P^i) \\ &= \sum_{k+l=i} (P^k_\beta \otimes P^l + P^k \otimes P^l_\beta). \end{aligned}$$

Recall that $E_{\mathcal{A}} \subset \mathcal{A}$ is the set of algebra generators with

$$E_{\mathcal{A}} = \begin{cases} \{Sq^1, Sq^2, \ldots\} & \text{for } p = 2, \\ \{\beta, P^1, P^1_\beta, P^2, P^2_\beta, \ldots\} & \text{for } p \text{ odd.} \end{cases}$$

For $\mathbb{G} = \mathbb{Z}/p^2$ let $T_{\mathbb{G}}(E_{\mathcal{A}})$ be the $\mathbb{G}$-tensor algebra generated by $E_{\mathcal{A}}$. We have the following canonical commutative diagram.

(11.1.4)
$$\begin{array}{ccc} \mathcal{A} & \xrightarrow{\delta} & \mathcal{A} \otimes \mathcal{A} \\ {\scriptstyle p}\uparrow & & \uparrow{\scriptstyle p\otimes p} \\ T_{\mathbb{G}}(E_{\mathcal{A}}) & \xrightarrow{\Delta} & T_{\mathbb{G}}(E_{\mathcal{A}}) \otimes T_{\mathbb{G}}(E_{\mathcal{A}}) \end{array}$$

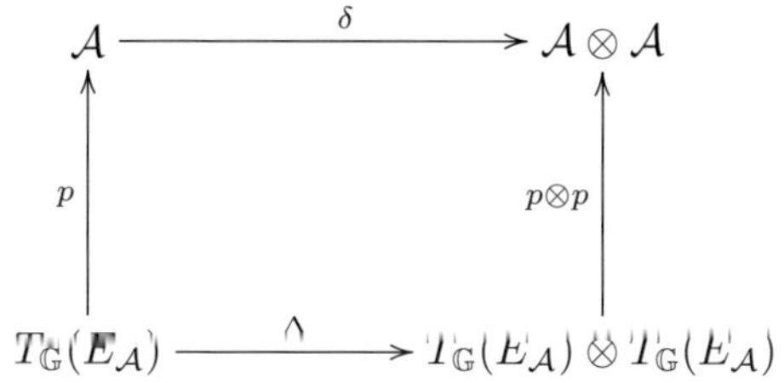

Here p is the surjective algebra map which is the identity on generators in $E_{\mathcal{A}}$. Moreover $T_{\mathbb{G}}(E_{\mathcal{A}}) \otimes T_{\mathbb{G}}(E_{\mathcal{A}})$ is an algebra with the even sign convention as in (11.1.1). The diagonal Δ is the unique algebra map defined on generators in $E_{\mathcal{A}}$ by the formulas

$$\begin{aligned} \Delta(Sq^i) &= \sum_{k+l=i} Sq^k \otimes Sq^l \text{ for } p = 2, \text{ and} \\ \Delta(\beta) &= \beta \otimes 1 + 1 \otimes \beta, \\ \Delta(P^i) &= \sum_{k+l=i} P^k \otimes P^l, \\ \Delta(P^i_\beta) &= \sum_{k+l=i} (P^k_\beta \otimes P^l + P^k \otimes P^l_\beta) \text{ for } p \text{ odd.} \end{aligned}$$

Here we have $k, l \geq 0$ and $Sq^0 = 1$, $P^0 = 1$ and $P^0_\beta = \beta$. It is clear that diagram (11.1.4) commutes and that $p \otimes p$ is an algebra map.

11.1.5 Lemma. *$(T_{\mathbb{G}}(E_{\mathcal{A}}), \Delta)$ is a Hopf algebra which is co-commutative for all primes p since we use the even sign convention.*

Moreover, the tensor algebra

$$T_{\mathbb{F}}(E_{\mathcal{A}}) = T_{\mathbb{G}}(E_{\mathcal{A}}) \otimes \mathbb{F}$$

is a co-commutative Hopf algebra $(T_{\mathbb{F}}(E_{\mathcal{A}}), \Delta)$ with Δ defined by the same formula as above. We write

$$\begin{aligned} \mathcal{B}_0 &= T_{\mathbb{G}}(E_{\mathcal{A}}), \\ \mathcal{F}_0 &= T_{\mathbb{F}}(E_{\mathcal{A}}) = \mathcal{B}_0/p\mathcal{B}_0, \end{aligned}$$

so that we have canonical algebra maps

$$q : \mathcal{B}_0 \longrightarrow \mathcal{F}_0 \longrightarrow \mathcal{A} \tag{11.1.6}$$

which are the identity on generators. These quotient maps are also Hopf algebra maps. If x is an element in a $\mathbb{G}$-module $\mathcal{M}$, then its image in $\mathcal{M} \otimes \mathbb{F}$ is denoted by $x_{\mathbb{F}}$ or also by x. Moreover for $x \in \mathcal{B}_0$ the image in $\mathcal{F}_0$, or in $\mathcal{A}$, is also denoted by x.

Remark. For $p = 2$ the Hopf algebra $(T_{\mathbb{G}}(E_{\mathcal{A}}), \Delta)$ is not co-commutative if we use the odd sign convention, since then we have

$$T\Delta(Sq^i) = \sum_{k+l=i} (-1)^{kl} Sq^l \otimes Sq^k.$$

Here the sign is non-trivial for kl odd since we work over $\mathbb{G} = \mathbb{Z}/4$.

11.2 Δ-tracks

In (10.8.2) we have seen that stable Künneth-Steenrod operations $s\alpha$ associated to $\alpha \in E_{\mathcal{A}}$ yield a function s as in the following commutative diagram.

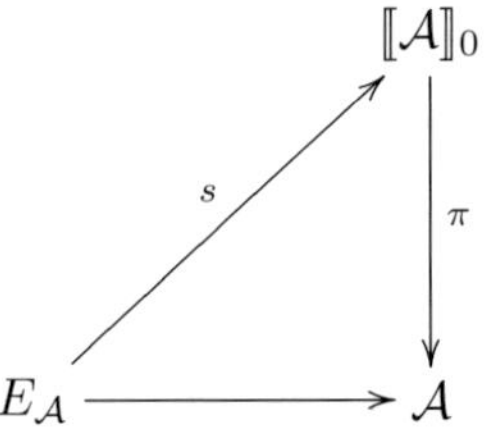

Moreover s induces as in (5.2.1) the function

$$s : T_{\mathbb{G}}(E_{\mathcal{A}}) \longrightarrow [\![\mathcal{A}]\!]_0 \tag{11.2.1}$$

which is a pseudo functor by (5.2.3). For $\alpha \in T_{\mathbb{G}}(E_{\mathcal{A}})$ we call $s\alpha$ also a (generalized) *Künneth-Steenrod operation*. If $|\alpha|= r$ and $x : X \to Z^q$ is a map in **Top***, then we obtain the composite

$$\alpha(x) = (s\alpha)_q \circ x : X \longrightarrow Z^q \longrightarrow Z^{q+r}.$$

Here $\alpha(x)$ is an element in the $\mathbb{F}$-vector space $[\![X, Z^{q+r}]\!]_0$. Since $\alpha(x)$ is linear in α we see that $\alpha(x)$ depends only on $\alpha \otimes 1$ in $T_{\mathbb{F}}(E_{\mathcal{A}}) = T_{\mathbb{G}}(E_{\mathcal{A}}) \otimes \mathbb{F}$. Hence for $p = 2$ we have $((-1) \cdot \alpha)(x) = \alpha(x)$ though $(-1) \cdot \alpha \neq \alpha$ in $T_{\mathbb{G}}(E_{\mathcal{A}})$.Now consider the diagonal Δ for $T_{\mathbb{G}}(E_{\mathcal{A}})$ in (11.1.4). For each $\alpha \in T_{\mathbb{G}}(E_{\mathcal{A}})$ we can find a family

(11.2.2) $$\bar{\alpha} = \{(\alpha_i', \alpha_i''); i \in I_\alpha\}$$

with $\alpha_i',\ \alpha_i'' \in T_{\mathbb{G}}(E_{\mathcal{A}})$ such that

$$\Delta\alpha = \sum_{i \in I_\alpha} \alpha_i' \otimes \alpha_i'' \in T_{\mathbb{G}}(E_{\mathcal{A}}).$$

We say that $\bar{\alpha}$ is a *Δ-family associated to* α. Since (11.1.4) commutes there exists a track

(11.2.3) $$H : \alpha(x \cdot y) \Longrightarrow \sum_i (-1)^{|x||\alpha_i''|} \alpha_i'(x) \cdot \alpha_i''(y)$$

where $x : X \to Z^q$, $y : X \to Z^{q'}$ and $x \cdot y : X \to Z^{q+q'}$. For $p = 2$ the signs on the right-hand side may be omitted. The track $H = H_\alpha^{x,y}$ can be chosen for all spaces X and all elements x, y of degree $|\, x \,| = q \geq 1$, $|\, y \,| = q' \geq 1$ such that $H_\alpha^{x,y}$ is natural in X. Here *naturality* means that a pointed map $f : Y \longrightarrow X$ induces the equation

$$H_\alpha^{x,y} f = H_\alpha^{xf,yf}.$$

We call a natural family $H_\alpha = (\alpha, H_\alpha^{x,y})$ a *Δ-track* associated to α. We point out that $\alpha(x, y)$ and $\alpha_i'(x), \alpha_i''(y)$ in (11.2.3) depend only on $\alpha, \alpha_i', \alpha_i'' \in T_{\mathbb{F}}(E_{\mathcal{A}})$. A track $H = H_\alpha$ as in (11.2.3), however, will be constructed below in such a way that H_α depends actually on $\alpha \in T_{\mathbb{G}}(E_{\mathcal{A}})$. Therefore we insist that a Δ-track H is associated to an element $\alpha \in T_{\mathbb{G}}(E_{\mathcal{A}})$ defined over $\mathbb{G}$ and not over $\mathbb{F}$.

11.2.4 Definition. Let $\xi \in T_{\mathbb{G}}(E_{\mathcal{A}}) \otimes T_{\mathbb{G}}(E_{\mathcal{A}})$ and let $x : X \to Z^q, y : X \to Z^{q'}$ be maps in **Top***. Then we define

$$\xi(x, y) : X \xrightarrow{(x,y)} Z^q \times Z^{q'} \longrightarrow Z^{q+q'+|\xi|}$$

as follows. We can write $\xi = \sum_i \xi_i' \otimes \xi_i''$ and we set

$$\xi(x, y) = \sum_i (-1)^{|x| \cdot |\xi_i''|} \xi_i'(x) \cdot \xi_i''(y).$$

Since $\alpha(x)\cdot\beta(y)$ is linear in α, resp. β, we see that $\xi(x,y)$ is well defined. In a similar way we define for $x,y,z : X \to Z^{q''}$ and $\xi \in T_{\mathbb{G}}(E_{\mathcal{A}})\otimes T_{\mathbb{G}}(E_{\mathcal{A}})\otimes T_{\mathbb{G}}(E_{\mathcal{A}})$ the map

$$\xi(x,y,z) : X \longrightarrow Z^{q+q'+q''+|\xi|}.$$

We point out that $\xi(x,y)$ or $\xi(x,y,z)$ depend only on $\xi_{\mathbb{F}}$, see (11.1.6). Using this notation a Δ-track is a family of tracks

(11.2.5) $$H_\alpha^{x,y} = \alpha(x,y) \Rightarrow (\Delta\alpha)(x,y)$$

which is natural in x,y. The universal example $H_\alpha^{\bar{x},\bar{y}} = H_\alpha^{q,q'}$ is given by the projections $\bar{x} : Z^q \times Z^{q'} \longrightarrow Z^q$ and $\bar{y} : Z^q \times Z^{q'} \longrightarrow Z^q$ with $q,q' \geq 1$. Hence we have by naturality $H_\alpha^{x,y} = H_\alpha^{q,q'}(x,y)$ for all x,y with $|\, x \,| = q$, $|\, y \,| = q'$. A Δ-*track* is thus determined by the family $(\alpha, H_\alpha^{q,q'}, q,q' \geq 1)$ where $H_\alpha^{q,q'}$ is a track as in the following diagram.

(11.2.6) $$\begin{array}{ccc} Z^q \times Z^{q'} & \xrightarrow{\bar{\alpha}} & \prod_i Z^{q+|\alpha_i'|} \times Z^{q+|\alpha_i''|} \\ \Big\downarrow \mu & \Longrightarrow & \Big\downarrow \mu_0 \\ Z^{q+q'} & \xrightarrow[s\alpha]{} & Z^{q+q'+|\alpha|} \end{array}$$

Here μ is the multiplication map and $\bar{\alpha}$ is defined by the coordinates $((-1)^{|x||\alpha_i''|}s\alpha_i') \times s\alpha_i''$ and μ_0 carries the tuple (x_i,y_i) to the sum $\sum\limits_i x_i \cdot y_i$.

11.3 Linearity tracks Γ_α and $\Gamma_{\alpha,\beta}$

In this section we fix some notation on linearity tracks. For $\alpha \in T_{\mathbb{G}}(E_{\mathcal{A}})$ we have the map $\alpha(x)$ as in (11.2.1). Here $\alpha(x)$ is linear in α, that is

$$(\alpha+\beta)(x) = \alpha(x) + \beta(x),$$

but not linear in x. There are well-defined *linearity tracks* (4.2.2)

(11.3.1) $$\Gamma_\alpha = \Gamma_\alpha^{x,y} : \alpha(x+y) \Rightarrow \alpha(x) + \alpha(y),$$

$$\Gamma_\alpha : \alpha(\sum_{i=1}^k n_i x_i) \Rightarrow \sum_{i=1}^k n_i \alpha(x_i),$$

for $n_i \in \mathbb{G}$, $x_i : X \longrightarrow Z^q$, $i = 1,\dots,k$. While source and target of Γ_α depend only on $(n_i)_{\mathbb{F}}$, the track Γ_α depends on $(n_i)_{\mathbb{G}}$. In fact, let

$$\varphi : \mathbb{G}^k \longrightarrow \mathbb{G}$$

be defined by the matrix $\varphi = (n_1, \dots, n_k)$, then $\Gamma_\alpha = \Gamma(\varphi)_\alpha^{x_1,\dots,x_k}$ is defined in (4.2.13). The track Γ_α is linear in α so that Γ_α depends only on $\alpha_{\mathbb{F}}$ and φ.

Let $\lambda : I \longrightarrow J$ be a function between finite index sets. Then the following diagram commutes

(11.3.2)
$$\begin{array}{ccc}
\alpha(\sum\limits_{i\in I} n_i x_{\lambda i}) & \xrightarrow{\Gamma_\alpha} & \sum\limits_{i\in I} n_i \alpha(x_{\lambda i}) \\
\| & & \| \\
\alpha(\sum\limits_{j\in J} m_j x_j) & \xrightarrow{\Gamma_\alpha} & \sum\limits_{j\in J} m_j \alpha(x_j)
\end{array}$$

where $m_j = \sum\limits_{i\in\lambda^{-1}(j)} n_i$ for $j \in J$. We derive the result from (4.2.15)(3).

Moreover recall that (s, Γ) in (11.2.1) is a pseudo functor so that Γ induces the following track ($q =| x |$).

(11.3.3)
$$\begin{array}{cccc}
\Gamma_{\alpha,\beta} = \Gamma(\alpha,\beta)x : & (\alpha\beta)(x) & \Longrightarrow & \alpha(\beta(x)) \\
& \| & & \| \\
& s(\alpha\beta)_q x & & (s\alpha)_{q+|\beta|}(s\beta)_q x
\end{array}$$

Here $\Gamma(\alpha, \beta)$ is defined in (5.2.2). We point out that $\Gamma_{\alpha,\beta}$ is the identity track if β is a monomial of generators in $E_{\mathcal{A}}$, moreover $\Gamma_{\alpha,\beta}$ is linear in α. Therefore $\Gamma_{\alpha,\beta}$ depends on $\alpha_{\mathbb{F}}$ and $\beta_{\mathbb{G}}$.

11.4 Sum and product of Δ-tracks

Given Δ-tracks
$$H_\alpha : \alpha(x \cdot y) \Longrightarrow (\Delta\alpha)(x, y),$$
$$H_\beta : \beta(x \cdot y) \Longrightarrow (\Delta\beta)(x, y),$$

with $\alpha, \beta \in T_{\mathbb{G}}(E_{\mathcal{A}})$ and $| \alpha |=| \beta |$ we obtain the *sum of Δ-tracks*

(11.4.1)
$$H_\alpha + H_\beta : (\alpha + \beta)(x \cdot y) \Longrightarrow \Delta(\alpha + \beta)(x, y)$$

which is a Δ-track associated to $\alpha + \beta$. If H_α and H_β are linear, then $H_\alpha + H_\beta$ is also linear. We therefore define the following graded $\mathbb{G}$-modules.

11.4.2 Definition. Let $\mathcal{T}_\Delta$ be the graded $\mathbb{G}$-module consisting of all pairs (α, H_α) with $\alpha \in T_\mathbb{G}(E_\mathcal{A})$ and H_α a Δ-track associated to α. The degree of (α, H_α) is the degree $|\alpha|$ of α and the sum in $\mathcal{T}_\Delta$ is given by the sum of Δ-tracks above, that is $(\alpha, H_\alpha) + (\beta, H_\beta) = (\alpha + \beta, H_\alpha + H_\beta)$. We call $\mathcal{T}_\Delta$ the *module of Δ-tracks.*

11.4.3 Definition. We introduce for Δ-tracks (α, H_α), $(\beta, H_\beta) \in \mathcal{T}_\Delta$ the *product of Δ-tracks* $(\alpha\beta, H_\alpha \odot H_\beta)$. Here

$$H_\alpha \odot H_\beta : (\alpha\beta)(x \cdot y) \Longrightarrow \Delta(\alpha\beta)(x, y) \tag{1}$$

is a Δ-track associated to $\alpha\beta$ where $\alpha\beta$ is the product of α and β in the algebra $T_\mathbb{G}(E_\mathcal{A})$. For the definition of $H_\alpha \odot H_\beta$ below it is crucial that α and β are defined over $\mathbb{G}$.

Let $M = \mathrm{Mon}(E_\mathcal{A})$ be the free graded monoid generated by $E_\mathcal{A}$ so that $T_\mathbb{G}(E_\mathcal{A})$ is the free $\mathbb{G}$-module generated by M and the tensor product $T_\mathbb{G}(E_\mathcal{A}) \otimes T_\mathbb{G}(E_\mathcal{A})$ is the free $\mathbb{G}$-module generated by the product $M \times M$. Here a pair $\xi = (\xi', \xi'') \in M \times M$ corresponds to the basis element $\xi' \otimes \xi''$. Each element $a \in T_\mathbb{G}(E_\mathcal{A}) \otimes T_\mathbb{G}(E_\mathcal{A})$ can be written uniquely as a sum

$$a = \sum_{\xi \in M \times M} \varphi_a(\xi)\xi' \otimes \xi'' \tag{2}$$

with $\varphi_a(\xi) \in \mathbb{G}$ and $|a| = |\xi|$. With the notation in (11.2.4) we have

$$a(x, y) = \sum_\xi (-1)^{|x||\xi''|} \varphi_a(\xi)\, \xi'(x) \cdot \xi''(y). \tag{3}$$

Now the Δ-track H_α induces the track

$$H_{\alpha,a} : \alpha(a(x, y)) \Longrightarrow ((\Delta\alpha) \cdot a)(x, y) \tag{4}$$

as follows. Let $H_{\alpha,a}$ be the composite of tracks as in the following commutative diagram.

$$\begin{array}{ccc}
\alpha(a(x,y)) & = & \alpha\big(\sum_\xi \epsilon_\xi^x \varphi_a(\xi)\xi'(x) \cdot \xi''(y)\big) \\
 & & \Big\downarrow {\scriptstyle \Gamma_\alpha = \Gamma_\alpha^a} \\
\Big\downarrow {\scriptstyle H_{\alpha,a}} & & \sum_\xi \epsilon_\xi^x \varphi_a(\xi)\alpha(\xi'(x) \cdot \xi''(y)) \\
 & & \Big\downarrow {\scriptstyle \sum_\xi \epsilon_\xi^x \varphi_a(\xi) H_\alpha^{\xi'(x),\xi''(y)}} \\
((\Delta\alpha) \cdot a)(x,y) & = & \sum_\xi \epsilon_\xi^x \varphi_a(\xi)\Delta(\alpha)(\xi'(x) \cdot \xi''(y))
\end{array} \tag{5}$$

Here Γ_α is defined by the elements $\epsilon_\xi^x \varphi_a(\xi) \in \mathbb{G}$ for $\xi \in M$ as in (11.3.1) with ε_ξ^x given by the *even sign convention*: $\varepsilon_\xi^x = (-1)^{p|x||\xi''|}$.

Remark. One can also use in the definition of Γ_α the *odd sign convention*: $\varepsilon_\xi^x = (-1)^{|x||\xi''|}$ for all primes $p \geq 2$.

In this case we obtain $(\Gamma_\alpha)_{\mathrm{odd}}$ with $\Gamma_\alpha = (\Gamma_\alpha)_{\mathrm{odd}}$ if p is odd and $\Gamma_\alpha \neq (\Gamma_\alpha)_{\mathrm{odd}}$ if p is even. In fact one can check for $p = 2$ the formula

(5a)
$$(\Gamma_\alpha)_{\mathrm{odd}} = \Gamma_\alpha + \Sigma\mathcal{L},$$

$$\mathcal{L} = |x| \cdot \kappa(\alpha) \cdot \sum_{\substack{\xi=(\xi',\xi'') \\ |\xi''| \text{ odd}}} \varphi_{\Delta\beta}(\xi)(\xi' x) \cdot (\xi'' y)$$

where $\mathcal{L} \in \tilde{H}^*(X)$. Here we use the following argument for the sign $(-1)^{|x||\xi''|} = -1$ in case $|x|$ and $|\xi''|$ are odd.

(5b)
$$\begin{aligned} \Gamma(-1)_\alpha &= \Gamma(3)_\alpha = \Gamma(2+1)_\alpha \\ &= (\Gamma(2)_\alpha + \Gamma(1)_\alpha)\Gamma(+_{\mathbb{G}})_\alpha, \quad \text{see (4.2.16)(2)} \\ &= (O^\square \oplus \kappa(\alpha))\Gamma(+_{\mathbb{G}})_\alpha, \quad \text{see (4.5.8)} \\ &= O^\square \oplus \kappa(\alpha). \end{aligned}$$

The equation in the bottom row of (5) is checked below. We are now ready to define the product $H_\alpha \odot H_\beta$ of Δ-tracks by the following commutative diagram, see (11.3.3).

(6)
$$\begin{array}{ccc} (\alpha\beta)(x \cdot y) & \xrightarrow{\Gamma_{\alpha,\beta}} & \alpha(\beta(x \cdot y)) \\ \Big\downarrow{\scriptstyle H_\alpha \odot H_\beta} & & \Big\downarrow{\scriptstyle \alpha H_\beta} \\ \Delta(\alpha\beta)(x,y) & \xleftarrow[H_{\alpha,\Delta\beta}]{} & \alpha(\Delta(\beta)(x,y)) \end{array}$$

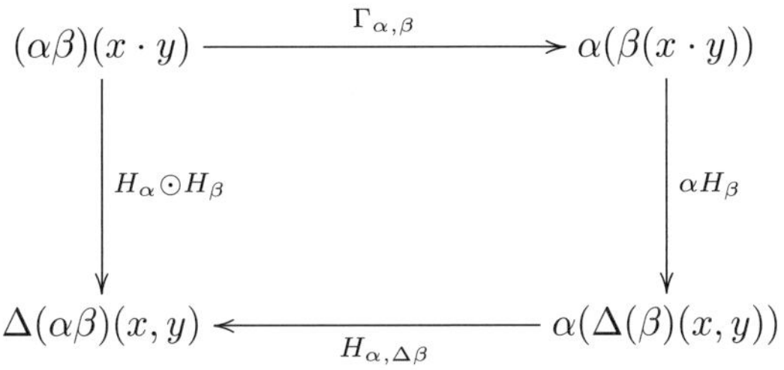

Remark. If we use the odd convention for Γ_α we get the product $(H_\alpha \odot H_\beta)_{\mathrm{odd}}$ with $(H_\alpha \odot H_\beta)_{\mathrm{odd}} = H_\alpha \odot H_\beta$ if p is odd and if p is even:

$$(H_\alpha \odot H_\beta)_{\mathrm{odd}} = (H_\alpha \odot H_\beta) + \Sigma\mathcal{L}$$

with $\mathcal{L}$ defined as in (5a) above.

We check the equation in the bottom row of (5) as follows. Let

$$\Delta(\alpha) = \sum_i \alpha_i' \otimes \alpha_i''.$$

Then we get

$$\begin{aligned}(\Delta\alpha)\cdot a &= (\sum_i \alpha_i'\otimes\alpha_i'')\cdot(\sum_\xi \varphi_a(\xi)\xi'\otimes\xi'')\\ &= \sum_{i,\xi}(-1)^{|\xi'||\alpha_i''|}\varphi_a(\xi)\alpha_i'\xi'\otimes\alpha_i''\xi''.\end{aligned}$$

Since ξ' , ξ'' are monomials we have $(\alpha_i'\xi')(x) = \alpha_i'(\xi'(x))$ and $(\alpha_i''\xi'')(y) = \alpha_i''(\xi''(y))$. Hence we get by (11.2.4)

$$\begin{aligned}((\Delta\alpha)\cdot a)(x,y) &= \sum_{i,\xi}\pm\varphi_a(\xi)\alpha_i'(\xi'x)\cdot\alpha_i''(\xi''y)\\ &= \sum_\xi \epsilon_\xi\varphi_a(\xi)\Delta(\alpha)(\xi'(x),\xi''(y)).\end{aligned}$$

The signs $\pm$ are achieved by the sign rules according to (11.1.1) and (11.2.4), that is, the sign $\pm$ in the first row is $(-1)^{|\xi'||\alpha_i''|}\cdot(-1)^{|x|\cdot|\alpha_i''\xi''|}$ and in the second row is $(-1)^{|x||\xi''|}$. If p is even, then we are in an $\mathbb{F}$-vector space so that in this case the signs can be omitted.

Remark. We have seen in section (10.5) that the interchange relation for Cartan tracks is of different form for p even and p odd respectively. This, in fact, forces us to use the even sign convention for the definition of $H_\alpha \odot H_\beta$.

11.5 The algebra $\mathcal{T}_\Delta$ of Δ-tracks

We show that the module of Δ-tracks (11.4.2) with the product (11.4.3) is an associative algebra over $\mathbb{G}$. For this we prove the following results.

11.5.1 Theorem. *The product $\odot$ on $\mathcal{T}_\Delta$ is bilinear, so that $\odot$ induces a well-defined multiplication map*

$$\odot : \mathcal{T}_\Delta\otimes\mathcal{T}_\Delta\to\mathcal{T}_\Delta.$$

The unit of the algebra $(\mathcal{T}_\Delta,\odot)$ is the Δ-track $(1,0^\square)$ where 1 is the unit of the tensor algebra $T_{\mathbb{G}}(E_{\mathcal{A}})$ with $\Delta(1)=1\otimes 1$ and

$$0^\square : 1(x\cdot y) = x\cdot y \Rightarrow (1\otimes 1)(x\cdot y) = x\cdot y$$

is the identity track which is a special Δ-track. We do not claim that the multiplication $\odot$ of τ_Δ is associative. Therefore $(\tau_\Delta,\odot)$ is only a "*magma algebra*".

Proof of (11.5.1). It suffices to consider the odd sign convention for the definition of $H_\alpha\odot H_\beta$ since $\mathcal{L}$ in (11.4.3)(5a) is linear in α and in β. We show that

$$(H_\alpha + H_{\alpha'})\odot H_\beta = H_\alpha\odot H_\beta + H_{\alpha'}\odot H_\beta.$$

This in fact is clear since $\Gamma_{\alpha,\beta}$ in (11.4.3)(6) is linear in α and Γ_α in (11.4.3)(5) is linear in α.

Next we show that

$$H_\alpha \odot (H_\beta + H_{\beta'}) = H_\alpha \odot H_\beta + H_\alpha \odot H_{\beta'}.$$

Both sides are tracks as in the following diagram.

$$\begin{array}{ccc}
(\alpha(\beta+\beta'))(x\cdot y) & \longrightarrow & \Delta(\alpha(\beta+\beta'))(x,y) \\
\| & & \| \\
(\alpha\beta+\alpha\beta')(x\cdot y) & & \| \\
\| & & \| \\
(\alpha\beta)(x\cdot y)+(\alpha\beta')(x\cdot y) & \longrightarrow & \Delta(\alpha\beta)(x,y)+\Delta(\alpha\beta')(x,y)
\end{array}$$

Now we observe that the following diagram commutes.

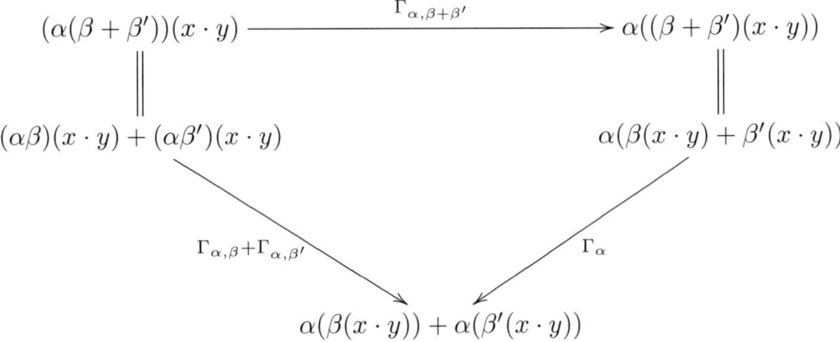

Moreover the next diagram commutes

$$\begin{array}{ccc}
\alpha(\beta(x\cdot y)+\beta'(x\cdot y)) & \xrightarrow{\Gamma_\alpha} & \alpha(\beta(x\cdot y))+\alpha(\beta'(x\cdot y)) \\
\Big\downarrow{\scriptstyle \alpha(H_\beta+H_{\beta'})} & & \Big\downarrow{\scriptstyle \alpha H_\beta+\alpha H_{\beta'}} \\
\alpha(\Delta\beta(x,y)+\Delta\beta'(x,y)) & \xrightarrow{\Gamma_\alpha} & \alpha(\Delta\beta(x,y))+\alpha(\Delta\beta'(x,y))
\end{array}$$

as follows from (4.2.5)(6). Hence it remains to check that the following diagram commutes.

$$\begin{array}{ccc}
\alpha(\Delta\beta(x,y)+\Delta\beta'(x,y)) & \xrightarrow{\Gamma_\alpha} & \alpha(\Delta\beta(x,y)+\alpha(\Delta\beta'(x,y)) \\
\| & & \Big\downarrow{\scriptstyle \Gamma_{\alpha,\Delta\beta}+\Gamma_{\alpha,\Delta\beta'}} \\
\alpha(\Delta(\beta+\beta')(x,y)) & \xrightarrow[\Gamma_{\alpha,\Delta(\beta+\beta')}]{} & \Delta(\alpha\beta)(x,y)+\Delta(\alpha\beta')(x,y)
\end{array}$$

For this we observe that by definition of the coordinate function φ in (11.4.3)(3) we have in $\mathbb{G}$ the equation

$$\varphi_{\Delta(\beta+\beta')}(\xi) = \varphi_{\Delta(\beta)}(\xi) + \varphi_{\Delta(\beta')}(\xi)$$

for all $\xi \in M \times M$ with $\mid \xi \mid = \mid \beta \mid$ so that the following diagram commutes with $w = \xi'(x) \cdot \xi''(y)$, see (11.3.1)(2).

$$\begin{array}{ccc}
\alpha(\sum_{\xi} \pm \varphi_{\Delta\beta+\Delta\beta'}(\xi)w) & \xrightarrow{\Gamma_\alpha} & \\
\Big\downarrow \Gamma_\alpha & & \alpha(\sum_{\xi} \pm \varphi_{\Delta\beta}(\xi)w) + \alpha(\sum_{\xi} \pm \varphi_{\Delta\beta'}(\xi)w) \\
\sum_{\xi} \pm \varphi_{\Delta\beta+\Delta\beta'}(\xi)\alpha(w) & \xleftarrow{\Gamma_\alpha+\Gamma_\alpha} &
\end{array}$$

This completes the proof of the theorem. □

11.6 The algebra of linear Δ-tracks

We first introduce the following linearity tracks $\Gamma \otimes 1$ and $1 \otimes \Gamma$. Let $\xi \in T_{\mathbb{G}}(E_{\mathcal{A}}) \otimes T_{\mathbb{G}}(E_{\mathcal{A}})$. Then we obtain linearity tracks.

(11.6.1)
$$\begin{aligned}\Gamma \otimes 1 &: \xi(x+x', y) \Longrightarrow \xi(x,y) + \xi(x',y),\\ 1 \otimes \Gamma &: \xi(x, y+y') \Longrightarrow \xi(x,y) + \xi(x,y')\end{aligned}$$

as follows. Let $\xi = \sum_i \xi_i' \otimes \xi_i''$ and let

$$\Gamma_{\xi_i'} : \xi_i'(x+x') \Longrightarrow \xi_i'(x) + \xi_i'(y)$$

be the linearity track. Then

$$\Gamma \otimes 1 = \sum_i (-1)^{|x||\xi_i''|} \Gamma_{\xi_i'} \cdot \xi_i''(y),$$

$$\begin{array}{ccc}
\Gamma \otimes 1 : \sum_i (-1)^{|x||\xi_i''|} \xi_i'(x+x') \cdot \xi_i''(y) & \Longrightarrow & \sum_i (-1)^{|x||\xi_i''|} (\xi_i'(x) + \xi_i'(x')) \cdot \xi_i''(y) \\
\Vert & & \Vert \\
\xi(x+x', y) & & \xi(x,y) + \xi(x',y)
\end{array}$$

Since $\Gamma_{\xi_i'}$ is linear in ξ_i' and since multiplication is bilinear, we see that $\Gamma \otimes 1$ is well defined by $\xi_{\mathbb{F}}$ and does not depend on the choice of ξ_i', ξ_i''. In a similar way we obtain $1 \otimes \Gamma$.

11.6.2 Definition. We say that a Δ-track H_α is a *linear Δ-track* (or satisfies the linearity relation) if for all x, x', y, y' the following diagrams of tracks commute.

$$\begin{array}{ccc}
\alpha((x+x')y) & \xrightarrow{H_\alpha} & \Delta(\alpha)(x+x',y) \\
\Big\downarrow{\scriptstyle \Gamma} & & \Big\downarrow{\scriptstyle \Gamma\otimes 1} \\
\alpha(xy)+\alpha(x'y) & \xrightarrow{H_\alpha+H_\alpha} & \Delta(\alpha)(x,y)+\Delta(\alpha)(x',y)
\end{array}$$

$$\begin{array}{ccc}
\alpha(x(y+y')) & \xrightarrow{H_\alpha} & \Delta(\alpha)(x,y+y') \\
\Big\downarrow{\scriptstyle \Gamma} & & \Big\downarrow{\scriptstyle 1\otimes\Gamma} \\
\alpha(xy)+\alpha(xy') & \xrightarrow{H_\alpha+H_\alpha} & \Delta(\alpha)(x,y)+\Delta(\alpha)(x,y')
\end{array}$$

Here $\Gamma \otimes 1$ and $1 \otimes \Gamma$ are defined in (11.6.1) above. By the linearity relations for Cartan tracks in section (10.7) we know that Cartan tracks C_α for $\alpha \in E_{\mathcal{A}}$ are linear Δ-tracks. Since Γ_α is linear in α we see that the sum $H_\alpha + H_\beta$ of linear tracks is again a linear Δ-track. Hence linear Δ-tracks yield a submodule

$$\mathcal{T}_\Delta(\mathrm{lin}) \subset \mathcal{T}_\Delta.$$

We now show that this submodule is actually a subalgebra. For this we prove:

11.6.3 Theorem. *If H_α and H_β are linear Δ-tracks, then $H_\alpha \odot H_\beta$ is a linear Δ-track.*

Proof of (11.6.3). It suffices to consider the odd sign convention for the definition of $H_\alpha \odot H_\beta$ since $\mathcal{L}$ in (11.4.3)(5a) is linear in x and in y. We show that all subdiagrams in the following diagram $(*)$ commute.

$$\begin{array}{ccccc}
(\alpha\beta)((x+x')y) & & \xrightarrow{\Gamma_{\alpha\beta}} & & (\alpha\beta)(xy)+(\alpha\beta)(x'y) \\
\Big\downarrow{\scriptstyle \Gamma_{\alpha,\beta}} & & & & \Big\downarrow{\scriptstyle \Gamma_{\alpha,\beta}+\Gamma_{\alpha,\beta}} \\
\alpha(\beta((x+x')y)) & \xrightarrow{\alpha\Gamma_\beta} & \bullet & \xrightarrow{\Gamma_\alpha} & \alpha(\beta(xy))+\alpha(\beta(x'y)) \\
\Big\downarrow{\scriptstyle \alpha H_\beta} & & \Big\downarrow{\scriptstyle \alpha(H_\beta+H_\beta)} & & \Big\downarrow{\scriptstyle \alpha H_\beta+\alpha H_\beta} \\
\alpha(\Delta\beta(x+x',y)) & \xrightarrow{\alpha(\Gamma\otimes 1)} & \bullet & \xrightarrow{\Gamma_\alpha} & \alpha(\Delta\beta(x,y))+\alpha(\Delta\beta(x',y)) \\
\Big\downarrow{\scriptstyle H_{\alpha,\Delta\beta}} & & & & \Big\downarrow{\scriptstyle H_{\alpha,\Delta\beta}+H_{\alpha,\Delta\beta}} \\
\Delta(\alpha\beta)(x+x',y) & & \xrightarrow[\Gamma\otimes 1]{} & & \Delta(\alpha\beta)(x,y)+\Delta(\alpha\beta)(x',y)
\end{array}$$

The top square commutes by properties of the pseudo functor (s, Γ). The square in the middle to the left commutes since we assume that H_β is a linear Δ-track; moreover the one to the right commutes by (3.1.3). Hence it remains to check that the bottom square commutes. For this let $\Delta\alpha = \sum_i \alpha_i' \otimes \alpha_i''$ as in (11.4.3)(7). Now $\Gamma \otimes 1$ in the bottom row is defined by

$$\Gamma^{x,x'}_{\alpha_i'\xi'} : \alpha_i'\xi'(x + x') \Longrightarrow \alpha_i'\xi'(x) + \alpha_i'\xi'(x').$$

Here we can use (4.2.5)(2) so that

$$\Gamma^{x,x'}_{\alpha_i'\xi'} = \Gamma^{\xi'x,\xi'x'}_{\alpha_i'} \square \alpha_i' \Gamma^{x,x'}_{\xi'}.$$

Now the following diagram commutes with $z = \xi''(y)$.

$$\begin{array}{ccccc}
\alpha(\xi'(x+x')\cdot z) & \xrightarrow{\alpha(\Gamma_{\xi'}\cdot z)} & \alpha((\xi'x+\xi'x')\cdot z) & \xrightarrow{\Gamma_\alpha} & \alpha(\xi'x\cdot z) + \alpha(\xi'x'\cdot z) \\
\Big\downarrow {\scriptstyle H_\alpha^{\xi'(x+x'),z}} & & & & \Big\downarrow {\scriptstyle H_\alpha^{\xi'x,z} + H_\alpha^{\xi'x',z}} \\
\Delta(\alpha)(\xi'(x+x'), z) & & \xrightarrow[\Gamma\otimes 1]{} & & \Delta(\alpha)(\xi'x, z) + \Delta(\alpha)(\xi'x', z)
\end{array}$$

This follows from (3.1.3) and the assumption that H_α is linear. Again using (3.1.3) for Γ_α in (11.4.3)(5) and for $\alpha(\Gamma_{\xi'} \cdot z)$ we see that the bottom square of diagram $(*)$ commutes. □

11.7 Generalized Cartan tracks and the associativity relation

The Cartan tracks C_α for $\alpha \in E_\mathcal{A}$ are linear Δ-tracks

(11.7.1) $$C_\alpha : \alpha(x \cdot y) \Longrightarrow (\Delta\alpha)(x, y).$$

We now define such tracks C_α for all $\alpha \in T_\mathbb{G}(E_\mathcal{A})$. For this we use the fact that the $\mathbb{G}$-module $\mathcal{T}_\Delta(\mathrm{lin})$ of linear Δ-tracks is an algebra and that $T_\mathbb{G}(E_\mathcal{A})$ is the free $\mathbb{G}$-algebra generated by $E_\mathcal{A}$. Hence there is a unique linear map

(11.7.2) $$C : T_\mathbb{G}(E_\mathcal{A}) \longrightarrow \mathcal{T}_\Delta(\mathrm{lin})$$

which on generators $\alpha \in E_\mathcal{A}$ is defined by $C(\alpha) = (\alpha, C_\alpha)$ where C_α is the Cartan track in (10.4.1) and which for monomials $\alpha \in \mathrm{Mon}(E_\mathcal{A})$ and $\beta \in E_\mathcal{A}$ satisfies

$$C(\alpha\beta) = (\alpha\beta, C_\alpha \odot C_\beta).$$

We point out that here $C_\alpha \odot C_\beta$ is defined by the even sign convention in (11.4.3). We get for $\alpha = \alpha_1 \cdots \alpha_r \in \mathrm{Mon}(E_{\mathcal{A}})$ with $\alpha_1 \cdots \alpha_r \in E_{\mathcal{A}}$ the Δ-track

$$C(\alpha) = (\alpha, C_\alpha)$$

with

$$C_\alpha = (\cdots(C_{\alpha_1} \odot C_{\alpha_2}) \odot \cdots \odot C_{\alpha_{r-1}}) \odot C_{\alpha_r}.$$

We call C_α the *generalized Cartan track*. We summarize the following properties of generalized Cartan tracks for $\alpha, \beta \in T_{\mathbb{G}}(E_{\mathcal{A}})$.

(1) $C_{\alpha+\beta} = C_\alpha + C_\beta,$

(2) $C_{\alpha\beta} = C_\alpha \odot C_\beta,$ for $\beta \in E_{\mathcal{A}},$

(3) C_α is a linear Δ–track,

(4) C_α is the Cartan track (10.4.1) if $\alpha \in \mathrm{E}_{\mathcal{A}}$.

Given $\xi \in T_{\mathbb{G}}(E_{\mathcal{A}}) \otimes T_{\mathbb{G}}(E_{\mathcal{A}})$ we define $\xi(x, y)$ as in (11.2.4). Now the generalized Cartan tracks induce tracks

$$\begin{aligned} C \otimes 1 &: \xi(x \cdot y, z) \Longrightarrow ((\Delta \otimes 1)(\xi))(x, y, z), \\ 1 \otimes C &: \xi(x, y \cdot z) \Longrightarrow ((1 \otimes \Delta)(\xi))(x, y, z) \end{aligned} \tag{11.7.3}$$

which are linear in ξ. Here the right-hand side is defined also in (11.2.4). We get for $\xi = \sum_i \xi_i' \otimes \xi_i''$ the track $C \otimes 1$ by the generalized Cartan tracks

$$C_i : \xi_i'(x \cdot y) \Longrightarrow (\Delta \xi_i')(x, y).$$

Namely we set

$$C \otimes 1 = \sum_i (-1)^{|xy| \cdot |\xi_i''|} C_i \cdot \xi_i''(z)$$

so that

$$\begin{array}{ccc} \sum_i (-1)^{|xy||\xi_i''|} \xi_i'(xy) \cdot \xi_i''(z) & = & \xi(xy, z) \\ \Big\downarrow{\scriptstyle C\otimes 1} & & \\ \sum_i (-1)^{|xy||\xi_i''|} (\Delta(\xi_i')(x, y)) \cdot \xi_i''(z) & = & ((\Delta \otimes 1)(\xi))(x, y, z). \end{array}$$

Since C_i is linear in ξ_i' and since multiplication is bilinear we see that $C \otimes 1$ is well defined. In a similar way we get $C \otimes 1$.

We prove that the associativity relation for Cartan tracks in (10.6) yields a corresponding relation for generalized Cartan tracks.

11.7.4 Theorem. *Let $\alpha \in T_{\mathbb{G}}(E_{\mathcal{A}})$ and let x, y, z be elements in $[\![X, Z^*]\!]_0$. Then the following diagram is commutative.*

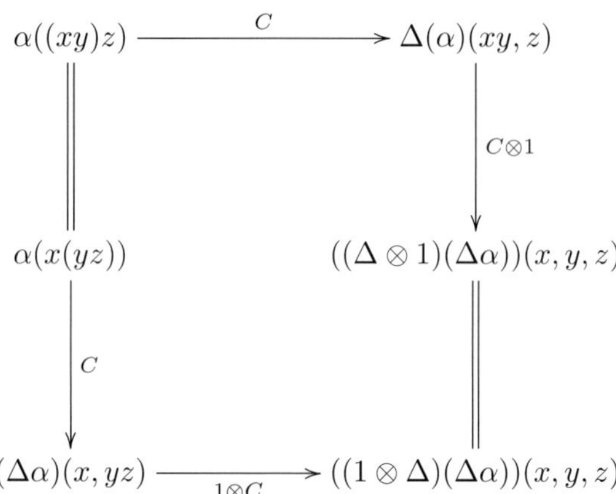

Proof. All morphisms and tracks are linear in α. Therefore we need to prove the result only for $\alpha \in \mathrm{Mon}(E_{\mathcal{A}})$. For $\alpha \in E_{\mathcal{A}}$ we know already that the diagram is commutative since it is equivalent to the commutative diagram in (10.6.4). We now use an induction argument.

Assume the proposition holds for all monomials $\alpha = \alpha_1 \dots \alpha_r \in \mathrm{Mon}(E_{\mathcal{A}})$ of length $\leq r$ and let $\beta \in E_{\mathcal{A}}$. Then we get for $C = C_{\alpha\beta} = C_\alpha \odot C_\beta$ the following diagram with $\bar{\Delta} = (\Delta \otimes 1)\Delta = (1 \otimes \Delta)\Delta$.

(1)
$$\begin{array}{ccccc}
(\alpha\beta)(xyz) & & & & \\
\downarrow{\scriptstyle \Gamma_{\alpha,\beta}} & & & & \\
\alpha(\beta(xyz)) & \xrightarrow{\alpha C_\beta} & \alpha(\Delta\beta(xy,z)) & \xrightarrow{C_{\alpha,\Delta\beta}} & \Delta(\alpha\beta)(xy,z)) \\
\downarrow{\scriptstyle \alpha C_\beta} & & \downarrow{\scriptstyle \alpha(C\otimes 1)} & & \\
\alpha(\Delta\beta(x,yz)) & \xrightarrow{\alpha(1\otimes C)} & \alpha(\bar{\Delta}\beta(x,y,z)) & & \downarrow{\scriptstyle C\otimes 1} \\
\downarrow{\scriptstyle C_{\alpha,\Delta\beta}} & & {\scriptstyle u}\searrow \searrow{\scriptstyle v} & & \\
\Delta(\alpha\beta)(x,yz) & & \xrightarrow{\quad 1\otimes C \quad} & & \bar{\Delta}(\alpha\beta)(x,y,z)
\end{array}$$

The square containing αC_β commutes by (10.6) since $\beta \in E_{\mathcal{A}}$. Hence it remains to check that the subdiagrams containing u, v are commutative. For this we consider the next diagram with

$$a = \Delta\beta = \sum_{\xi \in M \times M} \varphi_a(\xi)\xi' \otimes \xi''.$$

Here we have $\xi', \xi'' \in E_{\mathcal{A}}$ since $\beta \in E_{\mathcal{A}}$, compare the definition of Δ. Moreover let

$$\Delta\alpha = \sum_i \alpha_i' \otimes \alpha_i''.$$

We consider the following diagram.

(2)

$$\begin{array}{ccc}
\alpha(\sum_\xi \pm \varphi_a(\xi)\xi'(x) \cdot \xi''(yz)) & \xrightarrow{\alpha(1 \otimes C)} & \alpha(\sum_\xi \pm \varphi_a(\xi)\xi'(x) \cdot (\Delta\xi'')(y,z)) \\
\downarrow{\scriptstyle \Gamma_\alpha} & & \downarrow{\scriptstyle \Gamma_\alpha} \\
\sum_\xi \pm \varphi_a(\xi)\alpha(\xi'(x) \cdot \xi''(yz)) & \xrightarrow{\alpha(1 \cdot C)} & \sum_\xi \pm \varphi_a(\xi)\alpha(\xi'(x) \cdot (\Delta\xi'')(y,z)) \\
\downarrow{\scriptstyle C_\alpha^{\xi'x, \xi''(yz)}} & & \downarrow{\scriptstyle C_\alpha^{\xi'x, \Delta\xi''(y,z)}} \\
\sum_\xi \pm \varphi_a(\xi)\Delta\alpha(\xi'(x), \xi''(yz)) & \xrightarrow{\Delta\alpha(1, C_{\xi''})} & \sum_\xi \pm \varphi_a(\xi)(\Delta\alpha)(\xi'(x), \Delta\xi''(y,z)) \\
\| & & \downarrow{\scriptstyle 1 \cdot H_{\alpha_i'', \Delta\xi''}} \\
\sum_{i,\xi} \pm \varphi_a(\xi)(\alpha_i'\xi')x \cdot (\alpha_i''\xi'')(yz) & \xrightarrow{1 \cdot C_{\alpha_i''\xi''}} & \sum_{i,\xi} \pm \varphi_a(\xi)(\alpha_i'\xi')x \cdot \Delta(\alpha_i''\xi'')(y,z) \\
\| & & \| \\
\Delta(\alpha\beta)(x, yz) & \xrightarrow{1 \otimes C} & \bar{\Delta}(\alpha\beta)(x,y,z)
\end{array}$$

The squares containing $\alpha(1 \cdot C)$ commute by (3.1.3) and the square containing $1 \cdot H_{\alpha_i'', \Delta\xi''}$ commutes since $C_{\alpha_i''\xi''} = C_{\alpha_i''} \odot C_{\xi''}$. The bottom square commutes by definition of $1 \otimes C$. The column to the left-hand side yields $C_{\alpha,\Delta\beta}$ and the column to the right-hand side defines the track u. In the same way we get the track v in diagram (1). Hence it remains to check that the following diagram commutes.

(3)

$$
\begin{array}{ccccc}
\alpha(\sum_\xi \pm \varphi_a(\xi)\xi'(x)\cdot(\Delta\xi'')(y,z)) & & = & & \alpha(\sum_\xi \pm \varphi_a(\xi)\Delta(\xi')(x,y)\cdot\xi''(z)) \\
\downarrow{\scriptstyle \Gamma_\alpha} & & & & \downarrow{\scriptstyle \Gamma_\alpha} \\
\sum_\xi \pm \varphi_a(\xi)\alpha(\xi'(x)\cdot(\Delta\xi'')(y,z)) & \xrightarrow{\Gamma_\alpha} & A & \xleftarrow{\Gamma_\alpha} & \sum_\xi \pm \varphi_a(\xi)\alpha(\Delta(\xi')(x,y)\cdot\xi''(z)) \\
\downarrow{\scriptstyle C_\alpha^{\xi' x,\Delta\xi''(y,z)}} & & & & \downarrow{\scriptstyle C_\alpha^{\Delta(\xi')(x,y),\xi'' z}} \\
\sum_\xi \pm \varphi_a(\xi)\Delta\alpha(\xi' x,\Delta\xi''(y,z)) & & \downarrow & & \sum_\xi \pm \varphi_a(\xi)\Delta\alpha(\Delta\xi'(x,y),\xi'' z) \\
\downarrow{\scriptstyle \alpha'_i\xi' x\cdot C_{\alpha''_i,\Delta\xi''}} & & & & \downarrow{\scriptstyle C_{\alpha'_i,\Delta\xi'}\cdot\alpha''_i\xi'' z} \\
\bar{\Delta}(\alpha\beta)(x,y,z) & = & B & = & \bar{\Delta}(\alpha\beta)(x,y,z)
\end{array}
$$

Here the left-hand column is u and the right-hand column is v. We have for $\bar{\Delta} = (1\otimes\Delta)\Delta = (\Delta\otimes 1)\Delta$ the equation

$$
\begin{aligned}
\bar{\Delta}(\beta) &= \sum_{\eta\in M\times M\times M} \psi(\eta)\cdot\eta'\otimes\eta''\otimes\eta''' \\
&= \sum_\xi \varphi_{\Delta\beta}\xi'\otimes(\sum_\rho \varphi_{\Delta\xi''}(\rho)\rho'\otimes\rho'') \\
&= \sum_\xi \varphi_{\Delta\beta}(\sum_\rho \varphi_{\Delta\xi'}(\rho)\rho'\otimes\rho'')\otimes\xi''.
\end{aligned}
$$

Moreover let

$$\bar{\Delta}(\alpha) = \sum_j \alpha'_j\otimes\alpha''_j\otimes\alpha'''_j.$$

Now we define the track $A \to B$ in (3) by the following diagram.

(4)

$$
\begin{array}{c}
A = \sum_\eta \pm\psi(\eta)\alpha(\eta' x\cdot\eta'' y\cdot\eta''' z) \\
\downarrow{\scriptstyle \sum_\eta \pm\psi(\eta)\bar{C}_\alpha} \\
B = \sum_{\eta,j} \pm\psi(\eta)(\alpha'_j\eta')x\cdot(\alpha''_j\eta'')y\cdot(\alpha'''_j\cdot\eta''')z
\end{array}
$$

Here we set

$$\bar{C_\alpha} = (C \otimes 1)\Box C_\alpha = (1 \otimes C)\Box C_\alpha \tag{5}$$

since the proposition holds for α. Now the top square in (3) commutes by (11.3.2). Moreover the bottom squares of (3) commute by the linearity of C_α and by (5). For this recall that $C_{\alpha,\Delta\beta}$ is the composite of Γ_α and tracks defined by C. This completes the proof of the theorem. □

Remark. In the proof of (11.7.4) above we use the even convention for the definition of $C_\alpha \odot C_\beta$ so that one has to be aware that for p odd and p even different sign rules are used, see (11.4.3). It is also possible to go through all arguments of the proof of (11.7.4) using the odd convention for the definition of $C_\alpha \odot C_\beta$. Then alteration is necessary according to the correction term $\mathcal{L}$ in (11.4.3)(5a). One can show that all correction terms arising cancel. In fact, if we replace α in (11.7.4) by $\alpha\beta$ with $\beta = Sq^n$ we get the following four corrections with $i + j + k = n$:

$$\mathcal{L}_1 = \kappa(\alpha)|x| \sum_{j+k \text{ odd}} Sq^i x \cdot Sq^j y \cdot Sq^k z, \quad \text{for } x \cdot (yz),$$

$$\mathcal{L}_2 = \kappa(\alpha)|xy| \sum_{k \text{ odd}} Sq^i x \cdot Sq^j y \cdot Sq^k z, \quad \text{for } (xy) \cdot z,$$

$$\mathcal{L}_3 = \kappa(\alpha)|x| \sum_{j \text{ odd}} Sq^i x \cdot Sq^j y \cdot Sq^k z,$$

$$\mathcal{L}_4 = \kappa(\alpha)|y| \sum_{k \text{ odd}} Sq^i x \cdot Sq^j y \cdot Sq^k z.$$

Here $\mathcal{L}_3$ and $\mathcal{L}_4$ arise for $C \otimes 1$ and $1 \otimes C$ respectively. Now it is easy to see that $\mathcal{L}_1 + \mathcal{L}_2 + \mathcal{L}_3 + \mathcal{L}_4 = 0$.

11.8 Stability of Cartan tracks

The universal Cartan track C_α is stable with respect to partial loop operations. This property of Cartan tracks is the crucial argument in the construction of the relation diagonal in the next section.

Let $\alpha \in \mathcal{B}_0$ and let $\bar{\alpha} = \{(\alpha'_i, \alpha''_i),\ i \in I\}$ be a Δ-family associated to α. Then the universal Cartan track $C_\alpha = C_\alpha^{q,q'}$ is given by

(11.8.1)

$$\begin{array}{ccc} Z^q \times Z^{q'} & \xrightarrow{s\bar{\alpha}} & \prod_{i \in I} Z^{n_i} \times Z^{n'_i} \\ \Big\downarrow \mu & \overset{C_\alpha}{\Longrightarrow} & \Big\downarrow \mu_0 \\ Z^n & \xrightarrow[s\alpha]{} & Z^{n+m} \end{array}$$

with $n_i = q + \mid \alpha_i' \mid$, $n_i' = q' + \mid \alpha_i'' \mid$ and $s\bar{\alpha}$ defined with signs as in (11.2.6). Here C_α can be chosen to be a track under

$$Z^q \vee Z^{q'} \subset Z^q \times Z^{q'}$$

since $C_\alpha | Z^q \vee Z^{q'} = 0$ is the identity track of the trivial map 0. We therefore can apply the following partial loop operation to C_α.

11.8.2 Definition. Let A, B and U be pointed spaces and let

$$f : A \times B \longrightarrow U$$

be a pointed map with $f|A \vee B = 0 : A \vee B \to * \to U$. Then the *left partial loop*, resp. the *right partial loop*, are maps

$$\begin{aligned} Lf &: (\Omega A) \times B \longrightarrow \Omega U, \\ L'f &: A \times (\Omega B) \longrightarrow \Omega U, \end{aligned}$$

defined as follows. Let $t \in S^1$ and $\sigma \in \Omega A$, $\sigma' \in \Omega B$. Then we set $(Lf)(\sigma, b)(t) = f(\sigma(t), b)$ for $b \in B$, resp. we set $(Lf')(a, \sigma')(t) = f(a, \sigma'(t))$ for $a \in A$.

We apply the partial loop operation L to C_α in (11.8.1) and we get the following diagram.

(11.8.3)
$$\begin{array}{ccc} (\Omega(Z^q) \times Z^{q'} & \xrightarrow{\tilde{\Omega}(\bar{\alpha})} & \prod_{i \in I} (\Omega Z^{n_i}) \times Z^{n_i'} \\ {\scriptstyle L\mu_0} \downarrow & \overset{LC_\alpha}{\Longrightarrow} & \downarrow {\scriptstyle L^I \bar{\mu}_0} \\ \Omega Z^n & \xrightarrow[\Omega(s\alpha)]{} & \Omega Z^{n+m} \end{array}$$

Here we define the arrow $\tilde{\Omega}(\bar{\alpha})$ by the coordinates $\Omega(s\alpha_i') \times s\alpha_i''$ for $i \in I$. Moreover we have for the maps $(s\alpha)\mu_0$ and $\mu_0 s\bar{\alpha}$ in (11.8.1) the equations

(1)
$$\begin{aligned} L((s\alpha)\mu_0) &= (\Omega s\alpha) \circ (L\mu_0), \\ L(\bar{\mu_0}(s\bar{\alpha})) &= (L^I \bar{\mu_0}) \circ (\tilde{\Omega}\bar{\alpha}), \end{aligned}$$

so that diagram (11.8.3) is well defined. Here $L^I \bar{\mu}_0$ carries a tuple $z = ((\sigma_i, b_i), i \in I)$ to the sum in ΩZ^{n+m},

(2)
$$(L^I \bar{\mu_0})(z) = \sum_{i \in I} (L\mu_0)(\sigma_i, b_i)$$

where we use addition of loops induced by the vector space structure of Z^{n+m}.

Recall that $s\alpha$ is a stable map defined in (10.8.7) so that we have the diagram

(3)
$$\begin{array}{ccc} Z^{n-1} & \xrightarrow{s\alpha} & Z^{n+m-1} \\ \downarrow r & \overset{H}{\Longrightarrow} & \downarrow r \\ \Omega Z^n & \xrightarrow[\Omega(s\alpha)]{} & \Omega Z^{n+m} \end{array}$$

where we omit indices, see (10.8.6). A similar diagram is available for the stable map $s\alpha_i'$ for each $i \in I$. This shows that we have the following diagram.

(4)
$$\begin{array}{ccc} Z^{q-1} \times Z^{q'} & \xrightarrow{s\bar{\alpha}} & \prod_{i\in I} Z^{n_i-1} \times Z^{n_i'} \\ \downarrow (-1)^{q'} r\times 1 & \overset{\bar{H}}{\Longrightarrow} & \downarrow \prod_{i\in I}(-1)^{n_i'} r\times 1 \\ (\Omega Z^q) \times Z^{q'} & \xrightarrow{\tilde{\Omega}\bar{\alpha}} & \prod_{i\in I} (\Omega Z^{n_i}) \times Z^{n_i'} \end{array}$$

Here $\bar{H}$ has coordinates $H_i \times 1$ where H_i is the track defined by the stable map $s\alpha_i'$ as in (3) above. We point out that the definition of $s\bar{\alpha}$ in (11.2.6) involves signs which cancel with the signs $(-1)^{q'} \cdot (-1)^{n_i'} = (-1)^{|\alpha_i''|}$.

Moreover we have a canonical permutation track for the following diagram.

(5)
$$\begin{array}{ccc} Z^{q-1} \times Z^{q'} & \xrightarrow{(-1)^{q'} r\times 1} & (\Omega Z^q) \times Z^{q'} \\ \downarrow \mu_0 & \Longrightarrow & \downarrow L\mu_0 \\ Z^{n-1} & \xrightarrow[r]{} & \Omega Z^n \end{array}$$

This is easily seen by taking the adjoint t of r as in (10.8.4)(1),(3).

Similarly we get as in (5) the following track.

(6)
$$\begin{array}{ccc} \prod_{i\in I} Z^{n_i-1} \times Z^{n_i'} & \xrightarrow{\prod(-1)^{n_i'} r\times 1} & \prod_{i\in I} (\Omega Z^{n_i}) \times Z^{n_i'} \\ \downarrow \bar{\mu}_0 & \Longrightarrow & \downarrow L^I \bar{\mu}_0 \\ Z^{n+m-1} & \xrightarrow[r]{} & \Omega Z^{n+m} \end{array}$$

Now pasting tracks LC_α and (3), (4), (5) and (6) yields a track $\mathcal{L}C_\alpha$ as in the following diagram.

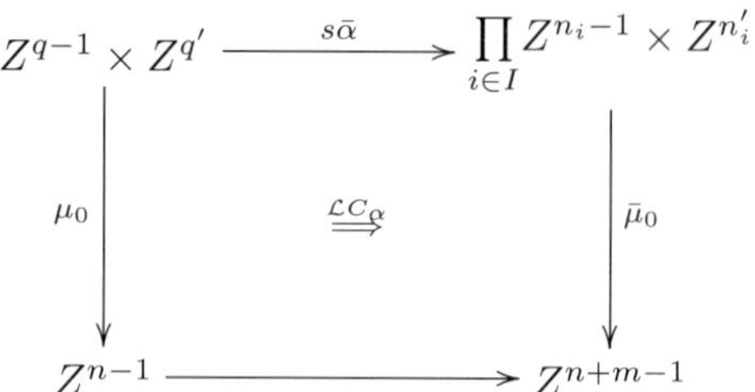

The *left stability of the Cartan track* is expressed in the following theorem.

11.8.4 Theorem. *Pasting the tracks* $LC_\alpha^{q,q'}$ *and the tracks* (3), (4), (5) *and* (6) *above yields a track* $\mathcal{L}C_\alpha$ *which coincides with* $C_\alpha^{q-1,q'}$.

Proof. We point out that H and $\bar{H}$ are defined by Cartan tracks as in (10.8.6). Therefore the theorem is a consequence of the associativity relation for Cartan tracks in section (11.7). □

A similar result as above yields the *right stability of the Cartan track* showing that the right partial loop $L'C_\alpha^{q,q'}$ yields a track $\mathcal{L}'C_\alpha$ which coincides with $C_{\alpha'}^{q,q'-1}$. The proof of this case is somewhat simpler since signs such as in (11.8.3)(4) do not arise.

11.9 The relation diagonal

According to the definition of the pair algebra $\mathcal{B}$ we have the pull back diagram

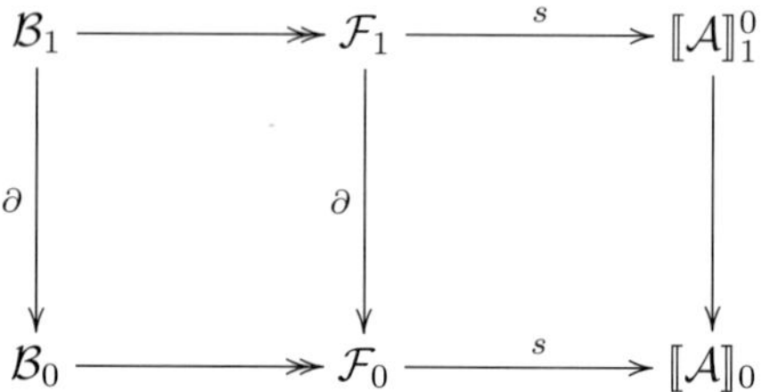

where $\mathcal{F}_0 = \mathcal{B}_0/p\mathcal{B}_0$ and $\mathcal{F}_1 = \mathcal{B}_1/[p]\mathcal{B}_0$ and $\twoheadrightarrow$ is the quotient map. Here $\mathcal{F}_0$ and $\mathcal{F}_1$ are $\mathbb{F}$-vector spaces, see also (11.1.6).

11.9.1 Lemma. *The diagram above is well defined and both squares are pull back diagrams*

Proof. The pair algebra $\mathcal{B}$ is derived from the secondary Steenrod algebra $[\![\mathcal{A}]\!]$ by the pull back diagram

(1)

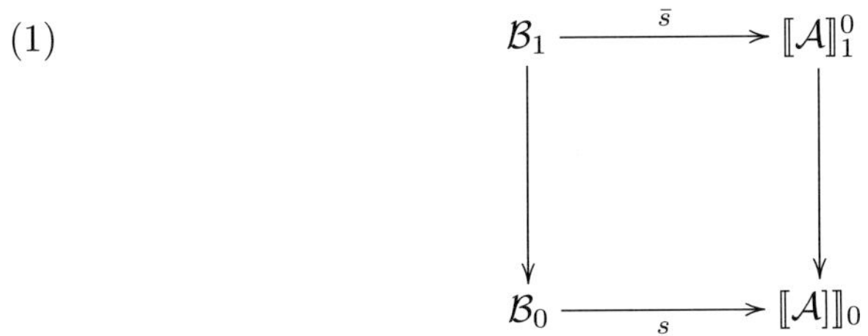

where the $\mathbb{G}$-linear map s is defined by Künneth-Steenrod operations. Here $[\![\mathcal{A}]\!]_1^0$ and $[\![\mathcal{A}]\!]_0$ are graded $\mathbb{F}$-vector spaces. Hence s carries $p\mathcal{B}_0$ to zero and, in fact, $\bar{s}$ carries $[p]\mathcal{B}_0$ to zero since $\bar{s}([p]\alpha)$ with $\alpha \in \mathcal{B}_0$ is defined by the track

(2)

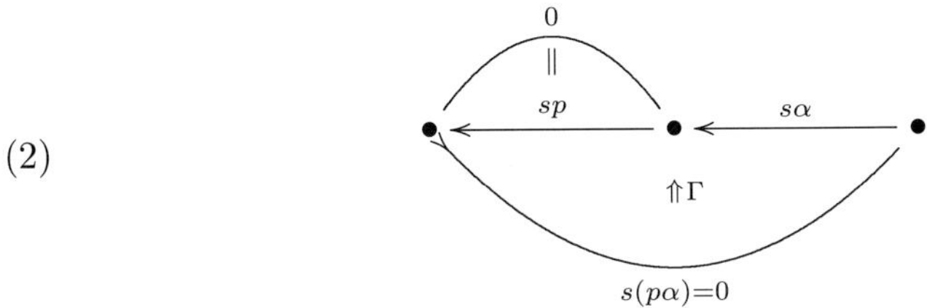

where $\Gamma = \Gamma(p, \alpha)$. Here $\Gamma(x, \alpha)$ is linear in x so that $\Gamma = 0$ and hence $\bar{s}([p]\alpha) = 0$. Now we get the induced diagram (11.9.1). The square at the left-hand side is a pull back since $\partial : [p] \cdot \mathcal{B}_0 \to p\mathcal{B}_0$ is an isomorphism. Moreover the pair $(s, \bar{s})$ is a pull back (1) by definition of $\mathcal{B}_1$. This shows that also the square at the right-hand side is a pull back and therefore $\mathcal{B}_1/[p]\mathcal{B}_0 = \mathcal{F}_1$ is an $\mathbb{F}$-vector space. $\square$

We say that $\alpha \in \mathcal{B}_0$, resp. $\alpha \in \mathcal{F}_0$, is a *relation* if α is in the image of ∂ or equivalently α is in the kernel of $q : \mathcal{B}_0 \to \mathcal{A}$, resp. $q : \mathcal{F}_0 \to \mathcal{A}$. Let

(11.9.2)
$$\begin{cases} \mathcal{R}_{\mathcal{B}} = \operatorname{im}(\partial) = \ker(q) \subset \mathcal{B}_0, \\ \mathcal{R}_{\mathcal{F}} = \operatorname{im}(\partial) = \ker(q) \subset \mathcal{F}_0 \end{cases}$$

be the submodules of relations. We can choose an $\mathbb{F}$-linear map ρ for which the following diagram commutes, we call ρ a *splitting of* $\partial_{\mathcal{F}}$.

(1)
$$\begin{array}{ccc} & & \mathcal{F}_1 \\ & \overset{\rho}{\nearrow} & \big\downarrow {\scriptstyle \partial = \partial_{\mathcal{F}}} \\ \mathcal{R}_{\mathcal{F}} & \subset & \mathcal{F}_0 \end{array}$$

In this section we associate to ρ a well-defined $\mathbb{G}$-linear map of degree -1, termed the *relation diagonal*,

(2)
$$\Theta_\rho : \mathcal{R}_{\mathcal{B}} \longrightarrow \mathcal{A} \otimes \mathcal{A}.$$

Since the kernel of $\partial_{\mathcal{F}}$ is $\Sigma\mathcal{A}$, each $\mathbb{F}$-linear map $t : R_{\mathcal{F}} \to \mathcal{A}$ of degree -1 yields the splitting $\rho + t : R_{\mathcal{F}} \to \mathcal{F}_1$ of ∂. We shall prove the formula

$$\text{(3)} \qquad \Theta_{\rho+t}(\alpha) = \Theta_\rho(\alpha) + \bar{t}(\Delta\alpha) - \delta t(\alpha).$$

Here $\delta : \mathcal{A} \to \mathcal{A}\otimes\mathcal{A}$ is the diagonal of the Steenrod algebra $\mathcal{A}$ and $\bar{t}(\Delta\alpha)$ is defined as follows. Since $\alpha \in R_{\mathcal{B}}$ is a relation we can write

$$\text{(4)} \qquad \Delta\alpha = \sum_{i\in I_0} \alpha_i' \otimes \alpha_i'' + \sum_{i\in I_1} \alpha_i' \otimes \alpha_i''$$

with $\alpha_i' \in R_{\mathcal{B}}$ for $i \in I_0$ and $\alpha_i'' \in R_{\mathcal{B}}$ for $i \in I_1$. Now we set

$$\text{(5)} \qquad \bar{t}(\Delta\alpha) = \sum_{i\in I_0} (t\alpha_i') \otimes q\alpha_i'' + \sum_{i\in I_1} (-1)^{|\alpha_i'|} (q\alpha_i') \otimes t\alpha_i''.$$

Hence formula (3) shows that the coset of Θ_ρ,

$$\{\Theta_\rho\} \in \mathrm{Hom}_{-1}(\mathcal{R}_{\mathcal{B}}, \mathcal{A}\otimes\mathcal{A})/\{\bar{t}\Delta - \delta t, t \in \mathrm{Hom}_{-1}(\mathcal{R}_{\mathcal{F}}, \mathcal{A})\},$$

does not depend on the choice of ρ. Hence this coset is an additional structure of the pair algebra $\mathcal{B}$. We point out that $R_{\mathcal{B}} \otimes \mathbb{F}$ does not coincide with $R_{\mathcal{F}}$, for example in degree 0 the module $(R_{\mathcal{F}})^0$ is trivial but $(R_{\mathcal{B}} \otimes \mathbb{F})^0$ is not trivial.

11.9.3 Definition. Using the map s in (11.2.1) we define for $x : X \to Z^q$ and $u \in \mathcal{F}_1$ with $\partial u = \alpha \in \mathcal{F}_0$ the track

$$u(x) : \alpha(x) \Rightarrow 0$$

with $\alpha(x) = (s\alpha)_q \circ x$ as in (11.2.1) and $u(x) = (su)_q \circ x$ accordingly.

For $\alpha \in R_{\mathcal{B}}$ we have the generalized Cartan track C_α as in the following commutative diagram of tracks.

$$\text{(11.9.4)} \qquad \begin{array}{ccc} \alpha(x\cdot y) & \xrightarrow{C_\alpha} & \Delta(\alpha)(x,y) \\ \downarrow{\scriptstyle (\rho\alpha)(x\cdot y)} & & \downarrow{\scriptstyle (\bar{\rho}\Delta\alpha)(x,y)} \\ 0 & \xrightarrow[\Theta_{x,y}]{} & 0 \end{array}$$

Here $(\rho\alpha)(x\cdot y)$ is defined by the notation (11.9.3) and we get by use of (11.9.1)(4),

$$\begin{aligned} (\bar{\rho}\Delta\alpha)(x,y) &= \sum_{i\in I_0} (-1)^{|x||\alpha_i''|} (\rho\alpha_i')(x)\cdot\alpha_i''(y) \\ &+ \sum_{i\in I_1} (-1)^{|x||\alpha_i''|} \alpha_i'(x)\cdot(\rho\alpha_i'')(y). \end{aligned}$$

Again we use the notation in (11.9.3).

11.9.5 Lemma. *The element $(\bar{\rho}\Delta\alpha)(x,y)$ only depends on ρ, α, x, y and does not depend on the choice of the decomposition (11.9.1)(4) of $\Delta\alpha$.*

Proof. We claim that each element $u \in (\mathcal{F}\bar{\otimes}\mathcal{F})_1$ with $\partial u = \xi \in \mathcal{F}_0 \otimes \mathcal{F}_0$ yields a well-defined track

$$u(x,y) : \xi(x,y) \Longrightarrow 0.$$

Here $\xi(x,y)$ is defined as in (11.2.4). Here $u(x,y)$ is linear in u so that we can define $u(x,y)$ by the special cases

$$\begin{aligned}(a \otimes \beta)(x,y) &= (-1)^{|x||\beta|} a(x) \cdot \beta(y),\\ (\alpha \otimes b)(x,y) &= (-1)^{|x||b|} \alpha(x) \cdot b(y),\end{aligned}$$

for $\alpha, \beta \in \mathcal{F}_0$, $a, b \in \mathcal{F}_1$. By (3.1.3) we see that the definition of $u(x,y)$ is compatible with the $\bar{\otimes}$ relation in (5.1.2). Let $\mathcal{R} = (R_{\mathcal{F}} \subset \mathcal{F}_0)$ be the pair given by $R_{\mathcal{F}}$ so that $\rho : \mathcal{R} \longrightarrow \mathcal{F}$ is a pair map over the identity of $\mathcal{F}_0$. Then we see that the following diagram commutes.

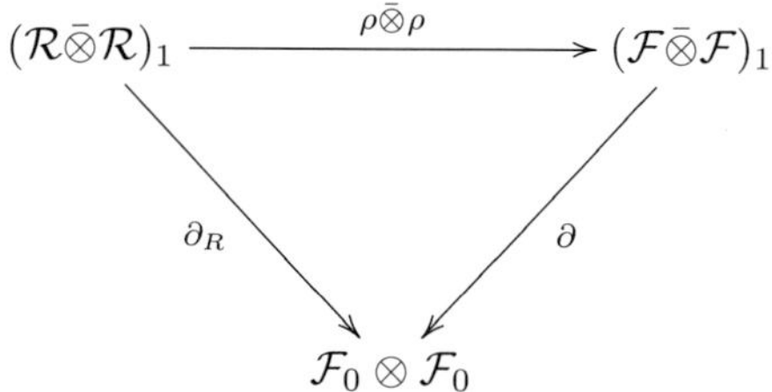

Here ∂_R is injective inducing an identification

$$(\mathcal{R}\bar{\otimes}\mathcal{R})_1 \cong \mathrm{im}(\partial)$$

so that $\bar{\rho} : \mathrm{im}(\partial) \longrightarrow (\mathcal{F}\bar{\otimes}\mathcal{F})_1$ is defined by $\rho\bar{\otimes}\rho$. For $\alpha \in R_{\mathcal{B}}$ we can consider $\Delta\alpha$ as an element in $\mathrm{im}(\partial)$ so that $\bar{\rho}(\Delta\alpha) \in (\mathcal{F}\bar{\otimes}\mathcal{F})_1$ is well defined. Now one can check that for $u = \bar{\rho}(\Delta\alpha)$ the track $u(x,y)$ above coincides with $(\bar{\rho}\Delta\alpha)(x,y)$ defined in (11.9.4). □

11.9.6 Theorem. *The composite of tracks $\theta_{x,y}$ in (11.9.4) is a track $0 \Longrightarrow 0$ which represents an element*

$$\Theta_{x,y} \in H^{|x|+|y|+|\alpha|-1}(X)$$

depending only on α and ρ and x and y by (11.9.5). There exists a unique element

$$\Theta_\rho(\alpha) \in (\mathcal{A} \otimes \mathcal{A})^{|\alpha|-1}$$

such that for all X, x, y the multiplication map $\mu : H^*(X) \otimes H^*(X) \longrightarrow H^*(X)$ satisfies

$$\mu(\Theta_\rho(\alpha) \cdot (x \otimes y)) = \Theta_{x,y}.$$

Moreover θ_ρ satisfies formula (11.9.2)(3).

This theorem defines the *relation diagonal* θ_ρ in (11.9.2(2)). Since diagram (11.9.4) is linear in α we see that θ_ρ is $\mathbb{G}$-linear.

Proof of (11.9.6). The theorem is a direct consequence of the left and right stability of Cartan tracks in (11.9). Formula (11.9.2)(3) is easily checked by use of diagram (11.9.4). □

Remark. The main result of Kristensen in [Kr2] can be interpreted as a corollary of Theorem (11.9.6). We also point out that the definition of $\Theta_{x,y}$ in (11.9.4) can be compared with the definition of the secondary products of Kock-Kristensen [KKr].

11.10 The right action on the relation diagonal

The module of relations $\mathcal{R}_\mathcal{B} \subset \mathcal{B}_0$ is an ideal so that for $\alpha \in \mathcal{R}_\mathcal{B}$ and $\beta \in \mathcal{B}_0$ also $\beta\alpha, \alpha\beta \in \mathcal{R}_\mathcal{B}$. In this section we describe the relation diagonal element $\Theta_\rho(\alpha\beta)$ in terms of $\Theta_\rho(\alpha)$. Let $\rho_\mathcal{B}$ be the splitting of $\partial_\mathcal{B}$ as in the diagram

(11.10.1)
$$\begin{array}{ccc} & & \mathcal{B}_1 \\ & \overset{\rho_\mathcal{B}}{\nearrow} & \big\downarrow {\scriptstyle \partial=\partial_\mathcal{B}} \\ \mathcal{R}_\mathcal{B} & \subset & \mathcal{B}_0 \end{array}$$

where $\rho_\mathcal{B}$ is induced by ρ via the pull back diagram (11.9.1). The splitting $\rho_\mathcal{B}$ is $\mathbb{G}$-linear but not a morphism of $\mathcal{B}_0$-bimodules. Since the kernel of $\partial_\mathcal{B}$ is $\Sigma\mathcal{A}$ we get elements $\nabla_\rho(\beta,\alpha)$, $\nabla'_\rho(\alpha,\beta) \in \mathcal{A}$ of degree $\mid \alpha \mid + \mid \beta \mid -1$ defined by the following equation in $\mathcal{B}_1$,

(11.10.2)
$$\begin{aligned} \rho_\mathcal{B}(\beta\alpha) &= \beta \cdot \rho_\mathcal{B}(\alpha) + \nabla_\rho(\beta,\alpha), \\ \rho_\mathcal{B}(\alpha\beta) &= \rho_\mathcal{B}(\alpha)\cdot\beta + \nabla'_\rho(\alpha,\beta). \end{aligned}$$

Here ∇_ρ and $\nabla^{'}_\rho$ are bilinear functions.

The elements $u = \beta \cdot \rho_\mathcal{B}(\alpha)$ or $v = \rho_\mathcal{B}(\alpha)\cdot\beta$ considered as elements in $\mathcal{F}_1$ by $\mathcal{B}_1 \to \mathcal{F}_1$ yield, as in (11.9.3), tracks

(11.10.3)
$$\begin{aligned} (\beta\cdot\rho_\mathcal{B}(\alpha))(x) &: (\beta\alpha)(x) \longrightarrow 0, \\ (\rho_\mathcal{B}(\alpha)\cdot\beta)(x) &: (\alpha\beta)(x) \longrightarrow 0. \end{aligned}$$

According to the definition of the $\mathcal{B}_0$-bimodule structure of $\mathcal{B}_1$ these tracks are obtained by the Γ-product • in (5.3.2) so that we get the composites

$$(\beta\cdot\rho_\mathcal{B}(\alpha))(x) : (\beta\alpha)(x) \xrightarrow{\Gamma_{\beta,\alpha}} \beta(\alpha(x)) \xrightarrow{\beta(\rho(\alpha)(x))} 0$$

$$(\rho_\mathcal{B}(\alpha)\cdot\beta)(x) : (\alpha\beta)(x) \xrightarrow{\Gamma_{\alpha,\beta}} \alpha(\beta(x)) \xrightarrow{\rho(\alpha)(\beta(x))} 0$$

Here $\Gamma_{\beta,\alpha}$, $\Gamma_{\alpha,\beta}$ are given by the pseudo functor (s,Γ) as in (11.3.3). The bilinear map ∇'_ρ in (11.10.2) induces the following operators where T is the interchange map (11.1.1) and μ is the multiplication map of the algebra $\mathcal{B}_0$.

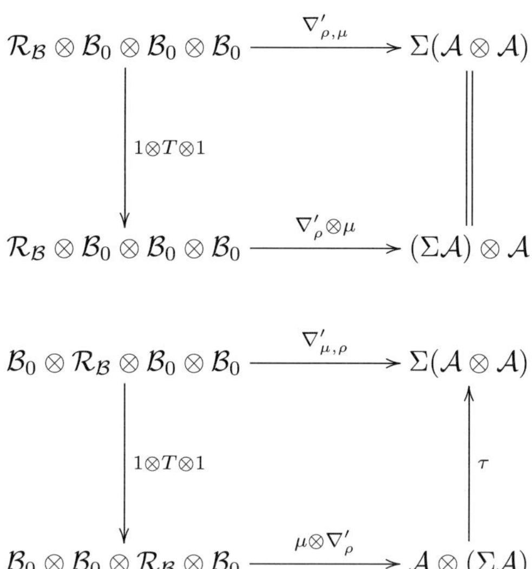

Here τ is the interchange of Σ, that is

$$\tau(a \otimes \Sigma b) = (-1)^{|a|}\Sigma a \otimes b.$$

11.10.4 Theorem. *Let $\alpha \in \mathcal{R}_\mathcal{B}$ and let $\Delta\alpha = \alpha_0 + \alpha_1$ with $\alpha_0 \in \mathcal{R}_\mathcal{B} \otimes \mathcal{B}_0$ and $\alpha_1 \in \mathcal{B}_0 \otimes \mathcal{R}_\mathcal{B}$ as in (11.9.2)(4). Then we have for $\beta \in \mathcal{B}_0$ the formula in $\mathcal{A} \otimes \mathcal{A}$,*

$$\Theta_\rho(\alpha\beta) = \Theta_\rho(\alpha)\cdot(\delta\beta) - \delta\nabla'_\rho(\alpha,\beta) + \nabla'_{\rho,\mu}(\alpha_0 \otimes \Delta\beta) + \nabla'_{\mu,\rho}(\alpha_1 \otimes \Delta\beta).$$

Here an element β in $\mathcal{B}_0$ represents also an element in $\mathcal{A}$ and $\delta : \mathcal{A} \to \mathcal{A} \otimes \mathcal{A}$ is the diagonal.

Proof of (11.10.4). Since $C_{\alpha\beta} = C_\alpha \odot C_\beta$ the element $\Theta_\rho(\alpha\beta)$ is determined by the composite of the following tracks.

(1)
$$\begin{array}{ccc}
(\alpha\beta)(x\cdot y) & \xrightarrow{C_\alpha \odot C_\beta} & \Delta(\alpha\beta)(x,y) \\
\downarrow{\scriptstyle \rho(\alpha\beta)(x\cdot y)} & & \downarrow{\scriptstyle (\bar\rho\Delta(\alpha\beta))(x,y)} \\
0 & & 0
\end{array}$$

Here $\rho(\alpha\beta)(x \cdot y)$ can be replaced by (11.10.2) and (11.10.3). Moreover we have for the decomposition of $\Delta\alpha$ in (11.9.2)(4) and for $\Delta\beta = \sum_\xi \varphi_{\Delta\beta}(\xi)\xi' \otimes \xi''$ the decomposition

(2) $$\Delta(\alpha\beta) = \sum_{i,\xi} (\pm\varphi_{\Delta\beta}(\xi)\alpha_i'\xi') \otimes \alpha_i''\xi''.$$

Now we can replace

$$\rho(\alpha_i'\xi')(x) \quad \text{for} \quad i \in I_0$$

and

$$\rho(\alpha_i''\xi'')(y) \quad \text{for} \quad i \in I_1$$

in $(\bar{\rho}\Delta(\alpha\beta))(x, y)$, see (11.9.4). Hence we compare $\Theta_\rho(\alpha\beta)$ defined by the composite in (1) with the element $\Theta_\rho(\alpha, \beta)$ defined by the composite

(3) $$\begin{array}{ccc} (\alpha\beta)(x \cdot y) & \xrightarrow{C_\alpha \odot C_\beta} & \Delta(\alpha\beta)(x, y) \\ \Big\downarrow{\scriptstyle (\rho_{\mathcal{B}}(\alpha)\cdot\beta)(x\cdot y)} & & \Big\downarrow{\scriptstyle u} \\ 0 & & 0 \end{array}$$

with

(4) $$\begin{aligned} u = &\sum_{i \in I_0, \xi} \pm\varphi_{\Delta\beta}(\xi)(\rho_{\mathcal{B}}(\alpha_i') \cdot \xi')(x) \cdot \alpha_i''(\xi'' y) \\ &+ \sum_{i \in I_1, \xi} \pm\varphi_{\Delta\beta}(\xi)\alpha_i'(\xi' x) \cdot (\rho_{\mathcal{B}}(\alpha_i'') \cdot \xi'')(y). \end{aligned}$$

Here we use the fact that ξ', ξ'' are monomials. Hence by (11.10.2) we get the equation in $\mathcal{A} \otimes \mathcal{A}$,

(5) $$\Theta_\rho(\alpha\beta) = \Theta_\rho(\alpha, \beta) - \delta\nabla_\rho'(\alpha, \beta) + v,$$

$$\begin{aligned} v = &\sum_{i \in I_0, \xi} \pm\varphi_{\Delta\beta}(\xi)\nabla_\rho'(\alpha_i', \xi') \otimes (\alpha_i''\xi'') \\ &+ \sum_{i \in I_1, \xi} \pm\varphi_{\Delta\beta}(\xi)(\alpha_i'\xi') \otimes \nabla_\rho'(\alpha_i'', \xi''). \end{aligned}$$

For the computation of $\Theta_\rho(\alpha, \beta)$ we use the composites in (11.10.3) and (11.4.3)(6).

Hence $\Theta_\rho(\alpha, \beta)$ is given by the following composite since $\Gamma_{\alpha,\beta}$ is cancelled.

(6)

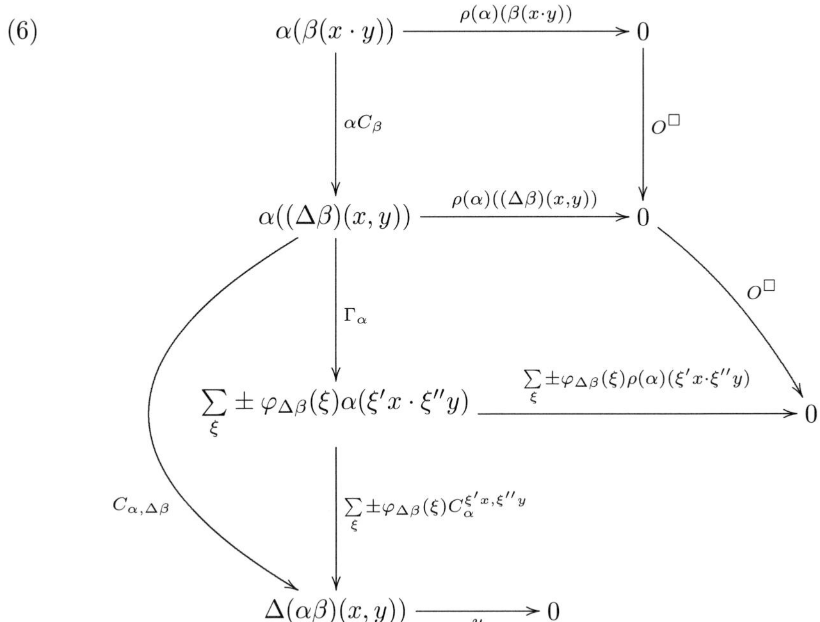

The top square and the square in the middle are commutative by (3.1.3). Since $\Gamma_{\alpha,\xi}$ is the identity track if ξ is a monomial, we can use (11.10.3) to show that u in (4) coincides with

(7)
$$u = \sum_\xi \pm\varphi_{\Delta\beta}(\xi)(\bar{\rho}(\Delta\alpha))(\xi' x, \xi'' y).$$

Hence diagram (6) yields the formula

(8)
$$\begin{aligned}\Theta_\rho(\alpha, \beta) &= \sum_\xi \varphi_{\Delta\beta}(\xi)\Theta_\rho(\alpha) \cdot (\xi' \otimes \xi'') \\ &= \Theta_\rho(\alpha) \cdot (\Delta\beta).\end{aligned}$$

□

Chapter 12

Secondary Hopf Algebras

We prove the crucial fact that the relation diagonal

$$\Theta_\rho : R_{\mathcal{B}} \longrightarrow \mathcal{A} \otimes \mathcal{A}$$

determines the secondary diagonal

$$\Delta : \mathcal{B} \longrightarrow \mathcal{B} \hat{\otimes} \mathcal{B}$$

where $\mathcal{B}\hat{\otimes}\mathcal{B}$ is the "folding product" of $\mathcal{B}$. Though Θ_ρ depends on the splitting ρ, it turns out that the secondary diagonal does not depend on the splitting and hence is the "invariant form" of the relation diagonal. Then the properties of generalized Cartan tracks imply that $(\mathcal{B}, \Delta)$ is a secondary Hopf algebra. This is a main result in this book, generalizing the fact that the Steenrod algebra $\mathcal{A}$ is a Hopf algebra.

12.1 The monoidal category of [p]-algebras

For a prime p we use the field $\mathbb{F} = \mathbb{Z}/p$ of p elements and the ring $\mathbb{G} = \mathbb{Z}/p^2$. An $\mathbb{F}$-vector space is also a $\mathbb{G}$-module via the ring homomorphism $\mathbb{G} \to \mathbb{F}$. For a graded module $M = \{M^n, n \in \mathbb{Z}\}$ we have the suspension ΣM which is the graded module given by

$$(\Sigma M)^n = M^{n-1}.$$

Let $\Sigma : M \to \Sigma M$ be the map of degree $+1$ given by the identity. In particular let $\mathbb{F}$ be concentrated in degree 0 so that $\Sigma\mathbb{F}$ is concentrated in degree 1. We have canonical isomorphisms

$$M \otimes (\Sigma N) \xrightarrow{\ \tau\ } \Sigma(M \otimes N) \xleftarrow{\ 1\ } (\Sigma M) \otimes N \tag{12.1.1}$$

for graded modules M, N. Here the left-hand side is the interchange of Σ and M given by $\tau(m \otimes \Sigma n) = (-1)^m \Sigma(m \otimes n)$. Let

$$\mathcal{A}^{\otimes n} = \mathcal{A} \otimes \cdots \otimes \mathcal{A}$$

be the n-fold tensor product of the Steenrod algebra $\mathcal{A}$ with $\mathcal{A}^{\otimes n} = \mathbb{F}$ for $n = 0$. Here $\mathcal{A}^{\otimes n}$ is an algebra over $\mathbb{F}$ and $\Sigma\mathcal{A}^{\otimes n}$ is a left and a right $\mathcal{A}^{\otimes n}$ module.

12.1.2 Definition. A $[p]$-*algebra* D of type $n \geq 0$ is given by an exact sequence of non-negatively graded $\mathbb{G}$-modules

$$(1) \qquad 0 \longrightarrow \Sigma\mathcal{A}^{\otimes n} \xrightarrow{i} D_1 \xrightarrow{\partial} D_0 \xrightarrow{q} \mathcal{A}^{\otimes n} \longrightarrow 0$$

such that D_0 is a free $\mathbb{G}$-module and an algebra over $\mathbb{G}$ and q is an algebra map. Moreover D_1 is a right D_0-module and ∂ is D_0 linear. Using q also $\Sigma\mathcal{A}^{\otimes n}$ is a right D_0-module and i is also D_0-linear. In degree 0 we have the unique element

$$(2) \qquad [p] \in D_1 \quad \text{with} \quad \partial[p] = p \cdot 1$$

where 1 is the unit of the algebra D_0. Let $[p] \cdot D_0$ be the $\mathbb{G}$-submodule of D_1 given by the right action of D_0 on the element $[p]$. As part of the definition of a $[p]$-algebra D we assume that the quotient

$$(3) \qquad D_1/[p] \cdot D_0$$

is a graded $\mathbb{F}$-vector space so that $D_1/[p] \cdot D_0$ is a right module over the algebra $D_0/p \cdot D_0 = D_0 \otimes \mathbb{F}$. Now let D and E be $[p]$-algebras of type n and m respectively. Then a morphism $f : D \to E$ is a commutative diagram

(4)

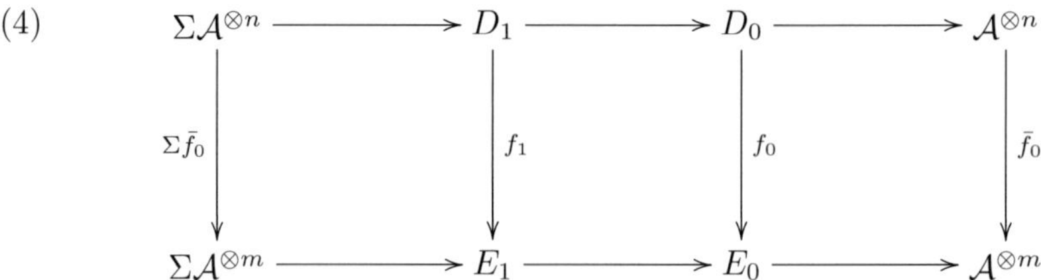

where f_0 is an algebra map and f_1 is an f_0-equivariant map of modules. We point out that f_1 induces $\Sigma\bar{f}_0$ where $\bar{f}_0$ is induced by f_0. Let $\mathcal{A}lg^{[p]}$ be the category of such $[p]$-algebras and maps.

12.1.3 Example. The *initial object* $\mathbb{G}^\Sigma$ in $\mathcal{A}lg^{[p]}$ is the $[p]$-algebra of type 0 given by the exact sequence

$$\begin{array}{ccccccccc} 0 \longrightarrow & \Sigma\mathbb{F} & \longrightarrow & \mathbb{G}_1^\Sigma & \xrightarrow{\partial} & \mathbb{G}_0^\Sigma & \longrightarrow & \mathbb{F} & \longrightarrow 0 \\ & & & \| & & \| & & & \\ & & & \mathbb{F} \oplus \Sigma\mathbb{F} & & \mathbb{G} & & & \end{array}$$

with $\partial|\mathbb{F}$ the inclusion and $\partial|\Sigma\mathbb{F} = 0$. The generator of $\mathbb{F} \subset \mathbb{G}_1^\Sigma$ is $[p]$. For each $[p]$-algebra D there is a unique morphism

$$\mathbb{G}_\Sigma \longrightarrow D$$

carrying $1 \in \mathbb{G}_0^\Sigma$ to $1 \in D_0$ and $[p] \in \mathbb{G}_1^\Sigma$ to $[p] \in D_1$. Therefore $\mathbb{G}_\Sigma$ is the initial object of $\mathcal{A}lg^{[p]}$. We call a morphism

$$\epsilon : D \longrightarrow \mathbb{G}_\Sigma$$

a *secondary augmentation* of D.

12.1.4 Proposition. *The pair algebra $\mathcal{B}$ of secondary cohomology operations is a $[p]$-algebra of type 1.*

This is a consequence of (11.9.1). We point out that $\mathcal{B}$ is also a crossed algebra, in particular $\mathcal{B}_1$ is a $\mathcal{B}_0$-bimodule. But only the right $\mathcal{B}_0$-module structure of $\mathcal{B}_1$ is used in the definition of a $[p]$-algebra.

12.1.5 Definition. For the $[p]$-algebra $\mathcal{B}$ of secondary cohomology operations we have the *secondary augmentation of* $\mathcal{B}$,

$$\epsilon : \mathcal{B} \longrightarrow \mathbb{G}_\Sigma,$$

in $\mathcal{A}lg^{[p]}$ defined as follows. Here ϵ is the diagram

(1)
$$\begin{array}{ccc} \mathcal{B}_1 & \xrightarrow{\epsilon_1} & \mathbb{F} \oplus \Sigma\mathbb{F} \\ \partial \downarrow & & \downarrow \\ \mathcal{B}_0 & \xrightarrow{\epsilon_0} & \mathbb{G} \end{array}$$

where ϵ_0 is the augmentation of the tensor algebra $\mathcal{B}_0 = T_{\mathbb{G}}(E_{\mathcal{A}})$. Moreover the $\mathbb{F}$-coordinate of ϵ_1 is given by the commutative diagram (1) and the $\Sigma\mathbb{F}$-coordinate of ϵ_1 is given by the retraction

(2)
$$\tilde{\epsilon} : \mathcal{B}_1 \longrightarrow \Sigma\mathbb{F}$$

defined in degree 1 as follows. An element $x \in \mathcal{B}_1$ with $|x| = 1$ is a pair $x = (a, \alpha)$ with $\alpha \in \mathcal{B}_0$, $|\alpha| = 1$, and $a : s\alpha \Rightarrow 0$. Here $|\alpha| = 1$ implies that α is a multiple of Sq^1 if $p = 2$ and of the Bockstein β if p is odd. This implies that $s\alpha = 0$ since $s\alpha \Rightarrow 0$ exists. Therefore $a : 0 \Rightarrow 0$ represents an element $\tilde{a}$ in $\mathbb{F}$. We set $\tilde{\epsilon}(x) = \tilde{a}$. The map $\tilde{\epsilon}$ is compatible with $\Sigma\epsilon : \Sigma\mathcal{A} \to \Sigma\mathbb{F}$. Moreover for the element $[p] = (p \cdot 1, 0) \in \mathcal{B}_1$ we have ($\beta \in \mathcal{B}_0$)

(3)
$$\tilde{\epsilon}([p] \cdot \beta) = \tilde{\epsilon}([p]) \cdot \epsilon(\beta) = 0.$$

For each $[p]$-algebra D we have an associated commutative diagram.

(12.1.6)

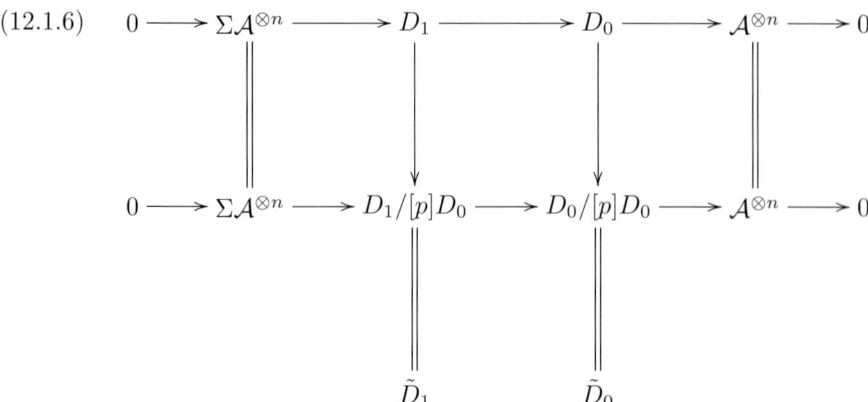

Here the rows are exact and all maps are morphisms of right D_0-modules. The square in the middle is a push out and a pull back diagram. This square defines the pair map

$$D \longrightarrow \tilde{D} = (\partial : \tilde{D}_1 \longrightarrow \tilde{D}_0) \tag{1}$$

where $\tilde{D}$ consists of graded $\mathbb{F}$-vector spaces. Hence there is an $\mathbb{F}$-isomorphism

$$\tilde{D}_1 \cong (\Sigma\mathcal{A}^{\otimes n}) \oplus kernel(\tilde{D}_0 \longrightarrow \mathcal{A}^{\otimes n}). \tag{2}$$

For $n, m \geq 0$ we define the *folding map* φ by the commutative diagram

$$\begin{array}{ccc} \mathcal{A}^{n,m} & = & (\Sigma\mathcal{A}^{\otimes n}) \otimes \mathcal{A}^{\otimes m} \oplus \mathcal{A}^{\otimes n} \otimes (\Sigma\mathcal{A}^{\otimes m}) \\ \downarrow{\scriptstyle \varphi} & & \downarrow{\scriptstyle (1,\tau)} \\ \Sigma\mathcal{A}^{\otimes(n+m)} & = & \Sigma\mathcal{A}^{\otimes n} \otimes \mathcal{A}^{\otimes m} \end{array} \tag{12.1.7}$$

where we use the maps in (12.1.1). Now we are ready to introduce the product of $[p]$-algebras.

12.1.8 Definition. Let D and E be $[p]$-algebras of type n and m respectively. Then the *folding product* $D\hat{\otimes}E$ is a $[p]$-algebra of type $n + m \geq 0$ defined as follows. Let $\tilde{D}$ and $\tilde{E}$ be the associated pairs as in (12.1.6). Since $\tilde{D}$ and $\tilde{E}$ are defined over $\mathbb{F}$ we get the exact top row in the following diagram where we use the tensor product $\bar{\otimes}$ of pairs in (5.1.2).

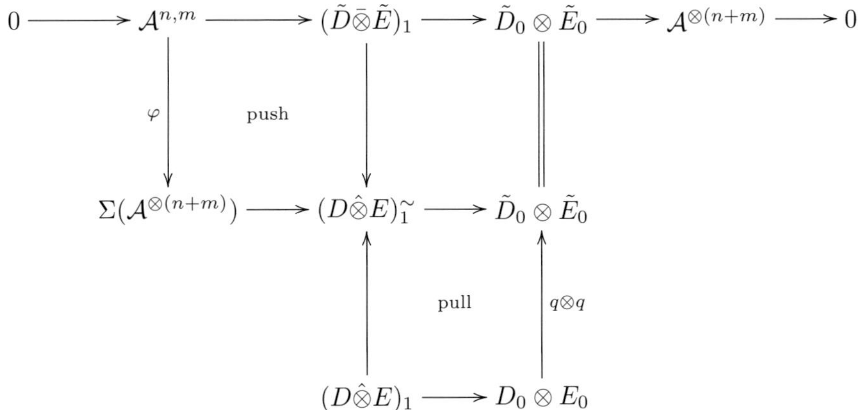

Here φ is the folding map in (12.1.7). Moreover the push out and pull back in the diagram define the bottom row which is the folding product $D\hat{\otimes}E$. The pull back in the diagram is also a push out. Hence the kernel of $D\hat{\otimes}E$ is $\Sigma(\mathcal{A}^{\otimes(n+m)})$ and the cokernel is $\mathcal{A}^{\otimes(n+m)}$.

One can check that

$$(D\hat{\otimes}E)_1^{\sim} = (D\hat{\otimes}E)_1/[p](D_0 \otimes E_0).$$

This shows that $D\hat{\otimes}E$ is a $[p]$-algebra of type $n+m$. The algebra $D_0\hat{\otimes}E_0$ acts on $(D\hat{\otimes}E)_1$ since D_0 acts on $\tilde{D}_1$ and E_0 acts on $\tilde{E}_1$ and φ is equivariant. Here we use the even sign convention depending on the prime p in (11.1.1).

12.1.9 Theorem. *The category $\mathcal{A}lg^{[p]}$ with the folding product $\hat{\otimes}$ is a symmetric monoidal category. The unit object is $\mathbb{G}_\Sigma$.*

In particular we have the natural isomorphisms in $\mathcal{A}lg^{[p]}$,

$$\mathbb{G}_\Sigma\hat{\otimes}D = D = D\hat{\otimes}\mathbb{G}_\Sigma,$$

$$(D\hat{\otimes}E)\hat{\otimes}F = D\hat{\otimes}(E\hat{\otimes}F),$$

$$T : D\hat{\otimes}E \cong E\hat{\otimes}D.$$

The interchange map T is induced by T in (11.1.1), see also (12.1.11)(4) below. There is a natural surjective map

(12.1.10) $$q : (D\bar{\otimes}E)_1 \twoheadrightarrow (D\hat{\otimes}E)_1$$

given by the following commutative diagram.

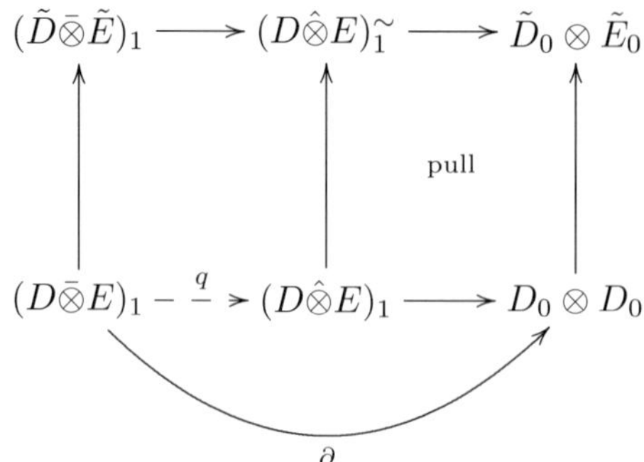

The arrow at the left-hand side is the $\bar{\otimes}$-product of $D \to \tilde{D}$ and $E \to \tilde{E}$ given by (12.1.6). The map q is a morphism of right $D_0 \otimes E_0$-modules. Here $D_0 \otimes E_0$ acts on $(D\bar{\otimes}E)_1$ by use of the interchange map T in (11.1.1).

As a special case of the map (12.1.10) we get the surjective map

$$q : (\mathcal{B}\bar{\otimes}\mathcal{B})_1 \longrightarrow (\mathcal{B}\hat{\otimes}\mathcal{B})_1$$

where $(\mathcal{B}\bar{\otimes}\mathcal{B})_1$ is a $(\mathcal{B}_0 \otimes \mathcal{B}_0)$-bimodule and q is a map of right $(\mathcal{B}_0 \otimes \mathcal{B}_0)$-modules.

12.1.11 Theorem. *Assume that a $[p]$-algebra $\mathcal{B}$ is also a pair algebra and that the derivation*

$$\Gamma[p] = \kappa : \mathcal{A} \longrightarrow \mathcal{A}$$

defined by $\kappa(\alpha) = [p] \cdot \alpha - \alpha \cdot [p]$ for $\alpha \in \mathcal{B}_0$ satisfies

$$(\kappa \otimes 1)\delta = \delta\kappa : \mathcal{A} \longrightarrow \mathcal{A} \otimes \mathcal{A}.$$

Then the folding product $(\mathcal{B}\hat{\otimes}\mathcal{B})_1$ is a left $\mathcal{B}_0$-module in such a way that the map q satisfies

$$q(\Delta(\alpha) \cdot x) = \alpha \cdot q(x)$$

for $\alpha \in \mathcal{B}_0$, $x \in (\mathcal{B}\bar{\otimes}\mathcal{B})_1$; that is, q is a map of left $\mathcal{B}_0$-modules with the left action of $\mathcal{B}_0$ on $(\mathcal{B}\bar{\otimes}\mathcal{B})_1$ induced by the diagonal $\Delta : \mathcal{B}_0 \longrightarrow \mathcal{B}_0 \otimes \mathcal{B}_0$.

We write for $x, y \in \mathcal{B}_1$ and $\xi, \eta \in \mathcal{B}_0$

(1)
$$\begin{aligned} x\hat{\otimes}\eta &= q(x \otimes \eta) \in (\mathcal{B}\hat{\otimes}\mathcal{B})_1, \\ \xi\hat{\otimes}y &= q(\xi \otimes y) \in (\mathcal{B}\hat{\otimes}\mathcal{B})_1. \end{aligned}$$

Hence for $\Delta\alpha = \sum_i \alpha_i' \otimes \alpha_i'' \in \mathcal{B}_0\hat{\otimes}\mathcal{B}_0$ we get the left action of α on $x\hat{\otimes}\eta$, $\xi\hat{\otimes}y$ by the formulas:

(2)
$$
\begin{aligned}
\alpha\cdot(x\hat{\otimes}\eta) &= q((\Delta\alpha)\cdot x\otimes\eta)\\
&= q(\sum_i(-1)^{|x||\alpha_i''|}\alpha_i' x\otimes\alpha_i''\eta)\\
&= \sum_i(-1)^{|x||\alpha_i''|}(\alpha_i' x)\hat{\otimes}(\alpha_i''\eta),\\
\alpha\cdot(\xi\hat{\otimes}y) &= q((\Delta\alpha)\cdot\xi\otimes y)\\
&= q(\sum_i(-1)^{|\xi||\alpha_i''|}(\alpha_i'\xi)\otimes(\alpha_i'' y))\\
&= \sum_i(-1)^{|\xi||\alpha_i''|}(\alpha_i'\xi)\hat{\otimes}(\alpha_i'' y).
\end{aligned}
$$

We point out, however, that $(\mathcal{B}\hat{\otimes}\mathcal{B})_1$ is not a left $\mathcal{B}_0\otimes\mathcal{B}_0$-module so that $(\alpha'\otimes\alpha'')\cdot(x\hat{\otimes}\eta)$ or $(\alpha'\otimes\alpha'')\cdot(\xi\hat{\otimes}y)$ are not defined for $\alpha',\alpha''\in\mathcal{B}_0$. But the right action of $\alpha'\otimes\alpha''$ is defined satisfying

(3)
$$
\begin{aligned}
(x\hat{\otimes}\eta)\cdot(\alpha'\otimes\alpha'') &= (-1)^{|\eta||\alpha'|}(x\cdot\alpha')\hat{\otimes}(\eta\cdot\alpha''),\\
(\xi\hat{\otimes}y)\cdot(\alpha'\otimes\alpha'') &= (-1)^{|y||\alpha'|}(\xi\cdot\alpha')\hat{\otimes}(y\cdot\alpha'').
\end{aligned}
$$

Though $(\mathcal{B}\bar{\otimes}\mathcal{B})_1$ is also a left $\mathcal{B}_0\otimes\mathcal{B}_0$-module, the folding product $(\mathcal{B}\hat{\otimes}\mathcal{B})_1$ is not a left $\mathcal{B}_0\otimes\mathcal{B}_0$-module, only a left $\mathcal{B}_0$-module. In fact, $\mathcal{B}\bar{\otimes}\mathcal{B}$ is a pair algebra since $\mathcal{B}$ is a pair algebra. Hence $(\mathcal{B}\bar{\otimes}\mathcal{B})_1$ is a $\mathcal{B}_0\otimes\mathcal{B}_0$-bimodule and for $x,y\in(\mathcal{B}\bar{\otimes}\mathcal{B})_1$ the equation

$$(\partial x)\cdot y = x\cdot(\partial y)$$

holds in $(\mathcal{B}\bar{\otimes}\mathcal{B})_1$. Such an equation is not available for $x,y\in(\mathcal{B}\hat{\otimes}\mathcal{B})_1$. However, using the surjective map $q:(\mathcal{B}\bar{\otimes}\mathcal{B})_1\to(\mathcal{B}\hat{\otimes}\mathcal{B})_1$ and Theorem (12.1.11) we still get for $x,y\in(\mathcal{B}\hat{\otimes}\mathcal{B})_1$,

(4)
$$\alpha\cdot y = x\cdot(\partial y)\ \text{ if }\ \partial x=\Delta(\alpha),\quad \alpha\in\mathcal{B}_0.$$

In fact $x=q(x')$, $y=q(y')$ with $\partial x=\partial x'=\Delta(\alpha)$ and $\partial y=\partial y'$ so that

$$
\begin{aligned}
x\cdot\partial y &= q(x'\cdot\partial y') = q(\partial x'\cdot y')\\
&= q(\Delta(\alpha)\cdot y') = \alpha\cdot q(y') = \alpha\cdot y.
\end{aligned}
$$

The *interchange operator*

$$T:(\mathcal{B}\hat{\otimes}\mathcal{B})_1\longrightarrow(\mathcal{B}\hat{\otimes}\mathcal{B})_1$$

is defined by

$$
\begin{aligned}
T(x\hat{\otimes}\eta) &= (-1)^{|x||\eta|}\eta\hat{\otimes}x,\\
T(\xi\hat{\otimes}y) &= (-1)^{|\xi||y|}y\hat{\otimes}\xi.
\end{aligned}
$$

Now formula (3) shows that T is a T-equivariant map of right $(\mathcal{B}_0 \hat{\otimes} \mathcal{B}_0)$-modules, that is, for $v \in (\mathcal{B} \hat{\otimes} \mathcal{B})_1$ and $a \in \mathcal{B}_0 \otimes \mathcal{B}_0$ we get

$$T(v \cdot a) = T(v) \cdot T(a). \tag{5}$$

This also follows from the fact that T is a map between $[p]$-algebras. For $a = \Delta\alpha$ and $v \cdot \alpha = v \cdot (\Delta\alpha)$ we get

$$T(\alpha \cdot v) = \alpha \cdot T(v). \tag{6}$$

In fact, by (2) we get for $v = x \hat{\otimes} \eta$ the formula

$$\begin{aligned} T(\alpha \cdot v) &= qT((\Delta\alpha) \cdot (x \otimes \eta)) \\ &= q((T\Delta\alpha) \cdot T(x \otimes \eta)) \\ &= q(\Delta\alpha \cdot T(x \otimes \eta)) \\ &= \alpha \cdot T(v). \end{aligned}$$

A similar computation holds for $v = \xi \hat{\otimes} y$.

For the proof of theorem (12.1.11) we need the next lemma.

12.1.12 Lemma. *For the derivation $\kappa = \Gamma[p] : \mathcal{A} \longrightarrow \mathcal{A}$ of degree -1 the composite*

$$\mathcal{A} \xrightarrow{\delta} \mathcal{A} \otimes \mathcal{A} \xrightarrow{\kappa_*} \mathcal{A} \otimes \mathcal{A}$$

is trivial with $\kappa_*(a \otimes b) = \kappa(a) \otimes b - (-1)^{|a|} a \otimes \kappa(b)$. Here κ_* is a derivation of degree -1.

Proof. The lemma is also a consequence of the equation $(\kappa \otimes 1)\delta = (1 \otimes \kappa)\delta = \delta\kappa$ in (12.1.11). □

Proof of (12.1.11). Consider the following diagram.

$$\begin{array}{ccccc} \mathcal{A}^{\otimes 4} & & \xrightarrow{\kappa\#} & & \Sigma\mathcal{A}^{\otimes 2} \\ \uparrow & & & & \downarrow \\ \mathcal{B}_0^{\otimes 4} & \xrightarrow{\psi} & (\mathcal{B} \bar{\otimes} \mathcal{B})_1 & \xrightarrow{q} & (\mathcal{B} \hat{\otimes} \mathcal{B})_1 \end{array} \tag{1}$$

Here the left-hand side is the quotient map and the right-hand side is the inclusion. Moreover we define ψ by

$$\begin{aligned} \psi(\alpha \otimes \beta \otimes \alpha' \otimes \beta') &= (\alpha \otimes \beta) \cdot ([p] \otimes 1 - 1 \otimes [p]) \cdot (\alpha' \otimes \beta') \\ &= ((\alpha[p]) \otimes \beta - \alpha \otimes (\beta[p])) \cdot (\alpha' \otimes \beta') \end{aligned}$$

We observe that $\partial\psi = 0$ so that $\partial q\psi = 0$ and hence there is a well-defined map $\kappa_\#$ for which the diagram commutes. Since $\kappa = \Gamma[p]$ we have the equations in $\mathcal{B}_1$,

$$\alpha \cdot [p] - [p] \cdot \alpha = \Sigma\kappa(\alpha),$$
$$\beta \cdot [p] - [p] \cdot \beta = \Sigma\kappa(\beta).$$

Therefore we obtain $\kappa_\#$ by the formula

$$\kappa_\#(\alpha \otimes \beta \otimes \alpha' \otimes \beta') = \varphi(A - B) \cdot (\alpha' \otimes \beta')$$

where

$$A = (\Sigma\kappa(\alpha)) \otimes \beta \;\text{ and }\; B = \alpha \otimes (\Sigma\kappa(\beta)).$$

Hence we get

$$\varphi(A) = \Sigma(\kappa(\alpha) \otimes \beta) \;\text{ and }\; \varphi(B) = (-1)^{|\alpha|}\Sigma(\alpha \otimes \kappa(\beta)).$$

This shows that

$$\begin{aligned}\kappa_\#(\alpha \otimes \beta \otimes \alpha' \otimes \beta') = \Sigma C \;\text{ with}&\\ C &= (\kappa(\alpha) \otimes \beta - (-1)^{|\alpha|}\alpha \otimes \kappa(\beta)) \cdot (\alpha' \otimes \beta')\\ &= \kappa_*(\alpha \otimes \beta) \cdot (\alpha' \otimes \beta').\end{aligned}$$

Here we use κ_* in the lemma above. Hence $\kappa_\#$ satisfies for $\xi \in \mathcal{B}_0$ the equations

$$\kappa_\#((\xi \cdot (\alpha \otimes \beta)) \otimes (\alpha' \otimes \beta')) = \Sigma(\kappa_*(\xi \cdot (\alpha \otimes \beta)) \cdot (\alpha' \otimes \beta')).$$

Now we get

$$\begin{aligned}\kappa_*(\xi \cdot (\alpha \otimes \beta)) &= \kappa_*(\delta(\xi) \cdot (\alpha \otimes \beta))\\ &= \kappa_*(\delta(\xi)) \cdot (\alpha \otimes \beta) + (-1)^{|\xi|}\delta(\xi) \cdot \kappa_*(\alpha \otimes \beta)\\ &= (-1)^{|\xi|}\delta(\xi) \cdot \kappa_*(\alpha \otimes \beta).\end{aligned}$$

since $\kappa_*(\delta) = 0$ by Lemma (12.1.12). Therefore we get

$$\kappa_\#((\xi \cdot (\alpha \otimes \beta) \otimes (\alpha' \otimes \beta')) = \xi \cdot \kappa_\#(\alpha \otimes \beta \otimes \alpha' \otimes \beta') \tag{2}$$

where the action of ξ is induced by the diagonal. Now we claim that the sequence

$$\mathcal{K} \xrightarrow{(\bar{i},\psi)} (\mathcal{B}\bar{\otimes}\mathcal{B})_1 \xrightarrow{q} (\mathcal{B}\hat{\otimes}\mathcal{B})_1 \longrightarrow 0 \tag{3}$$

is exact where $\mathcal{K}$ is the kernel of

$$\mathcal{A}^{1,1} \oplus \mathcal{B}_0^{\otimes 4} \xrightarrow{(\varphi,\kappa_\#)} \Sigma\mathcal{A}^{\otimes 2}.$$

Here $\bar{i} : \mathcal{A}^{1,1} \longrightarrow (\mathcal{B}\bar{\otimes}\mathcal{B})_1$ is induced by $\Sigma\mathcal{A} \subset \mathcal{B}_1$. One readily checks that $q(\bar{i}, \psi) = 0$. Since $\mathcal{K}$ is a left $\mathcal{B}_0$-module by (2) and $(\bar{i}, \psi)$ is a map of left $\mathcal{B}_0$-modules, also q is a map of left $\mathcal{B}_0$-modules. We derive the exactness of (3) from the exactness of

(4) $$\mathcal{B}_0^{\otimes 4} \xrightarrow{\psi} (\mathcal{B}\bar{\otimes}\mathcal{B})_1/\operatorname{image}(\bar{i}) \longrightarrow \mathcal{B}_0 \otimes \mathcal{B}_0.$$

In fact, since $\mathcal{B}$ is a pull back as in (12.1.4)(2) we know that $\mathcal{B}_1 = (\Sigma\mathcal{A}) \oplus R_{\mathcal{B}}$ where $R_{\mathcal{B}} = kernel(\mathcal{B}_0 \to \mathcal{A})$ with $[p] \in R_{\mathcal{B}}$.

Therefore we get

$$(\mathcal{B}\bar{\otimes}\mathcal{B})_1/\operatorname{image}(\bar{i}) = (R\bar{\otimes}R)_1$$

with $R = (R_{\mathcal{B}} \subset \mathcal{B}_0)$ the inclusion.

One now can check the exactness of

(5) $$\mathcal{B}_0^{\otimes 4} \xrightarrow{\psi} (R\bar{\otimes}R)_1 \longrightarrow \mathcal{B}_0 \otimes \mathcal{B}_0.$$

This completes the proof of Theorem (12.1.11). □

12.2 The secondary diagonal

For the pair algebra $\mathcal{B}$ of secondary cohomology operations there is a canonical *secondary diagonal* Δ which is a pair map

(12.2.1) $$\Delta : \mathcal{B} \longrightarrow \mathcal{B}\hat{\otimes}\mathcal{B}$$

where $\mathcal{B}\hat{\otimes}\mathcal{B}$ is the folding product. Here Δ corresponds to the commutative diagram

$$\begin{array}{ccc} \mathcal{B}_1 & \xrightarrow{\Delta_1} & (\mathcal{B}\hat{\otimes}\mathcal{B})_1 \\ \downarrow{\partial} & & \downarrow{\partial} \\ \mathcal{B}_0 & \xrightarrow{\Delta_0} & \mathcal{B}_0 \otimes \mathcal{B}_0 \end{array}$$

where Δ_0 is the Hopf algebra structure of the tensor algebra $T_{\mathbb{G}}(E_{\mathcal{A}})$ with the even sign convention in Section (11.1). We obtain Δ_1 using the pull back property of $(\mathcal{B}\hat{\otimes}\mathcal{B})_1$ by the commutative diagram

(12.2.2)

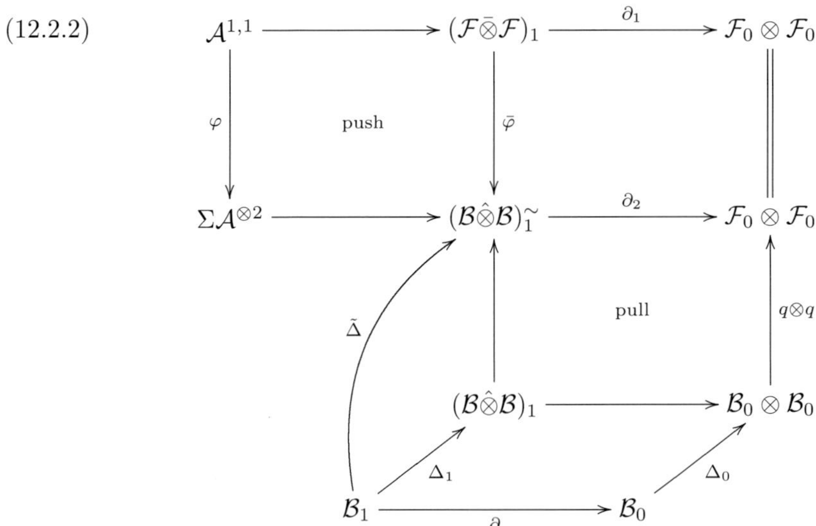

with $\mathcal{F} = (\mathcal{B}_1/[p]\mathcal{B}_0 = \mathcal{F}_1 \to \mathcal{B}_0/p\mathcal{B}_0 = \mathcal{F}_0) = \tilde{\mathcal{B}}$ as in (11.9.1) and (12.1.6). Below we define a map $\tilde{\Delta}$ as in the diagram. The pair of maps $(\tilde{\Delta}, \Delta_0\partial)$ yields Δ_1 by the pull back property. We define $\tilde{\Delta}$ by use of the relation diagonal Θ_ρ in Section (11.10).

We choose a splitting ρ of $\mathcal{F}$ as in (11.9.2).

(1)

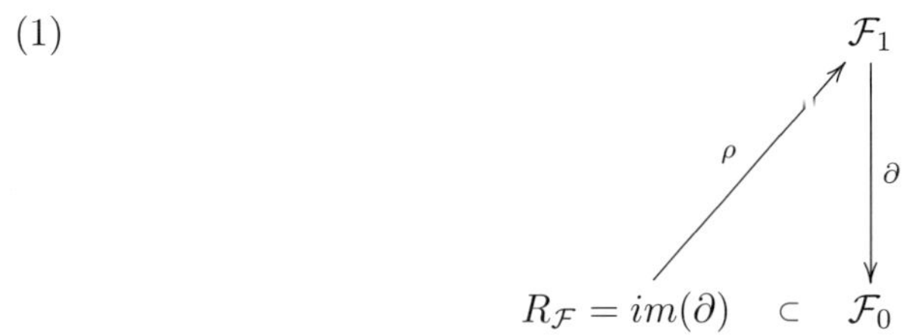

Then ρ induces a splitting $\bar{\rho} = \rho\bar{\otimes}\rho$ of $\mathcal{F}\bar{\otimes}\mathcal{F}$ as in the following commutative diagram.

(2)

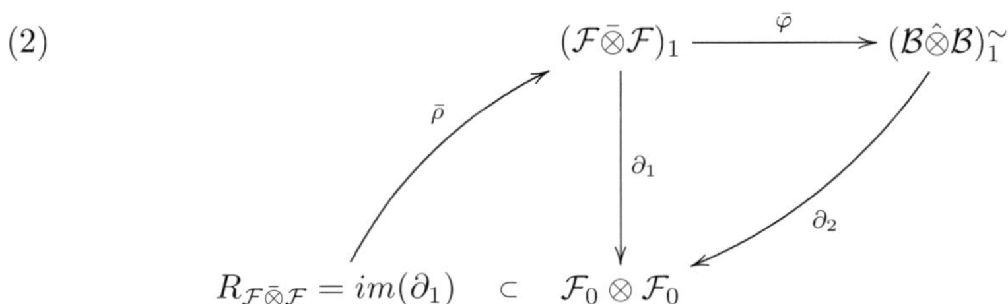

Here we have

(3) $$R_{\mathcal{F}\bar{\otimes}\mathcal{F}} = R_{\mathcal{F}} \otimes \mathcal{F}_0 + \mathcal{F}_0 \otimes R_{\mathcal{F}}$$

and $\bar{\rho}$ carries $x \otimes a$ and $b \otimes y$ with $x, y \in R_{\mathcal{F}}$ and $a, b \in \mathcal{F}_0$ to $(\rho x)\bar{\otimes}a$ and $b\bar{\otimes}(\rho y)$ respectively. Compare the proof of (11.9.5). Hence $\bar{\varphi}\bar{\rho}$ is a splitting of ∂_2 with $\bar{\varphi}$ defined by the push out in (12.2.1). We thus obtain the following diagram.

(4)

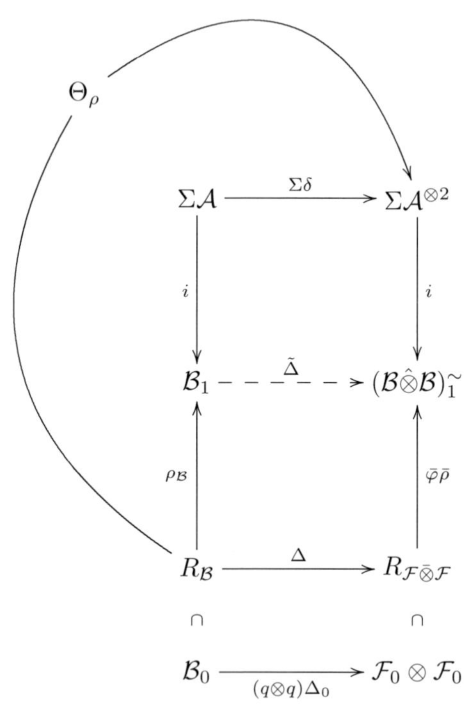

Here the splitting $\rho_{\mathcal{B}}$ of $\mathcal{B}$ is induced by ρ as in (11.10.1) and Θ_ρ is the relation diagonal in (11.9.6). We are now ready to define $\tilde{\Delta}$ by the following formula with $x \in \mathcal{B}_1$, $\xi = \partial x \in R_{\mathcal{B}}$,

(5) $$\tilde{\Delta}(x) = \bar{\varphi}\bar{\rho}(\Delta\xi) - \Theta_\rho(\xi) + \delta(x - \rho_{\mathcal{B}}(\xi)).$$

Equivalently $\tilde{\Delta}$ is the unique $\mathbb{G}$-linear map satisfying $\tilde{\Delta}i = i\Sigma\delta$ and

(6) $$\tilde{\Delta}(\rho_{\mathcal{B}}(\xi)) = \bar{\varphi}\bar{\rho}(\Delta\xi) - \Theta_\rho(\xi)$$

for $\xi \in R_{\mathcal{B}}$.

12.2.3 Lemma. *The $\mathbb{G}$-linear map $\tilde{\Delta}$ does not depend on the choice of the splitting ρ. Moreover $\partial_2\tilde{\Delta} = (q \otimes q)\Delta_0\partial$ holds.*

Proof. Let $\tilde{\Delta} = \tilde{\Delta}_\rho$ be defined by the formula above. As in (11.9.2)(3) we can alter ρ by $t : R_{\mathcal{F}} \longrightarrow \mathcal{A}$ of degree -1. Then we get

$$\begin{aligned}\tilde{\Delta}_{\rho+t}(x) &= \bar{\varphi}\overline{\rho + t}(\Delta\xi) - \Theta_{\rho+t}(\xi) + \delta(x - (\rho+t)_{\mathcal{B}}(\xi))\\ &= \bar{\varphi}\bar{\rho}(\Delta\xi) + \bar{t}(\Delta\xi) - \Theta_{\rho+t}(\xi) + \delta(x - \rho_{\mathcal{B}}(\xi)) - \delta(t\xi).\end{aligned}$$

Compare $\bar{t}(\Delta\xi)$ defined in (11.9.2)(5). Now formula (11.9.2)(3) shows $\tilde{\Delta}_{\rho+t}(x) = \tilde{\Delta}_\rho(x)$. Moreover we get

$$\begin{aligned}\partial_2\tilde{\Delta}(x) &= \partial_2(\bar{\varphi}\bar{\rho}(\Delta\xi) - \Theta_\rho(\xi) + \delta(x - \rho_{\mathcal{B}}(\xi)))\\ &= \partial_2(\bar{\varphi}\bar{\rho}(\Delta\xi)) = (q \otimes q)\Delta_0\xi.\end{aligned}$$

□

The lemma shows that the map $\tilde{\Delta}$ and hence the secondary diagonal in (12.2.1) is independent of the choice of ρ though $\tilde{\Delta}$ is defined in terms of the splitting ρ. Hence Δ is canonically defined, that is, Δ does not depend on choices. Therefore the secondary diagonal Δ can be considered as the "invariant form" of the relation diagonal. In fact, searching such an invariant form of the relation diagonal forces us to introduce the folding product of $[p]$-algebras.

12.3 The right action on the secondary diagonal

The pair algebra $\mathcal{B}$ of secondary cohomology operations is also a crossed algebra and therefore $\mathcal{B}_1$ is a $\mathcal{B}_0$-bimodule. Moreover the folding product $(\mathcal{B}\hat{\otimes}\mathcal{B})_1$ is a $\mathcal{B}_0$ bimodule with the right action of $\mathcal{B}_0$ given by Δ_0 and the right $\mathcal{B}_0 \otimes \mathcal{B}_0$-module structure of $(\mathcal{B}\hat{\otimes}\mathcal{B})_1$. The left action of $\mathcal{B}_0$ on $(\mathcal{B}\hat{\otimes}\mathcal{B})_1$ is described in (12.1.11).

12.3.1 Theorem. *The secondary diagonal*

$$\Delta_1 : \mathcal{B}_1 \longrightarrow (\mathcal{B}\hat{\otimes}\mathcal{B})_1$$

is a morphism of right $\mathcal{B}_0$-modules.

We describe in Section (11.10) the right action of $\mathcal{B}_0$ on the relation diagonal Θ_ρ. It turns out that the complicated formula (11.10.4) yields exactly the right equivariance of the $\mathcal{B}_0$ action on Δ_1.

Proof of (12.3.1). Let $\beta \in E_{\mathcal{A}}$ and $x \in \mathcal{B}_1$ with $\partial x = \xi \in R_{\mathcal{B}}$. In order to prove $\tilde{\Delta}(x \cdot \beta) = \tilde{\Delta}(x) \cdot \beta$ it suffices to show for $\rho = \rho_{\mathcal{B}}$,

(1) $$\tilde{\Delta}((\rho\xi) \cdot \beta) = (\tilde{\Delta}\rho\xi) \cdot \beta.$$

Here we have by (11.10.2) the equation

(2) $$(\rho\xi) \cdot \beta = \rho(\xi \cdot \beta) - \delta\nabla'(\xi, \beta).$$

Hence (1) is equivalent to

(3) $$\tilde{\Delta}\rho(\xi \cdot \beta) = (\tilde{\Delta}\rho\xi) \cdot \beta + \delta\nabla'(\xi, \beta).$$

Here we can use the formula (12.2.2)(6) with $\tilde{\rho} = \bar{\varphi}\bar{\rho}$, namely

(4) $$\tilde{\Delta}\rho\xi = \tilde{\rho}\Delta\xi - \Theta_\rho(\xi).$$

Hence we get by (11.10.4) the formula

(5) $$\begin{aligned}\tilde{\Delta}\rho(\xi \cdot \beta) &= \tilde{\rho}\Delta(\xi \cdot \beta) - \Theta_\rho(\xi \cdot \beta)\\ &= \tilde{\rho}\Delta(\xi \cdot \beta) - \Theta_\rho(\xi) \cdot \delta\beta + \delta\nabla'(\xi, \beta)\end{aligned}$$

(6) $$-\nabla'_{\rho,\mu}(\xi_0 \otimes \Delta\beta) - \nabla'_{\mu,\rho}(\xi_1 \otimes \Delta\beta)$$

(7) $$= \tilde{\rho}\Delta(\xi \cdot \beta) - (\tilde{\rho}\Delta\xi) \cdot \beta + (6)$$

(8) $$+(\tilde{\Delta}\rho\xi) \cdot \beta + \delta\nabla'(\xi, \beta).$$

Here (8) is the right-hand side of (2) so that (2) is equivalent to (7) $= 0$. Since $\Delta\xi = \xi_0 + \xi_1$ this is easily checked by the definition of $\tilde{\rho} = \bar{\varphi}\bar{\rho}$ and the definition of $\nabla'_{\rho,\mu}, \nabla'_{\mu,\rho}$. In fact, we have for $\eta \in R_{\mathcal{B}}$ and $\alpha, \beta', \beta'', \in \mathcal{B}_0$ the equation

(9) $$\begin{aligned}\nabla'_{\rho,\mu}(\eta \otimes \alpha \otimes \beta' \otimes \beta'') &= \pm\nabla'_\rho(\eta, \beta') \otimes \alpha \cdot \beta''\\ &= \pm(\rho(\eta\beta') - \rho(\eta) \cdot \beta') \otimes \alpha \cdot \beta''\\ &= \tilde{\rho}((\eta \otimes \alpha) \cdot (\beta' \otimes \beta'')) - \tilde{\rho}(\eta \otimes \alpha) \cdot (\beta' \otimes \beta'').\end{aligned}$$

A similar formula we get for $\nabla'_{\rho,\mu}$. This completes the proof of (1). □

12.4 The secondary Hopf algebra $\mathcal{B}$

In this section we describe a main result in this book. Almost all the arguments in this book are part of the proof of this result. We have seen that the pair algebra

(12.4.1) $$\mathcal{B} = (\partial : \mathcal{B}_1 \longrightarrow \mathcal{B}_0)$$

of secondary cohomology operations with $\mathcal{B}_0 = T_{\mathbb{G}}(E_{\mathcal{A}})$ admits an augmentation

(1) $$\varepsilon : \mathcal{B} \longrightarrow \mathbb{G}^\Sigma$$

and a secondary diagonal

(2) $$\Delta : \mathcal{B} \longrightarrow \mathcal{B}\hat{\otimes}\mathcal{B}$$

where $\mathcal{B}\hat{\otimes}\mathcal{B}$ is the folding product with the even sign convention in (11.1). Here ε and δ are maps in the category of $[p]$-algebras $\mathcal{A}lg^{[p]}$ which by $(\mathbb{G}^{\Sigma}, \hat{\otimes})$ is a monoidal category. More explicitly the augmentation ε and the diagonal Δ of $\mathcal{B}$ are given by the following commutative diagrams.

(3)

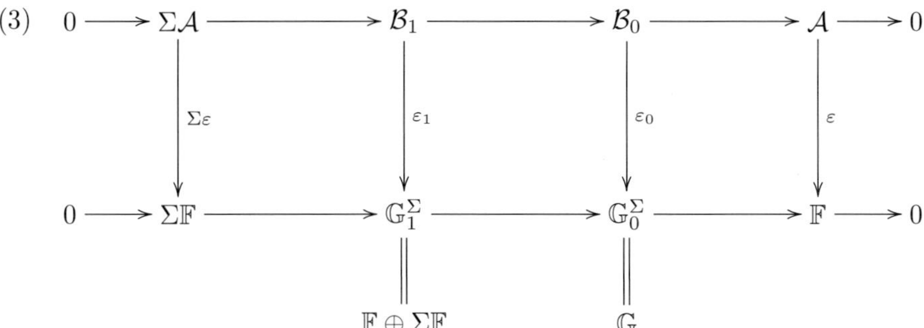

(4)

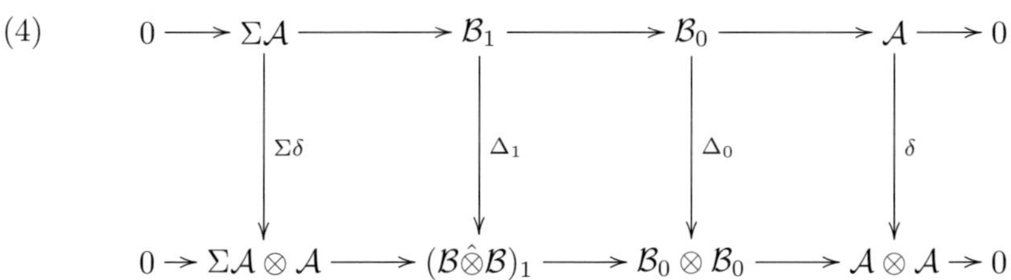

Here ε_0 is the augmentation of $\mathcal{B}_0$ and Δ_0 is the diagonal of $\mathcal{B}_0$ in Section (11.1). We know by (12.3.1) that Δ_1 is a map between $\mathcal{B}_0$-bimodules; that is, the diagram

(5)
$$\begin{array}{ccc} \mathcal{B}_0 \otimes \mathcal{B}_1 \otimes \mathcal{B}_0 & \xrightarrow{1\otimes\Delta_1\otimes\Delta_0} & \mathcal{B}_0 \otimes (\mathcal{B}\hat{\otimes}\mathcal{B})_1 \otimes \mathcal{B}_0 \otimes \mathcal{B}_0 \\ \downarrow{\scriptstyle \mu} & & \downarrow{\scriptstyle \mu} \\ \mathcal{B}_1 & \xrightarrow{\Delta_1} & (\mathcal{B}\hat{\otimes}\mathcal{B})_1 \end{array}$$

commutes where μ denotes the action map with the left action of $\mathcal{B}_0$ on $(\mathcal{B}\hat{\otimes}\mathcal{B})_1$ defined in (12.1.11).

In (11.1.2) we have seen that a Hopf algebra is a coalgebra in the monoidal category of algebras; in particular, $\mathcal{A}$ is such a Hopf algebra. We now obtain the corresponding result for the pair algebra $\mathcal{B}$.

12.4.2 Theorem. *The pair algebra $\mathcal{B}$ together with the augmentation ε and the diagonal Δ is a coalgebra in the monoidal category of $[p]$-algebras, that is, the* *following diagrams in $\mathcal{A}lg^{[p]}$ are commutative.*

(1)

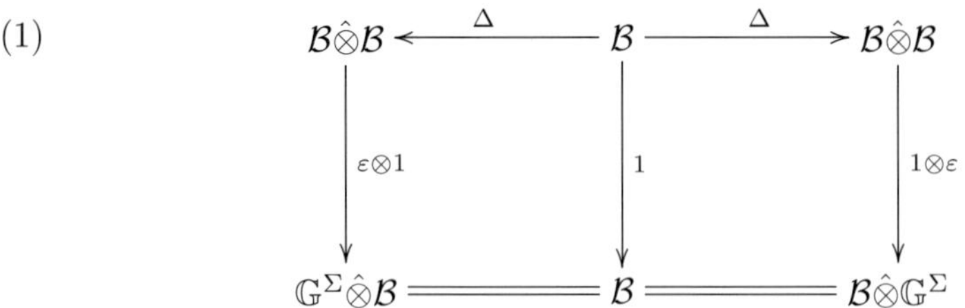

(2)

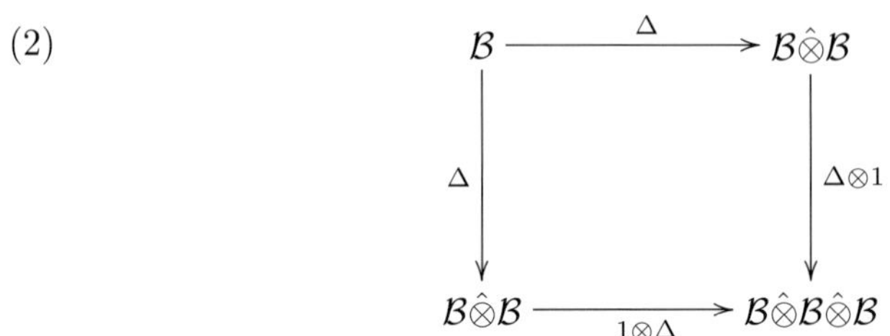

The theorem together with the properties in (12.4.1) describes the algebraic structure of the *secondary Hopf algebra* $\mathcal{B}$. We shall prove the theorem by using the action of the secondary Hopf algebra $\mathcal{B}$ on the strictified secondary cohomology in the next chapter. A somewhat more direct proof is also possible by using the definition of Δ_1 and the properties of the generalized Cartan tracks in Chapter 11, in particular (11.7.4).

Chapter 13

The Action of $\mathcal{B}$ on Secondary Cohomology

The pair algebra $\mathcal{B}$ of secondary cohomology operations is the strictification of the secondary Steenrod algebra $[\![\mathcal{A}]\!]$. We have seen that $\mathcal{B}$ is a secondary Hopf algebra generalizing the fact that the Steenrod algebra $\mathcal{A}$ is a Hopf algebra. We now consider the action of $\mathcal{A}$ on the cohomology $H^*(X)$ of a space and the corresponding action of $\mathcal{B}$ on the secondary cohomology. For this we use the strictification of secondary cohomology as defined in (5.6.2).

13.1 Pair algebras over the secondary Hopf algebra $\mathcal{B}$

The cohomology $H^*(X)$ of a space is an algebra and a module over the Steenrod algebra $\mathcal{A}$. Both structures are related by the Cartan formula which corresponds to the diagonal

$$\delta : \mathcal{A} \longrightarrow \mathcal{A} \otimes \mathcal{A}$$

of the Hopf algebra $\mathcal{A}$. The Cartan formula in terms of the diagonal is equivalent to the following commutative diagram with $H = H^*(X)$.

(13.1.1)

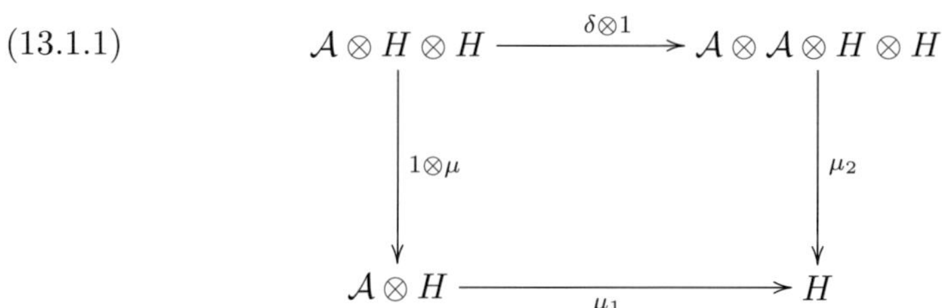

$$\begin{array}{ccc} \mathcal{A}\otimes H\otimes H & \xrightarrow{\delta\otimes 1} & \mathcal{A}\otimes\mathcal{A}\otimes H\otimes H \\ \downarrow{\scriptstyle 1\otimes\mu} & & \downarrow{\scriptstyle \mu_2} \\ \mathcal{A}\otimes H & \xrightarrow[\mu_1]{} & H \end{array}$$

Here μ_1 is given by the $\mathcal{A}$-module structure of H, that is, μ_1 carries $\alpha\otimes x$ to $\alpha(x)$. Moreover $1\otimes\mu$ carries $\alpha\otimes x\otimes y$ to $\alpha\otimes(x\cdot y)$ where $x\cdot y$ is the multiplication in H and μ_2 carries $\alpha \otimes \beta \otimes x \otimes y$ to $(-1)^{|\beta||x|}(\alpha x) \cdot (\beta y)$. One says that H is an *$\mathcal{A}$-algebra* or an *algebra over the Hopf algebra* $\mathcal{A}$ if the diagram commutes. This is equivalent to the condition $(\mathcal{K}1)$ in (1.1.7).

In this section we introduce a secondary analogue of an algebra over a Hopf algebra.

Let R be a commutative ring, for example $R = \mathbb{G} = \mathbb{Z}/p^2$. A pair X in the category of R-modules is an R-linear map $\partial : X_1 \to X_0$. The category of such pairs is a monoidal category with the tensor product $X \bar{\otimes} Y = \partial_{\otimes}$ in (5.1.2). A *graded pair* X is an R-linear map of degree 0,

$$X = (\partial : X_1 \longrightarrow X_0)$$

between (non-negatively) graded R-modules. For $n \in \mathbb{Z}$ we have the pair $X^n = (\partial^n : X_1^n \to X_0^n)$ in degree n given by X. The tensor product of graded pairs X, Y is defined by

$$(X \bar{\otimes} Y)^k = \bigoplus_{n+m=k} X^n \bar{\otimes} Y^m. \tag{13.1.2}$$

Compare (5.1.3). This is a monoidal structure of the category of graded pairs in **Mod**(R). A monoid B in this category is the same as a pair algebra, see (5.1.5). A module X over the pair algebra B is an action of the monoid B on X.

The category of graded pairs in **Mod**(R) is a track category in which homotopies or tracks are defined as follows. Let $f, g : X \to Y$ be maps between graded pairs, so that $\partial f_1 = f_0 \partial$ and $\partial g_1 = g_0 \partial$ as in the following diagram.

$$\begin{array}{ccc} X_1 & \xrightarrow{f_1, g_1} & Y_1 \\ {\scriptstyle\partial}\downarrow & \overset{H}{\nearrow} & \downarrow{\scriptstyle\partial} \\ X_0 & \xrightarrow[f_0, g_0]{} & Y_0 \end{array} \tag{13.1.3}$$

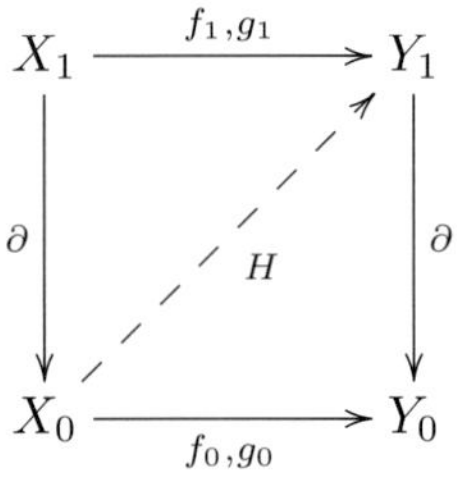

A *homotopy* $H : f \Rightarrow g$ is an R-linear map H as in the diagram such that

$$\begin{cases} \partial H = f_0 - g_0, \\ H\partial = f_1 - g_1. \end{cases}$$

Compare (6.4.2). The pasting of homotopies $H : f \Rightarrow g$ and $G : g \Rightarrow h$ is defined by

$$G \square H : f \Longrightarrow h$$

with $G \square H = G + H$ given by addition of maps.

We are now ready to define "pair algebras over the secondary Hopf algebra $\mathcal{B}$ " by replacing diagram (13.1.1) by a corresponding homotopy commutative diagram as follows. Recall that we have the secondary diagonal

$$\Delta : \mathcal{B} \longrightarrow \mathcal{B}\hat{\otimes}\mathcal{B}$$

where $\mathcal{B}\hat{\otimes}\mathcal{B}$ is the folding product for which we have the surjective map

$$q : \mathcal{B}\bar{\otimes}\mathcal{B} \twoheadrightarrow \mathcal{B}\hat{\otimes}\mathcal{B}.$$

Moreover we point out that we use the even sign convention in (11.1).

13.1.4 Definition. Let H be a pair algebra over $\mathbb{G}$ and a $\mathcal{B}$-module. We say that H is a *pair algebra over the secondary Hopf algebra* $\mathcal{B}$ if the following properties are satisfied. There is a commutative diagram

(1)

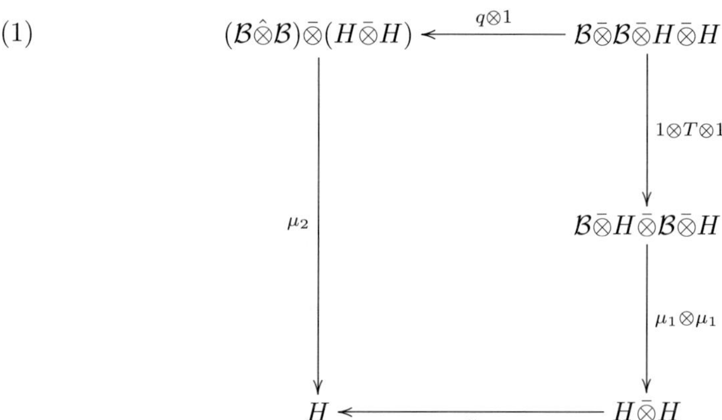

where μ_1 is the action of $\mathcal{B}_1$ on H and μ is the multiplication of H and T is the interchange map (11.1.1). Since q is surjective the map μ_2 is uniquely determined by μ and μ_1. Moreover there is given a homotopy C as in the diagram

(2)
$$\begin{array}{ccc}
\mathcal{B}\bar{\otimes}H\bar{\otimes}H & \xrightarrow{\Delta\otimes 1} & (\mathcal{B}\hat{\otimes}\mathcal{B})\bar{\otimes}(H\bar{\otimes}H) \\
{\scriptstyle 1\otimes\mu}\downarrow & \overset{C}{\Longrightarrow} & \downarrow{\scriptstyle \mu_2} \\
\mathcal{B}\bar{\otimes}H & \xrightarrow[\mu_1]{} & H
\end{array}$$

where C is a $\mathbb{G}$-linear map

$$C : \mathcal{B}_0 \otimes H_0 \otimes H_0 \longrightarrow H_1$$

satisfying the equations

$$
\begin{aligned}
\partial C &= (\mu_1(1 \otimes \mu))_0 - (\mu_2(\Delta \bar{\otimes} 1))_0 \quad \text{and} \\
C\partial &= (\mu_1(1 \otimes \mu))_1 - (\mu_2(\Delta \bar{\otimes} 1))_1.
\end{aligned} \tag{3}
$$

Compare (13.1.3). Moreover the homotopy C has the following property (4). Let $\alpha, \beta \in \mathcal{B}_0$ and $x, y, z \in H_0$. We write $(x, y) = x \otimes y \in H_0 \otimes H_0$. Then the *associativity formula* is satisfied:

$$
\begin{aligned}
&C(\alpha \otimes (x \cdot y, z)) + (C \otimes 1)(\Delta(\alpha) \otimes (x, y, z)) \\
&= C(\alpha \otimes (x, y \cdot z)) + (1 \otimes C)(\Delta(\alpha) \otimes (x, y, z)).
\end{aligned} \tag{4}
$$

Here the operators

$$C \otimes 1, 1 \otimes C : \mathcal{B}_0 \otimes \mathcal{B}_0 \otimes H_0 \otimes H_0 \otimes H_0 \longrightarrow H_1$$

are defined by

$$
\begin{aligned}
(C \otimes 1)(\alpha \otimes \beta \otimes (x, y, z)) &= (-1)^{p|\beta|(|x|+|y|)} C(\alpha \otimes (x, y)) \cdot \beta(z), \\
(1 \otimes C)(\alpha \otimes \beta \otimes (x, y, z)) &= (-1)^{p|\beta||x|} \alpha(x) \cdot C(\beta \otimes (y, z))
\end{aligned}
$$

with $(x, y, z) = x \otimes y \otimes z$.

Equation (4) can be expressed in terms of diagrams as follows. We set

$$
\begin{aligned}
\mathcal{B}^2 = \mathcal{B}\hat{\otimes}\mathcal{B}, \quad &\mathcal{B}^3 = \mathcal{B}\hat{\otimes}\mathcal{B}\hat{\otimes}\mathcal{B}, \\
H^2 = H\bar{\otimes}H, \quad &H^3 = H\bar{\otimes}H\bar{\otimes}H.
\end{aligned}
$$

Now we consider the following diagrams.

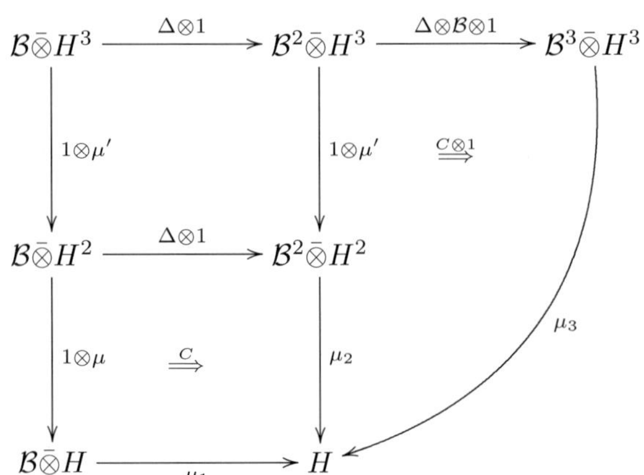

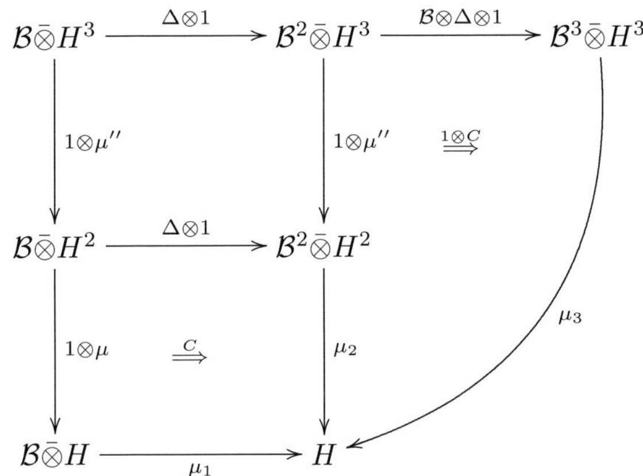

Here μ' and μ'' are defined by $\mu'(x,y,z) = (x\cdot y, z)$ and $\mu''(x,y,z) = (x, y\cdot z)$ and μ_3 is given by $\mu_3(\alpha\otimes\beta\otimes\gamma\otimes(x,y,z)) = \pm\alpha(x)\cdot\beta(y)\cdot\gamma(z)$ with the obvious sign. The associativity of $\mu_H : H\bar{\otimes}H \to H$ and $\Delta : \mathcal{B} \to \mathcal{B}\hat{\otimes}\mathcal{B}$ implies that

$$\begin{aligned}(1\otimes\mu)(1\otimes\mu') &= (1\otimes\mu)(1\otimes\mu''),\\ (\Delta\otimes\mathcal{B}\otimes 1)(\Delta\otimes 1) &= (\mathcal{B}\otimes\Delta\otimes 1)(\Delta\otimes 1).\end{aligned} \tag{13.1.5}$$

This shows that the boundaries of the two diagrams above coincide. Now equation (4) in (13.1.4) is equivalent to saying that the pasting of tracks in the two diagrams yields the same track, that is

$$(C\otimes 1)(\Delta\otimes 1)\square C(1\otimes\mu') = (1\otimes C)(\Delta\otimes 1)\square C(1\otimes\mu'').$$

One can check that the formulas for $C\otimes 1$ and $1\otimes C$ in (13.1.4)(5) yield well-defined homotopies for the diagrams above.

13.2 Secondary cohomology as a pair algebra over $\mathcal{B}$

Let X be a connected space and let $\mathcal{H}^*(X)$ be the secondary cohomology of X. Then $\mathcal{H}^*(X)$ is a pair algebra and its strictification defined in (5.6.2) is a $\mathcal{B}$-module. We now consider a strictification of $\mathcal{H}^*(X)$ which is a $\mathcal{B}$-module and also a pair algebra.

A *graded set* is a sequence of sets S^i, $i\in\mathbb{Z}$, such that $S^i = \emptyset$ is empty for $i<0$. The product of graded sets $S\times S'$ is the set of pairs (x,y) with $x\in S, y\in S'$ and degree $\mid (x,y)\mid = \mid x\mid + \mid y\mid$. Let $\mathbf{Set}^*$ be the category of graded sets and maps of degree 0. Then $(\mathbf{Set}^*, \times, *)$ is a monoidal category where the unit $*$ is the singleton concentrated in degree 0. A *graded monoid* M is a monoid object in $\mathbf{Set}^*$ given by $1\in M^0$ and by the associative multiplication

$$\mu^{n,k} : M^n\times M^k \longrightarrow M^{n+k}$$

which carries (x,y) to $x\cdot y$ with $1\cdot x = x = x\cdot 1$.

Let $\mathcal{M}$ be a graded monoid. Then $\mathcal{M}$ acts on the graded set S if an action map

$$\mathcal{M}^n \times S^k \longrightarrow S^{n+k}$$

is given which carries (x, u) to $x \cdot u$ such that $1 \cdot u = u$ and $(x \cdot y) \cdot u = x \cdot (y \cdot u)$. In this case we will call S an $\mathcal{M}$*-set*.

13.2.1 Definition. Recall that $E_{\mathcal{A}}$ denotes the set of generators of the Steenrod algebra $\mathcal{A}$. Let $\mathcal{M} = \mathrm{Mon}(E_{\mathcal{A}})$ be the free graded monoid generated by $E_{\mathcal{A}}$ and

$$s_X = M_X \longrightarrow \mathcal{H}^*(X)_0 = [\![X, Z^*]\!]_0$$

be a function with the following properties. Here M_X is a monoid and s_X is a morphism of monoids with the multiplication in $\mathcal{H}^*(X)_0$ induced by $Z^n \times Z^m \to Z^{n+m}$. Moreover M_X is a free $\mathcal{M}$-set and s_X is an $\mathcal{M}$-equivariant morphism of $\mathcal{M}$-sets with the action of $\alpha \in \mathcal{M}$ on $\xi : X \to Z^q \in \mathcal{H}^*(X)_0$ by composition

$$\alpha\xi : X \xrightarrow{\xi} Z^q \xrightarrow{(s\alpha)_q} Z^{q+|\alpha|}.$$

Then we obtain the *strictification* $\mathcal{H}^*(X, M_X, s_X)$ by the pull back diagram (compare (5.6.2))

$$\begin{array}{ccc} \mathcal{H}^*(X, M_X, s_X)_1 & \xrightarrow{\quad s \quad} & \mathcal{H}^*(X)_1 \\ \Big\downarrow \partial & \text{pull} & \Big\downarrow \partial \\ \mathcal{H}^*(X, M_X, s_X)_0 = \mathbb{G}[M_X] & \xrightarrow[\quad s_X \quad]{} & \mathcal{H}^*(X)_0 \end{array}$$

where $\mathbb{G}[M_X]$ is the free $\mathbb{G}$-module generated by the set M_X. Since M_X is a free $\mathcal{M}$-set we have a set $E_X \subset M_X$ of generators of the free $\mathcal{M}$-set M_X such that

$$\mathbb{G}[M_X] = T_{\mathbb{G}}(E_{\mathcal{A}}) \otimes \mathbb{G}E_X.$$

Moreover s_X is the free $\mathbb{G}$-linear extension of s_X above. The $\mathcal{B}$-module structure of $\mathcal{H}^*(X, M_X, s_X)$ is defined as in (5.6.2) by the Γ-product $\bullet$. Since s_X is a morphism of monoids we see that $\mathcal{H}^*(X, M_X, s_X)$ is a pair algebra so that the strictification $\mathcal{H}^*(X, M_X, s_X)$ is both, a $\mathcal{B}$-module and a pair algebra.

13.2.2 Example. Let

$$\mathcal{M}_X = \mathrm{Mon}(\mathcal{M} \times \mathcal{H}^*(X)_0)$$

be the free monoid generated by pairs $\alpha\xi = (\alpha, \xi)$ with $\alpha \in \mathcal{M} = \mathrm{Mon}(E_{\mathcal{A}})$ and $\xi \in \mathcal{H}^*(X)_0$. Then $\mathcal{M}_X$ is a monoid and a free $\mathcal{M}$-set and we have the natural map as in (13.2.1),

$$s_X : \mathcal{M}_X \longrightarrow \mathcal{H}^*(X)_0$$

which is the identity on the generating set $\mathcal{H}^*(X)_0$ of $\mathcal{M}_X$.

In this case $\mathcal{H}^*(X, \mathcal{M}_X, s_X)$ is a functor which carries a path connected pointed space X to a pair algebra which is also a $\mathcal{B}$-module.

13.2.3 Theorem. *The strictification $\mathcal{H}^*(X, M_X, s_X)$ of the secondary cohomology is a $\mathcal{B}$-module and a pair algebra in such a way that $\mathcal{H}^*(X, M_X, s_X)$ is a pair algebra over the secondary Hopf algebra $\mathcal{B}$ as defined in* (13.1.4). *Moreover $\mathcal{H}^*(X, \mathcal{M}_X, s_X)$ yields a functor from spaces to the category of pair algebras over the secondary Hopf algebra $\mathcal{B}$.*

This result is proved by the lemmas below.

We know that $\mathcal{H}^*(X)$ is not only a pair algebra but also a secondary permutation algebra. We shall consider this richer structure below.

13.2.4 Lemma. *For $H = \mathcal{H}^*(X, M_X, s_X)$ the map μ_2 in* (13.1.4) *is well defined.*

Proof. Since $q \otimes 1$ is an isomorphism at level 0 we only have to consider level 1. We have the commutative diagram

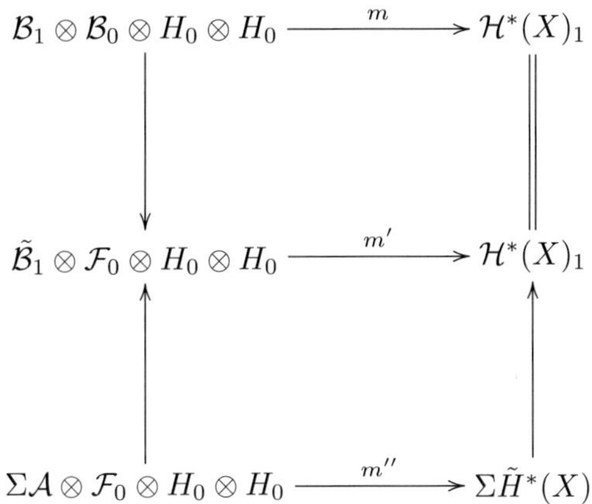

where we use $s : H_1 \longrightarrow \mathcal{H}^*(X)_1$ in (13.2.1) to define m by the formula

(1) $$m(a \otimes \beta \otimes x \otimes y) = \varepsilon s(a \cdot x) \cdot s(\beta \cdot y) \quad \text{with } \varepsilon = (-1)^{|\beta||x|}.$$

For $a = [p]$ we get $m([p] \otimes \beta \otimes x \otimes y) = 0$ so that m induces the map m'. Moreover we set

(2) $$m''((\Sigma\alpha) \otimes \beta \otimes x \otimes y) = \varepsilon\Sigma((\alpha x) \cdot (\beta y))$$

with $\Sigma\alpha \in \Sigma\mathcal{A}$ and $\alpha x, \beta y$ defined by the action of $\mathcal{A}$ on $\tilde{H}^*(X)$. The product $(\alpha x) \cdot (\beta y)$ is given by the multiplication in the algebra $H^*(X)$.

In a similar way we get the commutative diagram

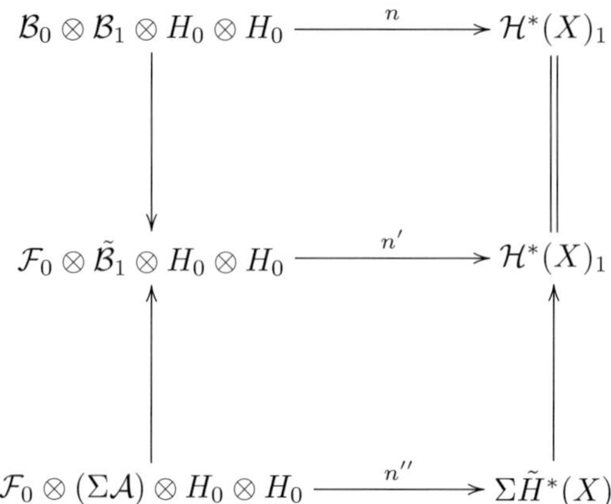

with

(3) $$n(\alpha \otimes b \otimes x \otimes y) = \varepsilon s(\alpha x) \cdot s(b \cdot y) \quad \text{with } \varepsilon = (-1)^{|b||x|},$$

(4) $$n''(\alpha \otimes (\Sigma\beta) \otimes x \otimes y) = \varepsilon \cdot (-1)^{|\alpha|}\Sigma(\alpha x) \cdot (\beta y).$$

Now (2) and (4) show that the map

(5) $$(\tilde{\mathcal{B}}\bar{\otimes}\tilde{\mathcal{B}})_1 \otimes H_0 \otimes H_0 \longrightarrow \mathcal{H}^*(X)_1$$

defined by (m', n') is compatible with the folding map φ. Therefore (5) induces

(6) $$(\mathcal{B}\hat{\otimes}\mathcal{B})_1^{\sim} \otimes H_0 \otimes H_0 \longrightarrow \mathcal{H}^*(X)_1.$$

Compare (12.2.2). The map

(7) $$(\mathcal{B}\hat{\otimes}\mathcal{B})_1 \otimes H_0 \otimes H_0 \longrightarrow H_1$$

is a map between pull backs induced by the following diagram.

$$\begin{array}{ccc}
(\mathcal{B}\hat{\otimes}\mathcal{B})_1^{\sim} \otimes H_0 \otimes H_0 & \longrightarrow & \mathcal{H}^*(X)_1 \\
\downarrow \partial & & \downarrow \partial \\
\mathcal{F}_0 \otimes \mathcal{F}_0 \otimes H_0 \otimes H_0 & \longrightarrow & \mathcal{H}^*(X)_0 \\
\uparrow & & \uparrow \\
\mathcal{B}_0 \otimes \mathcal{B}_0 \otimes H_0 \otimes H_0 & \longrightarrow & H_0
\end{array}$$

Moreover μ_2 is induced by (7) and

$$(8) \qquad \mathcal{B}_0 \otimes \mathcal{B}_0 \otimes (H \bar{\otimes} H)_1 \longrightarrow H_1$$

where (8) is given by the left action of $\mathcal{B}_0$ on H and the multiplication of the pair algebra H. This completes the proof that μ_2 is well defined. □

13.2.5 Definition. For the strictification $H = \mathcal{H}^*(X, M_X, s_X)$ we define the *Cartan homotopy*

$$(1) \qquad C : \mathcal{B}_0 \otimes H_0 \otimes H_0 \longrightarrow H_1$$

as follows. Let $\alpha \in \mathcal{B}_0$ and $x, y \in H_0$. We have the following pull back diagram.

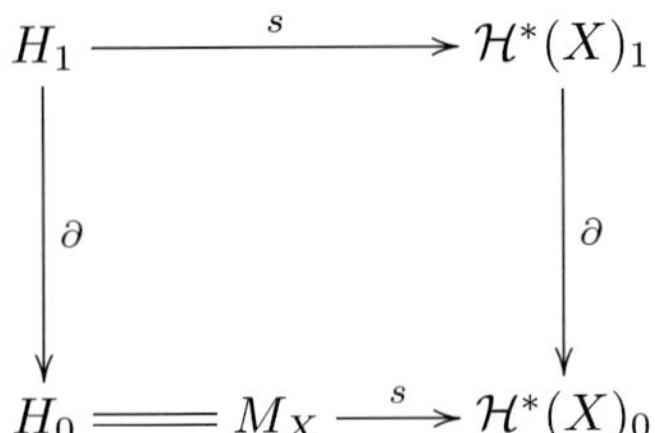

Hence $C(\alpha \otimes x \otimes y)$ is completely determined by (2) and (5). In H_0 we set

$$(2) \qquad \partial C(\alpha \otimes x \otimes y) = \alpha(x \cdot y) - \mu_2(\Delta(\alpha) \otimes x \otimes y).$$

Here $x \cdot y$ is the product in the algebra H_0 and $\alpha(x \cdot y)$ is defined by the action of $\mathcal{B}_0$ on H_0. Moreover $\Delta(\alpha) \in \mathcal{B}_0 \otimes \mathcal{B}_0$ is given by the diagonal Δ of $\mathcal{B}_0$ and μ_2 is defined in (13.1.4)(1).

Moreover using the generalized Cartan track $C_\alpha^{sx,sy}$ we get the commutative diagram of tracks in $[\![X, Z^*]\!]$.

$$(3) \qquad \begin{array}{ccc} \alpha(sx \cdot sy) & \xrightarrow{C_\alpha^{sx,sy}} & \Delta(\alpha)(sx, sy) \\ \Big\downarrow{\scriptstyle \Gamma(\alpha, x\cdot y)} & & \Big\downarrow{\scriptstyle \Gamma_{\Delta\alpha}} \\ s(\alpha(x \cdot y)) & \xrightarrow{\bar{C}_\alpha^{x,y}} & s\mu_2(\Delta(\alpha) \otimes x \otimes y) \end{array}$$

Here $\Gamma(\alpha, x \cdot y)$ is given as in (5.3.1).

For $\sum_i \alpha_i' \otimes \alpha_i''$ we get $\Gamma_{\Delta\alpha}$ by the following diagram.

(4)
$$\begin{array}{ccc} \Delta(\alpha)(sx,sy) & = & \sum_i \pm\alpha_i'(sx)\cdot\alpha_i''(sy) \\ \Big\downarrow{\scriptstyle \Gamma^{x,y}_{\Delta\alpha}=\Gamma_{\Delta\alpha}} & & \Big\downarrow{\scriptstyle \sum_i \pm\Gamma(\alpha_i',x)\cdot\Gamma(\alpha_i'',y)} \\ s\mu_2(\Delta(\alpha)\otimes x\otimes y) & = & \sum_i \pm s(\alpha_i'x)\cdot s(\alpha_i''y) \end{array}$$

Now we set in $\mathcal{H}^*(X)_1$,

(5)
$$sC(\alpha\otimes x\otimes y) = \bar{C}^{x,y}_\alpha - s\mu_2(\Delta(\alpha)\otimes x\otimes y).$$

Then we see that $C(\alpha\otimes x\otimes y)\in H_1$ is well defined by (2) and (5). Since all tracks in (3) are linear in α we see that $C(\alpha\otimes x\otimes y)$ is linear in α. Moreover since C_α is a linear Δ-track we see that $C(\alpha\otimes x\otimes y)$ is also linear in x and y. Therefore the $\mathbb{G}$-linear map C in (1) above is well defined.

13.2.6 Lemma. *The diagonal* $\Delta_1 : \mathcal{B}_1 \longrightarrow (\mathcal{B}\hat{\otimes}\mathcal{B})_1$ *satisfies the formula* ($a\in\mathcal{B}_1, x,y\in H_0$)

$$C(\alpha\otimes x\otimes y) = a(x\cdot y) - \Delta_1(a)(x,y) \ \in H_1$$

with $\alpha = \partial a\in\mathcal{B}_0$. *Here* $a(x\cdot y)$ *is defined by the left action of* $\mathcal{B}$ *on* H *and*

$$\Delta_1(a)(x,y) = \mu_2(\Delta_1(a)\otimes x\otimes y)$$

is defined by μ_2 *in* (13.2.4), (13.1.4)(1).

Hence (13.2.5)(1) and (13.2.6) show that the equations (13.1.4)(3) are satisfied.

Proof. Let ρ be a splitting of $\tilde{\mathcal{B}}$ and let $a=\rho_{\mathcal{B}}(\alpha)$. Then we define $\Delta_1(a)$ by $\tilde{\Delta}(a)$ with $\tilde{\Delta}$ in (12.2.2), namely

(1)
$$\tilde{\Delta}(\rho_{\mathcal{B}}(\alpha)) = \bar{\varphi}\bar{\rho}(\Delta_0\alpha) - \Theta_\rho(\alpha).$$

Compare formula (12.2.2)(6). According to (11.9.4) the element $\Theta_\rho(\alpha)$ is given by the track $\Theta_{x,y}$ in the following commutative diagram of tracks.

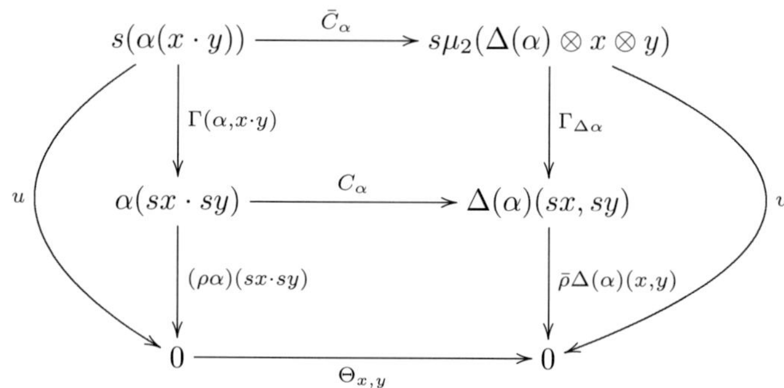

Here the composites u and v satisfy

$$u = s((\rho\alpha)\cdot(x\cdot y)),$$
$$v = s\mu_2(\bar{\rho}\Delta\alpha\otimes x\otimes y).$$

Moreover by definition of $C(\alpha\otimes x\otimes y)$ we have

(2) $$\bar{C}_\alpha = sC(\alpha\otimes x\otimes y) + s\mu_2(\Delta(\alpha)\otimes x\otimes y).$$

We get by (2.2.6) that

(3) $$\Theta_{x,y}\Box u = \Theta_{x,y} + u,$$

(4) $$\begin{aligned} v\Box\bar{C}_\alpha &= v + (\bar{C}_\alpha - s\mu_2(\Delta(\alpha)\otimes x\otimes y)) \\ &= v + sC(\alpha\otimes x\otimes y). \end{aligned}$$

Since we have (3)=(4) (see the diagram above) one gets

$$\begin{aligned} sC(\alpha\otimes x\otimes y) &= u + \Theta_{x,y} - v \\ &= s(\rho(\alpha)(x\cdot y)) + \Theta_{x,y} - s\mu_2(\bar{\rho}\Delta\alpha\otimes x\otimes y) \\ &= s(a(x\cdot y)) - s\Delta_1(a)(x,y). \end{aligned}$$

In fact, by (1) we have

$$\begin{aligned} s\Delta_1(\alpha)(x,y) &= s(\bar{\varphi}\bar{\rho}(\Delta_0\alpha) - \Theta_\rho(\alpha))(x,y) \\ &= s\mu_2(\bar{\rho}\Delta\alpha\otimes x\otimes y) - \Theta_{x,y}. \end{aligned}$$

□

13.2.7 Proposition. *The Cartan homotopy C satisfies the associativity formula (13.1.4)(4).*

Proof. We derive from (11.7.4) and the definition of $\bar{C}_\alpha = \bar{C}_\alpha^{x,y}$ that the following diagram commutes.

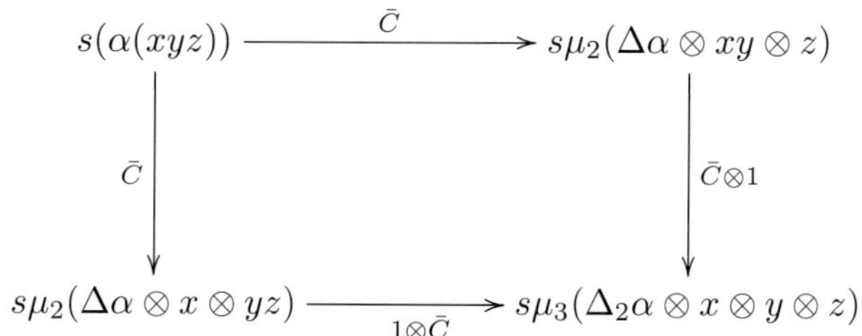

Here $\Delta_2 = (1\otimes\Delta)\Delta = (\Delta\otimes 1)\Delta$ and μ_3 is similarly defined as μ_2. Moreover $\bar{C}\otimes 1$ is similarly defined as $C\otimes 1$ in (11.7.4). Now (2.2.6) applied to the tracks in the diagram yields, for $\xi = s\mu_3(\Delta_2\alpha\otimes x\otimes y\otimes z)$,

$$(\bar{C}\otimes 1 - \xi) + C(\alpha\otimes xy\otimes z) = (1\otimes\bar{C} - \xi) + C(\alpha\otimes x\otimes yz)$$

where $\bar{C}\otimes 1 - \xi = (C\otimes 1)(\Delta\alpha\otimes(x,y,z))$ and $(1\otimes\bar{C}) - \xi = (1\otimes C)(\Delta\alpha\otimes(x,y,z))$. □

13.2.8 Lemma. *Let $\xi, \eta \in (\mathcal{B}\hat{\otimes}\mathcal{B})_1$ and let $\partial\xi = \partial\eta$ and $\xi(x,y) = \eta(x,y) \in H = \mathcal{H}^*[X]$ for all spaces X and elements $x, y \in H$. Then $\xi = \eta$.*

The lemma shows that there is at most one element $\Delta_1(a)$ satisfying (13.2.6) for all X and x, y.

Proof of (13.2.8). Exactness shows that there is $\Sigma(u) \in \Sigma(\mathcal{A} \otimes \mathcal{A})$ with $\xi = \eta + \hat{\imath}(\Sigma(u))$, $u \in \mathcal{A} \otimes \mathcal{A}$. Then

$$\xi(x,y) = \eta(x,y) + \Sigma(u(x,y))$$

where $u(x,y) \in \pi_0(H) = H^*(X)$ and we use (2.2.10). Hence $u(x,y) = 0$ for all X and x, y. But this implies $u = 0$. In fact, let $\mid u \mid < k$ and let $X = Y \times Y$ where

$$Y = Z^1 \times \cdots \times Z^1$$

is the k-fold product. For $p = 2$ let $u : Z^1 \to Z^1$ be the identity and for p odd let $u = u' \cdot \beta(u') : Z^1 \to Z^3$ where u' is the identity. Let

$$x = y = u \boxtimes \cdots \boxtimes u$$

be the k-fold $\boxtimes$-product of u, see (2.1.5). Then one can check that

$$\begin{array}{ccc} \mathcal{A}^{\leq k} & \longrightarrow & H^*(Y), \\ \alpha & \mapsto & \alpha(u) \end{array}$$

is injective, see 1.2 [Sch]. Hence the map

$$\mathcal{A}^{\leq k} \otimes \mathcal{A}^{\leq k} \longrightarrow H^*(Y) \otimes H^*(Y) = H^*(X)$$

which carries $\alpha \otimes \beta$ to $(-1)^{|\beta||x|}\alpha(x) \otimes \beta(y) = (\alpha \otimes \beta)((xp_1), (yp_2))$ is injective where p_1, p_2 are the projections $X \to Y$. □

13.2.9 Lemma. *The diagonal $\Delta = (\Delta_1, \Delta_0)$ of $\mathcal{B}$ is coassociative.*

Proof. We show that diagram (12.4.2)(2) commutes. Let $a \in \mathcal{B}_1$ with $\partial a = \alpha$ so that $\partial(\Delta_1 a) = \Delta_0\alpha$. Moreover let $x, y, z \in H$. Then we have the equations

$$\begin{array}{lrl} (1) & C(\alpha \otimes (x \cdot y, z)) = & a(x \cdot y \cdot z) - \Delta_1(a)(x \cdot y, z), \\ (2) & (C \otimes 1)((\Delta_0\alpha) \otimes (x,y,z)) = & (\Delta_1 a)(x \cdot y, z) - ((\Delta \otimes 1)_1\Delta_1(a))(x,y,z), \\ (3) & C(\alpha \otimes (x, y \cdot z)) = & a(x \cdot y \cdot z) - (\Delta_1 a)(x, y \cdot z), \\ (4) & (1 \otimes C)((\Delta_0\alpha) \otimes (x,y,z)) = & \Delta_1(a)(x, y \cdot z) - ((1 \otimes \Delta)_1\Delta_1(a))(x,y,z). \end{array}$$

Here we have $(1) + (2) = (3) + (4)$ by (13.2.7). This implies

$$((\Delta \otimes 1)_1\Delta_1(a))(x,y,z) = ((1 \otimes \Delta)_1\Delta_1(a))(x,y,z)$$

and this shows by uniqueness as in (13.2.8) that

$$(\Delta \otimes 1)_1\Delta_1 = (1 \otimes \Delta)_1\Delta_1$$

and hence Δ is coassociative. □

13.3 Secondary Instability

The cohomology $H^*(X)$ of a space is an *unstable* algebra over the Steenrod algebra $\mathcal{A}$ in the sense that for $\alpha \in \mathcal{A}$, $x \in H^*(X)$ we have $\alpha x = 0$ if $e(\alpha) > |x|$ where $e(\alpha)$ is the excess of α. Moreover

$$(13.3.1) \qquad \begin{aligned} Sq^{|x|} x &= x^2 \quad \text{if } p = 2, \\ P^{|x|/2} x &= x^p \quad \text{for } |x| \text{ even and } p \text{ odd.} \end{aligned}$$

In this section similar properties are also described for the strictified secondary cohomology

$$\mathcal{H}^*[X] = \mathcal{H}^*(X, \mathcal{P}_X, s_X)$$

defined in (5.6.2).

13.3.2 Definition. Let $\mathcal{M} = \mathrm{Mon}(E_{\mathcal{A}})$ be the free monoid generated by the set $E_{\mathcal{A}}$ and let

$$(1) \qquad \mathcal{E}(X) \subset \mathcal{M} \times \mathcal{H}^*[X]_0$$

be the *excess subset* given by all pairs (α, x) with $e(\alpha) < |x|$, $\alpha \in \mathcal{M}$ and $x \in \mathcal{H}^*[X]_0$. Then there is a well-defined *unstable structure map*

$$(2) \qquad v : \mathcal{E}(X) \longrightarrow \mathcal{H}^*[X]_1$$

which is natural in X. We define $v(\alpha, x)$ by the pull back

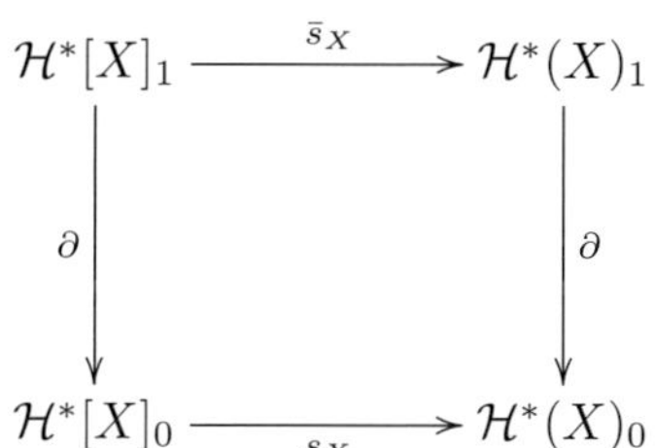

namely by $\partial v(\alpha, x) = \alpha \cdot x$ and by the track $\bar{s}_X v(\alpha, x) : s_X(\alpha x) \Longrightarrow 0$ in $\mathcal{H}^*(X)_1$ given by

$$(3) \qquad \Gamma(\alpha, x)^{\mathrm{op}} : s_X(\alpha x) \Longrightarrow s(\alpha) s_X(x) = 0.$$

See (5.3.2) and (5.4.3). Here the right-hand side is trivial by (5.5.1), (10.2.5) and (10.2.6) since $e(\alpha) < |x|$. The existence of $v(\alpha, x)$ with $\partial v(\alpha, x) = \alpha x$ implies that αx represents the trivial element in cohomology $H^*(X) = \textit{cokernel}(\partial)$.

In addition we get the following *unstable structure maps* corresponding to (13.3.1) above. We have

$$\begin{aligned} u : \mathcal{H}^*[X]_0 &\longrightarrow \mathcal{H}^*[X]_1 \quad \text{for } p = 2 \text{ and} \\ u : \mathcal{H}^*[X]_0^{\text{even}} &\longrightarrow \mathcal{H}^*[X]_1 \quad \text{for } p \text{ odd} \end{aligned}$$

with $|ux| = p|x|$ and

$$\begin{aligned} \partial(ux) &= Sq^{|x|}(x) - x^2 && \text{for } p = 2, \\ \partial(ux) &= P^{|x|/2}(x) - x^p && \text{for } |x| \text{ even and } p \text{ odd.} \end{aligned}$$

We define ux using the pull back above by the track $\bar{s}_X(ux) = \Gamma(\alpha, x)^{\mathrm{op}} - s_X(x^p)$ with $\alpha = Sq^{|x|}$ for $p = 2$ and $\alpha = P^{|x|/2}$ for p odd where

$$\Gamma(\alpha, x) : s_X(\alpha x) \Longrightarrow s(\alpha)s_X x = (s_X x)^p = s_X(x^p)$$

is a track in (5.4.3). Here the right-hand side holds by (5.5.1), (10.2.5), (10.2.6). Again the map u is natural in X and the existence of u implies the equations (13.3.1) in cohomology.

13.3.3 Theorem. *The unstable structure map v of $H = \mathcal{H}^*[X]$ has the following properties (provided both sides of the equations are defined),*

$$\begin{aligned} \alpha v(\beta, x) &= v(\alpha\beta, x), \\ v(\beta\gamma, x) &= v(\beta, \gamma x) \end{aligned}$$

for $\alpha, \beta, \gamma \in \mathcal{M}$, $x \in H_0$. Moreover

$$\begin{aligned} v(\alpha, \partial\bar{x}) &= \alpha\bar{x} \quad \text{for } \bar{x} \in H_1, \\ pv(\alpha, x) &= [p] \cdot (\alpha x), \\ \partial v(\alpha, x) &= \alpha x. \end{aligned}$$

For p odd the element $v(\alpha, x)$ is linear in x, but for $p = 2$ the element $v(\alpha, x)$ is quadratic in x with cross effect

$$v(\alpha, x + y) - v(\alpha, x) - v(\alpha, y) = v(\alpha, x|y) \in \Sigma\tilde{H}^*X$$

determined by the properties above and the delicate linearity track formula (5.5.1)(2d),

$$v(Sq^k, x|y) = \begin{cases} \Sigma(x \cdot y) & \text{for } k = |x| + 1, \\ 0 & \text{for } k > |x| + 1. \end{cases}$$

13.3.4 Theorem. *The unstable structure map u of $H = \mathcal{H}^*[x]$ has the following properties with $\alpha = Sq^{|x|}$, $x \in H_0$ for $p = 2$ and $\alpha = P^{|x|/2}$, $x \in H_0^{\mathrm{even}}$ for p odd:*

$$\begin{aligned} \partial(ux) &= \alpha x - x^p, \\ ux &= \alpha\bar{x} - \bar{x} \cdot x^{p-1} \quad \text{for } \bar{x} \in H_1 \text{ and } \partial\bar{x} = x, \\ p(ux) &= [p] \cdot (\alpha x - x^p). \end{aligned}$$

Chapter 14

Interchange and the Left Action

In this chapter we compute the symmetry operator S and the left action operator L associated to the secondary Hopf algebra $\mathcal{B}$ of secondary cohomology operations.

14.1 The operators S and L

Let $(\mathcal{B}, \Delta)$ be a secondary Hopf algebra as in (12.4) and assume that $\mathcal{B}$ is a pair algebra such that $\Gamma[p] = \kappa$ satisfies $(\kappa \otimes 1)\delta = \kappa\delta$. For example the algebra $\mathcal{B}$ of secondary cohomology operations has these properties. Then we define the *symmetry operator*

(14.1.1) $$S : R_{\mathcal{B}} \longrightarrow \tilde{\mathcal{A}} \otimes \tilde{\mathcal{A}} \text{ of degree } -1$$

as follows. Here $\tilde{\mathcal{A}}$ is the augmentation ideal in the Steenrod algebra $\mathcal{A}$ and $R_{\mathcal{B}} = kernel(\mathcal{B}_0 \to \mathcal{A})$ is the ideal of relations in $\mathcal{B}_0$. We define for $\xi \in R_{\mathcal{B}}$ with $\xi = \partial x$, $x \in \mathcal{B}_1$, the element $S(\xi) \in \tilde{\mathcal{A}} \otimes \tilde{\mathcal{A}}$ by the formula

$$T\Delta_1(x) = \Delta_1(x) + \Sigma S(\xi).$$

14.1.2 Lemma. *The operator S is a well-defined linear map.*

Proof. Since $T\Delta_0 = \Delta_0$ by (11.1) we see that there is a unique element $S(\xi)$ satisfying the equation. Moreover altering x by $\Sigma a \in \Sigma\mathcal{A}$ we get $\Delta_1(x + \Sigma a) = \Delta_1(x) + \Sigma\delta a$ and $T\delta = \delta$ so that $S(\xi)$ does not depend on the choice of x. Using the augmentation of $\mathcal{B}$ we see that $S(\xi) \in \tilde{\mathcal{A}} \otimes \tilde{\mathcal{A}}$. □

Next we define the *left action operator*

(14.1.3) $$L : \mathcal{A} \otimes R_{\mathcal{B}} \longrightarrow \tilde{\mathcal{A}} \otimes \tilde{\mathcal{A}} \text{ of degree } -1$$

as follows. By (12.1.11) we know that $(\mathcal{B}\hat{\otimes}\mathcal{B})_1$ is a left $\mathcal{B}_0$-module but the diagonal $\Delta_1 : \mathcal{B}_1 \longrightarrow (\mathcal{B}\hat{\otimes}\mathcal{B})_1$ need not be a map of left $\mathcal{B}_0$-modules. Therefore we define L by the formula

$$\Delta_1(\alpha \cdot x) = \alpha \cdot \Delta_1(x) + \Sigma L(\alpha \otimes \xi)$$

for $\alpha \in \mathcal{B}_0$, $x \in \mathcal{B}_1$ with $\partial x = \xi \in R_{\mathcal{B}}$.

14.1.4 Lemma. *The operator L is a well-defined linear map satisfying the equations* $(\alpha, \beta \in \mathcal{B}_0,\ \xi \in R_{\mathcal{B}})$

$$\begin{aligned} L(\alpha\beta \otimes \xi) &= L(\alpha \otimes \beta\xi) + (-1)^{|\alpha|}\delta(\alpha) \cdot L(\beta \otimes \xi), \\ L(\alpha \otimes \xi\beta) &= L(\alpha \otimes \xi) \cdot \delta(\beta), \\ L(\alpha \otimes p) &= 0. \end{aligned}$$

The equations show that the operator L is determined by the values $L(\alpha\otimes\xi)$ with $\alpha \in \mathcal{A}$ and $\xi \in E^1_{\mathcal{A}}$ where $E^1_{\mathcal{A}}$ is a set of generators of the ideal $R_{\mathcal{B}}$, for example, the set of generators given by Adem relations.

Proof. Since Δ_0 is an algebra map we see that there is a unique element $L(\alpha \otimes \xi)$ satisfying the equation. If $\alpha = \partial a$, $a \in \mathcal{B}_1$, then $\alpha \cdot x = a \cdot \partial x = a \cdot \xi$ and since Δ_1 is a map of right $\mathcal{B}_0$-modules we get

$$\begin{aligned} \Delta_1(\alpha \cdot x) = \Delta_1(a \cdot \xi) &= \Delta_1(a) \cdot \Delta_0\xi \\ &= \alpha \cdot \Delta_1 x \quad \text{see (12.1.11)(4)} \end{aligned}$$

so that $L(\partial a \otimes x) = 0$. Moreover $L(\alpha \otimes \xi)$ does not depend on the choice of x since $\delta : \mathcal{A} \to \mathcal{A} \otimes \mathcal{A}$ is an algebra map. Finally using the augmentation of $\mathcal{B}$ we see $L(\alpha \otimes \xi) \in \tilde{\mathcal{A}} \otimes \tilde{\mathcal{A}}$.

Finally we check that $L(\alpha \otimes p) = 0$ for the prime p considered as an element in $R_{\mathcal{B}}$. In fact we have for $L = \Sigma L(\alpha \otimes p)$,

$$\begin{aligned} \Delta_1(\alpha \cdot [p]) &= \alpha\Delta_1[p] + L \\ &= q((\Delta\alpha) \cdot ([p] \otimes 1)) + L \\ &= q\left(\sum_i \alpha_i'[p] \otimes \alpha_i''\right) + L \\ &= q\left(\sum_i ([p]\alpha_i' + \Sigma\kappa(\alpha_i')) \otimes \alpha_i''\right) + L \\ &= \Delta_1[p] \cdot \Delta_0(\alpha) + \Sigma(\kappa \otimes 1)\delta\alpha + L. \end{aligned}$$

On the other hand $\alpha \cdot [p] = [p] \cdot \alpha + \Sigma\kappa(\alpha)$ so that

$$\begin{aligned} \Delta_1(\alpha \cdot [p]) &= \Delta_1([p] \cdot \alpha) + \Sigma\delta\kappa(\alpha) \\ &= \Delta_1([p]) \cdot \Delta_0(\alpha) + \Sigma\delta\kappa(\alpha). \end{aligned}$$

Since $(\kappa \otimes 1)\delta = \delta\kappa$ we see that $L = 0$. □

14.1.5 Proposition. *For $\alpha \in \mathcal{B}_0$, $\xi \in R_{\mathcal{B}}$ the following formulas hold in $\mathcal{A} \otimes \mathcal{A}$.*

$$\begin{aligned} L(\alpha \otimes \xi) + S(\alpha \cdot \xi) &= TL(\alpha \otimes \xi) + (-1)^{|\alpha|} \delta\alpha \cdot S(\xi), \\ S(\xi \cdot \alpha) &= S(\xi) \cdot \delta\alpha. \end{aligned}$$

Proof. Let $\xi = \partial x$, $x \in \mathcal{B}_1$. Then we get:

$$\begin{aligned} T\Delta_1(\alpha \cdot x) &= \Delta_1(\alpha \cdot x) + \Sigma S(\alpha \cdot \xi) \\ &= \alpha \cdot \Delta_1(x) + \Sigma(L(\alpha \otimes \xi) + S(\alpha \cdot \xi)). \end{aligned}$$

On the other hand one has:

$$\begin{aligned} T\Delta_1(\alpha \cdot x) &= T(\alpha \cdot \Delta_1(x)) + \Sigma TL(\alpha \otimes \xi) \\ &= \alpha \cdot T\Delta_1(x) + \Sigma TL(\alpha \otimes \xi) \\ &= \alpha\Delta_1(x) + \delta\alpha \cdot \Sigma S(\xi) + \Sigma TL(\alpha \otimes \xi). \end{aligned}$$

Finally it is readily checked that $S(\xi \cdot \alpha) = S(\xi) \cdot \delta\alpha$ since Δ_1 is right equivariant. In fact,

$$\begin{aligned} T\Delta_1(x \cdot \alpha) &= T(\Delta_1(x) \cdot \alpha) \\ &= (T\Delta_1(x)) \cdot \alpha \\ &= (\Delta_1(x) + \Sigma S(\xi)) \cdot \alpha. \end{aligned}$$

□

14.2 The extended left action operator

We consider a pair algebra H over the secondary Hopf algebra $\mathcal{B}$ as defined in (13.1.4). Then the homotopy

$$C : \mathcal{B}_0 \otimes H_0 \otimes H_0 \longrightarrow H_1$$

leads to the *extended left action operator*

$$\mathcal{L} : \mathcal{B}_0 \otimes \mathcal{B}_0 \otimes H^* \otimes H^* \longrightarrow \tilde{H}^* \text{ of degree } -1 \tag{14.2.1}$$

as follows. Here $H^* = \operatorname{cokernel}(\partial : H_1 \to H_0)$ and $\Sigma\tilde{H}^* = \operatorname{kernel}(\partial : H_1 \to H_0)$. We define $\mathcal{L}$ for $\alpha, \beta \in \mathcal{B}_0$ and $u \in H_0 \otimes H_0$ by the equation

$$C(\alpha\beta \otimes u) = \alpha C(\beta \otimes u) + C(\alpha \otimes \Delta(\beta) \cdot u) + \Sigma\mathcal{L}(\alpha \otimes \beta \otimes u).$$

14.2.2 Lemma. *The map $\mathcal{L}$ is a well-defined linear map.*

Proof. We have

$$\partial C(\alpha\beta \otimes u) = \alpha\beta \cdot \mu(u) - \mu_2(\Delta(\alpha\beta) \otimes u) \tag{1}$$

with μ and μ_2 as in (13.1.4)(2). Moreover

$$\partial\alpha C(\beta\otimes u)=\alpha\cdot(\beta\cdot\mu(u)-\mu_2(\Delta\beta\otimes u)), \tag{2}$$
$$\partial C(\alpha\otimes\Delta\beta\cdot u)=\alpha\cdot\mu(\Delta\beta\cdot u)-\mu_2(\Delta\alpha\otimes\Delta\beta\cdot u). \tag{3}$$

Since $\mu_2(\Delta\beta\otimes u)=\mu(\Delta\beta\cdot u)$ we see that $(2)+(3)=(1)$. Therefore the element $\mathcal{L}(\alpha\otimes\beta\otimes u)$ is uniquely determined by the equation. It remains to check that for

$$u\in \operatorname{image}((H\bar{\otimes}H)_1\xrightarrow{\partial} H_0\otimes H_0)$$

we have $\mathcal{L}(\alpha\otimes\beta\otimes u)=0$. In fact let $x\in H_1$, $y\in H_0$ and $u=\partial x\otimes y$. Then we get

$$\begin{aligned}C(\alpha\otimes u)&=C\partial(\alpha\otimes x\otimes y)\\&=\alpha(x\cdot y)-\mu_2(\Delta\alpha\otimes x\otimes y).\end{aligned}$$

Hence we get:

$$\begin{aligned}C(\alpha\beta\otimes u)&=\alpha\beta(x\cdot y)-\mu_2(\Delta(\alpha\beta)\otimes x\otimes y),\\ \alpha C(\beta\otimes u)&=\alpha(\beta(x\cdot y)-\mu_2(\Delta\beta\otimes x\otimes y)),\\ C(\alpha\otimes\Delta\beta\cdot u)&=\alpha\cdot\mu(\Delta\beta\cdot(x\otimes y))-\mu_2(\Delta\alpha\otimes\Delta\beta\cdot(x\otimes y)).\end{aligned}$$

Since $\mu_2(\Delta\beta\otimes u)=\mu(\Delta\beta\cdot u)$ we see that $\mathcal{L}(\alpha\otimes\beta\otimes u)=0$ for $u=\partial(x\otimes y)$. □

14.2.3 Proposition. *For $\alpha\in\mathcal{B}_0$, $\xi\in R_{\mathcal{B}}$ and $u\in H_0\otimes H_0$ we have the equations*

$$\begin{aligned}&\mathcal{L}(\alpha\otimes\xi\otimes u)=-\mu_2(L(\alpha\otimes\xi)\otimes u),\\&\mathcal{L}(\xi\otimes\alpha\otimes u)=0.\end{aligned}$$

Proof. Let $\xi=\partial x$, $x\in\mathcal{B}_1$. Then

$$\begin{aligned}C(\alpha\xi\otimes u)&=(\alpha x)\cdot\mu(u)-\mu_2(\Delta_1(\alpha x)\otimes u),\\ \alpha C(\xi\otimes u)&=\alpha(x\cdot\mu(u)-\mu_2(\Delta_1 x\otimes u)),\\ C(\alpha\otimes\Delta\xi\cdot u)&=C\partial(\alpha\otimes\Delta_1(x)\cdot u)\\&=\alpha\cdot\mu_2(\Delta_1(x)\otimes u)-\mu_2(\Delta\alpha\otimes\Delta_1(x)\cdot u).\end{aligned}$$

Now $\Delta_1(\alpha x)=\alpha\Delta_1(x)+\Sigma L(\alpha\otimes x)$ and $(\alpha x)\cdot\mu(u)=\alpha(x\cdot\mu(u))$ and $\mu_2(\alpha\Delta_1(x)\otimes u)=\mu_2(\Delta\alpha\otimes\Delta_1(x)\cdot u)$ show the first equation. Next consider:

$$C(\xi\alpha\otimes u)=(x\cdot\alpha)\cdot\mu(u)-\mu_2(\Delta_1(x\cdot\alpha)\otimes u),$$

$$\begin{aligned}\xi C(\alpha\otimes u)&=x\cdot\partial C(\alpha\otimes u)\\&=x\cdot(\alpha\cdot\mu(u)-\mu_2(\Delta(\alpha)\otimes\mu)),\\ C(\xi\otimes\Delta\alpha\cdot u)&=x\cdot\mu(\Delta\alpha\cdot u)-\mu_2(\Delta_1 x\otimes\Delta\alpha\cdot u).\end{aligned}$$

This shows $\mathcal{L}(\xi\otimes\alpha\otimes u)=0$. □

The extended left action operator $\mathcal{L}$ in (14.2.1) satisfies the following (α, β, γ)-formula.

14.2.4 Proposition. *Let* $\alpha, \beta, \gamma \in \mathcal{B}_0$ *and* $u \in H^* \otimes H^*$. *Then*

$$\mathcal{L}(\alpha \otimes \beta\gamma \otimes u) + (-1)^{|\alpha|}\alpha\mathcal{L}(\beta \otimes \gamma \otimes u) = \mathcal{L}(\alpha\beta \otimes \gamma \otimes u) + \mathcal{L}(\alpha \otimes \beta \otimes \Delta(\gamma) \cdot u).$$

The formula shows that $\mathcal{L}$ is completely determined by the elements $\mathcal{L}(\alpha \otimes \beta \otimes u)$ with $\beta \in E_{\mathcal{A}}$ and $\alpha \in \mathrm{Mon}(E_{\mathcal{A}})$ and $u \in H^* \otimes H^*$.

Proof. We consider the defining equations as in (14.2.1):

$$\begin{aligned}
&(1) \quad C(\alpha(\beta\gamma) \otimes u) = \alpha C(\beta\gamma \otimes u)\\
&(2) \quad \qquad\qquad + C(\alpha \otimes \Delta(\beta\gamma) \cdot u) + \Sigma\mathcal{L}(\alpha \otimes \beta\gamma \otimes u)\\
&(3) \quad \qquad = \alpha(\beta C(\gamma \otimes u) + C(\beta \otimes \Delta(\gamma) \cdot u) + \Sigma\mathcal{L}(\beta \otimes \gamma \otimes u)) + (2),
\end{aligned}$$

$$\begin{aligned}
&(4) \quad C((\alpha\beta)\gamma \otimes u) = C(\alpha\beta \otimes \Delta(\gamma) \otimes u)\\
&(5) \quad \qquad\qquad + \alpha\beta C(\gamma \otimes u) + \Sigma\mathcal{L}(\beta \otimes \gamma \otimes u)\\
&(6) \quad \qquad = \alpha C(\beta \otimes \Delta\gamma \cdot u)\\
&\qquad\qquad\qquad + C(\alpha \otimes \Delta\beta \cdot \Delta\gamma \cdot u) + \Sigma\mathcal{L}(\alpha \otimes \beta \otimes \Delta\gamma \cdot u) + (5).
\end{aligned}$$

□

14.3 The interchange acting on secondary cohomology

In Section (13.2) we used the fact that secondary cohomology $\mathcal{H}^*(X)$ is a pair algebra and we constructed strictifications $\mathcal{H}^*(X, M_X, s_X)$ which are pair algebras and $\mathcal{B}$-modules. We here consider the additional structure of $\mathcal{H}^*(X)$ as a secondary permutation algebra and we choose $M_X = P_X$ in such a way that the strictification $\mathcal{H}^*(X, P_X, s_X)$ is also a secondary permutation algebra and a $\mathcal{B}$-module. We show that the interchange operator T in $\mathcal{H}^*(X, P_X, s_X)$ is compatible with the interchange operator T in $\mathcal{B}$.

We say that a graded monoid P is a *permutation monoid* if the permutation group σ_n acts on P^n for $n \geq 0$ and

$$\mu^{n,k} : P^n \times P^k \longrightarrow P^{n+k}$$

is $(\sigma_n \times \sigma_k \to \sigma_{n+k})$-equivariant and the equation

$$\tau_{x,y}(x \cdot y) = y \cdot x$$

holds for $x, y \in P$, see (6.1.3)(4). For example $\mathcal{H}^*(X)_0$ is a permutation monoid. Moreover, each permutation algebra is a permutation monoid and for a ring R the free R-module $R[P]$ generated by a permutation monoid P is a permutation algebra.

14.3.1 Definition. As in (13.2.1) let $\mathcal{M} = \mathrm{Mon}(E_{\mathcal{A}})$ be the free monoid generated by $E_{\mathcal{A}}$. Let

$$s_X = P_X \longrightarrow \mathcal{H}^*(X)_0 = [\![X, \mathbb{Z}^*]\!]_0$$

be a function with the following properties. Here P_X is a permutation monoid and s_X is a morphism of permutation monoids. Moreover P_X is a free $\mathcal{M}$-set and s_X is an $\mathcal{M}$-equivariant morphism between $\mathcal{M}$-sets with the action of $\mathcal{M}$ on $\mathcal{H}^*(X)_0$ as in (13.2.1). Then we obtain the *strictification* $\mathcal{H}^*(X, P_X, s_X)$ as in (13.2.1) by the following pull back diagram.

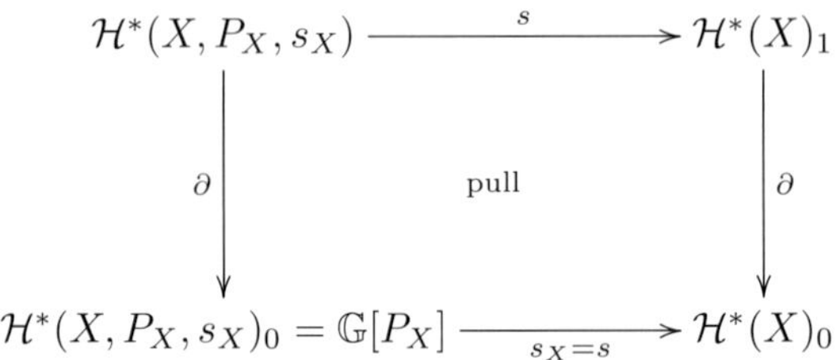

Since P_X is a permutation monoid we see that s_X induced by s_X above is a morphism between permutation algebras. Moreover, since P_X is a free $\mathcal{M}$-set with generating set $E_X \subset P_X$ we have

$$\mathbb{G}[P_X] = T_{\mathbb{G}}(E_{\mathcal{A}}) \otimes \mathbb{G}E_X.$$

Hence the $\mathcal{B}$-module structure of $\mathcal{H}^*(X, P_X, s_X)$ is defined as in (5.6.2) by the Γ-product $\bullet$.

14.3.2 Lemma. *The strictification $\mathcal{H}^*(X, P_X, s_X)$ is a secondary permutation algebra.*

Moreover we show by theorem (13.2.3) that $\mathcal{H}^*(X, P_X, s_X)$ is a pair algebra over the secondary Hopf algebra $\mathcal{B}$.

Proof of (14.3.2). Let $H = \mathcal{H}^*(X, P_X, s_X)$ and let R_* and $I(R_*)$ be given by $R_n = \mathbb{G}[\sigma_n]$ and $I(R_n) = \mathrm{kernel}(\varepsilon)$ where $\varepsilon : \mathbb{G}[\sigma_n] \to \mathbb{G}$ is the sign augmentation. Then we obtain the following commutative diagram.

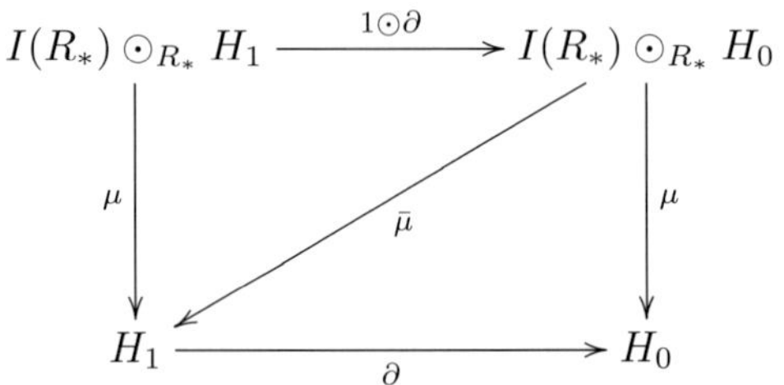

See (6.2.5). Here ∂ is a crossed permutation algebra as in (6.2.1). This structure is induced via the pull back over $s = s_X : H_0 = \mathbb{G}[P_X] \to \mathcal{H}^*(X)_0$ by the corresponding structure of $\mathcal{H}^*(X)$ in (6.3.2). More precisely the pull back H_1 is a left H_0-module by setting $\eta\cdot(\xi,x) = (\eta\cdot\xi, s(\eta)\cdot x)$ where $\eta,\xi\in H_0$ and $(\xi,x)\in H_1$ with $x\in\mathcal{H}^*(X)_1$ and $s\xi = \partial x$. Then we get for $(\xi,x),(\eta,y)\in H_1$ the formula

$$\begin{aligned}(\partial(\xi,x))\cdot(\eta,y) &= (\xi\eta, s(\xi)\cdot y)\\ &= (\xi\eta,(\partial x)\cdot y) = (\tau(\eta\xi),\tau(\partial y)\cdot x)\\ &= \tau(\eta\xi, s(\eta)\cdot x) = \tau(\partial(\eta,y))\cdot(\xi,x)\end{aligned}$$

where $\tau = \tau_{\partial y,x}$. Compare (6.2.1). Moreover we define $\bar{\mu}$ above by

$$\bar{\mu}(\sigma\otimes\xi) = (\sigma\cdot\xi, \bar{\mu}(\bar{\sigma}\otimes s(\xi))).$$

Here $\bar{\sigma}$ is the image of $\sigma\in I(\mathbb{G}[\sigma_n])$ under the following map.

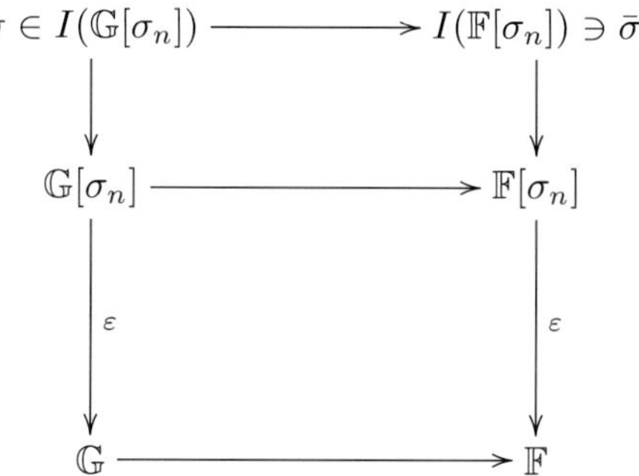

Moreover $\bar{\mu}(\bar{\sigma}\otimes s\xi)$ is given by the secondary permutation algebra $\mathcal{H}^*(X)$ as in (6.3.2). □

14.3.3 Example. For $\mathcal{M} = \mathrm{Mon}(E_{\mathcal{A}})$ let $\mathcal{M}$-**Perm** be the following category. Objects are graded sets S which have two (independent) structures, namely on the one hand S is a permutation monoid and on the other hand S is an $\mathcal{M}$-set. Morphisms are maps in **Set*** which preserves both structures. We have the forgetful functor

$$\mathcal{M}\text{-}\mathbf{Perm} \xrightarrow{\phi} \mathbf{Set}^*.$$

Let $F_{\mathcal{M}}$ be the left adjoint of this functor. We call $F_{\mathcal{M}}(S)$ the *free $\mathcal{M}$-permutation monoid generated by S*.

Now let

$$\mathcal{P}_X = F_{\mathcal{M}}(\mathcal{H}^*(X)_0)$$

be the free permutation monoid generated by the graded set $\mathcal{H}^*(X)_0$ and let

$$s_X : \mathcal{P}_X \longrightarrow \mathcal{H}^*(X)_0$$

be the morphism in $\mathcal{M}$-**Perm** extending the identity on $\mathcal{H}^*(X)_0$. Then $(\mathcal{P}_X, s_X)$ is an example of (14.3.1). By naturality of s_X we see that $\mathcal{H}^*(X,\mathcal{P}_X,s_X)$ yields a functor from path connected pointed spaces X to the category of secondary permutation algebras which are pair algebras over the secondary Hopf algebra $\mathcal{B}$.

14.3.4 Definition. Let H be a secondary permutation algebra as in (14.3.2). Then we obtain the following *interchange homotopy.*

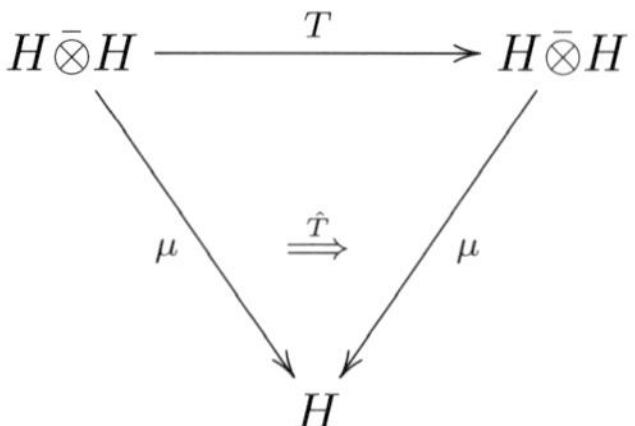

Here T is the interchange map induced by T with the even sign convention in (11.1.1) and μ is the multiplication of the pair algebra H. Moreover the homotopy $\hat{T} : \mu \Rightarrow \mu T$ is the $\mathbb{G}$-linear map

$$\hat{T} : H_0 \otimes H_0 \longrightarrow H_1$$

defined by the formula

$$\hat{T}(x \otimes y) = \bar{\mu}((\tau(y,x) - \varepsilon\tau(y,x)) \otimes y \cdot x)$$

for $x, y \in H_0$. Here $\varepsilon : \mathbb{G}[\sigma_n] \to \mathbb{G}$ with $n = |x| + |y|$ is the augmentation satisfying

$$\varepsilon\tau(y,x) = (-1)^{p|x||y|} \in \mathbb{G} \subset \mathbb{G}[\sigma_n]$$

and $\bar{\mu}$ is the map in (14.3.2). Moreover the element $y \cdot x$ is given by the multiplication of H_0, and $\tau(y,x)$ is the interchange permutation with

$$\tau(y,x)(y \cdot x) = x \cdot y$$

in the permutation algebra H_0. We have

$$\partial\hat{T}(x \otimes y) = x \cdot y - (-1)^{p|x||y|} y \cdot x$$

so that $\partial\hat{T} = (\mu - \mu T)_0$. Compare the even sign convention in (11.1.1). Moreover for $\bar{x} \in H_1$ with $\partial\bar{x} = x$ we get $\bar{x} \otimes y \in (H\bar{\otimes}H)_1$ such that

$$\begin{aligned} \hat{T}\partial(\bar{x} \otimes y) &= \bar{\mu}((\tau(y,x) - \varepsilon\tau(y,x)) \otimes \partial(y \cdot \bar{x})) \\ &= \tau(y,x)(y \cdot \bar{x}) - \varepsilon\tau(y,x) y \cdot \bar{x} \\ &= \bar{x} \cdot y - (-1)^{p|x||y|} y \cdot \bar{x} \end{aligned}$$

by the equation $\bar{\mu}(1 \odot \partial) = \mu$; see the diagram in (14.3.2). Hence we also get $\hat{T}\partial = (\mu - \mu T)_1$. Therefore $\hat{T} : \mu \Rightarrow \mu T$ is a well-defined homotopy.

Let $H = \mathcal{H}^*(X, P_X, s_X)$ be the strictification of $\mathcal{H}^*(X)$ in (14.3.1). Using the interchange homotopy $\hat{T}$ we define the *extended symmetry operator* with $H^* = H^*(X)$,

(14.3.5) $$\mathcal{S} : \mathcal{B}_0 \otimes H^* \otimes H^* \longrightarrow \tilde{H}^* \text{ of degree } -1$$

as follows. Let $\alpha \in \mathcal{B}_0$ and $u \in H_0 \otimes H_0$. Then $\mathcal{S}(\alpha \otimes u)$ is given by the *interchange formula* in H_1:

$$C(\alpha \otimes u) + \hat{T}(\Delta\alpha \cdot u) = \alpha\hat{T}(u) + C(\alpha \otimes Tu) + \Sigma\mathcal{S}(\alpha \otimes u).$$

14.3.6 Lemma. *The operator $\mathcal{S}$ is a well-defined linear map.*

Proof. We have the elements:

$$\begin{aligned}
\partial C(\alpha \otimes u) &= \alpha \cdot \mu(u) - \mu_2(\Delta\alpha \otimes u),\\
\partial \hat{T}(\Delta\alpha \cdot u) &= \mu_2(\Delta\alpha \otimes u) - \mu_2(T\Delta\alpha \otimes Tu),\\
\partial \alpha\hat{T}(u) &= \alpha(\mu(u) - \mu(Tu)),\\
\partial C(\alpha \otimes Tu) &= \alpha \cdot \mu(Tu) - \mu_2(\Delta\alpha \otimes Tu).
\end{aligned}$$

This shows that the element $\mathcal{S}(\alpha \otimes u)$ is uniquely determined. It remains to check that $\mathcal{S}(\alpha \otimes u) = 0$ if u is a boundary as in the proof of (14.1.5), that is, $u = \partial(x \otimes y)$ with $x \in H_1$ and $y \in H_0$. In this case we get as in (14.1.5):

$$\begin{aligned}
C(\alpha \otimes u) &= \alpha(x \cdot y) - \mu_2(\Delta\alpha \otimes x \otimes y),\\
C(\alpha \otimes Tu) &= \varepsilon\alpha(y \cdot x) - \varepsilon\mu_2(\Delta\alpha \otimes y \otimes x)
\end{aligned}$$

with $\varepsilon = (-1)^{p|x||y|}$. Moreover

$$\begin{aligned}
\alpha\hat{T}(u) &= \alpha(x \cdot y - \varepsilon y \cdot x) \quad \text{by (14.1.4)},\\
\hat{T}(\Delta\alpha \cdot u) &= \hat{T}\Big(\sum_i \pm \alpha_i' x \otimes \alpha_i'' y\Big)\\
&= \sum_i \pm (\alpha_i' x \cdot \alpha_i'' y - \pm\alpha_i'' y \cdot \alpha_i' x)\\
&= \mu_2(\Delta\alpha \otimes x \otimes y) - \varepsilon\mu_2(T\Delta\alpha \otimes y \otimes x).
\end{aligned}$$

This shows that $\mathcal{S}(\alpha \otimes u) = 0$ for $u = \partial(x \otimes y)$. □

The symmetry operator S in (14.1.1) has the following property:

14.3.7 Proposition. *For $\xi \in R_{\mathcal{B}}$ and $u \in H_0 \otimes H_0$ we get the formula in $\tilde{H}^*(X)$,*

$$\mathcal{S}(\xi \otimes u) = \mu_2(S(\xi) \otimes u).$$

Proof. Let $x \in \mathcal{B}_1$ with $\partial x = \xi$. Then we get

$$\begin{aligned}
C(\xi \otimes u) &= C\partial(x \otimes u) = x \cdot \mu(u) - \mu_2(\Delta_1 x \otimes u),\\
\hat{T}(\Delta\xi \cdot u) &= \hat{T}\partial(\Delta_1 x \cdot u) = \mu_2(\Delta_1 x \otimes u) - \mu_2(T\Delta_1 x \otimes Tu),\\
\xi \cdot \hat{T}(u) &= \partial x \cdot \hat{T}(u) = x \cdot \partial\hat{T}(u) = x \cdot (\mu(u) - \mu T(u)),\\
C(\xi \otimes Tu) &= C\partial(x \otimes Tu) = x \cdot \mu T(u) - \mu_2(\Delta_1 x \otimes Tu).
\end{aligned}$$

This yields the result. □

14.3.8 Proposition. *For $\alpha, \beta \in \mathcal{B}_0$ and $u \in H^* \otimes H^*$ we have the equation:*

$$\mathcal{S}(\alpha\beta \otimes u) = (-1)^{|\alpha|}\alpha \cdot \mathcal{S}(\beta \otimes u) + \mathcal{S}(\alpha \otimes \Delta\beta \cdot u) + \mathcal{L}(\alpha \otimes \beta \otimes (u - Tu)).$$

This is the extended version of (14.1.5).

Proof. Let $v = u - Tu$. Then (14.3.5) is equivalent to

$$C(\alpha \otimes v) = \alpha\hat{T}(u) - \hat{T}(\Delta\alpha \cdot u) + \Sigma\mathcal{S}(\alpha \otimes u). \tag{1}$$

Moreover we have by (14.2.1)

$$C(\alpha\beta \otimes v) = \alpha C(\beta \otimes v) + C(\alpha \otimes \Delta\beta \cdot v) + \Sigma\mathcal{L}(\alpha \otimes \beta \otimes v). \tag{2}$$

Here we get by (1) the following equations in H_1.

$$\begin{aligned} C(\alpha\beta \otimes v) &= \alpha\beta\hat{T}(u) - \hat{T}(\Delta(\alpha\beta) \cdot u) && (3)\\ &\quad + \Sigma\mathcal{S}(\alpha\beta \otimes u), && (4) \end{aligned}$$

$$\begin{aligned} \alpha C(\beta \otimes v) &= \alpha(\beta\hat{T}(u) - \hat{T}(\Delta\beta \cdot u)) && (5)\\ &\quad + \alpha\Sigma\mathcal{S}(\beta \otimes u). && (6) \end{aligned}$$

Finally we get

$$\begin{aligned} C(\alpha \otimes \Delta\beta \cdot v) &= C(\alpha \otimes (\Delta\beta \cdot u - \Delta\beta \cdot Tu)) && (7)\\ &= C(\alpha \otimes (\Delta\beta \cdot u - T(\Delta\beta \cdot u))). && (8) \end{aligned}$$

Moreover we get by (1):

$$\begin{aligned} (8) &= \alpha\hat{T}(\Delta\beta \cdot u) - \hat{T}(\Delta\alpha \cdot \Delta\beta \cdot u) && (9)\\ &\quad + \Sigma\mathcal{S}(\alpha \otimes \Delta\beta \cdot u). && (10) \end{aligned}$$

Now we observe (3) = (5) + (9). Hence the remaining terms yield the equation in (14.3.8). □

14.4 Computation of the extended left action

We first prove the following result from which we can derive the operator $\mathcal{L}$ completely by (14.2.3).

14.4.1 Theorem. *For $\alpha \in \mathcal{B}_0$ and $\beta \in E_{\mathcal{A}}$ and $x \otimes y \in H^* \otimes H^*$ the extended left action operator $\mathcal{L}$ is given as follows. One has*

$$\mathcal{L}(\alpha \otimes \beta \otimes x \otimes y) = 0 \text{ if } p \text{ is odd.}$$

Moreover, if p is even and $\beta = Sq^n$, one gets

$$\mathcal{L}(\alpha \otimes Sq^n \otimes x \otimes y) = |x| \sum_{i+j=n,\ j \text{ odd}} \kappa(\alpha)(Sq^i(x) \cdot Sq^j(y)).$$

Proof of 14.4.1. It suffices to consider the odd convention for the definition of $C_\alpha \odot C_\beta$. For $x, y \in H_0$ and $\alpha \in \mathcal{B}_0$ and $\beta \in E_{\mathcal{A}}$ we get the generalized Cartan track

$$C_{\alpha\beta}^{sx,sy} = C_\alpha \odot C_\beta = C_{\alpha,\Delta\beta} \square \alpha C_\beta \square \Gamma_{\alpha,\beta}$$

by (11.4.3). We therefore consider the following commutative diagram of tracks, compare (11.4.3)(6) and (13.2.5)(3).

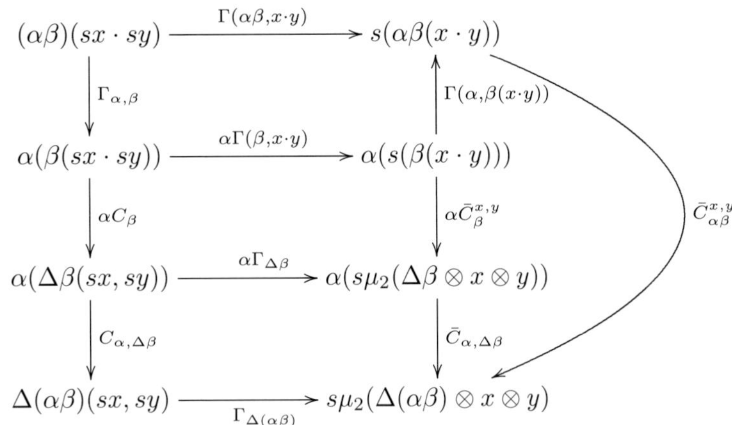

Let $\xi = \mu_2(\Delta(\alpha\beta) \otimes x \otimes y)$ and $\eta = \mu_2(\Delta(\beta) \otimes x \otimes y)$. Then we get by (2.2.6)(3) and (13.2.5)(5):

(1) $$sC(\alpha\beta \otimes x \otimes y) + s\xi = \bar{C}_{\alpha\beta}^{x,y} = H + G + s\xi,$$

$$H = \bar{C}_{\alpha,\Delta\beta} - s\xi,$$
$$G = (\alpha\bar{C}_\beta^{x,y}) \square \Gamma(\alpha, \beta(x \cdot y))^{\mathrm{op}} - \alpha s\eta.$$

On the other hand we get by (5.5.2)(2)

(2) $$\begin{aligned} s(\alpha C(\beta \otimes x \otimes y)) &= \alpha \bullet C(\beta \otimes x \otimes y) \\ &= (\alpha(\bar{C}_\beta^{x,y} - s\eta)) \square \Gamma(\alpha, \beta(x \cdot y) - \eta)^{\mathrm{op}}. \end{aligned}$$

This composite is the left-hand column of the following commutative diagram.

(3) $$\begin{array}{ccc} s(\alpha(\beta(x \cdot y) - \eta)) & = & s(\alpha\beta(x \cdot y)) - s(\alpha\eta) \\ \downarrow & & \uparrow {\scriptstyle \Gamma(\alpha,\beta(x\cdot y)) - \Gamma(\alpha,\eta)} \\ \alpha(s(\beta(x \cdot y) - s\eta)) & \xrightarrow{\Gamma} & \alpha s(\beta(x \cdot y)) - \alpha s\eta \\ \downarrow & & \downarrow {\scriptstyle \alpha\bar{C}_\beta^{x,y} - \alpha s\eta} \\ 0 & = & 0 \end{array}$$

Here the bottom square commutes by (4.2.5)(6). Now (2) and (3) imply

$$
(4)\qquad \begin{aligned} s(\alpha C(\beta\otimes x\otimes y)) &= G\square(s(\alpha\beta(x\cdot y)) - \Gamma(\alpha,\eta)^{\mathrm{op}}) \\ &= G - (\Gamma(\alpha,\eta)^{\mathrm{op}} - \alpha s\eta). \end{aligned}
$$

Therefore we get by (1)

$$
(5)\qquad sC(\alpha\beta\otimes x\otimes y) = H + G = F + s(\alpha C(\beta\otimes x\otimes y)) \quad \text{with}
$$

$$
(6)\qquad F = (\bar{C}_{\alpha,\Delta\beta} - s\xi) + (\Gamma(\alpha,\eta)^{\mathrm{op}} - \alpha s\eta).
$$

It remains to compare F with $sC(\alpha\otimes(\Delta\beta)\cdot(x\otimes y))$. Using (2.2.6)(3) we have by (6) the equation

$$
(7)\qquad F + s\xi = \bar{C}_{\alpha,\Delta\beta}\square\Gamma(\alpha,\eta)^{\mathrm{op}}.
$$

We now consider the following commutative diagram in which the left-hand column is $C_{\alpha,\Delta\beta}$ by (11.4.3)(5) and the sign is $\pm = (-1)^{|x||\rho''|}$.

$$
(8)\qquad \begin{array}{ccc}
\alpha((\Delta\beta)(sx,sy)) & \xrightarrow{\alpha\Gamma_{\Delta\beta}} & \alpha s\eta \\
\downarrow{\scriptstyle \Gamma_\alpha^{\Delta\beta}} & & \downarrow{\scriptstyle \Gamma(\alpha,\eta)} \\
\sum\limits_\rho \pm\varphi_{\Delta\beta}(\rho)\alpha(\rho' sx\cdot\rho'' sy) & \xrightarrow{\sum_\rho \pm\varphi_{\Delta\beta}(\rho)\Gamma^{(\rho)}} & s(\alpha\eta) \\
\downarrow{\scriptstyle \tilde{C}=\sum_\rho \pm\varphi_{\Delta\beta}(\rho)C_\alpha^{\rho' sx,\rho'' sy}} & & \downarrow{\scriptstyle F+s\xi} \\
(\Delta(\alpha\beta))(sx,sy) & \xrightarrow[\Gamma_{\Delta(\alpha\beta)}]{} & s\xi
\end{array}
$$

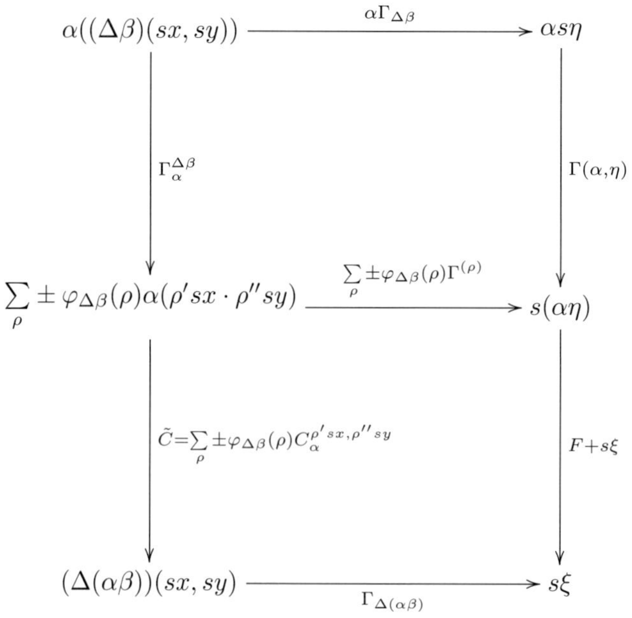

Here we obtain $\Gamma^{(\rho)}$ by the following commutative diagram.

$$
(9)\qquad \begin{array}{ccc}
\alpha(\rho' sx\cdot\rho'' sy) & \xrightarrow{\Gamma^{(\rho)}} & s(\alpha(\rho' x\cdot\rho'' y)) \\
\downarrow{\scriptstyle \alpha(\Gamma(\rho',x)\cdot\Gamma(\rho'',y))} & & \uparrow{\scriptstyle \Gamma(\alpha,\rho' x\cdot\rho'' y)} \\
\alpha(s(\rho' x)\cdot s(\rho'' y)) & = & \alpha(s(\rho' x\cdot\rho'' y))
\end{array}
$$

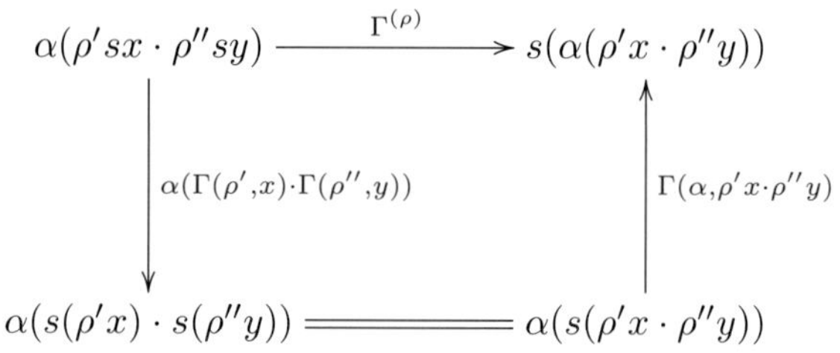

On the other hand we have by (3.1.3) the following commutative diagram for each ρ.

$$\begin{array}{ccc} \alpha(\rho' sx\cdot\rho'' sy) & \xrightarrow{\alpha(\Gamma(\rho',x)\cdot\Gamma(\rho'',y))} & \alpha(s(\rho' x)\cdot s(\rho'' y)) \\ \Big\downarrow C_\alpha^{\rho' sx,\rho'' sy} & & \Big\downarrow C_\alpha^{s(\rho' x),s(\rho'' y)} \\ (\Delta\alpha)(\rho' sx,\rho'' sy) & \xrightarrow{\Gamma^\rho_{\Delta\alpha}} & (\Delta\alpha)(s(\rho' x),s(\rho'' y)) \end{array} \tag{10}$$

$$\Gamma^\rho_{\Delta\alpha}=\sum_i \pm\alpha_i'\Gamma(\rho',x)\cdot\alpha_i''\Gamma(\rho'',y)$$

Moreover using the definition of $\Gamma^{x,y}_{\Delta(\alpha\beta)}$ in (13.2.5)(4) we get

$$\Gamma^{x,y}_{\Delta(\alpha\beta)}=\sum_{i,\rho}\pm\varphi_{\Delta\beta}(\rho)\Gamma(\alpha_i'\rho',x)\cdot\Gamma(\alpha_i''\rho'',y) \tag{11}$$

with the sign

$$\pm=(-1)^{|x|\cdot|\alpha_i''\rho''|}\cdot(-1)^{|\alpha_i''||\rho'|}\quad\text{and with}\quad \Gamma(\alpha_i'\rho',x)=\Gamma(\alpha_i,\rho' x)\Box\alpha_i\Gamma(\rho',x)$$

since ρ' is a monomial. This shows that

$$\begin{aligned}\Gamma^{x,y}_{\Delta(\alpha\beta)}&=\Big(\sum_{i,\rho}\pm\varphi_{\Delta\beta}(\rho)\Gamma(\alpha_i',\rho' x)\cdot\Gamma(\alpha_i'',\rho'' y)\Big)\\ &\quad\Box\Big(\sum_{i,\rho}\pm\varphi_{\Delta\beta}(\rho)\alpha_i'\Gamma(\rho',x)\cdot\alpha_i''\Gamma(\rho'',y)\Big)\\ &=\Big(\sum_\rho\pm\varphi_{\Delta\beta}(\rho)\Gamma^{\rho' x,\rho'' y}_{\Delta\alpha}\Big)\Box\Big(\sum_\rho\pm\varphi_{\Delta\beta}(\rho)\Gamma^\rho_{\Delta\alpha}\Big).\end{aligned} \tag{12}$$

Here in the first row the sign $\pm$ is the same as in (11) and in the second row the sign is $\pm=(-1)^{|x||\rho''|}$. Finally we get by multilinearity and the definition of μ_2 the equation:

$$\begin{aligned}sC(\alpha\otimes(\Delta\beta)\cdot(x\otimes y))+s\xi&=\sum_\rho\pm\varphi_{\Delta\beta}(\rho)sC(\alpha\otimes\rho' x\otimes\rho'' y)\\ &\quad+\sum_\rho\pm\varphi_{\Delta\beta}(\rho)s\mu_2(\Delta\alpha\otimes\rho' x\otimes\rho'' y)\\ &=\sum_\rho\pm\varphi_{\Delta\beta}(\rho)\bar C_\alpha^{\rho' x,\rho'' y},\end{aligned} \tag{13}$$

with the sign given by $\pm = (-1)^{|x||\rho''|}$. Here $\bar{C}_\alpha^{\rho' x,\rho'' y}$ is defined by the following commutative diagram (see (13.2.5)(4)).

$$\begin{array}{ccc} \alpha(s(\rho' x)\cdot s(\rho'' y)) & \xrightarrow{\Gamma(\alpha,\rho' x\cdot\rho'' y)} & s(\alpha(\rho' x\cdot\rho'' y)) \\ \Big\downarrow C_\alpha^{s(\rho' x),s(\rho'' y)} & & \Big\downarrow \bar{C}_\alpha^{\rho' x,\rho'' y} \\ \Delta\alpha(s(\rho' x),s(\rho'' y)) & \xrightarrow[\Gamma_{\Delta\alpha}^{\rho' x,\rho'' y}]{} & s\mu_2(\Delta\alpha\otimes\rho' x\otimes\rho'' y) \end{array} \tag{14}$$

Using (10) and (14) we see that the bottom square of (8) is subdivided into two squares corresponding to (10) and (14) respectively. This shows by (13) that $F = sC(\alpha\otimes(\Delta\beta)\cdot(x\otimes y))$ and the proof of (14.4.1) is complete if p is odd. If p is even we observe that all arguments above remain true for the odd convention. Hence for the even convention the result follows by use of formula (11.4.3)(5a). □

We now can compute the left action operator L by use of (14.2.3) since we know $\mathcal{L}$ by the result above. Since $L(\alpha\otimes p) = 0$ by (14.1.4) we know that L is given by the operator

$$L : \mathcal{A}\otimes R_{\mathcal{F}} \longrightarrow \tilde{\mathcal{A}}\otimes\tilde{\mathcal{A}} \text{ of degree } -1 \tag{14.4.2}$$

where $R_{\mathcal{F}} = \text{kernel}(q : \mathcal{F}_0 = T_{\mathbb{F}}(E_{\mathcal{A}}) \to \mathcal{A})$.

We have the *Adem relation* in $R_{\mathcal{F}} \subset \mathcal{F}_0$ given by the formula $(0 < a < 2b)$

$$[a,b] = Sq^a Sq^b + \sum_{k=0}^{[a/2]} \binom{b-k-1}{a-2k} Sq^{a+b-k} Sq^k.$$

Compare (1.1). we now define $(n,m \geq 0)$

$$L(Sq^n Sq^m) = \sum_{\substack{n_1+n_2=n\\ m_1+m_2=m\\ m_1,n_2 \text{ odd}}} Sq^{n_1}Sq^{m_1}\otimes Sq^{n_2}Sq^{m_2},$$

and we define $L[a,b]$ accordingly by

$$L[a,b] = L(Sq^a Sq^b) + \sum_{k=0}^{[a/2]} \binom{b-k-1}{a-2k} L(Sq^{a+b-k} Sq^k).$$

For $a+b \leq 9$ we have the following explicit formulas for $L[a,b]$.

2

$L[1,1] = \mathrm{Sq}^1 \otimes \mathrm{Sq}^1$

3

$L[1,2] = 0$

4

$L[1,3] = \mathrm{Sq}^1 \otimes \mathrm{Sq}^3 + \mathrm{Sq}^3 \otimes \mathrm{Sq}^1$
$L[2,2] = \mathrm{Sq}^2\,\mathrm{Sq}^1 \otimes \mathrm{Sq}^1 + \mathrm{Sq}^1 \otimes \mathrm{Sq}^3$

5

$L[1,4] = 0$
$L[2,3] = \mathrm{Sq}^3\,\mathrm{Sq}^1 \otimes \mathrm{Sq}^1$
$L[3,2] = \mathrm{Sq}^1 \otimes \mathrm{Sq}^3\,\mathrm{Sq}^1$

6

$L[1,5] = \mathrm{Sq}^1 \otimes \mathrm{Sq}^5 + \mathrm{Sq}^3 \otimes \mathrm{Sq}^3 + \mathrm{Sq}^5 \otimes \mathrm{Sq}^1$
$L[2,4] = \mathrm{Sq}^4\,\mathrm{Sq}^1 \otimes \mathrm{Sq}^1 + \mathrm{Sq}^2\,\mathrm{Sq}^1 \otimes \mathrm{Sq}^3 + \mathrm{Sq}^1 \otimes \mathrm{Sq}^5$
$L[3,3] = \mathrm{Sq}^1 \otimes \mathrm{Sq}^5 + \mathrm{Sq}^3 \otimes \mathrm{Sq}^3 + \mathrm{Sq}^5 \otimes \mathrm{Sq}^1$

7

$L[1,6] = 0$
$L[2,5] = \mathrm{Sq}^3\,\mathrm{Sq}^1 \otimes \mathrm{Sq}^3 + \mathrm{Sq}^5\,\mathrm{Sq}^1 \otimes \mathrm{Sq}^1$
$L[3,4] = \mathrm{Sq}^1 \otimes \mathrm{Sq}^5\,\mathrm{Sq}^1 + \mathrm{Sq}^3 \otimes \mathrm{Sq}^3\,\mathrm{Sq}^1$
$L[4,3] = \mathrm{Sq}^3\,\mathrm{Sq}^1 \otimes \mathrm{Sq}^3 + \mathrm{Sq}^5\,\mathrm{Sq}^1 \otimes \mathrm{Sq}^1 + \mathrm{Sq}^2\,\mathrm{Sq}^1 \otimes \mathrm{Sq}^3\,\mathrm{Sq}^1 + \mathrm{Sq}^1 \otimes \mathrm{Sq}^5\,\mathrm{Sq}^1$

8

$L[1,7] = \mathrm{Sq}^1 \otimes \mathrm{Sq}^7 + \mathrm{Sq}^3 \otimes \mathrm{Sq}^5 + \mathrm{Sq}^5 \otimes \mathrm{Sq}^3 + \mathrm{Sq}^7 \otimes \mathrm{Sq}^1$
$L[2,6] = \mathrm{Sq}^6\,\mathrm{Sq}^1 \otimes \mathrm{Sq}^1 + \mathrm{Sq}^4\,\mathrm{Sq}^1 \otimes \mathrm{Sq}^3 + \mathrm{Sq}^2\,\mathrm{Sq}^1 \otimes \mathrm{Sq}^5 + \mathrm{Sq}^1 \otimes \mathrm{Sq}^7$
$L[3,5] = 0$
$L[4,4] = \mathrm{Sq}^6\,\mathrm{Sq}^1 \otimes \mathrm{Sq}^1 + \mathrm{Sq}^4\,\mathrm{Sq}^1 \otimes \mathrm{Sq}^3 + \mathrm{Sq}^2\,\mathrm{Sq}^1 \otimes \mathrm{Sq}^5 + \mathrm{Sq}^1 \otimes \mathrm{Sq}^7 + \mathrm{Sq}^3\,\mathrm{Sq}^1 \otimes \mathrm{Sq}^3\,\mathrm{Sq}^1$
$L[5,3] = \mathrm{Sq}^5 \otimes \mathrm{Sq}^3 + \mathrm{Sq}^1 \otimes \mathrm{Sq}^5\,\mathrm{Sq}^2 + \mathrm{Sq}^5\,\mathrm{Sq}^2 \otimes \mathrm{Sq}^1 + \mathrm{Sq}^3 \otimes \mathrm{Sq}^5$

9

$L[1,8] = 0$
$L[2,7] = \mathrm{Sq}^3\,\mathrm{Sq}^1 \otimes \mathrm{Sq}^5 + \mathrm{Sq}^5\,\mathrm{Sq}^1 \otimes \mathrm{Sq}^3 + \mathrm{Sq}^7\,\mathrm{Sq}^1 \otimes \mathrm{Sq}^1$
$L[3,6] = \mathrm{Sq}^1 \otimes \mathrm{Sq}^7\,\mathrm{Sq}^1 + \mathrm{Sq}^3 \otimes \mathrm{Sq}^5\,\mathrm{Sq}^1 + \mathrm{Sq}^5 \otimes \mathrm{Sq}^3\,\mathrm{Sq}^1$
$L[4,5] = \mathrm{Sq}^4\,\mathrm{Sq}^1 \otimes \mathrm{Sq}^3\,\mathrm{Sq}^1 + \mathrm{Sq}^2\,\mathrm{Sq}^1 \otimes \mathrm{Sq}^5\,\mathrm{Sq}^1 + \mathrm{Sq}^1 \otimes \mathrm{Sq}^7\,\mathrm{Sq}^1$
$L[5,4] = \mathrm{Sq}^1 \otimes \mathrm{Sq}^7\,\mathrm{Sq}^1 + \mathrm{Sq}^3 \otimes \mathrm{Sq}^5\,\mathrm{Sq}^1 + \mathrm{Sq}^5 \otimes \mathrm{Sq}^3\,\mathrm{Sq}^1$

14.4.3 Theorem. *For p odd the left action operator L is trivial. For p even the operator L is the unique linear map of degree -1 satisfying the equations*

$$\begin{aligned} L(\alpha \otimes [a,b]) &= (\delta\kappa(\alpha)) \cdot L[a,b], \\ L(\alpha \otimes \beta\xi) &= L(\alpha\beta \otimes \xi) + \delta(\alpha) \cdot L(\beta \otimes \xi), \\ L(\alpha \otimes \xi\beta) &= L(\alpha \otimes \xi) \cdot \delta(\beta), \end{aligned}$$

with α, $\beta \in \mathcal{F}_0$ and $\xi \in R_{\mathcal{F}}$.

Proof. If p is odd then $\mathcal{L}$ is trivial by (14.4.2) and (14.2.4). Hence L is trivial by (14.2.3). If p is even we compute $\mathcal{L}(\alpha \otimes [a,b] \otimes x \otimes y)$ by considering, $u = x \otimes y$,

$$\begin{aligned} &\mathcal{L}(\alpha \otimes Sq^n Sq^m \otimes u) \\ &= \alpha\mathcal{L}(Sq^n \otimes Sq^m \otimes u) + \mathcal{L}(\alpha Sq^n \otimes Sq^m \otimes u) + \mathcal{L}(\alpha \otimes Sq^n \otimes \Delta(Sq^m) \cdot u), \end{aligned} \tag{1}$$

see (14.2.4). Here all terms are computed in (14.4.1). Hence we get (1) = (2) + (3) + (4).

$$\alpha|x| \sum_{\substack{i+j=m \\ j \text{ odd}}} Sq^{n-1}(Sq^i x \cdot Sq^j y), \tag{2}$$

$$|x| \sum_{\substack{i+j=m \\ j \text{ odd}}} \kappa(\alpha Sq^n)(Sq^i x \cdot Sq^j y) \tag{3}$$

with $\kappa(\alpha Sq^n) = \kappa(\alpha)Sq^n + \alpha Sq^{n-1}$, and

$$\sum_{r+s=m} |Sq^r x| \sum_{\substack{i+j=n \\ j \text{ odd}}} \kappa(\alpha)(Sq^i(Sq^r x) \cdot Sq^j(Sq^s y)). \tag{4}$$

Here we have $|Sq^r x| = r + |x|$. We now compute the sum of all summands in (2)+(3)+(4) containing the factor $|x|$. First we observe that

$$(2) + (3) = |x|\kappa(\alpha) \sum_{i+j=n} \sum_{\substack{r+s=m \\ s \text{ odd}}} Sq^i Sq^r x \cdot Sq^j Sq^s y. \tag{5}$$

On the other hand the part of (4) containing the factor $|x|$ is given by

$$|x|\kappa(\alpha) \sum_{r+s=m} \sum_{\substack{i+j=n \\ j \text{ odd}}} Sq^i Sq^r x \cdot Sq^j Sq^s y. \tag{6}$$

The summands (i,j,r,s) with j odd and s odd appear in (5) and (6) and hence cancel. Therefore we need only to consider in (5) j even and in (6) s even. Hence we get $(5)+(6)=|x|\kappa(\alpha)(7)$,

$$\lambda(Sq^nSq^m) = \sum_{\substack{r+s=m\\ i+j=n\\ j+s \text{ odd}}} Sq^iSq^rx\cdot Sq^jSq^sy \tag{7}$$
$$= \mu_2((1\otimes\varphi)\Delta(Sq^nSq^m)\otimes x\otimes y).$$

Here $\varphi:\mathcal{A}\longrightarrow\mathcal{A}$ is the function $\varphi(x)=x$ if $|x|$ is odd and $\varphi(x)=0$ if $|x|$ is even. Since $[a,b]$ is trivial in $\mathcal{A}$ we see that the sum $\lambda[a,b]$ defined by $\lambda(Sq^nSq^m)$ is trivial. Hence we need only to consider the part of (4) containing the factor r. This part is given by

$$\sum_{\substack{r+s=m\\ i+j=n\\ j \text{ odd}}} r\kappa(\alpha)(Sq^iSq^rx\cdot Sq^jSq^sy), \tag{8}$$
$$= \mu_2(\delta\kappa(\alpha)\cdot L(Sq^nSq^m)\otimes u).$$

Adding up such summands according to $[a,b]$ we see by (14.2.3) that

$$L(\alpha\otimes[a,b]) = \delta\kappa(\alpha)\cdot L[a,b].$$

Now the proof is complete. □

14.5 Computation of the extended symmetry

The extended symmetry operator $\mathcal{S}$ is determined by the extended left action operator $\mathcal{L}$ and the elements $\mathcal{S}(\alpha\otimes x\otimes y)$ with $\alpha\in E_{\mathcal{A}}$. This follows from (14.3.8) We now obtain the following result.

14.5.1 Theorem. *For $\alpha\in E_{\mathcal{A}}$ and $x\otimes y\in H^*\otimes H^*$ we have*

$$\mathcal{S}(\alpha\otimes x\otimes y)=0 \text{ if } p \text{ is odd.}$$

If p is even we get for $\alpha=Sq^n$ the formula

$$\mathcal{S}(Sq^n\otimes x\otimes y) = |x||y|\kappa(\alpha)(x\cdot y)+\mu_2(S_n\otimes x\otimes y)$$

with $S_n\in\tilde{A}\otimes\tilde{A}$ defined by

$$S_n = \sum_{\substack{i+j=n-1\\ i,j \text{ odd}}} Sq^i\otimes Sq^j.$$

In particular S_n is trivial if n is even.

Proof of (14.5.1). We first consider the case that p is odd. Let $H = \mathcal{H}^*(X, P_X, s_X)$ and let $x, y \in H_0$ and $\alpha \in E_A$. Then we claim that $\mathcal{S}(\alpha \otimes x \otimes y)$ is the primary element of the following diagram.

(1)
$$\begin{array}{ccc} \alpha(sx \cdot sy) & \xrightarrow{C_\alpha^{sx,sy}} & \Delta(\alpha)(sx, sy) \\ \Big\downarrow{\scriptstyle T_\varepsilon^{sx,sy}} & & \Big\downarrow{\scriptstyle T^{sx,sy}} \\ \varepsilon\alpha(sy \cdot sx) & \xrightarrow{\varepsilon C_\alpha^{sy,sx}} & \varepsilon\Delta(\alpha)(sx, sy) \end{array}$$

Here $s : H_1 \to \mathcal{H}^*(X)_1$ is defined in the pull back diagram (14.3.1) and T_ε and T are defined as follows. Recall that we have for pointed maps $x : X \longrightarrow Z^q, y : X \longrightarrow Z^{q'}$ the *interchange tracks* (6.3.1)(7)

$$\begin{cases} T(x, y) : x \cdot y \longrightarrow (-1)^{|x||y|} y \cdot x, \\ T(x, y) = \Gamma_{\tau(y,x)}(y \cdot x). \end{cases}$$

We therefore get for $\alpha \in T_{\mathbb{G}}(E_{\mathcal{A}})$ the interchange track

$$T_\epsilon : \alpha(x \cdot y) \longrightarrow \alpha(\epsilon y \cdot x) \longrightarrow \epsilon\alpha(y \cdot x) \quad \text{for } p \geq 2.$$

Here $\epsilon = (-1)^{|x||y|} \in \mathbb{G}$ and T_ϵ is the composite of $\alpha(T(x, y))$ and $\Gamma(\epsilon)_\alpha$. Moreover we get for

$$\xi = \sum_i \xi_i' \otimes \xi_i'' \ \in \ T_{\mathbb{G}}(E_A) \otimes T_{\mathbb{G}}(E_A)$$

the interchange track

$$T : \xi(x, y) \longrightarrow (-1)^{|x||y|}(T\xi)(y, x)$$

where $\xi(x, y)$ is defined in (11.2.4). Here $\xi(x, y)$ and $(T\xi)(y, x)$ depend only on $\xi_{\mathbb{F}}$ and T is linear in ξ.

We define T by the sum

$$T = \sum_i (-1)^{|x||\xi_i''|} T(\xi_i'(x), \xi_i''(y)).$$

We have to consider the interchange formula (14.3.5) under the operator $s : H_1 \to \mathcal{H}^*(X)_1$. We have (13.2.5)(5)

(2)
$$sC(\alpha \otimes x \otimes y) = \tilde{C}_\alpha^{x,y} - s\mu_2(\Delta\alpha \otimes x \otimes y).$$

Moreover by (6.3.1)(6) we get

$$\begin{aligned} s\hat{T}(x\otimes y) &= (\Gamma_\tau - \varepsilon(\tau))\circ s(y\cdot x) \\ &= \Gamma_\tau s(y)\cdot s(x) - \varepsilon(\tau)s(y)\cdot s(x) \end{aligned} \tag{3}$$

where $\tau = \tau(y,x)$ and $\Gamma_\tau : \tau \Rightarrow \varepsilon(\tau)$ with $\varepsilon = \varepsilon(\tau) = (-1)^{|x||y|}$. Here we have for $\Delta\alpha = \sum_i \alpha'_i \otimes \alpha''_i$

$$T^{sx,sy} = \sum_i \pm\Gamma_{\tau(\alpha''_i y, \alpha'_i x)}(\alpha''_i sy)\cdot(\alpha'_i sx). \tag{4}$$

Moreover we have

$$T^{sx,sy}_\varepsilon = \Gamma(\varepsilon)_\alpha \square \alpha(\Gamma_\tau s(y)\cdot s(x)), \tag{5}$$

where $\Gamma(\varepsilon)_\alpha : \alpha(\varepsilon sy\cdot sx) \Rightarrow \varepsilon\alpha(sy\cdot sx)$. Next we observe that the following diagram commutes:

$$\begin{array}{ccc} \Delta(\alpha)(sx,sy) & \xleftarrow{\Gamma^{y,x}_{\Delta\alpha}} & s\mu_2(\Delta\alpha\otimes x\otimes y) \\ \downarrow{\scriptstyle T^{sx,sy}} & & \downarrow{\scriptstyle s\hat{T}(\Delta(\alpha)\cdot(x\otimes y))+s\xi} \\ \varepsilon\Delta(sy\cdot sx) & \xleftarrow{\Gamma^{y,x}_{\Delta\alpha}} & s\mu_2(\Delta\alpha\otimes y\otimes x) \end{array} \tag{6}$$

where $\xi = \mu_2(\Delta\alpha\otimes y\otimes x)$. This follows from (4), the definition of $\Gamma^{y,x}_{\Delta\alpha}$ in (13.2.5)(4), and (2) applied to $s\hat{T}(\alpha'_i x\otimes\alpha''_i y)$, and (3.1.3). Also the following diagram commutes,

$$\begin{array}{ccc} s(\alpha(x\cdot y)) & \xrightarrow{\Gamma(\alpha,x\cdot y)} & \alpha(sx\cdot sy) \\ \downarrow{\scriptstyle s(\alpha\hat{T}(x\otimes y))+s\eta} & & \downarrow{\scriptstyle T^{sx,sy}_\varepsilon} \\ s(\varepsilon\alpha(y\cdot x)) & \xrightarrow[\varepsilon\Gamma(\alpha,y\cdot x)]{} & \varepsilon\alpha(sy\cdot sx) \end{array} \tag{7}$$

with $\eta = \varepsilon\alpha(y\cdot x)$. In fact, we have

$$s(\alpha\hat{T}(x\otimes y)) = \alpha\bullet\hat{T}(x\otimes y).$$

Here the right-hand side is the composite in the left column of the following commutative diagram.

(8)

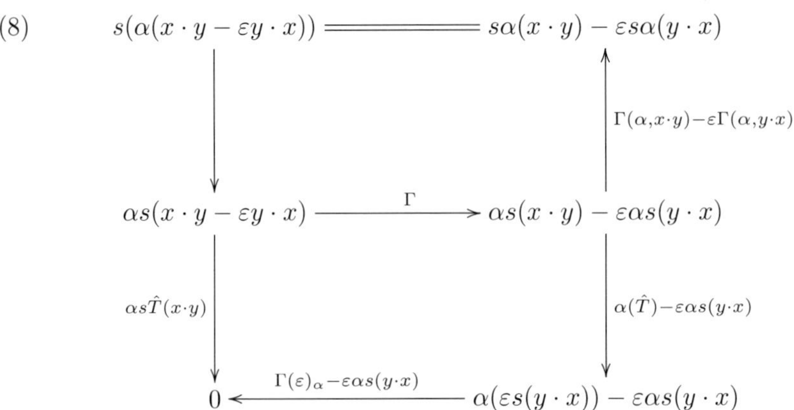

Here we set $\bar{T}=s\hat{T}(x\otimes y)+\varepsilon s(y\cdot x)=\Gamma_\tau s(y)\cdot s(x)$ by (2). This shows by (5) that (7) commutes.

The composite of top arrows in (1), (6) and (7) yields by (13.2.5)(4) the track $\bar{C}_\alpha^{x,y}$ and the composite of the bottom arrows in (1), (6) and (7) yields accordingly $\varepsilon\bar{C}_\alpha^{y,x}$. This shows by (2.2.6)(2) that the primary element of (1) is indeed $\mathcal{S}(\alpha\otimes x\otimes y)$. The primary element of (1), however, is trivial for p odd by the result in Section (10.5).

Now let $p=2$. Then the arguments above for p odd also hold if we use the odd sign convention for $p=2$. Now comparing the difference of the odd sign convention and the even sign convention yields the formula in (14.5.1) for $\mathcal{S}(\alpha\otimes x\otimes y)$ for $p=2$. □

Using (14.3.8) the extended symmetry operator $\mathcal{S}$ is completely determined by the formula in (14.5.1) above and by the operator $\mathcal{L}$ in Section (14.4). Hence we are able to compute the symmetry operator S by (14.3.7). For this we define the following elements in $\tilde{\mathcal{A}}\otimes\tilde{\mathcal{A}}$ with $n,m\geq 0$.

$$S(Sq^nSq^m)=(\delta Sq^n)\cdot S_m+S_n\cdot(\delta Sq^m)+\sum_{\substack{m_1+m_2=m\\ m_1,m_2\text{ odd}}}(\delta Sq^{n-1})\cdot(Sq^{m_1}\otimes Sq^{m_2})$$

Here S_n is defined in (14.5.1). Similarly as in (14.4.3) we define for $0<a<2b$ the element

$$S[a,b]=S(Sq^aSq^b)+\sum_{k=0}^{[a/2]}\binom{b-k-1}{a-2k}S(Sq^{a+b-k}Sq^k).$$

For $a+b\leq 9$ one gets the following explicit formulas for $S[a,b]$.

$$S[1,1] = 0$$

$$S[1,2] = 0$$

$$S[1,3] = 0$$
$$S[2,2] = 0$$

$$S[1,4] = 0$$
$$S[2,3] = \mathrm{Sq}^1 \otimes \mathrm{Sq}^2\,\mathrm{Sq}^1 + \mathrm{Sq}^2\,\mathrm{Sq}^1 \otimes \mathrm{Sq}^1 + \mathrm{Sq}^1 \otimes \mathrm{Sq}^3 + \mathrm{Sq}^3 \otimes \mathrm{Sq}^1$$
$$S[3,2] = \mathrm{Sq}^1 \otimes \mathrm{Sq}^2\,\mathrm{Sq}^1 + \mathrm{Sq}^2\,\mathrm{Sq}^1 \otimes \mathrm{Sq}^1 + \mathrm{Sq}^1 \otimes \mathrm{Sq}^3 + \mathrm{Sq}^3 \otimes \mathrm{Sq}^1$$

$$S[1,5] = 0$$
$$S[2,4] = \mathrm{Sq}^1 \otimes \mathrm{Sq}^3\,\mathrm{Sq}^1 + \mathrm{Sq}^3\,\mathrm{Sq}^1 \otimes \mathrm{Sq}^1$$
$$S[3,3] = 0$$

$$S[1,6] = 0$$
$$S[2,5] = \mathrm{Sq}^1 \otimes \mathrm{Sq}^4\,\mathrm{Sq}^1 + \mathrm{Sq}^4\,\mathrm{Sq}^1 \otimes \mathrm{Sq}^1 + \mathrm{Sq}^2\,\mathrm{Sq}^1 \otimes \mathrm{Sq}^3 + \mathrm{Sq}^3 \otimes \mathrm{Sq}^2\,\mathrm{Sq}^1 \\ + \mathrm{Sq}^1 \otimes \mathrm{Sq}^5 + \mathrm{Sq}^5 \otimes \mathrm{Sq}^1$$
$$S[3,4] = \mathrm{Sq}^1 \otimes \mathrm{Sq}^4\,\mathrm{Sq}^1 + \mathrm{Sq}^4\,\mathrm{Sq}^1 \otimes \mathrm{Sq}^1 + \mathrm{Sq}^2\,\mathrm{Sq}^1 \otimes \mathrm{Sq}^3 + \mathrm{Sq}^3 \otimes \mathrm{Sq}^2\,\mathrm{Sq}^1 \\ + \mathrm{Sq}^1 \otimes \mathrm{Sq}^5 + \mathrm{Sq}^5 \otimes \mathrm{Sq}^1$$
$$S[4,3] = 0$$

$$S[1,7] = 0$$
$$S[2,6] = \mathrm{Sq}^1 \otimes \mathrm{Sq}^5\,\mathrm{Sq}^1 + \mathrm{Sq}^5\,\mathrm{Sq}^1 \otimes \mathrm{Sq}^1 + \mathrm{Sq}^3 \otimes \mathrm{Sq}^3\,\mathrm{Sq}^1 + \mathrm{Sq}^3\,\mathrm{Sq}^1 \otimes \mathrm{Sq}^3$$
$$S[3,5] = 0$$
$$S[4,4] = \mathrm{Sq}^1 \otimes \mathrm{Sq}^5\,\mathrm{Sq}^1 + \mathrm{Sq}^5\,\mathrm{Sq}^1 \otimes \mathrm{Sq}^1 + \mathrm{Sq}^2\,\mathrm{Sq}^1 \otimes \mathrm{Sq}^3\,\mathrm{Sq}^1 + \mathrm{Sq}^3\,\mathrm{Sq}^1 \otimes \mathrm{Sq}^2\,\mathrm{Sq}^1$$
$$S[5,3] = \mathrm{Sq}^2\,\mathrm{Sq}^1 \otimes \mathrm{Sq}^3\,\mathrm{Sq}^1 + \mathrm{Sq}^3\,\mathrm{Sq}^1 \otimes \mathrm{Sq}^2\,\mathrm{Sq}^1 + \mathrm{Sq}^3 \otimes \mathrm{Sq}^3\,\mathrm{Sq}^1 + \mathrm{Sq}^3\,\mathrm{Sq}^1 \otimes \mathrm{Sq}^3$$

$$S[1,8] = 0$$
$$S[2,7] = \mathrm{Sq}^1 \otimes \mathrm{Sq}^6\,\mathrm{Sq}^1 + \mathrm{Sq}^6\,\mathrm{Sq}^1 \otimes \mathrm{Sq}^1 + \mathrm{Sq}^3 \otimes \mathrm{Sq}^4\,\mathrm{Sq}^1 + \mathrm{Sq}^4\,\mathrm{Sq}^1 \otimes \mathrm{Sq}^3 \\ + \mathrm{Sq}^2\,\mathrm{Sq}^1 \otimes \mathrm{Sq}^5 + \mathrm{Sq}^5 \otimes \mathrm{Sq}^2\,\mathrm{Sq}^1 + \mathrm{Sq}^1 \otimes \mathrm{Sq}^7 + \mathrm{Sq}^7 \otimes \mathrm{Sq}^1$$
$$S[3,6] = \mathrm{Sq}^1 \otimes \mathrm{Sq}^6\,\mathrm{Sq}^1 + \mathrm{Sq}^6\,\mathrm{Sq}^1 \otimes \mathrm{Sq}^1 + \mathrm{Sq}^3 \otimes \mathrm{Sq}^4\,\mathrm{Sq}^1 + \mathrm{Sq}^4\,\mathrm{Sq}^1 \otimes \mathrm{Sq}^3 \\ + \mathrm{Sq}^2\,\mathrm{Sq}^1 \otimes \mathrm{Sq}^5 + \mathrm{Sq}^5 \otimes \mathrm{Sq}^2\,\mathrm{Sq}^1 + \mathrm{Sq}^1 \otimes \mathrm{Sq}^7 + \mathrm{Sq}^7 \otimes \mathrm{Sq}^1$$
$$S[4,5] = \mathrm{Sq}^1 \otimes \mathrm{Sq}^6\,\mathrm{Sq}^1 + \mathrm{Sq}^6\,\mathrm{Sq}^1 \otimes \mathrm{Sq}^1 + \mathrm{Sq}^3 \otimes \mathrm{Sq}^4\,\mathrm{Sq}^1 + \mathrm{Sq}^4\,\mathrm{Sq}^1 \otimes \mathrm{Sq}^3 \\ + \mathrm{Sq}^2\,\mathrm{Sq}^1 \otimes \mathrm{Sq}^5 + \mathrm{Sq}^5 \otimes \mathrm{Sq}^2\,\mathrm{Sq}^1 + \mathrm{Sq}^1 \otimes \mathrm{Sq}^7 + \mathrm{Sq}^7 \otimes \mathrm{Sq}^1$$
$$S[5,4] = \mathrm{Sq}^1 \otimes \mathrm{Sq}^6\,\mathrm{Sq}^1 + \mathrm{Sq}^6\,\mathrm{Sq}^1 \otimes \mathrm{Sq}^1 + \mathrm{Sq}^4\,\mathrm{Sq}^1 \otimes \mathrm{Sq}^3 + \mathrm{Sq}^3 \otimes \mathrm{Sq}^4\,\mathrm{Sq}^1 \\ + \mathrm{Sq}^2\,\mathrm{Sq}^1 \otimes \mathrm{Sq}^5 + \mathrm{Sq}^5 \otimes \mathrm{Sq}^2\,\mathrm{Sq}^1 + \mathrm{Sq}^1 \otimes \mathrm{Sq}^7 + \mathrm{Sq}^7 \otimes \mathrm{Sq}^1$$

One can check that $S[1,n]=0$ for all $n\geq 1$. Here one has $[1,n]=\mathrm{Sq}^1\mathrm{Sq}^n$ if n is odd and $[1,n]=\mathrm{Sq}^1\mathrm{Sq}^n+\mathrm{Sq}^{n+1}$ if n is even.

14.5.2 Theorem. *For p odd the symmetry operator S is trivial. For p even the operator S is the unique linear map of degree -1*

$$S : R_{\mathcal{F}} \longrightarrow \tilde{A}\otimes\tilde{A}$$

satisfying the equations $(\alpha,\beta\in\mathcal{F}_0,\xi\in R_{\mathcal{F}})$

$$\begin{aligned} &S([a,b]) = S[a,b],\\ &S(\alpha\cdot\xi) = (\delta\alpha)\cdot S(\xi) + L(\alpha\otimes\xi) + TL(\alpha\otimes\xi),\\ &S(\xi\cdot\beta) = S(\xi)\cdot(\delta\beta).\end{aligned}$$

Proof. If p is odd we know that $\mathcal{L}=0$ so that by (14.3.8) and (14.5.1) also $\mathcal{S}=0$. Hence $S=0$ by (14.3.7). Now let p be even. We compute $\mathcal{S}([a,b]\otimes u)$ with $u=x\otimes y$ by considering (see (14.3.8))

$$\begin{aligned}&\mathcal{S}(\mathrm{Sq}^n\mathrm{Sq}^m\otimes u)\\ &= (\delta\,\mathrm{Sq}^n)\cdot\mathcal{S}(\mathrm{Sq}^m\otimes u) + \mathcal{S}(\mathrm{Sq}^n\otimes(\Delta\,\mathrm{Sq}^m)\cdot u) + \mathcal{L}(\mathrm{Sq}^n\otimes\mathrm{Sq}^m\otimes(u+Tu)).\end{aligned}\tag{1}$$

Hence by (14.5.1) we get (1) = (2) + (3) + (4):

$$\mathrm{Sq}^n(|x||y|\,\mathrm{Sq}^{m-1}(x\cdot y) + \mu_2(S_m\otimes u)),\tag{2}$$

$$\sum_{i+j=m}(|\mathrm{Sq}^i x||\mathrm{Sq}^j y|\,\mathrm{Sq}^{n-1}(\mathrm{Sq}^i x\cdot\mathrm{Sq}^j y) + \mu_2(S_n\otimes\mathrm{Sq}^i x\otimes\mathrm{Sq}^j y))\tag{3}$$

$$|x|\sum_{\substack{i+j=m\\ j\text{ odd}}}\mathrm{Sq}^{n-1}(\mathrm{Sq}^i x\cdot\mathrm{Sq}^j y) + |y|\sum_{\substack{i+j=m\\ j\text{ odd}}}\mathrm{Sq}^{n-1}(\mathrm{Sq}^i y\cdot\mathrm{Sq}^j x)\tag{4}$$

with $\mathrm{Sq}^i y\cdot\mathrm{Sq}^j x = \mathrm{Sq}^j x\cdot\mathrm{Sq}^i y$ since H^* is a commutative algebra. Hence we get by definition of $S(\mathrm{Sq}^n\mathrm{Sq}^m)$,

$$(2)+(3)+(4) = \mu_2(S(\mathrm{Sq}^n\mathrm{Sq}^m)\otimes u) + |x||y|\kappa(\mathrm{Sq}^n\mathrm{Sq}^m)(x\cdot y)$$

with $\kappa(\mathrm{Sq}^n\mathrm{Sq}^m) = \mathrm{Sq}^{n-1}\mathrm{Sq}^m + \mathrm{Sq}^n\mathrm{Sq}^{m-1}$. Since $\kappa[a,b]=0$ in $\mathcal{A}$ we see by (14.3.7) that $S([a,b]) = S[a,b]$. $\square$

In Section (15.2) below we show that there exist elements $\xi[a,b]$ satisfying

$$S[a,b] = \xi[a,b] + T\xi[a,b].\tag{14.5.3}$$

Remark. The existence of elements $\xi[a,b]$ satisfying equation (14.5.3) is obtained using the definition of S by the following result on *relations associated to Adem relations*. For $n,m \geq 0$ let

$$\hat{S}(\mathrm{Sq}^n\,\mathrm{Sq}^m) = \begin{cases} \mathrm{Sq}^i\,\mathrm{Sq}^j & \text{if } 2i = n-1,\ i \text{ odd},\ 2k = m,\ k \text{ even},\\ \mathrm{Sq}^i\,\mathrm{Sq}^k & \text{if } 2i = n-1,\ i \text{ even},\ 2k = m,\ k \text{ odd},\\ \mathrm{Sq}^i\,\mathrm{Sq}^k & \text{if } 2i = n,\ 2k = m-1,\ k \text{ odd},\\ 0 & \text{otherwise.} \end{cases}$$

Then we define

$$\hat{S}[a,b] = \hat{S}(\mathrm{Sq}^a\,\mathrm{Sq}^b) + \sum_{k=0}^{[a/2]} \binom{b-k-1}{a-2k} \hat{S}(\mathrm{Sq}^{a+b-k}\,\mathrm{Sq}^k).$$

One can check that $\hat{S}[a,b]$ is a relation, that is, $\hat{S}[a,b]$ considered as an element in $\mathcal{A}$ is trivial. Since $\hat{S}[a,b]$ is trivial one gets by definition of $S[a,b]$ elements $\xi[a,b]$ satisfying (14.5.3).

14.6 The track functor $\mathcal{H}^*[\,]$

In (13.1.4) we define a pair algebra over the secondary Hopf algebra $\mathcal{B}$ of secondary cohomology operations. We now refine this notion as follows.

14.6.1 Definition. A *secondary permutation algebra over* $\mathcal{B}$ is a pair algebra H over $\mathcal{B}$ as in (13.1.4) such that the pair algebra H has also the structure of a secondary permutation algebra as defined in section (6.2). Moreover the homotopy C satisfies the formulas (14.2.1) and (14.3.5) where the operators C and S are uniquely given by the formulas in Sections (14.4) and (14.5). We say that H is an *unstable* secondary permutation algebra over $\mathcal{B}$ if unstable structure maps with properties as in Section (13.3) are given:

$$\begin{aligned} v : \mathcal{E}(H) &\longrightarrow H_1, \\ u : H_0 &\longrightarrow H_1 \text{ if } p = 2 \text{ and} \\ u : H_0^{\text{even}} &\longrightarrow H_1 \text{ if } p \text{ is odd} \end{aligned}$$

Here $\mathcal{E} \subset \mathcal{M} \times H_0$ is the excess subset. These maps have properties as in (13.3.3) and (13.3.4).

Moreover for $p = 2$ the function u is quadratic with cross effect

$$u(x+y) - u(x) - u(y) = -\hat{T}(x \otimes y) - [p](n-1)y \cdot x$$

where $n = |x| = |y|$ and $\hat{T}$ is the interchange homotopy in (14.2.4). The cross effect is a consequence of the argument in the proof of (6.5.2).

We defined homotopies in the category of secondary permutation algebras in (6.4.2). We now consider such homotopies in the category of secondary permutation algebras over the secondary Hopf algebra $\mathcal{B}$.

Let A and B be secondary permutation algebras over $\mathcal{B}$ as defined in (14.6.1). Let $f = (f_0, f_1)$ and $g = (g_0, g_1)$ be maps

$$f, g : A \longrightarrow B$$

in the category of secondary permutation algebras over $\mathcal{B}$. Such maps are defined in the obvious way by compatibility with all structure maps.

14.6.2 Definition. A *homotopy* or *track* $H : f \Rightarrow g$ is a map

$$H : A_0 \longrightarrow B_1$$

which is R_*-linear as in (6.4.2) and which also is $\mathcal{B}_0$-linear such that the following properties hold

$$\partial H = f_0 - g_0, \tag{1}$$

$$H\partial = f_1 - g_1, \tag{2}$$

$$H(x \cdot y) = (Hx)(g_0 y) + (f_0 x) \cdot (Hy) \tag{3}$$

for $x, y \in A_0$. If A and B are unstable as is (14.6.1) and if f and g are compatible with the structure map u, v then we also assume

$$H(\alpha \cdot x) = (f_1 - g_1)v(\alpha, x) \tag{4}$$

for $\alpha \in \mathcal{M}$, $x \in A_0$ and $e(\alpha) < |x|$. Moreover

$$H(\alpha x - x^p) = (f_1 - g_1)u(x) \tag{5}$$

for $\alpha = Sq^{|x|}$, $x \in A_0$ if $p = 2$ and $\alpha = P^{|x|/2}$, $x \in A_0^{\mathrm{even}}$ if p odd.

Let $[\![\mathcal{K}_p^0]\!]$ be the category of unstable secondary permutation algebras over the secondary Hopf algebra $\mathcal{B}$, which is a track category given by such maps and homotopies.

14.6.3 Theorem. *There is a track functor*

$$\mathcal{H}^*[\,] : [\![Top_0^*]\!] \longrightarrow [\![\mathcal{K}_p^0]\!].$$

Proof. We define $\mathcal{H}^*[X]$ as in (14.2.1) by

$$\mathcal{H}^*[X] = \mathcal{H}^*(X, \mathcal{P}_X, s_X).$$

Moreover for a map $f : X \longrightarrow Y$ in Top_0^* let

$$f^* : \mathcal{H}^*[Y] \longrightarrow \mathcal{H}^*[X]$$

be given by

$$(f^*)_0 = F_{\mathcal{M}}(\mathcal{H}^*(f)_0)_*$$

in degree 0 and by $((f^*)_0, \mathcal{H}^*(f)_1)$ in degree 1 where we use the pull back property of $\mathcal{H}^*[X]_1$. For this we need the functor $\mathcal{H}^*$ of secondary cohomology in (6.3.1). Moreover for a track $H : f \Rightarrow g$ in Top_0^* we define the induced track

$$G = H^* : f^* \Longrightarrow g^*$$

in the category $[\![\mathcal{K}_p^0]\!]$ by the unique map

$$G : \mathcal{H}^*[Y]_0 = \mathbb{G}[F_{\mathcal{M}}(\mathcal{H}^*(Y)_0)] \longrightarrow \mathcal{H}^*[X]_1$$

which is a homotopy $G : f^* \Rightarrow g^*$ with the properties (1)–(4) in (14.6.2) and for which the following diagram commutes.

$$\begin{array}{ccc} \mathcal{H}^*(Y)_0 & \xrightarrow{\;H^*\;} & \mathcal{H}^*(X)_1 \\ \cap & & \uparrow \\ \mathcal{H}^*[Y]_0 & \xrightarrow[\;G\;]{} & \mathcal{H}^*[X]_1 \end{array}$$

Here H is the induced homotopy defined in (6.4.1). One readily checks by the freeness property of $F_{\mathcal{M}}(\mathcal{H}^*(Y)_0)$ that G is well defined and that G, in particular, satisfies (14.6.2)(4), (5). $\square$

Recall that $\mathcal{K}_p^0$ is the category of connected unstable algebras over the Steenrod algebra $\mathcal{A}$ which by (1.5.2) is isomorphic to the category of models of the theory $\mathbf{K}_p \subset \mathbf{Top}^*/\simeq$. Moreover the diagram

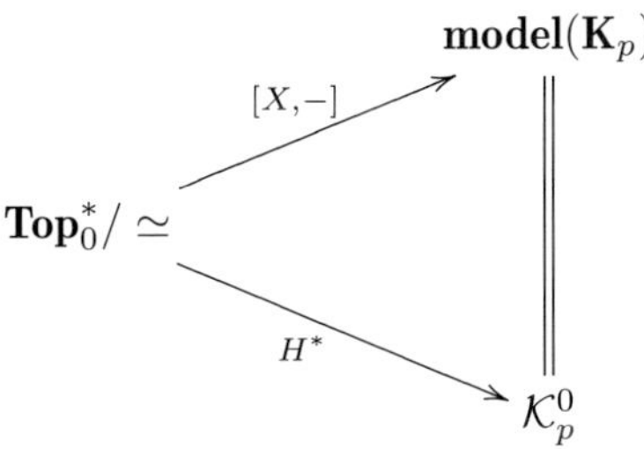

commutes, see (1.5.3). We now describe a similar diagram for the secondary theory. Let $[\![\mathcal{K}_p^0]\!]$ be the category of unstable secondary permutation algebras over the secondary Hopf algebra $\mathcal{B}$. This is a track category with tracks as in (14.6.2).

One gets the commutative diagram of track categories

(14.6.4)

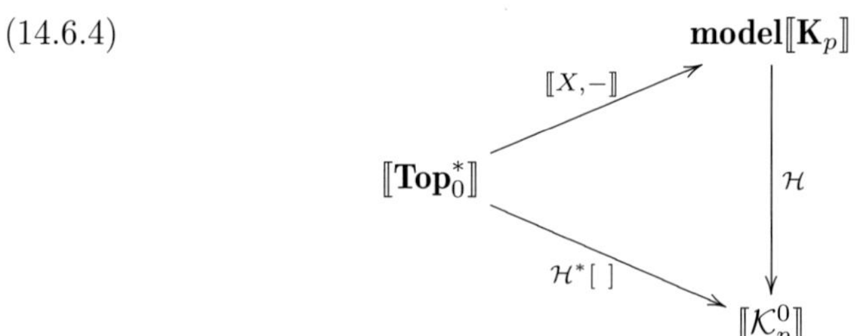

Here the functor $\mathcal{H}$ is similarly defined as the functor $\mathcal{H}^*[\,]$ in (14.6.3). The functor $\mathcal{H}$ induces on π_0 the isomorphism $\mathbf{model}(\mathbf{K}_p) = \mathcal{K}_p^0$. We conjecture that $\mathcal{H}$ is a weak equivalence of track categories.

Chapter 15

The Uniqueness of the Secondary Hopf Algebra $\mathcal{B}$

We show that the secondary Hopf algebra $\mathcal{B}$ of secondary cohomology operations is determined up to isomorphisms by the triple (κ, S, L) where $\kappa = \Gamma[p]$ is the derivation associated to $\mathcal{B}$ and S is the symmetry operator and L is the left action operator. We have seen in Chapter 14 that $S = 0$ and $L = 0$ for p odd.

15.1 The Δ-class of $\mathcal{B}$

The structure of the secondary Hopf algebra $\mathcal{B}$ leads to a Δ-class which can be expressed directly in terms of the Steenrod algebra $\mathcal{A}$. For this we have to choose a splitting u of $\mathcal{B}$.

Recall that for the prime $p \geq 2$ we have the field $\mathbb{F} = \mathbb{Z}/p$ and the ring $\mathbb{G} = \mathbb{Z}/p^2$. There is a canonical set $E_{\mathcal{A}}$ of generators of the Steenrod algebra $\mathcal{A}$ given by

$$
\begin{aligned}
E_{\mathcal{A}} &= \{Sq^i;\ i \geq 1\} && \text{for } p = 2,\\
E_{\mathcal{A}} &= \{\beta, P^i, P^i_\beta;\ i \geq 1\} && \text{for } p \text{ odd.}
\end{aligned}
$$

Let $\mathcal{B}_0 = T_{\mathbb{G}}(E_{\mathcal{A}})$ and $\mathcal{F}_0 = T_{\mathbb{F}}(E_{\mathcal{A}})$ be the tensor algebras over $\mathbb{G}$, resp. $\mathbb{F}$, generated by $E_{\mathcal{A}}$. We have the surjective algebra maps

$$\mathcal{B}_0 \longrightarrow \mathcal{F}_0 \longrightarrow \mathcal{A}$$

which are the identity on $E_{\mathcal{A}}$. Therefore we have the *ideals of relations*

$$
\begin{aligned}
R_{\mathcal{B}} &= \mathrm{kernel}(\mathcal{B}_0 \longrightarrow \mathcal{A}),\\
R_{\mathcal{F}} &= \mathrm{kernel}(\mathcal{F}_0 \longrightarrow \mathcal{A}).
\end{aligned}
$$

15.1.1 Definition. A *splitting* of $\mathcal{B}$ is a linear map u with the following properties. The diagram

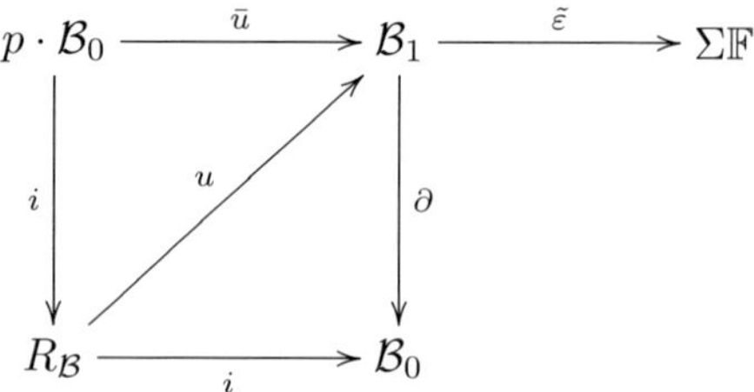

commutes and $\tilde{\varepsilon}u = 0$. Here $\tilde{\varepsilon}$ is the augmentation of $\mathcal{B}$, see (12.1.5), and $\bar{u}$ carries $p\alpha$ to $[p] \cdot \alpha$ for $\alpha \in \mathcal{B}_0$. The maps i in the diagram denote the inclusions.

15.1.2 Proposition. *A splitting u of $\mathcal{B}$ exists.*

Proof. Consider the short exact sequence of $\mathbb{F}$-vector spaces

$$0 \longrightarrow R_{\mathcal{F}} \longrightarrow \mathcal{F}_0 \longrightarrow \mathcal{A} \twoheadrightarrow 0.$$

We choose a basis N_R of $R_{\mathcal{F}}$ and we extend N_R by a basis N_S of a complement of $R_{\mathcal{F}}$ in $\mathcal{F}_0$ so that N_S maps bijectively to a basis of $\mathcal{A}$ and $1 \in N_S$. For example N_S is given by the set of admissible monomials. A lift b as in the commutative diagram

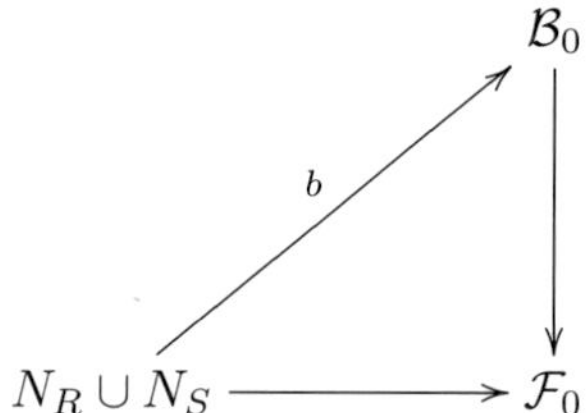

yields a basis $b(N_R \cup N_S)$ of the free $\mathbb{G}$-module $\mathcal{B}_0$. Hence we get the direct sum of free $\mathbb{G}$-modules

$$R \oplus S = \mathcal{B}_0$$

where R is generated by bN_R and S is generated by bN_S. Now one gets accordingly

$$R \oplus pS = R_{\mathcal{B}}.$$

Since R is a free $\mathbb{G}$-module we can choose a lift t as in the commutative diagram

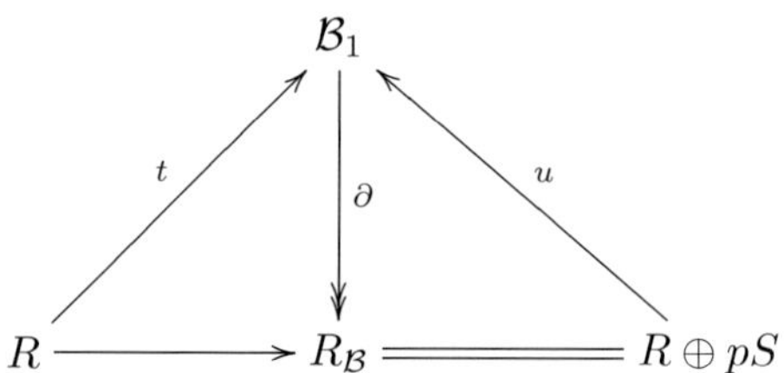

with $\tilde{\varepsilon}t = 0$, see (12.1.5)(2) and (3). We now define the splitting u by $u(x) = t(x)$ for $x \in R$ and $u(p\alpha) = [p] \cdot \alpha$ for $\alpha \in S$. Then for $\alpha \in R$ we have

$$u(p\alpha) = pu(\alpha) = pt(\alpha) = [p] \cdot \partial t(\alpha) = [p] \cdot \alpha.$$

This shows that u has the properties in (15.1.1). □

Now let

$$\begin{aligned} R^2_{\mathcal{B}} &= \mathrm{kernel}(\mathcal{B}_0 \otimes \mathcal{B}_0 \longrightarrow \mathcal{A} \otimes \mathcal{A}) \\ &= R_{\mathcal{B}} \otimes \mathcal{B}_0 + \mathcal{B}_0 \otimes R_{\mathcal{B}} \end{aligned}$$

and more generally let

$$R^n_{\mathcal{B}} = \mathrm{kernel}(\mathcal{B}_0^{\otimes n} \longrightarrow \mathcal{A}^{\otimes n})$$

for $n \geq 1$ so that $R^1_{\mathcal{B}} = R_{\mathcal{B}}$.

15.1.3 Proposition. *A splitting u of $\mathcal{B}$ induces via the following commutative diagram a splitting $u_\sharp$ of $\mathcal{B}\hat{\otimes}\mathcal{B}$.*

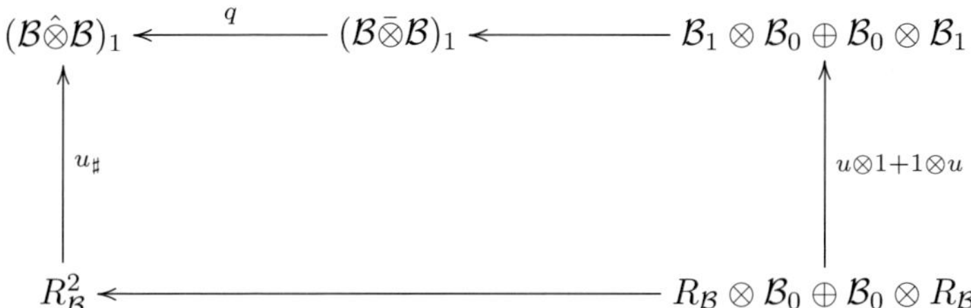

More generally one obtains in a similar way a splitting

$$u_\sharp : R^n_{\mathcal{B}} \longrightarrow (\mathcal{B}^{\hat{\otimes} n})_1$$

of $\mathcal{B}^{\hat{\otimes} n} = \mathcal{B}\hat{\otimes} \cdots \hat{\otimes}\mathcal{B}$.

The horizontal arrows in the diagram are the canonical quotient maps.

Proof of (15.1.3). We have

$$R^2_{\mathcal{B}} = R \otimes R \oplus R \otimes S \oplus S \otimes R \oplus p(S \otimes S)$$

where we use the direct sum in the proof of (15.1.2). We now define $u_\sharp$ for $x, y \in R$ and $a, b \in S$ by the equations

$$\begin{aligned} u_\sharp(x \otimes y) &= (tx)\hat{\otimes}y = x\hat{\otimes}(ty), \\ u_\sharp(a \otimes y) &= a\hat{\otimes}(ty), \\ u_\sharp(x \otimes b) &= (tx)\hat{\otimes}b, \\ u_\sharp(p \cdot (a \otimes b)) &= [p] \cdot (a \otimes b). \end{aligned}$$

Here $[p] = q(1 \otimes [p]) = q([p] \otimes 1)$ is defined by the quotient map q. Now one can check that $u_\sharp$ fits into the commutative diagram above. □

Next let $\tilde{\mathcal{A}} = \text{kernel}(\varepsilon : \mathcal{A} \longrightarrow \mathbb{F})$ be the augmentation ideal in the Steenrod algebra $\mathcal{A}$. Then one readily checks:

15.1.4 Proposition. *Let u, u' be splittings of $\mathcal{B}$ as in (15.1.1); then there is a unique linear map of degree -1,*

$$\alpha : R_{\mathcal{F}} \longrightarrow \tilde{\mathcal{A}},$$

such that $u' = u + \Sigma\alpha$ with $\Sigma\alpha$ being the composite

$$\Sigma\alpha : R_{\mathcal{B}} \twoheadrightarrow R_{\mathcal{F}} \xrightarrow{\alpha} \tilde{\mathcal{A}} \xrightarrow{\Sigma} \Sigma\mathcal{A} \subset \mathcal{B}_1.$$

We now obtain for a splitting u of $\mathcal{B}$ the following diagram.

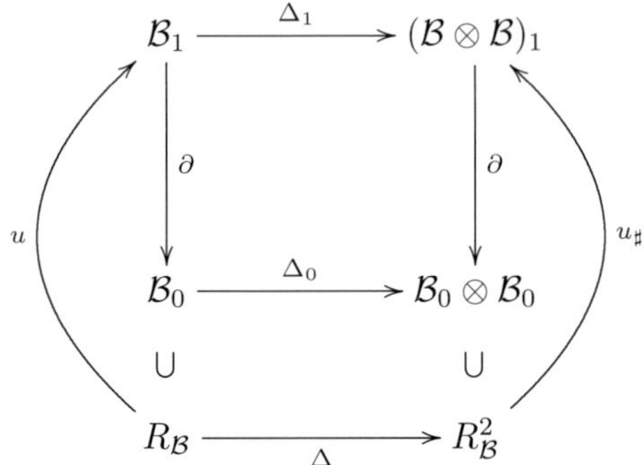

15.1.5 Proposition. *For each splitting u of $\mathcal{B}$ there is a unique linear map of degree -1,*

$$\nabla_u : R_{\mathcal{F}} \longrightarrow \tilde{\mathcal{A}} \otimes \tilde{\mathcal{A}},$$

satisfying

$$\Delta_1 u = u_\sharp \Delta + \Sigma \nabla_u.$$

We call ∇_u the Δ-*difference element* associated to the splitting u of $\mathcal{B}$. The Δ-class $\nabla_{\mathcal{B}}$ is the set of all Δ-difference elements so that $\nabla_{\mathcal{B}}$ is a subset of $Hom_{-1}(R_{\mathcal{F}}, \tilde{\mathcal{A}} \otimes \tilde{\mathcal{A}})$.

Proof of (15.1.5). We have

$$\partial \Delta_1 u = \Delta_0 \partial u = \Delta = \partial u_\sharp \Delta$$

so that $\partial(\Delta_1 u - u_\sharp \Delta) = 0$ and hence $F = \Delta_1 u - u_\sharp \Delta$ maps to $\Sigma\mathcal{A} \otimes \mathcal{A}$. Moreover

$$F(p \cdot \alpha) = 0 \text{ and } (\varepsilon \otimes 1)F = (1 \otimes \varepsilon)F = 0.$$

For this we need the fact that the augmentation ε of $\mathcal{B}$ satisfies

$$(\varepsilon \otimes 1)u_\sharp = u(\varepsilon \otimes 1),$$
$$(1 \otimes \varepsilon)u_\sharp = u(1 \otimes \varepsilon).$$

Hence F induces a unique map ∇_u as in the theorem. □

The diagonal Δ of $\mathcal{F}_0$ yields the commutative diagram

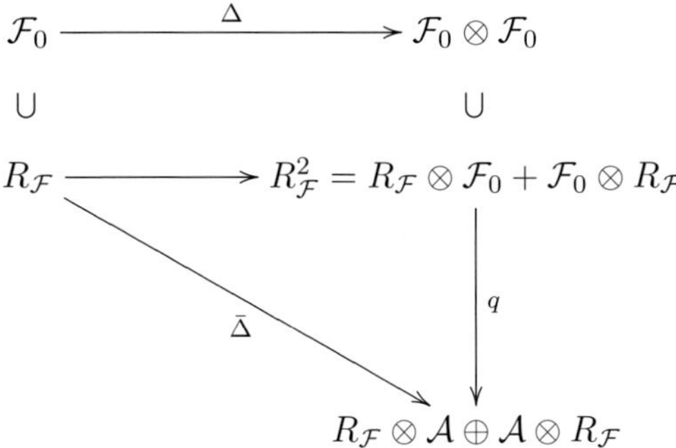

where q is the quotient map with kernel$(q) = R_{\mathcal{F}} \otimes R_{\mathcal{F}}$. We define the *differential*

$$d^1 : \mathrm{Hom}_{-1}(R_{\mathcal{F}}, \tilde{\mathcal{A}}) \longrightarrow \mathrm{Hom}_{-1}(R_{\mathcal{F}}, \tilde{\mathcal{A}} \otimes \tilde{\mathcal{A}})$$

by the formula

$$d^1(\alpha) = \delta\alpha - (\alpha \otimes 1, \tau(1 \otimes \alpha))\bar{\Delta}. \tag{15.1.6}$$

Here τ yields the sign which corresponds to the interchange of Σ in $\mathcal{A} \otimes \Sigma\mathcal{A}$. We point out that $d^1(\alpha)$ maps to $\tilde{\mathcal{A}} \otimes \tilde{\mathcal{A}}$ since

$$\begin{aligned}(\varepsilon \otimes 1)d^1(\alpha) &= (\varepsilon \otimes 1)\delta\alpha - (\varepsilon \otimes 1)(\alpha \otimes 1, \tau(1 \otimes \alpha))\bar{\Delta} \\ &= \alpha - \alpha = 0.\end{aligned}$$

Similarly one gets $(\varepsilon \otimes 1)d^1(\alpha) = 0$.

15.1.7 Proposition. *The Δ-difference element ∇_u satisfies the formula*

$$\nabla_{u+\Sigma\alpha} = \nabla_u + d^1(\alpha).$$

Hence the Δ-class $\nabla_{\mathcal{B}}$ is a well defined element in cokernel(d^1).

The proposition is readily checked by definition of $u_\sharp$ and properties of $(\mathcal{B}, \Delta)$. Next we define the *differential*

$$d^2 : \mathrm{Hom}_{-1}(R_{\mathcal{F}}, \tilde{\mathcal{A}} \otimes \tilde{\mathcal{A}}) \longrightarrow \mathrm{Hom}_{-1}(R_{\mathcal{F}}, \tilde{\mathcal{A}} \otimes \tilde{\mathcal{A}} \otimes \tilde{\mathcal{A}})$$

by the formula

$$d^2(\xi) = (\delta \otimes 1 - 1 \otimes \delta)\xi - (-\xi \otimes 1, \tau(1 \otimes \xi))\bar{\Delta}. \tag{15.1.8}$$

Here the right-hand side is given by composites in the following diagram.

$$\begin{array}{ccc} R_{\mathcal{F}} & \xrightarrow{\ \bar{\Delta}\ } & R_{\mathcal{F}} \otimes \mathcal{A} \oplus \mathcal{A} \otimes R_{\mathcal{F}} \\ \downarrow{\scriptstyle \xi} & & \downarrow{\scriptstyle (-\xi\otimes 1, \tau(1\otimes\xi))} \\ \Sigma\mathcal{A} \otimes \mathcal{A} & \xrightarrow{\ \Sigma(\delta\otimes 1 - 1\otimes\delta)\ } & \Sigma\mathcal{A} \otimes \mathcal{A} \otimes \mathcal{A} \end{array}$$

As in (15.1.6) one can check that $d^2(\xi)$ maps to $\tilde{\mathcal{A}} \otimes \tilde{\mathcal{A}} \otimes \tilde{\mathcal{A}}$.

15.1.9 Proposition. *Each element ξ in the Δ-class $\nabla_{\mathcal{B}}$ satisfies $d^2(\xi) = 0$.*

Since $d^2 d^1 = 0$ the proposition shows that the Δ-class is an element in the cohomology

$$\nabla_{\mathcal{B}} \in \operatorname{kernel}(d^2)/\operatorname{image}(d^1) = H_{\mathcal{A}}. \tag{15.1.10}$$

Proof of (15.1.9). We consider the following diagram where $\Delta' = \Delta \otimes 1 - 1 \otimes \Delta$ is given by $\mathcal{B}$.

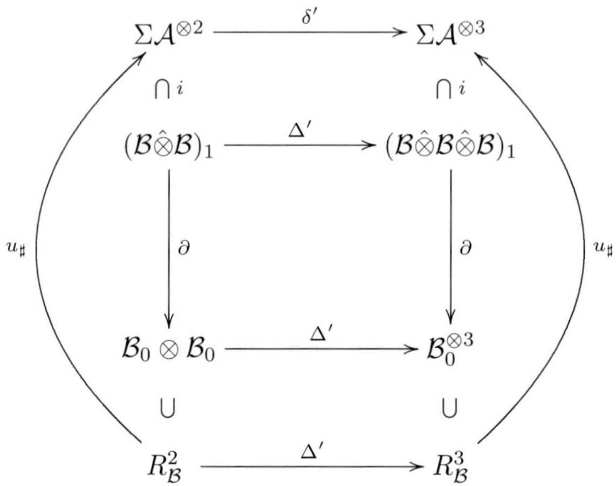

The diagram shows that there is a unique linear map of degree -1,

$$\mu : R^2_{\mathcal{B}} \longrightarrow \mathcal{A}^{\otimes 3},$$

satisfying

$$u_\sharp \Delta' = \Delta' u_\sharp + \Sigma\mu. \tag{1}$$

Now we get, for $\xi = \nabla_u \in \nabla_{\mathcal{B}}$ with $\delta' = \delta \otimes 1 - 1 \otimes \delta$,

$$\begin{aligned} i((\Sigma\delta')\xi - \mu\Delta) &= \Delta' i\xi - i\mu\Delta \\ &= \Delta'(\Delta u - u_\sharp\Delta) - (-\Delta' u_\sharp + u_\sharp\Delta')\Delta \\ &= 0. \end{aligned} \tag{2}$$

Here we use the fact that $\Delta'\Delta = 0$, as follows from the associativity of the diagonal Δ of $\mathcal{B}$. By (2) we see

$$\Sigma(\delta'\xi) = \mu\Delta. \tag{3}$$

Now the following diagram commutes.

This, in fact, proves (15.1.9) since $\Sigma d^2\xi = \Sigma(\delta'\xi) - \bar{\mu}\bar{\Delta}$ and $\bar{\mu}\bar{\Delta} = \mu\Delta$.

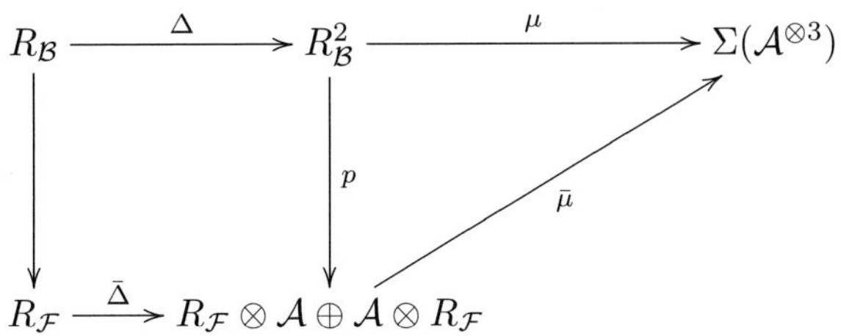

Here we set $\bar{\mu} = (-\xi \otimes 1, \tau(1 \otimes \xi))$ and p is the quotient map given by $R^2_{\mathcal{B}} \longrightarrow R^2_{\mathcal{F}}$. We have to check that $\bar{\mu}p = \mu$. Let $x, y \in R_{\mathcal{B}}$ and $a, b \in \mathcal{B}_0$. Then we get

$$\begin{aligned} i\mu(p \cdot (a \otimes b)) &= (-\Delta' u_\sharp + u_\sharp \Delta')(p \cdot (a \otimes b)) \\ &= (-\Delta \otimes 1 + 1 \otimes \Delta)[p] \cdot (a \otimes b) + u_\sharp p(\Delta a \otimes b - a \otimes \Delta b) \\ &= [p] \cdot (-\Delta a \cdot b + a \otimes \Delta b) + [p] \cdot (\Delta a \otimes b - a \otimes \Delta b) \\ &= 0, \end{aligned}$$

$$\begin{aligned} i\mu(x \otimes y) &= (-\Delta' u_\sharp + u_\sharp \Delta')(x \otimes y) \\ &= (-\Delta \otimes 1 + 1 \otimes \Delta) u_\sharp (x \otimes y) + u_\sharp(\Delta x \otimes y - x \otimes \Delta y) \\ &= -\Delta x \hat{\otimes} uy + ux \hat{\otimes} \Delta y + \Delta x \hat{\otimes} uy - ux \hat{\otimes} \Delta y \\ &= 0, \end{aligned}$$

$$\begin{aligned} i\mu(x \otimes b) &= (-\Delta' u_\sharp + u_\sharp \Delta')(x \otimes b) \\ &= (-\Delta \otimes 1 + 1 \otimes \Delta)(ux \otimes b) + u_\sharp(\Delta x \otimes b - x \otimes \Delta b) \\ &= -\Delta ux \hat{\otimes} b + ux \hat{\otimes} \Delta b + (u_\sharp \Delta x) \hat{\otimes} b - ux \hat{\otimes} \Delta b \\ &= -\Delta ux \hat{\otimes} b + u_\sharp \Delta x \hat{\otimes} b \\ &= -((\Delta u - u_\sharp \Delta) x) \hat{\otimes} b \\ &= -i(\xi \otimes 1)(x \otimes b). \end{aligned}$$

This shows $\mu(x \otimes b) = \bar{\mu}p(x \otimes b)$. Similarly one gets the equation $\mu(a \otimes y) = \bar{\mu}p(a \otimes y)$. This completes the proof of (15.1.9). □

Recall that we have the symmetry operator S which factorizes as a composite

$$R_{\mathcal{B}} \longrightarrow R_{\mathcal{F}} \xrightarrow{S} \tilde{\mathcal{A}} \otimes \tilde{\mathcal{A}} \quad \subset \quad \mathcal{A} \otimes \mathcal{A}.$$

Compare Section (14.1).

15.1.11 Proposition. *Each element ξ in the Δ-class $\nabla_{\mathcal{B}}$ satisfies the* symmetry formula

$$T\xi = \xi + S.$$

In the next section we show that there is exactly one cohomology class in $H_{\mathcal{A}}$, see (15.1.10), satisfying the symmetry formula. This is the zero class for the algebra $\mathcal{B}$ of secondary cohomology operations and p odd since we have seen that $\mathcal{B}$ has the trivial symmetry operator $S = 0$ for odd primes.

Proof of (15.1.10). The definition of $u_\sharp$ shows that $Tu_\sharp = u_\sharp T$ where $T : R^2_{\mathcal{B}} \cong R^2_{\mathcal{B}}$ is the restriction of T on $\mathcal{B}_0 \otimes \mathcal{B}_0$. Let $\xi = \nabla_u$, then we get:

$$\begin{aligned} i\,\Sigma(T\xi) &= Ti\xi = T(\Delta_1 u - u_\sharp \Delta_0) \\ &= (T\Delta_1)u - u_\sharp(T\Delta_0). \end{aligned}$$

Here we have $T\Delta_1 = \Delta_1 + \Sigma S(\partial_-)$ and $T\Delta_0 = \Delta_0$. Hence we obtain

$$i\Sigma(T\xi) = \Delta u - u_\sharp \Delta_0 + \Sigma S = \Sigma(\xi + S).$$

This completes the proof of (15.1.10). □

15.2 Computation of the Δ-class

We have seen that a secondary Hopf algebra $\mathcal{B}$ yields a Δ-class $\nabla_{\mathcal{B}}$ which is an element in the cohomology $H_{\mathcal{A}}$ in (15.1.10).

15.2.1 Theorem. *There is a unique element $\nabla \in H_{\mathcal{A}}$ such that all cocycles $\xi \in \nabla$ satisfy the symmetry formula*

$$T\xi = \xi + S$$

where S is the symmetry operator in (14.1).

By (15.1.10) the class ∇ in this theorem coincides with the Δ-class $\nabla_{\mathcal{B}}$. For the proof of the theorem we prove the following result on the cocycles of the cohomology $H_{\mathcal{A}}$.

15.2.2 Theorem. *Let $\xi : R_{\mathcal{F}} \to \tilde{A} \otimes \tilde{A}$ be a cocycle, i.e. $d^2\xi = 0$, and assume $T\xi = \xi$ holds. Then there is $\alpha : R_{\mathcal{F}} \to \tilde{A}$ with $d^1\alpha = \xi$, that is ξ is a coboundary.*

Proof of (15.2.1). Let ∇, $\nabla' \in H_{\mathcal{A}}$ such that for $\xi \in \nabla$, $\xi', \in \nabla'$ the symmetry formulas $T\xi = \xi + S$ and $T\xi' = \xi' + S$ hold. Then $\eta = \xi - \xi'$ satisfies

$$T\eta = T\xi - T\xi' = \xi - \xi' = \eta$$

so that $\eta = d^1\alpha$ is a coboundary by (15.2.2). Therefore $\nabla = \nabla'$. □

Theorem (15.2.2) corresponds to the 'dual' of a result of Penkava-Vanhaecke [PV]2.1. To see this we dualize the exact sequence

$$0 \longrightarrow R_{\mathcal{F}} \longrightarrow \mathcal{F}_0 \longrightarrow \mathcal{A} \longrightarrow 0$$

where $\mathcal{F}_0 \longrightarrow \mathcal{A}$ is a map between commutative coalgebras. The functor $\mathrm{Hom}(-, \mathbb{F})$ with $\mathrm{Hom}(V, \mathbb{F}) = V^*$ carries the sequence to the exact sequence

$$0 \longleftarrow R^*_{\mathcal{F}} \longleftarrow \mathcal{F}^*_0 \longleftarrow \mathcal{A}^* \longleftarrow 0.$$

Here $\mathcal{A}^* \longrightarrow \mathcal{F}^*_0$ is a morphism of commutative graded algebras so that $R^*_{\mathcal{F}} = \mathrm{Hom}(R_{\mathcal{F}}, \mathbb{F})$ is an $\mathcal{A}^*$-bimodule. This bimodule structure is also induced by dualizing $\bar{\Delta}$ in (15.1.6). For $x \in R^*_{\mathcal{F}}$ and $m \in \mathcal{A}^*$ we have the equation

$$x \cdot m = (-1)^{|x||m|} m \cdot x$$

since the diagonal Δ of $\mathcal{F}_0$ is cocommutative. We now consider the following normalized chain complex of Hochschild cohomology.

$$C_1 \xrightarrow{d_1} C_2 \xrightarrow{d_2} C_3. \tag{15.2.3}$$

Here C_i is the $\mathbb{F}$-vector space of all linear maps of degree $+1$,

$$C : A^{\otimes i} \longrightarrow M,$$

where $A = \mathcal{A}^*$, $M = R^*_{\mathcal{F}}$ satisfying the normalization condition $c(a_1 \otimes \cdots \otimes a_i) = 0$ if there is j with $0 \leq j \leq i$ and $a_j = 1$. Moreover the Hochschild differential is defined by

$$d_1(c)(x \otimes y) = -(-1)^{|x|} x \cdot c(y) + c(xy) - c(x) \cdot y$$

for $c \in C_1$ and

$$d_2(c)(x \otimes y \otimes z) = (-1)^{|x|} x \cdot c(y \otimes z) - c(xy \otimes z) + c(x \otimes yz) - c(x \otimes y)z$$

for $c \in C_2$. We now observe that we have the dualization isomorphism

$$C^i = \mathrm{Hom}_{-1}(R_{\mathcal{F}}, \tilde{\mathcal{A}}^{\otimes i}) \cong C_i$$

which carries ξ to the dual ξ^* of the composite $R_{\mathcal{F}} \xrightarrow{\xi} \tilde{\mathcal{A}}^{\otimes i} \subset \mathcal{A}^{\otimes i}$. Since the dual of the augmentation ε of $\mathcal{A}$ is the inclusion $1 : \mathbb{F} \to \mathcal{A}^*$, we see that ξ^* is normalized.

15.2.4 Lemma. *The differentials d_1 d_2 above are isomorphic to the differentials d^1, d^2 in* (15.1.10)*. that is, the following diagram commutes.*

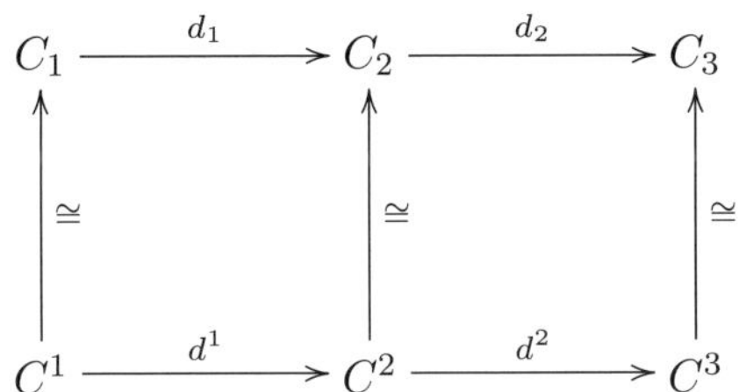

The lemma is readily checked by the definitions of the differential. We now prove (15.2.2) by the following result, see Penkava-Vanhaecke [PV]2.1.

15.2.5 Theorem. *Let $D \in C_2$ with $d_2 D = 0$ and $TD = D$; then there exists $\varphi \in C_1$ with $d_1\varphi = D$.*

Proof. To avoid signs we only consider the case of the even prime. Since D is a normalized cocycle we can consider the algebra extension with $\Sigma\bar{M} = M$,

$$\begin{array}{ccccccc} 0 & \longrightarrow & \bar{M} & \longrightarrow & E & \longrightarrow & A \longrightarrow 0 \\ & & & & \| & & \\ & & & & A \oplus \bar{M} & & \end{array}$$

with the multiplication

$$(a, m) \cdot (a', m') = (aa', am' + ma' + D(a, a'))$$

for (a, m), $(a', m') \in E$. Since $TD = D$ and since $am = ma$ we see that

$$(a, m) \cdot (a', m') = (a', m') \cdot (a, m).$$

Hence E is a commutative algebra. Now A is a free commutative algebra and therefore there exists a section $A \longrightarrow E$. Hence the cohomology class of the extension (represented by D) is trivial. Compare also section (16.2).

□

15.3 The multiplication class of $\mathcal{B}$

Let $\xi : R_{\mathcal{F}} \longrightarrow \tilde{A} \otimes \tilde{A}$ be a cocycle representing the Δ-class $\nabla_{\mathcal{B}}$ which is the element determined in (15.2.1) by the symmetry operator S. Let $\xi = 0$ be trivial if $S = 0$. Since $\xi \in \nabla_{\mathcal{B}}$ there exists a splitting u of $\mathcal{B}$ with

$$\nabla_u = \Delta_1 u - u_\sharp \Delta = \Sigma\xi.$$

In this case we call u a *ξ-splitting* of $\mathcal{B}$. If $\xi = 0$ then we call u a *Δ-splitting* of $\mathcal{B}$. Hence u is a Δ-splitting of $\mathcal{B}$ if u is a splitting as in (15.1.1) and the following diagram commutes.

(15.3.1)
$$\begin{array}{ccc} \mathcal{B}_1 & \xrightarrow{\Delta_1} & (\mathcal{B}\hat{\otimes}\mathcal{B})_1 \\ \uparrow u & & \uparrow u^\sharp \\ R_{\mathcal{B}} & \longrightarrow & R^2_{\mathcal{B}} \\ \cap & & \cap \\ \mathcal{B}_0 & \xrightarrow[\Delta_0]{} & \mathcal{B}_0 \otimes \mathcal{B}_0 \end{array}$$

Since the algebra $\mathcal{B}$ of secondary cohomology operations has a trivial symmetry operator $S = 0$ for odd primes p we see:

15.3.2 Theorem. *The algebra $\mathcal{B}$ of secondary cohomology operations has a Δ-splitting over odd primes p.*

A ξ-splitting u of $\mathcal{B}$ is not uniquely determined by ξ. According to (15.1.7) we get:

15.3.3 Lemma. *If u is a ξ-splitting of $\mathcal{B}$, then $u + \Sigma\varphi$ is a ξ-splitting if and only if $d^1(\varphi) = 0$.*

Here we have $d^1(\varphi) = 0$ if and only if the dual φ^* of φ satisfies $d_1(\varphi^*) = 0$ and this is the case if and only if $\varphi^* : \mathcal{A}^* \longrightarrow R^*_{\mathcal{F}}$ is a *derivation* of degree $+1$, that is

$$\varphi^*(xy) = \varphi^*(x) \cdot y + (-1)^{|x|} x \cdot \varphi^*(y)$$

for $x, y \in \mathcal{A}^*$. We point out that such a derivation is completely determined by its values on Milnor generators in the free commutative graded algebra $\mathcal{A}^*$, see [Mn].

We now consider the multiplication of the pair algebra $\mathcal{B}$ which is determined by the $\mathcal{B}_0$-bimodule structure of $\mathcal{B}_1$. The left and right action of $\mathcal{B}_0$ on $\mathcal{B}_1$ yield functions $A = A_u$ and $B = B_u$ by the formulas $(\alpha, \beta \in \mathcal{B}_0, x \in R_{\mathcal{B}})$

$$\begin{aligned} \Sigma A_u(\alpha \otimes x) &= \alpha \cdot u(x) - u(\alpha \cdot x) \in \Sigma\mathcal{A}, \\ \Sigma B_u(x \otimes \beta) &= u(x) \cdot \beta - u(x \cdot \beta) \in \Sigma\mathcal{A}. \end{aligned} \tag{15.3.4}$$

Here u is a ξ-splitting of $\mathcal{B}$. We call the pair (A_u, B_u) a *multiplication structure* of $\mathcal{B}$. Such a multiplication structure has the following properties.

15.3.5 Definition. A *$\mathcal{B}$-structure* (A, B) is given by a pair of $\mathbb{G}$-linear maps of degree -1,

$$A : \mathcal{B}_0 \otimes R_{\mathcal{B}} \longrightarrow \mathcal{A},$$

$$B : R_{\mathcal{B}} \otimes \mathcal{B}_0 \longrightarrow \tilde{\mathcal{A}},$$

satisfying the following properties with $\alpha, \alpha', \beta, \beta' \in \mathcal{B}_0$ and $x, y \in R_{\mathcal{B}}$.

$$A(\alpha, x\beta) + (-1)^{|\alpha|}\alpha B(x, \beta) = B(\alpha x, \beta) + A(\alpha, x)\beta, \tag{1}$$

$$A(x, y) = B(x, y), \tag{2}$$

$$A(\alpha\alpha', x) = A(\alpha, \alpha' x) + (-1)^{|\alpha|}\alpha A(\alpha', x), \tag{3}$$

$$B(x, \beta\beta') = B(x\beta, \beta') + B(x, \beta)\beta', \tag{4}$$

$$B(p\alpha, \beta) = 0. \tag{5}$$

Let M^2 be the set of all $\mathcal{B}$-structures (A, B). Hence M^2 is an $\mathbb{F}$-vector space by addition of maps. Moreover let

$$M^2_{\kappa_0} \subset M^2$$

be the subset of all $\mathcal{B}$-structures (A, B) satisfying

(6) $$A(\alpha, p\beta) = -\kappa_0(\alpha) \cdot \beta.$$

Here $\kappa_0 : \mathcal{A} \longrightarrow \mathcal{A}$ is a derivation of degree -1 of $\mathcal{A}$ satisfying $(\kappa_0 \otimes 1)\delta = \delta\kappa_0$.

For the trivial derivation $\kappa_0 = 0$ the subset $M_0^2 \subset M^2$ is a vector space. In general $M_{\kappa_0}^2$ is a coset in the quotient

(7) $$M_{\kappa_0}^2 \in M^2/M_0^2$$

so that M_0^2 acts transitively and effectively on the set $M_{\kappa_0}^2$. By (5) and (6) we see that pairs (A, B) in M_0^2 induce maps

(8) $$\begin{aligned} A &: \mathcal{F}_0 \otimes R_{\mathcal{F}} \longrightarrow \tilde{\mathcal{A}}, \\ B &: R_{\mathcal{F}} \otimes \mathcal{F}_0 \longrightarrow \tilde{\mathcal{A}}, \end{aligned}$$

also denoted by A and B, since $A(\alpha, p\beta) = 0$ for κ_0 by (6).

15.3.6 Lemma. *Let $\mathcal{B}$ be the algebra of secondary cohomology operations. Then a multiplication structure (A_u, B_u) of $\mathcal{B}$ is an element in M_κ^2 where $\kappa = \Gamma[p]$ is the derivation with $\kappa(Sq^n) = Sq^{n-1}$ for $p = 2$ and $\kappa(\beta) = 1$, $\kappa(P^n) = 0$ for p odd.*

The lemma is readily checked since $\Sigma\kappa(\alpha) = [p]\alpha - \alpha[p]$. We now define the differential

(15.3.7) $$M^1 \xrightarrow{\partial^1} M_0^2.$$

Here $M^1 = \text{kernel}(d^1 : C^1 \longrightarrow C^2)$ is defined by d^1 in (15.1.6). Hence we see as in (15.3.1) that $\varphi \in \text{Hom}_{-1}(R_{\mathcal{F}}, \tilde{\mathcal{A}}) = C^1$ is an element in M^1 if and only if φ^* is a derivation. We define $\partial^1(\varphi) = (A_\varphi, B_\varphi)$ by

$$\begin{aligned} A_\varphi(\alpha, x) &= (-1)^{|\alpha|}\alpha \cdot \varphi(x) - \varphi(\alpha x), \\ B_\varphi(x, \beta) &= \varphi(x) \cdot \beta - \varphi(x\beta). \end{aligned}$$

One readily checks by (15.3.2) that

15.3.8 Lemma. $(A_{u+\Sigma\varphi}, B_{u+\Sigma\varphi}) = (A_u, B_u) + \partial^1(\varphi)$.

Next we define the function

(15.3.9) $$\partial^2 : M_{\kappa_0}^2 \longrightarrow M^3,$$

$$M^3 = Hom_{-1}(\mathcal{B}_0 \otimes R_{\mathcal{B}} \oplus R_{\mathcal{B}} \otimes \mathcal{B}_0, \mathcal{A} \otimes \mathcal{A}).$$

For $(A, B) \in M_{\kappa_0}^2$ we consider the following diagrams.

(1) $$\begin{array}{ccc} \mathcal{A} & \xrightarrow{\delta} & \mathcal{A} \otimes \mathcal{A} \\ \uparrow A & & \uparrow A_\sharp \\ \mathcal{B}_0 \otimes R_{\mathcal{B}} & \xrightarrow{1 \otimes \Delta} & \mathcal{B}_0 \otimes R_{\mathcal{B}}^2 \end{array}$$

(2)
$$\begin{array}{ccc} \mathcal{A} & \xrightarrow{\delta} & \mathcal{A}\otimes\mathcal{A} \\ \uparrow{\scriptstyle B} & & \uparrow{\scriptstyle B_\sharp} \\ R_{\mathcal{B}}\otimes\mathcal{B}_0 & \xrightarrow{\Delta\otimes\Delta} & R^2_{\mathcal{B}}\otimes\mathcal{B}_0\otimes\mathcal{B}_0 \end{array}$$

Here the map $A_\sharp$ in (1) induced by A is defined by

(3)
$$\begin{aligned} A_\sharp(\alpha\otimes x\otimes\beta') &= \sum_i(-1)^{|\alpha_i''|\cdot|x|}A(\alpha_i',x)\otimes(\alpha_i''\cdot\beta'),\\ A_\sharp(\alpha\otimes\beta\otimes y) &= \sum_i(-1)^{\varepsilon_i}(\alpha_i'\cdot\beta)\otimes A(\alpha_i'',y), \end{aligned}$$

with $\varepsilon_i = |\alpha_i''||\beta| + |\alpha_i'| + |\beta|$ and

$$\Delta(\alpha) = \sum_i \alpha_i'\otimes\alpha_i'' \quad \in \mathcal{B}_0\otimes\mathcal{B}_0.$$

One can check that $A_\sharp$ is well defined, in particular, if restricted to $p(\mathcal{B}_0\otimes\mathcal{B}_0)\subset R^2_{\mathcal{B}}$. For this we use the assumption $(\kappa_0\otimes 1)\delta = \delta\kappa_0$ in (15.3.5)(6).

Moreover the map $B_\sharp$ in (2) induced by B is defined by

(4)
$$\begin{aligned} B_\sharp(x\otimes\alpha'\otimes\beta\otimes\beta') &= (-1)^{|\alpha'||\beta|}B(x,\beta)\otimes(\alpha'\cdot\beta'),\\ B_\sharp(\alpha\otimes y\otimes\beta\otimes\beta') &= \varepsilon(\alpha\cdot\beta),\otimes B(y,\beta') \end{aligned}$$

with $\varepsilon = (-1)^{|y||\beta|+|\alpha|+|\beta|}$. Using (15.3.3)(5) we see that $\mathcal{B}_\sharp$ is well defined.

Using (1) and (2) we define the function ∂^2 in (15.3.6) by

(5)
$$\begin{aligned} \partial^2(A,B) &= (A^\partial, B^\partial) \quad \text{with}\\ A^\partial &= \delta A - A_\sharp(1\otimes\Delta),\\ B^\partial &= \delta B - B_\sharp(\Delta\otimes\Delta). \end{aligned}$$

The cocycle $\xi : R_{\mathcal{F}} \longrightarrow \tilde{\mathcal{A}}\otimes\tilde{\mathcal{A}}\subset\mathcal{A}\otimes\mathcal{A}$ associated to the symmetry operator S (with $\xi = 0$ for $S = 0$) yields ∇^A_ξ, ∇^B_ξ by the formulas in $\Sigma(\mathcal{A}\otimes\mathcal{A})$,

$$\begin{aligned} \Sigma\nabla^A_\xi(\alpha\otimes x) &= (\delta\alpha)\cdot(\Sigma\xi(x)) - \Sigma\xi(\alpha\cdot x),\\ \Sigma\nabla^B_\xi(y\otimes\beta) &= (\Sigma\xi(y))\cdot(\delta\beta) - \Sigma\xi(y\cdot\beta). \end{aligned}$$

15.3.10 Theorem. *Let u be a ξ-splitting of $\mathcal{B}$; then the multiplication structure (A_u, B_u) is an element in M^2_κ which satisfies*

$$\partial^2(A_u,B_u) = (\nabla^A_\xi + L, \nabla^B_\xi)$$

where $L : \mathcal{B}_0\otimes R_{\mathcal{B}} \longrightarrow \mathcal{A}\otimes R_{\mathcal{B}} \longrightarrow \mathcal{A}\otimes\mathcal{A}$ is the left action operator of $\mathcal{B}$ in (14.1.3).

The theorem implies that the composite

$$M^1 \xrightarrow{\partial^1} M_0^2 \xrightarrow{\partial^2} M^3$$

is trivial, i.e., $\partial^2\partial^1 = 0$. In fact, we have for $\varphi \in M^1$,

$$\begin{aligned}\partial^2(A_{u+\Sigma\varphi}, B_{u+\Sigma\varphi}) &= (\nabla_\xi^A + L, \nabla_\xi^B)\\ &= \partial^2((A_u, B_u) + \partial^1\varphi)\\ &= (\nabla_\xi^A + L, \nabla_\xi^B) + \partial^2\partial^1\varphi.\end{aligned}$$

Let $M^2_{\kappa,\xi,L}$ be the subset of M^2_κ consisting of all pairs $(A, B) \in M^2_\kappa$ with $\partial^2(A, B) = (\nabla_\xi^A + L, \nabla_\xi^B)$. Then $\partial^1 M^1 = \text{image}(\partial^1)$ acts on this set by addition in M^2. Let $M^2_{\kappa,\xi,L}/\partial^1 M^1$ be the set of orbits of this action. This is a subset of $M^2/\partial^1 M^1$. If u is a ξ-splitting of $\mathcal{B}$, then (15.3.7) shows that (A_u, B_u) represents an element

$$\langle A, B\rangle \in M^2_{\kappa,\xi,L}/\partial^1 M^1 \tag{15.3.11}$$

which we call the *multiplication class* of $\mathcal{B}$ associated to the triple (κ, ξ, L). This class is independent of the choice of the ξ-splitting u. According to its construction we see:

15.3.12 Theorem. *The multiplication class determines the isomorphism type of the secondary Hopf algebra $\mathcal{B}$.*

We are now ready to prove

15.3.13 Theorem. (Uniqueness)*: For all primes $p \geq 2$ there exists up to isomorphism only one unique secondary Hopf algebra associated to the Steenrod algebra $\mathcal{A}$, the derivation κ, the symmetry operator S and the left action operator L (with $L = S = 0$ for p odd).*

Proof. The group M_0^2 acts transitively and effectively on M^2_κ, see (15.3.5)(7). Therefore the group $ker(\partial^2)/\partial^1 M^1 \subset M_0^2/\partial^1 M_1$ acts transitively and effectively on $M^2_{\kappa,\xi,L}$. In the next section we show that

$$\ker(\partial^2)/\partial^1 M^1 = 0$$

consists of a single element. Hence $M^2_{\kappa,\xi,L}/\partial^1 M^1$ consists of a single element. Thus the uniqueness theorem follows from (15.3.8). □

We can use $(A, B) \in M^2_{\kappa,\xi,L}$ for the computation of Massey products in the Steenrod algebra. Let $\alpha, \beta, \gamma \in \mathcal{A}$ with $\alpha \cdot \beta = 0$ and $\beta \cdot \gamma = 0$. Then the Massey product

$$\langle \alpha, \beta, \gamma\rangle \in \mathcal{A}^{|\alpha|+|\beta|+|\gamma|-1}/\mathcal{U}$$

is a coset of $\mathcal{U} = \alpha\mathcal{A}^{|\beta|+|\gamma|-1} + \mathcal{A}^{|\alpha|+|\beta|-1}\gamma$. An element representing the coset is obtained as follows:

15.3.14 Theorem. *Let* $(A, B) \in M^2_{\kappa,\xi,L}$ *and let* $\bar{\alpha}, \bar{\beta}, \bar{\gamma} \in \mathcal{B}_0$ *be elements representing* $\alpha, \beta, \gamma \in \mathcal{A}$. *Then*

$$A(\bar{\alpha}, \bar{\beta} \cdot \bar{\gamma}) - B(\bar{\alpha} \cdot \bar{\beta}, \bar{\gamma}) \in \langle \alpha, \beta, \gamma \rangle.$$

This result shows that Massey products in $\mathcal{A}$ are completely determined by an element $(A, B) \in M^2_{\kappa,\xi,L}$ which in turn can be obtained by solving equations in $\mathcal{A}$. Hence we get a computational method to determine $\langle \alpha, \beta, \gamma \rangle$ solving an old problem of Kristensen and Madsen [Kr4], [KrM2].

15.4 Proof of the uniqueness theorem

According to (15.3.9) the uniqueness theorem is a consequence of the following result.

15.4.1 Theorem. *The following sequence is exact:*

$$M^1 \xrightarrow{\partial^1} M^2_0 \xrightarrow{\partial^2} M^3.$$

Here M^2_0 is the $\mathbb{F}$-vector space of all pairs (A, B) with

$$A : \mathcal{F}_0 \otimes R_{\mathcal{F}} \longrightarrow \tilde{\mathcal{A}} \subset \mathcal{A},$$
$$B : R_{\mathcal{F}} \otimes \mathcal{F}_0 \longrightarrow \tilde{\mathcal{A}} \subset \mathcal{A}$$

satisfying equations (15.3.5)(1)... (4), see (15.3.5)(8). We have

$$\partial^2(A, B) = 0 \tag{15.4.2}$$

if and only if the dual maps

$$A^* : \mathcal{A}^* \longrightarrow \mathcal{F}^*_0 \otimes R^*_{\mathcal{F}},$$
$$B^* : \mathcal{A}^* \longrightarrow R^*_{\mathcal{F}} \otimes \mathcal{F}^*_0$$

are $(\mu^* : \mathcal{A}^* \longrightarrow \mathcal{A}^* \otimes \mathcal{A}^*)$-derivations of degree $+1$. That is , for $a, b \in \mathcal{A}^*$ we have

$$A^*(ab) = A^*(a) \cdot \mu^*(b) + (-1)^{|a|} \mu^*(a) \cdot A^*(b)$$

and the same formula holds if we replace A by B. Here we use the action of $\mathcal{A}^*$ on $\mathcal{F}^*_0$ and $R^*_{\mathcal{F}}$ defined in the exact sequences following (15.2.2) above.

Next we describe the equations in (15.3.5) in terms of commutative diagrams which can be easily dualized. Equation (15.4.3)(1) corresponds to the diagram

(15.4.3) (1)

$$\begin{array}{ccc}
\mathcal{A} & \xleftarrow{\quad\mu\quad} & \mathcal{A} \otimes \mathcal{A} \\
\uparrow{\scriptstyle (A,B)} & & \uparrow{\scriptstyle A\otimes q - \tau(q\otimes B)} \\
\mathcal{F}_0 \otimes R_{\mathcal{F}} \otimes R_{\mathcal{F}} \otimes \mathcal{F}_0 & \xleftarrow[(1\otimes\mu, -\mu\otimes 1)]{} & \mathcal{F}_0 \otimes R_{\mathcal{F}} \otimes \mathcal{F}_0
\end{array}$$

where τ corresponds to the interchange of Σ and μ is the multiplication map and $q : \mathcal{F} \to \mathcal{A}$ is the quotient map. Next (15.3.5)(2) corresponds to the diagram.

(2)

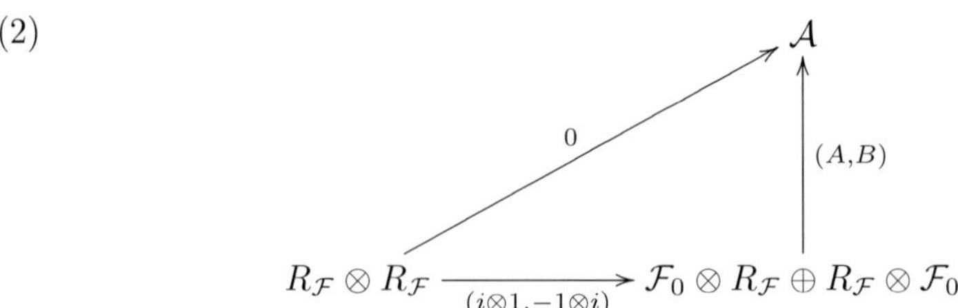

where $i : R_{\mathcal{F}} \subset \mathcal{F}_0$ is the inclusion. Moreover (15.3.5)(3), (4) corresponds to the diagram

(3)
$$\begin{array}{ccc} \mathcal{A} \otimes \mathcal{A} & \xrightarrow{\mu} & \mathcal{A} \\ \uparrow{\scriptstyle (\tau(q\otimes A), B\otimes q)} & & \uparrow{\scriptstyle (A,B)} \\ \mathcal{F}_0 \otimes \mathcal{F}_0 \otimes R_{\mathcal{F}} \oplus R_{\mathcal{F}} \otimes \mathcal{F}_0 \otimes \mathcal{F}_0 & \xrightarrow{\bar{\mu}} & \mathcal{F}_0 \otimes R_{\mathcal{F}} \oplus R_{\mathcal{F}} \otimes \mathcal{F}_0 \end{array}$$

with $\bar{\mu} = (\mu \otimes 1 - 1 \otimes \mu) \oplus (\mu \otimes 1 - 1 \otimes \mu)$. In the following definition we dualize the diagrams (1),(2),(3) above.

Let K_2 be the set of all $(\mu^* : \mathcal{A}^* \longrightarrow \mathcal{A}^* \otimes \mathcal{A}^*)$-derivations

(15.4.4) $$C = (A^*, B^*) : \mathcal{A}^* \longrightarrow \mathcal{F}_0^* \otimes R_{\mathcal{F}}^* \oplus R_{\mathcal{F}}^* \otimes \mathcal{F}_0^*$$

satisfying

(1) $$(1 \otimes \mu^*, -\mu^* \otimes 1)C = (A^* \otimes q^* - (q^* \otimes B^*)\tau^*)\mu^*,$$

(2) $$(i^* \otimes 1, -1 \otimes i^*)C = 0,$$

(3) $$\bar{\mu}C = ((q^* \otimes A^*)\tau^*, B^* \otimes q^*)\mu^*$$

with $\bar{\mu}^* = (\mu^* \otimes 1 - 1 \otimes \mu^*) \oplus (\mu^* \otimes 1 - 1 \otimes \mu^*)$. These equations correspond to the dualization of the diagram (15.4.3)(1), (2), (3). Hence we get the isomorphism

(15.4.5) $$ker(\partial^2 : M_0^2 \to M^3) \cong K_2$$

which carries (A, B) to $C = (A^*, B^*)$, see (15.3.5). Moreover the following diagram commutes.

(15.4.6)
$$\begin{array}{ccc} M^1 & \xrightarrow{\partial^1} & ker(\partial^2 : M_0^2 \to M^3) \\ \downarrow{\scriptstyle \cong} & & \downarrow{\scriptstyle \cong} \\ K_1 & \xrightarrow{\partial_1} & K_2 \end{array}$$

Here K_1 is the set of all derivations $\varphi : \mathcal{A}^* \longrightarrow R^*_{\mathcal{F}}$ of degree $+1$ and $M^1 \cong K^1$ carries γ to the dual γ^* and ∂_1 is the dual of ∂^1 in (15.3.6). That is, ∂_1 carries φ to the pair $(A^*_\varphi, B^*_\varphi)$ with

(15.4.7)
$$\begin{aligned} A^*_\varphi &= (q^* \otimes \varphi)\tau^*\mu^* - \mu^*\varphi, \\ B^*_\varphi &= (\varphi \otimes q^*)\mu^* - \mu^*\varphi. \end{aligned}$$

Here we use the maps in the following diagrams.

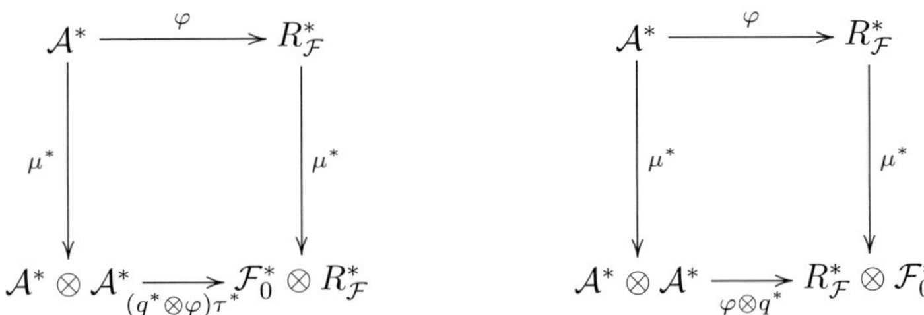

By a result of Milnor [Mn] the algebra $\mathcal{A}^*$ is a free commutative graded algebra generated by elements x_i, $i \geq 1$, of degree

(15.4.8)
$$n_i = |x_i| = \begin{cases} 2^i - 1 & \text{for } p \text{ even}, \\ 2p^j - 1 & \text{for } p \text{ odd}, \ i = 2j+1, \\ 2p^j - 2 & \text{for } p \text{ odd}, \ i = 2j. \end{cases}$$

We have $n_1 = 1 < n_2 < n_3 < \cdots$. Using the generators $x_1, x_2, \ldots$ of $\mathcal{A}^*$ we obtain the derivations $\varphi : \mathcal{A}^* \longrightarrow R^*_{\mathcal{F}}$ in K_1 as follows.

For each element $a \in R^*_{\mathcal{F}}$ with $|a| = n_i + 1$ there is a unique derivation

(15.4.9)
$$\varphi(a) : \mathcal{A}^* \longrightarrow R^*_{\mathcal{F}}$$

of degree $+1$ satisfying $\varphi(a)(x_i) = a$ and $\varphi(a)(x_j) = 0$ for $j \neq i$. Moreover each derivation $\varphi : \mathcal{A}^* \longrightarrow R^*_{\mathcal{F}}$ of degree $+1$ yields a sequence of elements $a_i = \varphi(x_i)$ such that

$$\varphi = \sum_{i=1}^{\infty} \varphi(a_i).$$

Here the infinite sum is well defined since $\varphi(a_i)(b) = 0$ for $b \in \mathcal{A}^*$ with $|b| < n_i$. Now Theorem (15.4.1) is a consequence of the following result.

15.4.10 Theorem. *∂_1 in (15.4.6) is surjective.*

Proof. We proceed inductively as follows. Let $C \in K_2$. We construct inductively elements $a_1, a_2, \ldots$ such that (see (15.4.9))

(1)
$$C_i = C - C_{\varphi_i} \ \text{ with } \ \varphi_i = \varphi(a_1) + \cdots + \varphi(a_i)$$

satisfies $C_i(\eta) = 0$ for $\eta \in \mathcal{A}^*$ with $|\eta| < n_{i+1}$. This shows that $C = C_\varphi$ so that $\partial_1(\varphi) = C$ and hence ∂_1 is surjective. We have $C_i(\eta) = 0$ for $|\eta| < n_{i+1}$ if and only if (see (15.4.8))

(2) $$C_i(x_j) = 0 \text{ for } j \leq i$$

since C_i is a derivation. For the case $i = 1$ we observe

(3) $$C \in K_1 \text{ satisfies } C(x_1) = 0.$$

Hence for any a_1, for example $a_1 = 0$, we get $C_1 = C - C_{\varphi_1}$ with $C_1(x_1) = 0$. Given C_{m-1} we obtain a_m and C_m in (1) by the following lemma. □

15.4.11 Lemma. *Suppose $C \in K_1$ satisfies $C(x_i) = 0$ for $i < m$. Then there exists $a = a_m$ with $|a| = n_m + 1$ such that $\bar{C} = C - C_{\varphi(a)}$ satisfies $\bar{C}(x_i) = 0$ for $i \leq m$.*

Proof of (15.4.11). We consider the element

$$C(x_m) = (x, y) \in \mathcal{F}_0^* \otimes R_\mathcal{F}^* \oplus R_\mathcal{F}^* \otimes \mathcal{F}_0^*.$$

The diagonal μ^* of $\mathcal{A}^*$ satisfies

(*) $$\mu^*(x_m) = 1 \otimes x_m + x_m \otimes 1 + \sum_t \xi_t' \otimes \xi_t''$$

with $|\xi_t'|, |\xi_t''| < |x_m|$. Therefore $C = (A^*, B^*)$ satisfies

$$\begin{aligned} A^*(\xi_t') = 0 = B^*(\xi_t''), \\ A^*(x_m) = x, \quad B^*(x_m) = y. \end{aligned}$$

Now (15.4.4)(1), (2), (3) yield the following equations.

(1) $$\begin{aligned} (1 \otimes \mu^*, -\mu^* \otimes 1)C(x_m) &= (1 \otimes \mu^*)x - (\mu^* \otimes 1)y \\ &= A^*(x_m) \otimes 1 - 1 \otimes B^*(x_m) \\ &= x \otimes 1 - 1 \otimes y, \end{aligned}$$

(2) $$(i^* \otimes 1)x - (1 \otimes i^*)y = 0,$$

(3) $$\begin{aligned} (\mu^* \otimes 1 - 1 \otimes \mu^*)x &= 1 \otimes A^*(x_m) = 1 \otimes x, \\ (\mu^* \otimes 1 - 1 \otimes \mu^*)y &= B^*(x_m) \otimes 1 = y \otimes 1. \end{aligned}$$

We shall show that the equations (1), (2), (3) on the pair (x, y) imply that there is $a = a_m$ with $(x, y) = C_{\varphi(a)}(x_m)$ so that $\bar{C} = C - C_{\varphi(a)}$ satisfies the proposition. We have by (15.4.7)

(4) $$C_{\varphi(a)}(x_m) = (-\mu^*(a) + 1 \otimes a, -\mu^*(a) + a \otimes 1).$$

Here again we use $(*)$ above. We associate to the equations (1)...(4) the morphisms in the following diagram.

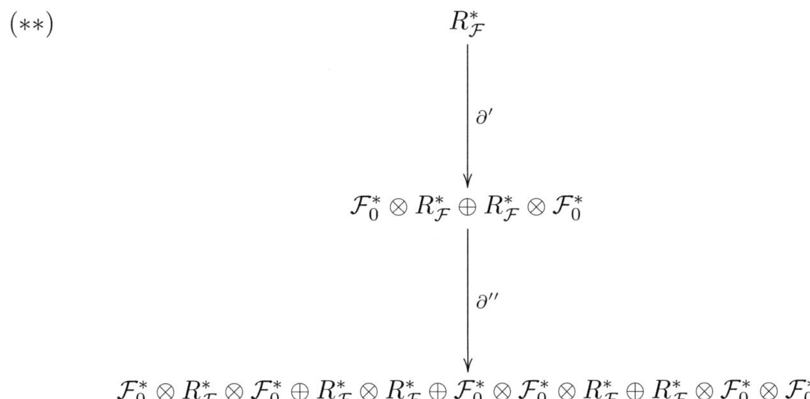

Recall that 1 denotes the identity of an object and also the unit $1 = \tilde{1}$ of an algebra. According to (4) we set

$$\partial'(a) = (-\mu^*(a) + \tilde{1} \otimes a, -\mu^*(a) + a \otimes \tilde{1}) \tag{5}$$

and we define the coordinates of $\partial''(x, y)$ as in (1), (2) and (3) respectively by

$$\partial''(x,y)_1 = (1 \otimes \mu^*)x - x \otimes \tilde{1} - (\mu^* \otimes 1)y + \tilde{1} \otimes y, \tag{6}$$

$$\partial''(x,y)_2 = (i^* \otimes 1)x - (1 \otimes i^*)y, \tag{7}$$

$$\partial''(x,y)_3 = (\mu^* \otimes 1)x - (1 \otimes \mu^*)x - \tilde{1} \otimes x, \tag{8}$$

$$\partial''(x,y)_4 = (\mu^* \otimes 1)y - (1 \otimes \mu^*)y - y \otimes \tilde{1}. \tag{9}$$

We have $\partial''(x, y) = 0$ if and only if (1), (2), (3) are satisfied. Moreover there exists $a = a_m$ with $(x, y) = C_{\varphi(a)}(x_m) = \partial'(a)$ if the sequence $(**)$ above is exact, that is image(∂') = kernel(∂''), in degree $n_m + 1$. The sequence $(**)$ is exact if and only if the following dual sequence $(***)$ is exact.

$(***)$

$$\begin{array}{c} R_{\mathcal{F}} \\ \uparrow {\scriptstyle d' = (\partial')^*} \\ \mathcal{F}_0 \otimes R_{\mathcal{F}} \oplus R_{\mathcal{F}} \otimes \mathcal{F}_0 \\ \uparrow {\scriptstyle d' = (\partial'')^*} \\ \mathcal{F}_0 \otimes R_{\mathcal{F}} \otimes \mathcal{F}_0 \oplus R_{\mathcal{F}} \otimes R_{\mathcal{F}} \oplus \mathcal{F}_0 \otimes \mathcal{F}_0 \otimes R_{\mathcal{F}} \oplus R_{\mathcal{F}} \otimes \mathcal{F}_0 \otimes \mathcal{F}_0 \end{array}$$

Here d' and d'' are dual to ∂' and ∂'' respectively and hence d' and d'' can be described as follows. We have the augmentation

$$\varepsilon : \mathcal{F}_0 \longrightarrow \mathbb{F}$$

which is dual to the inclusion $\mathbb{F} \subset \mathcal{F}_0^*$ given by the unit $\tilde{1} \in \mathcal{F}_0^*$ of the algebra $\mathcal{F}_0^*$. Therefore $\tilde{1}$ in (5)...(7) corresponds by dualization to ε. This yields the following formulas for d' and d'' with $\alpha, \alpha', \beta, \beta' \in \mathcal{F}_0$ and $x, y \in R_{\mathcal{F}}$.

(10) $d'(\alpha \otimes x + y \otimes \beta) = \varepsilon\alpha \otimes x - \alpha x + y \otimes \varepsilon\beta - y\beta,$

(11) $d''(\alpha \otimes x \otimes \beta) = (\alpha \otimes x\beta - (\alpha \otimes x) \cdot \varepsilon(\beta), -\alpha x \otimes \beta + \varepsilon(\alpha) \cdot (x \otimes \beta)),$

(12) $d''(x \otimes y) = (ix \otimes y, x \otimes iy),$

(13) $d''(\alpha \otimes \alpha' \otimes x) = \alpha\alpha' \otimes x - \alpha \otimes \alpha' x - \varepsilon(\alpha) \cdot (\alpha' \otimes x),$

(14) $d''(y \otimes \beta \otimes \beta') = y\beta \otimes \beta' - y \otimes \beta\beta' - (y \otimes \beta) \cdot \varepsilon(\beta').$

Now let

$$\tilde{\mathcal{F}}_0 = \text{kernel}(\varepsilon : \mathcal{F}_0 \longrightarrow \mathbb{F})$$

be the augmentation ideal. Then we have $\mathcal{F}_0 = \mathbb{F} \oplus \tilde{\mathcal{F}}_0$ and this shows that $(***)$ is exact in degree $n_m + 1$ if and only if the left-hand column in the following diagram is exact in degree $n_m + 1$.

$(****)$

$$\begin{array}{ccc}
R_{\mathcal{F}} & \longrightarrow & \tilde{\mathcal{F}}_0 \\
\uparrow d_2 & & \uparrow d_2 \\
(\tilde{\mathcal{F}}_0 \otimes R_{\mathcal{F}} \oplus R_{\mathcal{F}} \otimes \tilde{\mathcal{F}}_0)/\sim\ = \tilde{\mathcal{F}}_0 \otimes R_{\mathcal{F}} + R_{\mathcal{F}} \otimes \tilde{\mathcal{F}}_0 & \longrightarrow & \tilde{\mathcal{F}}_0 \otimes \tilde{\mathcal{F}}_0 \\
\uparrow d_3 & & \uparrow d_3 \\
\tilde{\mathcal{F}}_0 \otimes R_{\mathcal{F}} \otimes \tilde{\mathcal{F}}_0 + \tilde{\mathcal{F}}_0 \otimes \tilde{\mathcal{F}}_0 \otimes R_{\mathcal{F}} + R_{\mathcal{F}} \otimes \tilde{\mathcal{F}}_0 \otimes \tilde{\mathcal{F}}_0 & \longrightarrow & \tilde{\mathcal{F}}_0 \otimes \tilde{\mathcal{F}}_0 \otimes \tilde{\mathcal{F}}_0
\end{array}$$

Here the horizontal arrows are the inclusions and $d_2 = \mu$ is the multiplication and d_3 is given by

(15) $$d_3(\alpha \otimes \beta \otimes \gamma) = \alpha\beta \otimes \gamma - \alpha \otimes \beta\gamma.$$

The equivalence relation is defined by $(ix) \otimes y \sim x \otimes (iy)$. Now we observe that the right-hand column of the diagram is part of the bar construction $\bar{B}\mathcal{F}_0$ of the free algebra $\mathcal{F}_0$, compare for example page 32[A]. Moreover the projection $q : \mathcal{F}_0 \twoheadrightarrow \mathcal{A}$ induces the short exact sequence of chain complexes

(16) $$0 \longrightarrow \bar{K} \longrightarrow \bar{B}\mathcal{F}_0 \xrightarrow{q_*} \bar{B}A \longrightarrow 0$$

where K is the kernel of q_*. It is easy to see that the left-hand column of $(****)$ is part of the chain complex $\bar{K}$. The short exact sequence (16) induces the long exact sequence of homology group

$$H_3\bar{B}\mathcal{F}_0 \longrightarrow H_3\bar{B}\mathcal{A} \longrightarrow H_2\bar{K} \longrightarrow H_2\bar{B}\mathcal{F}_0$$

where $H_n \bar{B}\mathcal{F}_0 = 0$ for $n \geq 2$ since $\mathcal{F}_0$ is a free algebra. Hence we have the isomorphism

$$H_2 K \cong H_3 \bar{B}\mathcal{A} \tag{17}$$

so that the left-hand column of $(****)$ is exact if and only if (17) is trivial in degree $n_m + 1$. Now

$$(H_3 \bar{B}\mathcal{A})^t = Tor^{\mathcal{A}}_{s,t}(\mathbb{F}, \mathbb{F})$$

is dual to the cohomology

$$H^3(\bar{B}\mathcal{A}, \mathbb{F})^t = Ext^{\mathcal{A}}_{s,t}(\mathbb{F}, \mathbb{F})$$

of the Steenrod algebra $\mathcal{A}$, see page 28 of [A]. Therefore (17) is trivial in degree $n_m + 1$ since we can use the following result. This completes the proof of (15.4.10). □

15.4.12 Proposition. *The cohomology $H^3(\mathcal{A})$ of the Steenrod algebra $\mathcal{A}$ is trivial in degree $n_i + 1$ for $i \geq 1$ where n_i is defined in* (15.4.8).

Proof. Compare [A], [Ta], [Li], [ShY], [No]. In fact, Tangora describes in 1.2[Ta] a complete list of algebra generators which contribute to $H^3(\mathcal{A})$. The degree of these generators implies the proposition. Tangora proves the result for the prime $p = 2$. For odd primes p Liulevicius [Li] proves a result describing a similar list of generators contributing to $H^3(\mathcal{A})$. That this is a complete list needs an extension of Tangora's argument to the case of odd primes. □

15.5 Right equivariant cocycle of $\mathcal{B}$

A splitting

$$u : R_{\mathcal{B}} \longrightarrow \mathcal{B}_1$$

of $\mathcal{B}$ (as defined in (15.1.1)) is a *right equivariant splitting* if $u(x \cdot \beta) = u(x) \cdot \beta$ for $x \in R_{\mathcal{B}}, \beta \in \mathcal{B}_0$ or equivalently if $B_u(x \otimes \beta) = 0$, see (15.3.2). Moreover a cocycle

$$\xi : R_{\mathcal{F}} \longrightarrow \tilde{\mathcal{A}} \otimes \tilde{\mathcal{A}}$$

in the Δ-class is a *right equivariant cocycle* if $\xi(x \cdot \beta) = \xi(x) \cdot \delta(\beta)$ for $x \in R_{\mathcal{F}}, \beta \in \mathcal{F}_0$ or equivalently if $\nabla^B_\xi = 0$, see (15.3.10).

15.5.1 Theorem. *If the Δ-class in* (15.2.1) *contains a right equivariant cocycle ξ, then there exists a right equivariant ξ-splitting of $\mathcal{B}$.*

Proof of (15.5.1). We have to show that there exists

$$(A, B) \in M^2_{\kappa,\xi,L}$$

with $B = 0$. We construct (A, B) inductively. Since $M^2_{\kappa,\xi,L}$ is non-empty there is an element (A, B) with $B(x, \beta) = 0$ for $|x \otimes \beta| \leq 2$. (Here $\beta \in \mathbb{F}$ so that we can use (15.3.5)(4) for $\beta = \beta' = 1$.)

Next we assume inductively that there exists

$$(A, B) \in M^2_{\kappa,\xi,L} \text{ with}$$

(1) $$B(x, \beta) = 0 \text{ for } |x \otimes \beta| < N$$

with $N \geq 4$. Then we get for $x \otimes \beta\beta' \in R_{\mathcal{F}} \otimes \mathcal{F}_0$ with $|x \otimes \beta\beta'| = N$ the formula

(2) $$B(x, \beta\beta') = B(x\beta, \beta') \text{ for } |\beta'| > 0.$$

Therefore

(3) $$B : (R_{\mathcal{F}} \otimes \mathcal{F}_0)_N \longrightarrow \mathcal{A}$$

is uniquely determined by its restriction

(4) $$\bar{B} : (R_{\mathcal{F}} \otimes E)_N \longrightarrow \mathcal{A}$$

where $E \subset \mathcal{F}_0$ is the submodule generated by $E_{\mathcal{A}}$. We now observe by the following commutative diagram that the multiplication map μ with $\mu(x \otimes \beta) = x \cdot \beta$ is injective.

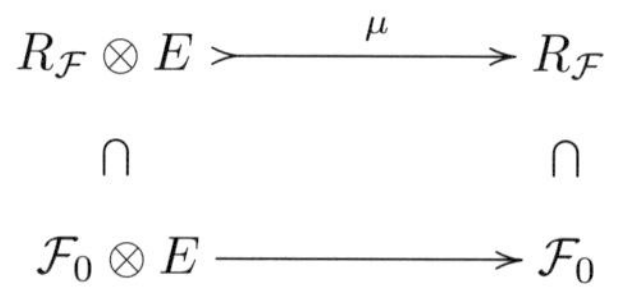

$$\begin{array}{ccc} R_{\mathcal{F}} \otimes E & \overset{\mu}{\rightarrowtail} & R_{\mathcal{F}} \\ \cap & & \cap \\ \mathcal{F}_0 \otimes E & \longrightarrow & \mathcal{F}_0 \end{array}$$

Let $K = \text{kernel}(\tilde{\delta} : \mathcal{A} \longrightarrow \mathcal{A} \otimes \mathcal{A})$ be the kernel of the reduced diagonal $\tilde{\delta}$. Then we can choose a map γ as in the following commutative diagram.

(5) $$\begin{array}{ccc} (R_{\mathcal{F}} \otimes E)_N & \overset{\bar{B}}{\longrightarrow} & K \subset \mathcal{A} \\ & {}_{\mu}\searrow \quad \nearrow_{\gamma} & \\ & (R_{\mathcal{F}})_N & \end{array}$$

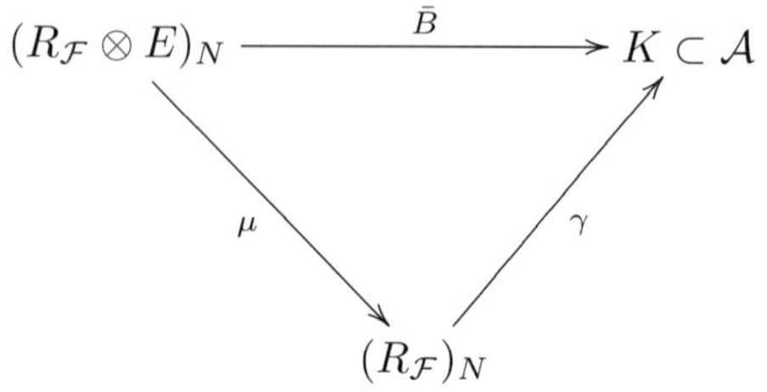

This is possible since μ being injective is a direct summand of the vector space. We define $\gamma(x) = 0$ for $|x| < N$ so that

(6) $$\bar{B}(x, \beta) = \gamma(x\beta)$$

for $|x \otimes \beta| = N$, $\beta \in E$. Here (6) is satisfied by (5). Now (2) and (6) show that

(7) $$B(x, \beta) = \gamma(x\beta) - \gamma(x) \cdot \beta$$

for $|x \otimes \beta| = N$ and $\beta \in \mathcal{F}_0$. The lemma in (15.5.2) below shows that there is a map

$$(8) \qquad \gamma : R_{\mathcal{F}} \longrightarrow \tilde{\mathcal{A}} \subset \mathcal{A}$$

satisfying $\gamma \in M^1$ such that γ is an extension of γ in (5). Now we define

$$\begin{aligned} A'(\alpha, x) &= A(\alpha, x) - (\gamma(\alpha x) - (-1)^{|\alpha|}\alpha\gamma(x)), \\ B'(x, \beta) &= B(x, \beta) - (\gamma(x\beta) - \gamma(x) \cdot \beta). \end{aligned}$$

Then (A', B') is also in $M^2_{\kappa,\xi,L}$ and (7) shows that $B'(x, \beta) = 0$ for $|x \otimes \beta| \leq N$. Hence we can proceed inductively to obtain an element $(A', B') \in M^2_{\kappa,\xi,L}$ with $B' = 0$. □

15.5.2 Lemma. *The map γ in* (8) *exists.*

Proof. We first observe that γ satisfies $\delta\gamma x = \gamma_\sharp \Delta x$ in degree $|x| \leq N$. Hence the dual of γ yields a map

$$\gamma^*_{<N} : \mathcal{A}^* \longrightarrow R^*_{\mathcal{F}}$$

defined in degree $< N$ such that $\gamma^*_{<N}$ is a derivation in degree $< N$. Now $\mathcal{A}^*$ is a free commutative graded algebra generated by the set M of Milnor generators. Hence there is a unique derivation γ^* defined for $\xi \in M$ by

$$\gamma^*(\xi) = \begin{cases} \gamma^*_{<N}(\xi) & \text{for } |\xi| < N, \\ 0 & \text{for } |\xi| \geq N. \end{cases}$$

Hence the dual of γ^* yields the map γ in (8). □

Chapter 16

Computation of the Secondary Hopf Algebra $\mathcal{B}$

We describe an algorithm which computes the secondary diagonal of $\mathcal{B}$ and the multiplication in the pair algebra $\mathcal{B}$. In the final section we give a multiplication table in low degrees. This yields a computation of triple Massey products.

16.1 Right equivariant splitting of $\mathcal{B}$

We first observe the following property of the ideal $R_{\mathcal{F}} \subset \mathcal{F}_0$.

16.1.1 Proposition. *The ideal $R_{\mathcal{F}}$ is a free right $\mathcal{F}_0$-module.*

Proof. One can find the result in the book of Cohn [Co] section 2,4. But it is also easy to see that a basis B of $R_{\mathcal{F}}$ as a free right $\mathcal{F}_0$-module is inductively obtained as follows. Let B_2 be the set which contains the unique element Sq^1Sq^1 or $\beta\beta$ of degree 2 and let $C_2 = \phi$ be the empty set and let $D_2 = B_2 \cup C_2$. Assume linearly independent subsets of elements of degree i,

$$B_i, D_i \subset \mathcal{F}_0 \text{ with } B_i \subset D_i$$

are defined for $i \leq n-1$, $n \geq 3$. Then let C_n be the union

$$C_n = \bigcup_{i=2}^{n-1} D_i \cdot Sq^{n-i}.$$

One can check that C_n is linearly independent and we can choose a basis D_n of all elements of degree n in $\mathcal{F}_0$, containing C_n. Let $B_n = D_n - C_n$ be the complement. Then $B = B_2 \cup B_3 \cup \cdots \cup B_n \cdots$ is a basis of the free right $\mathcal{F}_0$-module $R_{\mathcal{F}}$. We choose the elements of B_n as follows. Consider the set G_n of all elements $\alpha[a,b]$ of degree n with $\alpha \in \mathrm{Mon}(E_{\mathcal{A}})$. We choose lexicographical ordering of this set so that by this ordering $G_n = \{x_1, x_2, \ldots\}$. Let $k \geq 1$. If x_k is not a linear combination of $x_1, \ldots, x_{k-1}$ and of elements in C_n then x_k is an element in B_n and these are all elements of B_n. □

We now choose a basis B of the free right $\mathcal{F}_0$-module $R_{\mathcal{F}}$ and we choose a lift as in the following diagram.

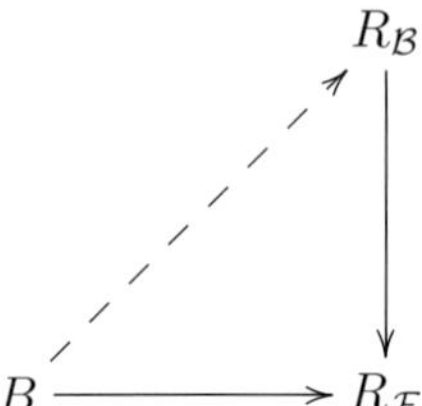

Then we get the induced equivariant injections

$$B \otimes \mathcal{B}_0 \longrightarrow R_{\mathcal{B}},$$

$$p\mathcal{B}_0 \longrightarrow R_{\mathcal{B}},$$

with $p\mathcal{B}_0 \cap B \otimes \mathcal{B}_0 = p(B \otimes \mathcal{B}_0)$. Hence we have the push out diagram

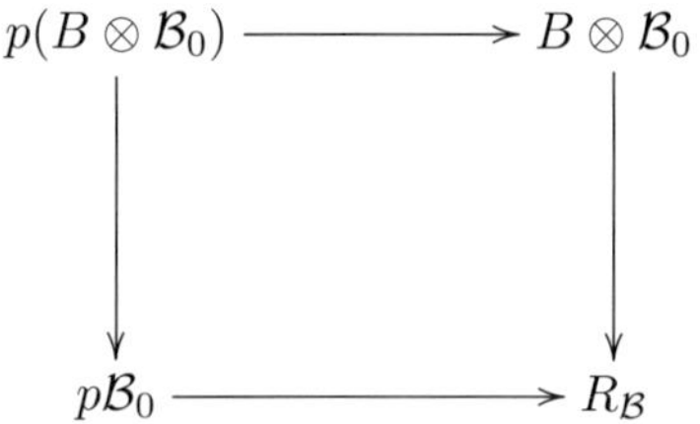

which is used for the construction of a splitting in the next result.

16.1.2 Theorem. *There is a splitting u of $\mathcal{B}$, see* (15.1.1), *which is right equivariant with respect to the action of $\mathcal{B}_0$.*

Proof. We choose a lift as in the diagram

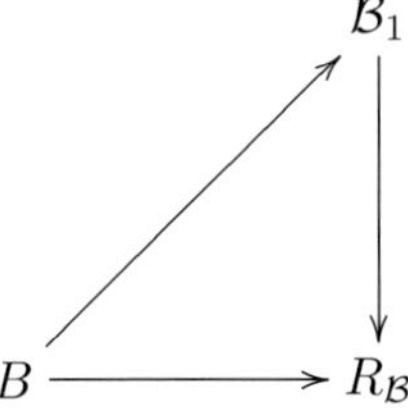

which defines the right equivariant map

$$u_1 : B \otimes \mathcal{B}_0 \longrightarrow \mathcal{B}_1.$$

Moreover we define

$$u_2 : p\mathcal{B}_0 \longrightarrow \mathcal{B}_1$$

as in (15.1.1) by $u(p\alpha) = [p] \cdot \alpha$. Then u_1 and u_2 coincide on the intersection $p(B \otimes \mathcal{B}_0)$ since for $x \in B$, $\alpha \in \mathcal{B}_0$,

$$u_1(p(x \otimes \alpha)) = p(u_1(x) \cdot \alpha) = [p] \cdot x \cdot \alpha = u_2(p(x \cdot \alpha)).$$

Therefore the section $u = u_1 \cup u_2$ is well defined and right equivariant. □

We consider the following diagram where u is a splitting of $\mathcal{B}$ as in (15.1.1) and $u_\sharp$ is defined as in (15.1.3).

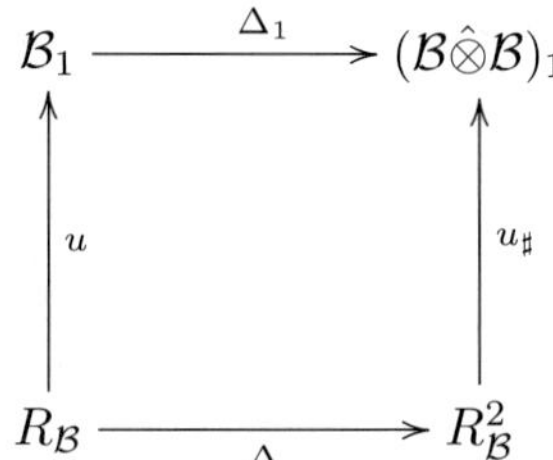

16.1.3 Theorem. *If the prime p is odd, there exists a splitting u of $\mathcal{B}$ which is right equivariant with respect to the action of $\mathcal{B}_0$ and for which the diagram commutes.*

Proof. If p is odd we know that the symmetry operator $S = 0$ is trivial. Hence the Δ-class of $\mathcal{B}$ is trivial by (15.2.1). Hence by definition of the Δ-class we obtain the result. Here we use (15.5.1). □

We now consider the case that the prime p is even. In this case the symmetry operator S is non-trivial and computed in (14.5.2). We consider the following diagram where ξ is a linear map of degree -1.

(16.1.4)

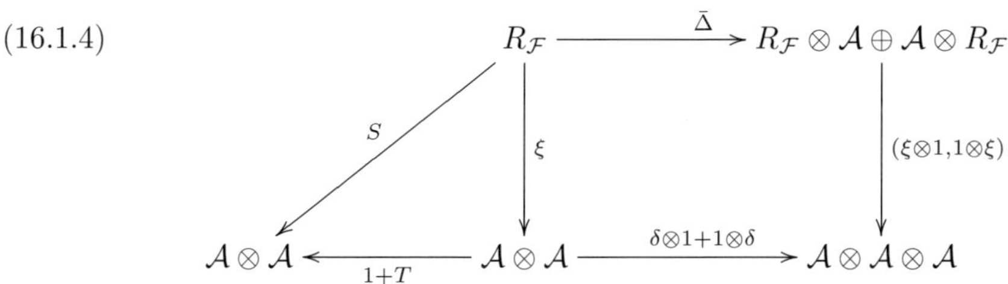

Here $\bar{\Delta}$ is induced by $\Delta : \mathcal{F}_0 \to \mathcal{F}_0 \otimes \mathcal{F}_0$.

16.1.5 Theorem. *Assume the prime p is even. Then there is a right $\mathcal{F}_0$-equivariant map of degree -1,*

$$\xi : R_\mathcal{F} \longrightarrow \tilde{\mathcal{A}} \otimes \tilde{\mathcal{A}} \subset \mathcal{A} \otimes \mathcal{A},$$

for which the diagram above commutes. Moreover for each such ξ there exists a splitting $u = u_\xi$ associated to ξ such that u is right $\mathcal{B}_0$-equivariant and

$$\Delta_1 u = u_\sharp \Delta_0 + \Sigma\xi$$

holds where we use $R_\mathcal{B} \twoheadrightarrow R_\mathcal{F}$.

Proof. Using the right equivariant splitting u in (16.1.2) we obtain a right equivariant ξ by the formula $\Delta_1 u = u_\sharp \Delta_0 + \Sigma\xi$. Moreover given ξ we obtain u_ξ by (15.5.1). □

The diagonal $\bar{\Delta}$ in (16.1.4) has a left and a right part Δ_L and Δ_R respectively, so that

$$\begin{aligned} \bar{\Delta} = (\Delta_R, \Delta_L) : R_\mathcal{F} &\longrightarrow R_\mathcal{F} \otimes \mathcal{A} \oplus \mathcal{A} \otimes R_\mathcal{F} \\ \text{with} \quad \Delta_R : R_\mathcal{F} &\longrightarrow R_\mathcal{F} \otimes \mathcal{A}, \\ \Delta_L : R_\mathcal{F} &\longrightarrow \mathcal{A} \otimes R_\mathcal{F} \end{aligned} \tag{16.1.6}$$

satisfying $T\Delta_L = \Delta_R$. We define

$$\begin{cases} \Delta' : R_\mathcal{F} \longrightarrow R_\mathcal{F} \otimes \tilde{\mathcal{A}} \\ \text{by} \quad \Delta_R(x) = x \otimes 1 + \Delta'(x) \end{cases}$$

so that the reduced diagonal $\tilde{\bar{\Delta}}$ is given by $(\Delta', T\Delta')$. A list of values $\Delta'[a, b]$ for the prime 2 is given as follows.

$$\Delta'[1,1] = 0$$

$$\Delta'[1,2] = [1,1] \otimes \mathrm{Sq}^1$$

$$\begin{aligned} \Delta'[1,3] &= [1,1] \otimes \mathrm{Sq}^2 + [1,2] \otimes \mathrm{Sq}^1 \\ \Delta'[2,2] &= [1,1] \otimes \mathrm{Sq}^2 + [1,2] \otimes \mathrm{Sq}^1 \end{aligned}$$

$$\begin{aligned} \Delta'[1,4] &= [1,1] \otimes \mathrm{Sq}^3 + [1,2] \otimes \mathrm{Sq}^2 + [1,3] \otimes \mathrm{Sq}^1 \\ \Delta'[2,3] &= [1,3] \otimes \mathrm{Sq}^1 + [2,2] \otimes \mathrm{Sq}^1 \\ \Delta'[3,2] &= [1,1] \otimes \mathrm{Sq}^2\,\mathrm{Sq}^1 + [1,2] \otimes \mathrm{Sq}^2 + [2,2] \otimes \mathrm{Sq}^1 \end{aligned}$$

$$\begin{aligned} \Delta'[1,5] &= [1,1] \otimes \mathrm{Sq}^4 + [1,2] \otimes \mathrm{Sq}^3 + [1,3] \otimes \mathrm{Sq}^2 + [1,4] \otimes \mathrm{Sq}^1 \\ \Delta'[2,4] &= [1,1] \otimes \mathrm{Sq}^4 + [1,2] \otimes \mathrm{Sq}^3 + [2,2] \otimes \mathrm{Sq}^2 + [1,4] \otimes \mathrm{Sq}^1 + [2,3] \otimes \mathrm{Sq}^1 \\ \Delta'[3,3] &= [1,1] \otimes \mathrm{Sq}^4 + [1,1] \otimes \mathrm{Sq}^3\,\mathrm{Sq}^1 + [1,2] \otimes \mathrm{Sq}^2\,\mathrm{Sq}^1 + [1,3] \otimes \mathrm{Sq}^2 + [2,3] \otimes \mathrm{Sq}^1 \\ &\quad + [3,2] \otimes \mathrm{Sq}^1 \end{aligned}$$

$$\begin{aligned} \Delta'[1,6] &= [1,1] \otimes \mathrm{Sq}^5 + [1,2] \otimes \mathrm{Sq}^4 + [1,3] \otimes \mathrm{Sq}^3 + [1,4] \otimes \mathrm{Sq}^2 + [1,5] \otimes \mathrm{Sq}^1 \\ \Delta'[2,5] &= [1,3] \otimes \mathrm{Sq}^3 + [2,2] \otimes \mathrm{Sq}^3 + [2,3] \otimes \mathrm{Sq}^2 + [1,5] \otimes \mathrm{Sq}^1 + [2,4] \otimes \mathrm{Sq}^1 \end{aligned}$$

$$
\begin{aligned}
\Delta'[3,4] &= [1,1]\otimes \mathrm{Sq}^5 + [1,1]\otimes \mathrm{Sq}^4\mathrm{Sq}^1 + [1,2]\otimes \mathrm{Sq}^3\mathrm{Sq}^1 + [1,3]\otimes \mathrm{Sq}^2\mathrm{Sq}^1 + [1,4]\otimes \mathrm{Sq}^2\\
&\quad + [2,2]\otimes \mathrm{Sq}^3 + [3,2]\otimes \mathrm{Sq}^2 + [2,4]\otimes \mathrm{Sq}^1 + [3,3]\otimes \mathrm{Sq}^1\\
\Delta'[4,3] &= [1,1]\otimes \mathrm{Sq}^4\mathrm{Sq}^1 + [1,2]\otimes \mathrm{Sq}^4 + [1,2]\otimes \mathrm{Sq}^3\mathrm{Sq}^1 + [1,3]\otimes \mathrm{Sq}^3\\
&\quad + [2,2]\otimes \mathrm{Sq}^3 + [2,2]\otimes \mathrm{Sq}^2\mathrm{Sq}^1 + [2,3]\otimes \mathrm{Sq}^2 + [3,2]\otimes \mathrm{Sq}^2 + [3,3]\otimes \mathrm{Sq}^1
\end{aligned}
$$

$$
\begin{aligned}
\Delta'[1,7] &= [1,1]\otimes \mathrm{Sq}^6 + [1,2]\otimes \mathrm{Sq}^5 + [1,3]\otimes \mathrm{Sq}^4 + [1,4]\otimes \mathrm{Sq}^3 + [1,5]\otimes \mathrm{Sq}^2 + [1,6]\otimes \mathrm{Sq}^1\\
\Delta'[2,6] &= [1,1]\otimes \mathrm{Sq}^6 + [1,2]\otimes \mathrm{Sq}^5 + [2,2]\otimes \mathrm{Sq}^4 + [1,4]\otimes \mathrm{Sq}^3 + [2,3]\otimes \mathrm{Sq}^3 + [2,4]\otimes \mathrm{Sq}^2\\
&\quad + [1,6]\otimes \mathrm{Sq}^1 + [2,5]\otimes \mathrm{Sq}^1\\
\Delta'[3,5] &= [1,1]\otimes \mathrm{Sq}^5\mathrm{Sq}^1 + [1,2]\otimes \mathrm{Sq}^5 + [1,2]\otimes \mathrm{Sq}^4\mathrm{Sq}^1 + [1,3]\otimes \mathrm{Sq}^3\mathrm{Sq}^1\\
&\quad + [1,4]\otimes \mathrm{Sq}^2\mathrm{Sq}^1 + [2,3]\otimes \mathrm{Sq}^3 + [3,2]\otimes \mathrm{Sq}^3 + [1,5]\otimes \mathrm{Sq}^2 + [3,3]\otimes \mathrm{Sq}^2\\
&\quad + [2,5]\otimes \mathrm{Sq}^1 + [3,4]\otimes \mathrm{Sq}^1\\
\Delta'[4,4] &= [1,1]\otimes \mathrm{Sq}^6 + [1,2]\otimes \mathrm{Sq}^5 + [1,3]\otimes \mathrm{Sq}^3\mathrm{Sq}^1 + [2,2]\otimes \mathrm{Sq}^4 + [2,2]\otimes \mathrm{Sq}^3\mathrm{Sq}^1\\
&\quad + [1,4]\otimes \mathrm{Sq}^3 + [2,3]\otimes \mathrm{Sq}^2\mathrm{Sq}^1 + [2,4]\otimes \mathrm{Sq}^2 + [3,4]\otimes \mathrm{Sq}^1 + [4,3]\otimes \mathrm{Sq}^1\\
\Delta'[5,3] &= [1,1]\otimes \mathrm{Sq}^4\mathrm{Sq}^2 + [1,2]\otimes \mathrm{Sq}^4\mathrm{Sq}^1 + [1,3]\otimes \mathrm{Sq}^4 + [2,2]\otimes \mathrm{Sq}^3\mathrm{Sq}^1 + [2,3]\otimes \mathrm{Sq}^3\\
&\quad + [3,2]\otimes \mathrm{Sq}^2\mathrm{Sq}^1 + [3,3]\otimes \mathrm{Sq}^2 + [4,3]\otimes \mathrm{Sq}^1
\end{aligned}
$$

$$
\begin{aligned}
\Delta'[1,8] &= [1,1]\otimes \mathrm{Sq}^7 + [1,2]\otimes \mathrm{Sq}^6 + [1,3]\otimes \mathrm{Sq}^5 + [1,4]\otimes \mathrm{Sq}^4 + [1,5]\otimes \mathrm{Sq}^3 + [1,6]\otimes \mathrm{Sq}^2\\
&\quad + [1,7]\otimes \mathrm{Sq}^1\\
\Delta'[2,7] &= [1,3]\otimes \mathrm{Sq}^5 + [2,2]\otimes \mathrm{Sq}^5 + [2,3]\otimes \mathrm{Sq}^4 + [1,5]\otimes \mathrm{Sq}^3 + [2,4]\otimes \mathrm{Sq}^3\\
&\quad + [2,5]\otimes \mathrm{Sq}^2 + [1,7]\otimes \mathrm{Sq}^1 + [2,6]\otimes \mathrm{Sq}^1\\
\Delta'[3,6] &= [1,1]\otimes \mathrm{Sq}^6\mathrm{Sq}^1 + [1,2]\otimes \mathrm{Sq}^6 + [1,2]\otimes \mathrm{Sq}^5\mathrm{Sq}^1 + [1,3]\otimes \mathrm{Sq}^5 + [1,3]\otimes \mathrm{Sq}^4\mathrm{Sq}^1\\
&\quad + [2,2]\otimes \mathrm{Sq}^5 + [1,4]\otimes \mathrm{Sq}^3\mathrm{Sq}^1 + [3,2]\otimes \mathrm{Sq}^4 + [1,5]\otimes \mathrm{Sq}^2\mathrm{Sq}^1 + [2,4]\otimes \mathrm{Sq}^3\\
&\quad + [3,3]\otimes \mathrm{Sq}^3 + [1,6]\otimes \mathrm{Sq}^2 + [3,4]\otimes \mathrm{Sq}^2 + [2,6]\otimes \mathrm{Sq}^1 + [3,5]\otimes \mathrm{Sq}^1\\
\Delta'[4,5] &= [1,1]\otimes \mathrm{Sq}^6\mathrm{Sq}^1 + [1,2]\otimes \mathrm{Sq}^6 + [1,2]\otimes \mathrm{Sq}^5\mathrm{Sq}^1 + [2,2]\otimes \mathrm{Sq}^4\mathrm{Sq}^1\\
&\quad + [1,4]\otimes \mathrm{Sq}^3\mathrm{Sq}^1 + [2,3]\otimes \mathrm{Sq}^3\mathrm{Sq}^1 + [3,2]\otimes \mathrm{Sq}^4 + [1,5]\otimes \mathrm{Sq}^3 + [2,4]\otimes \mathrm{Sq}^2\mathrm{Sq}^1\\
&\quad + [3,3]\otimes \mathrm{Sq}^3 + [2,5]\otimes \mathrm{Sq}^2 + [4,3]\otimes \mathrm{Sq}^2 + [3,5]\otimes \mathrm{Sq}^1 + [4,4]\otimes \mathrm{Sq}^1\\
\Delta'[5,4] &= [1,1]\otimes \mathrm{Sq}^5\mathrm{Sq}^2 + [1,1]\otimes \mathrm{Sq}^6\mathrm{Sq}^1 + [1,2]\otimes \mathrm{Sq}^6 + [1,2]\otimes \mathrm{Sq}^4\mathrm{Sq}^2\\
&\quad + [1,3]\otimes \mathrm{Sq}^4\mathrm{Sq}^1 + [2,2]\otimes \mathrm{Sq}^5 + [1,4]\otimes \mathrm{Sq}^4 + [2,3]\otimes \mathrm{Sq}^3\mathrm{Sq}^1 + [3,2]\otimes \mathrm{Sq}^4\\
&\quad + [3,2]\otimes \mathrm{Sq}^3\mathrm{Sq}^1 + [2,4]\otimes \mathrm{Sq}^3 + [3,3]\otimes \mathrm{Sq}^2\mathrm{Sq}^1 + [3,4]\otimes \mathrm{Sq}^2\\
&\quad + [4,4]\otimes \mathrm{Sq}^1 + [5,3]\otimes \mathrm{Sq}^1
\end{aligned}
$$

16.2 Computation of ξ and the diagonal Δ_1 of $\mathcal{B}$

We describe a cocycle $\xi = \xi_S$ in (16.1.5) only in terms of the symmetry operator S. We shall present an explicit formula for ξ_S and we show that ξ_S is right equivariant. For p odd we have $\xi = 0$ so that we only need to consider the case p even.

Let $A = \mathcal{A}^*$ be the dual of the Steenrod algebra and let $M = R^*_{\mathcal{F}}$ and $F = \mathcal{F}^*_0$. Then A and F are commutative algebras and M is an A-bimodule satisfying $a \cdot m = m \cdot a$ for $a \in A,\ m \in M$. We have the exact sequence

(16.2.1) $$0 \longrightarrow A \longrightarrow F \longrightarrow M \longrightarrow 0$$

as in Section (15.2). Given ξ as in (16.1.5) we obtain the dual map with suspension $\Sigma\bar{M} = M$,

(1) $$\xi^* = D : A \otimes A \longrightarrow \bar{M}$$

which is a normalized cocycle (see (15.2.4)) with

(2) $$D(a \otimes b) + D(b \otimes a) = C(a \otimes b).$$

Here C is the dual of the symmetry operator

(3) $$S^* = C : A \otimes A \longrightarrow \bar{M}.$$

We associate with D the algebra extension

(4) $$\begin{array}{ccccccc} 0 \longrightarrow & \bar{M} & \longrightarrow & E & \longrightarrow & A & \longrightarrow 0 \\ & & & \| & & & \\ & & & A \oplus \bar{M} & & & \end{array}$$

with the algebra structure in E defined by the formula ($a, a' \in A, m, m' \in \bar{M}$)

(5) $$(a, m) \cdot (a', m') = (aa', am' + ma' + D(a, a')).$$

Here we set $a \cdot (\Sigma\bar{m}) = \Sigma(a \cdot \bar{m})$ for $\bar{m} \in M$. The commutator in the algebra E satisfies

(6) $$(a, m) \cdot (a', m') - (a', m') \cdot (a, m) = (0, C(a, a'))$$

so that $S^* = C$ in (3) is the *commutator map* in E. Since commutators $[x, y] = xy - yx$ satisfy $[xx', y] = x[x', y] + [x, y]x'$ we see that C is a derivation in each variable, that is

(7) $$C(ab, a') = aC(b, a') + C(a, a')b.$$

16.2.2 Definition. The algebra A is a polynomial algebra generated by the Milnor generators ζ_i, $i \geq 1$, of degree $|\zeta_i| = 2^i - 1$. Hence a basis of A is given by the elements

(1) $$\zeta^{\underline{n}} = \zeta_1^{n_1} \cdot \zeta_2^{n_2} \cdots$$

with $\underline{n} = (n_1, n_2, \dots)$ and $n_i \geq 0$ and only finitely many $n_i \neq 0$. We define a linear section

(2) $$s : A \longrightarrow E$$

for the algebra extension (16.2.1)(4) as follows. Let $s(\zeta_i) = e_i = (\zeta_i, 0)$ and let

(3) $$s(\zeta^{\underline{n}}) = e_1^{n_1} \cdot e_2^{n_2} \cdot \dots = e^{\underline{n}}.$$

Here the right-hand side is given by multiplication in E. The section s, in general, does not coincide with the inclusion $A \subset A \oplus \bar{M}$. In terms of the section s we obtain a new cocycle

(4) $$D_S : A \otimes A \longrightarrow \bar{M},$$
$$D_S(a \otimes b) = s(ab) - s(a) \cdot s(b).$$

Now we define ξ_S to be the dual of D_S, namely

(5) $$\xi_S = D_S^* : R_{\mathcal{F}} \longrightarrow \mathcal{A} \otimes \mathcal{A} \text{ of degree } -1.$$

16.2.3 Theorem. *The map $\xi = \xi_S$ defined in (16.2.2) is completely determined by the symmetry operator S. Moreover diagram (16.1.4) commutes for $\xi = \xi_S$ and ξ_S is right equivariant with respect to the action of $\mathcal{F}_0$. Hence by (16.1.5) there is a right equivariant splitting $u = u_S$ of $\mathcal{B}$ associated to $\xi = \xi_S$.*

Remark. Theorem (16.2.3) is compatible with the main result of Kristensen (theorem 3.3 in [Kr4]). In fact, Kristensen defines elements in $\mathcal{A} \otimes \mathcal{A}$ of the form

$$K[a,b] = (\mathrm{Sq}^1 \otimes (\mathrm{Sq}^2 \mathrm{Sq}^1 + \mathrm{Sq}^3)) \cdot \delta(Y_{a,b})$$

with $Y_{a,b} \in \mathcal{A}$ given by the formula

$$\begin{aligned} Y_{a,b} =& \mathrm{Sq}^{a-3} \mathrm{Sq}^{b-2} + \mathrm{Sq}^{a-2} \mathrm{Sq}^{b-3} \\ &+ \sum_j \binom{b-1-j}{a-2j} (\mathrm{Sq}^{a+b-j-3} \mathrm{Sq}^{j-2} + \mathrm{Sq}^{a+b-j-2} \mathrm{Sq}^{j-3}). \end{aligned}$$

One can check that the Kristensen elements satisfy the formulas

$$S[a,b] = (1+T)K[a,b], \text{ and}$$
$$\xi_S[a,b] = TK[a,b].$$

Here S is described in (14.5.2) and ξ_S is the map above in (16.2.2).

One can compute the diagonal Δ_1 of $\mathcal{B}$ in terms of the splitting u_S, that is

$$\Delta_1 u_S = (u_S)_\sharp \Delta_0 + \Sigma\xi_S. \tag{16.2.4}$$

Compare (16.1.5). In order to compute ξ_S we first determine D_S in (16.2.2) by the following formula. For $\underline{n} = (n_1, n_2, \dots)$ and $\underline{m} = (m_1, m_2, \dots)$ let $\underline{n} + \underline{m} = (n_1 + m_1, n_2 + m_2, \dots)$. Then we have the product in the commutative algebra A given by

$$\zeta^{\underline{n}} \cdot \zeta^{\underline{n}} = \zeta^{\underline{n}+\underline{m}}$$

and hence we get

$$\begin{aligned}
D_S(\zeta^{\underline{n}} \otimes \zeta^{\underline{m}}) &= s(\zeta^{\underline{n}+\underline{m}}) - s(\zeta^{\underline{n}})s(\zeta^{\underline{m}}) \\
&= e^{\underline{n}+\underline{m}} + e^{\underline{n}} \cdot e^{\underline{m}} \\
&= e^{\underline{n}+\underline{m}} + e_1^{n_1} \cdot (e_2^{n_2} \dots) \cdot e_1^{m_1} \cdot (e_2^{m_2} \dots) \\
&= e^{\underline{n}+\underline{m}} + e_1^{n_1} \cdot e_1^{m_1} \cdot (e_2^{n_2} \dots) \cdot (e_2^{m_2} \dots) + e_1^{n_1} C(\zeta_2^{n_2} \dots, \zeta_1^{m_1}) e_2^{m_2} \dots .
\end{aligned}$$

Here we use the commutator rule (16.2.1)(6). Hence we get inductively the formula

$$D_S(\zeta^{\underline{n}} \otimes \zeta^{\underline{m}}) = \sum_{i \geq 1} \zeta_1^{n_1+m_1} \dots \zeta_{i-1}^{n_{i-1}+m_{i-1}} \zeta_i^{n_i} C(\zeta_{i+1}^{n_{i+1}} \dots, \zeta_i^{m_i}) \zeta_{i+1}^{m_{i+1}} \dots . \tag{16.2.5}$$

This formula shows that D_S is completely determined by C and hence by S. Therefore also ξ_S, the dual of D_S, is completely determined by S. We obtain a formula for ξ_S as follows. Let

$$\mathrm{Sq}^{\underline{n}} \in \mathcal{A}$$

be the Milnor generator dual to $\zeta^{\underline{n}} \in A = \mathcal{A}^*$. Then the elements $\mathrm{Sq}^{\underline{n}}$ form a basis of $\mathcal{A}$ and for $x \in \mathcal{A}$ we denote by $(x)_{\underline{n}} \in \mathbb{F}$ the coordinate of x at the basis element $\mathrm{Sq}^{\underline{n}}$ so that

$$x = \sum_{\underline{n}} (x)_{\underline{n}} \, \mathrm{Sq}^{\underline{n}}.$$

Similarly we denote coordinates of $x \in \mathcal{A} \otimes \mathcal{A}$ by $(x)_{\underline{n},\underline{m}}$ and coordinates of $x \in \mathcal{A} \otimes \mathcal{A} \otimes \mathcal{A}$ by $(x)_{\underline{n},\underline{m},\underline{k}}$.

Recall that we have the left coaction Δ_L in (16.1.6). Then the map

$$\xi_S : R_{\mathcal{F}} \longrightarrow \mathcal{A} \otimes \mathcal{A}$$

defined in (16.2.2) is given by the coordinates, $x \in R_{\mathcal{F}}$,

$$(\xi_S(x))_{\underline{n},\underline{m}} = \sum_{i \geq 1} ((1 \otimes S)\Delta_L(x))_{\underline{k}(i),\underline{a}(i),\underline{b}(i)}, \tag{16.2.6}$$

$$\begin{aligned}
\underline{k}(i) &= (n_1 + m_1, \dots, n_{i-1} + m_{i-1}, n_i, m_{i+1}, m_{i+2}, \dots), \\
\underline{a}(i) &= (\underbrace{0, \dots, 0}_{i}, n_{i+1}, n_{i+2}, \dots), \\
\underline{b}(i) &= (\underbrace{0, \dots, 0}_{i-1}, m_i, 0, 0 \dots).
\end{aligned}$$

This formula of ξ_S is easily checked to be the dual of D_S in (16.2.5). Thus ξ_S is completely determined by Δ_L and the symmetry operator S computed in Section (14.5).

Proof of (16.2.3). We use (16.2.6) and the fact that D_S is a cocycle representing the extension (16.2.1)(4). This shows that diagram (16.1.4) commutes for $\xi = \xi_S$. Hence it remains to check that ξ_S is right equivariant. This follows from the next lemma. □

16.2.7 Lemma. *The following diagram commutes.*

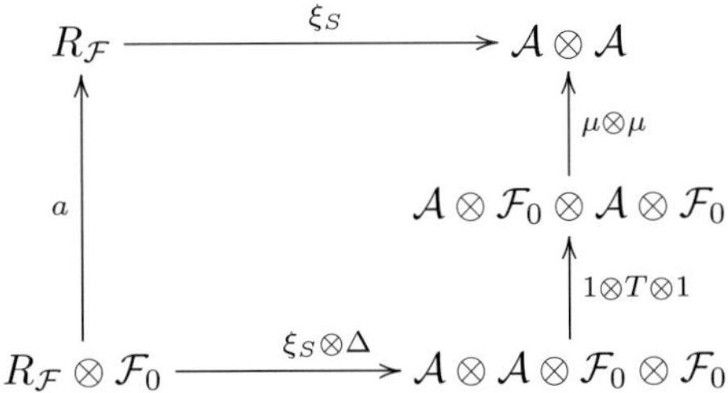

Here μ is given by the multiplication of $\mathcal{A}$ and a is the right action of $\mathcal{F}_0$ on $R_{\mathcal{F}}$.

Proof. We check that the following dual diagram commutes with $D = D_S$ dual to $\xi = \xi_S$.

(1)

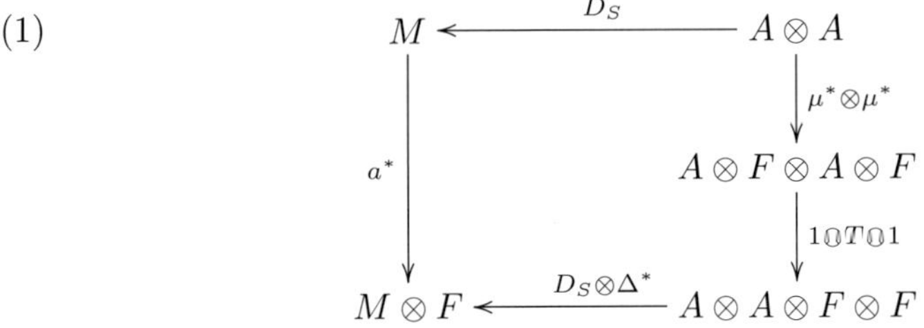

Compare the notation in (16.2.1). Using the extension (16.2.1)(4) we get the following diagram.

(2)

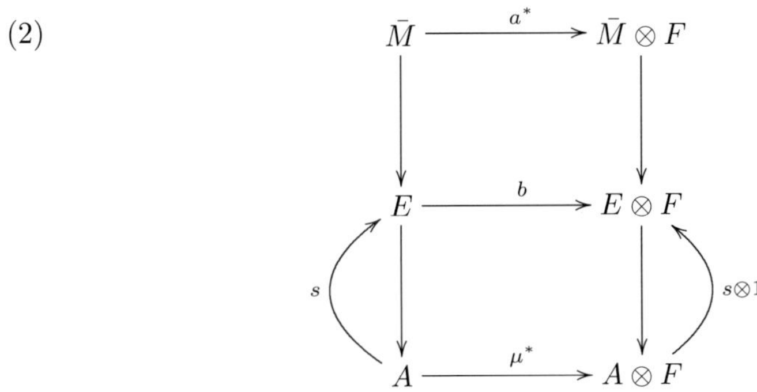

Here we have $E = A \oplus \bar{M}$ and $b = \mu^* \oplus a^*$. The map μ^* is given by the Milnor diagonal of A which is the unique algebra map satisfying

(3) $$\mu^*(\zeta_n) = \sum_{i=0}^{n} \zeta_{n-i}^{2^i} \otimes \zeta_i.$$

Here we use the inclusion $A \subset F$ which is dual to $\mathcal{F}_0 \twoheadrightarrow \mathcal{A}$. The multiplication of E is defined in (16.2.1)(5) in terms of D with $D = \xi^*$ where ξ is right equivariant as in (16.1.5). Since ξ is right equivariant we observe that $b = \mu^* \oplus a^*$ is an algebra map. The section s is defined as in (16.2.2). We claim that

(4) $$bs = (s \otimes 1)\mu^*.$$

Assuming (4) we see that by definition of D_S in (16.2.2)(4) diagram (1) commutes. Finally we obtain (4) as follows. Since the commutator map C is a derivation we get

$$C(\alpha^{2^j} \otimes a) = 2^j \alpha^{2^j - 1} C(\alpha \otimes a) = 0 \text{ for } j \geq 1.$$

Hence we have in E the equation

$$e^{2^j} \cdot x = x \cdot e^{2^j} \text{ for } j \geq 1.$$

This implies by (3) and the definition of s in (16.2.2) that (4) holds. □

16.3 The multiplication in $\mathcal{B}$

We know that $\mathcal{B}_1$ is a $\mathcal{B}_0$-bimodule. This bimodule is determined by a multiplication map A as follows. Let

$$L : \mathcal{A} \otimes R_{\mathcal{B}} \longrightarrow \mathcal{A} \otimes \mathcal{A}$$

be given by the left action operator L in (14.4.3) with $L = 0$ for p odd. Moreover let

$$\xi = \xi_S : R_{\mathcal{B}} \longrightarrow \mathcal{A} \otimes \mathcal{A}$$

be defined by ξ_S in (16.2.2) with $\xi = 0$ for p odd.

16.3.1 Definition. A *multiplication map* (associated to L and ξ) is a linear map of degree -1,

$$A : \mathcal{A} \otimes R_{\mathcal{B}} \longrightarrow \mathcal{A},$$

satisfying the following properties with $\alpha, \alpha', \beta, \beta' \in \mathcal{B}_0$ and $x, y \in R_{\mathcal{B}}$. Recall that via $\mathcal{B}_0 \to \mathcal{A}$ an element $\alpha \in \mathcal{B}_0$ yields the corresponding element in $\mathcal{A}$ also denoted by α.

(1) $$A(\alpha, x\beta) = A(\alpha, x)\beta,$$

(2) $$A(\alpha\alpha', x) = A(\alpha, \alpha' x) + (-1)^{|\alpha|}\alpha A(\alpha', x),$$

(3) $$A(\alpha, p\beta) = -\kappa(\alpha) \cdot \beta.$$

Next we consider the diagram

$$\begin{array}{ccc} \mathcal{A} & \xrightarrow{\delta} & \mathcal{A} \otimes \mathcal{A} \\ \uparrow{\scriptstyle A} & & \uparrow{\scriptstyle A_\sharp} \\ \mathcal{A} \otimes R_{\mathcal{B}} & \xrightarrow[1\otimes\Delta]{} & \mathcal{A} \otimes R_{\mathcal{B}}^2 \end{array}$$

where the induced map $A_\sharp$ is defined by

$$A_\sharp(\alpha \otimes x \otimes \beta') = \sum_i (-1)^{|\alpha_i''||x|} A(\alpha_i', x) \otimes (\alpha_i'' \beta'),$$

$$A_\sharp(\alpha \otimes \beta \otimes y) = \sum_i (-1)^{\varepsilon_i} (\alpha_i' \beta) \otimes A(\alpha_i'', y),$$

with $\varepsilon_i = |\alpha_i''||\beta| + |\alpha_i'| + |\beta|$ and

$$\delta(\alpha) = \sum_i \alpha_i' \otimes \alpha_i'' \ \in \mathcal{A} \otimes \mathcal{A}.$$

Then the following property holds:

$$\delta A = A_\sharp(1 \otimes \Delta) + L + \nabla_\xi. \tag{4}$$

Here L is the left action operator and ∇_ξ is defined as in (15.3.10) by

$$\nabla_\xi : \mathcal{B}_0 \otimes R_{\mathcal{B}} \to \mathcal{A} \otimes \mathcal{A},$$

$$\nabla_\xi(\alpha \otimes x) = (\delta\alpha) \quad \xi(x) \quad \xi(\alpha \quad x).$$

We have $L = 0$ and $\nabla_\xi = 0$ if p is odd. Recall that the reduced diagonals

$$\tilde{\delta} : \mathcal{A} \longrightarrow \mathcal{A} \otimes \mathcal{A},$$

$$\tilde{\Delta} : \mathcal{B}_0 \longrightarrow \mathcal{B}_0 \otimes \mathcal{B}_0$$

are defined by

$$\tilde{\delta}(\alpha) = \delta(\alpha) - (1 \otimes \alpha + \alpha \otimes 1),$$

$$\tilde{\Delta}(\alpha) = \Delta(\alpha) - (1 \otimes \alpha + \alpha \otimes 1).$$

We can rewrite formula (4) by the equivalent equation

$$\begin{aligned} \tilde{\delta} A(\alpha \otimes x) &= A_\sharp(\alpha \otimes \tilde{\Delta} x) + L(\alpha \otimes x) + \nabla_\xi(\alpha \otimes x) \\ &+ \sum_j ((-1)^{|\tilde{\alpha}_j|} \tilde{\alpha}_j \otimes A(\tilde{\tilde{\alpha}}_j, x) + (-1)^{|\tilde{\tilde{\alpha}}_j||x|} A(\tilde{\alpha}_j, x) \otimes \tilde{\tilde{\alpha}}_j)) \end{aligned} \tag{5}$$

for $\tilde{\Delta}(\alpha) = \sum_j \tilde{\alpha}_j \otimes \tilde{\tilde{\alpha}}_j$ with $|\tilde{\alpha}_j|$, $|\tilde{\tilde{\alpha}}_j| < |\alpha|$.

We say that two multiplication maps A and A' as in (16.3.1) are *equivalent* if there exists a linear map of degree -1,

$$\gamma : R_{\mathcal{F}} \longrightarrow \mathcal{A} \tag{16.3.2}$$

satisfying the properties

(1) $$A(\alpha \otimes x) - A'(\alpha \otimes x) = \gamma(\alpha x) - (-1)^{|\alpha|}\alpha \cdot \gamma(x).$$

Moreover γ is right equivariant with respect to the action of $\mathcal{F}_0$ and the diagram

(2)

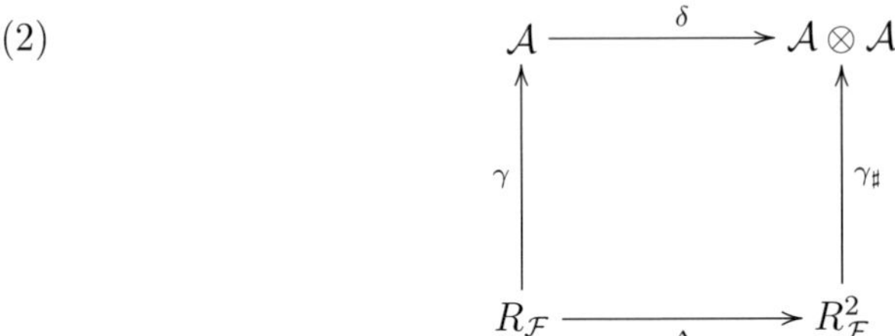

commutes with $R^2_{\mathcal{F}} = R_{\mathcal{F}} \otimes \mathcal{F}_0 + \mathcal{F}_0 \otimes R_{\mathcal{F}} \subset \mathcal{F}_0 \otimes \mathcal{F}_0$ and

$$\gamma_\sharp(x \otimes \beta) = \gamma(x) \otimes \beta,$$
$$\gamma_\sharp(\alpha \otimes y) = (-1)^{|\alpha|}\alpha \otimes \gamma(y),$$

for $x, y \in R_{\mathcal{F}}$ and $\alpha, \beta \in \mathcal{F}_0$. Commutativity of the diagram is equivalent to the condition that the dual map $\gamma^* : \mathcal{A}^* \to R^*_{\mathcal{F}}$ is a derivation, and also to the following equation corresponding to (14.6.1)(5):

(3) $$\tilde{\delta}\gamma(x) = \gamma_\sharp\tilde{\Delta}(x).$$

We now get the result:

16.3.3 Theorem. *There exists a multiplication map A and two such multiplication maps are equivalent.*

Proof. Using (15.3.5) we see that a $\mathcal{B}$-structure (A, B) with $B = 0$ is the same as a multiplication map. In fact, since $B = 0$, we see by (15.3.5)(2) that $A(x, y) = 0$ for $x, y \in R_{\mathcal{B}}$. Hence the exactness of the row in the following diagram shows that A induces a multiplication map also denoted by A.

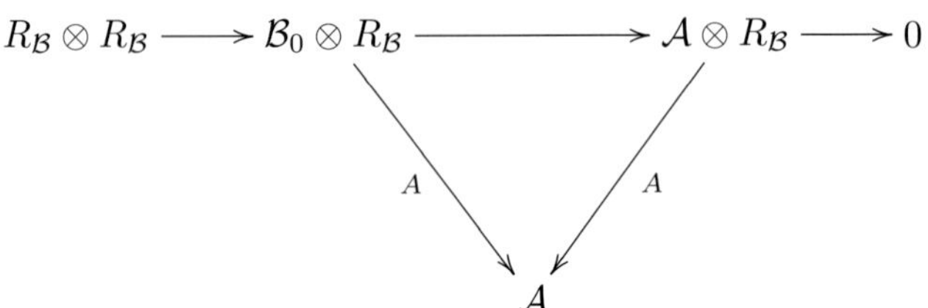

A $\mathcal{B}$-structure (A, B) with $B = 0$ exists by setting

$$\alpha \cdot u(x) - u(\alpha \cdot x) = \Sigma A(\alpha \otimes x)$$

where $u = u_\xi$ is the right equivariant splitting in (16.1.5). Two such sections yield equivalent multiplication maps. Now the uniqueness theorem (15.3.13) yields the result, see also (15.3.10). $\square$

We obtain by (15.3.14) the next result on triple Massey products.

16.3.4 Corollary. *Let A be a multiplication map and let $\langle \alpha, \beta, \gamma \rangle$ be defined for $\alpha, \beta, \gamma \in \mathcal{A}$, $\alpha\beta = 0$, $\beta\gamma = 0$. If $\bar{\beta}$, $\bar{\gamma} \in \mathcal{B}_0$ are elements representing β, γ, then $\bar{\beta} \cdot \bar{\gamma} \in R_{\mathcal{B}}$ and*

$$A(\alpha, \bar{\beta} \cdot \bar{\gamma}) \in \langle \alpha, \beta, \gamma \rangle.$$

Hence a computation of the multiplication map A yields the computation of all triple Massey products in the Steenrod algebra.

16.4 Computation of the multiplication map

We only consider the case $p = 2$ so that $\mathbb{F} = \mathbb{Z}/2\mathbb{Z}$ and $\mathbb{G} = \mathbb{Z}/4\mathbb{Z}$ leaving the case p= odd to the reader. Let

$$\chi : \mathbb{F} \longrightarrow \mathbb{G}$$

be the function with $\chi(0) = 0$ and $\chi(1) = 1$. We define for $0 < a < 2b$ the relation

$$[a, b] \in R_{\mathcal{B}} \subset T_{\mathbb{G}}(E_{\mathcal{A}}),$$

$$(16.4.1) \qquad [a, b] = \mathrm{Sq}^a \mathrm{Sq}^b + \sum_{k=0}^{[a/2]} \chi \binom{b-k-1}{a-2k} \mathrm{Sq}^{a+b-k} \mathrm{Sq}^k .$$

Let $E^1_{\mathcal{A}}$ be the subset of $R_{\mathcal{B}}$ consisting of the elements $p = p \cdot 1$ and $[a, b]$ for $0 < a < 2b$. Then the function

$$(16.4.2) \qquad \mathcal{B}_0 \otimes E^1_{\mathcal{A}} \otimes \mathcal{B}_0 \twoheadrightarrow R_{\mathcal{B}},$$

which carries $\alpha \otimes x \otimes \beta$ to $\alpha \cdot x \cdot \beta$, is surjective since $R_{\mathcal{B}}$ is the ideal in $\mathcal{B}_0$ generated by $E^1_{\mathcal{A}}$. The reduced diagonal

$$\tilde{\Delta} = \tilde{\Delta} : \mathcal{B}_0 \longrightarrow \mathcal{B}_0 \otimes \mathcal{B}_0$$

yields for $\tilde{\Delta}[a, b]$ the following result. Here we set $[a, b] = 0$ for $a = 0$ or $a \geq 2b$.

16.4.3 Proposition. *The element $\tilde{\Delta}[a, b] \in \mathcal{B}_0 \otimes \mathcal{B}_0$ is a linear combination*

$$\tilde{\Delta}[a, b] = p \cdot U[a, b] + \sum_t (\chi(n_t)[r, s] \otimes \mathrm{Sq}^u \mathrm{Sq}^v + \chi(m_t) \mathrm{Sq}^r \mathrm{Sq}^s \otimes [u, v])$$

where the p-term $U[a,b]$ *is a linear combination*

$$U[a,b] = \sum_t l_t \operatorname{Sq}^r \operatorname{Sq}^s \otimes \operatorname{Sq}^u \operatorname{Sq}^v .$$

The sums are taken over all $t = (r,s,u,v)$ *with* $r,s,u,v \geq 0$ *and* $n_t, m_t, l_t \in \mathbb{F}$.

Proof. We consider the following commutative diagram.

(1)
$$\begin{array}{ccc} \mathcal{B}_0 & \xrightarrow{\tilde{\Delta}_{\mathbb{G}}} & \mathcal{B}_0 \otimes \mathcal{B}_0 \\ \downarrow & & \downarrow \\ \mathcal{F}_0 & \xrightarrow{\tilde{\Delta}_{\mathbb{F}}} & \mathcal{F}_0 \otimes \mathcal{F}_0 \end{array}$$

It is easy to see that $\tilde{\Delta}_{\mathbb{F}}[a,b]$ is a linear combination

(2)
$$\tilde{\Delta}_{\mathbb{F}}[a,b] = \sum_t (n_t[r,s] \otimes \operatorname{Sq}^u \operatorname{Sq}^v + m_t \operatorname{Sq}^r \operatorname{Sq}^s \otimes [u,v])$$

with $n_t, m_t \in \mathbb{F}$. Let $(\tilde{\Delta}_{\mathbb{F}}[a,b])_\chi$ be the same sum with n_t, m_t replaced by $\chi(n_t)$, resp. $\chi(m_t)$. Then there is an element $U[a,b] \in \mathcal{B}_0$ such that

(3)
$$\tilde{\Delta}_{\mathbb{G}}[a,b] = (\tilde{\Delta}_{\mathbb{F}}[a,b])_\chi + pU[a,b].$$

This yields the canonical form of the p-term $U[a,b]$ in $\mathcal{F}_0$. □

Examples of the terms $U[a,b]$ considered as elements in $\mathcal{A} \otimes \mathcal{A}$ are given in the following table.

$U[1,1] = \operatorname{Sq}^1 \otimes \operatorname{Sq}^1$

$U[1,2] = \operatorname{Sq}^1 \otimes \operatorname{Sq}^2 + \operatorname{Sq}^2 \otimes \operatorname{Sq}^1$

$U[1,3] = \operatorname{Sq}^1 \otimes \operatorname{Sq}^3 + \operatorname{Sq}^2 \otimes \operatorname{Sq}^2 + \operatorname{Sq}^3 \otimes \operatorname{Sq}^1$
$U[2,2] = \operatorname{Sq}^1 \otimes \operatorname{Sq}^2 \operatorname{Sq}^1 + \operatorname{Sq}^2 \otimes \operatorname{Sq}^2 + \operatorname{Sq}^2 \operatorname{Sq}^1 \otimes \operatorname{Sq}^1$

$U[1,4] = \operatorname{Sq}^1 \otimes \operatorname{Sq}^4 + \operatorname{Sq}^2 \otimes \operatorname{Sq}^3 + \operatorname{Sq}^3 \otimes \operatorname{Sq}^2 + \operatorname{Sq}^4 \otimes \operatorname{Sq}^1$
$U[2,3] = \operatorname{Sq}^2 \otimes \operatorname{Sq}^3 + \operatorname{Sq}^2 \otimes \operatorname{Sq}^2 \operatorname{Sq}^1 + \operatorname{Sq}^3 \otimes \operatorname{Sq}^2 + \operatorname{Sq}^2 \operatorname{Sq}^1 \otimes \operatorname{Sq}^2$
$U[3,2] = 0$

$U[1,5] = \operatorname{Sq}^1 \otimes \operatorname{Sq}^5 + \operatorname{Sq}^2 \otimes \operatorname{Sq}^4 + \operatorname{Sq}^3 \otimes \operatorname{Sq}^3 + \operatorname{Sq}^4 \otimes \operatorname{Sq}^2 + \operatorname{Sq}^5 \otimes \operatorname{Sq}^1$
$U[2,4] = \operatorname{Sq}^3 \otimes \operatorname{Sq}^2 \operatorname{Sq}^1 + \operatorname{Sq}^2 \operatorname{Sq}^1 \otimes \operatorname{Sq}^3$
$U[3,3] = \operatorname{Sq}^2 \otimes \operatorname{Sq}^4 + \operatorname{Sq}^2 \otimes \operatorname{Sq}^3 \operatorname{Sq}^1 + \operatorname{Sq}^3 \otimes \operatorname{Sq}^3 + \operatorname{Sq}^4 \otimes \operatorname{Sq}^2 + \operatorname{Sq}^3 \operatorname{Sq}^1 \otimes \operatorname{Sq}^2$

$$
\begin{aligned}
U[1,6] &= \mathrm{Sq}^1\otimes\mathrm{Sq}^6+\mathrm{Sq}^2\otimes\mathrm{Sq}^5+\mathrm{Sq}^3\otimes\mathrm{Sq}^4+\mathrm{Sq}^4\otimes\mathrm{Sq}^3+\mathrm{Sq}^5\otimes\mathrm{Sq}^2+\mathrm{Sq}^6\otimes\mathrm{Sq}^1\\
U[2,5] &= \mathrm{Sq}^2\otimes\mathrm{Sq}^5+\mathrm{Sq}^4\otimes\mathrm{Sq}^2\,\mathrm{Sq}^1+\mathrm{Sq}^5\otimes\mathrm{Sq}^2+\mathrm{Sq}^2\,\mathrm{Sq}^1\otimes\mathrm{Sq}^4\\
U[3,4] &= \mathrm{Sq}^2\otimes\mathrm{Sq}^5+\mathrm{Sq}^2\otimes\mathrm{Sq}^4\,\mathrm{Sq}^1+\mathrm{Sq}^3\otimes\mathrm{Sq}^4+\mathrm{Sq}^4\otimes\mathrm{Sq}^3+\mathrm{Sq}^5\otimes\mathrm{Sq}^2+\mathrm{Sq}^4\,\mathrm{Sq}^1\otimes\mathrm{Sq}^2\\
U[4,3] &= \mathrm{Sq}^1\otimes\mathrm{Sq}^4\,\mathrm{Sq}^2+\mathrm{Sq}^3\otimes\mathrm{Sq}^4+\mathrm{Sq}^3\otimes\mathrm{Sq}^3\,\mathrm{Sq}^1+\mathrm{Sq}^4\otimes\mathrm{Sq}^3+\mathrm{Sq}^3\,\mathrm{Sq}^1\otimes\mathrm{Sq}^3\\
&\quad+\mathrm{Sq}^4\,\mathrm{Sq}^2\otimes\mathrm{Sq}^1
\end{aligned}
$$

$$
\begin{aligned}
U[1,7] &= \mathrm{Sq}^1\otimes\mathrm{Sq}^7+\mathrm{Sq}^2\otimes\mathrm{Sq}^6+\mathrm{Sq}^3\otimes\mathrm{Sq}^5+\mathrm{Sq}^4\otimes\mathrm{Sq}^4+\mathrm{Sq}^5\otimes\mathrm{Sq}^3+\mathrm{Sq}^6\otimes\mathrm{Sq}^2\\
&\quad+\mathrm{Sq}^7\otimes\mathrm{Sq}^1\\
U[2,6] &= \mathrm{Sq}^4\otimes\mathrm{Sq}^4+\mathrm{Sq}^5\otimes\mathrm{Sq}^2\,\mathrm{Sq}^1+\mathrm{Sq}^2\,\mathrm{Sq}^1\otimes\mathrm{Sq}^5\\
U[3,5] &= \mathrm{Sq}^2\otimes\mathrm{Sq}^5\,\mathrm{Sq}^1+\mathrm{Sq}^3\otimes\mathrm{Sq}^5+\mathrm{Sq}^4\otimes\mathrm{Sq}^3\,\mathrm{Sq}^1+\mathrm{Sq}^5\otimes\mathrm{Sq}^3+\mathrm{Sq}^3\,\mathrm{Sq}^1\otimes\mathrm{Sq}^4\\
&\quad+\mathrm{Sq}^5\,\mathrm{Sq}^1\otimes\mathrm{Sq}^2\\
U[4,4] &= \mathrm{Sq}^2\otimes\mathrm{Sq}^6+\mathrm{Sq}^2\otimes\mathrm{Sq}^4\,\mathrm{Sq}^2+\mathrm{Sq}^3\otimes\mathrm{Sq}^4\,\mathrm{Sq}^1+\mathrm{Sq}^4\otimes\mathrm{Sq}^4+\mathrm{Sq}^4\otimes\mathrm{Sq}^3\,\mathrm{Sq}^1\\
&\quad+\mathrm{Sq}^6\otimes\mathrm{Sq}^2+\mathrm{Sq}^3\,\mathrm{Sq}^1\otimes\mathrm{Sq}^4+\mathrm{Sq}^3\,\mathrm{Sq}^1\otimes\mathrm{Sq}^3\,\mathrm{Sq}^1+\mathrm{Sq}^4\,\mathrm{Sq}^1\otimes\mathrm{Sq}^3\\
&\quad+\mathrm{Sq}^4\,\mathrm{Sq}^2\otimes\mathrm{Sq}^2\\
U[5,3] &= \mathrm{Sq}^3\otimes\mathrm{Sq}^4\,\mathrm{Sq}^1+\mathrm{Sq}^3\,\mathrm{Sq}^1\otimes\mathrm{Sq}^3\,\mathrm{Sq}^1+\mathrm{Sq}^4\,\mathrm{Sq}^1\otimes\mathrm{Sq}^3
\end{aligned}
$$

$$
\begin{aligned}
U[1,8] &= \mathrm{Sq}^1\otimes\mathrm{Sq}^8+\mathrm{Sq}^2\otimes\mathrm{Sq}^7+\mathrm{Sq}^3\otimes\mathrm{Sq}^6+\mathrm{Sq}^4\otimes\mathrm{Sq}^5+\mathrm{Sq}^5\otimes\mathrm{Sq}^4\\
&\quad+\mathrm{Sq}^6\otimes\mathrm{Sq}^3+\mathrm{Sq}^7\otimes\mathrm{Sq}^2+\mathrm{Sq}^8\otimes\mathrm{Sq}^1\\
U[2,7] &= \mathrm{Sq}^2\otimes\mathrm{Sq}^7+\mathrm{Sq}^6\otimes\mathrm{Sq}^2\,\mathrm{Sq}^1+\mathrm{Sq}^7\otimes\mathrm{Sq}^2+\mathrm{Sq}^2\,\mathrm{Sq}^1\otimes\mathrm{Sq}^6\\
U[3,6] &= \mathrm{Sq}^2\otimes\mathrm{Sq}^6\,\mathrm{Sq}^1+\mathrm{Sq}^3\otimes\mathrm{Sq}^6+\mathrm{Sq}^4\otimes\mathrm{Sq}^4\,\mathrm{Sq}^1+\mathrm{Sq}^6\otimes\mathrm{Sq}^3+\mathrm{Sq}^4\,\mathrm{Sq}^1\otimes\mathrm{Sq}^4\\
&\quad+\mathrm{Sq}^6\,\mathrm{Sq}^1\otimes\mathrm{Sq}^2\\
U[4,5] &= \mathrm{Sq}^2\otimes\mathrm{Sq}^6\,\mathrm{Sq}^1+\mathrm{Sq}^3\otimes\mathrm{Sq}^4\,\mathrm{Sq}^2+\mathrm{Sq}^4\otimes\mathrm{Sq}^5+\mathrm{Sq}^5\otimes\mathrm{Sq}^4+\mathrm{Sq}^3\,\mathrm{Sq}^1\otimes\mathrm{Sq}^4\,\mathrm{Sq}^1\\
&\quad+\mathrm{Sq}^4\,\mathrm{Sq}^1\otimes\mathrm{Sq}^3\,\mathrm{Sq}^1+\mathrm{Sq}^4\,\mathrm{Sq}^2\otimes\mathrm{Sq}^3+\mathrm{Sq}^6\,\mathrm{Sq}^1\otimes\mathrm{Sq}^2\\
U[5,4] &= \mathrm{Sq}^2\otimes\mathrm{Sq}^5\,\mathrm{Sq}^2+\mathrm{Sq}^2\otimes\mathrm{Sq}^6\,\mathrm{Sq}^1+\mathrm{Sq}^4\otimes\mathrm{Sq}^5+\mathrm{Sq}^4\otimes\mathrm{Sq}^4\,\mathrm{Sq}^1+\mathrm{Sq}^5\otimes\mathrm{Sq}^4\\
&\quad+\mathrm{Sq}^5\otimes\mathrm{Sq}^3\,\mathrm{Sq}^1+\mathrm{Sq}^3\,\mathrm{Sq}^1\otimes\mathrm{Sq}^5+\mathrm{Sq}^4\,\mathrm{Sq}^1\otimes\mathrm{Sq}^4+\mathrm{Sq}^5\,\mathrm{Sq}^2\otimes\mathrm{Sq}^2\\
&\quad+\mathrm{Sq}^6\,\mathrm{Sq}^1\otimes\mathrm{Sq}^2
\end{aligned}
$$

The reduced diagonal $\tilde{\Delta}_{\mathbb{F}}$ of $\mathcal{F}_0$ induces the quotient map $\tilde{\tilde{\Delta}}$ in the diagram

$$
\begin{array}{ccc}
\mathcal{F}_0 & \xrightarrow{\tilde{\Delta}_{\mathbb{F}}} & \mathcal{F}_0\otimes\mathcal{F}_0\\
\cup & & \cup\\
R_{\mathcal{F}} & \longrightarrow & R^2_{\mathcal{F}}\\
& \searrow^{\tilde{\tilde{\Delta}}} & \downarrow\\
& & R_{\mathcal{F}}\otimes\mathcal{A}\oplus\mathcal{A}\otimes R_{\mathcal{F}}
\end{array}
\tag{4}
$$

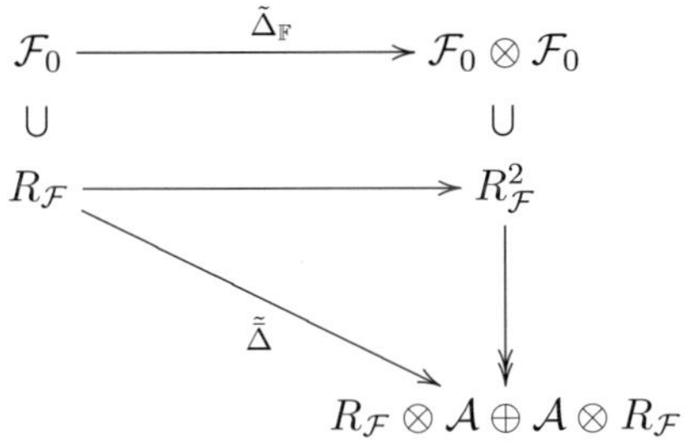

with $\tilde{\bar{\Delta}}[a,b]$ also given by

(5) $$\tilde{\bar{\Delta}}[a,b] = \sum_t (n_t[r,s] \otimes \mathrm{Sq}^u \mathrm{Sq}^v + m_t \mathrm{Sq}^r \mathrm{Sq}^s \otimes [u,v])$$

with n_t and m_t as in the proof of (16.3.3). The map $\tilde{\bar{\Delta}}$ in (4) has two coordinates

$$\tilde{\bar{\Delta}}(x) = (\Delta'(x), T\Delta'(x)) \;\text{ for } x \in R_{\mathcal{F}}$$

with Δ' as in (16.1.6).

Given a multiplication map (see (16.3.1))

$$A : \mathcal{A} \otimes R_{\mathcal{B}} \longrightarrow \mathcal{A}$$

we obtain for $0 < a < 2b$ the *multiplication function* (associated to L and ξ)

(16.4.4) $$A_{a,b} : \mathcal{A} \longrightarrow \mathcal{A}.$$

This is the linear map of degree $a + b - 1$ defined by

$$A_{a,b}(\alpha) = A(\alpha \otimes [a,b]).$$

The family of multiplication functions $\{A_{a,b}\}$ determines the multiplication map A uniquely by the following commutative diagram.

(1)

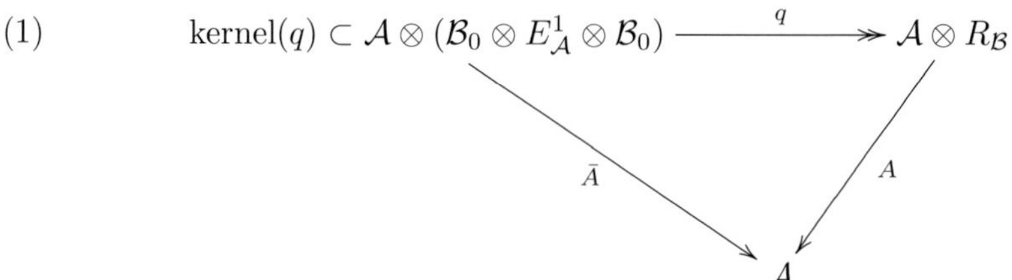

Here q is defined by (16.4.2). The map $\bar{A}$ is defined according to (16.3.1)(1)(2) by the formulas

(2) $$\begin{cases} \bar{A}(\alpha \otimes \alpha' \otimes p \otimes \beta) = \kappa(\alpha) \cdot \alpha'\beta, \\ \bar{A}(\alpha \otimes \alpha' \otimes [a,b] \otimes \beta) = (A_{a,b}(\alpha\alpha') + \alpha A_{a,b}(\alpha')) \cdot \beta. \end{cases}$$

Here we use the algebra map $\mathcal{B}_0 \twoheadrightarrow \mathcal{A}$ and the image of $\alpha \in \mathcal{B}_0$ in $\mathcal{A}$ is also denoted by α.

We express for $\alpha \in \mathcal{A}$ the reduced diagonal $\tilde{\delta}$ applied to α by the formula

(3) $$\tilde{\delta}(\alpha) = \sum_j \tilde{\alpha}_j \otimes \tilde{\tilde{\alpha}}_j$$

with $|\tilde{\alpha}_j|$, $|\tilde{\tilde{\alpha}}_j| > 0$. On the other hand the diagonal δ is written as

(4) $$\delta(\alpha) = \tilde{\delta}(\alpha) + 1 \otimes \alpha + \alpha \otimes 1 = \sum_i \alpha_i' \otimes \alpha_i''.$$

Moreover, recall that we have by (16.4.3) the elements

(5) $$U[a,b] \in \mathcal{A} \otimes \mathcal{A}$$

given by $U[a,b] \in \mathcal{F}_0 \otimes \mathcal{F}_0$ via the quotient map $\mathcal{F}_0 \to \mathcal{A}$.

Now we get by (16.3.1)(5) the following Δ-formula for the multiplication functions $A_{a,b}$.

16.4.5 Theorem. *The family of multiplication functions $A_{a,b}$ associated to L and $\xi = \xi_S$ satisfies the* Δ-formula *in $\tilde{\mathcal{A}} \otimes \tilde{\mathcal{A}}$:*

$$\begin{aligned}\tilde{\delta} A_{a,b}(\alpha) &= L(\alpha \otimes [a,b]) + \xi(\alpha \cdot [a,b]) + \alpha \cdot \xi([a,b]) \\ &\quad + (\delta\kappa(\alpha)) \cdot U[a,b] + W(\alpha \otimes [a,b])\end{aligned}$$

with

$$\begin{aligned}W(\alpha \otimes [a,b]) &= \sum_j \tilde{\alpha}_j \otimes A_{a,b}(\tilde{\tilde{\alpha}}_j) + A_{a,b}(\tilde{\alpha}_j) \otimes \tilde{\tilde{\alpha}}_j \\ &\quad + \sum_{t,i} (n_t A_{r,s}(\alpha_i') \otimes \alpha_i'' \operatorname{Sq}^u \operatorname{Sq}^v + m_t \alpha_i' \operatorname{Sq}^r \operatorname{Sq}^s \otimes A_{u,v}(\alpha_i'')).\end{aligned}$$

Here n_t, m_t are the coefficients in the formula (16.4.3)(5) *with $t = (r,s,u,v)$. We set $A_{a,b} = 0$ for $a = 0$ or $a \geq 2b$.*

The Δ-formula can be used for the inductive computation of $A_{a,b}(\alpha)$.

Proof. The formula in the theorem is a reformulation of the formula (16.3.1)(5). □

16.4.6 Theorem. *Consider a family of functions*

$$A_{a,b} : \mathcal{A} \longrightarrow \mathcal{A}$$

of degree $a + b - 1$ with $0 < a < 2b$ satisfying the Δ-formula (16.4.5) *and assume the map $\bar{A}$ in* (16.4.4)(2) *defined by the family $\{A_{a,b}\}$ satisfies the kernel condition*

$$\bar{A}(z) = 0 \text{ for } z \in \operatorname{kernel}(q)$$

as in (16.4.4)(1). *Then $\bar{A}$ induces via* (16.4.4)(1) *a multiplication map A and vice versa a multiplication map A yields a family $\{A_{a,b}\}$ with these properties.*

The theorem gives an inductive method for the computation of a multiplication map A. If $A_{a,b}(\alpha)$ is computed for $a+b+|\alpha| < n$, then we get for $a+b+|\alpha| = n$ the Δ-formula $\tilde{\delta} A_{a,b}(\alpha) =$ (terms already computed) and we can choose a solution $A_{a,b}(\alpha)$ of this formula.

We point out that the kernel of $\tilde{\delta} : \mathcal{A} \longrightarrow \mathcal{A} \otimes \mathcal{A}$ is generated by the elements Q_i of degree $2^i - 1$, $i \geq 1$, which are dual to the Milnor generators of the algebra $\mathcal{A}^*$. One obtains $Q_i = \mathrm{Sq}^{(0,\dots,0,1)}$ inductively by the formula (see [Mn])

(16.4.7)
$$\begin{cases} Q_i = \mathrm{Sq}^1, \\ Q_{i+1} = \mathrm{Sq}^{2^i} \cdot Q_i + Q_i \cdot \mathrm{Sq}^{2^i}, \quad i \geq 1. \end{cases}$$

Hence in the induction procedure above the element $A_{a,b}(\alpha)$ is uniquely determined for $a + b + |\alpha| \neq 2^i$.

16.4.8 Proposition. *Let $A_{a,b}(\alpha)$ be given for $a + b + |\alpha| \leq n = 2^i$ such that the Δ-formula holds and the kernel condition $\bar{A}(z) = 0$ for $z \in kernel(q)$ and $|z| \leq n$ is satisfied. Then there exists a multiplication map $\tilde{A}$ such that*

$$\tilde{A}(\alpha \otimes [a, b]) = A_{a,b}(\alpha)$$

for $a + b + |\alpha| \leq n$.

Proof. Assume the result holds for $n = 2^i$. Then the Δ-formula yields for $a + b + |\alpha| < m = 2^{i+1}$ unique elements $A_{a,b}(\alpha)$ and

$$\tilde{A}(\alpha \otimes [a, b]) = \tilde{A}_{a,b}(\alpha) = A_{a,b}(\alpha)$$

holds. We now choose for $a+b+|\alpha| = m$ elements $A_{a,b}(\alpha)$ such that the Δ-formula holds and such that the kernel condition

(1)
$$\bar{A}(z) = 0 \text{ for } |z| = m,\ z \in \mathrm{kernel}(q)$$

holds. Then $\bar{A}$ induces

(2)
$$A^m : (\mathcal{A} \otimes R_{\mathcal{B}})^{\leq m} \longrightarrow \mathcal{A}$$

and $A^m - \tilde{A} = \nabla$ is defined in degree $\leq m$ with $\nabla(\alpha, x) = 0$ for $|\alpha| + |x| < m$.

The argument that ∂_1 in (15.4.10) is surjective shows that there exists

(3)
$$\gamma : R_{\mathcal{F}}^{\leq m} \longrightarrow \mathcal{A} \text{ with}$$

$$\begin{cases} \gamma = 0 & \text{in degree} < m, \\ \delta\gamma = \gamma_{\sharp}\Delta & \text{in degree} \leq m, \\ \nabla(\alpha, x) = \gamma(\alpha x) + \alpha\gamma(x), & \\ \gamma(x\beta) = \gamma(x) \cdot \beta & \end{cases}$$

for $|\alpha|, |\beta| \leq m - |x|$.

The following lemma shows that there is an extension

(4) $$\gamma : R_{\mathcal{F}} \longrightarrow \mathcal{A}$$

of (3) in degree $> m$ such that the extension is an equivalence as in (16.3.2). Hence we obtain a multiplication map $\tilde{\tilde{A}}$ by

(5) $$\tilde{\tilde{A}}(\alpha, x) = \tilde{A}(\alpha, x) + \gamma(\alpha x) + \alpha\gamma(x)$$

and we have $\tilde{\tilde{A}} = A^m$ in degree $\leq m$. This completes the proof of (16.4.8). □

16.4.9 Lemma. *An equivalence γ extending* (3) *exists.*

Proof. We define the dual of γ,

(7) $$\gamma^* : \mathcal{A}^* \longrightarrow R^*_{\mathcal{F}},$$

in degree $\leq m - 1$ by the dual of (3). Let ζ_k be the Milnor generator in $\mathcal{A}^*$ of degree 2^{k-1}. Then γ^*, being a derivation, is determined by $\gamma^*(\zeta_k)$ for $k \geq 1$. We set $\gamma^*(\zeta_k) = 0$ for $2^k > m = 2^{i+1}$. Then γ^* is well defined and γ satisfies (16.3.2)(1). We have to check that also (16.3.2)(3) holds for γ. This is equivalent to the commutative diagram

(8) $$\begin{array}{ccc} \mathcal{A} \otimes \mathcal{A} & \xrightarrow{\mu} & \mathcal{A} \\ \uparrow{\scriptstyle \gamma\otimes q} & & \uparrow{\scriptstyle \gamma} \\ R_{\mathcal{F}} \otimes \mathcal{F}_0 & \xrightarrow{\nu} & R_{\mathcal{F}} \end{array}$$

where μ and ν are given by multiplication. The dual of the diagram is as follows.

(9) $$\begin{array}{ccc} \mathcal{A}^* \otimes \mathcal{A}^* & \xleftarrow{\mu^*} & \mathcal{A}^* \\ \downarrow{\scriptstyle \gamma^*\otimes q^*} & & \downarrow{\scriptstyle \gamma^*} \\ R^*_{\mathcal{F}} \otimes \mathcal{F}^*_0 & \xleftarrow{\nu^*} & R^*_{\mathcal{F}} \end{array}$$

According to Milnor the diagonal μ^* of $\mathcal{A}^*$ satisfies the formula

(10) $$\mu^*(\zeta_k) = \sum_{j=0}^{k} \zeta_{k-j}^{2^j} \otimes \zeta_j$$

We have to check that diagram (9) commutes on generators ζ_k. This is clear by (3) for $2^k \le m$. For $2^k > m = 2^{i+1}$, that is $k > i+1$, we have to check

(11) $$(\gamma^* \otimes q^*)\mu^*(\zeta_k) = 0$$

since $\gamma^*(\zeta_k) = 0$. Since also $\gamma^*(\zeta_k) = 0$ for $k < i+1$ we get by (10)

(12) $$(\gamma^* \otimes q^*)\mu^*(\zeta_k) = \gamma^*(\zeta_{k-j}^{2^j}) \otimes q^*\zeta_j$$

for $k - j = i+1$, $j > 0$. Now γ^* is derivation where the bimodule structure of $R_{\mathcal{F}}^*$ is given by the co-commutative diagonal of $\mathcal{F}_0$. Therefore we have for $\zeta = \zeta_{k-j}$ and $t = 2^j$ the formula

(13) $$\begin{aligned}\gamma^*(\zeta^t) &= \sum_{i=0}^{t-1} \zeta^t \gamma^*(\zeta)\zeta^{t-1-i} \\ &= \sum_{i=0}^{t-1} \zeta^{t-1}\gamma^*(\zeta) = t\zeta^{t-1}\gamma^*(\zeta) = 0\end{aligned}$$

since t is even. □

16.5 Admissible relations

We consider the case that the prime p is even. Then the admissible monomials

(16.5.1) $$\mathrm{Sq}^{a_1} \cdots \mathrm{Sq}^{a_k} \quad \text{with} \quad a_1 \ge 2a_2, \dots, a_{k-1} \ge 2a_k$$

form a basis of the Steenrod algebra $\mathcal{A}$. The basis yields the $\mathbb{F}$-linear section s of the algebra map q,

(1) $$\mathcal{A} \xrightarrow{s} \mathcal{F}_0 \xrightarrow{q} \mathcal{A}, \quad qs = 1$$

with $s(\mathrm{Sq}^{a_1} \cdots \mathrm{Sq}^{a_k}) = \mathrm{Sq}^{a_1} \cdots \mathrm{Sq}^{a_k}$ for each admissible word $\mathrm{Sq}^{a_1} \cdots \mathrm{Sq}^{a_k}$ in $\mathcal{A}$. In addition let $0 < a_k < 2a$. Then the *admissible relations* are the elements

(2) $$[a_1, \dots, a_k, a] = \mathrm{Sq}^{a_1} \cdots \mathrm{Sq}^{a_k}\,\mathrm{Sq}^a + sq(\mathrm{Sq}^{a_1} \cdots \mathrm{Sq}^{a_k}\,\mathrm{Sq}^a) \in R_{\mathcal{F}}.$$

For $k = 1$ this is the Adem relation $[a_1, a]$. Moreover the *preadmissible relations* are the elements

(3) $$\mathrm{Sq}^{a_1} \cdots \mathrm{Sq}^{a_{k-1}}[a_k, a] \in R_{\mathcal{F}}.$$

The next result was pointed out to the author by Mamuka Jibladze, see (16.1.1).

16.5.2 Proposition. *The set AR of all admissible relations and the set PAR of all preadmissible relations both are a basis of the free right $\mathcal{F}_0$-module $R_{\mathcal{F}}$. Moreover we have the formula in $R_{\mathcal{F}}$,*

$$[a_1, \dots, a_k, a] = \mathrm{Sq}^{a_1} \cdots \mathrm{Sq}^{a_{k-1}}[a_k, a] + \sum_{\alpha} \alpha \cdot \beta_\alpha,$$

where the sum is taken over all $\alpha \in AR$ with $\alpha < [a_1, \dots, a_k, a]$ and $\beta_\alpha \in \mathcal{F}_0$. Here the ordering of AR is the lexicographical ordering from the right.

The formula is easily checked by the definition of the Adem relation. The basis PAR of $R_{\mathcal{F}}$ as a right $\mathcal{F}_0$-module yields the section $\hat{s}$ of the multiplication q,

$$R_{\mathcal{F}} \xrightarrow{\hat{s}} \mathcal{F}_0 \otimes E^1 \otimes \mathcal{F}_0 \xrightarrow{q} R_{\mathcal{F}} \quad q\hat{s} = 1. \tag{16.5.3}$$

Here E^1 is in the set of Adem relations and q carries $\alpha \otimes [a, b] \otimes \beta$ to $\alpha[a, b]\alpha$. Moreover $\hat{s}$ is the unique map of right $\mathcal{F}_0$-modules satisfying

$$\hat{s}(\mathrm{Sq}^{a_1} \cdots \mathrm{Sq}^{a_{k-1}}[a_k, a]) = \mathrm{Sq}^{a_1} \cdots \mathrm{Sq}^{a_{k-1}} \otimes [a_k, a] \otimes 1$$

for all preadmissible relations in PAR. The elements

$$x - \hat{s}qx \ \text{ with } x = \alpha \otimes [a, b] \otimes \beta$$

and α , $\beta \in \mathrm{Mon}(E_{\mathcal{A}})$ generate the kernel of q.

Since the preadmissible relations form a basis of the free right $\mathcal{F}_0$-module $R_{\mathcal{F}}$, we can write for each monomial $\beta \in \mathrm{Mon}(E_{\mathcal{A}})$ in a unique way

$$\beta[a, b] - \Sigma \alpha_i [a_i, b_i] \beta_i \ \text{ in } \mathcal{F}_0 \tag{16.5.4}$$

with $\alpha_i[a_i, b_i]$ preadmissible. A list of examples for such equations is given in the following table.

$$\mathrm{Sq}^1[1, 1] = [1, 1]\,\mathrm{Sq}^1$$

$$\mathrm{Sq}^1[1, 2] = [1, 1]\,\mathrm{Sq}^2 + [1, 3]$$

$$\mathrm{Sq}^1[1, 3] = [1, 1]\,\mathrm{Sq}^3$$
$$\mathrm{Sq}^1[2, 2] = [1, 2]\,\mathrm{Sq}^2 + [1, 3]\,\mathrm{Sq}^1 + [3, 2]$$
$$\mathrm{Sq}^2\,\mathrm{Sq}^1[1, 1] = \mathrm{Sq}^2[1, 1]\,\mathrm{Sq}^1$$

$$\mathrm{Sq}^1[1, 4] = [1, 1]\,\mathrm{Sq}^4 + [1, 5]$$
$$\mathrm{Sq}^1[2, 3] = [1, 2]\,\mathrm{Sq}^3 + [1, 4]\,\mathrm{Sq}^1 + [1, 5] + [3, 3]$$
$$\mathrm{Sq}^1[3, 2] = [1, 3]\,\mathrm{Sq}^2$$

$$\begin{aligned}
\mathrm{Sq}^2[2,2] &= [2,2]\,\mathrm{Sq}^2 + [2,3]\,\mathrm{Sq}^1 + [3,3] + \mathrm{Sq}^4[1,1] + \mathrm{Sq}^3[1,2]\\
\mathrm{Sq}^2\,\mathrm{Sq}^1[1,2] &= \mathrm{Sq}^2[1,1]\,\mathrm{Sq}^2 + \mathrm{Sq}^2[1,3]
\end{aligned}$$

$$\begin{aligned}
\mathrm{Sq}^1[1,5] &= [1,1]\,\mathrm{Sq}^5\\
\mathrm{Sq}^1[2,4] &= [1,2]\,\mathrm{Sq}^4 + [1,5]\,\mathrm{Sq}^1 + [1,6] + [3,4]\\
\mathrm{Sq}^1[3,3] &= [1,3]\,\mathrm{Sq}^3 + [1,5]\,\mathrm{Sq}^1\\
\mathrm{Sq}^2[2,3] &= [2,2]\,\mathrm{Sq}^3 + [2,4]\,\mathrm{Sq}^1 + [2,5] + \mathrm{Sq}^5[1,1] + \mathrm{Sq}^3[1,3]\\
\mathrm{Sq}^2[3,2] &= [2,3]\,\mathrm{Sq}^2 + [4,3] + \mathrm{Sq}^4[1,2]\\
\mathrm{Sq}^3[2,2] &= [3,2]\,\mathrm{Sq}^2 + [3,3]\,\mathrm{Sq}^1 + \mathrm{Sq}^5[1,1]\\
\mathrm{Sq}^2\,\mathrm{Sq}^1[1,3] &= \mathrm{Sq}^2[1,1]\,\mathrm{Sq}^3\\
\mathrm{Sq}^2\,\mathrm{Sq}^1[2,2] &= [2,3]\,\mathrm{Sq}^2 + \mathrm{Sq}^2[1,2]\,\mathrm{Sq}^2 + \mathrm{Sq}^2[1,3]\,\mathrm{Sq}^1 + [4,3] + \mathrm{Sq}^4[1,2]
\end{aligned}$$

$$\begin{aligned}
\mathrm{Sq}^1[1,6] &= [1,1]\,\mathrm{Sq}^6 + [1,7]\\
\mathrm{Sq}^1[2,5] &= [1,2]\,\mathrm{Sq}^5 + [1,6]\,\mathrm{Sq}^1 + [3,5]\\
\mathrm{Sq}^1[3,4] &= [1,3]\,\mathrm{Sq}^4 + [1,7]\\
\mathrm{Sq}^1[4,3] &= [1,4]\,\mathrm{Sq}^3 + [1,5]\,\mathrm{Sq}^2 + [5,3]\\
\mathrm{Sq}^2[2,4] &= [2,2]\,\mathrm{Sq}^4 + [2,5]\,\mathrm{Sq}^1 + [2,6] + [3,5] + \mathrm{Sq}^6[1,1] + \mathrm{Sq}^3[1,4]\\
\mathrm{Sq}^2[3,3] &= [2,3]\,\mathrm{Sq}^3 + [2,5]\,\mathrm{Sq}^1 + [5,3] + \mathrm{Sq}^6[1,1] + \mathrm{Sq}^4[1,3]\\
\mathrm{Sq}^3[2,3] &= [3,2]\,\mathrm{Sq}^3 + [3,4]\,\mathrm{Sq}^1 + [3,5]\\
\mathrm{Sq}^3[3,2] &= [3,3]\,\mathrm{Sq}^2 + [5,3] + \mathrm{Sq}^5[1,2]\\
\mathrm{Sq}^2\,\mathrm{Sq}^1[1,4] &= \mathrm{Sq}^2[1,1]\,\mathrm{Sq}^4 + \mathrm{Sq}^2[1,5]\\
\mathrm{Sq}^2\,\mathrm{Sq}^1[2,3] &= [2,3]\,\mathrm{Sq}^3 + \mathrm{Sq}^2[1,2]\,\mathrm{Sq}^3 + [2,5]\,\mathrm{Sq}^1 + \mathrm{Sq}^2[1,4]\,\mathrm{Sq}^1 + [5,3] + \mathrm{Sq}^6[1,1]\\
&\quad + \mathrm{Sq}^4[1,3] + \mathrm{Sq}^2[1,5]\\
\mathrm{Sq}^2\,\mathrm{Sq}^1[3,2] &= \mathrm{Sq}^2[1,3]\,\mathrm{Sq}^2
\end{aligned}$$

$$\begin{aligned}
\mathrm{Sq}^1[1,7] &= [1,1]\,\mathrm{Sq}^7\\
\mathrm{Sq}^1[2,6] &= [1,2]\,\mathrm{Sq}^6 + [1,7]\,\mathrm{Sq}^1 + [3,6]\\
\mathrm{Sq}^1[3,5] &= [1,3]\,\mathrm{Sq}^5 + [1,7]\,\mathrm{Sq}^1\\
\mathrm{Sq}^1[4,4] &= [1,4]\,\mathrm{Sq}^4 + [1,6]\,\mathrm{Sq}^2 + [1,7]\,\mathrm{Sq}^1 + [5,4]\\
\mathrm{Sq}^1[5,3] &= [1,5]\,\mathrm{Sq}^3\\
\mathrm{Sq}^2[2,5] &= [2,2]\,\mathrm{Sq}^5 + [2,6]\,\mathrm{Sq}^1 + \mathrm{Sq}^7[1,1] + \mathrm{Sq}^3[1,5]\\
\mathrm{Sq}^2[3,4] &= [2,3]\,\mathrm{Sq}^4 + [2,7] + [4,5] + [5,4] + \mathrm{Sq}^4[1,4]\\
\mathrm{Sq}^2[4,3] &= [2,4]\,\mathrm{Sq}^3 + [2,5]\,\mathrm{Sq}^2 + \mathrm{Sq}^6[1,2] + \mathrm{Sq}^5[1,3]\\
\mathrm{Sq}^3[2,4] &= [3,2]\,\mathrm{Sq}^4 + [3,5]\,\mathrm{Sq}^1 + [3,6] + \mathrm{Sq}^7[1,1]\\
\mathrm{Sq}^3[3,3] &= [3,3]\,\mathrm{Sq}^3 + [3,5]\,\mathrm{Sq}^1 + \mathrm{Sq}^7[1,1] + \mathrm{Sq}^5[1,3]\\
\mathrm{Sq}^2\,\mathrm{Sq}^1[1,5] &= \mathrm{Sq}^2[1,1]\,\mathrm{Sq}^5
\end{aligned}$$

$$\begin{aligned}
\mathrm{Sq}^2\,\mathrm{Sq}^1[2,4] &= [2,3]\,\mathrm{Sq}^4 + \mathrm{Sq}^2[1,2]\,\mathrm{Sq}^4 + \mathrm{Sq}^2[1,5]\,\mathrm{Sq}^1 + [2,7] + [4,5] \\
&\quad + [5,4] + \mathrm{Sq}^4[1,4] + \mathrm{Sq}^2[1,6] \\
\mathrm{Sq}^2\,\mathrm{Sq}^1[3,3] &= \mathrm{Sq}^2[1,3]\,\mathrm{Sq}^3 + \mathrm{Sq}^2[1,5]\,\mathrm{Sq}^1 \\
\mathrm{Sq}^4[3,2] &= [4,3]\,\mathrm{Sq}^2 + [5,3]\,\mathrm{Sq}^1 + \mathrm{Sq}^5[2,2]
\end{aligned}$$

By (16.4.1) the equation (16.5.4) yields an element $\Theta \in \mathcal{B}_0$ such that

(16.5.5) $$\beta[a,b] = p\Theta + \sum_i \alpha_i[a_i,b_i]\beta_i \ \text{ in } \mathcal{B}_0.$$

Here Θ modulo $\mathbb{F}$ is well defined by β and $[a,b]$ so that Θ determines an element Θ in $\mathcal{A}$ by the algebra map $\mathcal{B}_0 \longrightarrow \mathcal{F}_0 \longrightarrow \mathcal{A}$.

16.5.6 Proposition. *Consider a family of functions* $A_{a,b}$ *as in* (16.4.6). *Then the associated map* $\bar{A}$ *in* (16.4.4)(1) *satisfies the kernel condition if and only if for all admissible monomials* α *and monomials* $\beta \in \mathrm{Mon}(E_{\mathcal{A}})$ *the following formula holds.*

$$A_{a,b}(\alpha\beta) + \alpha A_{a,b}(\beta) = \kappa(\alpha)\cdot\Theta + \sum_i (A_{a_i,b_i}(\alpha\alpha_i) + \alpha A_{a_i,b_i}(\alpha_i))\beta_i.$$

Proof. We observe that $\mathbb{F} \otimes R_{\mathcal{B}}$ has a basis consisting of the elements

(1) $$\begin{cases} p\alpha, & \alpha \in \mathrm{Mon}(E_{\mathcal{A}}),\ \text{and}\ \alpha\ \text{admissible}, \\ \alpha[a,b]\beta, & \alpha,\beta \in \mathrm{Mon}(E_{\mathcal{A}}),\ \text{and}\ \alpha[a,b]\ \text{preadmissible}. \end{cases}$$

Using this basis we obtain a section $\bar{s}$ of the multiplication map $\bar{q}$,

(2) $$\mathbb{F} \otimes R_{\mathcal{B}} \xrightarrow{\ \bar{s}\ } \mathbb{F} \otimes (\mathcal{B}_0 \otimes E^1_{\mathcal{A}} \otimes \mathcal{B}_0) \xrightarrow{\ \bar{q}\ } \mathbb{F} \otimes R_{\mathcal{B}}$$

with $\bar{s}p\alpha = p\otimes\alpha$ with $p \in E^1_{\mathcal{A}}$ and $\bar{s}(\alpha[a,b]\beta) = \alpha\otimes[a,b]\otimes\beta$. Therefore the kernel of $\bar{q}$ is generated by the elements $(\beta,\gamma \in \mathrm{Mon}(E_{\mathcal{A}}))$

(3) $$\beta\otimes p\otimes\gamma - s\bar{q}(\beta\otimes p\otimes\gamma),$$

(4) $$\beta\otimes[a,b]\otimes\gamma - s\bar{q}(\beta\otimes[a,b]\otimes\gamma).$$

Since the kernel of q in (16.4.4)(1) satisfies

(5) $$\mathrm{kernel}(q) = \mathcal{A}\otimes\mathrm{kernel}(\bar{q})$$

we have to check that for α admissible $\bar{A}$ vanishes on elements $\alpha\otimes(3)$ and $\alpha\otimes(4)$. Now definition of $\bar{A}$ on $\alpha\otimes(3)$ shows that $\bar{A}(\alpha\otimes(3)) = 0$ is always satisfied. Moreover by (16.4.4) we see that $\bar{A}(\alpha\otimes(4)) = 0$ holds if and only if

$$(A_{a,b}(\alpha\beta) + \alpha A_{a,b}(\beta))\gamma = \kappa(\alpha)\Theta\gamma + \sum_i (A_{a_i,b_i}(\alpha\alpha_i) + \alpha A_{a_i,b_i}(\alpha_i))\beta_i\gamma.$$

This equation is obtained by multiplying the equation in (16.5.6) by γ. □

In the next result we use the section $\bar{s}$ of the reduced diagonal $\tilde{\delta} : \tilde{\mathcal{A}} \longrightarrow \tilde{\mathcal{A}} \otimes \tilde{\mathcal{A}}$ defined as follows. For $x \in image(\tilde{\delta})$ there is a unique element $\tilde{s}(x) \in \tilde{\mathcal{A}}$ with $\delta\tilde{s}(x) = x$ such that $\tilde{s}(x)$ is a sum of non-primitive Milnor generators. Moreover we use the basis of preadmissible relations $\alpha[a, b]$ in the free right $\mathcal{F}_0$-module $R_{\mathcal{F}}$ in (16.5.1)(3). Let $PAR(n)$ be the set of all preadmissible relations of degree n.

16.5.7 Proposition. *Let $i \geq 1$. For each function*

$$\varepsilon^i : PAR(2^i) \longrightarrow \mathbb{F}$$

there is a unique map $\gamma = \gamma(\varepsilon^i) : R_{\mathcal{F}} \longrightarrow \tilde{\mathcal{A}}$ satisfying the properties in (16.3.2) *and satisfying*

$$\begin{aligned}
&\gamma(x) = 0 \ \text{ for } |x| < 2^i, \\
&\gamma(x) = \varepsilon^i(x) \cdot Q^i \ \text{ for } x \in PAR(2^i), \text{ see } (16.4.7), \\
&\gamma(x) = \text{sum of non} - \text{primitive Milnor generators for } x \in PAR(2^i),\ j > i.
\end{aligned}$$

Moreover each γ in (16.3.2) *determines well-defined functions ε^i, $i \geq 1$, such that*

$$\gamma = \sum_{i \geq 1} \gamma(\varepsilon^i).$$

Proof. Using (16.3.2)(3) we define $\gamma(x) = \gamma(\varepsilon^i)$ in degree $> 2^i$ by the formula $\gamma(x) = \tilde{s}\gamma_\sharp\tilde{\Delta}(x)$, $x \in PAR$. □

16.6 Computation of $\mathcal{B}$

We consider the case of the even prime p. The structure of $\mathcal{B}$ is completely determined by the function ξ_S and by the multiplication functions $A_{a,b}$ with $0 < a < 2b$. Since ξ_S is right equivariant the function ξ_S is well defined by the elements

$$\xi_S(\alpha[a, b]) \in \tilde{\mathcal{A}} \otimes \tilde{\mathcal{A}}$$

where $\alpha[a, b]$ is a preadmissible relation. A list of such elements in low degrees is given below. In this section we also describe well-defined elements

$$A_{a,b}(\alpha) \in \mathcal{A}$$

for all admissible α in $\mathcal{A}$. These elements determine $A_{a,b}$.

16.6.1 Theorem. *There exists a splitting u of $\mathcal{B}$,*

$$u : R_{\mathcal{B}} \longrightarrow \mathcal{B}_1$$

which is right equivariant with respect to the action of $\mathcal{B}_0$ and which satisfies

$$\Delta_1 u(x) = u_\sharp(\Delta_0 x) + \Sigma\xi_S(x)$$

for the diagonal $\Delta = (\Delta_1, \Delta_0)$ *of* $\mathcal{B}$. *Moreover*

$$\alpha \cdot u([a,b]) = u(\alpha[a,b]) + \Sigma A_{a,b}(\alpha)$$

for $[a,b] \in R_{\mathcal{B}}$ *as defined in* (16.4.1) *and* $\alpha \cdot u([a,b])$ *defined by the left action of* $\alpha \in \mathcal{B}_0$ *on* $\mathcal{B}_1$.

This result shows that $\mathcal{B}$ is completely determined by the elements $\xi(\alpha[a,b])$ and $A_{a,b}(\alpha)$. Theorem (16.6.1) allows the computation of *matrix Massey products* $\langle X, Y, Z\rangle$ as defined in (5.5.7).

16.6.2 Corollary. *Let* $X = (x^j)$, $Y = (y^i_j)$ *and* $Z = (z_i)$ *be matrices in* $\mathcal{A}$ *with* $XY = 0$ *and* $YZ = 0$. *Then we choose matrices* $\bar{Y} = (\bar{y}^i_j)$ *and* $\bar{Z} = (\bar{z}_i)$ *in* $\mathcal{B}_0$ *which map to* Y *and* Z *respectively. Then*

$$\sum_i \bar{y}^i_j \bar{z}_i \in R_{\mathcal{B}}$$

and we get

$$(*) \qquad -\sum_j A(x^j \otimes \sum_i \bar{y}^i_j \bar{z}_i) \in \langle X, Y, Z\rangle.$$

Here the function $A : \mathcal{A} \otimes R_{\mathcal{F}} \longrightarrow \mathcal{A}$ *of degree* -1 *is defined in terms of the multiplication functions* $A_{a,b} : \mathcal{A} \longrightarrow \mathcal{A}$ *as in* (16.4.4)(1).

Proof. Let also $\bar{X} = (\bar{x}^j)$ be a matrix mapping to X. Then the splitting u in (16.6.1) satisfies

$$(1) \qquad u(\bar{X}\bar{Y}) \cdot \bar{Z} - \bar{X}u(\bar{Y}\bar{Z}) \in \langle X, Y, Z\rangle.$$

Moreover since u is right equivariant (1) coincides with

$$(2) \qquad u(\bar{X}\,\bar{Y}\,\bar{Z}) - (u(\bar{X}\,\bar{Y}\,\bar{Z}) - \Sigma(*)) = \Sigma(*)$$

where $(*)$ is the term in the corollary. □

For the inductive determination of the elements $A_{a,b}$ we need the following splitting $\tilde{s}$ of the reduced diagonal $\tilde{\delta}$,

$$(16.6.3) \qquad \text{image}(\tilde{\delta}) \xrightarrow{\tilde{s}} \tilde{\mathcal{A}} \xrightarrow{\tilde{\delta}} \tilde{\mathcal{A}} \otimes \tilde{\mathcal{A}}$$

with $\tilde{\delta}\tilde{s}(x) = x$. The splitting $\tilde{s}$ is determined by the Milnor basis of $\tilde{\mathcal{A}}$ given by elements $\mathrm{Sq}^{\underline{n}}$ where $\underline{n} = (n_1, n_2, \dots)$ is a sequence of natural numbers $n_i \geq 0$ for $i \geq 1$ with $n_i = 0$ for almost all indices $i \geq 1$. For $\underline{n} = (n, 0, 0, \dots)$ we actually have $\mathrm{Sq}^{\underline{n}} = \mathrm{Sq}^n$. We have the formula

$$\tilde{\delta}\,\mathrm{Sq}^{\underline{n}} = \sum_{\substack{\underline{i},\underline{j} \neq 0 \\ \underline{i}+\underline{j}=\underline{n}}} \mathrm{Sq}^{\underline{i}} \otimes \mathrm{Sq}^{\underline{j}}.$$

The kernel of $\tilde{\delta}$ is generated by the primitive Milnor generators

$$Q_i = \mathrm{Sq}^{(0,\dots,0,1,0,\dots)}.$$

Therefore for $x \in \mathit{image}(\tilde{\delta})$ there is a unique element $\tilde{s}(x) \in \tilde{\mathcal{A}}$ with $\tilde{\delta}\tilde{s}(x) = x$ such that $\tilde{s}(x)$ is a sum of non-primitive Milnor generators. This defines the splitting $\tilde{s}$.

16.6.4 Definition. We introduce well-defined elements $A_{a,b}(\alpha)$ inductively as follows. In degree 3 we set

(1) $$A_{1,1}(\mathrm{Sq}^1) = 0.$$

Assume now $A_{a,b}(\alpha)$ is defined for degree $|\alpha| + a + b < n$, $n \geq 4$ and let $\alpha[a,b]$ be given where α is admissible and $|\alpha| + a + b = n$. Then the term $W(\alpha \otimes [a,b])$ in (16.4.5) is defined and we get the sum in $\tilde{\mathcal{A}} \otimes \tilde{\mathcal{A}}$:

(2) $$\begin{aligned} Z = {} & L(\alpha \otimes [a,b]) + \xi_S(\alpha[a,b]) + \alpha\xi_S([a,b]) \\ & + (\delta\kappa(\alpha))U(a,b) + W(\alpha \otimes [a,b]). \end{aligned}$$

This element is in the image of $\tilde{\delta}$ so that by the section $\tilde{s}$ in (16.6.2) also

(3) $$\tilde{s}(Z) \in \tilde{\mathcal{A}}$$

is defined.

We now introduce the *length function*. The function length_a carries a monomial α to the number ≥ 0 defined as follows. Let α' be the largest submonomial of α satisfying $\alpha = \alpha''\alpha'$ and $\alpha'[a,b]$ is preadmissible. Then $\mathrm{length}_a(\alpha)$ is the length of the monomial α''. We have $\mathrm{length}_a(\alpha) = 0$ if and only if $\alpha[a,b]$ is preadmissible.

If the degree n of $\alpha[a,b]$ is not a power of 2 we set

(4a) $$A_{a,b}(\alpha) = \tilde{s}(Z).$$

Moreover if the degree n of $\alpha[a,b]$ is a power of 2 with $n = 2^i$, then

(4b) $$A_{a,b}(\alpha) = \begin{cases} \tilde{s}(Z) & \text{if } \mathrm{length}_a(\alpha) = 0, \\ \tilde{s}(Z) + \varepsilon_{a,b}(\alpha) \cdot Q^i & \text{if } \mathrm{length}_a(\alpha) = 1. \end{cases}$$

Here $\varepsilon_{a,b}(\alpha) \in \mathbb{F}$ is a variable and Q^i is defined in (16.4.7).

If the degree n of $\alpha[a,b]$ is a power of 2 and if $\mathrm{length}_a(\alpha) \geq 2$, we define $A_{a,b}(\alpha)$ inductively as follows. Let Sq^m be the first factor of α, that is $\alpha = \mathrm{Sq}^m \beta$, with $\mathrm{length}_a(\beta) < \mathrm{length}_a(\alpha)$. Then we get $A_{a,b}(\alpha)$ by the formula:

(5) $$A_{a,b}(\alpha) = A_{a,b}(\mathrm{Sq}^m \beta) = \mathrm{Sq}^m A_{a,b}(\beta) + \mathrm{Sq}^{m-1}\Theta + \Psi.$$

Here Θ, $\Psi \in \mathcal{A}$ are defined as follows. Since the preadmissible relations form a basis of the right $\mathcal{F}_0$-module $R_{\mathcal{F}}$, we have the formula

$$\beta[a,b] = \sum_i \alpha_i[a_i,b_i]\beta_i \ \text{ in } \mathcal{F}_0$$

with $\alpha_i[a_i, b_i]$ preadmissible. As in (16.5.5) this formula yields an element $\Theta \in \mathcal{B}_0$ such that

(6) $$\beta[a,b] = p\Theta + \sum_i \alpha_i[a_i, b_i]\beta_i \text{ in } \mathcal{B}_0.$$

This determines the corresponding element Θ in $\mathcal{A}$. Moreover we define Ψ by the sum

(7) $$\Psi = \sum_i (A_{a_i,b_i}(\mathrm{Sq}^m \alpha_i) + \mathrm{Sq}^m A_{a_i,b_i}(\alpha_i)) \cdot \beta_i.$$

Here $\mathrm{length}_{a_i}(\mathrm{Sq}^m \alpha_i) \leq 1$ so that Ψ is already defined by (4b).

This shows that all elements $A_{a,b}(\alpha)$ (where α is a monomial and $a+b+|\alpha| = 2^i$) are defined in terms of the *ε-vectors*:

(8) $$\varepsilon_{a,b}(\alpha) \in \mathbb{F} \text{ with } \mathrm{length}_a(\alpha) = 1,\ a+b+|\alpha| = 2^i.$$

Now $A_{a,b}$ is a function on $\mathcal{A}$ so that for a relation $r \in R_{\mathcal{F}}$ we have $A_{a,b}(r) = 0$. The relations

$$\beta[s,t]\gamma \in R_{\mathcal{F}},\ \beta[s,t] \text{ preadmissible, } \gamma \text{ monomial}$$

with $|\alpha| + s + t + |\beta| = 2^i - a - b$ generate $R_{\mathcal{F}}$ in this degree. Therefore the Adem relation $[s,t]$ written in the form $[s,t] = \sum_j u_j$ with u_j a monomial of length 2 yields the equation

(9) $$\sum_j A_{a,b}(\beta u_j \gamma) = 0.$$

Here $A_{a,b}(\beta u_j \gamma)$ is defined in terms of (8) so that the equations (9) yield linear equations for the ε-vectors in (8). Any choice of ε-vector satisfying the equations defines $A_{a,b}(\alpha)$ in degree 2^i. This way a computer can compute a list of elements $A_{a,b}(\alpha)$ in low degrees. The list in the tables below is obtained by the choice of ε-vectors given as follows:

(10) $$\begin{aligned} \varepsilon_{3,3}(\mathrm{Sq}^2) &= 1, \\ \varepsilon_{3,3}(\mathrm{Sq}^4\,\mathrm{Sq}^6) &= 1, \\ \varepsilon_{3,3}(\mathrm{Sq}^8\,\mathrm{Sq}^{12}\,\mathrm{Sq}^6) &= 1, \end{aligned}$$

and $\varepsilon_{a,b}(\alpha) = 0$ otherwise, $a + b + |\alpha| \leq 63$.

We point out that the choices of ε satisfy the conditions above so that the elements $A_{a,b}(\alpha)$ are uniquely determined for $a + b + |\alpha| \leq 63$.

By induction the ε-vectors in degree $< 2^i$ are already chosen. Then the ε-vectors in degree 2^i satisfying the equations given by (9) form an affine subspace $V(2^i)$ in the vector space of all ε-vectors in degree 2^i. The dimension of this affine subspace can be controlled by the equivalences γ in (16.3.2) and (16.5.7). For this we define a linear map

$$L : \lambda\text{-vectors} \longrightarrow \varepsilon\text{-vectors} \tag{11}$$

follows. Here a λ-*vector* is a tuple

$$\lambda = (\lambda[a,b] \in \mathbb{F},\ 0 < a < 2b)$$

and L carries λ to the ε-vector defined by

$$\varepsilon_{a,b}(\alpha) = \sum_i \lambda[a_i, b_i].$$

Here the $[a_i, b_i]$ are all relations which appear in the sum decomposition (16.5.4) of the element $\alpha[a,b]$ without factors, that is, with $\alpha_i = 1$, $\beta_i = 1$. Then (16.5.7) shows that the image of L acts transitively and effectively on the affine space $V(2^i)$. For this we point out that we have the condition (16.6.3)(4b) for $\text{length}_a(\alpha) = 0$ where $\alpha \neq 1$.

The author is very grateful to Mamuka Jibladze for working out a computer program implementing the algorithm above for the computation of the multiplication table $A_{a,b}(\alpha)$. He implemented a Maple package on computations in the secondary Hopf algebra $\mathcal{B}$ which is based on Monk's package [Mo] on computations in the Steenrod algebra $\mathcal{A}$.

Using Mamuka's package one obtains the tables below which in particular show the multiplication table of the algebra $\mathcal{B}$ in degree ≤ 17. In degree 17 one can compute by (16.6.2) the triple matrix Massey product $\langle C, B, A\rangle$ with the matrices (see Harper [Ha], Section 6.2):

$$A = \begin{pmatrix} \mathrm{Sq}^1 \\ \mathrm{Sq}^2 \\ \mathrm{Sq}^4 \\ \mathrm{Sq}^8 \end{pmatrix},$$

$$B = \begin{pmatrix} \mathrm{Sq}^1 & 0 & 0 & 0 \\ \mathrm{Sq}^3 & \mathrm{Sq}^2 & 0 & 0 \\ \mathrm{Sq}^4 & \mathrm{Sq}^2\,\mathrm{Sq}^1 & \mathrm{Sq}^1 & 0 \\ \mathrm{Sq}^7 & \mathrm{Sq}^6 & \mathrm{Sq}^4 & 0 \\ \mathrm{Sq}^8 & \mathrm{Sq}^7 & \mathrm{Sq}^4\,\mathrm{Sq}^1 & \mathrm{Sq}^1 \\ \mathrm{Sq}^7\,\mathrm{Sq}^2 & \mathrm{Sq}^8 & \mathrm{Sq}^4\,\mathrm{Sq}^2 & \mathrm{Sq}^2 \\ \mathrm{Sq}^{15} & \mathrm{Sq}^{14} & \mathrm{Sq}^{12} & \mathrm{Sq}^8 \end{pmatrix},$$

$$\begin{aligned} C = (\,&\mathrm{Sq}^{15} + \mathrm{Sq}^{11}\,\mathrm{Sq}^4,\ \mathrm{Sq}^{11}\,\mathrm{Sq}^2,\ \mathrm{Sq}^{12} + \mathrm{Sq}^{11}\,\mathrm{Sq}^1, \\ &\mathrm{Sq}^8\,\mathrm{Sq}^1 + \mathrm{Sq}^6\,\mathrm{Sq}^3 + \mathrm{Sq}^6\,\mathrm{Sq}^2\,\mathrm{Sq}^1, \\ &\mathrm{Sq}^8 + \mathrm{Sq}^6\,\mathrm{Sq}^2,\ \mathrm{Sq}^7 + \mathrm{Sq}^4\,\mathrm{Sq}^2\,\mathrm{Sq}^1,\ \mathrm{Sq}^1). \end{aligned}$$

The multiplication table applied to $\langle C, B, A\rangle$ in (16.6.2) yields the matrix triple Massey product

$$\mathrm{Sq}^{16} \in \langle C, B, A\rangle \neq 0. \tag{16.6.5}$$

More precisely one gets by (16.6.2) and the tables the element

$$\begin{aligned}\mathrm{Sq}^{16} + \mathrm{Sq}^{2,0,2} + \mathrm{Sq}^{6,1,1} + \mathrm{Sq}^{1,0,0,1}& \\ = \mathrm{Sq}^{16} + \mathrm{Sq}^{14}\mathrm{Sq}^2 + \mathrm{Sq}^{13}\mathrm{Sq}^2\mathrm{Sq}^1 + \mathrm{Sq}^{11}\mathrm{Sq}^5 + \mathrm{Sq}^{11}\mathrm{Sq}^4\mathrm{Sq}^1& \\ + \mathrm{Sq}^{10}\mathrm{Sq}^5\mathrm{Sq}^1 + \mathrm{Sq}^{10}\mathrm{Sq}^4\mathrm{Sq}^2 + \mathrm{Sq}^9\mathrm{Sq}^4\mathrm{Sq}^2\mathrm{Sq}^1&\end{aligned}$$

which represents $\langle C, B, A\rangle$. This implies (16.6.5) by use of the indeterminacy of the triple Massey product. Using (16.6.5) it is easy to deduce the result of Adams [A] on secondary cohomology operations, see theorem 6.2.1 in [Ha]. We point out that the computation of the triple Massey product (16.6.5) is also a (non-immediate) consequence of the Adams result and of the fact that Sq^{2^n}, $n \geq 0$, generate the Steenrod algebra.

Moreover triple Massey products $\langle \alpha, \beta, \gamma\rangle$ with $\alpha, \beta, \gamma \in \mathcal{A}$ are computed in the tables below. It turns out that

$$\langle \alpha, \beta, \gamma\rangle = 0 \text{ for } |\alpha| + |\beta| + |\gamma| \leq 17 \tag{16.6.6}$$

and that for $\alpha = \mathrm{Sq}^{0,2}$ we have, in degree 18,

$$\mathrm{Sq}^{0,1,2} \in \langle \mathrm{Sq}^{0,2}, \mathrm{Sq}^{0,2}, \mathrm{Sq}^{0,2}\rangle \neq 0. \tag{16.6.7}$$

Here we use the Milnor generators

$$\mathrm{Sq}^{0,2} = \mathrm{Sq}^6 + \mathrm{Sq}^5\mathrm{Sq}^1 + \mathrm{Sq}^4\mathrm{Sq}^2$$

and

$$\begin{aligned}\mathrm{Sq}^{0,1,2} = \mathrm{Sq}^{17} + \mathrm{Sq}^{16}\mathrm{Sq}^1 + \mathrm{Sq}^{15}\mathrm{Sq}^2 + \mathrm{Sq}^{14}\mathrm{Sq}^3 + \mathrm{Sq}^{13}\mathrm{Sq}^4& \\ + \mathrm{Sq}^{12}\mathrm{Sq}^5 + \mathrm{Sq}^{11}\mathrm{Sq}^5\mathrm{Sq}^1 + \mathrm{Sq}^{11}\mathrm{Sq}^4\mathrm{Sq}^2 + \mathrm{Sq}^{10}\mathrm{Sq}^5\mathrm{Sq}^2.&\end{aligned}$$

This is the first example of a non-trivial triple Massey product in $\mathcal{A}$.

The computer calculations show that the inductive system of equations determining $A_{a,b}(\alpha)$ has indeed solutions. The author was very pleased by the calculations since they are a wonderful manifestation of the correctness of the new elaborate theory in this book.

Tables

Below are given all those triples $\langle\alpha,\beta,\gamma\rangle$ of homogeneous elements of degree $\leqslant 22$ in the Steenrod algebra with $\alpha\beta=\beta\gamma=0$, which are indecomposable in the sense that they cannot be presented in the form $\langle\alpha_1\alpha_2,\beta,\gamma\rangle$ with $\alpha_2\beta=0$, $\langle\alpha,\beta_1\beta_2,\gamma\rangle$ with $\alpha\beta_1=\beta_2\gamma=0$ or $\langle\alpha,\beta,\gamma_1\gamma_2\rangle$ with $\beta\gamma_1=0$.

All the corresponding triple Massey products contain 0 except for

$$\langle \mathrm{Sq}(0,2), \mathrm{Sq}(0,2), \mathrm{Sq}(0,2)\rangle$$

(where $\mathrm{Sq}(0,2)=\mathrm{Sq}^6+\mathrm{Sq}^5\,\mathrm{Sq}^1+\mathrm{Sq}^4\,\mathrm{Sq}^2$ is the Milnor basis element) which contains $\mathrm{Sq}(0,1,2)\notin \mathrm{Sq}(0,2)\mathscr{A}+\mathscr{A}\,\mathrm{Sq}(0,2)$.

Table 1. Triple Massey products in the mod 2 Steenrod algebra

Degree	α	β	γ
3	1	1	1
4			
5			
6	1	3	2
	2	2.1	1
7			
8	3	2	2.1
	3	3.1	1
	1	3.1	2.1
9	2	$5+4.1$	2
	$3+2.1$	$3+2.1$	$3+2.1$
	2.1	4.1	1
	1	5	3
10	5	3	2
	2	5.1	2
	1	5.2	2

	2	6.1	1
	2	2.1	4.1
11			
12	5	5.1	1
	1	7	4
	1	5.2.1	2.1
	3	7.1	1
	5	3.1	2.1
	3	3.1	4.1
	4 + 3.1	4.2.1	1
	1	5.1	4.1
13	2.1	8.1	1
	2	6.3	2
14	9 + 7.2	3	2
	5	3.1	4.1
	2	2.1	8.1
	2	10.1	1
15	1	5.2	7 + 4.2.1
	3	6	4.2
	6	4.2	2.1
	1	7.1	4.2
	7 + 4.2.1	6.1	1
	2	9 + 8.1	4
	4 + 3.1	9 + 8.1 + 7.2 + 6.2.1	2
	3	7.3	2
	2	6.3.1	2.1
	1	5	9 + 8.1 + 6.2.1
	1	9	5
	6	5.2.1	1
	3 + 2.1	7.2 + 9 + 8.1 + 6.3	3 + 2.1
16	5	7	4
	4	9.1 + 7.2.1	2
	1	9.2	3.1
	5	5.1	4.1
	3 + 2.1	7 + 6.1	4.2 + 5.1
	6 + 5.1	5.2 + 4.2.1	3 + 2.1
	1	5	10 + 8.2 + 6.3.1
	9.2	3.1	1
	2	9.2 + 8.3	3
	3	11.1	1
	9 + 7.2	3.1	2.1

	10 + 8.2 + 7.3	4.1	1
	4 + 3.1	4.2.1	4.1
	3	3.1	8.1
	1	9.2.1	2.1
	2.1	8.3	2
	2	9.1	4
	1	5.1	8.1
17	2.1	12.1	1
	9	5	3
	2	10.3	2
	2	9.3.1	2
	9	5.1	2
	1	9.4	3
	2	9.3	3
	2.1	4.1	8.1 + 6.2.1
18	9.2	3.1	2.1
	6	9.1 + 7.2.1	2
	1	11	6
	2.1	8.3.1	2.1
	2	2.1	12.1
	1	9.4.2	2
nonzero	**6 + 5.1 + 4.2**	**6 + 5.1 + 4.2**	**6 + 5.1 + 4.2**
	6 + 4.2	7.2 + 9 + 8.1 + 6.3	3 + 2.1
	9	5.2	2
	9 + 7.2	3.1	4.1
	2	6.1	8.1 + 6.2.1
	2	14.1	1
	5	3.1	8.1
	4	10 + 8.2 + 7.3 + 6.3.1	4 + 3.1
	2	6.3	7 + 4.2.1
	7 + 4.2.1	6.3	2
19	6.2	7.2.1	1
	6	5.2.1	4.1
	1	9.1	6.2
	1	9.3	4.2
	2	10.3.1	2.1
	1	7.2.1	6.2
	1	7.2	7.2
	3	11.3	2
	5	7.1	4.2
	10 + 8.2 + 7.3	5.2	2
	7.3	5.2.1	1

	1	5.2	8.2.1
	1	11.2	5
	6.2.1	6.2.1	1
	7.1	6.3.1	1
	5.2	8.2.1	1
	4	11.2.1	1
	1	7.1	6.3.1
	2	6.1	10 + 8.2 + 6.3.1
	4.1	10.2.1	1
	1	7.3	5.2.1
20	3	7.1	8.1 + 6.2.1
	4 + 3.1	4.2.1	8.1
	2	9 + 8.1	9 + 8.1
	9 + 7.2	7	4
	9.2	7.1	1
	6	9.2.1 + 8.3.1	2
	9	5.2.1	2.1
	6.2 + 5.2.1	8.2.1	1
	3 + 2.1	7.2 + 6.3	6.2 + 7.1 + 5.2.1
	7 + 4.2.1	6.3.1	2.1
	9.2	3.1	4.1
	1	9.4.2.1	2.1
	4	10 + 7.3 + 6.3.1 +8.2 + 7.2.1 + 9.1	6 + 5.1 + 4.2
	5	5.1	8.1
	7.3 + 7.2.1 + 6.3.1	5.2 + 4.2.1	3 + 2.1
	2.1	12.3	2
	6.2 + 7.1 + 5.2.1	6.3 + 6.2.1	3 + 2.1
	4 + 3.1	12.2.1	1
	9	5.1	4.1
	3	3.1	12.1
	1	9.4.1	4.1
	3	15.1	1
	1	9.1	8.1
	8.2.1 + 7.3.1 + 10.1	4.2	2.1
	6.1 + 5.2	8.3.1	1
	3	6	7.3.1 + 11 + 9.2 +8.2.1 + 10.1
	1	5.1	12.1
	3 + 2.1	7 + 6.1	7.3 + 7.2.1 + 6.3.1
	5	7.2	6 + 4.2
	7.3.1 + 11 + 9.2 +8.2.1 + 10.1	4.2	2.1

	5	13.1	1
	3	7.3	7 + 4.2.1
	2	9.1	6.2
	2	9.3	4.2
	3	6	11 + 9.2 + 7.3.1
	6 + 4.2	6.2.1	4.1
	3 + 2.1	7 + 6.1	7.3 + 6.3.1 + 9.1
	9	9.1	1
	2	10.5	3
21	2	10.5.2	2
	10.2	5.2	2
	1	9.4	4.2.1
	5.2	7 + 4.2.1	6.1
	7 + 6.1 + 5.2 + 4.2.1	7 + 6.1 + 5.2 + 4.2.1	7 + 6.1 + 5.2 + 4.2.1
	3	6	10.2 + 9.2.1
	7.2	7.2	2.1
	2.1	16.1	1
	2	13 + 12.1	6
	2	9.1	8.1
	7 + 4.2.1	6.1	4.2.1
	7.3	5.2.1	2.1
	3	11.3.1	2.1
	3	6.2	7.2.1
	9.1	6.2	2.1
	3	6.2	10 + 8.2 + 7.3
	4	8.2.1 + 7.3.1 + 10.1	4.2
	1	7.3.1	6.2.1
	1	7.3	7.2.1
	6 + 4.2	6.2 + 5.2.1	4.2.1
	9.2	6.2.1	1
	9	7.2	2.1
	1	13	7
	2	6.1	10.2
	5	7.2	4.2.1
	9.3	4.2	2.1
	3	6	9.2.1 + 8.3.1
	7	7.3.1	2.1
	11.4	4.1	1
	10.2 + 11.1	4.2	2.1
	4.2.1	8.4.1	1
	2	6.3	8.2
	7	4	10 + 8.2 + 7.3 + 6.3.1
	3	7.3	5.2.1

	2.1	4.1	10.2.1
	1	9.2	6.2.1
	9	5	4.2.1
	3	6	10.2 + 11.1
	10 + 8.2 + 7.3 + 6.3.1	4 + 3.1	4.2.1
	2	14.3	2
	1	11.4	4.1
	7	6.2.1	4.1
	3	7.1	6.3.1
	7.2	7.3.1	1
	7	4	9.1 + 7.2.1
	7	9.2 + 8.3	3
	7.1	6.3.1	2.1
	10.2 + 9.2.1	4.2	2.1
	4 + 3.1	11.2 + 10.2.1	4
	10 + 8.2 + 6.3.1	6.2	2.1
	7	9.3.1	1
	7.2.1	6.3.1	1
	1	9.3.1	4.2.1
	7	5.2	7 + 4.2.1
	9	7.3.1	1
	3	7.3.1	4.2.1
	6	11 + 9.2 + 7.3.1	4 + 3.1
	3	6	11.1 + 8.3.1
	7	4	10 + 7.3 + 6.3.1 +8.2 + 7.2.1 + 9.1
	11.2	5	3
	7	6.2 + 7.1	6 + 4.2
	7	4.1	8.1 + 6.2.1
	3	6.2.1	6.2.1
	10	6.3	2
	7.2.1	6.2	2.1
	4.2	9.4 + 8.4.1	2
22	7.2 + 6.3	7.3 + 7.2.1	3 + 2.1
	6.3	8.2	2.1
	2	18.1	1
	9 + 7.2	3.1	8.1
	3 + 2.1	7.2 + 9 + 8.1 + 6.3	10 + 8.2 + 7.2.1
	9 + 6.3	7.3 + 7.2.1	3 + 2.1
	6	10 + 8.2	4.2
	3	10	6.3
	9.1 + 6.3.1	6.3 + 6.2.1	3 + 2.1
	3 + 2.1	6.3.1 + 7.2.1	6.3 + 6.2.1

9.3 + 11.1 + 8.3.1	5.2 + 4.2.1	3 + 2.1
10 + 8.2 + 9.1	7.2 + 9 + 8.1 + 6.3	3 + 2.1
4	10.2 + 9.2.1	4.2
3	10.2	5.2
10 + 8.2 + 7.2.1	7.2 + 9 + 8.1 + 6.3	3 + 2.1
6.1	10.2	2.1
6.2 + 7.1	6 + 4.2	6.2 + 5.2.1
1	11.1	6.3
5	11	6
6.3.1	8.2.1	1
3 + 2.1	7 + 6.1	9.3 + 11.1 + 8.3.1
1	13.2	6
6.3	9.2.1	1
3 + 2.1	7.2 + 6.3	10 + 8.2 + 7.2.1
10 + 8.2 + 7.2.1	6.3 + 6.2.1	3 + 2.1
2	13.1	6
6	10.2 + 11.1	4 + 3.1
7 + 5.2	7.1 + 5.2.1	6.1 + 4.2.1
6.1 + 5.2	8.3.1	2.1
9 + 8.1	9 + 8.1	4
4.2	10.4.1	1
4.2	9.4.1	2
13 + 10.3	5.2	2
5	3.1	12.1
4	11.2.1	4
4.1	10.5 + 10.4.1	2
2	2.1	16.1
2	6.1	10.3 + 8.4.1
2.1	12.3.1	2.1
2	13.2 + 12.3	5
6.3.1 + 7.2.1	6.3 + 6.2.1	3 + 2.1
3 + 2.1	7.2 + 6.3	7.3 + 7.2.1
3	11.5	3
10 + 8.2 + 9.1	6.3 + 6.2.1	3 + 2.1
4	11.5 + 11.4.1	2
15.2 + 17 + 13.4 + 11.4.2	3	2

Table 2. The multiplication function $A_{a,b}$ on admissible monomials

$A_{1,2k+1}(\alpha) = 0$ for all k and α.

α	$A_{1,2}(\alpha)$	$A_{2,2}(\alpha)$	$A_{1,4}(\alpha)$	$A_{2,3}(\alpha)$	$A_{3,2}(\alpha)$	$A_{2,4}(\alpha)$	$A_{3,3}(\alpha)$
1	3	4 + 3.1	5	4.1 + 5	5	5.1	5.1
2	0	5	0	4.2 + 6	4.2 + 6 + 5.1	5.2 + 7	4.2.1 + 7 + 5.2 + 6.1
2.1	4.1 + 5	4.2 + 5.1	6.1	4.2.1 + 7 + 5.2 + 6.1	6.1	6.2 + 5.2.1	0
3	4.1 + 5	4.2 + 5.1	6.1	0	6.1	6.2 + 7.1	5.2.1 + 7.1
3.1	5.1	6.1 + 7	7.1	5.2.1 + 7.1	7.1	0	0
4	5.1	5.2	7.1	6.2 + 7.1	6.2	6.3 + 8.1 + 9 + 7.2	6.2.1 + 6.3 + 8.1 + 9
4.1	5.2	5.2.1 + 7.1	8.1 + 9 + 7.2	6.3 + 7.2	8.1 + 9 + 7.2	9.1 + 6.3.1 + 7.2.1 + 7.3	9.1 + 7.2.1
5	5.2	5.2.1	8.1 + 9 + 7.2	6.2.1 + 6.3	6.3	9.1 + 7.2.1	7.3
4.2	0	6.3 + 7.2	0	8.2 + 10	8.2 + 10 + 6.3.1 + 7.2.1	9.2 + 11 + 7.3.1	7.3.1 + 8.2.1 + 10.1 + 11 + 8.3
5.1	0	6.3 + 7.2	9.1	7.2.1 + 7.3 + 9.1	9.1	7.3.1 + 8.3 + 9.2	0

α	$A_{1,2}(\alpha)$	$A_{2,2}(\alpha)$	$A_{1,4}(\alpha)$	$A_{2,3}(\alpha)$	$A_{3,2}(\alpha)$	$A_{2,4}(\alpha)$	$A_{3,3}(\alpha)$
6	0	0	9.1	9.1 + 8.2 + 10 + 7.2.1 + 7.3	9.1 + 8.2 + 10 + 7.3	8.3 + 11	8.2.1 + 10.1 + 11 + 8.3
4.2.1	8.1 + 9 + 7.2 + 6.2.1	6.3.1 + 7.2.1	8.2.1 + 10.1	7.3.1 + 8.2.1 + 10.1 + 11 + 8.3	8.2.1 + 10.1	9.2.1 + 9.3 + 10.2	0
5.2	8.1 + 9 + 7.2 + 6.2.1	6.3.1 + 7.2.1 + 7.3	8.2.1 + 10.1	7.3.1 + 8.3 + 9.2	9.2 + 8.2.1 + 10.1 + 8.3	10.2 + 11.1 + 8.3.1 + 9.2.1 + 9.3	9.2.1 + 9.3 + 11.1
6.1	6.3	9.1 + 8.2 + 10 + 6.3.1	8.3 + 9.2	7.3.1 + 9.2 + 8.2.1 + 10.1 + 11	8.3 + 9.2	9.2.1 + 10.2	8.3.1 + 9.2.1
7	6.3	7.3 + 9.1 + 8.2 + 10 + 6.3.1	8.3 + 9.2	7.3.1	0	10.2 + 11.1 + 8.3.1 + 9.2.1	9.3 + 11.1 + 8.3.1
5.2.1	9.1 + 7.2.1	8.2.1 + 10.1 + 11 + 8.3	9.2.1 + 11.1	9.2.1 + 9.3 + 11.1	9.2.1 + 11.1	9.3.1	0
6.2	9.1 + 7.2.1	9.2 + 11 + 7.3.1	9.2.1 + 11.1	9.3 + 11.1 + 8.3.1	9.3 + 11.1 + 8.3.1	8.4.1 + 13 + 12.1 + 10.2.1 + 11.2 + 9.3.1 + 9.4	8.4.1 + 13 + 12.1 + 10.2.1 + 11.2 + 9.4
7.1	7.3	7.3.1	9.3	11.1 + 8.3.1	9.3	0	9.3.1

α	$A_{1,2}(\alpha)$	$A_{2,2}(\alpha)$	$A_{1,4}(\alpha)$	$A_{2,3}(\alpha)$	$A_{3,2}(\alpha)$	$A_{2,4}(\alpha)$	$A_{3,3}(\alpha)$
8	7.3	9.2 + 11 + 7.3.1	9.3	10.2 + 11.1	10.2 + 11.1	10.3 + 12.1 + 13 + 11.2 + 9.3.1	10.2.1 + 10.3 + 12.1 + 13 + 9.3.1
6.2.1	7.3.1 + 8.3 + 9.2	10.2 + 8.3.1	10.2.1 + 9.3.1	8.4.1 + 13 + 12.1 + 10.2.1 + 11.2 + 9.4	10.2.1 + 9.3.1	0	0
6.3	7.3.1 + 8.3 + 9.2	10.2 + 11.1 + 8.3.1 + 9.2.1 + 9.3	10.2.1 + 9.3.1	9.3.1 + 11.2 + 10.2.1	8.4.1 + 9.4 + 10.3	11.3 + 13.1 + 10.3.1 + 11.2.1 + 9.4.1	11.2.1 + 11.3 + 13.1
7.2	7.3.1 + 8.3 + 9.2	10.2 + 11.1 + 8.3.1 + 9.2.1 + 9.3	10.2.1 + 9.3.1	8.4.1 + 13 + 12.1 + 10.2.1 + 11.2 + 9.3.1 + 9.4	12.1 + 13 + 11.2 + 8.4.1 + 9.4	0	9.4.1 + 13.1 + 11.2.1
8.1	8.3	11.1 + 8.3.1 + 8.4 + 10.2	9.4	10.3 + 12.1 + 13 + 8.4.1 + 10.2.1 + 9.3.1 + 9.4	9.4	10.3.1 + 11.3 + 9.4.1	9.4.1
9	8.3	11.1 + 8.3.1 + 8.4 + 10.2	9.4	12.1 + 13 + 10.3 + 8.4.1 + 11.2 + 9.4	13 + 11.2 + 10.3 + 12.1 + 9.4	9.4.1	13.1 + 11.3 + 9.4.1 + 11.2.1
6.3.1	8.3.1 + 9.2.1	12.1 + 13 + 8.4.1 + 9.4	10.3.1 + 13.1	11.2.1 + 11.3 + 13.1	10.3.1 + 13.1	10.4.1 + 10.5	0

α	$A_{1,2}(\alpha)$	$A_{2,2}(\alpha)$	$A_{1,4}(\alpha)$	$A_{2,3}(\alpha)$	$A_{3,2}(\alpha)$	$A_{2,4}(\alpha)$	$A_{3,3}(\alpha)$
7.2.1	9.3	12.1 + 13 + 8.4.1 + 9.4	11.2.1	9.4.1 + 13.1 + 11.2.1	11.2.1	10.4.1 + 11.3.1 + 10.5 + 12.3 + 13.2	0
7.3	8.3.1 + 9.2.1 + 9.3	0	13.1 + 10.3.1 + 11.2.1	9.4.1 + 11.3	13.1 + 10.3.1 + 11.2.1	11.3.1 + 12.3 + 13.2	0
8.2	9.3	13 + 11.2 + 10.3 + 12.1 + 9.4 + 9.3.1	11.2.1	13.1 + 8.4.2 + 10.3.1 + 11.2.1 + 10.4 + 9.4.1	8.4.2 + 10.4	11.3.1 + 13.2 + 10.4.1 + 10.5 + 12.3 + 9.4.2 + 11.4	8.4.2.1 + 9.4.2 + 11.4 + 11.3.1 + 10.5
9.1	9.3	10.3 + 12.1 + 13 + 11.2 + 9.3.1	0	11.2.1 + 11.3 + 13.1	0	11.3.1 + 12.3 + 13.2	0
10	9.3	9.3.1 + 9.4 + 11.2	0	12.2 + 14 + 9.4.1 + 11.3	12.2 + 14 + 11.3	12.3 + 15	12.2.1 + 15 + 14.1 + 12.3
7.3.1	9.3.1	10.3.1 + 11.2.1 + 11.3	11.3.1	0	11.3.1	0	0
8.2.1	8.4.1 + 9.4	11.3 + 8.4.2	10.4.1 + 11.3.1	8.4.2.1 + 9.4.2 + 11.4 + 11.3.1 + 10.5	10.4.1 + 11.3.1	11.4.1 + 9.4.2.1 + 10.4.2 + 11.5	0
8.3	8.4.1 + 9.3.1 + 9.4	11.3 + 13.1 + 8.4.2 + 9.4.1	10.4.1	11.3.1 + 12.3 + 13.2	11.3.1 + 10.5	11.4.1 + 13.3 + 10.4.2 + 10.5.1	13.3 + 9.4.2.1 + 11.4.1
9.2	8.4.1 + 9.4	8.4.2 + 9.4.1	10.4.1 + 11.3.1	10.4.1 + 10.5	10.5	10.4.2 + 11.5 + 10.5.1	11.5 + 9.4.2.1

α	$A_{1,2}(\alpha)$	$A_{2,2}(\alpha)$	$A_{1,4}(\alpha)$	$A_{2,3}(\alpha)$	$A_{3,2}(\alpha)$	$A_{2,4}(\alpha)$	$A_{3,3}(\alpha)$
10.1	10.3	12.2 + 14 + 10.3.1 + 11.3 + 10.4	10.5	12.3 + 10.4.1 + 10.5 + 12.2.1 + 15 + 14.1	10.5	14.2 + 10.5.1 + 12.3.1	10.5.1
11	10.3	10.4 + 14 + 13.1 + 12.2 + 10.3.1	10.5	10.4.1 + 10.5 + 12.3 + 13.2	10.5 + 12.3 + 13.2	10.5.1 + 14.2 + 15.1	13.3 + 13.2.1 + 15.1 + 10.5.1
8.3.1	9.4.1	10.5 + 12.3 + 13.2 + 11.4 + 11.3.1	11.4.1 + 12.3.1 + 13.2.1	13.3 + 9.4.2.1 + 11.4.1	11.4.1 + 12.3.1 + 13.2.1	11.5.1	0
9.2.1	9.4.1	10.5 + 11.4	11.4.1	11.5 + 9.4.2.1	11.4.1	11.5.1	0
8.4	9.4.1	12.3 + 13.2 + 9.4.2	11.4.1 + 12.3.1 + 13.2.1	13.3 + 11.4.1 + 10.4.2	10.4.2 + 10.5.1 + 12.3.1 + 13.2.1	13.3.1 + 12.4.1 + 16.1 + 17 + 15.2 + 14.2.1 + 10.5.2 + 13.4 + 11.4.2	14.2.1 + 16.1 + 17 + 10.4.2.1 + 10.5.2 + 13.4 + 12.4.1 + 15.2 + 13.3.1
9.3	0	12.3 + 13.2	12.3.1 + 13.2.1	10.5.1 + 12.3.1 + 13.2.1	13.3 + 10.5.1 + 11.4.1 + 11.5	11.5.1 + 13.3.1	0
10.2	9.4.1	15 + 9.4.2 + 10.5 + 12.3	11.4.1	13.3 + 12.3.1 + 10.4.2 + 14.2 + 11.4.1	13.3 + 15.1 + 10.4.2 + 14.2 + 10.5.1	12.4.1 + 14.2.1 + 13.3.1 + 10.5.2 + 13.4 + 14.3 + 11.4.2 + 11.5.1	12.4.1 + 15.2 + 10.4.2.1 + 14.3 + 11.5.1 + 10.5.2 + 13.4

α	$A_{1,2}(\alpha)$	$A_{2,2}(\alpha)$	$A_{1,4}(\alpha)$	$A_{2,3}(\alpha)$	$A_{3,2}(\alpha)$	$A_{2,4}(\alpha)$	$A_{3,3}(\alpha)$
11.1	11.3	10.5 + 12.3 + 13.2 + 11.4 + 11.3.1	11.5	13.2.1 + 13.3 + 15.1 + 10.5.1 + 11.4.1 + 11.5	11.5	11.5.1 + 13.3.1	11.5.1
12	11.3	13.2 + 15 + 11.4 + 11.3.1	11.5	14.2 + 15.1 + 11.4.1 + 11.5 + 13.3	14.2 + 15.1 + 11.5 + 13.3	14.3 + 16.1 + 17 + 15.2 + 11.5.1	11.5.1 + 14.2.1 + 14.3 + 16.1 + 17
8.4.1	9.4.2	11.4.1 + 9.4.2.1 + 10.5.1	12.4.1 + 16.1 + 17 + 15.2 + 14.2.1 + 13.3.1 + 13.4 + 11.4.2	10.5.2 + 11.4.2	12.4.1 + 16.1 + 17 + 15.2 + 14.2.1 + 13.3.1 + 13.4 + 11.4.2	13.4.1 + 17.1 + 15.2.1 + 10.5.2.1 + 11.4.2.1 + 11.5.2	15.2.1 + 17.1 + 13.4.1 + 11.4.2.1
9.3.1	0	10.5.1 + 11.4.1 + 11.5	13.3.1	0	13.3.1	12.5.1 + 13.4.1 + 13.5	0
10.2.1	10.4.1 + 10.5	13.3 + 10.4.2 + 15.1 + 14.2 + 11.5 + 10.5.1	11.5.1	12.4.1 + 15.2 + 10.4.2.1 + 14.3 + 11.5.1 + 10.5.2 + 13.4	11.5.1	15.3 + 11.5.2 + 10.5.2.1 + 14.3.1	0
9.4	9.4.2	9.4.2.1	15.2 + 16.1 + 17 + 12.4.1 + 14.2.1 + 11.4.2 + 13.4	10.4.2.1 + 10.5.2	10.5.2 + 13.3.1 + 11.5.1	11.4.2.1 + 13.5 + 12.5.1 + 17.1 + 15.2.1	11.5.2

α	$A_{1,2}(\alpha)$	$A_{2,2}(\alpha)$	$A_{1,4}(\alpha)$	$A_{2,3}(\alpha)$	$A_{3,2}(\alpha)$	$A_{2,4}(\alpha)$	$A_{3,3}(\alpha)$
10.3	10.4.1 + 10.5	15.1 + 10.4.2 + 14.2 + 10.5.1	11.5.1 + 13.3.1	15.2 + 11.5.1 + 14.2.1	14.3 + 12.4.1 + 14.2.1 + 13.3.1 + 12.5	13.4.1 + 15.2.1 + 14.3.1 + 11.5.2	15.3 + 15.2.1 + 10.5.2.1 + 12.5.1 + 13.4.1 + 14.3.1
11.2	10.4.1 + 10.5	10.4.2 + 15.1 + 14.2 + 11.5 + 10.5.1	11.5.1	12.4.1 + 14.2.1 + 10.5.2 + 13.4 + 14.3 + 11.4.2	10.5.2 + 13.4 + 14.3 + 11.4.2 + 13.3.1 + 12.4.1 + 14.2.1	0	13.4.1 + 11.4.2.1 + 15.3 + 11.5.2
12.1	12.3	12.4 + 14.2 + 15.1 + 13.3 + 12.3.1 + 11.5	12.5	14.3 + 16.1 + 17 + 12.4.1 + 14.2.1 + 12.5 + 11.5.1	12.5	14.3.1 + 15.3 + 12.5.1	12.5.1
13	12.3	12.4 + 14.2 + 15.1 + 12.3.1	12.5	12.4.1 + 16.1 + 17 + 15.2 + 12.5 + 14.3	15.2 + 14.3 + 16.1 + 17 + 12.5	12.5.1	15.3 + 17.1 + 12.5.1 + 15.2.1
8.4.2	0	10.5.2 + 11.4.2	0	18 + 14.4 + 16.2 + 14.3.1 + 12.4.2	11.4.2.1 + 18 + 10.5.2.1 + 14.4 + 16.2 + 14.3.1 + 12.4.2	17.2 + 19 + 15.3.1 + 11.5.2.1 + 15.4 + 13.4.2	19 + 11.5.2.1 + 16.3 + 15.4 + 14.5 + 16.2.1 + 12.4.2.1 + 18.1 + 13.5.1 + 12.5.2

α	$A_{1,2}(\alpha)$	$A_{2,2}(\alpha)$	$A_{1,4}(\alpha)$	$A_{2,3}(\alpha)$	$A_{3,2}(\alpha)$	$A_{2,4}(\alpha)$	$A_{3,3}(\alpha)$
9.4.1	0	10.5.2 + 11.4.2 + 11.5.1	13.4.1 + 17.1 + 15.2.1	15.2.1 + 17.1 + 11.4.2.1 + 11.5.2 + 13.4.1	13.4.1 + 17.1 + 15.2.1	13.5.1 + 15.3.1 + 16.3 + 17.2 + 12.5.2 + 13.4.2 + 14.4.1 + 14.5 + 11.5.2.1	0
10.3.1	10.5.1	14.3 + 12.4.1 + 14.2.1 + 13.3.1 + 13.4	12.5.1 + 13.4.1	15.3 + 15.2.1 + 10.5.2.1 + 12.5.1 + 13.4.1 + 14.3.1	12.5.1 + 13.4.1	14.4.1 + 14.5	0
11.2.1	11.4.1 + 11.5	10.5.2 + 13.4 + 14.3 + 11.4.2 + 13.3.1 + 12.4.1 + 14.2.1	0	13.4.1 + 11.4.2.1 + 15.3 + 11.5.2	0	13.5.1 + 14.4.1 + 14.5 + 12.5.2 + 13.4.2 + 11.5.2.1	0
10.4	10.5.1	15.2 + 10.5.2	17.1 + 15.2.1 + 12.5.1	12.6 + 13.4.1 + 11.4.2.1 + 13.5 + 12.5.1 + 14.4 + 15.2.1 + 15.3	13.5 + 15.2.1 + 12.6 + 14.4 + 12.5.1	15.3.1 + 17.2 + 12.5.2 + 16.3 + 13.4.2 + 13.5.1 + 13.6 + 15.4	13.5.1 + 16.3 + 12.6.1 + 15.4 + 17.2 + 14.4.1 + 13.6 + 11.5.2.1
11.3	10.5.1 + 11.4.1 + 11.5	10.5.2 + 11.4.2 + 11.5.1	12.5.1 + 13.4.1	15.3 + 12.5.1	13.5	15.3.1 + 12.5.2 + 13.4.2	11.5.2.1 + 15.3.1 + 13.5.1
12.2	11.4.1 + 11.5	14.3 + 16.1 + 17 + 15.2 + 12.5 + 11.4.2 + 11.5.1	0	13.4.1 + 12.4.2 + 12.6 + 14.3.1 + 15.2.1 + 11.5.2	12.6 + 13.4.1 + 11.5.2 + 12.5.1 + 17.1 + 12.4.2 + 15.2.1	14.4.1 + 14.5 + 15.3.1 + 13.6 + 12.5.2	14.4.1 + 14.5 + 16.3 + 17.2 + 12.4.2.1 + 13.6 + 12.6.1 + 12.5.2

α	$A_{1,2}(\alpha)$	$A_{2,2}(\alpha)$	$A_{1,4}(\alpha)$	$A_{2,3}(\alpha)$	$A_{3,2}(\alpha)$	$A_{2,4}(\alpha)$	$A_{3,3}(\alpha)$
13.1	13.3	14.3 + 16.1 + 17 + 15.2 + 13.3.1 + 12.5 + 13.4	13.5	12.5.1 + 13.4.1 + 13.5 + 15.2.1 + 15.3 + 17.1	13.5	13.5.1 + 16.3 + 17.2 + 15.3.1	13.5.1
14	13.3	13.4 + 15.2 + 13.3.1	13.5	13.4.1 + 13.5 + 15.3 + 16.2 + 18	16.2 + 18 + 15.3 + 13.5	19 + 16.3 + 13.5.1	16.2.1 + 18.1 + 19 + 16.3 + 13.5.1

α	$A_{1,6}(\alpha)$	$A_{2,5}(\alpha)$	$A_{3,4}(\alpha)$	$A_{4,3}(\alpha)$	$A_{2,6}(\alpha)$	$A_{3,5}(\alpha)$	$A_{4,4}(\alpha)$	$A_{5,3}(\alpha)$
1	7	6.1 + 7	0	7 + 5.2 + 6.1	7.1	7.1	8 + 6.2 + 7.1	7.1
2	0	6.2	6.2 + 5.2.1	5.2.1	7.2	6.2.1 + 6.3 + 8.1 + 9	6.3 + 8.1 + 7.2	6.2.1 + 7.2
2.1	8.1 + 9	6.2.1 + 6.3	0	7.2 + 6.2.1 + 6.3 + 8.1 + 9	9.1 + 8.2 + 10 + 7.2.1 + 7.3	9.1	7.3 + 8.2 + 6.3.1	9.1
3	8.1 + 9	8.1 + 9	0	7.2	8.2 + 10	7.2.1 + 7.3	7.3 + 9.1 + 8.2	7.2.1 + 7.3
3.1	9.1	7.2.1 + 7.3 + 9.1	0	9.1 + 7.2.1	0	0	10.1 + 11 + 8.3 + 9.2 + 7.3.1	0

α	$A_{1,6}(\alpha)$	$A_{2,5}(\alpha)$	$A_{3,4}(\alpha)$	$A_{4,3}(\alpha)$	$A_{2,6}(\alpha)$	$A_{3,5}(\alpha)$	$A_{4,4}(\alpha)$	$A_{5,3}(\alpha)$
4	9.1	7.3	7.3 + 6.3.1	6.3.1 + 9.1	8.3 + 11	7.3.1	9.2	7.3.1 + 8.3 + 9.2
4.1	9.2 + 11	7.3.1 + 9.2 + 8.2.1 + 10.1 + 11	0	9.2 + 8.2.1 + 10.1 + 11	10.2 + 8.3.1	9.2.1 + 11.1	10.2 + 12 + 8.3.1 + 9.2.1	9.2.1 + 11.1
5	9.2 + 11	8.2.1 + 10.1 + 11 + 8.3	7.3.1 + 8.3 + 9.2	9.2 + 8.2.1 + 10.1 + 11	9.2.1 + 10.2	9.2.1 + 11.1	10.2 + 11.1 + 12 + 8.3.1 + 9.2.1	9.2.1 + 11.1
4.2	0	9.2.1 + 10.2	9.2.1 + 9.3 + 10.2	0	10.3 + 12.1 + 13 + 11.2 + 9.3.1	8.4.1 + 11.2 + 9.3.1 + 9.4 + 10.3	13 + 8.4.1 + 10.2.1 + 11.2 + 9.4	10.2.1 + 11.2
5.1	0	0	0	9.3	10.3 + 12.1 + 13 + 11.2 + 9.3.1	0	10.3 + 11.2	0
6	0	9.2.1 + 10.2	10.2 + 8.3.1	8.3.1 + 9.2.1	0	10.3 + 8.4.1 + 11.2 + 9.4	10.3 + 13 + 8.4.1 + 10.2.1 + 9.4	9.3.1 + 11.2 + 10.2.1
4.2.1	10.2.1 + 11.2	10.3 + 8.4.1 + 10.2.1 + 9.3.1 + 9.4	0	8.4.1 + 9.4 + 10.3	9.4.1 + 13.1 + 10.3.1 + 11.3	11.2.1	12.2 + 13.1 + 11.3	11.2.1
5.2	10.2.1 + 11.2	9.3.1 + 11.2 + 10.2.1	9.3.1	11.2 + 9.3.1	11.3 + 10.3.1	11.3 + 9.4.1 + 11.2.1	12.2 + 13.1 + 9.4.1 + 11.2.1 + 10.3.1	11.3 + 9.4.1 + 11.2.1

α	$A_{1,6}(\alpha)$	$A_{2,5}(\alpha)$	$A_{3,4}(\alpha)$	$A_{4,3}(\alpha)$	$A_{2,6}(\alpha)$	$A_{3,5}(\alpha)$	$A_{4,4}(\alpha)$	$A_{5,3}(\alpha)$
6.1	10.3 + 12.1 + 13	8.4.1 + 13 + 12.1 + 10.2.1 + 11.2 + 9.3.1 + 9.4	0	10.3 + 12.1 + 13 + 11.2 + 9.3.1	10.3.1 + 9.4.1	10.3.1 + 13.1	11.3 + 10.3.1 + 11.2.1 + 14 + 9.4.1	10.3.1 + 13.1
7	10.3 + 12.1 + 13	10.2.1 + 10.3 + 12.1 + 13 + 9.3.1	9.3.1	12.1 + 13 + 8.4.1 + 9.4	10.3.1 + 13.1	9.4.1 + 13.1 + 10.3.1 + 11.3	14	9.4.1 + 13.1 + 10.3.1 + 11.3
5.2.1	11.2.1	9.4.1 + 11.3	0	0	12.3 + 13.2	0	12.3 + 10.4.1 + 10.5 + 12.2.1	0
6.2	11.2.1	11.3 + 10.3.1	0	10.3.1 + 9.4.1	10.4.1 + 10.5	10.4.1 + 11.3.1 + 10.5 + 12.3 + 13.2	13.2 + 14.1 + 10.4.1 + 11.3.1 + 10.5	10.4.1 + 11.3.1 + 10.5 + 12.3 + 13.2
7.1	11.3 + 13.1	10.3.1 + 11.2.1 + 9.4.1	0	10.3.1 + 11.2.1 + 9.4.1	11.3.1 + 12.3 + 13.2	11.3.1	11.3.1 + 12.3 + 13.2 + 14.1 + 15	11.3.1
8	11.3 + 13.1	11.2.1	11.3 + 10.3.1	10.3.1 + 9.4.1	11.3.1 + 12.3 + 13.2	10.4.1 + 11.3.1 + 10.5 + 12.3 + 13.2	10.4.1 + 11.3.1 + 10.5 + 12.3 + 13.2 + 15	12.3 + 13.2
6.2.1	11.3.1 + 12.3 + 13.2	10.4.1 + 10.5	0	0	13.3	12.3.1 + 13.2.1	14.2 + 13.2.1	12.3.1 + 13.2.1
6.3	11.3.1 + 12.3 + 13.2	10.4.1 + 11.3.1 + 10.5 + 12.3 + 13.2	10.4.1 + 10.5	10.4.1 + 11.3.1 + 10.5 + 12.3 + 13.2	13.3 + 10.5.1 + 11.4.1 + 11.5	13.3 + 11.4.1 + 11.5 + 12.3.1 + 13.2.1	13.3 + 14.2 + 10.5.1 + 12.3.1 + 13.2.1	13.3 + 11.4.1 + 11.5 + 12.3.1 + 13.2.1

α	$A_{1,6}(\alpha)$	$A_{2,5}(\alpha)$	$A_{3,4}(\alpha)$	$A_{4,3}(\alpha)$	$A_{2,6}(\alpha)$	$A_{3,5}(\alpha)$	$A_{4,4}(\alpha)$	$A_{5,3}(\alpha)$
7.2	11.3.1 + 12.3 + 13.2	10.4.1 + 11.3.1 + 10.5	10.4.1 + 11.3.1 + 10.5 + 12.3 + 13.2	10.4.1 + 10.5 + 12.3 + 13.2	13.3	13.3 + 11.4.1 + 11.5 + 12.3.1 + 13.2.1	11.4.1 + 11.5 + 14.2 + 15.1	10.5.1
8.1	11.4 + 12.3 + 13.2	10.5 + 11.4	0	10.4.1 + 11.3.1 + 11.4 + 9.4.2	13.3 + 10.5.1 + 11.4.1 + 12.3.1 + 13.2.1	11.4.1 + 12.3.1 + 13.2.1	11.4.1 + 12.4 + 14.2 + 10.4.2 + 10.5.1	11.4.1 + 12.3.1 + 13.2.1
9	11.4 + 12.3 + 13.2	10.4.1 + 11.3.1 + 11.4	11.3.1 + 12.3 + 13.2	13.2 + 11.4 + 9.4.2 + 11.3.1 + 10.5 + 12.3	11.4.1 + 12.3.1 + 13.2.1	13.3 + 11.5 + 12.3.1 + 13.2.1	13.3 + 14.2 + 15.1 + 10.4.2 + 11.5 + 10.5.1 + 12.4	10.5.1 + 11.4.1
6.3.1	12.3.1 + 13.2.1	11.4.1 + 11.5 + 13.3	0	10.5.1 + 11.4.1 + 11.5	13.3.1 + 12.4.1 + 16.1 + 17 + 15.2 + 14.2.1 + 12.5 + 11.5.1	0	12.4.1 + 16.1 + 17 + 15.2 + 12.5 + 14.3	0
7.2.1	13.3	11.4.1 + 11.5 + 12.3.1 + 13.2.1	0	13.3 + 10.5.1 + 11.4.1 + 11.5	12.4.1 + 16.1 + 17 + 15.2 + 14.2.1 + 12.5	13.3.1	16.1 + 17 + 11.5.1 + 12.4.1 + 12.5	13.3.1
7.3	12.3.1 + 13.2.1 + 13.3	12.3.1 + 13.2.1 + 13.3	0	13.3	11.5.1 + 13.3.1	13.3.1	14.3 + 15.2 + 11.5.1	13.3.1

α	$A_{1,6}(\alpha)$	$A_{2,5}(\alpha)$	$A_{3,4}(\alpha)$	$A_{4,3}(\alpha)$	$A_{2,6}(\alpha)$	$A_{3,5}(\alpha)$	$A_{4,4}(\alpha)$	$A_{5,3}(\alpha)$
8.2	13.3	11.5 + 12.3.1 + 13.2.1 + 11.4.1 + 10.4.2	11.5 + 10.4.2 + 11.4.1 + 9.4.2.1	13.3 + 11.4.1 + 11.5 + 10.5.1 + 9.4.2.1	11.4.2 + 11.5.1	12.4.1 + 16.1 + 17 + 15.2 + 14.2.1 + 10.4.2.1 + 11.5.1 + 10.5.2 + 13.4	15.2 + 10.5.2 + 11.4.2 + 12.4.1 + 11.5.1 + 14.2.1	13.3.1 + 11.4.2 + 10.4.2.1
9.1	13.3	11.4.1 + 11.5 + 12.3.1 + 13.2.1	0	13.3 + 10.5.1	11.5.1 + 13.3.1	13.3.1	16.1 + 17 + 11.5.1 + 13.3.1	13.3.1
10	13.3	14.2 + 15.1 + 11.4.1 + 13.3	12.3.1 + 14.2	15.1 + 12.3.1 + 13.3 + 11.5	15.2 + 13.3.1	14.3 + 12.4.1 + 15.2 + 13.3.1 + 12.5	14.3 + 12.4.1 + 14.2.1 + 13.3.1 + 12.5 + 13.4 + 11.4.2 + 11.5.1	14.2.1 + 15.2 + 11.5.1 + 13.3.1

α	$A_{1,8}(\alpha)$	$A_{2,7}(\alpha)$	$A_{3,6}(\alpha)$	$A_{4,5}(\alpha)$	$A_{5,4}(\alpha)$
1	9	8.1	0	7.2	7.2 + 9
2	0	8.2 + 10	8.2 + 10 + 7.2.1 + 7.3	9.1 + 7.2.1 + 8.2 + 10	8.2 + 10 + 6.3.1
2.1	10.1	11 + 8.2.1 + 8.3	0	7.3.1	10.1 + 8.3 + 9.2
3	10.1	10.1	0	0	10.1 + 8.3 + 9.2 + 7.3.1

α	$A_{1,8}(\alpha)$	$A_{2,7}(\alpha)$	$A_{3,6}(\alpha)$	$A_{4,5}(\alpha)$	$A_{5,4}(\alpha)$
3.1	11.1	9.2.1 + 9.3	0	9.3	9.3 + 11.1
4	11.1	10.2 + 9.3	10.2 + 11.1 + 8.3.1 + 9.2.1	10.2 + 11.1 + 8.3.1 + 9.2.1 + 9.3	10.2
4.1	11.2 + 12.1	10.3 + 13 + 9.3.1	0	11.2	12.1 + 9.3.1
5	11.2 + 12.1	13 + 11.2 + 10.2.1 + 10.3	10.3 + 12.1 + 13 + 11.2 + 9.3.1	10.3 + 12.1 + 13	10.3 + 13 + 11.2 + 9.3.1
4.2	0	0	11.3 + 13.1 + 10.3.1 + 11.2.1 + 9.4.1	10.3.1 + 13.1	13.1 + 9.4.1
5.1	13.1	11.2.1 + 11.3	0	0	13.1
6	13.1	11.2.1 + 11.3	13.1 + 9.4.1	11.2.1 + 11.3 + 13.1	11.3 + 9.4.1 + 11.2.1 + 10.3.1
4.2.1	12.2.1	13.2 + 12.3 + 10.4.1 + 10.5 + 12.2.1	0	11.3.1	10.4.1 + 11.3.1 + 10.5 + 12.2.1
5.2	12.2.1	11.3.1 + 13.2 + 12.2.1 + 12.3	12.3 + 13.2	12.3 + 13.2	13.2 + 12.3 + 10.4.1 + 10.5 + 12.2.1

α	$A_{1,8}(\alpha)$	$A_{2,7}(\alpha)$	$A_{3,6}(\alpha)$	$A_{4,5}(\alpha)$	$A_{5,4}(\alpha)$
6.1	14.1 + 12.3 + 13.2	10.4.1 + 10.5 + 12.3 + 13.2 + 14.1	0	10.4.1 + 10.5 + 12.3 + 13.2	14.1
7	14.1 + 12.3 + 13.2	11.3.1 + 14.1 + 12.3 + 13.2	12.3 + 13.2	10.4.1 + 11.3.1 + 10.5	11.3.1 + 14.1 + 12.3 + 13.2
5.2.1	13.2.1	13.3 + 13.2.1 + 11.4.1 + 11.5	0	11.4.1 + 11.5	11.4.1 + 11.5 + 13.2.1
6.2	13.2.1	12.3.1	12.3.1 + 13.2.1 + 13.3	10.5.1	13.2.1 + 10.5.1
7.1	13.3 + 15.1	12.3.1 + 13.2.1 + 13.3 + 15.1 + 11.4.1 + 11.5	0	13.3	15.1
8	13.3 + 15.1	15.1	13.3 + 10.5.1	13.3 + 11.4.1 + 11.5 + 12.3.1 + 13.2.1	15.1 + 10.5.1 + 12.3.1 + 13.2.1

α	$A_{2,8}(\alpha)$	$A_{3,7}(\alpha)$	$A_{4,6}(\alpha)$	$A_{5,5}(\alpha)$	$A_{6,4}(\alpha)$
1	10 + 9.1	9.1	8.2 + 10 + 9.1	0	7.3 + 9.1 + 8.2
2	9.2	8.2.1 + 10.1 + 11 + 8.3	8.3 + 11	7.3.1 + 8.3 + 9.2	8.2.1 + 10.1 + 8.3 + 7.3.1
2.1	9.2.1 + 9.3 + 11.1 + 10.2	0	10.2 + 11.1 + 8.3.1 + 9.2.1	0	8.3.1
3	10.2	9.2.1 + 9.3 + 11.1	9.3 + 10.2 + 11.1	0	9.2.1 + 9.3
3.1	12.1 + 13	0	10.3 + 12.1 + 13 + 11.2 + 9.3.1	0	10.3 + 11.2 + 9.3.1
4	10.3 + 12.1 + 13 + 11.2	10.2.1 + 10.3 + 12.1 + 13 + 9.3.1	11.2	9.3.1	10.2.1
4.1	12.2 + 13.1 + 10.3.1 + 11.2.1 + 11.3	11.2.1	13.1 + 10.3.1 + 11.2.1	0	9.4.1 + 11.2.1 + 12.2
5	12.2 + 11.2.1	11.3 + 13.1	10.3.1 + 11.2.1 + 11.3	0	13.1 + 12.2 + 10.3.1 + 11.3 + 9.4.1
4.2	13.2 + 11.3.1	10.4.1 + 10.5 + 12.3 + 13.2	10.4.1 + 10.5 + 12.3 + 13.2	10.4.1 + 10.5	13.2 + 10.4.1 + 11.3.1 + 10.5

α	$A_{2,8}(\alpha)$	$A_{3,7}(\alpha)$	$A_{4,6}(\alpha)$	$A_{5,5}(\alpha)$	$A_{6,4}(\alpha)$
5.1	11.3.1 + 12.3 + 13.2	0	12.3 + 13.2	0	11.3.1
6	12.3	10.4.1 + 11.3.1 + 10.5 + 12.3 + 13.2	10.4.1 + 10.5	10.4.1 + 10.5	13.2 + 10.4.1 + 11.3.1 + 10.5
4.2.1	13.3 + 13.2.1 + 11.4.1 + 11.5	0	10.5.1 + 12.3.1 + 13.2.1	0	13.3 + 10.5.1 + 12.3.1
5.2	12.3.1 + 13.3	11.4.1 + 11.5 + 13.3	13.3 + 11.4.1 + 11.5 + 12.3.1 + 13.2.1	0	13.3 + 13.2.1 + 11.4.1 + 11.5
6.1	14.2 + 11.4.1 + 11.5 + 13.3	12.3.1 + 13.2.1	10.5.1 + 13.3 + 11.4.1 + 11.5 + 12.3.1 + 13.2.1	0	14.2 + 10.5.1 + 11.4.1 + 11.5 + 12.3.1 + 13.2.1
7	14.2 + 12.3.1 + 13.2.1 + 13.3	13.3 + 11.4.1 + 11.5 + 12.3.1 + 13.2.1	12.3.1 + 13.2.1	0	14.2

α	$A_{1,10}(\alpha)$	$A_{2,9}(\alpha)$	$A_{3,8}(\alpha)$	$A_{4,7}(\alpha)$	$A_{5,6}(\alpha)$	$A_{6,5}(\alpha)$	$A_{7,4}(\alpha)$
1	11	10.1	11	9.2 + 11	9.2 + 11	10.1 + 11 + 8.3 + 9.2	9.2
2	0	10.2	9.2.1 + 9.3 + 10.2	9.2.1 + 11.1	8.3.1 + 9.2.1 + 9.3	9.2.1 + 9.3 + 11.1 + 10.2	10.2 + 8.3.1
2.1	12.1 + 13	10.2.1 + 10.3 + 12.1 + 13	12.1 + 13	11.2 + 9.3.1 + 10.2.1 + 10.3 + 12.1 + 13	10.3 + 12.1 + 13	10.3 + 12.1 + 13 + 8.4.1 + 10.2.1 + 9.4	10.3
3	12.1 + 13	0	12.1 + 13	11.2	10.3 + 12.1 + 13 + 9.3.1	10.3 + 11.2	9.3.1 + 10.3
3.1	13.1	11.2.1 + 11.3	13.1	11.2.1	11.3 + 13.1	9.4.1 + 13.1 + 11.2.1	11.3
4	13.1	11.3 + 13.1	13.1 + 11.3 + 10.3.1	10.3.1 + 13.1	11.3 + 13.1	10.3.1	10.3.1
4.1	13.2	11.3.1 + 12.2.1	13.2	0	11.3.1 + 12.3 + 13.2	12.2.1 + 13.2	11.3.1 + 12.3
5	13.2	13.2 + 12.2.1 + 12.3	11.3.1 + 12.3	0	11.3.1 + 12.3 + 13.2	12.2.1 + 10.4.1 + 10.5 + 13.2	13.2
4.2	0	13.2.1	11.4.1 + 11.5 + 13.3	0	10.5.1 + 11.4.1 + 11.5	10.5.1 + 11.4.1 + 11.5 + 12.3.1	13.3 + 10.5.1
5.1	0	0	0	13.3	13.3	11.4.1 + 11.5 + 13.3	13.3

α	$A_{1,10}(\alpha)$	$A_{2,9}(\alpha)$	$A_{3,8}(\alpha)$	$A_{4,7}(\alpha)$	$A_{5,6}(\alpha)$	$A_{6,5}(\alpha)$	$A_{7,4}(\alpha)$
6	0	13.2.1	11.4.1 + 11.5 + 12.3.1 + 13.2.1	12.3.1 + 13.2.1	13.3 + 10.5.1 + 11.4.1 + 11.5	13.2.1 + 10.5.1	11.4.1 + 11.5

α	$A_{2,10}(\alpha)$	$A_{3,9}(\alpha)$	$A_{4,8}(\alpha)$	$A_{5,7}(\alpha)$	$A_{6,6}(\alpha)$	$A_{7,5}(\alpha)$
1	12 + 11.1	11.1	10.2 + 12	0	12 + 9.3	9.3 + 11.1
2	13 + 11.2	10.2.1 + 10.3 + 12.1 + 13	10.3 + 12.1 + 11.2	9.3.1 + 11.2 + 10.2.1	9.3.1 + 13 + 10.2.1	10.3 + 8.4.1 + 10.2.1 + 9.4
2.1	11.2.1 + 11.3 + 12.2	13.1	11.3 + 13.1 + 12.2 + 10.3.1	0	12.2 + 11.3 + 9.4.1	13.1
3	12.2 + 13.1	11.2.1 + 11.3	12.2 + 11.3	11.2.1 + 11.3 + 13.1	12.2 + 13.1 + 11.2.1 + 11.3	9.4.1 + 11.2.1
3.1	14.1 + 15	0	11.3.1 + 12.3 + 13.2 + 14.1 + 15	0	14.1 + 15	0
4	12.3	11.3.1	13.2	11.3.1 + 12.3 + 13.2	13.2 + 11.3.1	10.4.1 + 11.3.1 + 10.5
4.1	15.1 + 12.3.1	13.2.1	15.1 + 12.3.1	0	15.1 + 10.5.1 + 11.4.1 + 11.5	11.4.1 + 11.5 + 13.2.1
5	13.2.1	13.2.1	12.3.1	0	11.4.1 + 11.5	13.2.1 + 13.3

α	$A_{1,12}(\alpha)$	$A_{2,11}(\alpha)$	$A_{3,10}(\alpha)$	$A_{4,9}(\alpha)$	$A_{5,8}(\alpha)$	$A_{6,7}(\alpha)$	$A_{7,6}(\alpha)$	$A_{8,5}(\alpha)$
1	13	12.1 + 13	13	11.2 + 12.1	13 + 11.2	10.3 + 12.1 + 13 + 11.2	13	10.3 + 11.2 + 9.4
2	0	12.2 + 14	11.2.1 + 11.3 + 12.2 + 14	12.2 + 14 + 11.2.1	10.3.1 + 12.2 + 14	12.2 + 14 + 11.3	9.4.1 + 13.1 + 12.2 + 14 + 11.3	10.3.1 + 11.3 + 9.4.1
2.1	14.1	12.2.1 + 15 + 14.1 + 12.3	14.1	14.1 + 11.3.1	14.1 + 12.3 + 13.2	10.4.1 + 11.3.1 + 10.5 + 12.3 + 12.2.1 + 15 + 14.1	14.1	12.3 + 13.2 + 10.4.1
3	14.1	0	14.1	14.1	11.3.1 + 14.1 + 12.3 + 13.2	12.3 + 13.2	14.1	11.3.1 + 10.5
3.1	15.1	13.2.1 + 13.3 + 15.1	15.1	13.3 + 15.1	13.3 + 15.1	11.4.1 + 11.5 + 13.2.1 + 15.1	15.1	13.3 + 11.4.1
4	15.1	14.2 + 15.1 + 13.3	14.2 + 12.3.1 + 13.2.1	14.2 + 12.3.1 + 13.2.1 + 13.3	14.2	14.2 + 15.1	14.2 + 10.5.1	11.5 + 10.5.1

α	$A_{2,12}(\alpha)$	$A_{3,11}(\alpha)$	$A_{4,10}(\alpha)$	$A_{5,9}(\alpha)$	$A_{6,8}(\alpha)$	$A_{7,7}(\alpha)$	$A_{8,6}(\alpha)$	$A_{9,5}(\alpha)$
1	13.1	13.1	12.2 + 14	13.1	11.3 + 13.1	11.3 + 13.1	12.2 + 13.1 + 10.4 + 11.3	11.3
2	13.2 + 15	12.2.1 + 15 + 14.1 + 12.3	12.3 + 15	11.3.1 + 12.3 + 13.2	12.2.1 + 15 + 14.1 + 12.3 + 11.3.1	10.4.1 + 11.3.1 + 10.5 + 13.2 + 12.2.1 + 15 + 14.1	12.2.1 + 14.1 + 11.3.1 + 13.2 + 11.4	10.4.1 + 10.5
2.1	13.2.1 + 13.3 + 14.2	0	14.2 + 15.1 + 12.3.1 + 13.2.1	0	13.3 + 11.4.1 + 11.5 + 12.3.1 + 13.2.1	13.3	11.5 + 13.3 + 10.5.1 + 12.3.1	13.3
3	14.2 + 15.1	13.2.1 + 13.3 + 15.1	14.2 + 15.1 + 13.3	0	13.2.1 + 15.1	11.4.1 + 11.5 + 13.3 + 13.2.1 + 15.1	13.3 + 13.2.1 + 11.5	11.4.1 + 11.5

α	$A_{1,14}(\alpha)$	$A_{2,13}(\alpha)$	$A_{3,12}(\alpha)$	$A_{4,11}(\alpha)$	$A_{5,10}(\alpha)$	$A_{6,9}(\alpha)$	$A_{7,8}(\alpha)$	$A_{8,7}(\alpha)$	$A_{9,6}(\alpha)$
1	15	14.1 + 15	0	13.2 + 14.1	13.2 + 15	15 + 14.1 + 12.3	0	15 + 14.1 + 11.4	13.2 + 11.4
2	0	14.2	13.2.1 + 13.3 + 14.2	13.2.1	12.3.1 + 13.2.1 + 13.3	13.2.1 + 13.3	11.4.1 + 11.5 + 12.3.1 + 13.2.1	13.3 + 11.4.1 + 11.5 + 14.2 + 12.3.1 + 13.2.1	13.3 + 14.2 + 10.5.1 + 12.3.1

α	$A_{2,14}(\alpha)$	$A_{3,13}(\alpha)$	$A_{4,12}(\alpha)$	$A_{5,11}(\alpha)$	$A_{6,10}(\alpha)$	$A_{7,9}(\alpha)$	$A_{8,8}(\alpha)$	$A_{9,7}(\alpha)$	$A_{10,6}(\alpha)$
1	15.1	15.1	14.2 + 15.1	15.1	13.3 + 14.2	13.3 + 15.1	12.4 + 14.2 + 15.1 + 13.3 + 16	15.1	13.3 + 12.4 + 11.5

Table 3. The function $\xi = \xi_S$ on preadmissible relations

$\xi([1,1]) = 0$

$\xi([1,2]) = 0$

$\xi([1,3]) = 0$

$\xi([2,2]) = 0$

$\xi(2[1,1]) = 0$

$\xi([1,4]) = 0$

$\xi([2,3]) = 2.1 \otimes 1 + 3 \otimes 1$

$\xi([3,2]) = 2.1 \otimes 1 + 3 \otimes 1$

$\xi(2[1,2]) = 0$

$\xi(3[1,1]) = 0$

$\xi([1,5]) = 0$

$\xi([2,4]) = 3.1 \otimes 1$

$\xi([3,3]) = 0$

$\xi(2[1,3]) = 0$

$\xi(3[1,2]) = 0$

$\xi(4[1,1]) = 0$

$\xi([1,6]) = 0$

$\xi([2,5]) = 2.1 \otimes 3 + 3 \otimes 3 + 4.1 \otimes 1 + 5 \otimes 1$

$\xi([3,4]) = 2.1 \otimes 3 + 3 \otimes 3 + 4.1 \otimes 1 + 5 \otimes 1$

$\xi([4,3]) = 0$

$\xi(2[1,4]) = 0$

$\xi(3[1,3]) = 0$

$\xi(4[1,2]) = 0$

$\xi(5[1,1]) = 0$

$\xi([1,7]) = 0$

$\xi([2,6]) = 3.1 \otimes 3 + 5.1 \otimes 1$

$\xi([3,5]) = 0$

$\xi([4,4]) = 2.1 \otimes 3.1 + 3 \otimes 3.1 + 3.1 \otimes 3 + 5.1 \otimes 1$

$\xi([5,3]) = 2.1 \otimes 3.1 + 3 \otimes 3.1$

$\xi(2[1,5]) = 0$

$\xi(3[1,4]) = 0$

$\xi(4[1,3]) = 0$

$\xi(4[2,2]) = 3.1 \otimes 3 + 5.1 \otimes 1$

$\xi(5[1,2]) = 0$

$\xi(6[1,1]) = 0$

$\xi(4.2[1,1]) = 0$

$\xi([1,8]) = 0$

$\xi([2,7]) = 2.1 \otimes 5 + 3 \otimes 5 + 4.1 \otimes 3 + 5 \otimes 3 + 6.1 \otimes 1 + 7 \otimes 1$

$\xi([3,6]) = 2.1 \otimes 5 + 3 \otimes 5 + 4.1 \otimes 3 + 5 \otimes 3 + 6.1 \otimes 1 + 7 \otimes 1$

$\xi([4,5]) = 2.1 \otimes 5 + 3 \otimes 5 + 3.1 \otimes 3.1 + 4.1 \otimes 3 + 5 \otimes 3 + 6.1 \otimes 1 + 7 \otimes 1$

$\xi([5,4]) = 2.1 \otimes 5 + 3 \otimes 5 + 4.1 \otimes 3 + 5 \otimes 3 + 6.1 \otimes 1 + 7 \otimes 1$

$\xi(2[1,6]) = 0$

$\xi(3[1,5]) = 0$

$\xi(4[1,4]) = 0$

$\xi(4[2,3]) = 2.1 \otimes 5 + 3 \otimes 5 + 3.1 \otimes 3.1 + 4.1 \otimes 3 + 5 \otimes 3 + 4.2.1 \otimes 1 + 5.2 \otimes 1$

$\xi(5[1,3]) = 0$

$\xi(5[2,2]) = 2.1 \otimes 5 + 3 \otimes 5 + 4.1 \otimes 3 + 5 \otimes 3 + 4.2.1 \otimes 1 + 5.2 \otimes 1$

$\xi(6[1,2]) = 0$

$\xi(7[1,1]) = 0$

$\xi(4.2[1,2]) = 0$

$\xi(5.2[1,1]) = 0$

$\xi([1,9]) = 0$

$\xi([2,8]) = 3.1 \otimes 5 + 5.1 \otimes 3 + 7.1 \otimes 1$

$\xi([3,7]) = 0$

$\xi([4,6]) = 2.1 \otimes 5.1 + 3 \otimes 5.1 + 4.1 \otimes 3.1 + 5 \otimes 3.1$

$\xi([5,5]) = 2.1 \otimes 5.1 + 3 \otimes 5.1 + 4.1 \otimes 3.1 + 5 \otimes 3.1$

$\xi([6,4]) = 3.1 \otimes 5 + 5.1 \otimes 3 + 7.1 \otimes 1$

$\xi(2[1,7]) = 0$

$\xi(3[1,6]) = 0$

$\xi(4[1,5]) = 0$

$\xi(4[2,4]) = 2.1 \otimes 5.1 + 3 \otimes 5.1 + 3.1 \otimes 5 + 4.1 \otimes 3.1 + 5 \otimes 3.1 + 5.1 \otimes 3 + 5.2.1 \otimes 1$

$\xi(5[1,4]) = 0$

$\xi(5[2,3]) = 2.1 \otimes 5.1 + 3 \otimes 5.1 + 4.1 \otimes 3.1 + 5 \otimes 3.1$

$\xi(6[1,3]) = 0$

$\xi(6[2,2]) = 3.1 \otimes 5 + 5.1 \otimes 3 + 5.2.1 \otimes 1$

$\xi(7[1,2]) = 0$

$\xi(8[1,1]) = 0$

$\xi(4.2[1,3]) = 0$

$\xi(5.2[1,2]) = 0$

$\xi(6.2[1,1]) = 0$

$\xi([1,10]) = 0$

$\xi([2,9]) = 2.1 \otimes 7 + 3 \otimes 7 + 4.1 \otimes 5 + 5 \otimes 5 + 6.1 \otimes 3 + 7 \otimes 3 + 8.1 \otimes 1 + 9 \otimes 1$

$\xi([3,8]) = 2.1 \otimes 7 + 3 \otimes 7 + 4.1 \otimes 5 + 5 \otimes 5 + 6.1 \otimes 3 + 7 \otimes 3 + 8.1 \otimes 1 + 9 \otimes 1$

$\xi([4,7]) = 5.1 \otimes 3.1 + 3.1 \otimes 5.1$

$\xi([5,6]) = 0$

$\xi([6,5]) = 2.1 \otimes 5.2 + 3 \otimes 5.2 + 3.1 \otimes 5.1 + 4.1 \otimes 5 + 5 \otimes 5 + 5.1 \otimes 3.1 + 6.1 \otimes 3 + 7 \otimes 3 + 6.3 \otimes 1 + 7.2 \otimes 1$

$\xi([7,4]) = 2.1 \otimes 5.2 + 3 \otimes 5.2 + 4.1 \otimes 5 + 5 \otimes 5 + 6.1 \otimes 3 + 7 \otimes 3 + 6.3 \otimes 1 + 7.2 \otimes 1$

$\xi(2[1,8]) = 0$

$\xi(3[1,7]) = 0$

$\xi(4[1,6]) = 0$

$\xi(4[2,5]) = 2.1 \otimes 5.2 + 3 \otimes 5.2 + 4.1 \otimes 5 + 5 \otimes 5 + 4.2.1 \otimes 3 + 5.2 \otimes 3 + 6.2.1 \otimes 1 + 8.1 \otimes 1 + 9 \otimes 1 + 7.2 \otimes 1$

$\xi(5[1,5]) = 0$

$\xi(5[2,4]) = 2.1 \otimes 5.2 + 3 \otimes 5.2 + 4.1 \otimes 5 + 5 \otimes 5 + 4.2.1 \otimes 3 + 5.2 \otimes 3 + 6.2.1 \otimes 1 + 8.1 \otimes 1 + 9 \otimes 1 + 7.2 \otimes 1$

$\xi(6[1,4]) = 0$

$\xi(6[2,3]) = 2.1 \otimes 7 + 3 \otimes 7 + 4.1 \otimes 5 + 5 \otimes 5 + 5.1 \otimes 3.1 + 6.2.1 \otimes 1 + 6.3 \otimes 1 + 4.2.1 \otimes 2.1 + 7 \otimes 2.1 + 7 \otimes 3 + 6.1 \otimes 2.1 + 5.2 \otimes 2.1 + 6.1 \otimes 3 + 3.1 \otimes 5.1$

$\xi(6[3,2]) = 2.1 \otimes 7 + 3 \otimes 7 + 4.1 \otimes 5 + 5 \otimes 5 + 5.1 \otimes 3.1 + 6.2.1 \otimes 1 + 6.3 \otimes 1 + 4.2.1 \otimes 2.1 + 7 \otimes 2.1 + 7 \otimes 3 + 6.1 \otimes 2.1 + 5.2 \otimes 2.1 + 6.1 \otimes 3 + 3.1 \otimes 5.1$

$\xi(7[1,3]) = 0$

$\xi(7[2,2]) = 2.1 \otimes 7 + 3 \otimes 7 + 4.1 \otimes 5 + 5 \otimes 5 + 4.2.1 \otimes 2.1 + 7 \otimes 2.1 + 7 \otimes 3 + 6.1 \otimes 2.1 + 6.1 \otimes 3 + 5.2 \otimes 2.1 + 6.2.1 \otimes 1 + 6.3 \otimes 1$

$\xi(8[1,2]) = 0$

$\xi(9[1,1]) = 0$

$\xi(4.2[1,4]) = 0$

$\xi(5.2[1,3]) = 0$

$\xi(6.2[1,2]) = 0$

$\xi(6.3[1,1]) = 0$

$\xi(7.2[1,1]) = 0$

$\xi([1,11]) = 0$

$\xi([2,10]) = 9.1 \otimes 1 + 7.1 \otimes 3 + 5.1 \otimes 5 + 3.1 \otimes 7$

$\xi([3,9]) = 0$

$$\xi([4,8]) = 2.1 \otimes 7.1 + 3 \otimes 7.1 + 3.1 \otimes 7 + 4.1 \otimes 5.1 + 5 \otimes 5.1 + 5.1 \otimes 5 + 6.1 \otimes 3.1 + 7 \otimes 3.1 + 7.1 \otimes 3 + 9.1 \otimes 1$$

$\xi([5,7]) = 2.1 \otimes 7.1 + 3 \otimes 7.1 + 4.1 \otimes 5.1 + 5 \otimes 5.1 + 6.1 \otimes 3.1 + 7 \otimes 3.1$

$$\xi([6,6]) = 2.1 \otimes 7.1 + 3 \otimes 7.1 + 3.1 \otimes 5.2 + 4.1 \otimes 5.1 + 5 \otimes 5.1 + 5.1 \otimes 5 + 6.1 \otimes 3.1 + 7 \otimes 3.1 + 7.1 \otimes 3 + 7.3 \otimes 1$$

$\xi([7,5]) = 2.1 \otimes 7.1 + 3 \otimes 7.1 + 4.1 \otimes 5.1 + 5 \otimes 5.1 + 6.1 \otimes 3.1 + 7 \otimes 3.1$

$\xi(2[1,9]) = 0$

$\xi(3[1,8]) = 0$

$\xi(4[1,7]) = 0$

$\xi(4[2,6]) = 3.1 \otimes 5.2 + 3.1 \otimes 7 + 5.2.1 \otimes 3 + 7.1 \otimes 3 + 7.2.1 \otimes 1$

$\xi(5[1,6]) = 0$

$\xi(5[2,5]) = 0$

$\xi(6[1,5]) = 0$

$$\xi(6[2,4]) = 2.1 \otimes 7.1 + 3 \otimes 7.1 + 3.1 \otimes 5.2 + 3.1 \otimes 7 + 4.1 \otimes 5.1 + 5 \otimes 5.1 + 6.1 \otimes 3.1 + 7 \otimes 3.1 + 5.2.1 \otimes 2.1 + 5.2.1 \otimes 3 + 7.1 \otimes 2.1 + 7.1 \otimes 3 + 6.3.1 \otimes 1 + 7.2.1 \otimes 1 + 9.1 \otimes 1$$

$\xi(6[3,3]) = 0$

$\xi(7[1,4]) = 0$

$$\xi(7[2,3]) = 2.1 \otimes 7.1 + 3 \otimes 7.1 + 4.1 \otimes 5.1 + 5 \otimes 5.1 + 4.2.1 \otimes 3.1 + 5.2 \otimes 3.1 + 6.3.1 \otimes 1 + 7.2.1 \otimes 1 + 7.3 \otimes 1$$

$$\xi(7[3,2]) = 2.1 \otimes 7.1 + 3 \otimes 7.1 + 4.1 \otimes 5.1 + 5 \otimes 5.1 + 4.2.1 \otimes 3.1 + 5.2 \otimes 3.1 + 6.3.1 \otimes 1 + 7.2.1 \otimes 1 + 7.3 \otimes 1$$

$\xi(8[1,3]) = 0$

$$\xi(8[2,2]) = 3.1 \otimes 7 + 5.1 \otimes 5 + 4.2.1 \otimes 3.1 + 7 \otimes 3.1 + 6.1 \otimes 3.1 + 5.2 \otimes 3.1 + 5.2.1 \otimes 2.1 + 7.1 \otimes 2.1 + 7.1 \otimes 3 + 7.2.1 \otimes 1 + 7.3 \otimes 1$$

$\xi(9[1,2]) = 0$

$\xi(10[1,1]) = 0$

$\xi(4.2[1,5]) = 0$

$\xi(5.2[1,4]) = 0$

$\xi(6.2[1,3]) = 0$

$\xi(6.3[1,2]) = 0$

$\xi(7.2[1,2]) = 0$

$\xi(7.3[1,1]) = 0$

$\xi(8.2[1,1]) = 0$

$\xi([1,12]) = 0$

$\xi([2,11]) = 2.1 \otimes 9 + 3 \otimes 9 + 4.1 \otimes 7 + 5 \otimes 7 + 6.1 \otimes 5 + 7 \otimes 5 + 8.1 \otimes 3 + 9 \otimes 3 + 10.1 \otimes 1 + 11 \otimes 1$

$\xi([3,10]) = 2.1 \otimes 9 + 3 \otimes 9 + 4.1 \otimes 7 + 5 \otimes 7 + 6.1 \otimes 5 + 7 \otimes 5 + 8.1 \otimes 3 + 9 \otimes 3 + 10.1 \otimes 1 + 11 \otimes 1$

$\xi([4,9]) = 2.1 \otimes 9 + 3 \otimes 9 + 3.1 \otimes 7.1 + 4.1 \otimes 7 + 5 \otimes 7 + 5.1 \otimes 5.1 + 6.1 \otimes 5 + 7 \otimes 5 + 7.1 \otimes 3.1 + 8.1 \otimes 3 + 9 \otimes 3 + 10.1 \otimes 1 + 11 \otimes 1$

$\xi([5,8]) = 2.1 \otimes 9 + 3 \otimes 9 + 4.1 \otimes 7 + 5 \otimes 7 + 6.1 \otimes 5 + 7 \otimes 5 + 8.1 \otimes 3 + 9 \otimes 3 + 10.1 \otimes 1 + 11 \otimes 1$

$\xi([6,7]) = 2.1 \otimes 7.2 + 3 \otimes 7.2 + 2.1 \otimes 9 + 3 \otimes 9 + 4.1 \otimes 5.2 + 5 \otimes 5.2 + 6.1 \otimes 5 + 7 \otimes 5 + 6.3 \otimes 3 + 7.2 \otimes 3 + 8.3 \otimes 1 + 9.2 \otimes 1 + 10.1 \otimes 1 + 11 \otimes 1$

$\xi([7,6]) = 2.1 \otimes 7.2 + 3 \otimes 7.2 + 2.1 \otimes 9 + 3 \otimes 9 + 4.1 \otimes 5.2 + 5 \otimes 5.2 + 6.1 \otimes 5 + 7 \otimes 5 + 6.3 \otimes 3 + 7.2 \otimes 3 + 8.3 \otimes 1 + 9.2 \otimes 1 + 10.1 \otimes 1 + 11 \otimes 1$

$\xi([8,5]) = 3.1 \otimes 7.1 + 5.1 \otimes 5.1 + 7.1 \otimes 3.1$

$\xi(2[1,10]) = 0$

$\xi(3[1,9]) = 0$

$\xi(4[1,8]) = 0$

$\xi(4[2,7]) = 7 \otimes 5 + 2.1 \otimes 7.2 + 3.1 \otimes 7.1 + 5.1 \otimes 5.1 + 8.2.1 \otimes 1 + 11 \otimes 1 + 10.1 \otimes 1 + 3 \otimes 7.2 + 4.1 \otimes 5.2 + 5 \otimes 5.2 + 5 \otimes 7 + 6.2.1 \otimes 3 + 6.1 \otimes 5 + 5.2 \otimes 5 + 7.2 \otimes 3 + 9.2 \otimes 1 + 4.1 \otimes 7 + 7.1 \otimes 3.1 + 4.2.1 \otimes 5$

$\xi(5[1,7]) = 0$

$\xi(5[2,6]) = 2.1 \otimes 7.2 + 3 \otimes 7.2 + 4.1 \otimes 5.2 + 4.1 \otimes 7 + 5 \otimes 5.2 + 5 \otimes 7 + 4.2.1 \otimes 5 + 7 \otimes 5 + 6.1 \otimes 5 + 5.2 \otimes 5 + 6.2.1 \otimes 3 + 7.2 \otimes 3 + 8.2.1 \otimes 1 + 11 \otimes 1 + 10.1 \otimes 1 + 9.2 \otimes 1$

$\xi(6[1,6]) = 0$

$\xi(6[2,5]) = 9 \otimes 2.1 + 6.1 \otimes 5 + 7.2 \otimes 3 + 8.3 \otimes 1 + 9.2 \otimes 1 + 6.2.1 \otimes 3 + 4.1 \otimes 7 + 5.1 \otimes 5.1 + 4.2.1 \otimes 5 + 6.1 \otimes 4.1 + 7 \otimes 5 + 5 \otimes 7 + 6.3 \otimes 2.1 + 2.1 \otimes 7.2 + 8.1 \otimes 2.1 + 3 \otimes 7.2 + 4.2.1 \otimes 4.1 + 5.2 \otimes 4.1 + 3.1 \otimes 7.1 + 7 \otimes 4.1 + 6.2.1 \otimes 2.1 + 5 \otimes 5.2 + 5.2 \otimes 5 + 4.1 \otimes 5.2 + 7.3.1 \otimes 1 + 7.1 \otimes 3.1$

$\xi(6[3,4]) = 9 \otimes 2.1 + 6.1 \otimes 5 + 7.2 \otimes 3 + 8.3 \otimes 1 + 9.2 \otimes 1 + 6.2.1 \otimes 3 + 4.1 \otimes 7 + 5.1 \otimes 5.1 + 4.2.1 \otimes 5 + 6.1 \otimes 4.1 + 7 \otimes 5 + 5 \otimes 7 + 6.3 \otimes 2.1 + 2.1 \otimes 7.2 + 8.1 \otimes 2.1 + 3 \otimes 7.2$

$+ 4.2.1 \otimes 4.1 + 5.2 \otimes 4.1 + 3.1 \otimes 7.1 + 7 \otimes 4.1 + 6.2.1 \otimes 2.1 + 5 \otimes 5.2 + 5.2 \otimes 5 + 4.1 \otimes 5.2 + 7.3.1 \otimes 1 + 7.1 \otimes 3.1$

$\xi(7[1,5]) = 0$

$\xi(7[2,4]) = 3 \otimes 7.2 + 4.2.1 \otimes 4.1 + 7 \otimes 4.1 + 7 \otimes 5 + 4.2.1 \otimes 5 + 5 \otimes 5.2 + 5 \otimes 7 + 9.2 \otimes 1 + 6.1 \otimes 4.1 + 5.2.1 \otimes 3.1 + 7.1 \otimes 3.1 + 4.1 \otimes 5.2 + 6.1 \otimes 5 + 9 \otimes 2.1 + 4.1 \otimes 7 + 6.2.1 \otimes 3 + 6.3 \otimes 2.1 + 5.2 \otimes 4.1 + 8.1 \otimes 2.1 + 2.1 \otimes 7.2 + 6.2.1 \otimes 2.1 + 7.2 \otimes 3 + 5.2 \otimes 5 + 8.3 \otimes 1$

$\xi(7[3,3]) = 0$

$\xi(8[1,4]) = 0$

$\xi(8[2,3]) = 3 \otimes 9 + 4.1 \otimes 7 + 5 \otimes 7 + 5.1 \otimes 5.1 + 7 \otimes 4.1 + 6.1 \otimes 5 + 6.1 \otimes 4.1 + 4.2.1 \otimes 4.1 + 6.3 \otimes 2.1 + 8.2.1 \otimes 1 + 9 \otimes 2.1 + 7.1 \otimes 3.1 + 3.1 \otimes 7.1 + 9 \otimes 3 + 7 \otimes 5 + 8.1 \otimes 3 + 8.3 \otimes 1 + 5.2 \otimes 4.1 + 6.2.1 \otimes 2.1 + 8.1 \otimes 2.1 + 2.1 \otimes 9 + 7.3.1 \otimes 1$

$\xi(8[3,2]) = 3 \otimes 9 + 4.1 \otimes 7 + 5 \otimes 7 + 5.1 \otimes 5.1 + 7 \otimes 4.1 + 6.1 \otimes 5 + 6.1 \otimes 4.1 + 4.2.1 \otimes 4.1 + 6.3 \otimes 2.1 + 8.2.1 \otimes 1 + 9 \otimes 2.1 + 7.1 \otimes 3.1 + 3.1 \otimes 7.1 + 9 \otimes 3 + 7 \otimes 5 + 8.1 \otimes 3 + 8.3 \otimes 1 + 5.2 \otimes 4.1 + 6.2.1 \otimes 2.1 + 8.1 \otimes 2.1 + 2.1 \otimes 9 + 7.3.1 \otimes 1$

$\xi(9[1,3]) = 0$

$\xi(9[2,2]) = 6.2.1 \otimes 2.1 + 5.2.1 \otimes 3.1 + 7.1 \otimes 3.1 + 6.3 \otimes 2.1 + 8.1 \otimes 2.1 + 4.1 \otimes 7 + 5 \otimes 7 + 9 \otimes 3 + 5.2 \otimes 4.1 + 8.1 \otimes 3 + 8.2.1 \otimes 1 + 6.1 \otimes 4.1 + 6.1 \otimes 5 + 2.1 \otimes 9 + 3 \otimes 9 + 7 \otimes 4.1 + 7 \otimes 5 + 9 \otimes 2.1 + 4.2.1 \otimes 4.1 + 8.3 \otimes 1$

$\xi(10[1,2]) = 0$

$\xi(11[1,1]) = 0$

$\xi(4.2[1,6]) = 0$

$\xi(5.2[1,5]) = 0$

$\xi(6.2[1,4]) = 0$

$\xi(6.3[1,3]) = 0$

$\xi(7.2[1,3]) = 0$

$\xi(7.3[1,2]) = 0$

$\xi(8.2[1,2]) = 0$

$\xi(8.3[1,1]) = 0$

$\xi(9.2[1,1]) = 0$

$\xi([1,13]) = 0$

$\xi([2,12]) = 3.1 \otimes 9 + 5.1 \otimes 7 + 7.1 \otimes 5 + 9.1 \otimes 3 + 11.1 \otimes 1$

$\xi([3,11]) = 0$

$\xi([4,10]) = 2.1 \otimes 9.1 + 3 \otimes 9.1 + 4.1 \otimes 7.1 + 5 \otimes 7.1 + 6.1 \otimes 5.1 + 7 \otimes 5.1 + 8.1 \otimes 3.1 + 9 \otimes 3.1$

$\xi([5,9]) = 2.1 \otimes 9.1 + 3 \otimes 9.1 + 4.1 \otimes 7.1 + 5 \otimes 7.1 + 6.1 \otimes 5.1 + 7 \otimes 5.1 + 8.1 \otimes 3.1 + 9 \otimes 3.1$

$\xi([6,8]) = 3.1 \otimes 7.2 + 5.1 \otimes 5.2 + 5.1 \otimes 7 + 7.3 \otimes 3 + 9.1 \otimes 3 + 9.3 \otimes 1$

$\xi([7,7]) = 0$

$\xi([8,6]) = 7 \otimes 5.1 + 7.1 \otimes 5 + 7.3 \otimes 3 + 2.1 \otimes 7.3 + 3 \otimes 7.3 + 3.1 \otimes 9 + 4.1 \otimes 7.1 + 5.1 \otimes 5.2 + 11.1 \otimes 1 + 5 \otimes 7.1 + 9.3 \otimes 1 + 3.1 \otimes 7.2 + 6.1 \otimes 5.1 + 6.3 \otimes 3.1 + 7.2 \otimes 3.1$

$\xi([9,5]) = 2.1 \otimes 7.3 + 3 \otimes 7.3 + 4.1 \otimes 7.1 + 5 \otimes 7.1 + 6.1 \otimes 5.1 + 7 \otimes 5.1 + 6.3 \otimes 3.1 + 7.2 \otimes 3.1$

$\xi(2[1,11]) = 0$

$\xi(3[1,10]) = 0$

$\xi(4[1,9]) = 0$

$\xi(4[2,8]) = 2.1 \otimes 9.1 + 3 \otimes 9.1 + 3.1 \otimes 7.2 + 4.1 \otimes 7.1 + 5 \otimes 7.1 + 5.1 \otimes 5.2 + 5.1 \otimes 7 + 6.1 \otimes 5.1 + 7 \otimes 5.1 + 5.2.1 \otimes 5 + 7.1 \otimes 5 + 8.1 \otimes 3.1 + 9 \otimes 3.1 + 7.2.1 \otimes 3 + 9.2.1 \otimes 1 + 11.1 \otimes 1$

$\xi(5[1,8]) = 0$

$\xi(5[2,7]) = 2.1 \otimes 9.1 + 3 \otimes 9.1 + 4.1 \otimes 7.1 + 5 \otimes 7.1 + 6.1 \otimes 5.1 + 7 \otimes 5.1 + 8.1 \otimes 3.1 + 9 \otimes 3.1$

$\xi(6[1,7]) = 0$

$\xi(6[2,6]) = 5.2.1 \otimes 4.1 + 7 \otimes 5.1 + 2.1 \otimes 7.3 + 3 \otimes 7.3 + 6.3 \otimes 3.1 + 4.1 \otimes 7.1 + 5 \otimes 7.1 + 7.2.1 \otimes 2.1 + 7.3 \otimes 2.1 + 8.3.1 \otimes 1 + 7.1 \otimes 4.1 + 9.1 \otimes 2.1 + 6.3.1 \otimes 3 + 7.2.1 \otimes 3 + 11.1 \otimes 1 + 7.2 \otimes 3.1 + 7.3 \otimes 3 + 6.1 \otimes 5.1$

$\xi(6[3,5]) = 0$

$\xi(7[1,6]) = 0$

$\xi(7[2,5]) = 4.1 \otimes 7.1 + 2.1 \otimes 7.3 + 5 \otimes 7.1 + 7.2 \otimes 3.1 + 7.2.1 \otimes 3 + 5.2 \otimes 5.1 + 9 \otimes 3.1 + 8.3.1 \otimes 1 + 9.2.1 \otimes 1 + 9.3 \otimes 1 + 7.3 \otimes 3 + 3 \otimes 7.3 + 6.2.1 \otimes 3.1 + 6.3.1 \otimes 3 + 4.2.1 \otimes 5.1 + 8.1 \otimes 3.1$

$\xi(7[3,4]) = 4.1 \otimes 7.1 + 2.1 \otimes 7.3 + 5 \otimes 7.1 + 7.2 \otimes 3.1 + 7.2.1 \otimes 3 + 5.2 \otimes 5.1 + 9 \otimes 3.1 + 8.3.1 \otimes 1 + 9.2.1 \otimes 1 + 9.3 \otimes 1 + 7.3 \otimes 3 + 3 \otimes 7.3 + 6.2.1 \otimes 3.1 + 6.3.1 \otimes 3 + 4.2.1 \otimes 5.1 + 8.1 \otimes 3.1$

$\xi(8[1,5]) = 0$

$\xi(8[2,4]) = 9.1 \otimes 3 + 6.2.1 \otimes 3.1 + 7.2.1 \otimes 2.1 + 5 \otimes 7.1 + 2.1 \otimes 9.1 + 3 \otimes 9.1 + 3.1 \otimes 9 + 7.3 \otimes 2.1 + 7.2.1 \otimes 3 + 4.2.1 \otimes 5.1 + 6.3.1 \otimes 2.1 + 5.1 \otimes 5.2 + 3.1 \otimes 7.2 + 8.3.1 \otimes 1 + 9.3 \otimes 1 + 6.3 \otimes 3.1 + 5.2 \otimes 5.1 + 5.2.1 \otimes 5 + 4.1 \otimes 7.1$

$\xi(8[3,3]) = 0$

$\xi(9[1,4]) = 0$

$\xi(9[2,3]) = 4.2.1 \otimes 5.1 + 5 \otimes 7.1 + 7.3 \otimes 2.1 + 4.1 \otimes 7.1 + 8.3.1 \otimes 1 + 9.2.1 \otimes 1 + 9.3 \otimes 1 + 6.2.1 \otimes 3.1 + 7.2.1 \otimes 2.1 + 6.3.1 \otimes 2.1 + 2.1 \otimes 9.1 + 3 \otimes 9.1 + 6.3 \otimes 3.1 + 5.2 \otimes 5.1$

$\xi(9[3,2]) = 4.2.1 \otimes 5.1 + 5 \otimes 7.1 + 7.3 \otimes 2.1 + 4.1 \otimes 7.1 + 8.3.1 \otimes 1 + 9.2.1 \otimes 1 + 9.3 \otimes 1 + 6.2.1 \otimes 3.1 + 7.2.1 \otimes 2.1 + 6.3.1 \otimes 2.1 + 2.1 \otimes 9.1 + 3 \otimes 9.1 + 6.3 \otimes 3.1 + 5.2 \otimes 5.1$

$\xi(10[1,3]) = 0$

$$\xi(10[2,2]) = 9.1\otimes 3 + 9.2.1\otimes 1 + 9.3\otimes 1 + 6.3\otimes 3.1 + 8.1\otimes 3.1 + 9\otimes 3.1 + 7.3\otimes 2.1 + 7.1\otimes 4.1 + 5.1\otimes 7 + 4.2.1\otimes 5.1 + 7.1\otimes 5 + 7\otimes 5.1 + 6.1\otimes 5.1 + 5.2\otimes 5.1 + 3.1\otimes 9 + 9.1\otimes 2.1 + 7.2.1\otimes 2.1 + 6.2.1\otimes 3.1 + 5.2.1\otimes 4.1$$

$\xi(11[1,2]) = 0$

$\xi(12[1,1]) = 0$

$\xi(4.2[1,7]) = 0$

$\xi(5.2[1,6]) = 0$

$\xi(6.2[1,5]) = 0$

$\xi(6.3[1,4]) = 0$

$\xi(7.2[1,4]) = 0$

$\xi(7.3[1,3]) = 0$

$\xi(8.2[1,3]) = 0$

$\xi(8.3[1,2]) = 0$

$\xi(8.4[1,1]) = 0$

$\xi(9.2[1,2]) = 0$

$\xi(9.3[1,1]) = 0$

$\xi(10.2[1,1]) = 0$

$\xi([1,14]) = 0$

$$\xi([2,13]) = 2.1\otimes 11 + 3\otimes 11 + 4.1\otimes 9 + 5\otimes 9 + 6.1\otimes 7 + 7\otimes 7 + 8.1\otimes 5 + 9\otimes 5 + 10.1\otimes 3 + 11\otimes 3 + 12.1\otimes 1 + 13\otimes 1$$

$$\xi([3,12]) = 2.1\otimes 11 + 3\otimes 11 + 4.1\otimes 9 + 5\otimes 9 + 6.1\otimes 7 + 7\otimes 7 + 8.1\otimes 5 + 9\otimes 5 + 10.1\otimes 3 + 11\otimes 3 + 12.1\otimes 1 + 13\otimes 1$$

$$\xi([4,11]) = 3.1\otimes 9.1 + 5.1\otimes 7.1 + 7.1\otimes 5.1 + 9.1\otimes 3.1$$

$\xi([5,10]) = 0$

$$\xi([6,9]) = 2.1\otimes 9.2 + 2.1\otimes 11 + 3\otimes 9.2 + 3\otimes 11 + 3.1\otimes 9.1 + 4.1\otimes 7.2 + 5\otimes 7.2 + 5.1\otimes 7.1 + 6.1\otimes 5.2 + 6.1\otimes 7 + 7\otimes 5.2 + 7\otimes 7 + 7.1\otimes 5.1 + 6.3\otimes 5 + 8.1\otimes 5 + 9\otimes 5 + 7.2\otimes 5 + 9.1\otimes 3.1 + 8.3\otimes 3 + 9.2\otimes 3 + 10.3\otimes 1 + 12.1\otimes 1 + 13\otimes 1 + 11.2\otimes 1$$

$$\xi([7,8]) = 2.1\otimes 9.2 + 2.1\otimes 11 + 3\otimes 9.2 + 3\otimes 11 + 4.1\otimes 7.2 + 5\otimes 7.2 + 6.1\otimes 5.2 + 6.1\otimes 7 + 7\otimes 5.2 + 7\otimes 7 + 6.3\otimes 5 + 8.1\otimes 5 + 9\otimes 5 + 7.2\otimes 5 + 8.3\otimes 3 + 9.2\otimes 3 + 10.3\otimes 1 + 12.1\otimes 1 + 13\otimes 1 + 11.2\otimes 1$$

$$\xi([8,7]) = 3.1\otimes 7.3 + 7\otimes 5.2 + 5.1\otimes 7.1 + 11\otimes 3 + 8.3\otimes 3 + 6.3\otimes 5 + 7.1\otimes 5.1 + 3\otimes 9.2 + 4.1\otimes 7.2 + 5\otimes 7.2 + 4.1\otimes 9 + 5\otimes 9 + 9.2\otimes 3 + 2.1\otimes 9.2 + 10.3\otimes 1 + 7.2\otimes 5 + 11.2\otimes 1 + 7.3\otimes 3.1 + 10.1\otimes 3 + 6.1\otimes 5.2$$

$$\xi([9,6]) = 4.1\otimes 7.2 + 5\otimes 7.2 + 11.2\otimes 1 + 4.1\otimes 9 + 5\otimes 9 + 3\otimes 9.2 + 8.3\otimes 3 + 9.2\otimes 3 + 6.1\otimes 5.2 + 7\otimes 5.2 + 10.1\otimes 3 + 2.1\otimes 9.2 + 11\otimes 3 + 6.3\otimes 5 + 10.3\otimes 1 + 7.2\otimes 5$$

$\xi(2[1,12]) = 0$

$\xi(3[1,11]) = 0$

$\xi(4[1,10]) = 0$

$\xi(4[2,9]) = 10.2.1 \otimes 1 + 3 \otimes 9.2 + 3 \otimes 11 + 11.2 \otimes 1 + 2.1 \otimes 9.2 + 2.1 \otimes 11 + 6.1 \otimes 5.2 + 6.2.1 \otimes 5 + 7.2 \otimes 5 + 4.2.1 \otimes 7 + 5.2 \otimes 7 + 10.1 \otimes 3 + 9.2 \otimes 3 + 4.1 \otimes 7.2 + 5 \otimes 7.2 + 7 \otimes 5.2 + 8.2.1 \otimes 3 + 11 \otimes 3$

$\xi(5[1,9]) = 0$

$\xi(5[2,8]) = 10.2.1 \otimes 1 + 9.2 \otimes 3 + 2.1 \otimes 11 + 5 \otimes 7.2 + 11.2 \otimes 1 + 3 \otimes 9.2 + 3 \otimes 11 + 5.2 \otimes 7 + 4.2.1 \otimes 7 + 2.1 \otimes 9.2 + 6.2.1 \otimes 5 + 7.2 \otimes 5 + 6.1 \otimes 5.2 + 4.1 \otimes 7.2 + 7 \otimes 5.2 + 8.2.1 \otimes 3 + 11 \otimes 3 + 10.1 \otimes 3$

$\xi(6[1,8]) = 0$

$\xi(6[2,7]) = 8.2.1 \otimes 2.1 + 10.2.1 \otimes 1 + 11 \otimes 2.1 + 10.3 \otimes 1 + 10.1 \otimes 2.1 + 12.1 \otimes 1 + 3.1 \otimes 9.1 + 4.2.1 \otimes 6.1 + 13 \otimes 1 + 6.2.1 \otimes 4.1 + 8.3 \otimes 2.1 + 6.1 \otimes 6.1 + 7.3 \otimes 3.1 + 8.1 \otimes 4.1 + 9.3.1 \otimes 1 + 9.1 \otimes 3.1 + 5.2 \otimes 6.1 + 7 \otimes 6.1 + 3.1 \otimes 7.3 + 6.3 \otimes 4.1 + 9 \otimes 4.1 + 7.3.1 \otimes 3$

$\xi(6[3,6]) = 8.2.1 \otimes 2.1 + 10.2.1 \otimes 1 + 11 \otimes 2.1 + 10.3 \otimes 1 + 10.1 \otimes 2.1 + 12.1 \otimes 1 + 3.1 \otimes 9.1 + 4.2.1 \otimes 6.1 + 13 \otimes 1 + 6.2.1 \otimes 4.1 + 8.3 \otimes 2.1 + 6.1 \otimes 6.1 + 7.3 \otimes 3.1 + 8.1 \otimes 4.1 + 9.3.1 \otimes 1 + 9.1 \otimes 3.1 + 5.2 \otimes 6.1 + 7 \otimes 6.1 + 3.1 \otimes 7.3 + 6.3 \otimes 4.1 + 9 \otimes 4.1 + 7.3.1 \otimes 3$

$\xi(7[1,7]) = 0$

$\xi(7[2,6]) = 10.3 \otimes 1 + 10.2.1 \otimes 1 + 13 \otimes 1 + 12.1 \otimes 1 + 4.2.1 \otimes 6.1 + 7.1 \otimes 5.1 + 6.2.1 \otimes 4.1 + 6.1 \otimes 6.1 + 7 \otimes 6.1 + 5.2.1 \otimes 5.1 + 8.3 \otimes 2.1 + 8.1 \otimes 4.1 + 7.3 \otimes 3.1 + 9 \otimes 4.1 + 5.2 \otimes 6.1 + 7.2.1 \otimes 3.1 + 9.1 \otimes 3.1 + 8.2.1 \otimes 2.1 + 10.1 \otimes 2.1 + 11 \otimes 2.1 + 6.3 \otimes 4.1$

$\xi(7[3,5]) = 0$

$\xi(8[1,6]) = 0$

$\xi(8[2,5]) = 9.2 \otimes 3 + 8.2.1 \otimes 3 + 9.4 \otimes 1 + 4.1 \otimes 7.2 + 9.1 \otimes 3.1 + 5 \otimes 7.2 + 8.4.1 \otimes 1 + 7.2 \otimes 5 + 5.2 \otimes 5.2 + 3.1 \otimes 7.3 + 3.1 \otimes 9.1 + 6.2.1 \otimes 5 + 7.3.1 \otimes 2.1 + 8.1 \otimes 5 + 9 \otimes 5 + 7.3 \otimes 3.1 + 4.2.1 \otimes 5.2 + 9.3.1 \otimes 1 + 4.1 \otimes 9 + 5 \otimes 9 + 3 \otimes 9.2 + 2.1 \otimes 9.2 + 7.3.1 \otimes 3$

$\xi(8[3,4]) = 9.2 \otimes 3 + 8.2.1 \otimes 3 + 9.4 \otimes 1 + 4.1 \otimes 7.2 + 9.1 \otimes 3.1 + 5 \otimes 7.2 + 8.4.1 \otimes 1 + 7.2 \otimes 5 + 5.2 \otimes 5.2 + 3.1 \otimes 7.3 + 3.1 \otimes 9.1 + 6.2.1 \otimes 5 + 7.3.1 \otimes 2.1 + 8.1 \otimes 5 + 9 \otimes 5 + 7.3 \otimes 3.1 + 4.2.1 \otimes 5.2 + 9.3.1 \otimes 1 + 4.1 \otimes 9 + 5 \otimes 9 + 3 \otimes 9.2 + 2.1 \otimes 9.2 + 7.3.1 \otimes 3$

$\xi(8[4,3]) = 3.1 \otimes 7.2.1 + 3.1 \otimes 7.3 + 5.1 \otimes 5.2.1 + 5.2.1 \otimes 5.1 + 6.3.1 \otimes 3.1 + 7.3.1 \otimes 2.1 + 7.3.1 \otimes 3$

$\xi(9[1,5]) = 0$

$\xi(9[2,4]) = 7.3 \otimes 3.1 + 7.2 \otimes 5 + 8.2.1 \otimes 3 + 8.1 \otimes 5 + 9 \otimes 5 + 4.1 \otimes 9 + 3 \otimes 9.2 + 4.1 \otimes 7.2 + 4.2.1 \otimes 5.2 + 9.3.1 \otimes 1 + 9.2 \otimes 3 + 5.2 \otimes 5.2 + 9.4 \otimes 1 + 8.4.1 \otimes 1 + 5 \otimes 9 + 6.2.1 \otimes 5 + 5 \otimes 7.2 + 6.3.1 \otimes 3.1 + 7.2.1 \otimes 3.1 + 2.1 \otimes 9.2$

$\xi(9[3,3]) = 0$

$\xi(10[1,4]) = 0$

$\xi(10[2,3]) = 5.2 \otimes 6.1 + 6.3.1 \otimes 3.1 + 9 \otimes 4.1 + 9 \otimes 5 + 4.1 \otimes 9 + 8.1 \otimes 4.1 + 7 \otimes 7 + 6.3 \otimes 4.1 + 3 \otimes 11 + 10.1 \otimes 2.1 + 10.1 \otimes 3 + 8.2.1 \otimes 2.1 + 10.2.1 \otimes 1 + 7.3.1 \otimes 2.1 + 6.1 \otimes 6.1 + 5 \otimes 9 + 6.2.1 \otimes 4.1 + 3.1 \otimes 9.1 + 11 \otimes 2.1 + 5.1 \otimes 7.1 + 11 \otimes 3 + 2.1 \otimes 11 + 7.1 \otimes 5.1 + 7.2.1 \otimes 3.1 + 8.1 \otimes 5 + 8.3 \otimes 2.1 + 10.3 \otimes 1 + 7 \otimes 6.1 + 7.3 \otimes 3.1 + 9.1 \otimes 3.1 + 6.1 \otimes 7 + 9.3.1 \otimes 1 + 4.2.1 \otimes 6.1$

$\xi(10[3,2]) = 5.2 \otimes 6.1 + 6.3.1 \otimes 3.1 + 9 \otimes 4.1 + 9 \otimes 5 + 4.1 \otimes 9 + 8.1 \otimes 4.1 + 7 \otimes 7 + 6.3 \otimes 4.1 + 3 \otimes 11 + 10.1 \otimes 2.1 + 10.1 \otimes 3 + 8.2.1 \otimes 2.1 + 10.2.1 \otimes 1 + 7.3.1 \otimes 2.1 + 6.1 \otimes 6.1 + 5 \otimes 9 + 6.2.1 \otimes 4.1 + 3.1 \otimes 9.1 + 11 \otimes 2.1 + 5.1 \otimes 7.1 + 11 \otimes 3 + 2.1 \otimes 11 + 7.1 \otimes 5.1 + 7.2.1 \otimes 3.1 + 8.1 \otimes 5 + 8.3 \otimes 2.1 + 10.3 \otimes 1 + 7 \otimes 6.1 + 7.3 \otimes 3.1 + 9.1 \otimes 3.1 + 6.1 \otimes 7 + 9.3.1 \otimes 1 + 4.2.1 \otimes 6.1$

$\xi(11[1,3]) = 0$

$\xi(11[2,2]) = 8.1 \otimes 5 + 10.1 \otimes 3 + 7 \otimes 7 + 6.1 \otimes 6.1 + 10.2.1 \otimes 1 + 10.1 \otimes 2.1 + 5.2.1 \otimes 5.1 + 6.1 \otimes 7 + 5.2 \otimes 6.1 + 7 \otimes 6.1 + 11 \otimes 2.1 + 10.3 \otimes 1 + 6.3 \otimes 4.1 + 7.2.1 \otimes 3.1 + 8.2.1 \otimes 2.1 + 7.1 \otimes 5.1 + 8.3 \otimes 2.1 + 5 \otimes 9 + 2.1 \otimes 11 + 7.3 \otimes 3.1 + 8.1 \otimes 4.1 + 4.1 \otimes 9 + 9 \otimes 4.1 + 3 \otimes 11 + 9.1 \otimes 3.1 + 6.2.1 \otimes 4.1 + 4.2.1 \otimes 6.1 + 11 \otimes 3 + 9 \otimes 5$

$\xi(12[1,2]) = 0$

$\xi(13[1,1]) = 0$

$\xi(4.2[1,8]) = 0$

$\xi(5.2[1,7]) = 0$

$\xi(6.2[1,6]) = 0$

$\xi(6.3[1,5]) = 0$

$\xi(7.2[1,5]) = 0$

$\xi(7.3[1,4]) = 0$

$\xi(8.2[1,4]) = 0$

$\xi(8.3[1,3]) = 0$

$\xi(8.4[1,2]) = 0$

$\xi(9.2[1,3]) = 0$

$\xi(9.3[1,2]) = 0$

$\xi(9.4[1,1]) = 0$

$\xi(10.2[1,2]) = 0$

$\xi(10.3[1,1]) = 0$

$\xi(11.2[1,1]) = 0$

$\xi([1,15]) = 0$

$\xi([2,14]) = 3.1 \otimes 11 + 5.1 \otimes 9 + 7.1 \otimes 7 + 9.1 \otimes 5 + 11.1 \otimes 3 + 13.1 \otimes 1$

$\xi([3,13]) = 0$

$$\xi([4,12]) = 2.1 \otimes 11.1 + 3 \otimes 11.1 + 3.1 \otimes 11 + 4.1 \otimes 9.1 + 5 \otimes 9.1 + 5.1 \otimes 9 + 6.1 \otimes 7.1 + 7 \otimes 7.1 + 7.1 \otimes 7 + 8.1 \otimes 5.1 + 9 \otimes 5.1 + 9.1 \otimes 5 + 10.1 \otimes 3.1 + 11 \otimes 3.1 + 11.1 \otimes 3 + 13.1 \otimes 1$$

$$\xi([5,11]) = 2.1 \otimes 11.1 + 3 \otimes 11.1 + 4.1 \otimes 9.1 + 5 \otimes 9.1 + 6.1 \otimes 7.1 + 7 \otimes 7.1 + 8.1 \otimes 5.1 + 9 \otimes 5.1 + 10.1 \otimes 3.1 + 11 \otimes 3.1$$

$$\xi([6,10]) = 2.1 \otimes 11.1 + 3 \otimes 11.1 + 3.1 \otimes 9.2 + 3.1 \otimes 11 + 4.1 \otimes 9.1 + 5 \otimes 9.1 + 5.1 \otimes 7.2 + 6.1 \otimes 7.1 + 7 \otimes 7.1 + 7.1 \otimes 5.2 + 7.1 \otimes 7 + 8.1 \otimes 5.1 + 9 \otimes 5.1 + 7.3 \otimes 5 + 9.1 \otimes 5 + 10.1 \otimes 3.1 + 11 \otimes 3.1 + 9.3 \otimes 3 + 11.3 \otimes 1 + 13.1 \otimes 1$$

$$\xi([7,9]) = 2.1 \otimes 11.1 + 3 \otimes 11.1 + 4.1 \otimes 9.1 + 5 \otimes 9.1 + 6.1 \otimes 7.1 + 7 \otimes 7.1 + 8.1 \otimes 5.1 + 9 \otimes 5.1 + 10.1 \otimes 3.1 + 11 \otimes 3.1$$

$$\xi([8,8]) = 4.1 \otimes 7.3 + 10.1 \otimes 3.1 + 2.1 \otimes 9.3 + 3 \otimes 11.1 + 13.1 \otimes 1 + 7.2 \otimes 5.1 + 3 \otimes 9.3 + 6.3 \otimes 5.1 + 9.2 \otimes 3.1 + 11 \otimes 3.1 + 5 \otimes 7.3 + 6.1 \otimes 7.1 + 9.1 \otimes 5 + 7 \otimes 7.1 + 11.1 \otimes 3 + 7.1 \otimes 7 + 5.1 \otimes 9 + 8.3 \otimes 3.1 + 3.1 \otimes 11 + 2.1 \otimes 11.1$$

$$\xi([9,7]) = 2.1 \otimes 9.3 + 3 \otimes 9.3 + 2.1 \otimes 11.1 + 3 \otimes 11.1 + 4.1 \otimes 7.3 + 5 \otimes 7.3 + 6.1 \otimes 7.1 + 7 \otimes 7.1 + 6.3 \otimes 5.1 + 7.2 \otimes 5.1 + 8.3 \otimes 3.1 + 9.2 \otimes 3.1 + 10.1 \otimes 3.1 + 11 \otimes 3.1$$

$\xi([10,6] = 3.1 \otimes 9.2 + 5.1 \otimes 7.2 + 5.1 \otimes 9 + 7.1 \otimes 5.2 + 7.3 \otimes 5 + 9.3 \otimes 3 + 11.1 \otimes 3 + 11.3 \otimes 1$

$\xi(2[1,13]) = 0$

$\xi(3[1,12]) = 0$

$\xi(4[1,11]) = 0$

$$\xi(4[2,10]) = 7.2.1 \otimes 5 + 9.2.1 \otimes 3 + 5.1 \otimes 9 + 9.1 \otimes 5 + 11.2.1 \otimes 1 + 13.1 \otimes 1 + 5.1 \otimes 7.2 + 3.1 \otimes 9.2 + 5.2.1 \otimes 7 + 7.1 \otimes 7 + 7.1 \otimes 5.2$$

$\xi(5[1,10]) = 0$

$\xi(5[2,9]) = 0$

$\xi(6[1,9]) = 0$

$$\xi(6[2,8]) = 7.1 \otimes 5.2 + 9.3 \otimes 3 + 7.2 \otimes 5.1 + 6.3.1 \otimes 5 + 9 \otimes 5.1 + 5.2.1 \otimes 7 + 9.1 \otimes 4.1 + 5 \otimes 9.1 + 7.3 \otimes 5 + 11.1 \otimes 2.1 + 7.3 \otimes 4.1 + 7.2.1 \otimes 4.1 + 5 \otimes 7.3 + 9.2.1 \otimes 2.1 + 5.1 \otimes 7.2 + 5.2.1 \otimes 6.1 + 8.3.1 \otimes 3 + 13.1 \otimes 1 + 3.1 \otimes 9.2 + 3.1 \otimes 11 + 2.1 \otimes 9.3 + 4.1 \otimes 9.1 + 6.3 \otimes 5.1 + 7.1 \otimes 6.1 + 10.3.1 \otimes 1 + 8.3 \otimes 3.1 + 4.1 \otimes 7.3 + 9.3 \otimes 2.1 + 3 \otimes 9.3 + 8.1 \otimes 5.1 + 11.2.1 \otimes 1 + 11.1 \otimes 3 + 9.2 \otimes 3.1$$

$\xi(6[3,7]) = 0$

$\xi(7[1,8]) = 0$

$$\xi(7[2,7]) = 5 \otimes 7.3 + 10.1 \otimes 3.1 + 5 \otimes 9.1 + 7.2.1 \otimes 5 + 9.2.1 \otimes 3 + 8.3.1 \otimes 3 + 6.1 \otimes 7.1 + 9.3 \otimes 3 + 7 \otimes 7.1 + 5.2 \otimes 7.1 + 11.2.1 \otimes 1 + 7.3 \otimes 5 + 10.3.1 \otimes 1 + 2.1 \otimes 9.3 + 7.2 \otimes 5.1 + 4.1 \otimes 7.3 + 3 \otimes 9.3 + 11 \otimes 3.1 + 8.2.1 \otimes 3.1 + 4.1 \otimes 9.1 + 6.3.1 \otimes 5 + 4.2.1 \otimes 7.1 + 6.2.1 \otimes 5.1 + 9.2 \otimes 3.1 + 11.3 \otimes 1$$

$$\xi(7[3,6]) = 5\otimes 7.3 + 10.1\otimes 3.1 + 5\otimes 9.1 + 7.2.1\otimes 5 + 9.2.1\otimes 3 + 8.3.1\otimes 3 + 6.1\otimes 7.1 + 9.3\otimes 3 + 7\otimes 7.1 + 5.2\otimes 7.1 + 11.2.1\otimes 1 + 7.3\otimes 5 + 10.3.1\otimes 1 + 2.1\otimes 9.3 + 7.2\otimes 5.1 + 4.1\otimes 7.3 + 3\otimes 9.3 + 11\otimes 3.1 + 8.2.1\otimes 3.1 + 4.1\otimes 9.1 + 6.3.1\otimes 5 + 4.2.1\otimes 7.1 + 6.2.1\otimes 5.1 + 9.2\otimes 3.1 + 11.3\otimes 1$$

$$\xi(8[1,7]) = 0$$

$$\xi(8[2,6]) = 6.3.1\otimes 4.1 + 5.2.1\otimes 6.1 + 6.3.1\otimes 5 + 7.3\otimes 5 + 13.1\otimes 1 + 6.2.1\otimes 5.1 + 7.1\otimes 6.1 + 9.3\otimes 3 + 11.2.1\otimes 1 + 8.2.1\otimes 3.1 + 9.1\otimes 5 + 3\otimes 9.3 + 8.3.1\otimes 3 + 11.1\otimes 2.1 + 9.2\otimes 3.1 + 10.1\otimes 3.1 + 2.1\otimes 9.3 + 11\otimes 3.1 + 9.4.1\otimes 1 + 5\otimes 7.3 + 5.1\otimes 9 + 11.3\otimes 1 + 4.1\otimes 9.1 + 4.2.1\otimes 7.1 + 5.1\otimes 7.2 + 7\otimes 7.1 + 3.1\otimes 9.2 + 5\otimes 9.1 + 8.3.1\otimes 2.1 + 5.2\otimes 7.1 + 9.1\otimes 4.1 + 7.2\otimes 5.1 + 4.1\otimes 7.3 + 6.1\otimes 7.1 + 5.2.1\otimes 5.2$$

$$\xi(8[3,5]) = 0$$

$$\xi(8[4,4]) = 9.2\otimes 3.1 + 5\otimes 7.2.1 + 4.2.1\otimes 5.2.1 + 9.1\otimes 5 + 11.2.1\otimes 1 + 9.4.1\otimes 1 + 5.2\otimes 7.1 + 3.1\otimes 9.2 + 8.3\otimes 3.1 + 5.1\otimes 7.2 + 7.2.1\otimes 5 + 5.2.1\otimes 5.2 + 5.2.1\otimes 6.1 + 7\otimes 7.1 + 13.1\otimes 1 + 4.1\otimes 7.2.1 + 5.1\otimes 9 + 5\otimes 9.1 + 2.1\otimes 9.2.1 + 7.2\otimes 5.1 + 7.3\otimes 4.1 + 4.2.1\otimes 7.1 + 7.2.1\otimes 4.1 + 10.1\otimes 3.1 + 6.3\otimes 5.1 + 3.1\otimes 7.3.1 + 9.2.1\otimes 2.1 + 11\otimes 3.1 + 7.1\otimes 6.1 + 9.1\otimes 4.1 + 3\otimes 9.2.1 + 9.2.1\otimes 3 + 11.1\otimes 2.1 + 6.1\otimes 7.1 + 9.3\otimes 2.1 + 11.3\otimes 1 + 4.1\otimes 9.1 + 5.2\otimes 5.2.1$$

$$\xi(9[1,6]) = 0$$

$$\xi(9[2,5]) = 4.1\otimes 9.1 + 5\otimes 9.1 + 9.3\otimes 2.1 + 9\otimes 5.1 + 8.3\otimes 3.1 + 7.3.1\otimes 3.1 + 9.2\otimes 3.1 + 8.1\otimes 5.1 + 2.1\otimes 9.3 + 9.2.1\otimes 3 + 3\otimes 9.3 + 4.1\otimes 7.3 + 9.2.1\otimes 2.1 + 7.2.1\otimes 5 + 7.3\otimes 4.1 + 7.2.1\otimes 4.1 + 8.3.1\otimes 2.1 + 9.3\otimes 3 + 7.3\otimes 5 + 6.3\otimes 5.1 + 6.3.1\otimes 5 + 5\otimes 7.3 + 7.2\otimes 5.1 + 8.3.1\otimes 3 + 6.3.1\otimes 4.1$$

$$\xi(9[3,4]) = 4.1\otimes 9.1 + 5\otimes 9.1 + 9.3\otimes 2.1 + 9\otimes 5.1 + 8.3\otimes 3.1 + 7.3.1\otimes 3.1 + 9.2\otimes 3.1 + 8.1\otimes 5.1 + 2.1\otimes 9.3 + 9.2.1\otimes 3 + 3\otimes 9.3 + 4.1\otimes 7.3 + 9.2.1\otimes 2.1 + 7.2.1\otimes 5 + 7.3\otimes 4.1 + 7.2.1\otimes 4.1 + 8.3.1\otimes 2.1 + 9.3\otimes 3 + 7.3\otimes 5 + 6.3\otimes 5.1 + 6.3.1\otimes 5 + 5\otimes 7.3 + 7.2\otimes 5.1 + 8.3.1\otimes 3 + 6.3.1\otimes 4.1$$

$$\xi(9[4,3]) = 9.3\otimes 2.1 + 8.3\otimes 3.1 + 7.3.1\otimes 3.1 + 8.2.1\otimes 3.1 + 2.1\otimes 9.3 + 4.2.1\otimes 5.2.1 + 9.2.1\otimes 3 + 3\otimes 9.3 + 4.1\otimes 7.3 + 9.2.1\otimes 2.1 + 7.2.1\otimes 5 + 7.3\otimes 4.1 + 4.1\otimes 7.2.1 + 5\otimes 7.2.1 + 7.2.1\otimes 4.1 + 2.1\otimes 9.2.1 + 8.3.1\otimes 2.1 + 9.3\otimes 3 + 3\otimes 9.2.1 + 7.3\otimes 5 + 6.3\otimes 5.1 + 6.3.1\otimes 5 + 5\otimes 7.3 + 6.2.1\otimes 5.1 + 8.3.1\otimes 3 + 6.3.1\otimes 4.1 + 5.2\otimes 5.2.1$$

$$\xi(10[1,5]) = 0$$

$$\xi(10[2,4]) = 8.1\otimes 5.1 + 9.1\otimes 4.1 + 3.1\otimes 11 + 6.1\otimes 7.1 + 10.3.1\otimes 1 + 5.1\otimes 7.2 + 9\otimes 5.1 + 7.1\otimes 6.1 + 2.1\otimes 11.1 + 7.2.1\otimes 5 + 5.2.1\otimes 5.2 + 7\otimes 7.1 + 11\otimes 3.1 + 5.2.1\otimes 6.1 + 4.1\otimes 9.1 + 7.1\otimes 7 + 11.1\otimes 2.1 + 10.1\otimes 3.1 + 9.2.1\otimes 3 + 11.1\otimes 3 + 3.1\otimes 9.2 + 8.3.1\otimes 2.1 + 6.3.1\otimes 4.1 + 5\otimes 9.1 + 9.4.1\otimes 1 + 3\otimes 11.1$$

$$\xi(10[3,3]) = 0$$

$$\xi(11[1,4]) = 0$$

$$\xi(11[2,3]) = 9.3\otimes 2.1 + 8.3.1\otimes 2.1 + 4.2.1\otimes 7.1 + 6.3\otimes 5.1 + 6.2.1\otimes 5.1 + 8.2.1\otimes 3.1 + 10.3.1\otimes 1 + 7.3.1\otimes 3.1 + 2.1\otimes 11.1 + 6.3.1\otimes 4.1 + 4.1\otimes 9.1 + 9.2.1\otimes 2.1 + 11.2.1\otimes 1 + 7.2.1\otimes 4.1 + 5.2\otimes 7.1 + 8.3\otimes 3.1 + 5\otimes 9.1 + 7.3\otimes 4.1 + 3\otimes 11.1 + 11.3\otimes 1$$

$$\xi(11[3,2]) = 9.3 \otimes 2.1 + 8.3.1 \otimes 2.1 + 4.2.1 \otimes 7.1 + 6.3 \otimes 5.1 + 6.2.1 \otimes 5.1 + 8.2.1 \otimes 3.1 + 10.3.1 \otimes 1 + 7.3.1 \otimes 3.1 + 2.1 \otimes 11.1 + 6.3.1 \otimes 4.1 + 4.1 \otimes 9.1 + 9.2.1 \otimes 2.1 + 11.2.1 \otimes 1 + 7.2.1 \otimes 4.1 + 5.2 \otimes 7.1 + 8.3 \otimes 3.1 + 5 \otimes 9.1 + 7.3 \otimes 4.1 + 3 \otimes 11.1 + 11.3 \otimes 1$$

$\xi(12[1,3]) = 0$

$$\xi(12[2,2]) = 11.1 \otimes 3 + 8.3 \otimes 3.1 + 11.3 \otimes 1 + 7 \otimes 7.1 + 8.2.1 \otimes 3.1 + 11.2.1 \otimes 1 + 5.2.1 \otimes 6.1 + 5.2 \otimes 7.1 + 10.1 \otimes 3.1 + 7.1 \otimes 6.1 + 5.1 \otimes 9 + 6.3 \otimes 5.1 + 3.1 \otimes 11 + 9.2.1 \otimes 2.1 + 9.1 \otimes 5 + 7.3 \otimes 4.1 + 8.1 \otimes 5.1 + 9.1 \otimes 4.1 + 7.1 \otimes 7 + 6.1 \otimes 7.1 + 7.2.1 \otimes 4.1 + 4.2.1 \otimes 7.1 + 9.3 \otimes 2.1 + 11 \otimes 3.1 + 9 \otimes 5.1 + 11.1 \otimes 2.1 + 6.2.1 \otimes 5.1$$

$\xi(13[1,2]) = 0$

$\xi(14[1,1]) = 0$

$\xi(4.2[1,9]) = 0$

$\xi(5.2[1,8]) = 0$

$\xi(6.2[1,7]) = 0$

$\xi(6.3[1,6]) = 0$

$\xi(7.2[1,6]) = 0$

$\xi(7.3[1,5]) = 0$

$\xi(8.2[1,5]) = 0$

$\xi(8.3[1,4]) = 0$

$\xi(8.4[1,3]) = 0$

$$\xi(8.4[2,2]) = 5 \otimes 9.1 + 9.2 \otimes 3.1 + 4.1 \otimes 9.1 + 4.1 \otimes 7.3 + 5.1 \otimes 7.2 + 7.3 \otimes 5 + 9.3 \otimes 3 + 9.2.1 \otimes 2.1 + 8.3.1 \otimes 2.1 + 5.1 \otimes 9 + 2.1 \otimes 9.3 + 7.2.1 \otimes 4.1 + 9.1 \otimes 5 + 9.4.1 \otimes 1 + 7.3 \otimes 4.1 + 8.3.1 \otimes 3 + 7.2 \otimes 5.1 + 6.3.1 \otimes 4.1 + 8.3 \otimes 3.1 + 9.3 \otimes 2.1 + 3 \otimes 9.3 + 6.3.1 \otimes 5 + 9 \otimes 5.1 + 6.3 \otimes 5.1 + 5.2.1 \otimes 5.2 + 5 \otimes 7.3 + 3.1 \otimes 9.2 + 8.1 \otimes 5.1$$

$\xi(9.2[1,4]) = 0$

$\xi(9.3[1,3]) = 0$

$\xi(9.4[1,2]) = 0$

$\xi(10.2[1,3]) = 0$

$\xi(10.3[1,2]) = 0$

$\xi(10.4[1,1]) = 0$

$\xi(11.2[1,2]) = 0$

$\xi(11.3[1,1]) = 0$

$\xi(12.2[1,1]) = 0$

$\xi(8.4.2[1,1]) = 0$

$\xi([1,16]) = 0$

$\xi([2,15]) = 2.1\otimes 13+3\otimes 13+4.1\otimes 11+5\otimes 11+6.1\otimes 9+7\otimes 9+8.1\otimes 7+9\otimes 7+10.1\otimes 5 + 11\otimes 5 + 12.1\otimes 3 + 13\otimes 3 + 14.1\otimes 1 + 15\otimes 1$

$\xi([3,14]) = 2.1\otimes 13+3\otimes 13+4.1\otimes 11+5\otimes 11+6.1\otimes 9+7\otimes 9+8.1\otimes 7+9\otimes 7+10.1\otimes 5 + 11\otimes 5 + 12.1\otimes 3 + 13\otimes 3 + 14.1\otimes 1 + 15\otimes 1$

$\xi([4,13]) = 2.1\otimes 13+3\otimes 13+3.1\otimes 11.1+4.1\otimes 11+5\otimes 11+5.1\otimes 9.1+6.1\otimes 9+7\otimes 9 +7.1\otimes 7.1+8.1\otimes 7+9\otimes 7+9.1\otimes 5.1+10.1\otimes 5+11\otimes 5+11.1\otimes 3.1+12.1\otimes 3 + 13\otimes 3 + 14.1\otimes 1 + 15\otimes 1$

$\xi([5,12]) = 2.1\otimes 13+3\otimes 13+4.1\otimes 11+5\otimes 11+6.1\otimes 9+7\otimes 9+8.1\otimes 7+9\otimes 7+10.1\otimes 5 + 11\otimes 5 + 12.1\otimes 3 + 13\otimes 3 + 14.1\otimes 1 + 15\otimes 1$

$\xi([6,11]) = 2.1\otimes 11.2 + 3\otimes 11.2 + 4.1\otimes 9.2 + 4.1\otimes 11 + 5\otimes 9.2 + 5\otimes 11 + 6.1\otimes 7.2 +7\otimes 7.2+6.3\otimes 7+7.2\otimes 7+8.1\otimes 5.2+9\otimes 5.2+8.3\otimes 5+9.2\otimes 5+10.3\otimes 3 + 12.1\otimes 3 + 13\otimes 3 + 11.2\otimes 3 + 12.3\otimes 1 + 13.2\otimes 1$

$\xi([7,10]) = 2.1\otimes 11.2 + 3\otimes 11.2 + 4.1\otimes 9.2 + 4.1\otimes 11 + 5\otimes 9.2 + 5\otimes 11 + 6.1\otimes 7.2 +7\otimes 7.2+6.3\otimes 7+7.2\otimes 7+8.1\otimes 5.2+9\otimes 5.2+8.3\otimes 5+9.2\otimes 5+10.3\otimes 3 + 12.1\otimes 3 + 13\otimes 3 + 11.2\otimes 3 + 12.3\otimes 1 + 13.2\otimes 1$

$\xi([8,9]) = 2.1\otimes 13+3\otimes 13+3.1\otimes 9.3+4.1\otimes 11+5\otimes 11+5.1\otimes 7.3+5.1\otimes 9.1+6.1\otimes 9 +7\otimes 9+8.1\otimes 7+9\otimes 7+7.3\otimes 5.1+9.1\otimes 5.1+10.1\otimes 5+11\otimes 5+9.3\otimes 3.1 + 12.1\otimes 3 + 13\otimes 3 + 14.1\otimes 1 + 15\otimes 1$

$\xi([9,8]) = 2.1\otimes 13+3\otimes 13+4.1\otimes 11+5\otimes 11+6.1\otimes 9+7\otimes 9+8.1\otimes 7+9\otimes 7+10.1\otimes 5 + 11\otimes 5 + 12.1\otimes 3 + 13\otimes 3 + 14.1\otimes 1 + 15\otimes 1$

$\xi([10,7] = 10.1\otimes 5+11\otimes 5+6.1\otimes 9+6.3\otimes 5.2+7.3\otimes 5.1+7\otimes 9+5\otimes 9.2+11.4\otimes 1 +11.1\otimes 3.1+7.2\otimes 5.2+5.1\otimes 7.3+6.1\otimes 7.2+10.3\otimes 3+3.1\otimes 11.1+8.3\otimes 5 + 3.1\otimes 9.3 + 7.1\otimes 7.1 + 9.2\otimes 5 + 7\otimes 7.2 + 2.1\otimes 9.4 + 11.2\otimes 3 + 10.5\otimes 1 + 4.1\otimes 9.2 + 9.3\otimes 3.1 + 3\otimes 9.4$

$\xi([11,6] = 5\otimes 9.2+11.2\otimes 3+7.2\otimes 5.2+6.1\otimes 7.2+6.3\otimes 5.2+10.5\otimes 1+8.3\otimes 5+9.2\otimes 5 +7\otimes 7.2+10.3\otimes 3+6.1\otimes 9+2.1\otimes 9.4+11.4\otimes 1+10.1\otimes 5+7\otimes 9+3\otimes 9.4 + 11\otimes 5 + 4.1\otimes 9.2$

$\xi(2[1,14]) = 0$

$\xi(3[1,13]) = 0$

$\xi(4[1,12]) = 0$

$\xi(4[2,11]) = 5.1\otimes 9.1 + 9\otimes 5.2 + 12.2.1\otimes 1 + 4.1\otimes 9.2 + 5\otimes 9.2 + 11.1\otimes 3.1 + 6.1\otimes 7.2 + 8.2.1\otimes 5 + 2.1\otimes 13 + 3\otimes 13 + 8.1\otimes 5.2 + 11.2\otimes 3 + 7.1\otimes 7.1 + 3.1\otimes 11.1 + 10.2.1\otimes 3 + 2.1\otimes 11.2 + 6.2.1\otimes 7 + 9.2\otimes 5 + 7\otimes 7.2 + 12.1\otimes 3 + 13.2\otimes 1 + 7.2\otimes 7 + 4.2.1\otimes 9 + 13\otimes 3 + 3\otimes 11.2 + 5.2\otimes 9 + 9.1\otimes 5.1$

$\xi(5[1,11]) = 0$

$\xi(5[2,10]) = 4.1\otimes 9.2 + 2.1\otimes 13 + 3\otimes 13 + 3\otimes 11.2 + 6.2.1\otimes 7 + 6.1\otimes 7.2 + 7.2\otimes 7 + 4.2.1\otimes 9 + 7\otimes 7.2 + 5\otimes 9.2 + 8.2.1\otimes 5 + 8.1\otimes 5.2 + 10.2.1\otimes 3 + 9.2\otimes 5 + 2.1\otimes 11.2 + 9\otimes 5.2 + 5.2\otimes 9 + 13\otimes 3 + 12.1\otimes 3 + 11.2\otimes 3 + 13.2\otimes 1 + 12.2.1\otimes 1$

$\xi(6[1,10]) = 0$

$$\begin{aligned}\xi(6[2,9]) = {} & 11 \otimes 4.1 + 9.2 \otimes 5 + 11 \otimes 5 + 10.1 \otimes 5 + 6.2.1 \otimes 7 + 11.3.1 \otimes 1 + 6.1 \otimes 8.1 \\ & + 3.1 \otimes 9.3 + 3 \otimes 11.2 + 10.2.1 \otimes 3 + 9.3.1 \otimes 3 + 9 \otimes 7 + 4.1 \otimes 9.2 + 4.2.1 \otimes 9 \\ & + 9 \otimes 5.2 + 5 \otimes 11 + 9 \otimes 6.1 + 5.2 \otimes 9 + 13.2 \otimes 1 + 12.1 \otimes 2.1 + 9.1 \otimes 5.1 + \\ & 7.3 \otimes 5.1 + 6.3 \otimes 6.1 + 11.2 \otimes 3 + 6.1 \otimes 9 + 8.3 \otimes 4.1 + 9.3 \otimes 3.1 + 8.1 \otimes 5.2 \\ & + 13 \otimes 2.1 + 8.1 \otimes 7 + 10.3 \otimes 2.1 + 7.2 \otimes 7 + 5 \otimes 9.2 + 4.2.1 \otimes 8.1 + 7.3.1 \otimes 5 \\ & + 6.1 \otimes 7.2 + 5.1 \otimes 7.3 + 5.1 \otimes 9.1 + 7 \otimes 8.1 + 4.1 \otimes 11 + 7 \otimes 9 + 2.1 \otimes 11.2 \\ & + 6.2.1 \otimes 6.1 + 5.2 \otimes 8.1 + 8.1 \otimes 6.1 + 10.2.1 \otimes 2.1 + 8.2.1 \otimes 5 + 10.1 \otimes 4.1 \\ & + 12.3 \otimes 1 + 8.2.1 \otimes 4.1 + 7 \otimes 7.2\end{aligned}$$

$$\begin{aligned}\xi(6[3,8]) = {} & 11 \otimes 4.1 + 9.2 \otimes 5 + 11 \otimes 5 + 10.1 \otimes 5 + 6.2.1 \otimes 7 + 11.3.1 \otimes 1 + 6.1 \otimes 8.1 \\ & + 3.1 \otimes 9.3 + 3 \otimes 11.2 + 10.2.1 \otimes 3 + 9.3.1 \otimes 3 + 9 \otimes 7 + 4.1 \otimes 9.2 + 4.2.1 \otimes 9 \\ & + 9 \otimes 5.2 + 5 \otimes 11 + 9 \otimes 6.1 + 5.2 \otimes 9 + 13.2 \otimes 1 + 12.1 \otimes 2.1 + 9.1 \otimes 5.1 + \\ & 7.3 \otimes 5.1 + 6.3 \otimes 6.1 + 11.2 \otimes 3 + 6.1 \otimes 9 + 8.3 \otimes 4.1 + 9.3 \otimes 3.1 + 8.1 \otimes 5.2 \\ & + 13 \otimes 2.1 + 8.1 \otimes 7 + 10.3 \otimes 2.1 + 7.2 \otimes 7 + 5 \otimes 9.2 + 4.2.1 \otimes 8.1 + 7.3.1 \otimes 5 \\ & + 6.1 \otimes 7.2 + 5.1 \otimes 7.3 + 5.1 \otimes 9.1 + 7 \otimes 8.1 + 4.1 \otimes 11 + 7 \otimes 9 + 2.1 \otimes 11.2 \\ & + 6.2.1 \otimes 6.1 + 5.2 \otimes 8.1 + 8.1 \otimes 6.1 + 10.2.1 \otimes 2.1 + 8.2.1 \otimes 5 + 10.1 \otimes 4.1 \\ & + 12.3 \otimes 1 + 8.2.1 \otimes 4.1 + 7 \otimes 7.2\end{aligned}$$

$$\xi(7[1,9]) = 0$$

$$\begin{aligned}\xi(7[2,8]) = {} & 4.1 \otimes 11 + 8.1 \otimes 5.2 + 5.2.1 \otimes 7.1 + 10.1 \otimes 4.1 + 8.2.1 \otimes 5 + 7 \otimes 9 + 11 \otimes 4.1 \\ & + 5 \otimes 11 + 7 \otimes 7.2 + 9 \otimes 5.2 + 6.3 \otimes 6.1 + 12.1 \otimes 2.1 + 8.2.1 \otimes 4.1 + 5 \otimes 9.2 \\ & + 5.2 \otimes 9 + 10.3 \otimes 2.1 + 9.2.1 \otimes 3.1 + 13 \otimes 2.1 + 10.2.1 \otimes 3 + 13.2 \otimes 1 + 11 \otimes 5 \\ & + 5.2 \otimes 8.1 + 9.3 \otimes 3.1 + 9.2 \otimes 5 + 6.2.1 \otimes 7 + 6.1 \otimes 7.2 + 4.2.1 \otimes 8.1 + 8.1 \otimes 7 \\ & + 3 \otimes 11.2 + 7.2 \otimes 7 + 11.2 \otimes 3 + 7.1 \otimes 7.1 + 8.1 \otimes 6.1 + 9 \otimes 6.1 + 9 \otimes 7 + 2.1 \otimes 11.2 \\ & + 11.1 \otimes 3.1 + 6.2.1 \otimes 6.1 + 6.1 \otimes 8.1 + 10.2.1 \otimes 2.1 + 9.1 \otimes 5.1 + 4.1 \otimes 9.2 \\ & + 7.3 \otimes 5.1 + 7 \otimes 8.1 + 8.3 \otimes 4.1 + 10.1 \otimes 5 + 12.3 \otimes 1 + 4.2.1 \otimes 9 + 7.2.1 \otimes 5.1 \\ & + 6.1 \otimes 9\end{aligned}$$

$$\xi(7[3,7]) = 0$$

$$\xi(8[1,8]) = 0$$

$$\begin{aligned}\xi(8[2,7]) = {} & 5.2 \otimes 8.1 + 8.3 \otimes 4.1 + 12.3 \otimes 1 + 10.1 \otimes 4.1 + 8.4.1 \otimes 3 + 9.4 \otimes 3 + 8.1 \otimes 6.1 \\ & + 9 \otimes 6.1 + 5.2 \otimes 7.2 + 10.2.1 \otimes 2.1 + 6.1 \otimes 8.1 + 8.2.1 \otimes 5 + 12.1 \otimes 2.1 + 5.2 \otimes 9 \\ & + 4.2.1 \otimes 9 + 9.2 \otimes 5 + 2.1 \otimes 9.4 + 8.2.1 \otimes 4.1 + 4.2.1 \otimes 7.2 + 7.2 \otimes 5.2 + 6.3 \otimes 6.1 \\ & + 6.2.1 \otimes 5.2 + 13 \otimes 2.1 + 9 \otimes 5.2 + 9.3.1 \otimes 3 + 8.1 \otimes 5.2 + 6.2.1 \otimes 6.1 + 9.3.1 \otimes 2.1 \\ & + 13.2 \otimes 1 + 4.1 \otimes 9.2 + 10.4.1 \otimes 1 + 4.2.1 \otimes 8.1 + 7 \otimes 8.1 + 11.4 \otimes 1 + 11 \otimes 4.1 \\ & + 5 \otimes 9.2 + 10.3 \otimes 2.1 + 3 \otimes 9.4 + 7.3.1 \otimes 4.1\end{aligned}$$

$$\begin{aligned}\xi(8[3,6]) = {} & 5.2 \otimes 8.1 + 8.3 \otimes 4.1 + 12.3 \otimes 1 + 10.1 \otimes 4.1 + 8.4.1 \otimes 3 + 9.4 \otimes 3 + 8.1 \otimes 6.1 \\ & + 9 \otimes 6.1 + 5.2 \otimes 7.2 + 10.2.1 \otimes 2.1 + 6.1 \otimes 8.1 + 8.2.1 \otimes 5 + 12.1 \otimes 2.1 + 5.2 \otimes 9 \\ & + 4.2.1 \otimes 9 + 9.2 \otimes 5 + 2.1 \otimes 9.4 + 8.2.1 \otimes 4.1 + 4.2.1 \otimes 7.2 + 7.2 \otimes 5.2 + 6.3 \otimes 6.1 \\ & + 6.2.1 \otimes 5.2 + 13 \otimes 2.1 + 9 \otimes 5.2 + 9.3.1 \otimes 3 + 8.1 \otimes 5.2 + 6.2.1 \otimes 6.1 + 9.3.1 \otimes 2.1 \\ & + 13.2 \otimes 1 + 4.1 \otimes 9.2 + 10.4.1 \otimes 1 + 4.2.1 \otimes 8.1 + 7 \otimes 8.1 + 11.4 \otimes 1 + 11 \otimes 4.1 \\ & + 5 \otimes 9.2 + 10.3 \otimes 2.1 + 3 \otimes 9.4 + 7.3.1 \otimes 4.1\end{aligned}$$

$$\begin{aligned}\xi(8[4,5]) = {} & 7.3 \otimes 5.1 + 5.2.1 \otimes 5.2.1 + 8.3.1 \otimes 3.1 + 5.1 \otimes 9.1 + 3.1 \otimes 9.2.1 + 5.2 \otimes 8.1 \\ & + 6.3.1 \otimes 5.1 + 8.3 \otimes 4.1 + 12.3 \otimes 1 + 10.1 \otimes 4.1 + 8.4.1 \otimes 3 + 4.1 \otimes 7.3.1 + 9.4 \otimes 3 \\ & + 8.1 \otimes 6.1 + 9 \otimes 6.1 + 5.2 \otimes 7.2 + 2.1 \otimes 9.3.1 + 10.2.1 \otimes 2.1 + 6.1 \otimes 8.1 + 3 \otimes 9.3.1 \\ & + 8.2.1 \otimes 5 + 12.1 \otimes 2.1 + 9.3 \otimes 3.1 + 5.2 \otimes 9 + 4.2.1 \otimes 9 + 9.2 \otimes 5 + 2.1 \otimes 9.4 \\ & + 8.2.1 \otimes 4.1 + 4.2.1 \otimes 7.2 + 7.2 \otimes 5.2 + 6.3 \otimes 6.1 + 6.2.1 \otimes 5.2 + 13 \otimes 2.1 + 9 \otimes 5.2\end{aligned}$$

$+5\otimes 7.3.1+9.3.1\otimes 3+8.1\otimes 5.2+6.2.1\otimes 6.1+9.3.1\otimes 2.1+9.1\otimes 5.1+13.2\otimes 1 + 4.1\otimes 9.2+10.4.1\otimes 1+4.2.1\otimes 8.1+7\otimes 8.1+11.4\otimes 1+11\otimes 4.1+5\otimes 9.2 + 10.3\otimes 2.1+3\otimes 9.4+5.1\otimes 7.2.1+7.3.1\otimes 4.1$

$\xi(9[1,7]) = 0$

$\xi(9[2,6]) = 10.3\otimes 2.1+9\otimes 5.2+4.1\otimes 9.2+5\otimes 9.2+9.2\otimes 5+10.4.1\otimes 1+9.3.1\otimes 3 +4.2.1\otimes 8.1+11.3.1\otimes 1+13\otimes 2.1+9.4\otimes 3+7.2\otimes 5.2+3\otimes 9.4+4.2.1\otimes 7.2 +8.1\otimes 5.2+11.4\otimes 1+12.1\otimes 2.1+9.1\otimes 5.1+12.3\otimes 1+5.2.1\otimes 7.1+5.2\otimes 9 +8.3\otimes 4.1+5.2\otimes 7.2+8.2.1\otimes 5+11.1\otimes 3.1+13.2\otimes 1+11\otimes 4.1+8.3.1\otimes 3.1 +8.1\otimes 6.1+6.1\otimes 8.1+6.2.1\otimes 6.1+7\otimes 8.1+6.3.1\otimes 5.1+7.1\otimes 7.1+8.2.1\otimes 4.1 +5.2\otimes 8.1+8.4.1\otimes 3+10.1\otimes 4.1+4.2.1\otimes 9+6.2.1\otimes 5.2+9\otimes 6.1+6.3\otimes 6.1 + 2.1\otimes 9.4+10.2.1\otimes 2.1$

$\xi(9[3,5]) = 0$

$\xi(9[4,4]) = 5.2\otimes 7.2+7.2\otimes 5.2+8.3\otimes 4.1+9.1\otimes 5.1+4.2.1\otimes 9+5.2\otimes 9+10.1\otimes 4.1 +12.3\otimes 1+9.4\otimes 3+9.3\otimes 3.1+8.1\otimes 5.2+2.1\otimes 9.4+6.3\otimes 6.1+9.2.1\otimes 3.1+ 8.2.1\otimes 4.1+9.3.1\otimes 3+10.3\otimes 2.1+9.2\otimes 5+13.2\otimes 1+5.2.1\otimes 7.1+6.2.1\otimes 6.1 +5\otimes 9.2+4.1\otimes 7.3.1+5.2\otimes 8.1+11.4\otimes 1+7.1\otimes 7.1+8.2.1\otimes 5+11.3.1\otimes 1 + 7.2.1\otimes 5.1+4.2.1\otimes 7.2+10.2.1\otimes 2.1+7.3\otimes 5.1+3\otimes 9.3.1+8.1\otimes 6.1 +9\otimes 5.2+11\otimes 4.1+8.4.1\otimes 3+5\otimes 7.3.1+12.1\otimes 2.1+6.2.1\otimes 5.2+4.2.1\otimes 8.1 +7\otimes 8.1+13\otimes 2.1+6.1\otimes 8.1+9\otimes 6.1+10.4.1\otimes 1+4.1\otimes 9.2+11.1\otimes 3.1 + 2.1\otimes 9.3.1+3\otimes 9.4$

$\xi(10[1,6]) = 0$

$\xi(10[2,5]) = 11.2\otimes 2.1+6.3\otimes 7+7.1\otimes 7.1+4.1\otimes 9.2+7.2\otimes 7+11\otimes 4.1+7\otimes 6.3 +8.2.1\otimes 5+4.2.1\otimes 6.3+10.5\otimes 1+6.2.1\otimes 6.1+5.2\otimes 6.3+5\otimes 9.2+2.1\otimes 11.2 +3.1\otimes 9.3+11\otimes 5+9.4\otimes 2.1+9.3.1\otimes 2.1+10.4.1\otimes 1+8.3\otimes 4.1+7\otimes 7.2 + 8.2.1\otimes 4.1+3.1\otimes 11.1+9\otimes 6.1+9.2\otimes 5+5\otimes 11+7.3.1\otimes 5+6.3\otimes 5.2 + 8.4.1\otimes 2.1+10.3\otimes 2.1+6.1\otimes 6.3+10.1\otimes 4.1+5.1\otimes 7.3+6.1\otimes 7.2+ 7.2.1\otimes 5.1+6.2.1\otimes 5.2+9.3.1\otimes 3+3\otimes 11.2+4.1\otimes 11+10.2.1\otimes 3+9.2.1\otimes 3.1 +6.3\otimes 6.1+8.3.1\otimes 3.1+8.1\otimes 6.1+11.2\otimes 3+10.1\otimes 5+11.1\otimes 3.1+6.3.1\otimes 5.1$

$\xi(10[3,4]) = 11.2\otimes 2.1+6.3\otimes 7+7.1\otimes 7.1+4.1\otimes 9.2+7.2\otimes 7+11\otimes 4.1+7\otimes 6.3 +8.2.1\otimes 5+4.2.1\otimes 6.3+10.5\otimes 1+6.2.1\otimes 6.1+5.2\otimes 6.3+5\otimes 9.2+2.1\otimes 11.2 +3.1\otimes 9.3+11\otimes 5+9.4\otimes 2.1+9.3.1\otimes 2.1+10.4.1\otimes 1+8.3\otimes 4.1+7\otimes 7.2 + 8.2.1\otimes 4.1+3.1\otimes 11.1+9\otimes 6.1+9.2\otimes 5+5\otimes 11+7.3.1\otimes 5+6.3\otimes 5.2 + 8.4.1\otimes 2.1+10.3\otimes 2.1+6.1\otimes 6.3+10.1\otimes 4.1+5.1\otimes 7.3+6.1\otimes 7.2+ 7.2.1\otimes 5.1+6.2.1\otimes 5.2+9.3.1\otimes 3+3\otimes 11.2+4.1\otimes 11+10.2.1\otimes 3+9.2.1\otimes 3.1 +6.3\otimes 6.1+8.3.1\otimes 3.1+8.1\otimes 6.1+11.2\otimes 3+10.1\otimes 5+11.1\otimes 3.1+6.3.1\otimes 5.1$

$\xi(10[4,3]) = 6.3.1\otimes 5.1+8.3.1\otimes 3.1+7.3.1\otimes 4.1+7.3.1\otimes 5+9.3.1\otimes 2.1+3.1\otimes 9.3 + 9.3.1\otimes 3+5.2.1\otimes 5.2.1+5.1\otimes 7.2.1+5.1\otimes 7.3+3.1\otimes 9.2.1$

$\xi(11[1,5]) = 0$

$\xi(11[2,4]) = 11.1\otimes 3.1+8.3.1\otimes 3.1+6.2.1\otimes 6.1+8.3\otimes 4.1+8.4.1\otimes 2.1+6.1\otimes 6.3 + 6.3\otimes 5.2+6.3\otimes 7+7.2\otimes 7+11\otimes 4.1+9.4\otimes 2.1+11\otimes 5+6.2.1\otimes 5.2 +9\otimes 6.1+3\otimes 11.2+5.2\otimes 6.3+2.1\otimes 11.2+11.3.1\otimes 1+5\otimes 9.2+6.3\otimes 6.1 +9.1\otimes 5.1+10.4.1\otimes 1+10.1\otimes 4.1+10.1\otimes 5+8.2.1\otimes 5+7.1\otimes 7.1+7\otimes 6.3 + 6.3.1\otimes 5.1+8.2.1\otimes 4.1+10.2.1\otimes 3+11.2\otimes 2.1+4.2.1\otimes 6.3+6.1\otimes 7.2$

$+ 11.2 \otimes 3 + 9.2 \otimes 5 + 8.1 \otimes 6.1 + 10.3 \otimes 2.1 + 4.1 \otimes 11 + 5 \otimes 11 + 7 \otimes 7.2 + 5.2.1 \otimes 7.1 + 9.3.1 \otimes 2.1 + 4.1 \otimes 9.2 + 10.5 \otimes 1$

$\xi(11[3,3]) = 0$

$\xi(12[1,4]) = 0$

$\xi(12[2,3]) = 12.1 \otimes 3 + 7.3 \otimes 5.1 + 4.1 \otimes 11 + 8.3.1 \otimes 3.1 + 3 \otimes 13 + 5 \otimes 11 + 8.3 \otimes 4.1 + 6.1 \otimes 9 + 5.1 \otimes 9.1 + 9.3.1 \otimes 2.1 + 3.1 \otimes 11.1 + 7.3.1 \otimes 4.1 + 10.3 \otimes 2.1 + 11.3.1 \otimes 1 + 6.3 \otimes 6.1 + 8.2.1 \otimes 4.1 + 11.1 \otimes 3.1 + 7.1 \otimes 7.1 + 9.1 \otimes 5.1 + 6.3.1 \otimes 5.1 + 12.1 \otimes 2.1 + 10.1 \otimes 5 + 12.3 \otimes 1 + 13 \otimes 3 + 9.2.1 \otimes 3.1 + 12.2.1 \otimes 1 + 7 \otimes 9 + 4.2.1 \otimes 8.1 + 7 \otimes 8.1 + 11 \otimes 5 + 8.1 \otimes 7 + 5.2 \otimes 8.1 + 6.1 \otimes 8.1 + 11 \otimes 4.1 + 9 \otimes 7 + 9 \otimes 6.1 + 2.1 \otimes 13 + 7.2.1 \otimes 5.1 + 6.2.1 \otimes 6.1 + 10.1 \otimes 4.1 + 10.2.1 \otimes 2.1 + 13 \otimes 2.1 + 8.1 \otimes 6.1 + 9.3 \otimes 3.1$

$\xi(12[3,2]) = 12.1 \otimes 3 + 7.3 \otimes 5.1 + 4.1 \otimes 11 + 8.3.1 \otimes 3.1 + 3 \otimes 13 + 5 \otimes 11 + 8.3 \otimes 4.1 + 6.1 \otimes 9 + 5.1 \otimes 9.1 + 9.3.1 \otimes 2.1 + 3.1 \otimes 11.1 + 7.3.1 \otimes 4.1 + 10.3 \otimes 2.1 + 11.3.1 \otimes 1 + 6.3 \otimes 6.1 + 8.2.1 \otimes 4.1 + 11.1 \otimes 3.1 + 7.1 \otimes 7.1 + 9.1 \otimes 5.1 + 6.3.1 \otimes 5.1 + 12.1 \otimes 2.1 + 10.1 \otimes 5 + 12.3 \otimes 1 + 13 \otimes 3 + 9.2.1 \otimes 3.1 + 12.2.1 \otimes 1 + 7 \otimes 9 + 4.2.1 \otimes 8.1 + 7 \otimes 8.1 + 11 \otimes 5 + 8.1 \otimes 7 + 5.2 \otimes 8.1 + 6.1 \otimes 8.1 + 11 \otimes 4.1 + 9 \otimes 7 + 9 \otimes 6.1 + 2.1 \otimes 13 + 7.2.1 \otimes 5.1 + 6.2.1 \otimes 6.1 + 10.1 \otimes 4.1 + 10.2.1 \otimes 2.1 + 13 \otimes 2.1 + 8.1 \otimes 6.1 + 9.3 \otimes 3.1$

$\xi(13[1,3]) = 0$

$\xi(13[2,2]) = 10.3 \otimes 2.1 + 9 \otimes 6.1 + 12.2.1 \otimes 1 + 5 \otimes 11 + 7 \otimes 9 + 4.2.1 \otimes 8.1 + 6.1 \otimes 8.1 + 10.1 \otimes 5 + 12.3 \otimes 1 + 9.2.1 \otimes 3.1 + 7 \otimes 8.1 + 9 \otimes 7 + 12.1 \otimes 3 + 7.2.1 \otimes 5.1 + 11.1 \otimes 3.1 + 6.2.1 \otimes 6.1 + 7.3 \otimes 5.1 + 6.3 \otimes 6.1 + 8.2.1 \otimes 4.1 + 3 \otimes 13 + 2.1 \otimes 13 + 8.1 \otimes 7 + 8.3 \otimes 4.1 + 13 \otimes 2.1 + 9.3 \otimes 3.1 + 5.2.1 \otimes 7.1 + 8.1 \otimes 6.1 + 6.1 \otimes 9 + 5.2 \otimes 8.1 + 4.1 \otimes 11 + 12.1 \otimes 2.1 + 11 \otimes 4.1 + 11 \otimes 5 + 10.1 \otimes 4.1 + 7.1 \otimes 7.1 + 10.2.1 \otimes 2.1 + 13 \otimes 3 + 9.1 \otimes 5.1$

$\xi(14[1,2]) = 0$

$\xi(15[1,1]) = 0$

$\xi(4.2[1,10]) = 0$

$\xi(5.2[1,9]) = 0$

$\xi(6.2[1,8]) = 0$

$\xi(6.3[1,7]) = 0$

$\xi(7.2[1,7]) = 0$

$\xi(7.3[1,6]) = 0$

$\xi(8.2[1,6]) = 0$

$\xi(8.3[1,5]) = 0$

$\xi(8.4[1,4]) = 0$

$\xi(8.4[2,3]) = 9.4.2 \otimes 1 + 5.1 \otimes 9.1 + 5.2 \otimes 7.2 + 8.1 \otimes 5.2 + 9.2 \otimes 5 + 8.4.1 \otimes 3 + 4.1 \otimes 7.3.1 + 5 \otimes 7.3.1 + 2.1 \otimes 9.3.1 + 3 \otimes 9.3.1 + 9 \otimes 5.2 + 6.2.1 \otimes 5.2 + 3.1 \otimes 9.2.1 + 9.3.1 \otimes 3$

$+ 5.1 \otimes 7.2.1 + 8.2.1 \otimes 5 + 4.2.1 \otimes 9 + 7.2.1 \otimes 5.1 + 5.2.1 \otimes 5.2.1 + 5.2 \otimes 9 + 2.1 \otimes 9.4 + 9.4 \otimes 3 + 4.2.1 \otimes 7.2 + 4.1 \otimes 9.2 + 9.1 \otimes 5.1 + 5 \otimes 9.2 + 3 \otimes 9.4 + 9.2.1 \otimes 3.1 + 7.2 \otimes 5.2 + 8.4.2.1 \otimes 1$

$\xi(9.2[1,5]) = 0$

$\xi(9.3[1,4]) = 0$

$\xi(9.4[1,3]) = 0$

$\xi(9.4[2,2]) = 9.4.2 \otimes 1 + 5.2 \otimes 7.2 + 9.2 \otimes 5 + 5 \otimes 9.2 + 8.2.1 \otimes 5 + 7.3 \otimes 5.1 + 5.2 \otimes 9 + 3 \otimes 9.4 + 7.2 \otimes 5.2 + 8.3.1 \otimes 3.1 + 4.2.1 \otimes 7.2 + 6.3.1 \otimes 5.1 + 8.1 \otimes 5.2 + 6.2.1 \otimes 5.2 + 4.1 \otimes 9.2 + 9 \otimes 5.2 + 9.3.1 \otimes 3 + 9.2.1 \otimes 3.1 + 9.4 \otimes 3 + 8.4.1 \otimes 3 + 8.4.2.1 \otimes 1 + 4.2.1 \otimes 9 + 2.1 \otimes 9.4 + 9.3 \otimes 3.1 + 7.2.1 \otimes 5.1$

$\xi(10.2[1,4]) = 0$

$\xi(10.3[1,3]) = 0$

$\xi(10.4[1,2]) = 0$

$\xi(10.5[1,1]) = 0$

$\xi(11.2[1,3]) = 0$

$\xi(11.3[1,2]) = 0$

$\xi(11.4[1,1]) = 0$

$\xi(12.2[1,2]) = 0$

$\xi(12.3[1,1]) = 0$

$\xi(13.2[1,1]) = 0$

$\xi(8.4.2[1,2]) = 0$

$\xi(9.4.2[1,1]) = 0$

Bibliography

[A] Adams, J.F.: On the non-existence of elements of Hopf invariant one. Annals of Math. 72 (1960) 20–104

[BAH] Baues, H.-J.: Algebraic homotopy. Cambridge studies in advanced math. 15, Cambridge University Press 1989

[BD] Baues, H.-J. and Dreckmann, W.: The cohomology of homotopy categories and the general linear group. K-theory 3 (1989) 307–338

[BJ1] Baues, H.-J. and Jibladze, M.: Classification of abelian track categories. K-theory 486 (2002) 1–13.

[BJ2] Baues, H.-J. and Jibladze, M.: Suspension and loop objects and representability of tracks. Georgian Math. J. 8 (2001) 683–696

[BJ3] Baues, H.-J. and Jibladze, M.: Suspension and loop objects in theories and cohomology. Georgian Math. J. 8 (2001) 697–712

[BJ4] Baues,H.-J. and Jibladze, M.: The Steenrod algebra and theories associated to Hopf algebras. Preprint

[BJ5] Baues, H.-J. and Jibladze, M.: Secondary derived functors and the Adams spectral sequence, MPIM preprint 2004-43 (arXiv:math.AT/0407031, to apperar in Topology).

[BJ6] Baues,H.-J. and Jibladze, M.: Computation of the E_3-term of the Adams spectral sequence, MPIM preprint 2004-53 (arXiv:math.AT/0407045).

[BJ7] Baues,H.-J. and Jibladze, M.: The algebra of secondary cohomology operations and its dual, MPIM preprint 2004-111.

[Bl] Blanc, D.: Realising coalgebras over the Steenrod algebra. Topology 40 (2001) 993–1016.

[BM] Baues, H.-J. and Minian, E.G.: Crossed extensions of Algebras and Hochschild Cohomology. Preprint

[Bo] Borceux, F.: Handbook of categorical algebra 2. Encyclopedia of Math. and its Appl. 51 Cambridge University Press 1994

[BOT] Baues, H.-J.: Obstruction theory. Lecture Notes in Math. 628 Springer 1977, 387 pages

[BP] Baues, H.-J. and Pirashvili, T.: On the third MacLane cohomology. Preprint.

[BSC] Baues, H.-J.: Secondary cohomology and the Steenrod square. Preprint MPI für Mathematik 2000 (122)

[BT] Baues, H.-J. and Tonks A.: On sum normalized cohomology of categories, twisted homotopy pairs and universal Toda brackets. Quart. J. Math. Oxford (2) 47 (1996) 405–433

[BUT] Baues, H.-J.: On the cohomology of categories, universal Toda brackets and homotopy pairs. K-theory 11 (1997) 259–285

[BW] Baues, H.-J. and Wirsching, G.: The cohomology of small categories. J. Pure Appl. Algebra 38 (1985) 187–211

[C] Cartan, H.: Sur les groupes d'Eilenberg-MacLane II. Proc. Nat. Acad. Sci. USA 40 (1954) 704–707

[Co] Cohn, P.M. : Free rings and their relations. London Math. Soc. monographs 19, Acad. Press London 1985, 588p.

[EML] Eilenberg, S. and MacLane, S.: On the groups $H(\pi, n)$ I. Ann. of Math. 58 (1953) 55–106

[FM] Fantham, P.H.H. and Moore, E.J.: Groupoid enriched categories and homotopy theory. Can. J. Math. 35 (1983) 385–416

[G] Gray, B.: Homotopy Theory. Academic Press 1975, 378 pages

[GJ] Goerss, P. and Jardine, J.F.: Simplicial homotopy theory. Birkhäuser Verlag, Progress in Math. 174, (1999), 510 pages

[Ha] Harper, J.R.: Secondary cohomology operations.

[HFT] Félix, Y. and Halperin. S. and Thomas, J.C.: Rational Homotopy theory. Graduate Texts in Math., Springer 205, 2000.

[HS] Hovey, M. and Shipley, B. and Smith, J.: Symmetric spectra. Journal of the AMS 13 (2000) 149–208

[JP] Jibladze, M. and Pirashvili, T.: Cohomology of algebraic theories. J. of Algebra 137(2) (1991) 253–296

[Ka1] Karoubi, M.: Formes différentielles non commutatives et cohomologie a coefficients arbitraires. Transactions of the AMS 347 (1995) 4227–4299

[Ka2] Karoubi, M.: Formes différentielles non commutatives et opération de Steenrod. Topology 34 (1995) 699–715

[Ke] Kelly, G.M.: Basic concepts of enriched category theory. Volume 64 of London Math. Soc. Lecture Note Series. Cambridge University Press 1982

[KKr] Kock, A and Kristensen, L.: A secondary product structure in cohomology theory. Math. Scand. 17 (1965) 113–149

[Kr1] Kristensen, L.: On secondary cohomology operations, Math. Scand. 12 (1963) 57–82

[Kr2] Kristensen, L.: On a Cartan formula for secondary cohomology operations. Math. Scand. 16 (1965) 97–115

[Kr3] Kristensen, L.: On secondary cohomology operations II. Conf. on Algebraic Topology 1969 (Univ. of Illinois at Chicago Circle, Chicago, Ill. 1968) pp 117–133

[Kr4] Kristensen, L.: Massey products in Steenrod's algebra. Proc. Adv. Study Inst. Aarhus 1970, 240–255

[KrM1] Kristensen, L. and Madsen, Ib: Note on Whitehead products in spheres. Math. Scand. 21 (1967) 301–314

[KrM2] Kristensen, L. and Madsen, Ib: On the structure of the operation algebra for certain cohomology theories. Conf. Algebraic topology. Univ. of Illinois at Chicago Circle (1968) 134–160

[KrP] Kristensen, L. and Pedersen, E.K.: The $\mathcal{A}$-module structure for the cohomology of two stage spaces. Math. Scand. 30 (1972) 95–106

[L] Lawvere, F.W.: Functorial semantics of algebraic theories. Proc. Nat. Acad. Sci. USA 50 (1963) 869–873

[Li] Liulevicius, A.: The factorization of cyclic reduced powers by secondary cohomology operations. Memoirs AMS 42 (1962).

[Ma] Mandell, M.A.:E_∞-algebras and p-adic homotopy theory. Topology 40 (2001) 43-94.

[MaP] Massey, W.S. and Petersen, F.P.: Cohomology of certain fibre spaces I. Topology 4 (1965) 47–65

[May] May, J.P.: A general algebraic approach to Steenrod operations. Steenrod algebra and its Applications: A conference to Celebrate N.E. Steenrod's Sixtieth Birthday, Springer Lecture Notes 168 (1970) 153–231

[MC] McClure, J.E.: Power operations in H_∞ ring theories. Springer Lecture Notes 1176 (1986)

[MLC] MacLane, S.: Categories for the working mathematician. GTM, 5, Springer (1971)

[MLH] MacLane, S.: Homology. Grundlehren 114, Springer (1967)

[Mn] Milnor, J.: The Steenrod algebra and its dual. Annals of Math.(2) 81 (1965) 211–264

[Mo] Monks, K.G. : STEENROD: a Maple package for computing with the Steenrod algebra, (1995).

[No] Novikov, S.P.: Cohomology of the Steenrod algebra. Dokl. Akad. Nauk SSSR 128 (1959) 893–895.

[P] Pirashvili, T.: Spectral sequence for MacLane homology. J. Algebra 170 (1994) 422–428

[PV] Penkava, M. and Vanhaecke, P.: Hochschild cohomology of polynomial algebras. Communications in Contemporary Math. 3 (2001) 393–402

[PW] Pirashvili, T. and Waldhausen, F.: MacLane homology and topological Hochschild homology. J. Pure Appl. Algebra 82 (1992) 81–98

[S] Sullivan, D.: Infinitesimal computation in topology. Publ. Math. IHES 47 (1978) 269–331.

[S1] Serre, J.-P.: Cohomologie modulo 2 des complexes d'Eilenberg-MacLane. Commentari Math. Helv. 27 (1953) 198–232

[S2] Serre, J.-P.: Représentations linéaires des groupes finis. Hermann, Paris 1971

[Sch] Schwartz, L.: Unstable Modules over the Steenrod algebra and Sullivan's fixed point set construction. Chicago Lectures in Math. Series, the University of Chicago Press 1994, 230 pages

[SE] Steenrod, N.E. and Epstein, D.B.A.: Cohomology Operations. Annals of Math. Studies 50, Princeton

[ShY] Shimada, N. and Yamanoshita, T.: On triviality of the mod p Hopf invariant, Japan Jr. Math. 31 (1961) 1–25

[St] Stover, C.R.: The equivalence of certain categories of twisted Lie and Hopf algebras over a commutative ring. J. Pure Appl. Algebra 86 (1993) 289–326

[Ta] Tangora, M.C.: On the Cohomology of the Steenrod algebra. Math. Z. 116 (1970) 18–64

[To] Toda, H.: Compositional methods in homotopy groups of spheres, Ann. of Math. Studies 49, Princeton Univ. Press, (1962).

[W] Wood, R.M.W.: Problems in the Steenrod algebra. Bull. London Math. Soc. 30 (1998) 449–517

Index